Proceedings

"WASCOM 2007"

14th Conference on

Waves and Stability in Continuous Media

Proceedings
"WASCOM 2007"
14th Conference on
Waves and Stability in Continuous Media

Baia Samuele, Sicily, Italy

30 June – 7 July 2007

Editors

Natale Manganaro
Università di Messina, Italy

Roberto Monaco
Politecnico di Torino, Italy

Salvatore Rionero
Università di Napoli, Italy

World Scientific

NEW JERSEY • LONDON • SINGAPORE • BEIJING • SHANGHAI • HONG KONG • TAIPEI • CHENNAI

Published by

World Scientific Publishing Co. Pte. Ltd.
5 Toh Tuck Link, Singapore 596224
USA office: 27 Warren Street, Suite 401-402, Hackensack, NJ 07601
UK office: 57 Shelton Street, Covent Garden, London WC2H 9HE

British Library Cataloguing-in-Publication Data
A catalogue record for this book is available from the British Library.

WAVES AND STABILITY IN CONTINUOUS MEDIA
Proceedings of the 14th Conference on WASCOM 2007

ISBN-13 978-981-277-234-3
ISBN-10 981-277-234-0

Printed in Singapore by World Scientific Printers

PREFACE

In 1981 a group of Italian researchers planned an International Conference devoted to a fruitful exchange of ideas and views between researchers from several countries, interested in the fields of wave propagation and nonlinear stability in Continuous Media. Thus the cycle of WASCOM Conferences (Waves and Stability in Continuous Media) began. After its first edition, the WASCOM Conference was regularly held every two years. The meeting received an increasing interest and participation by the international scientific community and nowadays it is a well known and appreciated forum. The last conference, the XIV edition (WASCOM 2007), was held in Sampieri (Ragusa), from June 30 to July 7, 2007. The previous editions took place in Catania (1981), Arcavacata di Rende (Cosenza, 1983), Giovinazzo (Bari, 1985), Taormina (Messina, 1987), Sorrento (Napoli, 1989), Acireale (Catania, 1991), Bologna (1993), Altavilla Milicia (Palermo, 1995), Capitolo di Monopoli (Bari, 1997), Vulcano (Messina, 1999), Porto Ercole (Grosseto, 2001), Villasimius (Cagliari, 2003), Acireale (Catania, 2005). Each edition has provided a book of proceedings documenting the updated research work and progress in the area. The 14th edition of the International Conference on Waves and Stability in Continuous Media, has been promoted by the Research Project of National Interest "Nonlinear Propagation and Stability in Thermodynamical Processes of Continuous Media" (PRIN 2005, national coordinator Prof. Tommaso Ruggeri). The research group working on this project belongs to Local Units of Bologna, Messina, Naples, Palermo, Turin and came from 12 different Italian universities. The Local Unit of Messina was in charge of organizing the meeting. The members of the Organizing Committee of the Conference belong to the Department of Mathematics of Universities of Bologna, Catania, Messina and Naples. One hundred and twenty-three participants coming from 14 different countries attended the XIV edition of the Conference. The main topics discussed were:

- Discontinuity and shock waves;
- Linear and non-linear stability in fluid dynamics;
- Small parameter problems;
- Kinetic theories and comparison with continuum models;
- Wave propagation and non equilibrium thermodynamics;
- Group analysis and reduction techniques;
- Numerical and technical applications.

This volume contains 80 papers which have been presented at the WASCOM 2007 as invited lectures and communications. The Editors of this volume would like to thank the Scientific Committee who carefully suggested the invited lectures and selected the contributed papers, as well as the member of the Organizing Committee. A special thank is addressed to all the participants to whom ultimately the success of the conference has been ascribed to. A thank as well to Fiammetta Conforto (Università di Messina) and Sandra Pieraccini (Politecnico di Torino) for their help in preparing the final version of the manuscript. Lastly the Editors are especially indebted to Fondazione CRT of Turin which has partially supported the publishing expenses of this volume.

The Editors

Natale Manganaro
Roberto Monaco
Salvatore Rionero

January 2008

Professor Tommaso Ruggeri
Università di Bologna

Dedicated to Tommaso Ruggeri

The International Conference WASCOM 2007 has been dedicated to Professor Tommaso Ruggeri in the occasion of his 60th birthday. In what follows, for brevity, we limit ourselves to explain the main points of his scientific and academic career. More information can be found in the web site http://www.ciram.unibo.it/~ruggeri/.

Born in Messina on 31 July 1947, had his degree in Physics at the University of Messina on 1969 and soon began his academic career at the Institute of Mathematics of Messina. In 1973 he moved to the University of Bologna as assistant Professor. Since 1980, Tommaso Ruggeri is Full Professor of Rational Mechanics at the Faculty of Engineering of the University of Bologna.

He is:

- member of the National Accademia dei Lincei;
- member of the Director Board of Istituto Nazionale di Alta Matematica (INdAM);
- Director of Gruppo Nazionale per la Fisica Matematica (GNFM) of INdAM.

Since 2000, he was national coordinator of Research Projects of National Interest (PRIN) involving more than 50 researchers from several Italian universities. His scientific activity was developed along different lines of research on Mathematical Physics and Applied Mathematics. In particular, his main contributions were attributed to Nonlinear Wave Propagation theory as well as to the Non-Equilibrium Thermodynamics. Within the first research framework we would like to mention the very important results he obtained on symmetrization of hyperbolic systems of balance laws and in the theory of shock waves. In the field of non equilibrium thermodynamics, he has been one of the founders of the modern Extended Thermodynamics and has written, jointly with Ingo Müller, the well known book *Rational Extended Thermodynamics* (Springer-Verlag, 1988, 2nd edition). He is author of more than 150 scientific papers, several books and monographs. His high scientific value is widely recognized by the international scientific community where he is greatly appreciated. He contributed to the scientific training of many qualified researchers from different countries.

Apart from his outstanding position in the scientific community, Tommaso Ruggeri is also very well appreciated for his likableness, friendly character and playful nature. Tommaso is not only a distinguished scientist and an esteemed colleague but for all of us he is above all a great and sincere friend.

Natale Manganaro
Roberto Monaco
Salvatore Rionero

January, 2008

CONFERENCE DATA

WASCOM 2007
14th International Conference on Waves and Stability in Continuous Media

Sampieri (RG), Italy, June 30–July 7, 2007

Scientific Committee

Chairmen: **N. Manganaro** (Messina), **S. Rionero** (Naples),

G. Boillat (France), **Y. Choquet-Bruhat** (France),
L. Desvillettes (France), **J. N. Flavin** (Ireland), **D. Fusco** (Messina),
H. Gouin (France), **A. M. Greco** (Palermo),
R. Monaco (Turin), **I. Müller** (Germany), **G. Mulone** (Catania),
C. Rogers (Australia), **B. Straughan** (UK),
A. Strumia (Bari), **M. Sugiyama** (Japan), **C. Tebaldi** (Turin)

Organizing Committee

Chairmen: **N. Manganaro** (Messina), **S. Rionero** (Napoli)

F. Conforto (Messina), **S. Iacono** (Messina), **G. Valenti** (Messina),
B. Buonomo (Naples), **F. Capone** (Naples), **M. Gentile** (Naples),
A. Mentrelli (Bologna), **A. Valenti** (Catania)

Sponsors

- Research Project of National Interest "Nonlinear Propagation and Stability in Thermodynamical Processes of Continuous Media" (MIUR-COFIN 2005–07, national coordinator Prof. Tommaso Ruggeri);
- INdAM & GNFM;
- Department of Mathematics, University of Messina;
- Department of Mathematics , University of Naples;
- Comune di Scicli;
- Banca Agricola Popolare di Ragusa.

CONTENTS

RECIPROCAL TRANSFORMATIONS AND INTEGRABLE HAMILTONIAN HYDRODYNAMIC TYPE SYSTEMS

S. ABENDA

C.I.R.A.M. and Department of Mathematics, Bologna University, I-40126 Bologna BO, Italy
E-mail: abenda@ciram.unibo.it

Following our approach in Ref. 2, I present sufficient conditions so that the reciprocal Hamiltonian structure to a DN system is local.

Keywords: Hamiltonian systems of hydrodynamic type, reciprocal transformations, zero curvature equations

1. Dubrovin-Novikov systems and zero curvature equations

Equations of hydrodynamic type

$$u_t^i = \sum_{k=1}^{n} v_k^i(u) u_x^k, \qquad u = (u^1, \dots, u^n), \quad u_x^i = \frac{\partial u^i}{\partial x}, \quad u_t^i = \frac{\partial u^i}{\partial t}. \tag{1}$$

naturally arise in applications such as gas dynamics, hydrodynamics, chemical kinetics, the Whitham averaging procedure, differential geometry and topological field theory.[4,7,8,16,17] Dubrovin and Novikov[7] showed that Eq. (1) is a Hamiltonian (DN) system with Hamiltonian $H[u] = \int h(u)dx$, if there exists a flat contravariant metric $g^{il}(u)$ in $\mathbb{R}^n$ with Christoffel symbols $\Gamma^i_{jk}(u)$, such that the matrix $v^i_k(u)$ can be represented in the form

$$v_k^i(u) = \sum_{k=1}^{n} \left(g^{il}(u) \frac{\partial^2 h}{\partial u^l \partial u^k}(u) - \sum_{s=1}^{n} g^{ik}(u) \Gamma^l_{sk}(u) \frac{\partial h}{\partial u^l}(u) \right). \tag{2}$$

If $n = 2$, Eq. (1) are the Euler equations for ideal compressible fluids, the system can be put in diagonal form and is integrable by the hodograph method. For arbitrary n, Tsarev[16] proved that a DN system as in Eqs. (1), (2) can be integrated by a generalized hodograph method only if it may be transformed to the diagonal form

$$u_t^i = v^i(u) u_x^i, \qquad i = 1, \dots, n. \tag{3}$$

In the latter case, moreover the flat metric is diagonal, the Hamiltonian satisfies

$$\frac{\partial^2 h}{\partial u^i \partial u^j} = \Gamma^i_{ij}(u)\frac{\partial h}{\partial u^i} + \Gamma^j_{ji}(u)\frac{\partial h}{\partial u^j}, \tag{4}$$

and each solution $p(u)$ to Eq. (4) generates a conserved quantity for the DN system (1), (2) and all of these symmetries commute.

From a differential geometric point of view, a non-degenerate flat diagonal metric in $\mathbb{R}^n$ is associated to an orthogonal coordinate system $u^i = u^i(x_1, \ldots, x_n)$. Upon introducing the Lamé coefficients

$$H_i^2(u) = \sum_k \left(\frac{\partial x^i}{\partial u^k}\right)^2,$$

the metric tensor in the coordinate system u^i is diagonal $ds^2 = \sum_{i=1}^{n} H_i^2(u)(du^i)^2$, and the zero curvature conditions $R_{il,im}(u) = 0$ $(i \neq l \neq m \neq i)$ and $R_{il,il}(u) = 0$ $(i \neq l)$ form an overdetermined system:

$$\frac{\partial^2 H_i}{\partial u^l \partial u^m} = \frac{1}{H_l}\frac{\partial H_l}{\partial u^m}\frac{\partial H_i}{\partial u^l} + \frac{1}{H_m}\frac{\partial H_m}{\partial u^l}\frac{\partial H_i}{\partial u^m}, \tag{5}$$

$$\frac{\partial}{\partial u^l}\frac{\partial H_i}{H_l \partial u^l} + \frac{\partial}{\partial u^i}\frac{\partial H_l}{H_i \partial u^i} + \sum_{m \neq i,l} \frac{1}{H_m^2}\frac{\partial H_i}{\partial u^m}\frac{\partial H_l}{\partial u^m} = 0. \tag{6}$$

Bianchi and Cartan showed that a general solution to the zero curvature equations (5), (6) can be parametrized locally by $n(n-1)/2$ arbitrary functions of two variables. If the Lamé coefficients $H_i(u)$ are known, one can find $x^i(u^1, \ldots, u^n)$ solving the linear undetermined problem (embedding equations)

$$\frac{\partial^2 x^i}{\partial u^k \partial u^l} = \Gamma^k_{kl}(u)\frac{\partial x^i}{\partial u^k} + \Gamma^l_{lk}(u)\frac{\partial x^i}{\partial u^l}, \qquad \frac{\partial^2 x^i}{\partial (u^l)^2} = \sum_k \Gamma^k_{ll}(u)\frac{\partial x^i}{\partial u^k}. \tag{7}$$

Comparison of Eqs. (4) and (7) implies that the flat coordinates for the metric $g_{ii}(u) = (H^i(u))^2$ are the Casimirs of the corresponding Hamiltonian operator. Finally, Zakharov[19] showed that the dressing method may be used to determine the solutions to the zero curvature equations up to Combescure transformations.

It then follows that the classification of flat diagonal metrics $ds^2 = g_{ii}(u)(du^i)^2$ is an important preliminary step in the classification of integrable Hamiltonian systems of hydrodynamic type. An important technical point is that all known examples of integrable Hamiltonian systems of hydrodynamic type possess a pair of compatible flat metrics and have been obtained in the framework of semisimple Frobenius manifolds (axiomatic theory of integrable Hamiltonian systems).[4–6] In the latter case, one of the metrics is Egorov (*i.e.* its rotation coefficients are symmetric).

It is then of a certain interest to find and classify new examples of integrable Hamiltonian systems of hydrodynamic type which do not belong to the above class. Reciprocal transformations act non trivially on Hamiltonian structures; following our approach in Refs. 1,2, I explain below their possible role in the above picture.

2. Reciprocal transformations and integrable DN systems

Reciprocal transformations have been introduced by Rogers and Shadwick[14] and are an important class of nonlocal transformations which act on hydrodynamic–type systems.[1–3,10–15,18] These transformations originate from gas dynamics - the simplest example being the passage from Eulerian to Lagrangian coordinates in one–dimensional gas dynamics - and they change the independent variables of a system.

Let the integrable DN system in Riemann invariant form

$$u^i_t = v^i(u)u^i_x, \qquad i = 1, \dots, n, \tag{8}$$

admits conservation laws

$$B(u)_t = A(u)_x, \quad N(u)_t = M(u)_x \tag{9}$$

with $B(u)M(u) - A(u)N(u) \neq 0$. In the new independent variables $\tilde{x}$ and $\tilde{t}$ defined by

$$d\hat{x} = B(u)dx + A(u)dt, \qquad d\hat{t} = N(u)dx + M(u)dt, \tag{10}$$

the reciprocal system is still diagonal and takes the form

$$u^i_{\hat{t}} = \frac{B(u)v^i(u) - A(u)}{M(u) - N(u)v^i(u)} u^i_{\hat{x}} = \hat{v}^i(u)u^i_{\hat{x}}. \tag{11}$$

Moreover, the metric of the initial systems $g_{ii}(u)$ transforms to

$$\hat{g}_{ii}(u) = \left(\frac{M(u) - N(u)v^i(u)}{B(u)M(u) - A(u)N(u)}\right)^2 g_{ii}(u) \tag{12}$$

and all conservations laws and commuting flows of the original system (8) may be recalculated in the new independent variables.

If the reciprocal transformation is linear (*i.e.* A, B, N, M are constant functions), then the reciprocal to a flat metric is still flat and locality and compatibility of the associated Hamiltonian structures are preserved (see Refs. 13,17,18).

Under a general reciprocal transformation, the Hamiltonian structure does not behave trivially and a thorough study of reciprocal Hamiltonian structures is still an open problem. Ferapontov[10] takes a reciprocal transformation where the conservation laws in Eq. (11) are a linear combination of the Casimirs, momentum and Hamiltonian densities and gives sufficient conditions so that the reciprocal to the flat metric $g_{ii}(u)$ in Eq. (12) is a constant curvature metric. Finally, Ferapontov and Pavlov[12] construct the Riemann curvature tensor and the nonlocal Hamiltonian operator associated to the reciprocal to Eq. (2).

The classification of the reciprocal Hamiltonian structures is complicated by the fact that a DN system as in Eqs. (1)-(2) also possesses an infinite number of nonlocal Hamiltonian structures. It is then possible that two DN systems are linked by a reciprocal transformation and that the flat metrics of the first system are not reciprocal to the flat metrics of the second. In Ref. 1, we constructed such an example: the genus one modulation (Whitham-CH) equations associated to Camassa-Holm in Riemann invariant form ($n = 3$ in Eq. (8)). We proved that the Whitham-CH equations are a DN-system and possess a pair of compatible flat metrics (none of the metrics is Egorov). We also proved the connection via a reciprocal transformation of the Whitham-CH equations to the modulation equations associated to the first negative flow of the Korteweg de Vries hierarchy (Whitham-KdV$_{-1}$). In Ref. 1, finally we also found the relation between the Poisson structures of the Whitham-KdV$_{-1}$ and the Whitham-CH equations: both systems possess a pair of compatible flat metrics, and the two flat metrics of the first system are respectively reciprocal to the constant curvature and conformally flat metrics of the second (and vice versa).

In view of the above results, in Ref. 2 we have started to classify the reciprocal transformations which transform a DN system to a DN system, under the condition that the flat metric tensor $\tilde{g}(u)$ of the transformed system is reciprocal to a metric tensor $g(u)$ of the initial system, which is either flat or of constant Riemannian curvature or conformally flat.

In particular, in Ref. 2, we classify which conservation laws may preserve the flatness of the metric when the transformation changes only one independent variable and we get the following theorem:

Theorem 2.1. *Assume that the $u^i_t = v^i(u)u^i_x$, $i = 1,\dots,n$, is a DN system with flat diagonal metric $g_{ii}(u)$ and conservation laws $B(u)_t = A(u)_x$, $N(u)_t = M(u)_x$ such that $B(u)M(u) - A(u)N(u) \neq 0$. Let $(\nabla B)^2(u) = \sum\limits_{m=1}^{n} g^{mm}(u)(\partial_m B)^2$. Let the reciprocal transformation and the* *reciprocal metric tensor $\tilde{g}$ be as in Eqs. (11) and (12). Then*

(i) If $N \equiv 0$, $M \equiv 1$, the reciprocal metric tensor $\tilde{g}$ is flat if and only if:

- *(a) B and A are constant functions;*
- *(b) $B(u)$ is a Casimir for the metric $g_{ii}(u)$ and $(\nabla B)^2(u) = 0$;*
- *(c) $B(u)$ is a density of momentum for the metric $g_{ii}(u)$ and $(\nabla B)^2(u) = 2B(u)$;*

(ii) If $B = 1$, $A = 0$, the reciprocal metric tensor $\tilde{g}$ is flat if and only if:

- *(a) N and M are constant functions;*
- *(b) $N(u)$ is a Casimir for the metric $g_{ii}(u)$ and $(\nabla N)^2(u) = 0$;*
- *(c) $N(u) = \kappa H(u)$, where $H(u)$ is a density of Hamiltonian associated to the metric g^{ii}) and $(\nabla N(u))^2 = 2\kappa M(u)$.*

To prove the above theorem we express the zero curvature equations (5) and (6) of the reciprocal metric tensor (12) in function of the initial metric $g(u)$ and of the conservation laws in the reciprocal transformation (10).

We use the same technique also to produce sufficient conditions for a flat reciprocal metric $\tilde{g}$ in the case of reciprocal transformations of both independent variables: assuming that $B(u)$ is either a Casimir or a momentum density or a Hamiltonian associated to the initial metric $g_{ii}(u)$, we[2] prove that $N(u)$ is a linear combination of the Casimirs, momentum and Hamiltonian density for the initial metric $g_{ii}(u)$.

To construct explicit examples, we use Dubrovin[5] classification of flat metrics on Hurwitz spaces (spaces of meromorphic functions on Riemann surfaces): the genus g-KdV modulation equations correspond to the case in which the Riemann invariants $(u_1, \dots, u_{2g+1})$ of the integrable DN system (8) are the ramification points of genus g hyperelliptic Riemann surfaces. Indeed, we[2] find new examples of flat pencils of metrics associated to the reciprocal to genus g-KdV modulation equations ($n = 2g + 1$ in Eq. (8)), plugging the explicit expressions of the Casimirs, the conservation laws and the flat metrics associated to the genus g-KdV modulation equations into the zero curvature conditions for the reciprocal metrics. In particular, we show that the genus g Camassa-Holm modulation equations are a DN system with a pair of compatible flat non-Egorov metrics, generalizing our results in Ref. 1 to any genus $g \geq 1$.

There are of course still many open problems connected to the classification of reciprocal Hamiltonian structures. The results in Refs. 2,10 suggest that Casimirs, momentum and Hamiltonian densities have a privileged role in reciprocal transformations which preserve locality of the Hamiltonian structure. So a natural question is: do there exist two DN systems connected by a reciprocal transformation not in the above class? What about other types of transformations among hydrodynamic systems?

Finally, several systems of evolutionary PDEs arising in physics may be written as perturbations of hyperbolic systems of PDEs and their classification in case of Hamiltonian perturbations has recently been started by Dubrovin, Liu and Zhang.[9] It would also be interesting to investigate the role of reciprocal transformations in this perturbation scheme.

Acknowledgements

It is a pleasure to dedicate this paper to prof. Tommaso Ruggeri in occasion of His 60th birthday.

This research is partially supported by ESF Programme MISGAM, by RTN ENIGMA, by PRIN2006 "Metodi geometrici nella teoria delle onde non lineari ed applicazioni" and by GNFM-INdAM.

References

1. S. Abenda, T. Grava, *Ann. Inst. Fourier* **55**, 1803–1834 (2005).
2. S. Abenda, T. Grava, *J. Phys. A: Math. Theor.* **40**, 10769–10790 (2007).
3. K. W. Chow, C. C. Mak, C. Rogers, W. K. Schief. *J. Comput. Appl. Math.* **190**, 114–126 (2006).
4. B. A. Dubrovin, in *Lecture Notes in Math.* **1620**, 120–348, Springer (1996).
5. B. A. Dubrovin, in *Surv. Differ. Geom.*, IV, 213–238, Int. Press, Boston, MA, (1998).
6. B. A. Dubrovin, in *Integrable systems and algebraic geometry (Kobe/Kyoto, 1997)*, 47–72, World Sci. Publ. (1998).
7. B. A. Dubrovin, S. P. Novikov, *Sov. Math. Dokl.* **270**, 665–669 (1983).
8. Dubrovin, B.A.; Novikov, S.P. *Russian Math. Surveys* **44**, 35–124 (1989).
9. B. A. Dubrovin, Si-Qi Liu, Y. Zhang, *Comm. Pure Appl. Math.* **59**, 559–615 (2006).
10. E. V. Ferapontov, *Amer. Math. Soc. Transl. Ser.2* **170**, 33–58 (1995).
11. E. V. Ferapontov, C. Rogers, W. K. Schief. *J. Math. Anal. Appl.* **228**, 365–376 (1998).
12. E. V. Ferapontov, M. V. Pavlov, *J. Math. Phys.* **44**, 1150–1172 (2003).
13. M. V. Pavlov. *Math. Notes* **57**, 489–495 (1995).
14. C. Rogers, W. F. Shadwick. Bäcklund transformations and their applications, xiii+334 pp. Academic Press, Inc., New York-London, (1982).
15. C. Rogers, T. Ruggeri, *Lett. Nuovo Cimento* **44**, 289–296 (1985).
16. S. P. Tsarev. *Dokl. Akad. Nauk. SSSR* **282**, 534–537 (1985).
17. S. P. Tsarev. *Math, USSR Izv.* **37**, 397–419 (1991).
18. T. Xue, Y Zhang. *Lett. Math. Phys.* **75**, 79–92 (2006).
19. V. E. Zakharov, *Duke Math. Journal* **94**, 103–139 (1998).

MODELING AND NUMERICAL ANALYSIS OF WAVE PROPAGATION IN PARTIALLY SATURATED POROUS MEDIA*

B. ALBERS

Institute for Geotechnical Engineering and Soil Mechanics
Technical University of Berlin, Germany
E-mail: albers@grundbau.tu-berlin.de
www.mech-albers.de

The propagation of sound waves in partially saturated soils is investigated using a macroscopic linear model for a deformable skeleton and two compressible pore fluids which was constructed by a micro-macro transition. Four body waves exist: three longitudinal waves, $P1$, $P2$, $P3$, and one shear wave, S, whose phase velocities and attenuations are given in dependence on the frequency and the initial saturation. The combination of velocities of $P2$ and $P3$ waves in the unsaturated porous medium is similar to this of an air-water-mixture: for a certain degree of saturation there appears a minimum in the sonic velocity.

Keywords: Sound waves; Partially saturated porous media.

1. Introduction

In this work we investigate the propagation of sound waves in partially saturated poroelastic media by means of a new model [1] which is an extension of both the classical BIOT Model (e.g. [2]) for two-component saturated media and the Simple Mixture Model by WILMAŃSKI (e.g. [3]). The three-component model which describes linear processes in unsaturated poroelastic materials contains features of both models: the interaction between components by partial volume changes through an additional contribution to partial stresses like in Biot's model and an own balance equation for the field of porosity like in the Simple Mixture Model. The modeling included the application of a systematic method of derivation of relations between macroscopic (average) material parameters (compressibilities and coupling parameters) and their counterparts for true (microscopic) materials.

*Dedicated to Professor Tommaso Ruggeri on the occasion of his 60th birthday.

This paper is a summary of [1,4]. In the latter it is shown that due to the existence of the second pore fluid (the gas) in the unsaturated medium besides the two compressional waves ($P1$, $P2$) and the shear wave (S) which appear in the saturated porous medium an additional compressional wave ($P3$) emerges. The speeds and attenuations of all these waves are shown in dependence on the frequency ω and on the initial saturation S_0. The speeds are compared to those of sound waves in air-water-mixures.

2. Linear model

The linear thermodynamical model without memory effects can be described by the essential fields $\{\mathbf{v}^S, \mathbf{v}^F, \mathbf{v}^G, \mathbf{e}^S, \varepsilon^F, \varepsilon^G\}$ which have to satisfy the field equations

$$
\begin{aligned}
\rho_0^S \frac{\partial \mathbf{v}^S}{\partial t} &= \operatorname{div}\left\{\lambda^S e \mathbf{1} + 2\mu^S \mathbf{e}^S + Q^F \varepsilon^F \mathbf{1} + Q^G \varepsilon^G \mathbf{1}\right\} + \\
&\quad + \pi^{FS}\left(\mathbf{v}^F - \mathbf{v}^S\right) + \pi^{GS}\left(\mathbf{v}^G - \mathbf{v}^S\right), \\
\rho_0^F \frac{\partial \mathbf{v}^F}{\partial t} &= \operatorname{grad}\left\{\rho_0^F \kappa^F \varepsilon^F + Q^F e + Q^{FG} \varepsilon^G\right\} - \pi^{FS}\left(\mathbf{v}^F - \mathbf{v}^S\right), \qquad (1) \\
\rho_0^G \frac{\partial \mathbf{v}^G}{\partial t} &= \operatorname{grad}\left\{\rho_0^G \kappa^G \varepsilon^G + Q^G e + Q^{FG} \varepsilon^F\right\} - \pi^{GS}\left(\mathbf{v}^G - \mathbf{v}^S\right), \\
\frac{\partial \mathbf{e}^S}{\partial t} &= \operatorname{sym}\operatorname{grad} \mathbf{v}^S, \quad \frac{\partial \varepsilon^F}{\partial t} = \operatorname{div} \mathbf{v}^F, \quad \frac{\partial \varepsilon^G}{\partial t} = \operatorname{div} \mathbf{v}^G, \quad e \equiv \operatorname{tr} \mathbf{e}^S.
\end{aligned}
$$

This set of equations coincides with the classical Biot model, if we neglect the third component, i.e. the gas.

The quantities $\mathbf{v}^S, \mathbf{v}^F, \mathbf{v}^G$ are macroscopic velocity fields of the components, $\mathbf{e}^S$ is the macroscopic deformation tensor. Quantities with subindex zero are initial values of the corresponding current quantity.

Instead of the partial mass densities of the components, ρ^S, ρ^F, ρ^G, the equations depend on the volume changes of the components $e, \varepsilon^F, \varepsilon^G$ for which hold

$$
e = \frac{\rho_0^S - \rho^S}{\rho_0^S}, \quad \varepsilon^F = \frac{\rho_0^F - \rho^F}{\rho_0^F}, \quad \varepsilon^G = \frac{\rho_0^G - \rho^G}{\rho_0^G}. \qquad (2)
$$

In principle, the porosity, n, also is a field and satisfies an own balance equation. However, if we neglect memory effects, the balance equation can be solved and its consideration is not longer necessary to solve the problem. The current saturation of the fluid, S, i.e. the fraction of fluid in the voids, is not included in the series of fields. Instead, a constitutive law is used for this quantity.

The momentum sources include relative resistances π^{FS} and π^{GS} (fluid-skeleton and gas-skeleton). They depend on the one hand on the resistance parameter π which appears both in the Biot model and in the Simple Mixture Model and describes the resistance of the flow of the fluid through the channels of the skeleton. On the other hand they are related to the relative permeabilities k_f, k_g which were proposed by van Genuchten [5].

Material parameters λ^S and μ^S are Lamé constants and κ^F, κ^G are compressibility coefficients of an ideal fluid and of the gas, respectively. Q^F, Q^G and Q^{FG} reflect the couplings between the partial stresses of the components (F: fluid-skeleton, G: gas-skeleton, FG: fluid-gas). These material parameters have to be specified according to the material, they are functions of an initial porosity n_0 and an initial saturation S_0. Therefore a transition procedure from the micro- to the macro-scale is applied which is described in [1]. Several macroscopic, microscopic and mixed conditions (dynamical and geometrical compatibility conditions) have to be combined. Moreover, a microscopic relation for the capillary pressure is needed. We use the one proposed by van Genuchten [5] in 1980.

For the solution of the problem we still need 6 boundary condtions. We obtain them from three Gedankenexperiments which first have been proposed by Biot and Willis [6] for the saturated porous medium. In this case they yield 4 conditions (see: Wilmanski [7]). For the unsaturated medium we obtain 6 conditions (see: Albers [1]). Using those boundary conditions we are able to determine the six material parameters in dependence on the initial porosity n_0 and on the initial saturation S_0 and, hence, the model is complete.

3. Wave analysis

We proceed with the investigation of the propagation of sound waves in partially saturated soils by means of the above introduced model (for detailed information see: [1]).

We assume that the fields satisfy the following relations

$$\begin{aligned} \varepsilon^F = E^F \mathcal{E}, \quad \varepsilon^G = E^G \mathcal{E}, \quad \mathbf{e}^S = \mathbf{E}^S \mathcal{E}, \\ \mathbf{v}^F = \mathbf{V}^F \mathcal{E}, \quad \mathbf{v}^G = \mathbf{V}^G \mathcal{E}, \quad \mathbf{v}^S = \mathbf{V}^S \mathcal{E}, \end{aligned} \tag{3}$$

where $\mathcal{E} := \exp i(\mathbf{k} \cdot \mathbf{x} - \omega t)$ and $\mathbf{E}^S, E^F, E^G, \mathbf{V}^S, \mathbf{V}^F, \mathbf{V}^G$ are constant amplitudes, ω is a given frequency, $\mathbf{k}$ is the, possibly complex, wave vector. This means that $\mathbf{k} = k\mathbf{n}$, where k is the complex wave number and $\mathbf{n}$ is a unit vector in the direction of propagation. Such a solution describes the propagation of plane monochromatic waves in an infinite medium whose fronts are perpendicular to $\mathbf{n}$.

Separation of the transversal and normal contributions in the resulting equations yields the phase speeds in the two limits of the frequency. We achieve similar results to the well-known results for two-component media (e.g. [3], [8]). For the transversal wave we obtain

$$\lim_{\omega\to\infty} c_{ph} = \sqrt{\frac{\mu^S}{\rho_0^S}}, \qquad \lim_{\omega\to 0} c_{ph} = \sqrt{\frac{\mu^S}{\rho_0^S+\rho_0^F+\rho_0^G}}. \tag{4}$$

Secondly, we investigate the longitudinal waves. The problem for the unsaturated porous medium yields the existence of three longitudinal modes of propagation: $P1$-, $P2$- and $P3$-waves. In the frequency limits they possess the following phase velocities

$$\lim_{\omega\to 0} c_{ph} = \begin{cases} \sqrt{\dfrac{\lambda^S+2\mu^S+\rho_0^F\kappa^F+\rho_0^G\kappa^G+2(Q^F+Q^G+Q^{FG})}{\rho_0^S+\rho_0^F+\rho_0^G}} & \text{for } P1\text{-waves} \\ 0 & \text{for } P2\text{-waves} \\ 0 & \text{for } P3\text{-waves} \end{cases}. \tag{5}$$

For the high frequency limit we distinguish also the limits of the initial saturation. Results are given in [4], qualitatively holds

$$\lim_{\omega\to\infty} c_{ph} = \begin{cases} \begin{matrix} S_0=0 & S_0=1 \\ c_{P1}^{\infty,0} & c_{P1}^{\infty,1} \\ 0 & c_{P2}^{\infty,1} \\ c_{P3}^{\infty,0} & 0 \end{matrix} & \begin{matrix} \\ \text{for } P1\text{-waves} \\ \text{for } P2\text{-waves} \\ \text{for } P3\text{-waves} \end{matrix} \end{cases}. \tag{6}$$

4. Numerical analysis of the wave propagation

We numerically investigate the dispersion relations for the shear wave and for the longitudinal waves. The following data corresponding approximately to sandstones filled with an air water mixture are used

$$\rho_0^S = 2500 \text{ kg/m}^3, \quad \rho_0^F = 250 \text{ kg/m}^3, \quad \rho_0^G = 1.2 \text{ kg/m}^3$$
$$n_0 = 0.25, \quad \pi = 10^7 \text{ kg/m}^3\text{s}.$$

The phase speeds and attenuations of the waves have been investigated in dependence on the frequency and on the saturation. In [4] the results for several values of S_0 and ω are shown. Here, we limit our attention to the values $S_0 = 0.9$ and $\omega = 1000$ Hz (see Fig. 1). The dispersion relations have been solved for the complex wave number k. Results for k yield the phase speeds $c = \omega/(\operatorname{Re} k)$ and the attenuations $\operatorname{Im} k$.

For $S_0 = 0.9$ which is close to saturation, the change of speeds of the $P1$-, S- and $P3$-waves with frequency is small. The $P2$-wave increases with increasing frequency but also reaches an asymptotic value for high frequencies. The behavior of the $P2$- and $P3$-waves in dependence on the saturation is stronger. Their existence is ascribed to the pore fluids and therefore their speeds run in the opposite direction: While the speed of the $P2$-wave starts with zero for $S_0 = 0$ and then increases with increasing saturation, the speed of the $P3$-wave is decreasing and ends up with zero for $S_0 = 1$. Hence, the $P3$-wave does not appear in the saturated medium while the $P2$-wave does not exist in an air-filled medium. For medium values of the saturation both waves appear but, of course, the $P2$-wave is faster than the $P3$-wave for almost all values of saturation due to the faster sound velocity in water than in air. This is illustrated in the right panel of Fig. 2 which shows the combination of the speeds of the $P2$- and $P3$-waves. The bottom row of Fig. 1 shows that these waves are much stronger attenuated than the other two waves.

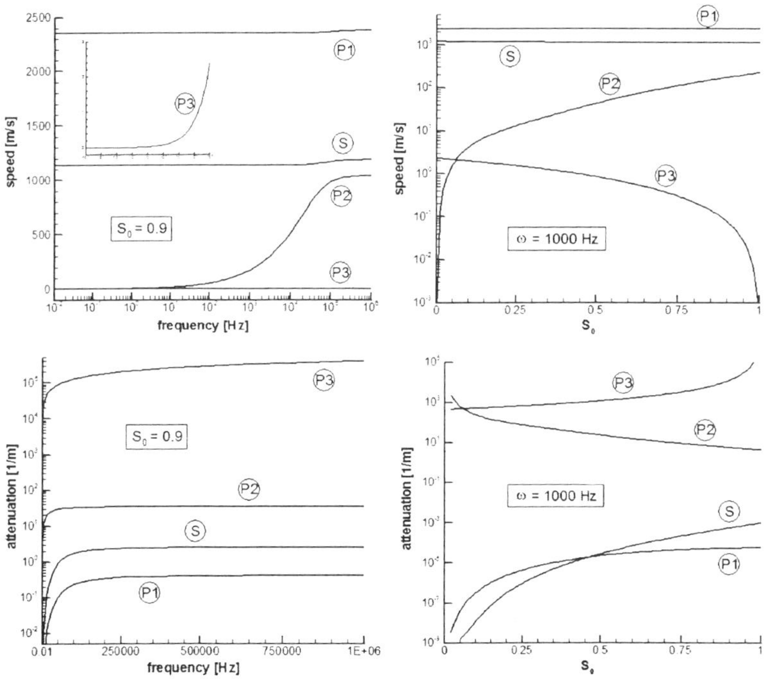

Fig. 1. Phase speeds (top row) and attenuations (bottom row) of the four waves appearing in partially saturated sandstones in dependence on the frequency (for $S_0 = 0.9$) and on the initial saturation (for $\omega = 1000$ Hz).

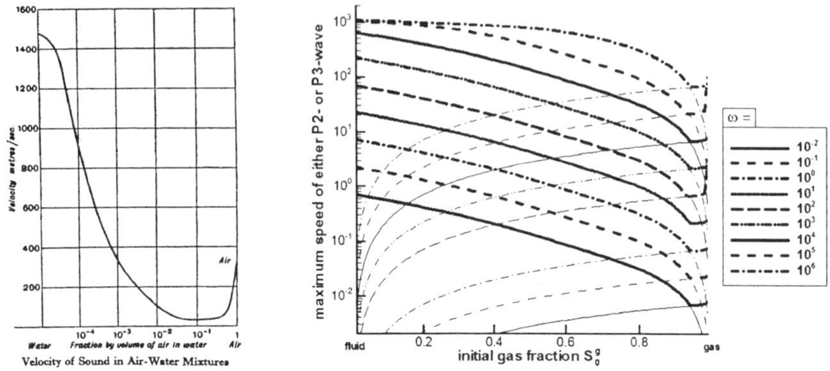

Fig. 2. Left: Sound velocity in air-water-mixtures [9], right: maximum speed either of the $P2$ or the $P3$ wave in dependence on the gas fraction S_0^g.

The left panel of Fig. 2 shows the sound velocity in a water-air-mixture. From the theory of suspensions it is well known that the existence of air bubbles in water results in a minimum in the sound velocity. The right hand side of Fig. 2 makes clear that this effect also arises in the partially saturated sandstone. In this case it appears in the combination of the speeds of the $P2$- and $P3$-waves.

Acknowledgement

I appreciate the financial support of the German Research Foundation (DFG).

References

1. B. Albers, *submitted to Géotechnique* (2007).
2. I. Tolstoy (ed.), *Acoustics, elasticity and thermodynamics of porous media: Twenty-one papers by M. A. Biot* (American Institute of Physics, 1992).
3. K. Wilmanski, Waves in porous and granular materials, in *Kinetic and continuum theories of granular and porous media*, eds. K. Hutter and K. Wilmanski, CISM Courses and Lectures, Vol. 400 (Springer, Wien, 1999) pp. 131–186.
4. B. Albers, *submitted to Proc. Royal Soc. A* (2007).
5. M. van Genuchten, *Soil Sci. Soc. Am. J.* **44**, 892 (1980).
6. M. Biot and D. Willis, *J. Appl. Mech.* **24**, 594 (1957).
7. K. Wilmanski, *Géotechnique* **54**, 593 (2004).
8. K. Wilmanski and B. Albers, Acoustic waves in porous solid-fluid mixtures, in *Dynamic response of granular and porous materials under large and catastrophic deformations*, eds. K. Hutter and N. Kirchner, Lecture Notes in Appl. and Computational Mechanics, Vol. 11 (Springer, Berlin, 2003) pp. 285–314.
9. A. B. Wood, *A Textbook of Sound* (G. Bell and Sons, London, 1957).

APPLICATION OF GENERALIZED OBSERVABLES TO STOCHASTIC QUANTUM MODELS IN PHASE SPACE

G. ALÌ, R. BENEDUCI, G. MASCALI

Dipartimento di Matematica
Università della Calabria and INFN-Gruppo c. Cosenza, 87036 Cosenza, Italy
g.ali@mat.unical.it, rbeneduci@unical.it,g.mascali@unical.it

Recently, quantum models of carrier transport in semiconductors have been proposed on the basis of the Wigner formulation of quantum mechanics. Such models are demanded by the semiconductor industry due to the fast transition from microelectronics to nanoelectronics. However, the Wigner function cannot be interpreted as a quantum analog of a distribution function, since it is not a probability density. In this work we show how this difficulty could be overcome by introducing a non standard quantum theory, based on the concept of Positive-Operator-Valued (POV) measures [3–5,7].

1. Notations and preliminaries

In what follows $\mathcal{B}(\mathbb{R}_q^N \times \mathbb{R}_p^N)$ will denote the Borel σ-algebra of the phase space $\Gamma := \mathbb{R}_q^N \times \mathbb{R}_p^N$. $\mathbf{0}$ and $\mathbf{1}$ respectively are the null and the identity operators. $\mathcal{B}_s(\mathcal{H})$ is the space of all bounded self-adjoint linear operators on a Hilbert space $\mathcal{H}$ with scalar product $\langle \cdot, \cdot \rangle$, $\mathcal{F}(\mathcal{H}) \subset \mathcal{B}_s(\mathcal{H})$ is the subspace of all positive, bounded, operators, and $\mathcal{L}^2(\mathbb{R}^N)$ is the space of Hilbert-Schmidt operators [2]. A key role will be played by the positive operator valued (POV) measures [2–5,7,8] which generalize the concept of quantum observables.

Definition 1.1. A POV measure is a map $F : \mathcal{B}(\mathbb{R}^{2N}) \to \mathcal{F}(\mathcal{H})$, such that $F\big(\bigcup_{n=1}^{\infty} \Delta_n\big) = \sum_{n=1}^{\infty} F(\Delta_n)$ for every countable family of disjoint sets $\{\Delta_n \in \mathcal{B}(\mathbb{R}^{2N})\}$, the series means the limit of the partial sums in the weak operator topology. A POV is said normalized if $F(\mathbb{R}^{2N}) = \mathbf{1}$, orthogonal if $F(\Delta_1)F(\Delta_2) = \mathbf{0}$ whenever $\Delta_1 \cap \Delta_2 = \emptyset$. In the latter case the values of the map are projectors and the measure is said spectral.

In what follows, POV measures will be normalized and the term "measurable", referred to functions, will mean Borel measurable functions.

2. Classical Representations of Quantum Mechanics

The Wigner formulation [10] of non-relativistic Quantum Mechanics is based on the following scheme:
1) there exists an affine, one-to-one map $\mathcal{W}: \mathcal{L}^2(\mathbb{R}^N) \to L^2(\Gamma)$ that assigns to each density operator ρ a square integrable function $f_\rho^W(q,p) := \mathrm{Tr}(\mathcal{P}_{q,p}\rho)$, said Wigner distribution function [8]. $\mathcal{P}_{q,p} = \frac{1}{\pi} V_{q,p} \mathcal{P} V_{q,p}^{-1}$, with $\mathcal{P}$ the parity operator and $V_{q,p} = e^{i(-q\cdot P + p\cdot Q)}$ the Weyl operator. $Q = (Q_1, \ldots, Q_N)$ and $P = (P_1, \ldots, P_N)$ are the position and the momentum operators.
2) The marginals of the Wigner distribution function are such that

$$\int_\Delta d^N q \int f_\rho^W(q,p)\, d^N p = \int_\Delta \langle q|\rho q\rangle d^N q = \mathrm{Tr}(\rho Q(\Delta)),$$
$$\int_\Delta d^N p \int f_\rho^W(q,p)\, d^N q = \int_\Delta \langle p|\rho p\rangle d^N p = \mathrm{Tr}(\rho P(\Delta)), \tag{1}$$

for every $\Delta \in \mathcal{B}(\mathbb{R}^N)$, where $Q(\Delta)$ and $P(\Delta)$ are the spectral measures corresponding to the position and momentum operators respectively.
3) there exists a map $\mathcal{W}^{-1}: L^2(\Gamma) \to \mathcal{L}^2(\mathbb{R}^N)$ (Weyl transform) that assigns to each $f(q,p) \in L^2(\Gamma)$ a self-adjoint operator $A_f = \int f(q,p)\mathcal{P}_{q,p}\, d^N q\, d^N p$ such that:

$$\mathrm{Tr}(A_f \rho) = \int f(q,p) f_\rho^W(q,p)\, d^N q\, d^N p, \quad \text{(duality)}.$$

By the Wigner theorem [1,9], $f_\rho^W(q,p)$ cannot be positive definite for each ρ. This drawback can be overcome by resorting to a stochastic approach to Quantum Statistical Mechanics, whose aim is to find a map $\rho \mapsto f_\rho^S(q,p)$, said classical representation (CR), such that $f_\rho^S(q,p) \geq 0$, at difference with the Wigner formulation. More precisely, a CR of quantum statistical mechanics is defined as follows [6]:

Definition 2.1. A CR is an injective, affine map

$$\mathcal{W}^S : \mathscr{T}_s(\mathcal{H}) \to L^1(\Gamma) \tag{2}$$

from the space of self-adjoint trace class operators to the space of integrable functions, which assigns to each density operator $\rho \in \mathscr{T}_s^1(\mathcal{H}) \subset \mathscr{T}_s(\mathcal{H})$ a density distribution function $\mathcal{W}^S(f) := f_\rho^S(q,p)$ such that
i)

$$f_\rho^S(q,p) \geq 0.$$

The positivity of $f_\rho^S(q,p)$ forbids that its marginals can satisfy relations like (1). In fact, the property of the marginals in item 2) is replaced by the following one

ii)

$$\int_\Delta d^N q \int f_\rho^S(q,p)\, d^N p = \mathrm{Tr}(\rho F^Q(\Delta)) \neq \int_\Delta \langle q|\rho q\rangle d^N q,$$
$$\int_\Delta d^N p \int f_\rho^S(q,p)\, d^N q = \mathrm{Tr}(\rho F^P(\Delta)) \neq \int_\Delta \langle p|\rho p\rangle d^N p,$$

where, $F^Q(\Delta)$ and $F^P(\Delta)$ define two POV measures, respectively named approximate position and momentum observables (see section 3).

Moreover, it is required that

iii) there exists a map $A \longmapsto f_A(q,p)$ assigning a real measurable function $f_A(q,p)$ to each self-adjoint operator A in $\mathcal{H}$ in such a way that:

$$Tr(A\rho) = \int f_A(q,p) f_\rho^S(q,p) d^N q d^N p.$$

2.1. *Classical Representations and POV Measures*

The aim of the present section is to show that there exists a one-to-one correspondence between classical representations of quantum mechanics and a particular class of POV measures, the informationally complete and absolutely continuous (ICAC) POV measures.

Definition 2.2. A POV measure F is said to be absolutely continuous if, for every $\rho \in \mathcal{T}_s^1(\mathcal{H})$, the measure $Tr(\rho F(\cdot))$ is absolutely continuous with respect to the Lebesgue measure. Moreover, it is said to be informationally complete if $Tr(\rho F(\Delta)) = 0$, for every $\Delta \in \mathcal{B}(\mathbb{R}^{2N})$, implies that $\rho = 0$.

Theorem 2.1. *There exists a one-to-one correspondence between ICAC POV measures* $F : \mathcal{B}(\mathbb{R}^{2N}) \to \mathcal{F}(H)$ *and classical representations* $\mathcal{W}^S : \mathcal{T}_s(\mathcal{H}) \mapsto L^1(\Gamma)$. *If* $\rho \in \mathcal{T}_s^1(\mathcal{H})$, *the function* $f_\rho^S(q,p) := \mathcal{W}^S(\rho)$ *is the probability distribution associated to the probability measure* $Tr(\rho F(\cdot))$ *and* $\int_\Delta f_\rho^S(q,p) d^N q\, d^N p = Tr(\rho F(\Delta))$.

In the following, $\mathcal{W}_F^S$ denote the CR associated to the POV measure F.

3. Stochastic Representations of Quantum Mechanics

In this section we give a meaningful example of a Classical Representation of Quantum Mechanics and analyze the problem of the marginals. Let us consider $(q,p) \mapsto U_{qp} := \prod_{j=1}^N e^{ip_jQ_j} \prod_{j=1}^N e^{iq_jP_j}$, which is an irreducible, strongly continuous projective unitary representation on $\mathcal{H} = L^2(\mathbb{R}^N)$ of the additive group $\mathbb{R}^{2N}$, see [2] .

Definition 3.1. A classical representation of quantum mechanics $\mathcal{W}_F^S$ is covariant with respect to the group $\mathbb{R}_q^N \times \mathbb{R}_p^N$ if

$$U_{qp}F(\Delta)U_{qp}^{-1} = F(\Delta + (q,p)), \quad \Delta \in \mathcal{B}(\mathbb{R}_q^N \times \mathbb{R}_p^N)$$

Theorem 3.1. *If F is a POV measure covariant with respect to the group $\mathbb{R}_q^N \times \mathbb{R}_p^N$, then F is absolutely continuous.*

Theorem 3.2. *Let $\gamma = |u\rangle\langle u| \in \mathcal{T}_s^1(\mathcal{H})$ and $\gamma_{qp} := U_{qp}\gamma U_{qp}^{-1}$. Then the map $(q,p) \to \gamma_{qp}$ is trace-norm continuous [9] and*

$$F(\Delta) := \frac{1}{(2\pi)^N}\int_\Delta \gamma_{qp} d^N q\, d^N p$$

defines a covariant POV measure $F : \mathcal{B}(\mathbb{R}^{2N}) \to \mathcal{F}(H)$, integration is intented in the Bochner sense [2].

Theorem 3.3. *The POV measure F of theorem 3.2 is informationally complete if and only if $\gamma_W := Tr(\gamma U_{qp}) \neq 0$, a. e.*

Example 3.1. If $u = u^\sigma(x) := (\pi\sigma^2)^{-\frac{N}{4}} e^{-\frac{x^2}{2\sigma^2}}$ then,

$$f_\rho^S(q,p) = (\pi\sigma^2)^{-\frac{N}{2}} \int e^{\left[-\frac{(q'-q)^2}{\sigma^2} - \sigma^2(p'-p)^2\right]} f_\rho^W(q',p')d^Nq'd^Np'$$

which coincides with the **Husimi transform.**

Let us analyze the marginals $F^P(\Delta) = F(\mathbb{R}^N \times \Delta)$ and $F^Q(\Delta) = F(\Delta \times \mathbb{R}^N)$, $\Delta \in \mathcal{B}(\mathbb{R})^N$ of $F : \mathcal{B}(\mathbb{R})^{2N} \to \mathcal{F}(H)$.

Theorem 3.4. *The marginals of the informationally complete observable F are:*

$$F^Q(\Delta) = \int \chi_\Delta(x) * |u(x)|^2 dE_x^Q, \quad F^P(\Delta) = \int \chi_\Delta(k) * |\widetilde{u}(k)|^2 dE_k^P,$$

where E_x^Q and E_k^P are the spectral resolutions of the standard position and momentum operators respectively, $*$ denotes the operation of convolution and $\widetilde{u}$ the Fourier transform of u.

The POV measures F^Q and F^P are the approximate position and momentum observables.

Definition 3.2. Two observables E and F are said to be informationally equivalent if and only if

$$Tr(\rho E(\Delta)) = 0, \quad \forall \Delta \in \mathcal{B}(\Gamma) \Leftrightarrow Tr(\rho F(\Delta)) = 0, \quad \forall \Delta \in \mathcal{B}(\Gamma)$$

The following theorem ensures that using the approximate position and momentum observables there is no loss of physical information.

Theorem 3.5. *The approximate position and momentum observables F^Q and F^P are informationally equivalent to the sharp position and momentum observables E^Q and E^P respectively.*

From (3.4) it is possible to obtain

$$\int_\Delta d^N q \int f^S_\rho(q,p)\, d^N p = \int_\Delta d^N q \int \chi_q(x)\langle x|\rho x\rangle\, d^N x,$$

$$\int_\Delta d^N p \int f^S_\rho(q,p)\, d^N q = \int_\Delta d^N p \int \chi'_p(k)\langle k|\rho k\rangle\, d^N k,$$

where $\chi_0(x-q) = \chi_q(x) = (2\pi\hbar)^N\langle x|\gamma_{q,p}|x\rangle$, $\chi'_0(k-p) = \chi'_p(k) = (2\pi\hbar)^N\langle k|\gamma_{q,p}|k\rangle$ are said confidence functions of the instrument. The confidence functions satisfy the Heisenberg inequality [4] $\mathrm{Var}(\chi_q)\mathrm{Var}(\chi'_p) \geq \frac{1}{4}$. The unavoidable presence of measure errors makes the concept of sharp points $(q,p) \in \Gamma$ untenable [2]. This suggests to interpret the confidence functions as giving the probability densities that the result q (respectively p) is obtained when the actual sharp value is x (respectively k). Therefore the concept of stochastic phase space $\Gamma^S := \{(q,\chi_q(x)) \times (p,\chi'_p(x)) \mid (q,p) \in \Gamma\}$ is introduced, where sharp points are replaced by stochastic points. As regards the interpretation of $f^S_\rho(q,p)$, certainly the integral $\int_\Delta f^S_\rho(q,p) d^N q\, d^N p$ cannot be interpreted as the probability that the position and the momentum of a particle are in the element $d^N q d^N p$ centered in (q,p), but as the probability that a measurement of position and momentum gives a result in the above element. In other words [3], $f^S_\rho(q,p)$ is a probability density on the stochastic phase space Γ^S.

3.1. *Classical representation on phase space and Wigner formulation*

Let us consider the classical representation $\mathcal{W}^S : \mathcal{T}_s(\mathcal{H}) \to L^1(\Gamma)$ induced by a POV measure F, generated by a spherically symmetric [2,9] density operator γ such that $\gamma_W := Tr(\gamma U_{qp}) \neq 0$, a.e. Now let us show the relationship between the CR and the Wigner representation of quantum mechanics and how the quantum observables are represented in the phase space. Let $f_\rho^S(q,p)$ be the probability density corresponding to the state ρ and $\widetilde{f}_\rho^W(q',p')$ the symplectic Fourier transform [2] of f_ρ^W.

Theorem 3.6. *There exists a one-to-one mapping $f_\rho^S(q,p) \mapsto f_\rho^W(q,p)$ of $\mathcal{W}^S(\mathscr{T}_s^1(\mathcal{H}))$ onto the space $W(T_s^1(\mathcal{H}))$ such that:*

$$\widetilde{f}_\rho^S(q',p') = \bar{\widetilde{\gamma}}_W(q',p')\widetilde{f}_\rho^W(q',p')$$

where the bar denotes complex conjugation.

Theorem 3.7. *To every symmetric operator A on $L^2(\mathbb{R}^N)$ with domain $D(A)$ containing the Schwartz space [10] $\mathscr{L}(\mathbb{R}^N)$, there corresponds a unique real generalized function $A_\gamma(q,p)$ such that*

$$Tr(\rho A) = \int A_\gamma(q,p) f_\rho^S(q,p) d^N q d^N p,$$

$$\widetilde{A}_\gamma(q',p') = [\widetilde{\gamma}_W(q',p')]^{-1}\widetilde{A}_W(q',p').$$

In the case of Example 3.1, $Q \mapsto q$ and $P \mapsto p$.

In a future paper we will treat the dynamics of the functions f_ρ^S and try to specialize it to the case of electron transport in semiconductors.

References

1. E. P. Wigner, *Quantum Mechanical Distribution Functions Revisited*, in Perspectives in Quantum Theory, W. Yourgrau and A. van der Merwe (eds.), 25, MIT Press, Cambridge, Mass (1971).
2. S.T. Ali, E. Prugovečki, *Int. J. Theor. Phys.*, **16**, 689 (1977).
3. E. Prugovečki, *Stochastic Quantum Mechanics and Quantum Space Time*, D. Reidel Publishing Corporation, Dordrecht, 2nd edition, (1986).
4. E.B.Davies, J.T. Lewis, *Commun. Math. Phys.*, **17**, 239 (1970).
5. A. S. Holevo, *Probabilistics and statistical aspects of quantum theory*, North Holland, Amsterdam (1982).
6. W. Guz, *Int. J. Theo. Phys.* **23**, 157 (1984).
7. G. Ludwig, *Foundations of quantum mechanics I*, Springer-Verlag, Neww York-Heidelberg-Berlin (1983).
8. P. Busch, M. Grabowski, P. Lahti, *Operational quantum physics*, LNP m, 31, Springer-Verlag, Berlin (1995).
9. W. Stulpe, *Classical Representations of Quantum Mechanics Related to Statistically Complete Observables*, arXiv:quant-ph/0610122 v1 16 Oct 2006.
10. J. C. T. Pool, *J. Math. Phys.*, **7**, 66 (1966).

A HYDRODYNAMIC MODEL FOR SEMICONDUCTORS WITH FIELD-DEPENDENT MOBILITY

G. ALÌ AND M. CARINI

Department of Mathematics
University of Calabria, Cosenza and INFN-Gruppo c. Cosenza
E-mail: giuseppe.ali@unical.it;carini@unical.it

We study the relaxation limit for a one-dimensional, isentropic, hydrodynamical model for semiconductors, with field-dependent mobility. We find a range where no solution can be built which satisfies usual shock and boundary layer conditions.

1. Introduction

We consider a scaled 1-d, isentropic, hydrodynamical model for a unipolar semiconductor:

$$\begin{cases} n_t + j_x = 0, \\ j_t + \left(\dfrac{j^2}{n} + p(n) \right)_x = nE - \dfrac{j}{\varepsilon\tau(n,j,E)}, \qquad (x,t) \in (0,\nu) \times \mathbb{R}^+ \\ E_x = n - N(x). \end{cases} \tag{1}$$

where $n(x,t)$ is the electron number density, $j(x,t)$ is the electron flux density, $E(x,t)$ is the electric field, $p(n)$ is the pressure and $\tau(n,j,E)$ is the momentum relaxation time, which we assume to depend also on the electric field, and small parameter ε is a typical value for the relaxation time with respect to the reference time. The device domain is the x-interval $(0,\nu)$, with $\nu = L/\lambda_{\text{db}}$ (L device length, λ_{db} Debye length), and $N = N(x)$ is the number density of the background ions (doping profile), with $N(x) > 0$.

This system is related to the drift-diffusion equations by the diffusive scaling $t \to t/\varepsilon$, $j \to \varepsilon j$, which leads to the system:

$$\begin{cases} n_t + j_x = 0, \\ \varepsilon^2 j_t + \left(\varepsilon^2 \dfrac{j^2}{n} + p(n) \right)_x = nE - \dfrac{j}{\tau(n,\varepsilon j,E)}, \\ E_x = n - N(x). \end{cases} \tag{2}$$

Formally, as ϵ tends to zero, the solutions of (2) tend to the solutions of the drift diffusion equations

$$\begin{cases} \mathcal{N}_t + \mathcal{J}_x = 0, \quad \mathcal{J} = -\mu(\mathcal{N},\mathcal{E})\left(p(\mathcal{N})_x - \mathcal{N}\mathcal{E}\right), \\ \mathcal{E}_x = \mathcal{N} - N(x), \end{cases} \tag{3}$$

where $\mu(\mathcal{N},\mathcal{E}) = \tau(\mathcal{N},0,\mathcal{E})$.

For τ = constant, this result was proved by Marcati and Natalini [5] for an infinitely extended domain, and we expect it to hold also for a finite device domain.

The simple above argument shows also the link between the relaxation time τ and the mobility μ. Actually, the existence of a wide literature on field dependent mobility models suggests to explore the possibility of a similar dependence also for the relaxation time.

We investigate the singular limit just discussed, in the stationary case.

2. The model

We assume N = constant and consider the scaled one-dimensional system

$$\left(\frac{\varepsilon^2 j^2}{n} + p(n)\right)_x = nE - \frac{j}{\mu(n,E)}, \qquad E_x = n - 1, \tag{4}$$

with $x \in (0,\nu)$, and boundary data

$$n(0) = n_1, \quad n(\nu) = n_2, \qquad \int_0^\nu E(x)\,dx = \left[V_{\rm bi}\right]_0^\nu + V, \tag{5}$$

where $V_{\rm bi}(n) = h(n)$, $h'(n) = \frac{1}{n}p'(n)$ and the second equation of (5) represents the applied voltage. It is possible to use the current-voltage relation [2] to express V as a function of j. Thus, we can assign the current j instead of the voltage V. Moreover, we can choose $j > 0$, otherwise we make the transformation: $x \to -x$, $E \to -E$, $j \to -j$.

We assume that the pressure $p(n) \in C^2([0,\infty[)$ satisfies the properties:

$$p(0) = 0, \quad q(n) := n^2 p'(n) > 0, \quad q'(n) > 0, \quad \lim_{n\to\infty} \frac{n^2 q'(n)}{q(n)^2} < \infty. \tag{6}$$

For the mobility we assume: $\mu = \mu(E) \in C^1(\mathbb{R})$ satisfies the properties:

$$\begin{aligned} &\mu(0) > 0, \quad \mu(-E) = \mu(E), \quad E\mu'(E) < 0, \quad Ev'(E) > 0 \quad (E \neq 0), \\ &\lim_{E\to+\infty} v(E) = v_S > 0, \qquad \lim_{E\to+\infty} |E|^\beta E v'(E) = a_S > 0 \quad (\beta > 0), \end{aligned} \tag{7}$$

where $v(E) := E\mu(E)$ is the drift velocity and v_S is the saturation velocity.

The assumption (6) for the pressure is satisfied by the polytropic state equation, $p(n) = p_0 n^\gamma$, $\gamma \geq 1$, with $q(n) = \gamma p_0 n^{\gamma+1}$. The assumption

(7) for the mobility models is satisfied by the Caughey-Thomas model [1], $\mu(E) = \frac{\mu_0}{[1+(\mu_0|E|/v_S)^\kappa]^{1/\kappa}}$, where $\kappa = 1$ for holes and $\kappa = 2$ for elcctrons, with $\beta = \kappa$, $a_S = \frac{v_S^\kappa}{\mu_0^\kappa} v_S$. For definiteness, we will consider the above models.

For smooth solutions, the system (4) can be written in the form

$$\left(p'(n) - \frac{\varepsilon^2 j^2}{n^2}\right) n_x = nE - \frac{j}{\mu(E)}, \qquad E_x = n - 1. \tag{8}$$

The first equation in (8) becomes singular when

$$q(n) \equiv n^2 p'(n) = \varepsilon^2 j^2 \quad \Leftrightarrow \quad n = n_s(\varepsilon, j) \equiv \left(\varepsilon^2 j^2/\gamma p_0\right)^{\frac{1}{\gamma+1}}.$$

The singularity line $\mathcal{C}_s := \{n = n_s\}$ is called sonic line. It separates the supersonic flows $(0 < n < n_s)$ from the subsonic flows $(n > n_s)$. A smooth solution can cross the sonic line only if

$$v(E) \equiv E\mu(E) = j/n_s(\varepsilon, j) \Leftrightarrow E = E_s(\varepsilon, j)$$

which is defined only if $n_s(\varepsilon, j) > j/v_S$. Thus the crossing can occur only if $n_s(\varepsilon, j) > j/v_S$. A non smooth solution can have a jump at $x = x_s \in (0, \nu)$ from a left state density n_L to a right state density n_R only if the jump satisfies the shock condition $\frac{\varepsilon^2 j^2}{n_L} + p(n_L) = \frac{\varepsilon^2 j^2}{n_R} + p(n_R)$. The jump is admissible if it satisfies the Lax condition (with $s = 0$, $j > 0$)

$$\frac{\varepsilon j}{n_L} - \sqrt{p'(n_L)} > 0 > \frac{\varepsilon j}{n_R} - \sqrt{p'(n_R)} \quad \Leftrightarrow \quad n_L < n_s(\varepsilon, j) < n_R. \tag{9}$$

Therefore, a jump can occur only from a supersonic to a subsonic state.

A non smooth solution can have a jump at the boundary (boundary layer) only in the following two cases [4]:

$$n_s(\varepsilon, j) < \lim_{x \to 0^+} n(x) < n_1, \qquad \lim_{x \to \nu^-} n(x) < n_2^c < n_s(\varepsilon, j) < n_2, \tag{10}$$

with $\frac{\varepsilon^2 j^2}{n_2^c} + p(n_2^c) = \frac{\varepsilon^2 j^2}{n_2} + p(n_2)$.

3. The method

We consider the (4) as a dynamical system for $(n(x), E(x))$.

A solution to (4)-(5) is represented by a path which: (a) starts from a state (n_1, E_1), at $x = x_1$, (b) and reaches a state (n_2, E_2), at $x = x_2$, (c) with $x_2 - x_1 = \nu$. To determine this (these) path(s), we need to:

i. determine all trajectories which cross the lines $\{n = n_1\}$, $\{n = n_2\}$;
ii. if a trajectory enters the supersonic region, consider possible jumps;
iii. evaluate the length of travel between the two lines along each trajectory (possibly with jumps).

To evaluate the length of travel, we consider the identities

$$\frac{dE}{dn} = \frac{(n-1)\,(q(n)-q(n_s))}{n^2\,(nE - j/\mu(E))}, \qquad E(n_1) = E_1, \qquad \text{if } n \neq j/v(E),$$

$$\frac{dn}{dE} = \frac{n^2\,(nE - j/\mu(E))}{(n-1)\,(q(n)-q(n_s))}, \qquad n(E_1) = n_1, \qquad \text{if } n \neq 1, n_s,$$

which can be recovered from (4). Then we have

$$x_2 - x_1 = \int_{n_1}^{n_2} \frac{p'(n) - \varepsilon^2 j^2/n^2}{nE(n) - j/\mu(E(n))}\,dn, \qquad \text{if } n \neq j/v(E),$$

$$x_2 - x_1 = \int_{E_1}^{E_2} \frac{1}{n(E)-1}\,dE, \qquad \text{if } n \neq 1, n_s.$$

In case of several intersections with $\{n = j/v(E)\}$, $\{n = 1\}$, $\{n = n_s\}$, we can combine the previous formulas. Both formulas depend on a single parameter (for example, E_1) and the monotonicity properties imply existence (or non existence) of E_1 such that $x_2 - x_1 = \nu$.

This method has been used by Ascher et al. [3] to study steady-state transonic solutions of the isentropic hydrodynamic model for semiconductors, and later extended by Rosini [6,7] to the non-isentropic case.

As an example, we apply this method to an n^+-n-n^+ MOSFET channel, corresponding to $n_1 = n_2 \equiv n^+ > 1$. To study the phase portrait of (8), we regularize the system:

$$\begin{cases} n_X = n^2 \left(nE - \dfrac{j}{\mu(E)} \right), \\ E_X = \left(q(n) - \varepsilon^2 j^2\right)(n-1), \end{cases} \qquad dX = \frac{dx}{n^2 p'(n) - \varepsilon^2 j^2}. \tag{11}$$

The transformation $x \to X$ changes the orientation of the trajectories in the supersonic region $0 < n < n_s$. The system (11) has (at most) two critical

$0 < j/v_S < 1$	P_d	P_s	P_S^∞
$1 < n_s$	RK/RF	S	S
$j/v_S < n_s < 1$	S	RK/RF	S
$0 < n_s < j/v_S$	S	–	RK

$1 < j/v_S$	P_d	P_s	P_S^∞
$j/v_S < n_s$	–	S	RK
$0 < n_s < j/v_S$	–	–	S

P_-^∞	P_+^∞	P_{0-}^∞	P_{0+}^∞
RK	AK	S	AK

Fig. 1. Character of the critical points at variance with the parameters n_s, j. S=saddle, K=knot, F=focus, A=attractive, R=repulsive,

points,

$$P_d = (1, E_d) \quad (\text{if } 1 > j/v_S), \qquad P_s = (n_s, E_s) \quad (\text{if } n_s > j/v_S),$$

and (at most) five critical points at infinity,

$$P_{0\pm}^{\infty} \simeq \{(n,E) \to (0,\pm\infty)\}, \qquad P_{S}^{\infty} \simeq \{(n,E) \to (j/v_S,\infty)\},$$
$$P_{\pm}^{\infty} \simeq \{(n,E/p(n)^{1/2}) \to (\infty,\pm\sqrt{2})\}.$$

The character of the critical points is shown in Fig. 1.

Using the above results, we can reconstruct qualitatively the phase portrait of system (11). The phase portrait of the original system (4) can be recovered by simply reverting the orientation of the trajectories in the region $0 < n < n_S$.

4. Discussion

We can distinguish two cases, at variance with the parameters j and n_s. **Case I:** $j/v_S < 1$. In the first two figures in Fig. 2, P_s is a saddle and can

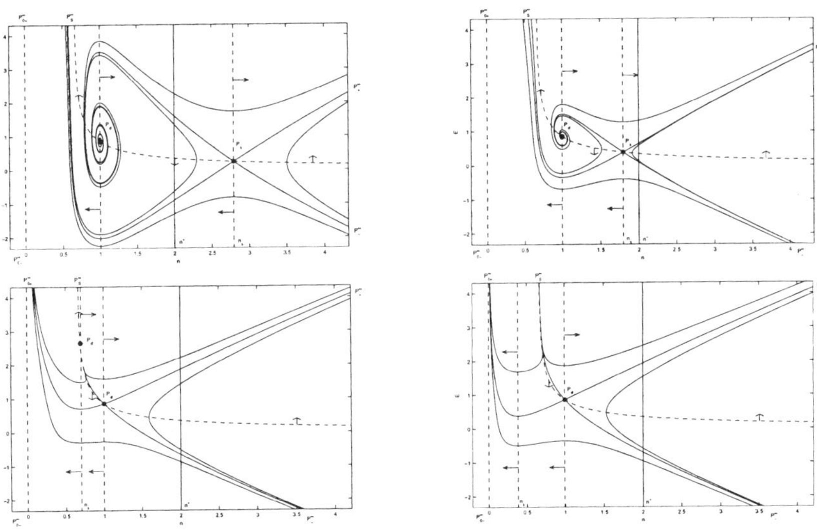

Fig. 2. Case I: $j/v_S < 1$. We use Caughey-Thomas mobility model ($\mu_0 = 1$, $v_S = 1$, $\kappa = 2$), and $j = 0.65$, $\epsilon = 6.3, 3.5, 0.985, 0.45$.

be crossed along two characteristics, while P_d is an attractive point. The solution is represented by the trajectory intersected by $\{n = n^+\}$ (entirely subsonic in the first case, supersonic in the second one), if its length of travel is ν. Otherwise, the solution becomes transonic and exhibits a shock [3]. As ε decreases, first, the character of P_s and P_d is exchanged, and eventually P_s disappears. In these cases the solution is entirely subsonic, since the intersect formed by the separatrices through P_d and the line $\{n = n^+\}$ has an infinite length of travel.

Case II: $1 < j/v_S$. In this case, for boundary values $n^+ > j/v_S > 1$

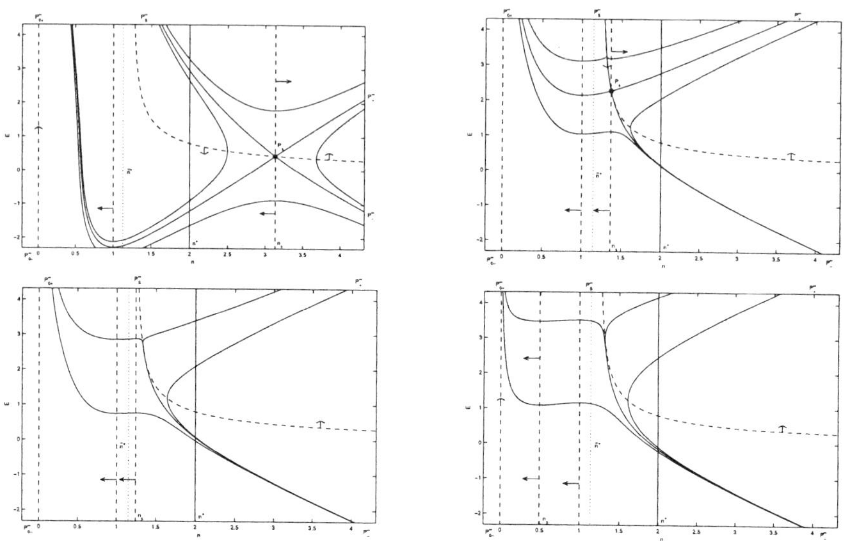

Fig. 3. Case II: $j/v_S > 1$. We use Caughey-Thomas mobility model ($\mu_0 = 1$, $v_S = 1$, $\kappa = 2$), and $j = 1.25$, $\epsilon = 5.3, 1.75, 1.55, 0.45$.

the discussion is similar to the previous case, the only difference being the disappearance of P_d, replaced by P_S^∞. However, for boundary values $j/v_S > \bar{n}^+ > 1$, it is not possible to built a solution which satisfies the admissibility conditions for shocks and/or boundary layers. Thus in this case the existence of solutions needs further study.

References

1. D. Caughey and R. Thomas, *Proc. IEEE* **55**, 2192 (1967).
2. P. Degond and P. A. Markowich, *Appl. Math. Letters* **3** (3), 25 (1990).
3. U. M. Ascher, P. A. Markowich, P. Pietra and C. Schmeiser, *M3AS* **1** (3), 347 (1991).
4. I. M. Gamba, *Commun. in Partial Differential Eqs.* **17** (3 & 4), 553 (1992).
5. P. Marcati and R. Natalini, *Arch. Rational Mech. Anal.* **129**, 129 (1995).
6. M. D. Rosini, *J. Differential Equations* **199** (2), 326 (2004).
7. M. D. Rosini, *Quarterly of Applied Mathematics* **63** (2), 251 (2005).

STOCK MARKETS AND THE OPERATOR NUMBER REPRESENTATION*

F. BAGARELLO

Dipartimento di Metodi e Modelli Matematici
Facoltà di Ingegneria,
Università di Palermo, I - 90128 Palermo, Italy
E-mail: bagarell@unipa.it

We review here some results concerning a recent approach to the description of stock markets which is based on the so-called *operator number representation* of quantum mechanics.

1. Introduction

In some recent papers, [1–3], we have shown how second quantization and Heisemberg dynamics, which are typically related to quantum mechanics, can be used in the description of stock markets as well. This point of view takes origin from few simple remarks: the total number of shares and the total amount of money in a closed market does not change in time, and the price of a single share does not change continuously, but for integer multiples of a certain minimal quantity, the *monetary unit.* The *number representation* of quantum mechanics provides a natural framework in which these features can be taken into account. Moreover it also provides natural tools to discuss the existence of conserved quantities and to find the differential equations of motion which drive the portfolio of each single trader, as we will see. In this note we essentially review few results concerning the most recent model we have proposed so far, which belongs to the same general area of research discussed in [4,5] and partially in [6].

*This work has been financially supported in part by M.U.R.S.T., within the project *Problemi Matematici Non Lineari di Propagazione e Stabilità nei Modelli del Continuo,* coordinated by Prof. T. Ruggeri.

2. The model

Our attention is focused here on a single trader, τ, which interacts with an ensemble of other traders. In other words we divide our toy stock market, which is defined in terms of the number of a **single type** of shares, the cash, the price of the shares and the supply of the market [1], in two main *ingredients*: we call *system*, S, all the dynamical quantities which refer to trader τ: its *shares number operators*, a, $a^\dagger$ and $\hat{n} = a^\dagger\, a$, the *cash operators* of τ, c, $c^\dagger$ and $\hat{k} = c^\dagger\, c$ as well as the *price operators* of the shares, p, $p^\dagger$ and $\hat{P} = p^\dagger\, p$. On the other hand, we associate to the *reservoir*, R, all the other quantities, that is first of all, the *shares number operators*, A_k, $A_k^\dagger$ and $\hat{N}_k = A_k^\dagger\, A_k$ and the *cash operators*, C_k, $C_k^\dagger$ and $\hat{K}_k = C_k^\dagger\, C_k$ of the other traders. Here $k \in \Lambda$ and Λ is a subset of $\mathbb{N}$ which labels the traders of the market (other than τ). Moreover we associate to the reservoir also the supply of the market, which is described by the operators o_k, $o_k^\dagger$ and $\hat{O}_k = o_k^\dagger\, o_k$, $k \in \Lambda$. The stock market is given by the union of S and R, and the hamiltonian is taken to be, see [2],

$$\begin{cases} H = H_0 + \lambda\, H_I, \text{ where} \\ H_0 = \omega_a\, \hat{n} + \omega_c\, \hat{k} + \omega_p \hat{P} + \sum_{k\in\Lambda} \left(\Omega_A(k)\, \hat{N}_k + \Omega_C(k)\, \hat{K}_k + \Omega_O(k)\, \hat{O}_k\right) \\ H_I = \left(z^\dagger\, Z(f) + z\, Z^\dagger(\overline{f})\right) + \left(p^\dagger\, o(g) + p\, o^\dagger(\overline{g})\right) \end{cases} \tag{1}$$

Here ω_a, ω_c and ω_p are positive real numbers and $\Omega_A(k)$, $\Omega_C(k)$ and $\Omega_O(k)$ are real valued non negative functions, whose interpretation was first discussed in [1]. We have also introduced the following *smeared fields* of the reservoir:

$$\begin{cases} Z(f) = \sum_{k\in\Lambda} Z_k\, f(k) = \sum_{k\in\Lambda} A_k\, {C_k^\dagger}^{\hat{P}}\, f(k), \\ Z^\dagger(\overline{f}) = \sum_{k\in\Lambda} Z_k^\dagger\, \overline{f(k)} = \sum_{k\in\Lambda} A_k^\dagger\, {C_k}^{\hat{P}}\, \overline{f(k)} \\ o(g) = \sum_{k\in\Lambda} o_k\, g(k) \\ o^\dagger(\overline{g}) = \sum_{k\in\Lambda} o_k^\dagger\, \overline{g(k)}, \end{cases} \tag{2}$$

as well as the operators $z = a\, {c^\dagger}^{\hat{P}}$, $Z_k = A_k\, {C_k^\dagger}^{\hat{P}}$ and their conjugates. They are relevant because, for instance, A_k and C_k appear always in this combination both in H_I and in all the computations we will perform in the following. This is natural because of the *economical* meaning of, e.g., z: the action of z on a fixed vector number destroys a share in the portfolio of τ and, at the same time, creates as many monetary units as $\hat{P}$ prescribes! Of course, in H_I such an operator is associated to $Z^\dagger(\overline{f})$ which acts exactly in the opposite way on the traders of the reservoir: one share is created in the cumulative portfolio of R while $\hat{P}$ *quanta* of money are destroyed, since they are used to pay for the share. The following non trivial commutation rules are assumed:

$$[c, c^\dagger] = [p, p^\dagger] = [a, a^\dagger] = 1\!\!1, \qquad [o_i, o_j^\dagger] = [A_i, A_j^\dagger] = [C_i, C_j^\dagger] = \delta_{i,j} 1\!\!1 \quad (3)$$

which implies

$$[\hat{K}_k, C_q^{\hat{P}}] = -\hat{P}\, C_q^{\hat{P}}\, \delta_{k,q}, \quad [\hat{K}_k, {C_q^\dagger}^{\hat{P}}] = \hat{P}\, {C_q^\dagger}^{\hat{P}}\, \delta_{k,q} \qquad (4)$$

Finally, the functions $f(k)$ and $g(k)$ in (1) and (2) are sufficiently regular to allows for the sums in (2) to be well defined (this becomes relevant if we are interested in considering *very large* markets).

Remark:– Of course, since τ can be chosen arbitrarily, changing τ we will be able, in principle, to describe the behavior of each trader of the market.

The interpretation suggested above concerning z and $Z(f)$ are also based on the following fact: let

$$\hat{N} := \hat{n} + \sum_{k\in\Lambda} \hat{N}_k, \quad \hat{K} := \hat{k} + \sum_{k\in\Lambda} \hat{K}_k, \quad \hat{\Gamma} := \hat{P} + \sum_{k\in\Lambda} \hat{O}_k \qquad (5)$$

Of course $\hat{N}$ is associated to the total number of shares in our closed market and, therefore, is called the *total number operator*. $\hat{K}$ is the total amount of money present in the market and is called the *total cash operator*. $\hat{\Gamma}$ has not a direct interpretation so far, since is just the sum of the price and the *total supply* operators, $\hat{O} = \sum_{k\in\Lambda} \hat{O}_k$. We have:

Prop 2.1. The operators $\hat{N}$, $\hat{K}$ and $\hat{\Gamma}$ are constants of motion.

The proof of this proposition is a simple exercise based on the commutation rules above. Indeed, it is not hard to check that H commutes with $\hat{N}$, $\hat{K}$ and with $\hat{\Gamma}$. This proves that our main motivation for introducing the hamiltonian in (1) is correct: with this choice we are constructing a closed market in which the total amount of money and the total number of shares are preserved and in which, if the total supply increases, then the price of the share must decrease in order for $\hat{\Gamma}$ to stay constant. Indeed, this is the simple mechanism which is assumed in the toy model discussed in this paper. Of course, it would be interesting to relate the changes of $\hat{O}$ to other (maybe external) conditions, but this problem will not be considered here.

The next step of our analysis should be to recover the equations of motion for the portfolio of the trader τ, defined as

$$\hat{\Pi}(t) = \hat{P}(t)\, \hat{n}(t) + \hat{k}(t). \qquad (6)$$

It is not surprising that this cannot be done exactly so that some perturbative technique is needed. This is what we are going to do next.

We first remind that given a generic operator X its time evolution, in the Heisemberg representation, is (formally) given by $X(t) = e^{iHt} X e^{-iHt}$. $X(t)$ satisfies the following Heisemberg equation of motion: $\dot{X}(t) = ie^{iHt}[H, X]e^{-iHt} = i[H, X(t)]$. In the attempt of deducing the analytic expression for $\hat{\Pi}(t)$, we get the following *semiclassical* equations [2]

$$\left\{\begin{array}{l} \frac{d\hat{n}(t)}{dt} = i\lambda\left(-z^\dagger(t)\, Z(f,t) + z(t)\, Z^\dagger(\overline{f},t)\right), \\ \frac{d\hat{k}(t)}{dt} = i\lambda\, P_c(t)\, \left(z^\dagger(t)\, Z(f,t) - z(t)\, Z^\dagger(\overline{f},t)\right), \\ \frac{dz(t)}{dt} = i\left(P_c(t)\omega_c - \omega_a\right)\, z(t) + i\lambda[z^\dagger(t), z(t)]\, Z(f,t), \\ \frac{dZ(f,t)}{dt} = i\, Z\left((P_c(t)\Omega_C - \Omega_A)f, t\right) + i\lambda\, z(t)\, [Z^\dagger(\overline{f},t), Z(f,t)], \end{array}\right. \tag{7}$$

where

$$P_c(t) = \omega(\hat{P}(t)) = \frac{1}{2}\left[(M+O) + (M-O)\cos(2\lambda t)\right] \tag{8}$$

As widely discussed in [1,2] taking mean values of the *observable* operators on a number state ω allows us to get a *classical* time evolution and to fix the initial conditions (initial number of shares, the initial cash and so on).

In order to simplify further the analysis of this system, it is also convenient to assume that both $\Omega_C(k)$ and $\Omega_A(k)$ are constant for $k \in \Lambda$. Indeed, under this assumption, the last two equation in (7) forms by themselves a closed system of differential equations in the non commuting variables $z(t)$ and $Z(f,t)$:

$$\left\{\begin{array}{l} \frac{dz(t)}{dt} = i\left(P_c(t)\omega_c - \omega_a\right)\, z(t) + i\lambda\, Z(f,t)\, [z^\dagger(t), z(t)], \\ \frac{dZ(f,t)}{dt} = i\left(P_c(t)\Omega_C - \Omega_A\right) Z(f,t) + i\lambda\, z(t)\, [Z^\dagger(\overline{f},t), Z(f,t)]. \end{array}\right. \tag{9}$$

Getting the exact solution of the system (7), with (9) as the two last equations, is an hard job. We refer to [2] for the details of our perturbation scheme which, roughly speaking, consists in constructing iteratively a solution of system (9), stopping at the first non trivial order. Then we first define $z_0(t)$ and $Z_0(f,t)$ as

$$z_0(t) = z\, e^{i\chi(t)}, \qquad Z_0(f,t) = Z(f)\, e^{i\tilde{\chi}(t)}, \tag{10}$$

where $\chi(t) = \alpha t + \beta\sin(2\lambda t)$, $\tilde{\chi}(t) = \tilde{\alpha} t + \tilde{\beta}\sin(2\lambda t)$ and

$$\left\{\begin{array}{ll} \alpha = \frac{1}{2}((M+O)\omega_c - 2\omega_a), & \beta = \frac{\omega_c}{4\lambda}(M-O) \\ \tilde{\alpha} = \frac{1}{2}((M+O)\Omega_C - 2\Omega_A), & \tilde{\beta} = \frac{\Omega_C}{4\lambda}(M-O). \end{array}\right. \tag{11}$$

$Z_0(t)$ and $z_0(f,t)$ correspond to the solution of (9) for $\lambda = 0$. Then we look for the next approximation of (9) by considering the following system:

$$\begin{cases} \frac{dz_1(t)}{dt} = i\,(P_c(t)\omega_c - \omega_a)\; z_0(t) + i\lambda\, Z_0(f,t)\,[z_0^\dagger(t), z_0(t)], \\ \frac{dZ_1(f,t)}{dt} = i\,(P_c(t)\Omega_C - \Omega_A)\, Z_0(f) + i\lambda\, z_0(t)\,[Z_0^\dagger(\overline{f},t), Z_0(f,t)]. \end{cases}$$

which can be still written as

$$\begin{cases} \frac{dz_1(t)}{dt} = i\,(P_c(t)\omega_c - \omega_a)\; z_0(t) + i\lambda\, Z_0(f,t)\,[z_0^\dagger, z_0], \\ \frac{dZ_1(f,t)}{dt} = i\,(P_c(t)\Omega_C - \Omega_A)\, Z_0(f) + i\lambda\, z_0(t)\,[Z_0^\dagger(\overline{f}), Z_0(f)]. \end{cases} \tag{12}$$

These equations can be solved and the solution can be written as

$$\begin{cases} z_1(t) = z\,\eta_1(t) + Z(f)\,[z^\dagger, z]\,\eta_2(t), \\ Z_1(f,t) = Z(f)\,\tilde{\eta}_1(t) + z\,[Z(\overline{f})^\dagger, Z(f)]\,\tilde{\eta}_2(t), \end{cases} \tag{13}$$

where we have introduced the following functions

$$\begin{cases} \eta_1(t) = 1 + i\int_0^t (P_c(t')\omega_c - \omega_a)\, e^{i\chi(t')}\, dt', \quad \eta_2(t) = i\lambda\int_0^t e^{i\tilde{\chi}(t')}\, dt' \\ \tilde{\eta}_1(t) = 1 + i\int_0^t (P_c(t')\Omega_C - \Omega_A)\, e^{i\tilde{\chi}(t')}\, dt', \quad \tilde{\eta}_2(t) = i\lambda\int_0^t e^{i\chi(t')}\, dt' \end{cases} \tag{14}$$

It is not a big surprise that this approximated solution does not share with $z(t)$ and $Z(f,t)$ all their properties. In particular, while for instance $[z(t), Z(f,t)] = 0$ for all t, $z_1(t)$ and $Z_1(f,t)$ do not commute.

It is now easy to deduce from the first two equations in (7) that

$$\begin{cases} \dot{n}(t) = \frac{dn(t)}{dt} = -2\lambda\Im\left\{\omega\left(z(t)\, Z^\dagger(\overline{f},t)\right)\right\}, \\ \dot{k}(t) = \frac{dk(t)}{dt} = 2\lambda\, P_c(t)\Im\left\{\omega\left(z(t)\, Z^\dagger(\overline{f},t)\right)\right\}, \end{cases} \tag{15}$$

where $n(t) = \omega(\hat{n}(t))$ and $k(t) = \omega(\hat{k}(t))$. Equation (15) in particular implies the identity $P_c(t)\dot{n}(t) + \dot{k}(t) = 0$ for all t, which in turns implies that $\dot{\Pi}(t) = \dot{P}_c(t)\, n(t)$. It should be remarked that, because of this relation, since $M = O$ implies $P_c(t) = P_c(0) = M$, see (8), then when $M = O$ the dynamics of the portfolio of τ is trivial, $\Pi(t) = \Pi(0)$, even if both $n(t)$ and $k(t)$ may change in time.

If we now put (13) in (15) and we introduce $\omega(1) := \omega\left(zz^\dagger\,[Z^\dagger(\overline{f}), Z(f)]\right\}$, $\omega(2) := \omega\left(Z(f)Z^\dagger(\overline{f})\,[z^\dagger, z]\right\}$, and the function $r(t) = \omega(1)\,\eta_1(t)\,\overline{\tilde{\eta}_2(t)} + \omega(2)\,\eta_2(t)\,\overline{\tilde{\eta}_1(t)}$, then we get

$$\begin{cases} n(t) = n - 2\lambda\Im\left\{\int_0^t r(t')\, dt'\right\}, \\ k(t) = k + 2\lambda\Im\left\{\int_0^t P_c(t')\, r(t')\, dt'\right\}. \end{cases} \tag{16}$$

Needless to say, some estimates on the validity of our approximation would be interesting. However this will not considered here. The time dependence of the portfolio can now be written as

$$\Pi(t) = \Pi(0) + \delta\Pi(t), \tag{17}$$

with

$$\delta\Pi(t) = n(O-M)\sin^2(\lambda t) + 2\lambda\,\Im\left\{\int_0^t P_c(t')\,r(t')\,dt'\right\} + \\ -2\lambda\,\Im\left\{\int_0^t r(t')\,dt'\right\}\left(M+(O-M)\sin^2(\lambda t)\right), \tag{18}$$

which gives the variation of the portfolio of τ in time. We observe that, as it is expected, $\delta\Pi(t) = 0$ if $\lambda = 0$.

We avoid here a detailed analysis of this solution, which can be found in [2], together with many numerical results. We just want to stress that this analysis suggested to interpretate the parameters appearing in (1) as a sort of *information* reaching the traders, in analogy with the interpretation discussed in [1]. This is because one expects that in a real market a crucial role in the behavior of the different traders is linked to *what the trader knows*! This is clearly a very delicate subject for which a much deeper analysis is still needed.

Acknowledgments

It is a pleasure to thank Prof. Manganaro for his hard job in Scicli.

References

1. F. Bagarello, *An operatorial approach to stock markets*, J. Phys. A, **39**, 6823-6840 (2006)
2. F. Bagarello, *An operatorial approach to stock markets*, Physica A, in press
3. F. Bagarello, *The Heisenberg picture in the analysis of stock markets and in other sociological contexts*, Proceedings del Workshop *How can Mathematics contribute to social sciences*, Bologna 2006, Italia, in *Quality and Quantity*, 10.1007/s11135-007-9076-4.
4. M. Schaden, *A Quantum Approach to Stock Price Fluctuations*, Physica A **316**, 511, (2002)
5. B.E. Baaquie, *Quantum Finance*, Cambridge University Press (2004)
6. R.M. Mantegna, E. Stanley, *Introduction to Econophysics*, Cambridge University Press (1999)

AN APPLICATION OF EXTENDED THERMODYNAMICS AND FLUCTUATION PRINCIPLE TO STATIONARY HEAT CONDUCTION IN RADIAL SYMMETRY

ELVIRA BARBERA

Dipartimento di Scienze per l'Ingegneria e per l'Architettura,
University of Messina, Messina, Italy
E-mail:ebarbera@unime.it

1. Introduction

Extended Thermodynamics[1] is a field theory for gases. The fields are moments of the distribution function. The field equations are obtained from the Boltzmann equation, which leads to an infinite hierarchy of balance equations for the moments. In this paper we close this hierarchy by use of Consistent-Order Extended Thermodynamics (COET)[2] where the closure is a consequence of the assignment of the order of magnitude.

We use the field equations of COET with a view to describing a gas at rest between two infinite coaxial cylinders maintained at two different temperatures. The same problem has been investigated[3] in the context of Extended Thermodynamics with thirteen moments, representing the 1st order in COET. Even with this simple theory, they have found that, while the pressure remains constant, the pressure deviator does not vanish and it affects the relation between the heat flux and the temperature gradient. The Fourier law is no longer valid and the temperature field is affected in the neighborhood of the inner cylinder.

In addition, we expect that the pressure and the temperature exhibit boundary layers near the two boundaries that are not present in[3] because they do not appear in a 1st order theory. Indeed, earlier work[4] suggests that, if we wish to find boundary layers, we have to employ COET at least of the 4th order. The 4th order COET implies a system containing 19 field equations, and not all boundary values can be controlled. In fact, we need 19 boundary conditions in order to obtain the solution and realistically we can control only two of them. We calculate the remaining boundary values by use of the fluctuation principle.[4]

As expected, we find a solution for the temperature with boundary layers superposed on the classical Fourier temperature. The pressure is no longer constant and it also exhibits boundary layers. Moreover, we obtain as in[3] two normal components of the traceless part of the stress tensor.

2. Field equations

The fields of Extended Thermodynamics[1] of a monatomic gas are moments of the distribution function $f\left(\underline{x}, \underline{C}, t\right)$ of the gas, defined as

$$F_{i_1 i_2 \ldots i_k} = m \int C_{i_1} C_{i_2} \ldots C_{i_k} f \mathrm{d}\underline{C}, \tag{1}$$

where m is the atomic mass and $\underline{C}$ is the velocity of the atoms relative to the velocity of the gas. Here we consider only gases at rest. Some moments can be expressed in terms of more common variables, such that

$$F = \rho, \quad F_i = 0, \quad F_{ll} = 3p = 3\rho \tfrac{k}{m} T, \quad F_{<ij>} = -t_{<ij>}, \quad F_{ill} = 2q_i, \tag{2}$$

where ρ is the density of the gas, p its pressure, T its temperature, k the Boltzmann constant, $t_{<ij>}$* the traceless part of the stress tensor and q_i the heat flux. In the following we will use $\theta = \frac{k}{m} T$ instead of T.

The balance equations for the moments are obtained[1] from the Boltzmann equation and in the BGK approximations[5] they read

$$\frac{\partial F_{i_1 i_2 \ldots i_A}}{\partial t} + \frac{\partial F_{i_1 i_2 \ldots i_A k}}{\partial x_k} = -\frac{F_{i_1 i_2 \ldots i_A} - F^{\mathrm{E}}_{i_1 i_2 \ldots i_A}}{\tau}, \qquad A = 0, 1, \ldots \quad . \tag{3}$$

τ is the constant relaxation time of the order of magnitude of the mean time of free flight of an atom. The symbol "E" denotes equilibrium.

In this paper we study stationary heat conduction in a gas at rest between two coaxial cylinders. Therefore, we use cylindrical coordinates (r, ϑ, z) and the physical cylindrical components $\tilde{\mathbf{F}}'s$ of the moments. Then, we assume that the fields depend only on r, and that $\tilde{F}_{<r\vartheta>}$, $\tilde{F}_{<rz>}$, $\tilde{F}_{<\vartheta z>}$, $\tilde{F}_{\vartheta ll}$ and $\tilde{F}_{zll}$ vanish. Although these assumptions, Eqs.(3) form a very complicate hierarchy of balance equations for the $\tilde{\mathbf{F}}$'s. We close it using the methods of COET.[2,6]

The closed system is linearized around $p = p_0$, $\theta = \theta_0$, $\tilde{F}_{llss} = 15 p_0 \theta_0$, $\tilde{F}_{llsskk} = 105 p_0 \theta_0^2$ and vanishing $\tilde{\mathbf{F}}'s$ and written in terms of the dimension-

*Angular brackets denote the traceless part of the tensor.

less variables, defined as

$$\bar{p} = \frac{p}{p_0}, \qquad \bar{\theta} = \frac{\theta}{\theta_0}, \qquad \bar{r} = \frac{r}{r_e}, \quad \bar{F}_{i_1 i_2 \ldots i_k} = \frac{\tilde{F}_{i_1 i_2 \ldots i_k}}{p_0 \sqrt{\theta_0}^{k-2}},$$

$$\bar{\Delta} = \frac{\tilde{F}_{llss} - 15p\theta}{p_0\theta_0}, \quad \bar{\Gamma} = \frac{\tilde{F}_{llsskk} - 105p\theta^2}{p_0\theta_0^2}, \qquad \text{and} \qquad Kn = \frac{\tau\sqrt{\theta_0}}{r_e}. \tag{4}$$

The Knudsen number, Kn, gives a measure of the degree of rarefaction of the gas. r_e is the radius of the external cylinder. The linearized system is

$$\begin{aligned}
&\frac{\mathrm{d}\left[\bar{p}+\bar{F}_{<rr>}\right]}{\mathrm{d}\bar{r}} + \frac{\bar{F}_{<rr>}-\bar{F}_{<\vartheta\vartheta>}}{\bar{r}} = 0,\\
&\frac{\mathrm{d}\bar{F}_{rll}}{\mathrm{d}\bar{r}} + \frac{\bar{F}_{rll}}{\bar{r}} = 0 \Rightarrow \bar{F}_{rll} = \frac{D}{\bar{r}},\\
&\frac{\mathrm{d}\left[\bar{F}_{<rrr>}+\frac{4}{15}\bar{F}_{rll}\right]}{\mathrm{d}\bar{r}} + \frac{\bar{F}_{<rrr>}-2\bar{F}_{<r\vartheta\vartheta>}-\frac{2}{15}\bar{F}_{rll}}{\bar{r}} = -\frac{\bar{F}_{<rr>}}{Kn},\\
&\frac{\mathrm{d}\left[\bar{F}_{<r\vartheta\vartheta>}-\frac{2}{15}\bar{F}_{rll}\right]}{\mathrm{d}\bar{r}} + \frac{3\bar{F}_{<r\vartheta\vartheta>}+\frac{4}{15}\bar{F}_{rll}}{\bar{r}} = -\frac{\bar{F}_{<\vartheta\vartheta>}}{Kn},\\
&\frac{\mathrm{d}\left[5\bar{p}+5\bar{\theta}+\frac{1}{3}\bar{\Delta}+\bar{F}_{<rr>ll}\right]}{\mathrm{d}\bar{r}} + \frac{\bar{F}_{<rr>ll}-\bar{F}_{<\vartheta\vartheta>ll}}{\bar{r}} = -\frac{\bar{F}_{rll}}{Kn},\\
&\frac{9}{35}\frac{\mathrm{d}\bar{F}_{<rr>ll}}{\mathrm{d}\bar{r}} - \frac{6}{35}\frac{\bar{F}_{<rr>ll}-\bar{F}_{<\vartheta\vartheta>ll}}{\bar{r}} = -\frac{\bar{F}_{<rrr>}}{Kn},\\
&\frac{\mathrm{d}}{\mathrm{d}\bar{r}}\left[-\frac{2}{35}\bar{F}_{<rr>ll}+\frac{1}{7}\bar{F}_{<\vartheta\vartheta>ll}\right] + \frac{8}{35}\frac{\bar{F}_{<rr>ll}-\bar{F}_{<\vartheta\vartheta>ll}}{\bar{r}} = -\frac{\bar{F}_{<r\vartheta\vartheta>}}{Kn},\\
&\frac{\mathrm{d}\bar{F}_{rllss}}{\mathrm{d}\bar{r}} + \frac{\bar{F}_{rllss}}{\bar{r}} = \frac{\bar{\Delta}}{Kn},\\
&\frac{\mathrm{d}}{\mathrm{d}\bar{r}}\left[\bar{F}_{<rrr>ll}+\frac{4}{15}\bar{F}_{rllss}\right] + \frac{\bar{F}_{<rrr>ll}-2\bar{F}_{<r\vartheta\vartheta>ll}-\frac{2}{15}\bar{F}_{rllss}}{\bar{r}} = -\frac{\bar{F}_{<rr>ll}}{Kn},\\
&\frac{\mathrm{d}}{\mathrm{d}\bar{r}}\left[\bar{F}_{<r\vartheta\vartheta>ll}-\frac{2}{15}\bar{F}_{rllss}\right] + \frac{3\bar{F}_{<r\vartheta\vartheta>ll}+\frac{4}{15}\bar{F}_{rllss}}{\bar{r}} = -\frac{\bar{F}_{<\vartheta\vartheta>ll}}{Kn},\\
&\frac{\mathrm{d}}{\mathrm{d}\bar{r}}\left[35\bar{p}+70\bar{\theta}+\frac{1}{3}\bar{\Gamma}+\bar{F}_{<rr>llss}\right] + \frac{\bar{F}_{<rr>llss}-\bar{F}_{<\vartheta\vartheta>llss}}{\bar{r}} = -\frac{\bar{F}_{rllss}}{Kn},\\
&\frac{9}{35}\frac{\mathrm{d}\bar{F}_{<rr>llss}}{\mathrm{d}\bar{r}} - \frac{6}{35}\frac{\bar{F}_{<rr>llss}-\bar{F}_{<\vartheta\vartheta>llss}}{\bar{r}} = -\frac{\bar{F}_{<rrr>ll}}{Kn},\\
&\frac{\mathrm{d}}{\mathrm{d}\bar{r}}\left[-\frac{2}{35}\bar{F}_{<rr>llss}+\frac{1}{7}\bar{F}_{<\vartheta\vartheta>llss}\right] + \frac{8}{35}\frac{\bar{F}_{<rr>llss}-\bar{F}_{<\vartheta\vartheta>llss}}{\bar{r}} = -\frac{\bar{F}_{<r\vartheta\vartheta>ll}}{Kn},\\
&\frac{\mathrm{d}\bar{F}_{rllsskk}}{\mathrm{d}\bar{r}} + \frac{\bar{F}_{rllsskk}}{\bar{r}} = \frac{\bar{\Gamma}}{Kn},\\
&\frac{\mathrm{d}\left[\bar{F}_{<rrr>llss}+\frac{4}{15}\bar{F}_{rllsskk}\right]}{\mathrm{d}\bar{r}} + \frac{\bar{F}_{<rrr>llss}-2\bar{F}_{<r\vartheta\vartheta>llss}-\frac{2}{15}\bar{F}_{rllsskk}}{\bar{r}} = -\frac{\bar{F}_{<rr>llss}}{Kn},\\
&\frac{\mathrm{d}\left[\bar{F}_{<r\vartheta\vartheta>llss}-\frac{2}{15}\bar{F}_{rllsskk}\right]}{\mathrm{d}\bar{r}} + \frac{3\bar{F}_{<r\vartheta\vartheta>llss}+\frac{4}{15}\bar{F}_{rllsskk}}{\bar{r}} = -\frac{\bar{F}_{<\vartheta\vartheta>llss}}{Kn},\\
&\frac{\mathrm{d}\left[315\bar{p}+945\bar{\theta}+12\bar{\Gamma}-126\bar{\Delta}+\bar{F}_{<rr>llsskk}\right]}{\mathrm{d}\bar{r}} + \frac{\bar{F}_{<rr>llsskk}-\bar{F}_{<\vartheta\vartheta>llsskk}}{\bar{r}} = -\frac{\bar{F}_{rllsskk}}{Kn},\\
&\frac{9}{35}\frac{\mathrm{d}\bar{F}_{<rr>llsskk}}{\mathrm{d}\bar{r}} - \frac{6}{35}\frac{\bar{F}_{<rr>llsskk}-\bar{F}_{<\vartheta\vartheta>llsskk}}{\bar{r}} = -\frac{\bar{F}_{<rrr>llss}}{Kn},\\
&\frac{\mathrm{d}\left[-\frac{2}{35}\bar{F}_{<rr>llsskk}+\frac{1}{7}\bar{F}_{<\vartheta\vartheta>llsskk}\right]}{\mathrm{d}\bar{r}} + \frac{8}{35}\frac{\bar{F}_{<rr>llsskk}-\bar{F}_{<\vartheta\vartheta>llsskk}}{\bar{r}} = -\frac{\bar{F}_{<r\vartheta\vartheta>llss}}{Kn},
\end{aligned} \tag{5}$$

with

$$\begin{aligned}\bar{F}_{<rr>llsskk} &= 33\bar{F}_{<rr>llss} - 297\bar{F}_{<rr>ll} + 693\bar{F}_{<rr>},\\ \bar{F}_{<\vartheta\vartheta>llsskk} &= 33\bar{F}_{<\vartheta\vartheta>llss} - 297\bar{F}_{<\vartheta\vartheta>ll} + 693\bar{F}_{<\vartheta\vartheta>}.\end{aligned} \tag{6}$$

Equations $(5)_{1,2}$ are the conservations laws of momentum and energy, while the conservation law of mass is identically satisfied. The other equations are the balance laws for $\bar{F}_{<rr>}$, $\bar{F}_{<\vartheta\vartheta>}$, $\bar{q}_r = \frac{\bar{F}_{rll}}{2}$, and for the other moments which do not have a suggestive physical meaning.

Equation $(5)_5$ relates $\bar{F}_{rll}$ with the gradient of temperature. In classical thermodynamics this relation is described by the Fourier law which reads

$$-5\bar{p}Kn\frac{\mathrm{d}\bar{\theta}}{\mathrm{d}\bar{r}} = \bar{F}_{rll}. \tag{7}$$

From $(5)_5$, we may say that in COET this relation is no longer valid because of the fields $\bar{\Delta}$, $\bar{F}_{<rr>ll}$ and $\bar{F}_{<\vartheta\vartheta>ll}$.

3. Boundary conditions and integration

Integration of system (5) needs 19 boundary conditions and this is a problem. In fact, in an experiment we can assign or control only the temperatures of the gas at the two cylinders, i.e.,

$$\bar{\theta}(\bar{r}_i) = \bar{\theta}_i \quad \text{and} \quad \bar{\theta}(\bar{r}_e) = \bar{\theta}_e. \tag{8}$$

We can neither impose nor control any other boundary value. Thus, it seems clear that the gas itself chooses the remaining values. The problem is to understand how the gas makes its choice. A possible solution of this problem has been proposed in[4] where the authors consider the gas subject to thermal fluctuations, so that the fields fluctuate freely together with their boundary values. But, these fluctuations are very rapid, so that the gas cannot adjust to the ever changing fluctuating boundary values. Therefore, they propose that "the gas adjusts the non-controllable boundary values to the mean values of the fluctuating boundary data". They calculate these mean values by use of the Boltzmann formula

$$S = k \ln W, \tag{9}$$

that relates the entropy S to W, the number of possibilities to realize a fluctuation. We proceed to explain how to calculate the mean values.

Firstly, we assign to Eqs.(5) the formal boundary values problem†

$$\begin{array}{lll}
\bar{p}(\bar{r}_e) = 1, & \bar{\theta}(\bar{r}_e) = \bar{\theta}_e, & \bar{F}_{<rr>}(\bar{r}_e) = c_1, \\
\bar{F}_{<\vartheta\vartheta>}(\bar{r}_e) = c_2, & \bar{F}_{<rrr>}(\bar{r}_e) = c_3, & \bar{F}_{<r\vartheta\vartheta>}(\bar{r}_e) = c_4, \\
\bar{\Delta}(\bar{r}_e) = c_5, & \bar{F}_{<rr>ll}(\bar{r}_e) = c_6, & \bar{F}_{<\vartheta\vartheta>ll}(\bar{r}_e) = c_7, \\
\bar{F}_{rllss}(\bar{r}_e) = c_8, & \bar{F}_{<rrr>ll}(\bar{r}_e) = c_9, & \bar{F}_{<r\vartheta\vartheta>ll}(\bar{r}_e) = c_{10}, \\
\bar{\Gamma}(\bar{r}_e) = c_{11}, & \bar{F}_{<rr>llss}(\bar{r}_e) = c_{12}, & \bar{F}_{<\vartheta\vartheta>llss}(\bar{r}_e) = c_{13}, \\
\bar{F}_{rllsskk}(\bar{r}_e) = c_{14}, & \bar{F}_{<rrr>llss}(\bar{r}_e) = c_{15}, & \bar{F}_{<r\vartheta\vartheta>llss}(\bar{r}_e) = c_{16}.
\end{array} \tag{10}$$

One of the two boundary temperatures (8) is directly used as boundary value, while the second has to be still imposed. This will be done by using the constant D in $(5)_2$ related to the heat flux as shooting parameter. That is, D will be determined by setting the temperature $\bar{\theta}$ at $\bar{r} = \bar{r}_i$ equal to $\bar{\theta}_i$.

In principle, we can determine the numerical solution of (5) for every 16-tuple of values c_1,...,c_{16}, in this way our solution is a function of the radial coordinate $\bar{r}$ and of c_1,...,c_{16}. Then, we insert this solution in the entropy density ρs, that in our case reads

$$\rho s = \rho s|_E - \frac{\rho_0}{2}\frac{k}{m}\left[\frac{3\bar{F}^2_{<rr>}}{4} + \frac{3\bar{F}^2_{<\vartheta\vartheta>}}{4} + \frac{\bar{F}^2_{rll}}{10} + \frac{\bar{\Delta}^2}{120} + \frac{3\left(\bar{F}_{<rr>ll} - 7\bar{F}_{<rr>}\right)^2}{56} + \right.$$
$$+\frac{3\left(\bar{F}_{<\vartheta\vartheta>ll} - 7\bar{F}_{<\vartheta\vartheta>}\right)^2}{56} + \frac{\left(\bar{F}_{<rr>llss} - 18\bar{F}_{<rr>ll} + 63\bar{F}_{<rr>}\right)^2}{672} +$$
$$\left. + \frac{\left(\bar{\Gamma} - 21\bar{\Delta}\right)^2}{5040} + \frac{\left(\bar{F}_{<\vartheta\vartheta>llss} - 18\bar{F}_{<\vartheta\vartheta>ll} + 63\bar{F}_{<\vartheta\vartheta>}\right)^2}{672}\right]. \tag{11}$$

In this way, ρs becomes a function of $\bar{r}, c_1, ..., c_{16}$. The entropy S follows by integration of ρs over the whole range of $\bar{r}$. So S is a function of $c_1, ..., c_{16}$. Inserting S in (9) we get $W(c_1, ..., c_{16})$, the number of possibilities of realizing a set of these constants $c_1, ..., c_{16}$. Then, the probability of the occurrence of an 16-tupla $(c_1, ..., c_{16})$ during a thermal fluctuation is

$$P(c_1, ..., c_{16}) = \frac{e^{\frac{S(c_1,...,c_{16})}{k}}}{\int_{-\infty}^{+\infty} e^{\frac{S(c_1,...,c_{16})}{k}} \mathrm{d}c_1 ... \mathrm{d}c_{16}} \tag{12}$$

and, consequently, the mean values of the 16 boundary data $c_1, ..., c_{16}$ are

$$< c_i > = \frac{\int_{-\infty}^{+\infty} c_i e^{\frac{S(c_1 ... c_{16})}{k}} \mathrm{d}c_1 ... \mathrm{d}c_{16}}{\int_{-\infty}^{+\infty} e^{\frac{S(c_1 ... c_{16})}{k}} \mathrm{d}c_1 ... \mathrm{d}c_{16}} \qquad i = 1, ..., 16. \tag{13}$$

†The choice for the boundary condition for the pressure at $\bar{r} = \bar{r}_e$ is not restrictive. In fact it affects only an additive constant in the pressure. We have set the pressure at $\bar{r} = \bar{r}_e$ equal to one in order to compare our pressure with the classical pressure and we have imposed that they are at least equal at $\bar{r} = \bar{r}_e$.

We have calculated these values for $Kn = 0.1$ and $Kn = 0.2$ and for $\bar{r}_i = 0.1$, $\bar{r}_e = 1$, $\bar{\theta}_i = 0.9$ and $\bar{\theta}_e = 1.1$. The obtained values are listed in Table 1. For completeness we also present in the table, the corresponding values of the parameter D. By following[4] the values in Table 1 are the boundary values that the body will choose.

Table 1. Mean values of the fluctuating boundary data.

	$Kn = 0.1$	$Kn = 0.2$		$Kn = 0.1$	$Kn = 0.2$
D	−0.0318032	−0.0113814			
c_1	−0.02417	−0.15312	c_9	−0.130742	−0.374756
c_2	0.186488	0.54062	c_{10}	−0.332508	−0.601461
c_3	0.0000165242	0.0317326	c_{11}	−8.98546	−39.8025
c_4	−0.110418	−0.262847	c_{12}	−0.00328976	0.507892
c_5	−0.540982	−2.81013	c_{13}	2.34161	3.7628
c_6	−0.0642759	−0.319328	c_{14}	0.799742	20.615
c_7	0.667051	1.5836	c_{15}	−2.1447	−6.29904
c_8	−0.0319872	1.63248	c_{16}	−0.334195	0.343621

4. Results

The solution of Eqs.(5), obtained with the boundary values predicted by the fluctuation principle, is illustrated in Figs.1,2.

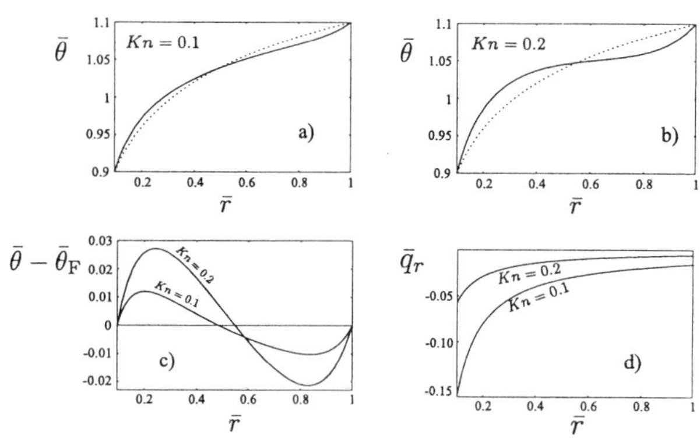

Fig. 1. Fig.1a,b: Temperature together with $\bar{\theta}_F$, the temperature predicted by the Fourier law (dashed lines). Fig.1c: Differences between the two temperatures and $\bar{\theta}_F$. Fig.1d: Radial heat flux.

In particular, Fig.1 shows the temperature together with the temperature predicted by the Fourier law $\bar{\theta}_F$ (dashed line). Our temperature exhibits boundary layers superposed on $\bar{\theta}_F$. The boundary layers are less pronounced for the denser gas (Fig.1a) and become more noticeable for the more rarefied gas (Fig.1b). This is what we expect, in fact for a dense gas, near the range of validity of the classical thermodynamics, there must be no great difference to the classical solution, whereas the difference must increase for a more rarefied gas. The difference between $\bar{\theta}$ and $\bar{\theta}_F$ is also illustrated in Fig.1c, while Fig.1d shows the radial component of the heat flux.

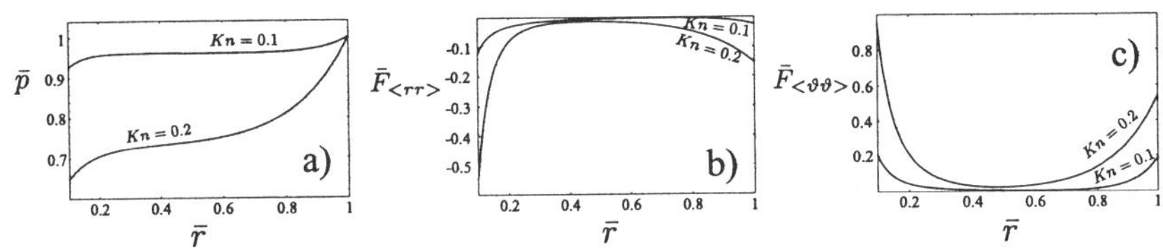

Fig. 2. Fig.1a: pressure. Fig.1b: rr−component of the deviatoric stress. Fig.1c: $\vartheta\vartheta$−component of the deviatoric stress for $Kn = 0.1$ and for $Kn = 0.2$.

Then, Fig.2a shows that the pressure is not longer constant, contrary to the pressure predicted by the classical theory. Our pressure exhibits boundary layers near the two boundaries and they depend on the Knudsen numbers. Furthermore, as in[3] we have two non-vanishing components of the traceless part of the stress tensor, although the gas is at rest, see Fig.2b,c.

References

1. Müller, I. and Ruggeri, T., *Rational Extended Thermodynamics,* 2nd ed., Tracts in Natural Philosophy **37**, Springer, New York, 1998.
2. Müller, I. Reitebuch, D. and Weiss, W., "Extended Thermodynamics – consistent in order of magnitude", *Cont.Mech.Thermodyn.* **15**, 113-146 (2002).
3. Müller I. and Ruggeri T., "Stationary Heat Conduction in Radially Symmetric Situations - An Application of Extended Thermodynamics". *J. Non Newtonian Fluid Mech.* **119**, 139-143 (2004).
4. Barbera, E. Müller, I. Reitebuch, D. and Zhao, N., "Determination of the Boundary Conditions in Extended thermodynamics via Fluctuation Theory", *Cont.Mech.Thermodyn.* **16**, 411-425 (2004).
5. Bhatnagar, P.L. Gross, E.P. and Krook, M., "A model for collision processes in gases. I. Small amplitude processes in charged and neutral one-component systems" *Phys.Rev.* **94** (1954).
6. Barbera, E., "Consistently Ordered Extended Thermodynamics – a proposal for an alternative method", *Cont.Mech.Thermodyn.* **17**, 61-81 (2005).

QUANTITATIVE ESTIMATES FOR THE LARGE TIME BEHAVIOR OF A REACTION-DIFFUSION EQUATION WITH RATIONAL REACTION TERM

M. BISI*, L. DESVILLETTES† and G. SPIGA‡

*‡*Dep. of Mathematics, Parma University,*
Viale G.P. Usberti 53/A, I-43100 Parma, Italy

†*CMLA, ENS Cachan, CNRS & IUF, PRES UniverSud,*
61, Av. du Pdt. Wilson, 94235 Cachan Cedex, France

**E-mail: marzia.bisi@unipr.it* †*E-mail: desville@cmla.ens-cachan.fr*
‡*E-mail: giampiero.spiga@unipr.it*

We show that the entropy method can be used to prove exponential convergence towards equilibrium with explicit constants when one considers a reaction-diffusion system corresponding to an irreversible mechanism of dissociation/recombination, for which no natural entropy is available.

Keywords: Reaction-diffusion equations; Entropy/entropy dissipation method.

1. Introduction

We consider a diatomic gas with dissociation/recombination reactions, made up by atoms A with mass m_1 and molecules A_2 with mass $m_2 = 2\,m_1$. According to a common kinetic model,[1] the gas is described as a mixture of three species, with an additional component, representing unstable molecules $A_3 \equiv A_2^*$ (with mass $m_3 = m_2$) and playing the role of a transition state. The mixture is then taken to diffuse in a much denser medium, whose evolution is not affected by the collisions going on, assumed in local thermodynamical equilibrium, namely with distribution function $f_0 = n_0 M_0$, where M_0 stands for the normalized Maxwellian with temperature T_0 (constant) and vanishing mass velocity. According to the model, both atoms A_1 and stable molecules A_2 may undergo elastic collisions with other atoms, stable molecules and background particles. Moreover, atoms A_1 may form a stable molecule A_2 passing through the transition state A_2^*, while, on the other hand, both stable and unstable diatomic molecules may dissociate into two atoms. More precisely, the recombination process occurs in two steps:

$$(R) \quad A_1 + A_1 \quad \rightarrow \quad A_2^*, \qquad\qquad (I) \quad A_2^* + P \quad \rightarrow \quad A_2 + P,$$

where $P = A,\, A_2$, while dissociation occurs via two possible reactions:

$$(D1) \quad A_2 + P \quad \rightarrow \quad 2\,A_1 + P, \qquad\qquad (D2) \quad A_2^* + P \quad \rightarrow \quad 2\,A_1 + P.$$

All above interactions, modelling the chemical reactions at the kinetic level, have to be understood as irreversible processes. The host medium is assumed here only elastically scattering, and not chemically reacting. A physical situation in which the background is actually involved in chemical processes is extensively discussed in Refs. 2, 3.

Kinetic Boltzmann equations relevant to species A_1, A_2, A_2^* have been scaled[2] in terms of the typical relaxation times, a small parameter defining the dominant process(es) has been introduced, and the formal asymptotic limit when this parameter vanishes has been consistently investigated. This leads to the derivation (for the stable species) of hydrodynamic limiting equations, whose nature varies considerably according to the relative importance of the various processes and to the corresponding pertinent scaling, but which are typically of reaction-diffusion type as long as the scattering with the background plays an important role. We shall deal here with one of the asymptotic limits which seems more realistic in practice, and leads to

$$\partial_t n_i - d_i\, \Delta_{\mathbf{x}} n_i = \mathcal{Q}_i(n_1, n_2), \qquad\qquad i = 1, 2, \tag{1}$$

where d_i are the diffusion coefficients

$$d_1 = \frac{m_1 + m_0}{2\, m_1\, m_0}\,\frac{T_0}{\bar{\nu}_{10}^s\, n_0}\,, \qquad\qquad d_2 = \frac{2\, m_1 + m_0}{4\, m_1\, m_0}\,\frac{T_0}{\bar{\nu}_{20}^s\, n_0}\,,$$

and $\mathcal{Q}_i$ are the reaction contributions

$$\mathcal{Q}_2 = \frac{1}{\nu_{31}^t\, n_1 + \nu_{32}^t\, n_2}\Big[A(n_1)^3 + B(n_1)^2 n_2 - C n_1 (n_2)^2 - D(n_2)^3\Big] \tag{2}$$

with $A = \nu_{11}^r\, \nu_{31}^i > 0$, $B = \nu_{11}^r\, \nu_{32}^i - \nu_{21}^d\, \nu_{31}^t$, $C = \nu_{21}^d\, \nu_{32}^t + \nu_{22}^d\, \nu_{31}^t > 0$, $D = \nu_{22}^d\, \nu_{32}^t > 0$, and $\mathcal{Q}_1 = -2\,\mathcal{Q}_2$ (preservation of total number of atoms $\bar{n}^0$). ν_{ij}^{k} are total microscopic collision frequencies, where the superscript k takes the values s, r, i, d, corresponding to elastic scattering, recombination R, inelastic scattering I, dissociations $D1$, $D2$, respectively, and $\nu_{3j}^t = \nu_{3j}^i + \nu_{3j}^d$. Just on the basis of the sign of the coefficients, it can be checked that the cubic function (in the variable n_1) into the square brackets in (2) has a positive root $n_1 = \gamma\, n_2$ (with $\gamma > 0$), while the other two roots are negative or complex conjugate with negative real part. Therefore

$$\mathcal{Q}_2 = (n_1 - \gamma\, n_2)\, \mathcal{P}(n_1, n_2) \tag{3}$$

where $\qquad \mathcal{P}(n_1, n_2) = \dfrac{(n_1)^2 + (\alpha + \beta) n_1\, n_2 + \alpha\, \beta\, (n_2)^2}{\lambda\, n_1 + \mu\, n_2} > 0$

with $\lambda,\ \mu > 0$ and $\alpha,\ \beta > 0$ or complex conjugate with positive real part. Taking into account conservation of $\bar{n}^0$, system (1) with Neumann boundary conditions on a bounded domain of unit measure admits a unique global equilibrium state:

$$n_1^* = \frac{\gamma}{2+\gamma}\,\bar{n}^0\,, \qquad n_2^* = \frac{1}{2+\gamma}\,\bar{n}^0\,. \tag{4}$$

2. Existence and uniqueness of a strong solution

Theorem 2.1. *Let $d_1, d_2 > 0$ and Ω be a bounded regular (C^2) open set of $\mathbb{R}^N$. We consider initial data in $C^2(\bar{\Omega})$, compatible with Neumann boundary conditions, and satisfying the bounds (for some strictly positive constants c_1, c_2, $\mathcal{C}_1$ and $\mathcal{C}_2$):*

$$0 < c_1 < n_1(0,\mathbf{x}) < \mathcal{C}_1\,, \qquad 0 < c_2 < n_2(0,\mathbf{x}) < \mathcal{C}_2\,. \tag{5}$$

Then, there exists a unique (strong) solution $n_1(t,\mathbf{x})$, $n_2(t,\mathbf{x})$ in $C^2(\mathbb{R}^+\times\bar{\Omega})$ to system (1) with homogeneous Neumann boundary conditions such that, for $(t,\mathbf{x}) \in \mathbb{R}^+ \times \Omega$,

$$k_1 \le n_1(t,\mathbf{x}) \le \mathcal{K}_1\,, \qquad k_2 \le n_2(t,\mathbf{x}) \le \mathcal{K}_2\,, \tag{6}$$

where

$$k_1 = \min\Big\{c_1\,, \gamma\, c_2\Big\}\,, \qquad k_2 = \min\Big\{\gamma^{-1}\, c_1\,, c_2\Big\}\,, \tag{7}$$

$$\mathcal{K}_1 = \max\Big\{\mathcal{C}_1\,, \gamma\, \mathcal{C}_2\Big\}\,, \qquad \mathcal{K}_2 = \max\Big\{\gamma^{-1}\, \mathcal{C}_1\,, \mathcal{C}_2\Big\}\,. \tag{8}$$

Proof. At first we shall prove that the "maximum principle" holds for $(t,\mathbf{x}) \in [0,T] \times \Omega$, for each $T > 0$, following the same lines as in Ref. 4. Let $\varepsilon > 0$ be fixed and let us consider the functions

$$n_1^\varepsilon(t,\mathbf{x}) = n_1(t,\mathbf{x})\ \mathrm{e}^{-\,\varepsilon\, t}\,, \qquad n_2^\varepsilon(t,\mathbf{x}) = n_2(t,\mathbf{x})\ \mathrm{e}^{-\,\varepsilon\, t}\,; \tag{9}$$

we prove that

$$n_1^\varepsilon(t,\mathbf{x}) < \mathcal{K}_1\,, \qquad n_2^\varepsilon(t,\mathbf{x}) < \mathcal{K}_2\,. \tag{10}$$

From equations (1) it follows that the evolution of n_1^ε, n_2^ε is governed by the system

$$\begin{aligned} \partial_t n_1^\varepsilon - d_1\ \Delta_{\mathbf{x}} n_1^\varepsilon &= -\,2\,(n_1^\varepsilon - \gamma\, n_2^\varepsilon)\,\mathcal{P}(n_1^\varepsilon, n_2^\varepsilon)\,\mathrm{e}^{\varepsilon\, t} - \varepsilon\, n_1^\varepsilon\,, \\ \partial_t n_2^\varepsilon - d_2\ \Delta_{\mathbf{x}} n_2^\varepsilon &= (n_1^\varepsilon - \gamma\, n_2^\varepsilon)\,\mathcal{P}(n_1^\varepsilon, n_2^\varepsilon)\,\mathrm{e}^{\varepsilon\, t} - \varepsilon\, n_2^\varepsilon\,. \end{aligned} \tag{11}$$

Suppose that inequalities (10) do not hold for all $(t,\mathbf{x}) \in [0,T] \times \Omega$, and define the set $B^\varepsilon = \big\{\tau > 0\ :\ n_1^\varepsilon(t,\mathbf{x}) < \mathcal{K}_1,\ \ n_2^\varepsilon(t,\mathbf{x}) < \mathcal{K}_2\ \ \forall\ (t,\mathbf{x}) \in [0,\tau) \times \Omega\big\}$. If we denote $\tilde{t} = \sup B^\varepsilon$, there must exist $\tilde{\mathbf{x}} \in \bar{\Omega}$ such that $(n_1^\varepsilon(\tilde{t},\tilde{\mathbf{x}}), n_2^\varepsilon(\tilde{t},\tilde{\mathbf{x}})) \in \partial B^\varepsilon$, hence one of the following equalities holds:

$$n_1^\varepsilon(\tilde{t},\tilde{\mathbf{x}}) = \mathcal{K}_1 \qquad \text{or} \qquad n_2^\varepsilon(\tilde{t},\tilde{\mathbf{x}}) = \mathcal{K}_2\,.$$

If $n_1^\varepsilon(\tilde{t},\tilde{\mathbf{x}}) = \mathcal{K}_1$, by definitions of $\tilde{t}$ and $\tilde{\mathbf{x}}$ we have $n_1^\varepsilon(\tilde{t},\tilde{\mathbf{x}}) \geq n_1^\varepsilon(\tilde{t},\mathbf{x})$ $\forall \mathbf{x} \in \Omega$, thus $d_1 \Delta_{\mathbf{x}} n_1^\varepsilon(\tilde{t},\tilde{\mathbf{x}}) \leq 0$. Moreover, by evaluating the chemical contributions in the first of (11) at $(\tilde{t},\tilde{\mathbf{x}})$ we get

$$-2\left(n_1^\varepsilon - \gamma n_2^\varepsilon\right)(\tilde{t},\tilde{\mathbf{x}})\,\mathcal{P}(n_1^\varepsilon,n_2^\varepsilon)(\tilde{t},\tilde{\mathbf{x}}) \leq -2\left(\mathcal{K}_1 - \gamma\,\mathcal{K}_2\right)\mathcal{P}(\mathcal{K}_1, n_2^\varepsilon(\tilde{t},\tilde{\mathbf{x}})) = 0$$

(the inequality holds since $n_2^\varepsilon(\tilde{t},\tilde{\mathbf{x}}) \geq \mathcal{K}_2$ and $\mathcal{P}(n_1^\varepsilon,n_2^\varepsilon) > 0$, and the last term vanishes since $\mathcal{K}_1 = \gamma\,\mathcal{K}_2$). Consequently, the equation (11) for n_1^ε implies that $\partial_t n_1^\varepsilon(\tilde{t},\tilde{\mathbf{x}}) \leq -\varepsilon\, n_1^\varepsilon(\tilde{t},\tilde{\mathbf{x}}) < 0$, hence $n_1^\varepsilon(t,\tilde{\mathbf{x}}) > n_1^\varepsilon(\tilde{t},\tilde{\mathbf{x}}) = \mathcal{K}_1$ for $t < \tilde{t}$, contradicting the definition of $\tilde{t}$. The case $n_2^\varepsilon(\tilde{t},\tilde{\mathbf{x}}) = \mathcal{K}_2$ may be treated in a similar way. Consequently, the set B^ε is unbounded, hence $n_1^\varepsilon(t,\mathbf{x}) < \mathcal{K}_1$ and $n_2^\varepsilon(t,\mathbf{x}) < \mathcal{K}_2$ for all $\mathbf{x} \in \Omega$ and for all $t \in [0,T]$. This means that $n_1(t,\mathbf{x}) < \mathcal{K}_1\, e^{\varepsilon t}$ and $n_2(t,\mathbf{x}) < \mathcal{K}_2\, e^{\varepsilon t}$, thus, passing to the limit $\varepsilon \to 0$, we have $n_1(t,\mathbf{x}) \leq \mathcal{K}_1$ and $n_2(t,\mathbf{x}) \leq \mathcal{K}_2$.

The "minimum principle" for n_1 and n_2 may be recovered analogously, by studying the evolution of the auxiliary functions $n_{1,\varepsilon}(t,\mathbf{x}) = n_1(t,\mathbf{x})\, e^{\varepsilon t}$ and $n_{2,\varepsilon}(t,\mathbf{x}) = n_2(t,\mathbf{x})\, e^{\varepsilon t}$.

Boundedness from above of n_1 and n_2 allows to prove existence and uniqueness of a strong solution on $[0,T]$ to the system (1) (with Neumann boundary conditions) by resorting to a suitable fixed point argument.[5] By sticking together the solutions on $[0,T]$ (for $T \in \mathbb{R}^+$), we obtain a (unique) solution in $C^2(\mathbb{R}^+ \times \bar{\Omega})$. □

3. Entropy functional and convergence to equilibrium

Large time behavior of reaction-diffusion systems has attracted a considerable interest in scientific literature,[6] and we show here that the "entropy / entropy dissipation method", already successfully used in the frame of reversible chemistry,[7] may be extended to the present irreversible situation. A crucial feature of our system (1) is that it admits a unique collision equilibrium (n_1^*, n_2^*), given explicitly in (4). Notice that $\mathcal{Q}_2 \geq 0 \Leftrightarrow n_1 \geq \gamma\, n_2$, and conversely for $\mathcal{Q}_1$. This suggests that a suitable entropy could be given by the quadratic functional

$$E(n_1,n_2) = \int_\Omega \left(\frac{1}{4}\,(n_1)^2 + \frac{\gamma}{2}\,(n_2)^2\right) d\mathbf{x}\,. \tag{12}$$

Lemma 3.1 (Relative entropy). *A direct computation shows that the relative entropy with respect to the equilibrium state (n_1^*, n_2^*) is related to the L^2-distance from the equilibrium itself:*

$$E(n_1,n_2) - E(n_1^*,n_2^*) = \frac{1}{4}\,\|n_1 - n_1^*\|_2^2 + \frac{\gamma}{2}\,\|n_2 - n_2^*\|_2^2\,, \tag{13}$$

hence the entropy $E(n_1, n_2)$ takes its minimum for $(n_1, n_2) = (n_1^*, n_2^*)$.

Lemma 3.2 (Entropy dissipation). *The entropy dissipation $D(n_1, n_2) = -\partial_t E(n_1, n_2)$ fulfils the inequality*

$$D(n_1, n_2) \geq \mathcal{C}\Big[E(n_1, n_2) - E(n_1^*, n_2^*)\Big], \tag{14}$$

with

$$\mathcal{C} = \min\left\{\min\left\{1, \frac{2+\gamma}{6}\right\}\frac{d_1}{2\,P(\Omega)}\,,\ \min\left\{\frac{1}{2}, \frac{2+\gamma}{6\gamma}\right\}\frac{d_2}{P(\Omega)}\,,\ \frac{2+\gamma}{6}\,d_3\right\}, \tag{15}$$

where $P(\Omega)$ is the Poincaré constant of Ω, and d_3 is a positive lower bound for $\mathcal{P}(n_1, n_2)$.

Proof. By direct computation, the entropy dissipation reads as

$$D(n_1, n_2) = \frac{d_1}{2}\int_\Omega |\nabla_{\mathbf{x}} n_1|^2 d\mathbf{x} + d_2\,\gamma\int_\Omega |\nabla_{\mathbf{x}} n_2|^2 d\mathbf{x} + \int_\Omega (n_1 - \gamma\, n_2)^2\, \mathcal{P}(n_1, n_2)\, d\mathbf{x}.$$

Thanks to Poincaré's inequality and to the lower and upper bounds for n_1 and n_2 given in Theorem 2.1, we have

$$D(n_1, n_2) \geq \frac{d_1}{2\,P(\Omega)}\|n_1 - \bar{n}_1\|_2^2 + \frac{d_2\,\gamma}{P(\Omega)}\|n_2 - \bar{n}_2\|_2^2 + d_3\|n_1 - \gamma\, n_2\|_2^2\,,$$

where $\bar{n}_i$ denotes the total number of atoms/molecules of species i in the domain Ω. Thanks to the inequality $|n_i - n_i^*|^2 \leq 2\Big[|n_i - \bar{n}_i|^2 + |\bar{n}_i - n_i^*|^2\Big]$, in order to prove Lemma 3.2 it suffices to show that

$$\begin{aligned} I &:= \int_\Omega \left[\frac{d_1}{2\,P(\Omega)}|n_1 - \bar{n}_1|^2 + \frac{d_2\,\gamma}{P(\Omega)}|n_2 - \bar{n}_2|^2 + d_3|n_1 - \gamma\, n_2|^2\right] d\mathbf{x} \\ &\geq \mathcal{C}\int_\Omega \left[\frac{1}{2}\,|n_1 - \bar{n}_1|^2 + \gamma\,|n_2 - \bar{n}_2|^2 + \frac{1}{2}\,|\bar{n}_1 - n_1^*|^2 + \gamma\,|\bar{n}_2 - n_2^*|^2\right] d\mathbf{x}\,. \end{aligned}$$

It obviously holds

$$\frac{1}{2}I \geq \mathcal{C}_1\int_\Omega \left[\frac{1}{2}\,|n_1 - \bar{n}_1|^2 + \gamma\,|n_2 - \bar{n}_2|^2\right] d\mathbf{x} \quad \text{with} \quad \mathcal{C}_1 = \frac{\min\{d_1, d_2\}}{2\,P(\Omega)}\,.$$

It remains to prove that

$$\frac{1}{2}I \geq \mathcal{C}_2\left[\frac{1}{2}\,|\bar{n}_1 - n_1^*|^2 + \gamma\,|\bar{n}_2 - n_2^*|^2\right], \tag{16}$$

and to take $\mathcal{C} = \min\{\mathcal{C}_1, \mathcal{C}_2\}$. It can be easily checked that $|\bar{n}_1 - n_1|^2 + |n_1 - \gamma\, n_2|^2 + \gamma^2\,|n_2 - \bar{n}_2|^2 \geq \frac{1}{3}\,|\bar{n}_1 - \gamma\,\bar{n}_2|^2$; moreover, bearing in mind the expressions of (n_1^*, n_2^*) together with the fact that $n_1^* + 2\,n_2^* = \bar{n}_1 + 2\,\bar{n}_2 = \bar{n}^0$, we get

$$\bar{n}_1 - \gamma\,\bar{n}_2 = \frac{2+\gamma}{2}\,(\bar{n}_1 - n_1^*) = -\,(2+\gamma)\,(\bar{n}_2 - n_2^*)\,.$$

Therefore (16) becomes

$$|\bar{n}_1 - n_1^*|^2 \geq \frac{6}{2+\gamma} \frac{1}{\min\left\{\frac{d_1}{2\,P(\Omega)}, \frac{d_2}{\gamma\,P(\Omega)}, d_3\right\}}\, \mathcal{C}_2\, |\bar{n}_1 - n_1^*|^2\,,$$

that is true once we put

$$\mathcal{C}_2 = \frac{2+\gamma}{6}\, \min\left\{\frac{d_1}{2\,P(\Omega)}, \frac{d_2}{\gamma\,P(\Omega)}, d_3\right\}.$$

Taking $\mathcal{C} = \min\{\mathcal{C}_1, \mathcal{C}_2\}$ concludes the proof of Lemma 3.2. □

Theorem 3.1 (Exponential convergence to equilibrium).
Let $d_1, d_2 > 0$, and Ω be a bounded regular (C^2) open set of $\mathbb{R}^N$. Let $(n_1(t,\mathbf{x}), n_2(t,\mathbf{x}))$ be a strong solution (that is, in $C^2(\mathbb{R}^+ \times \bar{\Omega})$) to system (1) with homogeneous Neumann boundary conditions and with initial conditions (5). Then, this solution satisfies the following property of exponential decay towards equilibrium:

$$\frac{1}{4}\,\|n_1 - n_1^*\|_2^2 + \frac{\gamma}{2}\,\|n_2 - n_2^*\|_2^2 \leq \Big(E(n_1^0, n_2^0) - E(n_1^*, n_2^*)\Big)\, \mathrm{e}^{-\mathcal{C}\,t}, \qquad (17)$$

where $\mathcal{C}$ is given explicitly in (15).

Proof. Thanks to Lemma 3.2 we have

$$\partial_t\Big[E(n_1, n_2) - E(n_1^*, n_2^*)\Big] \leq -\mathcal{C}\Big[E(n_1, n_2) - E(n_1^*, n_2^*)\Big],$$

thus, by applying Gronwall's lemma and bearing in mind the explicit relative entropy (12), we get the sought inequality (17). □

Unfortunately, in the case of chemically reacting background, the equilibrium state (still unique) is not available in explicit form.[3] However, the strategy presented in this paper, with some additional technicality, allows again explicit estimates on the large time behavior of the relevant solutions.

Acknowledgments

This work was sponsored by MIUR, by INDAM, by GNFM, by the University of Parma (Italy) and by the Franco-italian GdR GREFI-MEFI.

References

1. M. Groppi, A. Rossani and G. Spiga, *J. Phys. A* **33**, 8819 (2000).
2. M. Bisi and G. Spiga, *J. Math. Phys.* **46**, 113301 (2005).
3. M. Bisi, L. Desvillettes and G. Spiga, *Preprint n.11 CMLA*, ENS Cachan, France (2007).
4. M. Kirane, *Bull. Inst. Math. Acad. Sinica* **18**, 369 (1990).
5. L. Desvillettes, *Riv. Mat. Univ. Parma* (7) **7**, 81 (2007).
6. H. Gajewski and K. Gröger, *Math. Nachr.* **177**, 109 (1996).
7. L. Desvillettes and K. Fellner, *J. Math. Anal. Appl.* **319**, 157 (2006).

LINEARIZED EULER'S VARIATIONAL EQUATIONS IN LAGRANGIAN COORDINATES*

GUY BOILLAT AND YUE-JUN PENG

Laboratoire de Mathématiques, CNRS UMR 6620
Université Blaise Pascal (Clermont-Ferrand 2), 63177 Aubière cedex, France
peng@math.univ-bpclermont.fr

We show that in some special case, one-dimensional augmented Euler's variational equations can be written as a linear system in Lagrangian coordinates. This yields the well-posedness of its Cauchy problem for entropy solutions and provides a natural way to solve the corresponding Cauchy problem for Euler's variational equations in this special case. The Born-Infeld system is a typical example.

Keywords: Euler's variational equations, Born-Infeld system.

1. Introduction

Let $L(q_\alpha, q)$ be a Lagrangian depending on $q = (q^1, \cdots, q^N)$ and $q_\alpha = \partial_\alpha q$, the first order derivatives of q with respect to x^α ($\alpha = 0, 1, 2, 3$; $x^0 = t$). The Euler's variational equations read (see [9]) :

$$\partial_\alpha \Big(\frac{\partial L}{\partial q_\alpha} \Big) = \frac{\partial L}{\partial q}. \tag{1}$$

Here, as usual, the summation occurs for repeated indices. It is known that a classical solution of (1) also satisfies four additional conservation laws :

$$\partial_\alpha T^{\alpha\beta} = 0, \quad \beta = 0, 1, 2, 3, \tag{2}$$

where the tensor $T^{\alpha\beta}$ is defined by

$$T^\alpha{}_\beta = \frac{\partial L}{\partial q_\alpha} \cdot q_\beta - L\delta^\alpha_\beta, \quad T^\alpha{}_\beta = g_{\beta\gamma} T^{\alpha\gamma}, \tag{3}$$

with $T^{00} = h$ being the density, $T^{0i} = \sigma^i$ the momentum, $g_{\beta\gamma}$ the Minkowski metric tensor and δ^α_β its mixed components.

*To Tommaso Ruggeri: "Chi trova un amico trova un tesoro".

Let us introduce the field (the breve ˘ denotes the transpose)

$$\breve{u} = \left(\frac{\partial L}{\partial q_0}, q_i, q\right), \quad i = 1, 2, 3. \tag{4}$$

Then h and σ^i are functions of u and (1) can be rewritten as a symmetric Friedrichs-Lax-Godunov system of $5N$ equations

$$\partial_t u + A'^i \partial_i u' = B' u', \tag{5}$$

where u' is the main field defined by

$$\breve{u}' = \frac{\partial h(u)}{\partial u} = \left(q_0, -\frac{\partial L}{\partial q_i}, -\frac{\partial L}{\partial q}\right). \tag{6}$$

The constant symmetric matrices A'^i and the constant skew-symmetric matrix B' satisfy the Duffin-Kemmer-Petiau relations

$$A'^j A'^i A'^k + A'^k A'^i A'^j = \delta^{ik} A'^j + \delta^{ij} A'^k, \quad \forall\, i, j, k = 1, 2, 3 \tag{7}$$

and

$$B' A'^i B' = 0, \quad A'^i A'^j B' + B' A'^j A'^i = \delta^{ij} B',$$

where δ^{ij} is the symbol of Kronecker. A direct consequence of (7) is

$$A'^j A'^i A'^j - \delta^{ij} A'^j = 0. \tag{8}$$

Moreover, the solution satisfies the differential constraints

$$\partial_i c^{ij}(u) = B' A'^j u \ \text{ with } \ c^{ij}(u) = (A'^i A'^j - \delta^{ij} I) u. \tag{9}$$

Note that the matrix B' may be set equal to zero when L only depends on q_α, i.e. is independent of q. In this case, (5) reduces to a system of $4N$ conservation laws without source terms. We refer the reader to [1–5] for details.

2. Augmented Euler's equations

Equations (2) are rewritten (see [2,3]) :

$$\partial_t h + \partial_i h^i = 0, \tag{10}$$

$$\partial_t \sigma^j + \partial_i T^{ij} = 0, \tag{11}$$

with

$$\sigma^j = T^{0j} = \frac{1}{2}\, \breve{u} A'^j u, \tag{12}$$

$$h^i = T^{i0} = \frac{1}{2}\,\breve{u}'A'^i u',$$

$$T^{ij} = \breve{u}'A'^i A'^j u - \delta^{ij} h.$$

As in [4], introducing the energy density $\phi(u, \sigma) > 0$ (which is supposed to be a convex function) and its corresponding main field

$$v' = \frac{\partial \phi}{\partial u}, \quad \sigma'_j = \frac{\partial \phi}{\partial \sigma^j}, \tag{13}$$

which are functions of u and σ, then an additional conservation law holds :

$$\partial_t \phi + \partial_i \left[\frac{1}{2}\,\breve{v}'A'^i v' + \breve{v}'c^{ij}\sigma'_j + (\breve{v}'u + \sigma'_j \sigma^j)\sigma'_i \right] = 0. \tag{14}$$

Now (5) and (11) can be rewritten as :

$$\partial_t u + \partial_i (A'^i v' + c^{ij}\sigma'_j + \sigma'_i u) = B'(v' + A'^j \sigma'_j u), \tag{15}$$

$$\partial_t \sigma^j + \partial_i (\breve{v}'c^{ij} + \sigma'_i \sigma^j + \delta^{ij}\phi') = 0, \tag{16}$$

where the Legendre transform $\phi \longmapsto \phi'$ is defined by

$$\phi' = \breve{v}'u + \sigma'_j \sigma^j - \phi. \tag{17}$$

Equations (10) and (14) coincide when $\sigma = \sigma(u)$ defined by (12). Furthermore, differentiating the relation $\phi(u, \sigma(u)) = h(u)$ yields

$$u' = v' + A'^j \sigma'_j u. \tag{18}$$

By regarding u, σ^j and ϕ as independent variables, (14)-(16) forms a system of $5N + 4$ equations. called augmented Euler's equations.. It is clear that each classical solution of the Euler's equations (5) satisfies (14)-(16). Conversely, define

$$\psi^j = \sigma^j - \frac{1}{2}\,\breve{u}A'^j u.$$

Then (14)-(16) implies that ψ^j is a solution of the linear system :

$$\partial_t \psi^j + \partial_i (\sigma'_i \psi^j) + \psi^i \partial_j \sigma'_i = 0.$$

From the uniqueness of smooth solutions of the Cauchy problem to linear systems, we deduce that $\psi^j = 0$ for all $t > 0$, provided that $\psi^j = 0$ at $t = 0$. Thus, if (u, σ^j, ϕ) is a solution of (14)-(16) and $\sigma^j = \frac{1}{2}\,\breve{u}A'^j u$ at $t = 0$, then u is a solution of the Euler's equations (5). In this sense, the Euler's equations (5) and its augmented system (14)-(16) are equivalent for smooth solutions.

3. Linearized system in one space dimension

In relativity it is convenient to have a symmetric tensor $T^{\alpha\beta}$. It was shown that this condition is equivalent to $T^{0i} = T^{i0}$ (see [5]). Then we may replace equation (14) by

$$\partial_t \phi + \partial_i \sigma^i = 0.$$

Suppose now $B' = 0$. Then u is reduced to a vector in $\mathbb{R}^{4N}$. In one space variable x^1, the set of the augmented Euler's equations leads to :

$$\partial_t \phi + \partial_1 \sigma^1 = 0, \tag{19}$$

$$\partial_t u + \partial_1 (A'^1 v' + c^{1j} \sigma'_j + \sigma'_1 u) = 0, \tag{20}$$

$$\partial_t \sigma^j + \partial_1 (\breve{v}' c^{1j} + \sigma'_1 \sigma^j + \delta^{1j} \phi') = 0. \tag{21}$$

We first show that c^{1j} is constant. Indeed, (9) gives

$$\partial_1 c^{1j} = 0, \quad \partial_t c^{1j} = (A'^1 A'^j - \delta^{1j} I) \partial_t u.$$

Using (5) and the Duffin-Kemmer-Petiau relation (8), we obtain

$$\begin{aligned} \partial_t c^{1j} &= -(A'^1 A'^j - \delta^{1j} I) A'^1 \partial_1 u' \\ &= -(A'^1 A'^j A'^1 - \delta^{1j} A'^1) \partial_1 u' = 0. \end{aligned}$$

By (19), we may introduce an Euler-Lagrange change of coordinates $(t, x) \longmapsto (s, y)$ defined by (see [8,15]) :

$$s = t, \quad dy = \phi dx_1 - \sigma^1 dt. \tag{22}$$

Then in Lagrangian coordinates (s, y), the system (19)-(21) is equivalent to :

$$\partial_s \left(\frac{1}{\phi} \right) - \partial_y \left(\frac{\sigma^1}{\phi} \right) = 0, \tag{23}$$

$$\partial_s \left(\frac{u}{\phi} \right) + \partial_y \left(A'^1 v' + \sigma'_j c^{1j} + \sigma'_1 u - \frac{\sigma^1 u}{\phi} \right) = 0, \tag{24}$$

$$\partial_s \left(\frac{\sigma^j}{\phi} \right) + \partial_y \left(\breve{v}' c^{1j} + \delta^{1j} \phi' + \sigma'_1 \sigma^j - \frac{\sigma^1 \sigma^j}{\phi} \right) = 0. \tag{25}$$

We look for conditions such that the system (23)-(25) is linear with respect to v', σ'_j and ϕ'. In view of the nonlinear terms in the y-derivative of the equations, one necessary condition is

$$\sigma^1 = \phi \sigma'_1. \tag{26}$$

Then the system (23)-(25) is linear if and only if

$$\frac{1}{\phi} = a_* \phi' + \breve{u}_* v' + \sigma_*^k \sigma_k', \tag{27}$$

$$\frac{u}{\phi} = u_*' \phi' + K_* v' + w_*^k \sigma_k', \tag{28}$$

$$\frac{\sigma^j}{\phi} = \sigma_{j*} \phi' + \breve{w}_{j*} v' + b_*^{jk} \sigma_k', \tag{29}$$

where the asterisk stands for constant quantities, a_* and b_*^{jk} are scalars, u_*, u_*', w_*^j and w_{j*} vectors in $\mathbb{R}^{4N}$, σ_*^j and σ_{j*} vectors in $\mathbb{R}^3$ and K_* a square matrix of order $4N$. These quantities are not completely arbitrary. In fact from (17), we have

$$u = \frac{\partial \phi'}{\partial v'}, \quad \sigma^i = \frac{\partial \phi'}{\partial \sigma_i'}. \tag{30}$$

Then differentiating (28) with respect to σ_j' and using (27) yields

$$\frac{1}{\phi} \frac{\partial^2 \phi'}{\partial \sigma_j' \partial v'} + \frac{\partial \phi'}{\partial v'} \left(\sigma_*^j + a_* \frac{\partial \phi'}{\partial \sigma_j'} \right) = u_*' \frac{\partial \phi'}{\partial \sigma_j'} + w_*^j.$$

Similarly,

$$\frac{1}{\phi} \frac{\partial^2 \phi'}{\partial v' \partial \sigma_j'} + \frac{\partial \phi'}{\partial \sigma_j'} \left(u_* + a_* \frac{\partial \phi'}{\partial v'} \right) = \sigma_{j*} \frac{\partial \phi'}{\partial v'} + w_{j*}.$$

These equations give :

$$\sigma_*^j \frac{\partial \phi'}{\partial v'} - u_* \frac{\partial \phi'}{\partial \sigma_j'} = u_*' \frac{\partial \phi'}{\partial \sigma_j'} - \sigma_{j*} \frac{\partial \phi'}{\partial v'} + w_*^j - w_{j*},$$

or equivalently, in view of (30),

$$(\sigma_*^j + \sigma_{j*}) u - (u_* + u_*') \sigma_j = w_*^j - w_{j*}.$$

Since u and σ^j are independent variables, we obtain

$$\sigma_*^j + \sigma_{j*} = 0, \quad u_* + u_*' = 0, \quad w_*^j = w_{j*}. \tag{31}$$

On the other hand, from (26) and (29) we have the relations :

$$\sigma_{1*} = 0, \quad w_{1*} = 0, \quad b_*^{1k} = \delta^{1k}. \tag{32}$$

Let us give an explicit relation between ϕ' and (v', σ'). From (17), we have

$$\frac{\phi'}{\phi} = \breve{v}' \left(\frac{u}{\phi} \right) + \sigma_j' \left(\frac{\sigma^j}{\phi} \right) - 1$$

which together with (27)-(29) yields

$$\phi'(a_*\phi' + \breve{u}_* v' + \sigma^j_* \sigma'_j) = \breve{v}'(u_*\phi' + K_* v' + w^j_* \sigma'_j)$$
$$+(\sigma_{j*}\phi' + \breve{w}_{j*} v' + b^{jk}_* \sigma'_k)\sigma'_j - 1.$$

This implies the following relation

$$a_*\phi'^2 + 2\phi'(\breve{u}_* v' + \sigma_{j*}\sigma'_j) - \breve{v}' K_* v' - 2\,\breve{w}_{j*} v' \sigma'_j + 1 - b^{jk}_* \sigma'_k \sigma'_j = 0. \quad (33)$$

Thus, (27)-(29) and (31)-(33) determine a linear system of the augmented Euler's variational equations in Lagrangian coordinates.

4. An example and remarks

An example of (33) is given by the Born-Infeld Lagrangian [3,6,10] :

$$L = \sqrt{(k - |E|^2)(k + |B|^2) + |E \times B|^2}$$

and the energy density

$$\phi(u, \sigma) = \sqrt{k^2 + k|u|^2 + |\sigma|^2},$$

where $k > 0$ is a constant, $|\cdot|$ denotes the Euclidean norm and

$$u = (D, B), \quad \sigma = D \times B, \quad E = \frac{kD + B \times \sigma}{\phi}.$$

We deduce from (13) and (17) that

$$\phi'(v', \sigma') = -k\sqrt{1 - |v'|^2/k - |\sigma'|^2}, \quad (34)$$

i.e.

$$\phi\,\phi' = -k^2 \quad (35)$$

(see [4]). This example corresponds to (33) with

$$a_* = -k^{-2}, \;\; u_* = 0, \;\; w_{i*} = 0, \;\; K_* = I/k, \;\; b^{jk}_* = \delta^{jk}.$$

The augmented Born-Infeld equations are introduced by Brenier in [7] where many interesting properties of the equations are shown, such as fully linear degeneracy, Galilean invariance etc. A different augmentation of the equations is proposed by Serre [14] for general nonlinear Maxwell equations. The one dimensional Born-Infeld equations are studied by the second author in [11–13] where explicit entropy solutions are constructed for general bounded initial data.

Besides the restrictions (26)-(29), (31) and (32), suppose now that the linear system (23)-(25) is hyperbolic. This requires more restrictions on the constant quantities. In this case, it has been shown that the original system (19)-(21) is hyperbolic, fully exceptional (linearly degenerate) and its Cauchy problem with initial data in $L^\infty(\mathbb{R})$ admits a unique entropy solution (u, σ, ϕ) satisfying all entropy equality (see [13]). It is expected that u is a unique entropy solution of the Euler's variational equations in this special case, as proved for the Born-Infeld equations.

5. General case

In this section, we consider the linearization of the system without the symmetric condition $T^{\alpha\beta} = T^{\beta\alpha}$. In this case, equation (19) needs not be satisfied. Inspired by the treatment in section 3, suppose now that $J \neq 0$ is an independent variable satisfying an additional conservation law :

$$\partial_t \left(\frac{1}{J}\right) + \partial_1 \left(\frac{\sigma_1'}{J}\right) = 0. \tag{36}$$

Then the system of equations (36) and (20)-(21) forms an augmented Euler's equations in this general case for the field (u, σ, J). With the change of variables

$$s = t, \quad dy = \frac{1}{J}\, dx_1 - \frac{\sigma_1'}{J}\, dt, \tag{37}$$

replacing (22) this system becomes :

$$\partial_s J - \partial_y \sigma_1' = 0, \tag{38}$$

$$\partial_s (Ju) + A'^1 \partial_y v' + c^{1i} \partial_y \sigma_i' = 0, \tag{39}$$

$$\partial_s (J\sigma^j) + \partial_y v' c^{1j} + \delta^{1j} \partial_y \phi' = 0, \quad j = 1, 2, 3. \tag{40}$$

Remark that ϕ' appears in (40) only for $j = 1$.

As in section 3, suppose that

$$Ju = K_* v' + u_*^i \sigma_i' + u_* \phi' + v_*, \quad J\sigma^j = h_*^{ji} \sigma_i' + \breve{u}_*^j v' + \sigma_*^j \phi' + a_*^j, \tag{41}$$

with symmetric matrices K_* and (h_*^{ij}). Define

$$F = \frac{1}{2} \breve{v}' K_* v' + \breve{u}_*^i v' \sigma_i' + \frac{1}{2} h_*^{ij} \sigma_i' \sigma_j' + (\breve{u}_* v' + \sigma_*^i \sigma_i') \phi' + \breve{v}_* v' + a_*^i \sigma_i' + a_*. \tag{42}$$

From (17), since

$$d\phi' = u dv' + \sigma^i d\sigma_i', \tag{43}$$

a straightforward calculation shows that

$$dF = (J + \breve{u}_* v' + \sigma_*^i \sigma_i') d\phi'.$$

Hence,

$$F = F(\phi'), \quad F'(\phi') = J + \breve{u}_* v' + \sigma_*^i \sigma_i'. \tag{44}$$

Using (41) and (44), system (38)-(40) can be rewritten as

$$F''(\phi')\partial_s \phi' - \partial_y(\breve{u}_* v' + \sigma_*^i \sigma_i') - \partial_y \sigma_1' = 0,$$

$$\partial_s(K_* v' + u_*^i \sigma_i' + u_* \phi') + A'^1 \partial_y v' + c^{1i} \partial_y \sigma_i' = 0,$$

$$\partial_s(h_*^{ji} \sigma_i' + \breve{u}_*^j v' + \sigma_*^j \phi') + \partial_y v' c^{1j} + \delta^{1j} \partial_y \phi' = 0, \quad j = 1, 2, 3.$$

This is a linear system of unknown variables (v', σ', ϕ') only if $F(\phi')$ is a second order polynomial :

$$F(\phi') = \frac{1}{2}\phi'^2 + b\phi' + c. \tag{45}$$

Together with (44), this implies that

$$J = \phi' - \breve{u}_* v' - \sigma_*^i \sigma_i' + b.$$

Finally, (42) and (45) determine ϕ' as a function of (v', σ'). This function can be seen as a state equation of the Euler's variational equations. It is important to note that system (38)-(40) is not overdetermined since (38) is compatible with (39). Indeed, using (8) and the second relation in (9), we have

$$c^{1j}(\partial_s J - \partial_y \sigma_1') = (A'^1 A'^j - \delta^{1j} I)[\partial_s(Ju) + A'^1 \partial_y v' + c^{1i} \partial_y \sigma_i'], \; j = 1, 2, 3.$$

Thus, (38) is satisfied as soon as $c^{1j} \neq 0$ for one indice j.

6. Conclusion

From (42) and (45), we deduce that

$$\phi'(U') = P_1 \pm \sqrt{P_2}, \tag{46}$$

where P_1 and P_2 are two arbitrary polynomials respectively of first and second order in the variables v' and σ' :

$$P_1(U') = \breve{U}_* U' + a_*, \quad P_2(U') = \breve{U}' A_* U' + 2\breve{V}_* U' + b_*, \quad U' = \begin{pmatrix} v' \\ \sigma_i' \end{pmatrix} \quad i = 1, 2, 3. \tag{47}$$

By (44), (45) and (46), we see that

$$J = \phi' - P_1 = \pm\sqrt{P_2}.$$

With these notations equations (38)-(40) can be simply rewritten

$$\partial_s(\phi' - \breve{U}_* U') - \partial_y \sigma'_1 = 0,$$

$$A_* \partial_s U' + C_* \partial_y U' + U_* \partial_y \sigma'_1 + W_* \partial_y \phi' = 0,$$

where

$$C_* = \begin{pmatrix} A'^1 & c^{1i} \\ \breve{c}^{1i} & 0_3 \end{pmatrix}, \quad W_* = \begin{pmatrix} 0 \\ \delta^{1i} \end{pmatrix}, \quad i = 1, 2, 3.$$

It is also easy to show that $\phi(U)$ and $\phi'(U')$ have the same form. Equation (43) gives

$$U(u,\sigma) = \frac{\partial \phi'}{\partial U'} = U_* \pm \frac{1}{\sqrt{P_2}}(A_* U' + V_*),$$

which implies that

$$U' = \pm\sqrt{P_2}\, A_*^{-1}(U - U_*) - A_*^{-1} V_*. \tag{48}$$

Substituting into (47) results in

$$P_2\{1 - (\breve{U} - \breve{U}_*) A_*^{-1}(U - U_*)\} = b_* - \breve{V}_* A_*^{-1} V_* = c_*. \tag{49}$$

Now $\phi(U)$ and $\phi'(U')$ are linked by (17) and by (46)-(48). Then

$$\begin{aligned}\phi(U) &= \breve{U}'U - \phi' = \breve{U}'(U - U_*) - a_* \mp \sqrt{P_2} \\ &= \mp\sqrt{c_*\{1 - (\breve{U} - \breve{U}_*) A_*^{-1}(U - U_*)\}} - (\breve{U} - \breve{U}_*) A_*^{-1} V_* - a_*.\end{aligned}$$

In the special case of the Born-Infeld equations (see section 4),

$$a_* = U_* = V_* = 0, \quad b_* = k^2, \quad A_* = -\begin{pmatrix} kI_6 & 0 \\ 0 & k^2 I_3 \end{pmatrix}.$$

Also, according to (19) and (36)

$$J = \phi', \quad \sigma^i = \phi \sigma'_i.$$

By (49),

$$\phi = \mp\sqrt{k^4/P_2} = -k^2/\phi'$$

in agreement with (35).

References

1. G. Boillat, Chocs caractéristiques, *C.R. Acad. Sci. Paris, Série A*, 274 (1972), 1018-1021.
2. G. Boillat, Chocs dans les champs qui dérivent d'un principe variationnel : équation de Hamilton-Jacobi pour la fonction génératrice, *C.R. Acad. Sci. Paris, Série A*, 283 A (1976), 539-542.
3. G. Boillat, Non linear hyperbolic fields and waves, Lecture Notes Math., Vol. 1640, 1-47, Springer-Verlag, 1996.
4. G. Boillat, Euler's variational equations with independent momentum, *Il Nuovo Cimento*, 119 B (2004), 839-847.
5. G. Boillat and T. Ruggeri, Energy momentum, wave velocities and characteristic shocks in Euler's variational equations with application to the Born-Infeld theory, *J. Math. Phys.* 45 (2004), 3468-3478.
6. M. Born and L. Infeld, Foundation of the new field theory, *Proc. Roy. Soc. London*, A144 (1934), 425-451.
7. Y. Brenier, Hydrodynamic structure of the augmented Born-Infeld equations, *Arch. Rat. Mech. Anal.* 172 (2004), 65-91.
8. R. Courant and K.O. Friedrichs, Supersonic Flow and Shock Waves, Interscience publishers, New-York, 1948.
9. L.C. Evans, Partial Differential Equations, American Math. Society, 1998.
10. G.W. Gibbons and C.A.R. Herdeiro, Born-Infeld theory and stringy causality, *Phys. Rev.* D63 (2001), 064006.
11. Y.J. Peng, Explicit solutions for 2×2 linearly degenerate systems, *Appl. Math. Letters*, 11 (1998), 75-78.
12. Y.J. Peng, Entropy solutions of Born-Infeld systems in one space dimension, *Rend. Circ. Mat. Palermo.* Serie II, 78 (2006), 259-271.
13. Y.J. Peng, Euler-Lagrange change of variables in conservation laws, *Nonlinearity*, 20 (2007), 1927-1953.
14. D. Serre, Hyperbolicity of the nonlinear models of Maxwell's equations, *Arch. Rat. Mech. Anal.* 172 (2004), 309-331.
15. D. Wagner, Equivalence of the Euler and Lagrangian equations of gas dynamics for weak solutions, *J. Diff. Eqns.* 68 (1987), 118-136.

A 3-DIMENSIONAL INVERSE PROBLEM OF GEOMETRICAL OPTICS FOR CONTINUOUS ISOTROPIC INHOMOGENEOUS MEDIA

F. BORGHERO [1] and G. BOZIS [2]

[1] *Dipartimento di Matematica Università di Cagliari (Italy); e-mail: borghero@unica.it*
[2] *Department of Physics, University of Thessaloniki (Greece); e-mail: gbozis@auth.gr*

In the framework of Geometrical Optics we consider the general inverse problem consisting in obtaining refractive-index distributions $n = n(x, y, z)$, of a 3-dimensional transparent inhomogeneous isotropic medium, compatible with a given two-parametric family of monochromatic light rays $f(x, y, z) = c_1, g(x, y, z) = c_2$. We establish a system of two first order linear $PDEs$ relating the assigned family of light rays with all possible refractive-index distributions compatible with this family. We prove that, in order for this system to have solutions, the family must be a normal congruence. We conclude with some appropriate examples.

1. Introduction

In this work we shall be concerned with the propagation of light in a continuous transparent inhomogeneous and isotropic medium, dispersive or not, from the point of view of geometrical optics which is characterized by the neglect of the wavelength ($\lambda \to 0$). In the past only occasional attempts were made to study artificial inhomogeneous media and such investigations presented for a long time a merely academic interest. This was mainly due to the difficulty, or even impossibility, of constructing a medium of variable refractive index with very high degree of accuracy, as required in Optics. Recently the situation has considerably changed due to the advent of microwave techniques and their applications and also because inhomogeneous media are now playing an important role in integrated and fiber optics[1].[2] Taking into account these considerations, it is desirable to know what are the explicit equations that relate refractive-index distributions with compatible families of light rays. The aim of this paper is to study the following version of the general 3-dimensional inverse problem of geometrical optics: *we are given a congruence, that is a two-parametric family of regular spatial curves whose equations*

$$(\Gamma) \quad f(x,y,z) = c_1, \quad g(x,y,z) = c_2, \tag{1}$$

involve two independent parameters, c_1, c_2, *in such a way that only one curve passes through each point* (x_0, y_0, z_0). *We suppose that* Γ *is lying in a 3-dimensional inhomogeneous isotropic medium and we want to find all possible refractive-index distributions* $n(x,y,z)$ *allowing for the creation of the given congruence as family of monochromatic light rays.*
We establish, as our main result, a system of two linear first order $PDEs$ (the system (19) below) in the unique unknown function $N(x,y,z) = logn(x,y,z))$. The solution of the system (19), if it exists, is generally unique and gives the refractive-index profile compatible with the assigned family (1). Moreover, we prove analytically that, for the system (19) to admit of a nontrivial solution, the given two-parametric family (1) needs not merely be a simple congruence but must be a *normal congruence* Γ. In other words, the curves belonging to Γ must constitute an *orthotomic system of rays*, as we know from electromagnetic theory of light.[3]
In particular, if the medium is 2-dimensional, the system (19) reduces to a single first order linear PDE, as we already established it in our previous paper.[4] Media of such type are of great interest in the applications, for instance: (i) in optical systems with spherical or cylindrical symmetry (e.g. Luneburg lens, or Fletcher lens); (ii) in applications to microwaves constrained to travel along a given surface. In fact, it is well known that a wave guide can be formed by two surface-parallel metal plates where the spacing between the plates is sufficiently small. We conclude the paper with two applications of the theory (two exact solutions of the system (19)).

2. Some useful geometrical tools

In our study we need consider two useful tangent vectors to the curves (1), $\vec{\delta}$ and $\vec{\epsilon}$ and two scalars $\alpha(x,y,z)$, $\beta(x,y,z)$, connected with the differential geometry of the light rays. To this end we introduce the two gradients $gradf = (f_x, f_y, f_z)$, $gradg = (g_x, g_y, g_z)$, and their vector product

$$\vec{\delta} = gradf \times gradg, \tag{2}$$

which is tangent in a typical point to a generic curve of the family (1).
If $\delta_1, \delta_2, \delta_3$ are the Cartesian components of this tangent vector (under the assumption that $\delta_1 \neq 0$) we have

$$\vec{\delta} = \delta_1 \mathbf{i} + \delta_2 \mathbf{j} + \delta_3 \mathbf{k} = \delta_1 \Big(\mathbf{i} + \frac{\delta_2}{\delta_1} \mathbf{j} + \frac{\delta_3}{\delta_1} \mathbf{k} \Big) \tag{3}$$

where, **i**, **j**, **k** are the unit vectors along the perpendicular axes Ox, Oy, Oz and

$$\delta_1 = \begin{vmatrix} f_y & f_z \\ g_y & g_z \end{vmatrix}, \qquad \delta_2 = \begin{vmatrix} f_z & f_x \\ g_z & g_x \end{vmatrix}, \qquad \delta_3 = \begin{vmatrix} f_x & f_y \\ g_x & g_y \end{vmatrix}. \tag{4}$$

We consider also another tangent vector

$$\vec{\epsilon} = \frac{\vec{\delta}}{\delta_1} = \mathbf{i} + \frac{\delta_2}{\delta_1}\mathbf{j} + \frac{\delta_3}{\delta_1}\mathbf{k} \tag{5}$$

whose components are $\{1, \alpha, \beta\}$. The functions $\alpha(x, y, z)$, $\beta(x, y, z)$, in view of (5), are defined as

$$\alpha(x, y, z) = \frac{\delta_2}{\delta_1}, \qquad \beta(x, y, z) = \frac{\delta_3}{\delta_1}. \tag{6}$$

If we parametrize any curve of the family (1) by x, and if we differentiate with respect to x the two equations (1), we have

$$f_x + f_y y' + f_z z' = 0, \qquad g_x + g_y y' + g_z z' = 0 \tag{7}$$

where primes denote differentiation with respect to x. Thus, if we solve the system (7) with respect to y' and z', in view of (4) and (6) we obtain

$$y' = \alpha(x, y, z), \qquad z' = \beta(x, y, z). \tag{8}$$

We shall refer to α and β (which are essentially geometrical elements of the rays) as the *slope functions* of the family (1).
Now we can express the unit tangent vector **t** in terms of $\vec{\epsilon}$

$$\mathbf{t} = \frac{\vec{\epsilon}}{|\vec{\epsilon}|} = \frac{\vec{\epsilon}}{(1 + \alpha^2 + \beta^2)^{1/2}}. \tag{9}$$

3. Refractive-index distribution system (RIDS)

Suppose that, for a transparent isotropic inhomogeneous 3-dimensional medium, we are given a congruence Γ. We shall find a system of $PDEs$, in the unique unknown function $n(x, y, z)$, whose solutions give all possible refractive-index distributions allowing for the creation of the given family of curves as monochromatic light rays. We start with Fermat's principle of the shortest optical path

$$\delta \int_{P_0}^{P_1} n ds = 0, \tag{10}$$

where the line element ds, in Cartesian orthogonal coordinates, is

$$ds = \sqrt{dx^2 + dy^2 + dz^2}. \tag{11}$$

Equation (10) reads

$$\delta \int_{x_0}^{x_1} n(x,y,z)\sqrt{1+y'^2+z'^2}dx = 0 \tag{12}$$

where primes denote derivatives with respect to the variable x. The variational equation (12) is equivalent to the Euler-Lagrange's system of three $ODEs$ (see Born-Wolf,[3] Appendix 1, pag. 869). Due to our parametrisation, the first equation becomes an identity, so we have only two equations

$$\begin{aligned} n_y\sqrt{1+y'^2+z'^2} - \tfrac{d}{dx}\left[\tfrac{ny'}{\sqrt{1+y'^2+z'^2}}\right] &= 0 \\ n_z\sqrt{1+y'^2+z'^2} - \tfrac{d}{dx}\left[\tfrac{nz'}{\sqrt{1+y'^2+z'^2}}\right] &= 0 \end{aligned} \tag{13}$$

where subscripts denote partial derivatives of the function $n(x, y, z)$.
This system of $ODEs$ for the light rays, as far as we know, has been interpreted, up to now, only from the point of view of the direct problem of geometrical optics, that is: given the refractive index $n(x, y, z)$ of the medium, to solve (13) as a system of two ordinary nonlinear differential equations of the second order in the two unknown functions $y(x)$ and $z(x)$ with given initial conditions (y_0, z_0, y'_0, z'_0) at $x = x_0$.
We want now to transform the above system to make it suitable for inverse problem considerations that is: given a congruence Γ to find compatible refractive-index profiles. To this end we proceed as follows: taking into account that $\vec{\epsilon} = \{1, \alpha, \beta\}$, we have $|\vec{\epsilon}| = (1+\alpha^2+\beta^2)^{1/2}$ and in view of (8) we obtain

$$y'' = \vec{\epsilon}\cdot grad\alpha, \qquad z'' = \vec{\epsilon}\cdot grad\beta. \tag{14}$$

So, after some algebra, we can rewrite the system (13) as

$$\begin{aligned} \alpha n_x - (1+\beta^2)n_y + \alpha\beta n_z &= \left[\tfrac{\alpha(\alpha\alpha_0+\beta\beta_0)}{1+\alpha^2+\beta^2} - \alpha_0\right]n \\ \beta n_x + \alpha\beta n_y - (1+\alpha^2)n_z &= \left[\tfrac{\beta(\alpha\alpha_0+\beta\beta_0)}{1+\alpha^2+\beta^2} - \beta_0\right]n \end{aligned} \tag{15}$$

where, for the sake of simplicity, we have set

$$\alpha_0 = \vec{\epsilon}\cdot grad\alpha, \; \beta_0 = \vec{\epsilon}\cdot grad\beta. \tag{16}$$

The system (15) may be also written in the form

$$\begin{aligned} \alpha n_x - n_y + \Omega_1 n &= 0 \\ \beta n_x - n_z + \Omega_2 n &= 0 \end{aligned} \tag{17}$$

where

$$\Omega_1 = \frac{\vec{\epsilon} \cdot grad\alpha}{1 + \alpha^2 + \beta^2}, \qquad \Omega_2 = \frac{\vec{\epsilon} \cdot grad\beta}{1 + \alpha^2 + \beta^2}. \tag{18}$$

or, setting $N(x, y, z) = logn(x, y, z)$, in the simpler form

$$\begin{aligned} \alpha N_x - N_y + \Omega_1 &= 0 \\ \beta N_x - N_z + \Omega_2 &= 0. \end{aligned} \tag{19}$$

4. Discussion on the RIDS

The previous system (19), may called *refractive-index distribution system* (RIDS) and can be written also in the form:

$$F \equiv \alpha N_x - N_y + \Omega_1 = 0, \quad G \equiv \beta N_x - N_z + \Omega_2 = 0. \tag{20}$$

In general, a system as (20), of two *PDEs* in the unique unknown function $N = N(x, y, z)$ is not expected to be compatible. In fact, it is known that, if the two equations (20) are satisfied, then a third linear *PDE* (which is neither trivial nor an algebraic combination of the two equations (20)) must also be satisfied. If we set temporarily $x = x_1$, $y = x_2$, $z = x_3$ and $N_x = p_1$, $N_y = p_2$, $N_z = p_3$, this third equation reads

$$\sum_{i=1}^{3} \left(\frac{\partial F}{\partial p_i} \frac{\partial G}{\partial x_i} - \frac{\partial F}{\partial x_i} \frac{\partial G}{\partial p_i} \right) = 0$$

where the functions $F(x_1, x_2, x_3, p_1, p_2, p_3)$ and $G(x_1, x_2, x_3, p_1, p_2, p_3)$ are given by (20). So we obtain the new *PDE*

$$(\alpha\beta_x - \beta\alpha_x + \alpha_z - \beta_y)N_x = \beta\Omega_{1x} - \alpha\Omega_{2x} + \Omega_{2y} - \Omega_{1z}. \tag{21}$$

The system (19) or (20) admits nontrivial solution if and only if the equation (21) becomes an identity. But this is the case when

$$\alpha\beta_x - \beta\alpha_x + \alpha_z - \beta_y = 0. \tag{22}$$

In this case, the second member of (21) also becomes zero. It is easy to verify that the condition (22) can be written in vectorial form, i.e.

$$\vec{\epsilon} \cdot curl\vec{\epsilon} = 0, \ \textit{or equivalently} \ \mathbf{t} \cdot curl\mathbf{t} = 0. \tag{23}$$

So we have also established an analytical proof that the family of rays defined from Fermat's principle must be a *normal congruence*,[3] that is not obvious. We shall refer to equation (22) or (23) as the *normality condition* of the family. Therefore we can state the following two propositions:

Proposition 1: *In order that the system of PDE (19), derived from Fermat's principle, is compatible, i.e. admits nontrivial solutions, the family of curves (1) must be a normal congruence.*

Proposition 2: *In a continuous 3-dimensional transparent medium, inhomogeneous and isotropic, all the refractive-index distributions $n(x, y, z)$ that allow for the creation, as light rays, of a given normal congruence must satisfy the system of PDE (19).*

Remark: If either $\alpha = 0$ (or $\beta = 0$), the refractive index depends merely on two (not three) independent variables, so the problem becomes 2-dimensional, and the two equations (19) reduce to one.[4]

We conclude with some examples.

Example 1: Consider, in the 3-dimensional space, the two-parametric family of conics

$$f \equiv y - z = c_1, \qquad g \equiv (y+z)^2 - 4x = c_2; \tag{24}$$

it is easy to verify that this family is a normal congruence, so it can be generated as an orthotomic system of light rays inside a medium with a refractive index solution of the system of *PDEs* (17). One can check that

$$n(x, y, z) = e^x \sqrt{(y+z)^2 + 2} \tag{25}$$

is a compatible refractive index with the given family of light rays.

Example 2: The family of circles

$$f \equiv x + y + z = c_1, \qquad g \equiv x^2 + y^2 + z^2 = c_2 \tag{26}$$

is a normal congruence and therefore can be generated as an orthotomic system of light rays in a medium with a refractive index solution of the system of *PDE* (17). The pertinent refractive index

$$n(x, y, z) = \frac{1}{(x-z)^2} \sqrt{x^2 + y^2 + z^2 - (xy + xz + yz)} \tag{27}$$

is compatible with the assigned family of circles.

References

1. Kravtsov, Yu. A., and Orlov, Yu. I., *Geometrical Optics of inhomogeneous media*, Springer Series on Wave phenomena, 6., Springer-Verlag, Berlin, 1990.
2. Hindy, A., Refractive-index profile in fiber optics, Microwave and Opt. Tech. Letters, 29, 252-256 (2001).
3. Born, M. and Wolf, E., *Principles of Optics*, 7th edn.(revised), Cambridge University Press, 2002.
4. Borghero, F. and Bozis, G., A two-dimensional inverse problem of geometrical optics, *J. Phys. A: Math. Gen.*, **38** (2005) 175-184.

DERIVATION OF A QUANTUM HYDRODYNAMIC MODEL IN THE HIGH-FIELD CASE

G. BORGIOLI*, G. FROSALI†, C. MANZINI†

* *Dip. di Elettronica e Telecomunicazioni,* † *Dip. di Matematica Applicata, Università di Firenze, Via S.Marta 3, I-50139 Firenze, Italy*
**E-mail: giovanni.borgioli@unifi.it*

A fluid-dynamical set of equations is derived starting from a quantum kinetic description of transport in high-field regime.

1. Introduction and formulation of the scaled problem

We derive a hydrodynamical model for a quantum system evolving in high-field regime, namely when advection and dissipative terms are dominant and have the same order of magnitude. To this aim we consider a rescaled version of the Wigner equation with unknown the quasi-distribution function $w = w(x, v, t), (x, v) \in \mathbb{R}^{2d}$, $t > 0$, describing the time-evolution of a quantum system with d degrees of freedom, under the effect of an external potential $V = V(x), x \in \mathbb{R}^d$ and a "collisional" term $Q(w)$. It reads

$$\epsilon\partial_t w + \epsilon v \cdot \nabla_x w - \Theta[V]w = Q(w)\,, \quad (x, v) \in \mathbb{R}^{2d}, \quad t > 0. \tag{1}$$

The potential V enters through the pseudo-differential operator $\Theta[V]$ defined by

$$(\Theta[V]w)(x, v, t) = \frac{i}{(2\pi)^{d/2}} \int_{\mathbb{R}^d} \delta V(x, \eta)\mathbf{F}w(x, \eta, t)e^{iv\cdot\eta}\, d\eta\,, \tag{2}$$

where

$$\delta V(x, \eta) := \frac{1}{\hbar}\left[V\left(x + \frac{\hbar\eta}{2m}\right) - V\left(x - \frac{\hbar\eta}{2m}\right)\right]$$

and ϵ is a parameter corresponding to the Knudsen number. $\mathbf{F}f(\eta) \equiv [\mathbf{F}_{v\to\eta}f](\eta)$ denotes the Fourier transform of w from v to η. In the Fourier-transformed space $\mathbb{R}^d_x \times \mathbb{R}^d_\eta$ the operator $\Theta[V]$ is the multiplication operator by the function $i\,\delta V$; in symbols,

$$\mathbf{F}\left(\Theta[V]w\right)(x, \eta) = i\,\delta V(x, \eta)\mathbf{F}w(x, \eta)\,. \tag{3}$$

We choose the collisional term in the relaxation-time BGK form, i.e. $Q(w) = -\nu(w - w_{\rm eq})$. It describes the dissipative mechanism due to the interaction with the environment, which leads the system to a state $w_{\rm eq}$ of thermodynamical equilibrium with temperature $1/k\beta$. Explicitly,

$$w_{\rm eq}(x,v,t) = n(x,t)\left(\frac{\beta m}{2\pi}\right)^{d/2} e^{-\beta m v^2/2} \times \left\{1 + \hbar^2 \frac{\beta^2}{24}\left[-\frac{1}{m}\Delta V(x) + \beta \sum_{r,s=1}^{d} v_r v_s \frac{\partial^2 V(x)}{\partial x_r x_s}\right] + O(\hbar^4)\right\}. \quad (4)$$

This is obtained by inserting in the Wigner thermodynamical equilibrum function [8] the parameter $c = c(x,t)$ and then by assuming

$$\int w_{\rm eq}(x,v,t)\,dv = \int w(x,v,t)\,dv =: n[w](x,t) \equiv n(x,t). \quad (5)$$

The version of Wigner equation (1) under examination corresponds to the case in which a "strong" external potential is included. Accordingly the potential characteristic time t_V and the mean free time t_C between interactions of the system with the background are assumed to be comparable and small. The Knudsen number ϵ is inserted in order to identify terms of the same order of magnitude. Since dissipative interaction and advection terms coexist during the evolution, the high-field relaxation-time state shall be determined by considering the joint actions of collisions and external field. The corresponding distribution function is the solution of Eq.1 for $\epsilon = 0$ and it shall be adopted to close the system for the fluid-dynamical moments of the Wigner function

$$n(x,t) := \int_{R_v^d} w(x,v,t)\,dv\,, \quad nu(x,t) := \int_{R_v^d} v\,w(x,v,t)\,dv\,,$$
$$e(x,t) := \int_{R_v^d} \frac{1}{2} m|v|^2\, w(x,v,t)\,dv\,. \quad (6)$$

Let us rewrite the right-hand side part of Eq. (1) as

$$Q(w) := -(\nu\, w - \Omega\, w)\,,$$

where the operator Ω is defined by

$$\Omega w(x,v) = \nu\, n[w](x)\left[F(v) + \hbar^2 F^{(2)}(x,v)\right],$$

the function $F(v)$ is the normalized Maxwellian $F(v) := \left(\frac{\beta m}{2\pi}\right)^{d/2} e^{-\beta m v^2/2}$ and $F^{(2)}$ is the $O(\hbar^2)-$coefficient in w_{eq}

$$F^{(2)}(x,v) \equiv F^{(2)}[V](x,v) = \frac{\beta^2}{24}\left[-\frac{1}{m}\Delta V + \beta \sum_{r,s=1}^{d} v_r v_s \frac{\partial^2 V}{\partial x_r x_s}\right] F(v)\,.$$

2. The high-field relaxation-time function

Let us denote by $M = M(x,v)$ the solution of the following equation

$$\theta[V]w + Qw = 0 \tag{7}$$

In Ref. 7 it is proved that exists a unique solution M with $\int M(x,v,t)\,dv = \int w(x,v,t)\,dv$, by setting the problem in the Hilbert space $L^2(\mathbb{R}^{2d}; 1+|v|^{2k})$ with $2k > d$, provided V is regular enough ($V \in H^{k+2}$, e.g.). By taking formally the moments $(mv, mv\otimes v, mv|v|^2)$ of (7), it is possible to compute the corresponding moments of the relaxation-time function M. For brevity, we omit the proof here.

Lemma 2.1. *Let M be the solution of Eq.* (7) *such that* $\int M(x,v,t)\,dv = \int w(x,v,t)\,dv = n[w](x,t)$. *Then*

$$\int vM\,dv = -n\frac{\nabla V}{\nu m} := nu_M\,, \tag{8}$$

$$\int v\otimes mvM\,dv = n\frac{\mathcal{I}}{\beta} + 2mnu_M\otimes u_M + \frac{\beta\hbar^2}{12m}n\nabla\otimes\nabla V\,, \tag{9}$$

$$\int \frac{1}{2}m|v|^2 M\,dv = n\frac{d}{2\beta} + mnu_M^2 + \frac{\beta\hbar^2}{24m}n\Delta V\,, \tag{10}$$

$$\int \frac{1}{2}mv|v|^2 M\,dv = -\frac{\hbar^2}{8m}n\Delta u_M + u_M\int v\otimes mvM\,dv + u_M\int\frac{1}{2}m|v|^2 M\,dv\,.$$

The moments of the function M can be compared with the corresponding ones of the shifted-Maxwellian [3]. To the high-field relaxation-time distribution function is associated the fluid velocity in (8), accordingly the expression for the energy density $e = e(x,t)$ calculated at relaxation-time is

$$\int\frac{1}{2}m|v|^2 M\,dv = \frac{1}{2}mnu_M^2 - \frac{1}{2}\mathrm{tr}\mathbf{P}_M\,,$$

since it is natural to define the pressure tensor $\mathbf{P}_M$ as

$$\mathbf{P}_M := -n\frac{\mathcal{I}}{\beta} - mnu_M\otimes u_M - \frac{\beta\hbar^2}{12m}n\nabla\otimes\nabla V\,, \tag{11}$$

in analogy with Gardner [3]. The high-field assumption yields the additional tensor

$$-mnu_M \otimes u_M = -mn\frac{\nabla V}{\nu m} \otimes \frac{\nabla V}{\nu m}$$

in the pressure term. Moreover, by comparing the expression (11) with the one computed with the shifted-Maxwellian, the heat-flux term q can be defined as

$$q := -\frac{\hbar^2}{8m} n\Delta u_M + mnu_M \left(u_M \otimes u_M\right) . \tag{12}$$

Again, the heat-flux term, in the high-field case, consists of the expected quantum term and an additional one which is cubic in the fluid velocity. Next, we shall give another motivation for the definitions of the pressure and the heat-flux terms in the high-field case.

3. Derivation of the high-field QHD system

In this section we shall derive the QHD system, in the high field case. Starting from Eq. (1) with $\epsilon > 0$ and multiplying by $1, mv, \frac{1}{2}m|v|^2$ and integrating in dv, one gets

$$\begin{aligned} \partial_t n + \nabla_x \cdot (nu) &= \tfrac{1}{\epsilon} \textstyle\int Q(w)\, dv \\ \partial_t(mnu) + \nabla_x \cdot \textstyle\int mv \otimes vw\, dv + \tfrac{1}{\epsilon} \textstyle\int mv\Theta[V]w\, dv &= \tfrac{1}{\epsilon} \textstyle\int mvQ(w)\, dv \\ \partial_t e + \nabla_x \cdot \textstyle\int \tfrac{1}{2} mv|v|^2 w\, dv + \tfrac{1}{\epsilon} \textstyle\int \tfrac{1}{2} m|v|^2 \Theta[V] w\, dv &= \tfrac{1}{\epsilon} \textstyle\int \tfrac{1}{2} m|v|^2 Q(w)\, dv \end{aligned} \tag{13}$$

where n, nu, e are defined in (6). We use the relaxation-time distribution function M to close the $\mathcal{O}(\frac{1}{\epsilon})$ terms. Since M satisfies Eq. (7), that is the Wigner equation with $\epsilon = 0$, Eqs. (13) reduce immediately to

$$\begin{aligned} \partial_t n + \nabla_x \cdot (nu) &= 0 , \\ \partial_t(mnu) + \nabla_x \cdot \int mv \otimes vw\, dv &= 0 , \\ \partial_t e + \nabla_x \cdot \int \frac{1}{2} mv|v|^2 w\, dv &= 0 , \end{aligned} \tag{14}$$

where n, u and e are the unknown functions. Then we still need to close the integral terms. Eq. (8) indicates the fluid velocity computed at relaxation time. Accordingly, the velocity momentum can be expressed in terms of the deviation from the relaxation-time velocity, that is,

$$\begin{aligned} n\, u = \int vw\, dv \;&=\; \int \left(v - \frac{1}{n}\int vM\right) w\, dv + \frac{1}{n}\int vM\, dv \int w\, dv \\ &= \int (v - u_M)\, w\, dv + nu_M . \end{aligned}$$

Analogously, the tensor $\int mv \otimes vw\,dv$ can be written as

$$\int mv \otimes vw\,dv = \int m\,(v-u_M)\otimes(v-u_M)\,w\,dv + \int \left(mv\otimes v - m\,(v-u_M)\otimes(v-u_M)\right)w\,dv$$

and, by some manipulations in the same spirit of the calculations in Ref. 5, it becomes

$$\int mv \otimes vw\,dv = \int m\,(v-u_M)\otimes(v-u_M)\,w\,dv + mnu\otimes u - \frac{m}{n}\int (v-u_M)\,w\,dv \otimes \int (v-u_M)\,w\,dv\,. \tag{15}$$

We shall use the function M to close the previous expression: let us define

$$\begin{aligned}\mathbf{P}_M := -\int m\,(v-u_M)\otimes(v-u_M)\,M dv &= -\int mv\otimes vM(x,v)\,dv + mnu_M\otimes u_M \\ = -n\frac{\mathcal{I}}{\beta} - mnu_M\otimes u_M - \frac{\beta\hbar^2}{12m}n\nabla\otimes\nabla V\,,\end{aligned} \tag{16}$$

where the last equality is obtained by (9). The definition of the tensor $\mathbf{P}_M$ is consistent with classical kinetic theory [6] (apart from the sign that is a matter of convention) and with the quantum case [4]. Moreover, it coincides with (11), given in analogy with Gardner [3]. Thus (15) reads

$$\int mv\otimes vw\,dv = mnu\otimes u - \mathbf{P}_M\,, \tag{17}$$

and (16) has to be compared with the corresponding results obtained in Ref. 1 and Ref. 2, respectively. In the same spirit,

$$\begin{aligned}\int \frac{1}{2}mv|v|^2 w\,dv &= \int \frac{1}{2}m\,(v-u_M)\,|v-u_M+u_M|^2\,w\,dv + u_M\int\frac{1}{2}m|v|^2 w\,dv \\ &= \int \frac{1}{2}m\,(v-u_M)\,|v-u_M|^2\,w\,dv + (e\mathcal{I}-\mathbf{P}_M)u \\ &+ |u_M|^2\int\frac{1}{2}m\,(v-u_M)\,w\,dv\,.\end{aligned} \tag{18}$$

By using M and defining

$$q_M := \int \frac{1}{2}m\,(v-u_M)\,|v-u_M|^2\,M\,dv\,, \tag{19}$$

(18) becomes

$$\int \frac{1}{2}mv|v|^2 w\,dv = q_M + (e\mathcal{I}-\mathbf{P}_M)u\,. \tag{20}$$

The expression (19) can be rewritten as

$$q_M = \frac{1}{2}\left[\int mv|v|^2 M\, dv - 2u_M \int v \otimes mvM\, dv + |u_M|^2 \int mvM\, dv \right. \tag{21}$$
$$\left. - u_M \int m|v|^2 M\, dv + 2u_M \otimes u_M \int mvM\, dv - mnu_M|u_M|^2 \right]$$

and, by using Lemma 2.1,

$$q_M = -\frac{\hbar^2}{8m} n\Delta u_M + mnu_M\left(u_M \otimes u_M\right). \tag{22}$$

The definition (19) of the heat flux is consistent with the classical and quantum literature, moreover it coincides with (12), previously introduced in analogy with Gardner [3].
Finally, by using in Eqs. (14) the expressions (17) for the moment $\int v \otimes mvw\, dv$ and (20) for $\int \frac{1}{2} mv|v|^2 w\, dv$, we derive

$$\begin{aligned} \partial_t n + \nabla_x \cdot (nu) &= 0\,, \\ \partial_t (mnu) + \nabla_x \cdot mnu \otimes u - \nabla_x \cdot \mathbf{P}_M &= 0\,, \\ \partial_t e + \nabla_x \cdot (eu) - \nabla_x \cdot (\mathbf{P}_M u) + \nabla_x \cdot q_M &= 0\,. \end{aligned} \tag{23}$$

In conclusion, we remark that, unlike Ref. 3, the high-field relaxation-time function is adopted for the closure procedure. Accordingly, pressure tensor and heat-flux differ from the quantum (standard) ones in terms that are quadratic and cubic, respectively, in the ϵ^0 order velocity field u_M.

References

1. P. Degond, F. Méhats, C. Ringhofer, Quantum energy-transport and drift-diffusion models, *J. Stat. Phys.* **118**, 625–665 (2005).
2. P. Degond, C. Ringhofer, Quantum moment hydrodynamics and the entropy pronciple, *J. Stat. Phys.* **112**, 587–628 (2003).
3. C. Gardner, The Quantum Hydrodynamic Model for Semiconductor Devices, *SIAM J. App. Math.* **54(2)**, 409-427 (1994).
4. C. Gardner, Resonant Tunneling in the Quantum Hydrodynamic Model, *VLSI Design* **3**, 201-210 (1995).
5. A. Jüngel, D. Matthes, J.P. Milišić, A derivation of new quantum hydrodynamic equations using entropy minimization, to appear in *SIAM J. Appl. Math.* (2006).
6. C.D. Levermore, Moment Closure Hierarchies for Kinetic Theories, *J. Stat. Phys.* **83**, 1021–1065 (1996).
7. C. Manzini, G. Frosali, Rigorous drift-diffusion asymptotics of a strong-field quantum transport equation, submitted (2006)
8. E. Wigner, On the quantum correction for thermodynamic equilibrium, *Phys. Rev.* **40**, 749-759 (1932).

ON THE MAGNETIC RAYLEIGH-BÉNARD PROBLEM FOR COMPRESSIBLE FLUIDS

F. BRINI

Dipartimento di Matematica and C.I.R.A.M.
University of Bologna,
via Saragozza, 8 40123 Bologna, Italy,
E-mail: brini@ciram.unibo.it

G. MULONE AND M. TROVATO

Dipartimento di Matematica e Informatica,
Città Universitaria, Viale A. Doria, 6, 95125, Catania, Italy
E-mail: mulone@dmi.unict.it, trovato@dmi.unict.it

The linear stability of the magnetic Rayleigh Bénard problem for a general compressible horizontal layer of an electrically conducting Newtonian fluid is studied and a comparison of the stability results with the classical magnetic Bénard problem in the Boussinesq approximation is presented. In particular, the physical cases of mercury, liquid sodium and solar plasma have been considered. We find a stabilizing effect of the magnetic field, a dependence of the critical instability parameters on the depth of the layer (as in the case of fluids without magnetic field) and a good agreement with the experiments and the classical results of Chandrasekhar.

Keywords: magnetic Rayleigh-Bénard problem, linear stability, compressible fluids, stabilizing effect.

1. Introduction

The magnetic Rayleigh-Bénard problem concerns the onset of thermal instability in a horizontal layer of an electrically conducting fluid heated from below in the presence of an external magnetic fluid. This case was studied by many authors both from a theoretical and from an experimental point of view. With regard to the theoretical aspects we recall in particular, among the others, the work by Thompson[1] who first described the problem, the papers by Chandrasekhar[2] who proposed a linear stability analysis for incompressible fluids taking into account the Boussinesq approximation, the papers by Rionero,[3] Galdi,[4] Rionero and Mulone[5] who presented a nonlinear stability analysis for incompressible fluids. Instead, experiments were carried out for mercury in the fifties by Nakagawa,[6] Jirlow,[7] Lehnert and Little[8] and, more recently, by Cioni *et al.*[9] The theoretical studies and the experiments suggest the idea that, at least for fluids with negligible compressibility, there is a stabilizing effect of the magnetic field. In this paper we will consider a compressible model in order to verify such a property when the compressibility is no more neglected. A linear stability analysis is preformed and a comparison of the results with the ones known in the literature is presented. To this aim we consider first an application to fluids with small compressibility and pass then to the case of a solar plasma. We have to recall that for the classical Rayleigh-Bénard problem without magnetic effects other authors have already considered the compressible case, we mention in particular Spiegel,[10] Bormann,[11] Padula[12] and Barbera.[13] For more details about these studies we refer also to the review by Straughan,[14] Chap. 15.

2. A linear stability analysis of the compressible model

The model for compressible Newtonian fluids that we have considered is obtained adding the effect of a magnetic field together with the fluid electrical conductivity to the equation system used by Bormann:[11]

$$\begin{gathered}\frac{\partial \rho}{\partial t}+\frac{\partial \rho v_i}{\partial x_i}=0,\\ \frac{\partial \rho v_j}{\partial t}+\frac{\partial(\rho v_j v_i+\sigma_{ij})}{\partial x_i}=\rho f_j+\mu\left(H_i\frac{\partial H_j}{\partial x_i}-H_i\frac{\partial H_i}{\partial x_j}\right),\\ \frac{\partial \rho\epsilon}{\partial t}+\frac{\partial(\rho\epsilon v_i+q_i)}{\partial x_i}=-\sigma_{ji}\frac{\partial v_j}{\partial x_i},\\ \frac{\partial H_j}{\partial t}+\frac{\partial}{\partial x_i}(v_iH_j-v_jH_i)=\eta\frac{\partial^2 H_j}{\partial x_i\partial x_i},\quad \frac{\partial H_i}{\partial x_i}=0,\end{gathered}\tag{1}$$

with

$$\sigma_{ji}=p\delta_{ji}-\chi(\frac{\partial v_i}{x_j}+\frac{\partial v_j}{\partial x_i})+\frac{2}{3}\chi\frac{\partial v_k}{\partial x_k}\delta_{ji};\ q_i=-\kappa\frac{\partial T}{\partial x_i};\ f=(-g,0,0);\ \eta=\frac{1}{\mu\sigma},$$

where t and x_i are respectively the time and the spatial variable along the i-direction; while ρ, p, T, v_j, ϵ, H_j denote respectively the mass density, the pressure, the temperature, the j-component of the velocity, the internal energy and the j-component of the magnetic field, while the parameters χ, κ, g, σ and μ are respectively the dynamic viscosity, the heat conductivity, the gravitational acceleration, the electrical conductivity and the magnetic permeability. Furthermore, denoting by d the thickness of the layer, by κ_T

the isothermal compressibility, by α the thermal expansion coefficient, by c_p the specific heat, it is possible to introduce the following non-dimensional quantities

$$Ra = \frac{g\alpha(T_0 - T_1)\rho_1^2 c_p d^3}{\kappa\chi}, \; Pr = \frac{\chi c_p}{\kappa}, \; A = \alpha(T_1 - T_0), \; B = \frac{\chi^2 \kappa_T}{d^2 \rho_1}, \; z = \frac{x_1}{d};$$

where T_0, T_1 are the values of the temperature at $z = 0$, $z = 1$ and ρ_1 is the density at $z = 1$. In particular, the Rayleigh number Ra will be used as instability parameter for the linear stability analysis.[2,11] To system (1) we add the usual boundary conditions on a perfect conducting rigid surface, on an electrically non-conducting rigid surface, on electrically non-conducting free surface (see,[2] Chap. 2,4) for the velocity and magnetic fields, and we assign the temperature at the boundaries. By assuming that $p = p(\rho, T)$, by considering the previous quantities and the boundary conditions, the stationary basic solution of the system (1) reads:

$$\begin{aligned} &\bar{v}_i = 0, \quad \bar{H}_1 = H, \quad \bar{T} = (T_1 - T_0)\, z + T_0, \\ &\bar{\rho} = \rho_1 \frac{A}{\left(A - \frac{B\,Ra}{Pr\,A}\right)\exp(A(z-1)) + \frac{B\,Ra}{Pr\,A}}, \end{aligned} \tag{2}$$

where the magnetic field H is supposed to be constant, directed along the vertical z-axis and antiparallel to the gravitational acceleration. In order to study the linear stability of the basic solution, we take independent infinitesimal disturbance denoted by

$$\delta\mathbf{X} = (\delta\rho, \; \delta T, \; \delta v_1, \; \delta v_2, \; \delta v_3, \; \delta H_1, \; \delta\xi_1),$$

where $\delta\rho$, δT, δv_j, δH_1 and $\delta\xi_1 = \epsilon_{1jk}\partial\delta H_k/\partial x_j$ are the perturbations of the mass density, of the temperature, the j-component of the velocity, of the magnetic field along the vertical axis, of the current density. Introducing suitable non-dimensional field variables, along with the non-dimensional constants $D = \mu\kappa_T H^2$, $L = (\eta\rho_1)/\chi$, and supposing, as usual, that

$$\delta\mathbf{X}(x_1, x_2, x_3, t) = \delta\hat{\mathbf{X}}(x_1)\exp\left(i(\lambda_2 x_2 + \lambda_3 x_3) + st\right),$$

the linearized equation system reads

$$\begin{aligned} &\delta v_1 \frac{d\bar{\rho}}{dz} + \bar{\rho}\,\frac{d\delta v_1}{dz} + i\,(\lambda_2\,\delta v_2 + \lambda_3\,\delta v_3)\,\bar{\rho} + s\,\delta\rho = 0 \\ &\delta\rho\left(3\,A^2\,Pr\,\frac{d\bar{T}}{dz} - 6\,B\,Ra\,\bar{\rho}\right) + 3\,A\,B\,Pr\,\bar{\rho}\,\delta v_1\,(\lambda_2^2 + \lambda_3^2 + s\,\bar{\rho}) + \\ &A\,Pr\left\{3\,\frac{d\bar{\rho}}{dz} + \bar{\rho}\left[3\,A\,\frac{d\delta T}{dz} - i\,B\left(\lambda_2\,\frac{d\delta v_2}{dz} + \lambda_3\,\frac{d\delta v_3}{dz} - 4\,i\,\frac{d^2\delta v_1}{dz^2}\right)\right]\right\} = 0 \end{aligned} \tag{3}$$

$$i3\lambda_2\,\delta\rho + i3A\,\lambda_2\,\bar\rho\,\delta T + B\,\bar\rho\left[(4\lambda_2^2+3\lambda_3^2+3s\,\bar\rho)\,\delta v_2 + \lambda_2\lambda_3\delta v_3 - i\lambda_2\frac{d\delta v_1}{dz}\right.$$
$$\left.-3\,\frac{d^2\delta v_2}{dz^2}\right] = i3\,D\,\bar\rho\left[-\lambda_2\,\delta H_1 + \frac{1}{(\lambda_2^2+\lambda_3^2)}\left(\lambda_2\frac{d^2\delta H_1}{dz^2}+\lambda_3\frac{d\delta\xi_1}{dz}\right)\right]$$
$$i3\lambda_3\delta\rho + i3A\lambda_3\bar\rho\delta T + B\bar\rho\left[\left(4\lambda_3^2+3\lambda_2^2+3s\bar\rho\right)\,\delta v_3 + \lambda_2\lambda_3\delta v_2 - i\lambda_3\frac{d\delta v_1}{dz}\right.$$
$$\left.-3\,\frac{d^2\delta v_3}{dz^2}\right] = i\,3\,D\,\bar\rho\left[-\lambda_3\,\delta H_1 + \frac{1}{(\lambda_2^2+\lambda_3^2)}\left(\lambda_3\frac{d^2\delta H_1}{dz^2}-\lambda_2\frac{d\delta\xi_1}{dz}\right)\right]$$
$$\left(\frac{d\delta v_1}{dz} - A\,\frac{d\bar T}{dz}\,\delta v_1 + i\lambda_2\delta v_2 + i\lambda_3\delta v_3\right)\bar T + C\,Pr\,\bar\rho\,\frac{d\bar T}{dz}\delta v_1 - C\,\frac{d^2\delta T}{dz^2}$$
$$+\left[C\left(\lambda_2^2+\lambda_3^2+sPr\bar\rho\right) - sA\bar T\right]\delta T = 0$$
$$i\,(\lambda_2\,\delta v_2 + \lambda_3\,\delta v_3) + L\left[\left(\lambda_2^2+\lambda_3^2\right)\,\delta H_1 - \frac{d^2\delta H_1}{dz^2}\right] + s\,\delta H_1 = 0$$
$$i\left(\lambda_3\frac{d\delta v_2}{dz} - \lambda_2\frac{d\delta v_3}{dz}\right) + L\left[\left(\lambda_2^2+\lambda_3^2\right)\,\delta\xi_1 - \frac{d^2\delta\xi_1}{dz^2}\right] + s\,\delta\xi_1 = 0,$$

where $a=\sqrt{\lambda_2^2+\lambda_3^2}$ is the wave number. Apparently, the set of differential equation is of the thirteenth order; but inserting in $(3)_2$ the expression of $d\delta v_1/dz$ deduced from $(3)_1$ and, recalling that $d\bar\rho/dz$ is a known function of z, it is possible to obtain an equivalent system of the twelfth order.

3. Numerical Methods and boundary data assignment

In order to carry on the linear stability analysis and to determine the stability curves and the critical instability parameters, we made use of a numerical approach applying the Chebychev collocation[15] method and the Chebychev tau method.[16] Particular attention was devoted to the form of the equations and to the assignment of boundary conditions in order to try to avoid spurious eigenvalues,[14,16–18] but it was not always possible to succeed. Therefore, eventual spurious eigenvalues were detected and discarded from the spectrum. On a perfect conducting rigid surface we fix six boundary data as $\delta u_1=0$, $\delta u_2=0$, $\delta u_3=0$, $\delta T=0$, $\delta H_1=0$, $d\delta\xi_1/dz=0$; while on an electrically non-conducting rigid surface the six boundary data are assigned as $\delta u_1=0$, $\delta u_2=0$, $\delta u_3=0$, $\delta T=0$, $\delta\xi_1=0$, $d\delta H_1/dz=0$ and, finally, on an electrically non-conducting free surface the six boundary data are assigned as $\delta u_1=0$, $d\delta u_2/dz=0$, $d\delta u_3/dz=0$, $\delta T=0$, $\delta\xi_1=0$, $d\delta H_1/dz=0$. The main numerical difficulties we have met with are related to the requirement of multiple precision calculation for mercury and liquid sodium and to the presence of spurious eigenvalues for liquid sodium and solar plasma.

4. Numerical results and comparison with experiments

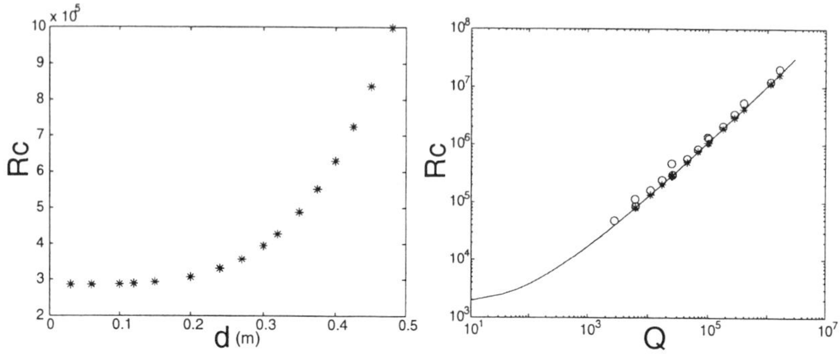

Fig. 1. On the left we report the critical Rayleigh number R_c as a function of the layer thickness d when the value of Q is fixed (for this example $Q \simeq 2.45 \times 10^4$). To this aim the case of mercury is taken into account. On the right we report the Rayleigh number as a function of Q for mercury: a comparison between classical results by Chandrasekhar (continuous line), experimental data by Nakagawa (circles) and the results obtained with the present compressible model (asterisks).

The numerical results for mercury, liquid sodium and solar plasma clearly show the stabilizing effect of the magnetic field; this fact is in accordance with the results known in the literature[2,5,6] for incompressible fluids. For the model of incompressible fluids in the Boussinesq approximation, introduced by Chandrasekhar and studied by many authors, the critical instability parameters depend on the non-dimensional number $Q = \mu^2 H^2 d^2 \sigma/\chi$. The linear stability analysis of our model shows an explicit dependence of the critical instability parameters on Q, but also on d, as shown on the left of Figure 1; in fact for a fixed value of Q different estimates of the critical Rayleigh number are obtained when d is changed. Another important question is the dependence of the critical Rayleigh number on the boundary conditions. In all the cases we have analyzed we found that the critical parameter is greater if the surfaces of the layer are rigid, with respect to both the cases of two free surfaces or of one free and one rigid surfaces. On the contrary, there is no relevant difference between the case of electrically non-conducting surfaces and the perfect conducting surfaces. These results are in perfect agreement with the case of incompressible fluids.
In the case of mercury it is also possible a comparison with experimental data[6–9] and Chandrasekhar's results.[2] In general, as for the incompressible

fluid model, instability sets in as stationary convection and a good agreement is found between our results, the classical ones for incompressible fluids and the experimental data, see on the right of Figure 1 . On the contrary, for solar plasma[19,20] we found that instability may set in as overstability and the critical wave number may depend on the layer thickness also in the absence of the magnetic field, in contrast with classical results.[2,21] More studies on this subject will appear in a forthcoming paper.[22]

Acknowledgments. This paper is dedicated to Tommaso Ruggeri on the occasion of his birthday. The paper was supported by funds MIUR Progetto di interesse Nazionale "Non linear Propagation and Stability in Thermodynamical Processes of Continuous Media" Coordinatore T. Ruggeri, and by the GNFM-INDAM.

References

1. W. B. Thompson, *Phil. Mag. Ser. A* **42**, 1417–1432 (1951).
2. S. Chandrasekhar, *Phil. Mag. Ser. A* **43**, 501–532 (1952); *ibidem* **45** (1954); *Hydrodynamic and Hydromagnetic stability*, Oxford Univ. Press (1961).
3. S. Rionero, *Ann. Mat. Pura App.* **76**, 75–92 (1967).
4. G.P. Galdi, *Arch. Rational Mech. Anal.* **87**, 167-186 (1985).
5. S. Rionero, G. Mulone, *Arch. Rational Mech. Anal.* **103**, 347–368 (1988); G. Mulone, S. Rionero, *ibidem* **166**, 197-218 (2003).
6. Y. Nakagawa, *Nature* **175**, 417-419 (1955); *Proc. Roy. Soc. (London) A* **240**, 108-113 (1957); *ibidem* **249**, 138–145 (1959).
7. K. Jirlow, *Tellus*, **8** 252–253 (1956).
8. B. Lehnert and N. C. Little, *Tellus* **9**, 97–103 (1957).
9. S. Cioni, S. Chaumat, J. Sommeria, *Physical Review E* **62**, 4520–4523 (2000).
10. E. Spiegel, *Astrophys. J.* **141** 1068–1090 (1965).
11. A. Bormann, *Continuum Mech.Thermodyn.* **13**, 9–23(2001).
12. M. Padula, *Boll. Un. Mat. Ital. B* **5**, 581–602 (1986).
13. E. Barbera, *Contin. Mech. Thermodyn.* **16**, 337–346 (2004).
14. B. Straughan, *The energy method, stability, and nonlinear convection* Springer-Verlag (2004).
15. L.N. Trefthen, *Spectral Methods in MATLAB*, SIAM (2000).
16. J.J. Dongarra, B. Straughan, D.W. Walker, *Appl. Numer. Math.* **22**, 399–435 (1996).
17. L.E. Payne, B. Straughan, *Int. J. Num. Anal. Meth. Geomech.* **24**, 815–836 (2000).
18. D. Bourne, *Continuum Mech. Thermodyn.* **15**, 571–579 (2003).
19. L. Spitzer, *Physics of fully ionized gases*, John Wiley & Sons (1962).
20. E. Priest, *Solar Magneto-Hydrodynamic*, Reidel Pub. Comp.(1987).
21. J. Jeng, B. Hassard, *Non-linear Mechanics* **34**, 221–229 (1999).
22. F. Brini, G. Mulone and M. Trovato, *Magnetic Rayleigh-Bénard problem for compressible fluids* (in preparation).

SOLUTIONS OF SOME COUPLED KORTEWEG-DE-VRIES EQUATIONS IN TERMS OF HYPERELLIPTIC FUNCTIONS OF GENUS TWO

T. BRUGARINO and M. SCIACCA

Dipartimento di Metodi e Modelli Matematici, Università di Palermo, Palermo, 90128, Italia

E-mail: brugarin@unipa.it (T. Brugarino), msciacca@unipa.it (M. Sciacca)

We suggest how one can obtain exact solutions of some type of coupled Korteweg-de Vries equations by means of hyperelliptic functions of genus two.

Keywords: Nonlinear equations, Exact solutions, Hyperelliptic functions.

1. Introduction

The construction of exact solutions of nonlinear evolution equations have aroused considerable interest in mathematical physics. To achieve this goal, many important methods have been established such as inverse scattering transform method, Darboux and Bäcklund transformations, Hirota method, Lie groups method, Painlevé analysis, auxiliary equation method and so on. Regarding the last method, let

$$U_t = F(U, U_x, U_{xx}, ...)$$

be a given a nonlinear evolution equation, then one can seek solution $U(x,t)$ having the following expansion $U(u) = \sum_i U_i \phi(u)^i$, where U_i are suitable constant, $u = x - \omega t$ and $\phi(u)$ is a solution of an auxiliary equation $\phi'(u) = E(\phi)$, with E being a regular function of $\phi(u)$. This auxiliary equation usually is the Riccati equation[1,2] $\phi'(u) = 1 + \phi(u)^2$. But, a more general form can be assumed such as[3,4]

$$\phi'(u)^2 = 4\phi(u)^3 - g_2\phi(u) - g_3 \tag{1}$$

with g_2 and g_3 suitable constant, which can be associated to the elliptic curve of genus one

$$y^2 = 4x^3 - g_2 x - g_3$$

It is to note that the equation (1) has the Weirestrass elliptic function $\wp(u)$ as solution.[5] The Weirestrass function can be obtained by the odd $\sigma(u)$ function, which has no singularities in the finite domain and whose only zero is the pole of the $\wp(u)$. The function $\sigma(u)$ is defined by the series[6]

$$\sigma(u) = u - \alpha_1 u^5 - \alpha_2 u^7 \ldots$$

with α_1 and α_2, ... dependent on g_2 and g_3, and linked to $\wp(u)$ by

$$\wp(u) = -\frac{\mathrm{d}^2}{\mathrm{d}u^2} \ln \sigma(u)$$

Here, we will deal with auxiliary equations which are associated to an hyperelliptic curve of genus two given by the nondegenerate algebraic equation

$$y^2 = 4x^5 + \lambda_4 x^4 + \lambda_3 x^3 + \lambda_2 x^2 + \lambda_1 x + \lambda_0 = 4\prod_{k=1}^{5}(x - a_k)$$

which represents a double torus as Riemann surface of genus two. The solutions of these auxiliary equations are the hyperelliptic functions $\wp_{11}$, $\wp_{12}$, $\wp_{22}$ which are the natural generalization of the Weierstrass elliptic function $\wp$ to the genus two.[7–9] Note that the hyperelliptic functions are related to the Θ function[10,11] and have four period ω_1, ω_2, ω_3 and ω_4 as property, which are linked to the four period of the double torus.

In a similar way to the case of genus 1, the hyperelliptic functions can be obtained from a function $\mathfrak{S}(\mathbf{u})$, the natural generalization of the $\sigma(u)$ function, which can be defined by its expansion near $\mathbf{u} = \mathbf{0}$[12]

$$\mathfrak{S}(\mathbf{u}) = \tilde{\alpha}_1 u_1 + \tilde{\alpha}_2 u_1^3 - \tilde{\alpha}_3 u_2^3 + \ldots$$

where $\tilde{\alpha}_1, \tilde{\alpha}_2, \tilde{\alpha}_3$, ... depend on $\lambda_0, \ldots \lambda_4$ and $\mathbf{u}^T = (u_1, u_2) \in \mathbb{C}^2$.

The functions $\wp_{11}$, $\wp_{12}$, $\wp_{22}$ are defined as

$$\wp_{11}(\mathbf{u}) = -\frac{\partial^2 \ln \mathfrak{S}(\mathbf{u})}{\partial u_1^2}, \quad \wp_{21}(\mathbf{u}) = -\frac{\partial^2 \ln \mathfrak{S}(\mathbf{u})}{\partial u_1 \partial u_2}, \quad \wp_{22}(\mathbf{u}) = -\frac{\partial^2 \ln \mathfrak{S}(\mathbf{u})}{\partial u_2^2}$$

with u_1 and u_2 independent variable.

As the auxiliary equation (1) puts in relation the Weierstrass elliptic function $\wp$ to its derivative $\wp'$, also here the auxiliary equations relate the hyperelliptic functions $\wp_{11}$, $\wp_{12}$, $\wp_{22}$ to their derivatives. They are[8,9]

$$\begin{cases} \wp_{222}^2 = 4\wp_{11} + \lambda_3\wp_{22} + 4\wp_{22}^3 + 4\wp_{12}\wp_{22} + \lambda_4\wp_{22}^2 + \lambda_2 \\ \wp_{122}^2 = \lambda_0 - 4\wp_{11}\wp_{12} + \lambda_4\wp_{12}^2 + \lambda_4\wp_{22}\wp_{12}^2 \\ \wp_{211}^2 = \lambda_0\wp_{22}^2 - \lambda_1\wp_{22}\wp_{12} + \lambda_2\wp_{12}^2 + 4\wp_{11}\wp_{12}^2 \\ \wp_{111}^2 = \lambda_0\wp_{12}^2 + \lambda_1\wp_{12}\wp_{11} + \lambda_2\wp_{11}^2 + 4\lambda_0\wp_{22}\wp_{11} + 4\wp_{11}^3 + \frac{1}{2}\lambda_0\lambda_3\wp_{12} + \\ \frac{1}{16}\left(\lambda_1^2\lambda_4 + \lambda_0\lambda_3^2 - 4\lambda_0\lambda_3\lambda_4\right) + \left(\frac{1}{4}\lambda_1^2 - \lambda_0\lambda_2\right)\wp_{22} + \left(\frac{1}{4}\lambda_1\lambda_3 - \lambda_0\lambda_4\right)\wp_{11} \end{cases}$$

Some explicit expressions for the derivative functions $\wp_{ijkh}$ can be found such as

$$\begin{cases} \wp_{2222} = 6\wp_{22}^2 + \frac{1}{2}\lambda_3 + \lambda_4\wp_{22} + 4\wp_{12} \\ \wp_{2221} = 6\wp_{22}\wp_{12} + \lambda_4\wp_{12} - 2\wp_{11} \\ \wp_{2211} = 2\wp_{22}\wp_{11} + 4\wp_{12}^2 + \frac{1}{2}\lambda_3\wp_{12} \\ \wp_{2111} = 6\wp_{12}\wp_{11} + \lambda_2\wp_{12} - \frac{1}{2}\lambda_1\wp_{22} - \lambda_0 \\ \wp_{1111} = 6\wp_{11}^2 - 3\lambda_0\wp_{22} + \lambda_1\wp_{12} + \lambda_2\wp_{11} - \frac{1}{2}\lambda_0\lambda_4 + \frac{1}{8}\lambda_1\lambda_3 \end{cases}$$

2. The description of the method

Now, we describe the method used to find wave solutions. Suppose to have a system of two coupled Korteweg-de Vries equations

$$A_i \frac{\partial U_i(x,t)}{\partial t} + \sum_{h,k} A_i^{hk} U_h(x,t) \frac{\partial U_k(x,t)}{\partial x} + \sum_k A_i^k \frac{\partial^3 U_k(x,t)}{\partial x^3} = 0 \qquad (2)$$

$(i,k,h = 1,2)$ then, after introducing the new variables $u_1 = Kx - \Omega t$ and $u_2 = kx - \omega t$, we obtained solutions $U_1(u_1,u_2)$, $U_2(u_1,u_2)$ that satisfied the differential equations.

The method for solving eqn.s (2) can be summarized by the following steps

Step 1. Assume that the solutions of the eqn.s (2) are of the form:

$$\begin{cases} U_1(x,t) = U(u_1,u_2) = U_{10} + U_{122}\ \wp_{22}(u_1,u_2) \\ U_2(x,t) = V(u_1,u_2) = U_{20} + U_{222}\ \wp_{22}(u_1,u_2) \end{cases} \qquad (3)$$

where $K, \Omega, k, \omega, U_{10}, U_{122}, U_{20}, U_{222}$ are constants to be determined.

Step 2. Inserting the expansion (3) in eqn.s (2), and using expressions of the derivatives $\wp_{ijk}$, $\wp_{ijkh}$ defined above, and their consequences, eqn.s (2) becomes polynomials in $\wp_{22}$, $\wp_{12}$ and $\wp_{11}$.

Step 3. Setting each coefficient of the polynomials to zero yields equations for $K, \Omega, k, \omega, U_{10}, U_{122}, U_{20}, U_{222}$.

Step 4. The equations obtained should be solved.

Step 5. Substituting these results into eqn.s (3) and recalling that $u_1 = Kx - \Omega t$ and $u_2 = kx - \omega t$, wave solutions of eqn.s (2) are obtained.

3. Applications of the method

3.1. *Korteweg–de Vries equation*

The first equation which we take into account is the well-known Korteveg-de Vries equation[13]

$$U_t(x,t) + \alpha U(x,t) U_x(x,t) + U_{xxx}(x,t) = 0 \qquad (4)$$

which, after the transformation $u_1 = Kx - \Omega t$ and $u_2 = kx - \omega t$, becomes

$$U_1(u_1,u_2)\,(k\alpha U(u_1,u_2) - \omega) + k^3 U_{222}(u_1,u_2) - \Omega U_1(u_1,u_2) + K\alpha U(u_1,u_2)$$
$$U_1(u_1,u_2) + 3k^2 K U_{122}(u_1,u_2) + 3kK^2 U_{112}(u_1,u_2) + K^3 U_{111}(u_1,u_2) = 0$$

where here $U_{i\ldots j}$ stands for $(\partial/\partial u_i \ldots \partial/\partial u_j)\,U$.

Now, follow the procedure introduced above: substitute (step 1) $U(u_1,u_2) = U_{10} + U_{122}\wp_{22}(u_1,u_2)$ into the previous equation, use successive derivatives of $\wp_{ij}$ (step 2) and, at least, set the coefficients of $\wp_{22}$, $\wp_{12}$ and $\wp_{11}$ to zero, then an algebraic system in $K, \Omega, k, \omega, U_{10}, U_{122}$ and α is obtained, whose solution are $K = 0$, $\Omega = 4k^2$, $U_{10} = \frac{\omega - k^2\lambda_4}{k\alpha}$ and $U_{122} = -\frac{12k^2}{\alpha}$. Now, substituting the above values in KdV equation and using $u_1 = Kx - \Omega t$ and $u_2 = kx - \omega t$, we get the exact soliton solution of genus 2 of Korteweg-De Vries equation

$$U = \frac{\omega - k^3\lambda_4}{k\alpha} - 12\frac{k^2}{\alpha}\,\wp_{22}(-4k^3 t, kx - \omega t)$$

3.2. *Complex version of Korteweg–de Vries equation*

Now, take into account the following complex version of Korteweg–de Vries equation[14]

$$\mathcal{U}_t(x,t) + \alpha\mathcal{U}(x,t)\mathcal{U}_x(x,t) + \mathcal{U}_{xxx}(x,t) = 0$$

with $\alpha = a_1 + \mathrm{i}a_2$ complex constant. Inserting $\mathcal{U}(x,t) = U_1(x,t) + \mathrm{i}U_2(x,t)$ into the above equation and separating the real and imaginary part, after some calculations the following solution is obtained

$$\begin{cases} U_1 = \frac{a_1(\omega - k^3\lambda_4)}{(a_1^2+a_2^2)k} - 12\frac{a_1 k^2}{a_1^2+a_2^2}\,\wp_{22}(-4k^3 t, kx - \omega t) \\[2ex] U_2 = -\frac{a_2(\omega - k^3\lambda_4)}{(a_1^2+a_2^2)k} + 12\frac{a_2 k^2}{a_1^2+a_2^2}\,\wp_{22}(-4k^3 t, kx - \omega t) \end{cases}$$

3.3. *Generalized Drinfeld-Sokolov equations*

We now apply the method to the generalized Drinfeld-Sokolov equations[15,16] (with $\alpha = 2$)

$$\begin{cases} U_{1t}(x,t) + a_1 U_{1xxx}(x,t) + a_2 U_1(x,t)U_{1x}(x,t) + a_3 U_2(x,t)U_{2x}(x,t) = 0 \\ U_{2t}(x,t) + b_1 U_{2xxx}(x,t) + b_2 U(x,t)U_{2x}(x,t) = 0 \end{cases}$$

Following the same procedure adopted for the previous equations and using the transformation $u_1 = Kx - \Omega t$ and $u_2 = kx - \omega t$, the solutions of the investigated equation are

$$\begin{cases} U_1 = \frac{\omega - b_1 k^3 \lambda_4}{b_2 k} - 12 \frac{b_1 k^2}{b_2} \wp_{22}(-4b_1 k^3 t, kx - \omega t) \\ U_2 = \mp \left(\omega - b_1 k^3 \lambda_4\right) \frac{\sqrt{b_2 - a_2}}{b_2 k \sqrt{a_3}} \pm 12 \frac{b_1 \sqrt{b_2 - a_2} k^2}{b_2 \sqrt{a_3}} \wp_{22}(-4b_1 k^3 t, kx - \omega t) \end{cases}$$

3.4. *Two-wave modes in stratified liquid equations*

The equations derived from a two-wave modes in stratified liquid[17,18] are

$$\begin{cases} U_{1t}(x,t) + a_1 U_{1x} + a_2 U_{1xxx}(x,t) + a_3 U_1(x,t) U_{1x}(x,t) + a_4 U_{2x}(x,t) = 0 \\ U_{2t}(x,t) + b_1 U_{2x} + b_2 U_{2xxx}(x,t) + b_3 U_2(x,t) U_{2x}(x,t) + b_4 U_{1x}(x,t) = 0 \end{cases}$$

The corresponding solution is

$$\begin{cases} U_1 = \frac{\omega}{a_3 k} - \frac{a_1}{a_3} - \frac{a_4}{b_3} - \frac{b_2 k^2 \lambda_4}{a_3} - 12 \frac{b_2 k^2}{a_3} \wp_{22}(-4b_2 k^3 t, kx - \omega t) \\ U_2 = \frac{\omega}{b_3 k} - \frac{b_1}{b_3} - \frac{b_4}{a_3} - \frac{b_2 k^2 \lambda_4}{b_3} - 12 \frac{b_2 k^2}{b_3} \wp_{22}(-4b_2 k^3 t, kx - \omega t) \end{cases}$$

with $a_2 = b_2$.

3.5. *Two-component Bose-Einstein condensates equations*

The equations derived to describe two-component Bose-Einstein condensates[19] are

$$\begin{cases} U_{1t}(x,t) + e_1 U_{1x} + U_{1xxx}(x,t) + U_{2xxx}(x,t) + a_1 U_1(x,t) U_{1x}(x,t) + \\ \quad b_1 U_1(x,t) U_{2x}(x,t) + c_1 U_2(x,t) U_{2x}(x,t) + d_1 U_2(x,t) U_{1x}(x,t) = 0 \\ U_{2t}(x,t) + e_2 U_{2x} + U_{1xxx}(x,t) + U_{2xxx}(x,t) + c_2 U_1(x,t) U_{1x}(x,t) + \\ \quad d_2 U_1(x,t) U_{2x}(x,t) + a_2 U_2(x,t) U_{2x}(x,t) + b_2 U_2(x,t) U_{1x}(x,t) = 0 \end{cases}$$

The obtained solutions are

$$\begin{cases} U_1 = \frac{(c_1+d_1)(e_2 k + 2k^3 \lambda_4 - \omega) - (a_2+b_2)(e_1 k + 2k^3 \lambda_4 - \omega)}{(a_2+b_2-c_1-d_1)(a_2+b_2+c_2+d_2)k} \\ \quad -24 \frac{k^2}{a2+b2+c2+d2} \wp_{22}(-8k^3 t, kx - \omega t) \\ U_2 = \frac{c_2(e_1-e_2)k + d_2(e_1-e_2)k - (a_2+b_2-c_1-d_1)(e_2 k + 2k^3 \lambda_4 - \omega)}{(a_2+b_2-c_1-d_1)(a_2+b_2+c_2+d_2)k} \\ \quad -24 \frac{k^2}{a_2+b_2+c_2+d_2} \wp_{22}(-8k^3 t, kx - \omega t) \end{cases}$$

with $a_1 = a_2 - b_1 + b_2 - c_1 + c_2 - d_1 + d_2$.

4. Conclusions

In conclusion, here we have proposed a simple and direct method to obtain solutions of some coupled Korteweg-de Vries equations; of course, the above method can be applied to other equations.

We want to underline that further studies on algebraic curves of genus higher than 2 could propose a method more general than that proposed here.

At least, all the proposed results are obtained applying the *MATHEMATICA* package of symbolic computation to the boring algebraic and analytical operation.

References

1. R. Conte, Phys. Lett. A **134**, (1988) 100.
2. W. Malfliet, Am. J. Phys. **60**, (1992) 650.
3. E G Fan, J. Phys. A **35**, (2002) 6853.
4. T. Brugarino and M. Sciacca, "Exact Travelling Wave Solutions of Nonlinear Equations Using the Auxiliary Equation Method", (pre-print).
5. E. T. Whittaker and G N Watson, *A Course of Modern Analysis* Cambridge U P (1973).
6. F. Tricomi, *Funzioni Ellittiche* Zanichelli (1951).
7. H. F. Baker, *Multiply Periodic Functions* Cambridge U P (1907).
8. E. D. Belokolos, V. Z. Enolskii, **I** J. Math. Sci. **106**, 6 (2001) 3395.
9. E. D. Belokolos, V. Z. Enolskii, **II** J. Math. Sci. **108**, (2002) 295.
10. B. A. Dubrovin, Russian Math. Survey **36** (1981) 11.
11. B. Deconinck, M. Heily, A. Bobenkoy, M. van Hoeijz, M. Schmiesy *Computing Riemann Theta Functions* 10 Jun 2002 `http://arxiv.org/abs/nlin.SI/0206099/`
12. P. L. Christiansen, J. C. Eilbeck, V. Z. Enolskii, and N. A. Kostov, *Quasi-Periodic and Periodic Solutions for Systems of Coupled Nonlinear Schrödinger Equations* 16 Apr 1999 `http://arXiv:solv-int/9904017 v1`.
13. D. J. Korteweg, F. de Vries, Phil. Mag. **39** (1895) 422.
14. L. Yang, K. Yang, H. Luo, Phys. Lett. A **267** (2000) 331.
15. V. G. Drinfeld, V. V. Sokolov, Sov. J. Math. **30**, (1985) 1975.
16. M. Gurses, A. Karus, Phys. Lett. A **251**, (1999) 2475.
17. J. A. Gear, R. Grimshaw, Stud. Appl, Math. **70**, (1984) 235.
18. J. A. Gear, Stud. Appl, Math. **72**, (1985) 95.
19. V. A. Brazhnyi, V. V. Konotop, Phys. Rev. E **72**, (2005) 026616.

A GLOBAL STABILITY RESULT FOR A CERTAIN BILINEAR SYSTEM

BRUNO BUONOMO

Department of Mathematics and Applications, University of Naples Federico II
via Cintia, I-80126 Naples (Italy)
E-mail: buonomo@unina.it

DEBORAH LACITIGNOLA

Department of Mathematics, University of Salento
via Provinciale Lecce-Arnesano, I-73100 Lecce (Italy)
E-mail: deborah.lacitignola@unile.it

A bilinear three dimensional ODE system is considered, which generalizes many mathematical models in epidemiology. The global stability problem is investigated through a geometrical approach, due to M. Li and J. Muldowney [8], and based on the use of a higher order generalization of the well-known Bendixson criterion. Global dynamics for the system is completely determined.

1. The general system

We consider a bilinear three dimensional ODE system, whose structure generalizes mathematical models of interest in epidemiology, like SEIS with recruitment [4], SEIR model [7] and model for HTLV-I infection and ATL progression [9]:

$$\begin{aligned} \dot{z}_1 &= -a_{13}z_1z_3 - e_1z_1 - b_{12}z_2 + b_{13}z_3 + c_1 \\ \dot{z}_2 &= -e_2z_2 + a_{13}z_1z_3 \\ \dot{z}_3 &= b_{32}z_2 - e_3z_3. \end{aligned} \tag{1}$$

We assume that all the parameters appearing in (1) are positive constants. In particular, the parameters a_{13}, b_{32} and c_1 are assumed to be strictly positive. In analogy with the *basic reproductive number* of the epidemic systems, we define the positive quantity $R_0 = \dfrac{c_1a_{13}b_{32}}{e_1e_2e_3}$. Then, we consider the set:

$$\Gamma = \left\{ (z_1, z_2, z_3) \in \mathbf{R}_3^+ : \sum_{i=1}^{3} z_i \leq \frac{c_1}{\Lambda} \right\}, \tag{2}$$

where,

$$\Lambda = \min\{e_1,\ e_2 + b_{12} - b_{32},\ e_3 - b_{13}\}, \tag{3}$$

which is strictly positive if:

$$e_2 + b_{12} - b_{32} > 0, \qquad e_3 - b_{13} > 0. \tag{4}$$

The region Γ is invariant. In fact, by (1), one gets:

$$\frac{d(z_1 + z_2 + z_3)}{dt} = -e_1 z_1 - (b_{12} + e_2 - b_{32}) z_2 - (e_3 - b_{13}) z_3 + c_1,$$

so that: $\limsup_{t \to \infty} (z_1 + z_2 + z_3) \leq \dfrac{c_1}{\Lambda}$.

Lemma 1.1. *Under the assumptions (4), if $R_0 > 1$, then there exists $\bar{t} > 0$ such that, for all $t > \bar{t}$: $a_{13} z_1(t) > b_{13}$, provided that:*

$$e_2 e_3 > b_{13} b_{32}, \qquad b_{12} < \Lambda \left(\frac{e_2 e_3 - b_{13} b_{32}}{e_2 e_3} \right). \tag{5}$$

Proof. From $(1)_1$ we get: $\dot{z}_1 = c_1 - e_1 z_1 - (a_{13} z_1 - b_{13}) z_3 - b_{12} z_2$. If $a_{13} z_1(t) \leq b_{13}$, then: $\dot{z}_1 \geq c_1 - e_1 z_1 - b_{12} z_2$. This implies:

$$\dot{z}_1 \geq c_1 - e_1 \frac{b_{13}}{a_{13}} - b_{12} \frac{c_1}{\Lambda}. \tag{6}$$

The right hand side is positive if:

$$c_1 a_{13} b_{32} \left(1 - \frac{b_{12}}{\Lambda} \right) > e_1 b_{13} b_{32}. \tag{7}$$

On the other hand, the hypothesis $R_0 > 1$ implies $c_1 a_{13} b_{32} > e_1 e_2 e_3$. The inequality (7) is satisfied if:

$$e_2 e_3 > \frac{b_{13} b_{32} \Lambda}{\Lambda - b_{12}},$$

which is equivalent to $(5)_2$. Hence, from (6) it follows $\dot{z} > 0$. This means that a finite time $\bar{t}$ exists such that $z_1(t)$ crosses the line $z_1(t) = \dfrac{b_{13}}{a_{13}}$, and remains above for all $t > \bar{t}$ #.

2. Global dynamics

We begin by recalling the following Theorem [8]:

Theorem 2.1. *Consider the autonomous dynamical system*

$$\dot{x} = f(x), \tag{8}$$

where $f : D \to \mathbf{R}^n$, $D \subset \mathbf{R}^n$ open set and $f \in C^1(D)$. Assume that the following hypotheses hold:
(H1) D is simply connected;
(H2) there exists a compact absorbing set $\Gamma \subset D$;
(H3) the equation (8) has a unique equilibrium x^* in D.
Then x^* is globally asymptotically stable in D provided that a function $P(x)$ and a Lozinskiĭ measure μ exist such that condition:

$$\limsup_{t\to\infty} \sup_{x_0\in\Gamma} \frac{1}{t}\int_0^t \mu(B(x(s,x_0)))ds < 0, \tag{9}$$

is satisfied.

Now we will state a theorem which gives the full dynamics of system (1) in terms of the critical parameter R_0. The global stability will be proved by applying Theorem 2.1.

Theorem 2.2. *Assume that*

$$\sigma = \min\{e_1, e_2 - b_{12}\} > 0 \tag{10}$$

and that (4)-(5) hold. If $R_0 < 1$, then system (1) admits only the globally stable trivial equilibrium $E_0 = \left(\dfrac{c_1}{e_1}, 0, 0\right)$. If $R_0 > 1$, then E_0 is unstable and there exists a unique globally asymptotically stable nontrivial equilibrium E in the interior of Γ.

Proof. *(i) Existence of equilibria.*
System (1) admits the *trivial* equilibrium $E_0 = \left(\dfrac{c_1}{e_1}, 0, 0\right) \in \partial\Gamma$. We search for *nontrivial* equilibria, i.e. $E = (z_1^*, z_2^*, z_3^*)$ with positive components. From (1), one has:

$$z_1^* = \frac{e_2 e_3}{a_{13}b_{32}}, \quad z_2^* = \frac{e_3(c_1 - e_1 z_1^*)}{e_3(e_2 + b_{12}) - b_{32}b_{13}}, \quad z_3^* = \frac{b_{32}(c_1 - e_1 z_1^*)}{e_3(e_2 + b_{12}) - b_{32}b_{13}}.$$

In order to ensure positivity of E, we observe that $R_0 > 1$ is equivalent to: $c_1 - e_1 z_1^* > 0$. Thus, if $R_0 > 1$, by (4) it follows that z_2^* and z_3^* are positive.

(ii) Uniform persistence.
Consider: $L = \overline{p}z_2 + \overline{q}z_3$, where $\overline{p}$ and $\overline{q}$ are positive constants to be chosen later. Along the solutions to (1) one has:

$$\frac{dL}{dt} = \overline{p}a_{13}z_1z_3 + (\overline{q}b_{32} - \overline{p}e_2)\, z_2 - \overline{q}e_3z_3.$$

Set $\overline{p}e_2 = \overline{q}b_{32}$. After some manipulations:

$$\frac{dL}{dt} = \frac{\overline{p}a_{13}}{R_0}\left(R_0 z_1 - \frac{c_1}{e_1}\right) z_3.$$

Because of $z_1 \leq \dfrac{c_1}{e_1}$, it follows that $R_0 \leq 1$ implies $L' \leq 0$ and hence the global asymptotic stability of E_0. On the other hand, one has $L' > 0$ for $R_0 > 1$ and z_1 sufficiently close to $\dfrac{c_1}{e_1}$, except when $z_3 = 0$. The instability of E_0 implies the uniform persistence [5] which, together with boundedness of Γ, is equivalent to the existence of a compact absorbing set in the interior of Γ [6].

(iii) Bendixson criterion.
Straightforward calculations reveal that the second additive compound matrix [8] $J^{[2]}(z_1, z_2, z_3)$ is:

$$J^{[2]} = \begin{pmatrix} -a_{13}z_3 - e_1 - e_2 & a_{13}z_1 & a_{13}z_1 - b_{13} \\ b_{32} & -a_{13}z_3 - e_1 - e_3 & -b_{12} \\ 0 & a_{13}z_3 & -e_2 - e_3 \end{pmatrix}.$$

We take $P(z_1, z_2, z_3) = \operatorname{diag}\left\{1, \dfrac{z_2}{z_3}, \dfrac{z_2}{z_3}\right\}$, so that:

$$P_f P^{-1} = \operatorname{diag}\left\{0, \frac{\dot{z}_2}{z_2} - \frac{\dot{z}_3}{z_3}, \frac{\dot{z}_2}{z_2} - \frac{\dot{z}_3}{z_3}\right\},$$

and the matrix $PJ^{[2]}P^{-1}$ is then given by:

$$\begin{bmatrix} -a_{13}z_3 - e_1 - e_2 & \dfrac{z_3}{z_2}a_{13}z_1 & \dfrac{z_3}{z_2}(a_{13}z_1 - b_{13}) \\ \dfrac{z_2}{z_3}b_{32} & -a_{13}z_3 - e_1 - e_3 & -b_{12} \\ 0 & a_{13}z_3 & -e_2 - e_3 \end{bmatrix}.$$

Defining $B = P_f P^{-1} + PJ^{[2]}P^{-1}$, one has:

$$B = \begin{bmatrix} B_{11} & B_{12} \\ B_{21} & B_{22} \end{bmatrix},$$

where,

$$B_{11} = -a_{13}z_3 - e_1 - e_2; \qquad B_{12} = \left[\frac{z_3}{z_2} a_{13}z_1, \ \frac{z_3}{z_2}(a_{13}z_1 - b_{13}) \right],$$

$$B_{21} = \left[\frac{z_2}{z_3} b_{32}, \ 0 \right]^T; \qquad B_{22} = \begin{bmatrix} \varphi(z_2, z_3) & -b_{12}, \\ \\ a_{13}z_3 & \psi(z_2, z_3) \end{bmatrix},$$

$$\begin{aligned} \varphi(z_2, z_3) &= \frac{\dot{z}_2}{z_2} - \frac{\dot{z}_3}{z_3} - a_{13}z_3 - e_1 - e_3, \\ \psi(z_2, z_3) &= \frac{\dot{z}_2}{z_2} - \frac{\dot{z}_3}{z_3} - e_2 - e_3. \end{aligned}$$

Consider now the norm in $\mathbf{R}^3$ as $|(u, v, w)| = \max\{|u|, |v| + |w|\}$, where (u, v, w) denotes the vector in $\mathbf{R}^3$. Denote by μ the Lozinskiĭ measure with respect to this norm. It follows:

$$\mu(B) \le \sup\{g_1, g_2\} \equiv \sup\{\mu_1(B_{11}) + |B_{12}|, \ \mu_1(B_{22}) + |B_{21}|\}, \tag{11}$$

where $|B_{21}|$, $|B_{12}|$ are matrix norms with respect to the L^1 vector norm and μ_1 denotes the Lozinskiĭ measure with respect to the L^1 norm.
We have,

$$\mu_1(B_{11}) = -a_{13}z_3 - e_1 - e_2; \quad |B_{12}| = \max\left\{ \left| \frac{z_3}{z_2} a_{13}z_1 \right|, \left| \frac{z_3}{z_2}(a_{13}z_1 - b_{13}) \right| \right\}$$

and, because of Lemma 1.1, for $t > \bar{t}$, it follows: $|B_{12}| = \dfrac{z_3}{z_2} a_{13}z_1$.
Furthermore, $|B_{21}| = \dfrac{z_2}{z_3} b_{32}$, and:

$$\mu_1(B_{22}) = \max\{\frac{\dot{z}_2}{z_2} - \frac{\dot{z}_3}{z_3} - e_1 - e_3, \ \frac{\dot{z}_2}{z_2} - \frac{\dot{z}_3}{z_3} - e_2 - e_3 + b_{12}\}.$$

Rearranging, from (11), we have:

$$\begin{aligned} g_1 &= -a_{13}z_3 - e_1 - e_2 + \frac{z_3}{z_2} a_{13}z_1, \\ g_2 &= \frac{\dot{z}_2}{z_2} - \frac{\dot{z}_3}{z_3} - e_3 + \max\{-e_1, -e_2 + b_{12}\} + \frac{z_2}{z_3} b_{32}. \end{aligned} \tag{12}$$

From system (1) we get: $\frac{\dot{z}_2}{z_2} = -e_2 + \frac{z_3}{z_2}a_{13}z_1$, and $\frac{\dot{z}_3}{z_3} = b_{32}\frac{z_2}{z_3} - e_3$.

Hence,

$$g_1 \le \frac{\dot{z}_2}{z_2} - e_1; \qquad g_2 \le \frac{\dot{z}_2}{z_2} + \max\{-e_1, -e_2 + b_{12}\},$$

and thus we conclude that for $t > \bar{t}$, one has $\mu(B) \le \frac{\dot{z}_2}{z_2} - \sigma$, where σ is given by (10). The Bendixson condition (9) is thus satisfied.#

We conclude by stressing that the global dynamics of a more general system, which include more epidemic models than (1), will be the subject of a forthcoming paper [3].

Acknowledgments

The present work has been performed under the auspices of the italian National Group for the Mathematical Physics (GNFM-Indam). Granted scientific project entitled *"Analisi di due metodi per la stabilità ed applicazioni a modelli ODE e PDE in dinamica delle popolazioni"*.

References

1. B.Buonomo and D.Lacitignola, *Nonlinear Analysis RWA* **5**, 749-762 (2004).
2. B.Buonomo and D.Lacitignola, *Proc. Dynamic Sys.Appl.* **4**, 53-57 (2004).
3. B.Buonomo and D.Lacitignola, On the use of the geometric approach to global stability for three-dimensional ODE systems: a bilinear case. Submitted (2007).
4. M.Fan, M.Y.Li and K.Wang, *Math. Biosci.* **170**, 199-208 (2001).
5. H.I.Freedman, S.Ruan and M.Tang, *J.Diff.Equations* **6**, 583-600 (1994).
6. V.Hutson and K.Schmitt, *Math.Biosci.* **111**, 1-71 (1992).
7. M.Y.Li and J.S.Muldowney, *Math. Biosci.* **125**, 155-164 (1995).
8. M.Y.Li and J.S.Muldowney, *SIAM J. Math. Anal.* **27**, 1070-1083 (1996).
9. L.Wang, M.Y.Li and D.Kirschner, *Math. Biosci.* **179**, 207-217 (2002).

RESTABILIZING FORCING FOR A DIFFUSIVE PREY-PREDATOR MODEL

BRUNO BUONOMO AND SALVATORE RIONERO

Department of Mathematics and Applications, University of Naples Federico II, via Cintia, 80126 Naples, Italy.
buonomo@unina.it, rionero@unina.it.

We consider a diffusive prey-predator model and find conditions under which a relevant non trivial equilibrium undergoes to Turing bifurcation. Then, a forcing is applied to the model and values of forcing able to regain the (nonlinear) stability are detected. A maximal restabilizing region is derived.

1. Introduction

Patterned distributions in the density of species may be crucial in Ecology and their importance refers also to management questions. An illustrative example coming from forestry management is the following: different species of trees react to and interact with fire in different ways. Hence, the ways that forestry managers deal with fire in forest strongly depends on spatial variations in species' densities.[1] The manager may use *forcing* (stocking or harvesting) to influence the long term dynamics and the spatial distribution. In particular, forestry manager may aim to restabilize an unstable equilibrium by forcing, so that the populations approach a stable equilibrium instead.

Diffusional (Turing) instability may be used as a mathematical model of pattern formation.[2,3] Hence, the following general question arises: find conditions under which diffusion rates can cause the system to undergo a Turing bifurcation and find also forcing rates which can '*reverse*' the bifurcation. For the so-called *pioneer-climax* interaction model, this probem has been studied, by using different methodologies, by Buchanan[5] and Buonomo and Rionero.[6] In particular, in Ref. 6 a new approach to nonlinear L^2-stability (see e.g. Ref. 7) is used. The same method is applied here, in order to find the restabilizing values of forcing for a diffusive prey-predator model. The basic model is due to Levin and Segel[8] and it has been considered also in Ref. 9.

2. The model and preliminaries

Let $\Omega \subset \mathbf{R}^3$ be a bounded smooth domain, and consider the model:

$$\begin{cases} u_t = \gamma_1 \Delta u + u(h_1 - d_1 v + e_1 u) + F \\ v_t = \gamma_2 \Delta v + d_2 uv - e_2 v^2, \end{cases} \tag{1}$$

where u and v denote a measure of a prey and a predator species, respectively. They are functions: $\psi : (\mathbf{x}, t) \in \Omega \times \mathbf{R}_+ \rightarrow \psi(\mathbf{x}, t) \in \mathbf{R}$. The parameters h_1, d_i, e_i and γ_i, $i = 1, 2$, are positive constants, and we assume that: $d_1 d_2 > e_1 e_2$. The term F represents the constant forcing. The term $+e_1 u$ in $(1)_1$ makes the per capita growth rate of prey to be an increasing function of the prey itself. This is called an *autocatalytic* effect (see Ref 4, p. 208). System (1), which is completed by non negative initial conditions and homogeneous Neumann (zero-flux) boundary conditions, admits the constant equilibrium $E = (u^*, v^*)$, whose components depend on the forcing F:

$$u^*(F) = \frac{h_1 e_2 + \sqrt{h_1^2 e_2^2 + 4(d_1 d_2 - e_1 e_2) e_2 F}}{2(d_1 d_2 - e_1 e_2)}, \quad v^*(F) = \frac{d_2}{e_2} u^*(F). \tag{2}$$

This equilibrium admits positive components if $F \geq -\Phi$, where $\Phi = h_1^2 e_2 / 4(d_1 d_2 - e_1 e_2)$. System (1) may admit other constant equilibria, but (2) is the only one that for $F = 0$ reduces to an equilibrium which is still internal to the positive cone. So, (2) is the relevant equilibrium for study here.

Once set $w_1 = u - u^*$, $w_2 = v - v^*$, we get from (1) the perturbation system:

$$\begin{cases} (w_1)_t = \gamma_1 \Delta w_1 + a_{11} w_1 + a_{12} w_2 + f(w_1, w_2) \\ (w_2)_t = \gamma_2 \Delta w_2 + a_{21} w_1 + a_{22} w_2 + g(w_1, w_2), \end{cases} \tag{3}$$

where

$$\begin{aligned} &a_{11} = e_1 u^* - (u^*)^{-1} F, \;\; a_{12} = -d_1 u^*, \;\; a_{21} = d_2 v^*, \;\; a_{22} = -e_2 v^*; \\ &f(w_1, w_2) = -d_1 w_1 w_2 + e_1 w_1^2; \quad g(w_1, w_2) = d_2 w_1 w_2 - e_2 w_2^2. \end{aligned} \tag{4}$$

3. Nonlinear stability

We will study the L^2-stability of the equilibrium E, with respect to the perturbations (w_1, w_2) belonging, for all $t \in \mathbf{R}_+$, to the Sobolev space $[H(\Omega)]^2$: $\varphi \in H(\Omega) \rightarrow \left\{ \varphi^2 + (\nabla \varphi)^2 \in L(\Omega), \; \frac{d\varphi}{d\mathbf{n}} = 0 \text{ on } \partial\Omega \right\}$. Let λ_1 be the positive constant appearing in the Poincaré-Wirtinger inequality,[10] $\|\nabla \varphi\|^2 \geq \lambda_1 \|\varphi\|^2$, holding in the space $H(\Omega)$ above. As it is well known, $\lambda_1 = \lambda_1(\Omega)$ is the lowest eigenvalue α of the problem, in $H(\Omega)$: $\Delta \phi = -\alpha \phi$. In other words, it is the principal eigenvalue of $-\Delta$.

We remark that in order to avoid that λ_1 is zero, one should consider the perturbations with zero mean value on Ω. However, the conditional nature of the nonlinear stability result that we will obtain in the next, allows us to restrict the initial data of the admissible perturbations to a neighborhood of E not including others equilibria. This restriction allows us to consider the perturbations in the space $H(\Omega)$ above (for more details about this matter, see Ref. 6).

Denote now by k_1 and k_2 two positive rescaling constants, and set

$$\begin{aligned}
&w_1 = k_1 u_1, \quad w_2 = k_2 u_2,\\
&b_{11} = a_{11} - \lambda_1\gamma_1, \quad b_{12} = \tfrac{k_2}{k_1}a_{12}, \quad b_{21} = \tfrac{k_1}{k_2}a_{21}, \quad b_{22} = a_{22} - \lambda_1\gamma_2,\\
&\overline{f} = k_1^{-1}\, f(w_1, w_2)|_{w_1=k_1u_1, w_2=k_2u_2},\\
&\overline{g} = k_2^{-1}\, g(w_1, w_2)|_{w_1=k_1u_1, w_2=k_2u_2},\\
&\overline{f}^* = \gamma_1\left(\Delta u_1 + \lambda_1 u_1\right), \quad \overline{g}^* = \gamma_2\left(\Delta u_2 + \lambda_1 u_2\right).
\end{aligned} \tag{5}$$

System (3) may be written:

$$\mathbf{u}_t = B\mathbf{u} + \overline{\mathbf{G}}(\mathbf{u}) + \overline{\mathbf{G}}^*(\mathbf{u}) \tag{6}$$

where $\mathbf{u} = (u_1, u_2)^T$; $B = [b_{ij}]$ and its entries are given by $(5)_2$; $\overline{\mathbf{G}} = (\overline{f}, \overline{g})^T$, where $\overline{f}$ and $\overline{g}$ are given by $(5)_{3,4}$; $\overline{\mathbf{G}}^* = (\overline{f}^*, \overline{g}^*)^T$, where $\overline{f}^*$ and $\overline{g}^*$ are given by $(5)_5$.

Observe that: $det(B) = b_{11}b_{22} - a_{12}a_{21}$, and $tr(B) = b_{11} + b_{22}$, so that setting,

$$T = \begin{pmatrix} b_{11} & a_{12}\\ a_{21} & b_{22}\end{pmatrix}, \tag{7}$$

it follows: $det(T) = det(B)$, and $tr(T) = tr(B)$. Taking into account of (4) and $(5)_2$, we have:

$$T(F, \gamma_1, \gamma_2) = \begin{pmatrix} e_1u^* - (u^*)^{-1}F - \lambda_1\gamma_1 & -d_1u^*\\ d_2v^* & -e_2v^* - \lambda_1\gamma_2\end{pmatrix}. \tag{8}$$

Denote by $\{\lambda_n, \phi_n\}$ $(n = 1, 2, \ldots)$ the sequence of eigenvalues (with associated eigenfunctions in $H(\Omega)$) of $\Delta\phi_n = -\alpha_n\phi_n$. The eigenvalues λ_n are also referred as *wavenumbers* and the eigenfunction ϕ_n as *modes.*

Set M_n the matrix obtained by T replacing λ_1 with λ_n (observe that: $M_1 = T$). It is well known that if the matrices M_n, for each n, are stable (i.e. both their eigenvalues have negative real parts), then the null solution of (3) is linearly stable. The next theorem gives sufficient conditions allowing to deduce the L^2-asymptotic exponential stability of the null solution of (3), only in view of the matrix T, corresponding to the first eigenvalue λ_1.

Introduce the following quantities:

$$\alpha_1 = det(T) + b_{21}^2 + b_{22}^2, \quad \alpha_2 = det(T) + b_{11}^2 + b_{12}^2,$$
$$\alpha_3 = det(T) + b_{11}b_{21} + b_{12}b_{22}.$$

Let $\gamma_1 \neq \gamma_2$, and assume, for the sake of concreteness, $\gamma_1 < \gamma_2$. Then, the following two Theorems hold (for the proofs and all the details, see Ref. 7):

Theorem 3.1. *Let be $\gamma_1 < \gamma_2$, and assume that one of the following conditions holds true:*

$$\begin{aligned} &(i)\ b_{11}b_{12}b_{21}b_{22} < 0; \quad (ii)\ (\gamma_1 + \gamma_2)\,|\alpha_3| \leq 2\sqrt{\gamma_1\gamma_2\alpha_1\alpha_2};\\ &(iii)\ (\gamma_1 + \gamma_2)\,|\ b_{22}| \leq 2\sqrt{\gamma_1\gamma_2\,(b_{22}^2 + det(T))};\\ &(iv)\ (\gamma_1 + \gamma_2)\,|\ b_{11}| \leq 2\sqrt{\gamma_1\gamma_2\,(b_{11}^2 + det(T))}, \end{aligned} \tag{9}$$

If $tr(T) < 0$ and $det(T) > 0$, then the null solution of (3) *is (locally) L^2-asymptotically exponentially stable.*

Theorem 3.2. *If either $tr(T) > 0$ or $det(T) < 0$, then the null solution of* (3) *is linearly unstable.*

4. Restabilizing forcing

Define the following matrices:

$$T_0 = T(0,0,0), \quad T_1 = T(F,0,0), \quad T_{0D} = T(0,\gamma_1,\gamma_2),$$

where T is the matrix given by (8).
Observe that:

$$det(T_0) = (d_1d_2 - e_1e_2)u^*(0)v^*(0) > 0, \tag{10}$$

so that only the sign of $tr(T_0)$ rules the stability - instability of matrix T_0. Further, because of: $tr(T_0) = e_1u^*(0) - e_2v^*(0)$, and taking into account of (2), it follows that $tr(T_0) < 0$ is equivalent to:

$$e_1 < d_2. \tag{11}$$

Theorem 4.1. *Under the hypotheses of* Theorem 3.1, *let be*

$$\mathcal{F} = \{F \in [-\Phi, +\infty) : tr(T) < 0, \;\; det(T) > 0\},$$

and, $tr(T_0) < 0$, $det(T_{0D}) < 0$, i.e. Turing instability for system (3) in absence of forcing. Then, for $F \in \mathcal{F}$ the null solution of (3) is L^2-asymptotic exponentially stable, i.e. the system is restabilized by forcing F for values in the region $\mathcal{F}$.

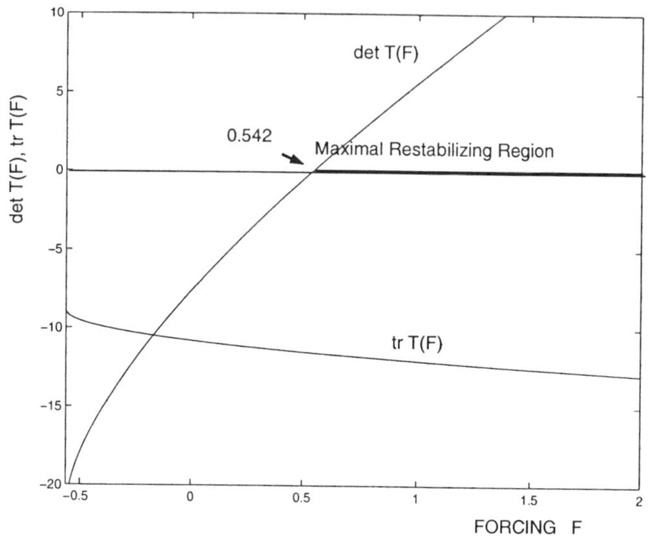

Fig. 1. The functions $trT(F)$ and $\det T(F)$, when $\gamma_1 = 0.05$. The forcing F may restabilize the system for values in the region (maximal restabilizing region): $\mathcal{F} = \{F : tr(T) < 0,\ det(T) > 0\} = [0.542, +\infty)$.

The proof is a direct consequence of Theorems 3.1 and 3.2.

In view of the direct application of Theorem 4.1, we first remark that because of (10), the condition (11) guarantees the local stability of the equilibrium, when $\gamma_1 = \gamma_2 = 0$. Then, in order to have Turing instability for system (3) when the forcing is absent ($F = 0$), by standard approach (see, e.g. Ref. 3, p. 383), we get the following Turing region: $I = \{\gamma_2/\gamma_1 : \gamma_2/\gamma_1 > \Gamma\}$, where:

$$\Gamma = \frac{d_2}{e_2} \cdot max\left\{\frac{e_2}{e_1}, \frac{2d_1d_2 - e_1e_2 + 2\sqrt{d_1d_2(d_1d_2 - e_1e_2)}}{2e_1^2}\right\}. \tag{12}$$

The conditions ensuring the Turing instability are (11) and $det(T_{0D}) < 0$, i.e., in terms of the diffusion coefficient γ_1:

$$\gamma_1 < \frac{e_1u^*(0)\gamma_2\lambda_1 - (d_1d_2 - e_1e_2)u^*(0)v^*(0)}{\gamma_2\lambda_1^2 + e_2\lambda_1v^*(0)} \equiv \Gamma_T. \tag{13}$$

We now provide a numerical example which shows the practical application of the results obtained above. The theoretical values chosen for the parameters are the following:

$$h_1 = 3, \quad d_1 = 2, \quad d_2 = 3, \quad e_1 = 2, \quad e_2 = 1, \quad \gamma_2 = 10. \qquad (14)$$

The diffusion coefficient of the prey species, γ_1, is taken as a bifurcation parameter. The spatial domain is assumed to be a one-dimensional domain $[0, \pi]$. In this case the minimum eigenvalue λ_1 is $\lambda_1 = 1$ (see, e.g. Ref. 10). System (1) with values (14) admits the equilibrium $E(F) \equiv (u^*(F), v^*(F))$, where:

$$u^*(F) = \frac{3}{8}\left(1 + \frac{1}{3}\sqrt{9 + 16F}\right), \quad v^*(F) = \frac{9}{8}\left(1 + \frac{1}{3}\sqrt{9 + 16F}\right).$$

This equilibrium is real positive provided that $F > -0.562$. From (12) it follows: $\Gamma = 7.424$ and, from (13), $\Gamma_T = 0.673$, so the Turing region for the system without forcing is $I_T = \{\gamma_1 : 0 < \gamma_1 < 0.673\}$. Further, it follows:

$$\det T(F) = 4u^*(F)v^*(F) - 20u^*(F) + \frac{Fv^*(F)}{u^*(F)} + 10\frac{F}{u^*(F)} + \gamma_1 v^*(F) + 10\gamma_1,$$

and,

$$trT(F) = 2u^*(F) - \frac{F}{u^*(F)} - \gamma_1 - v^*(F) - 10.$$

We take $\gamma_1 = 0.05$ inside the Turing region. The region $\mathcal{F}$, introduced in Theorem 3.1, can be numerically evaluated. For the example above we get the following maximal restabilizing region: $\mathcal{F} = [0.542, +\infty)$. Hence, in this case only prey stocking may restabilize the equilibrium (see Fig. 1).

Acknowledgments

The present work has been performed under the auspices of the italian National Group for the Mathematical Physics (GNFM-Indam). Granted scientific project entitled *"Analisi di due metodi per la stabilità ed applicazioni a modelli ODE e PDE in dinamica delle popolazioni"*.

References

1. C. Miller and D. L. Urban, *Can. J. For. Res.*, **29**, 202-212 (1999).
2. A. M. Turing, *Phil. Trans. Roy. Soc B.*, **237**, 5-72 (1952).
3. J. D. Murray, *Mathematical Biology*, 2nd edition (Springer, Berlin, 1993).
4. A. Okubo, *Diffusion and Ecological Problems: Mathematical Models* (Springer, Berlin, 1980).
5. J. R. Buchanan, *Math. Biosci.*, **194**, 199-216 (2005).
6. B. Buonomo and S. Rionero, Linear and nonlinear stability thresholds for a diffusive model of pioneer and climax species interaction. Submitted (2007).
7. S. Rionero, *Rend. Mat. Acc. Lincei*, s. 9, **16**, 227-238 (2005).
8. S. A. Levin and L. A. Segel, *Nature.*, **259**, 659 (1976).
9. S. Rionero, *Math. Biosci. Eng.*, **3**, n.1, 189-204 (2006).
10. J. N. Flavin and S. Rionero, *Qualitative Estimates for Partial Differential Equations. An Introduction* (CRC Press, Boca Raton, Florida, 1995).

MATHEMATICAL STUDY OF THE PLANAR OSCILLATIONS OF A HEAVY, ALMOST HOMOGENEOUS LIQUID IN A CONTAINER

P. CAPODANNO

2B, Rue des Jardins,
Besançon, 2500, France
E-mail: pierre.capodanno@wanadoo.fr

D. VIVONA

Dipartimento di Metodi e Modelli Matematici per le Scienze Applicate,
16, Via A.Scarpa, Roma, 00161, Italy
E-mail: vivona@dmmm.uniroma1.it

The authors study the planar small oscillations of a heavy almost homogeneous liquid partially filling an arbitrary container. The main object of the paper is to prove that the spectrum is real and decomposed in two parts: an essential spectrum which fills a bounded interval and a point spectrum formed by a sequence of eigenvalues, which tends to infinity.

Keywords: Small oscillations, Variational method.

1. The equation of motion

We restrict ourselves to planar motions. The liquid fills partially an arbitrary container.In the equilibrium position, it occupies a domain Ω bounded by the wetted wall S of the container and the horizontal free line $\Gamma : x_2 = 0; Ox_1$ lies on Γ and Ox_2 is directed vertically upwards.

We suppose that the liquid is *almost homogeneous in the fluid domain* Ω, i.e. its density in its equilibrium position can be written in the form $\rho_0(x_2) = \rho(1 - \beta x_2)$, $\rho > 0$, $\beta > 0$ such that, if h is the mean depth of the liquid, βh is sufficiently small, so $(\beta h)^2, (\beta h)^3, \ldots$ are negligible with respect to βh. Then, the Euler equation and the continuity equation can replaced by an approximate equation *analogous to Boussinesq equation of the theory of the convective fluid motions.*

If $\vec{u}(x_1, x_2, t) = (u_1, u_2)$ is the small displacement of a particle with respect to its equilibrium position, $p(x_1, x_2, t)$ the dynamic pressure, we obtain easily:

$$\ddot{\vec{u}} = \frac{1}{\rho} \vec{grad}\, p - \beta\, g\, u_2\, \vec{x_2} \qquad \text{in } \Omega\ ; \tag{1}$$

$$div\ \vec{u} = 0 \qquad \text{in } \Omega\ ; \tag{2}$$

$$u_n = \vec{u} \cdot \vec{n} = 0 \qquad \text{on } S\ ; \tag{3}$$

$$p\,(x_1, 0, t) = \rho\ g\ u_2(x_1, 0, t)\ ; \tag{4}$$

$$\int_\Gamma u_2(x_1, 0, t) d\Gamma = 0\ . \tag{5}$$

2. Transformation of the equations of the motion

We make use of the method of the orthogonal projection [2].

We introduce the following spaces:

$$\mathcal{J}_0(\Omega) = \{\vec{u} \in \mathcal{L}^2(\Omega) = [L^2(\Omega)]^2\ ;\ div\, \vec{u} = 0\ ;\ u_n = 0 \ \text{ in }\ H^{-1/2}(\partial\Omega)\}\ ;$$

$$\mathcal{G}(\Omega) = \{\vec{u} = \vec{grad}\, p\ ;\ p \in H^1(\Omega)\ ;\ \int_\Gamma p_{|\Gamma}\ d\Gamma = 0\}\ ;$$

$$\mathcal{G}_{0,\Gamma}(\Omega) = \{\vec{u} = \vec{grad}\, p\ ;\ p \in H^1(\Omega)\ ;\ p_{|\Gamma} = 0\}\ ;$$

$$H^1_{h,S}(\Omega) = \{p \in H^1(\Omega)\ ;\ \Delta p = 0 \ \text{ in }\ \mathcal{D}'(\Omega)\ ;\ \frac{\partial p}{\partial n} = 0 \ \text{ in }\ [H^{1/2}_{00}(S)]'\ ;$$
$$\int_\Gamma p_{|\Gamma}\ d\Gamma = 0\}\ ;$$

$$\mathcal{G}_{h,S}(\Omega) = \{\vec{u} = \vec{grad}\, p\ ;\ p \in H^1_{h,S}(\Omega)\}\ ;$$

and we use the orthogonal decompositions

$$\mathcal{L}^2(\Omega) = \mathcal{J}_0(\Omega) \oplus \mathcal{G}_{h,S}(\Omega) \oplus \mathcal{G}_{0,\Gamma}(\Omega) \quad ; \quad \mathcal{G}(\Omega) = \mathcal{G}_{h,S}(\Omega) \oplus \mathcal{G}_{0,\Gamma}(\Omega)\ .$$

Let be

$$\vec{u} = \vec{v} + \vec{grad}\, \Phi\ , \ \text{ with } \vec{v} \in \mathcal{J}_0(\Omega)\ ,\ \vec{grad}\, \Phi \in \mathcal{G}_{h,S}(\Omega)\ ;$$
$$\vec{grad}\, p = \vec{grad}\, \varphi + \vec{grad}\, k\ , \ \text{ with } \vec{grad}\, \varphi \in \mathcal{G}_{h,S}(\Omega)\ ,\ \vec{grad}\, k \in \mathcal{G}_{0,\Gamma}(\Omega)\ .$$

Denoting by P_0 and P_S the orthogonal projectors of $\mathcal{L}^2(\Omega)$ on $\mathcal{J}_0(\Omega)$ and $\mathcal{G}_{h,S}$ (the projection on $\mathcal{G}_{0,\Gamma}(\Omega)$ gives only k), we have

$$\frac{\partial^2 \vec{v}}{\partial t^2} = -\beta\ g\ P_0 \left[(v_2 + \frac{\partial \Phi}{\partial x_2}) \vec{x}_2 \right]\ , \tag{6}$$

$$\vec{grad} \left(\frac{\partial^2 \Phi}{\partial t^2} \right) = -\frac{1}{\rho}\ \vec{grad}\, \varphi - \beta\ g\ P_S \left[(v_2 + \frac{\partial \Phi}{\partial x_2}) \vec{x}_2 \right]\ . \tag{7}$$

But, we can write

$$P_S(v_2\vec{x_2}) = \vec{grad}\,\psi\ ; \quad P_S\left(\frac{\partial\Phi}{\partial x_2}\vec{x_2}\right) = \vec{grad}\ \Psi\ , \qquad \psi, \Psi \in H^1_{h,S}(\Omega)\ ,$$

so that, from (7), we can deduce the first integral

$$\frac{\partial^2\Phi}{\partial t^2} = \frac{\varphi}{\rho} - \beta\ g(\varphi + \Psi) \qquad \text{in } \Omega\ . \tag{8}$$

On the other hand, we have immeditely:

$$p|_\Gamma = \varphi|_\Gamma\ ,\ u_n|_\Gamma = u_2(x_1, 0, t) = \left(\frac{\partial\Phi}{\partial n}\right)|_\Gamma$$

and, consequently,

$$\varphi|_\Gamma = \rho\ g\ \left(\frac{\partial\Phi}{\partial n}\right)|_\Gamma\ .$$

Writing (8) on Γ, we obtain

$$\frac{\partial^2\Phi|_\Gamma}{\partial t^2} = -g\ \left(\frac{\partial\Phi}{\partial n}\right)|_\Gamma - \beta\ g\left(\varphi + \Psi\right)|_\Gamma\ . \tag{9}$$

It can be shown that the equations (6) and (9) permit to determine all the unknowns of the problem.

3. Operatorial equation of the problem

From the classical study of Neumann and Zaremba problems, it can be proved that exists a bounded linear operator C from $[\widetilde{H}^{1/2}_{00}(\Gamma)]'$ onto $\widetilde{H}^{1/2}_{00}(\Gamma) = \{f \in H^{1/2}_{00}(\Gamma);\ \int_\Gamma f\ d\Gamma = 0\}$, such that $\Phi|_\Gamma\ =\ C\ \left(\partial\Phi/\partial n\right)|_\Gamma$, $\Phi \in H^1_{h,S}(\Omega)$.

Setting

$$\eta = C^{1/2}\left(\frac{\partial\Phi}{\partial n}\right)|_\Gamma = C^{-1/2}\ \Phi|_\Gamma \in \widetilde{L}^2(\Gamma) = \{f \in L^{1/2}(\Gamma);\ \int_\Gamma f\ d\Gamma = 0\}\ ,$$

we can replace the equation (9) by

$$\frac{d^2\eta}{dt^2} = -g\ C^{-1}\eta - \beta g\ (C^{-1/2}\psi|_\Gamma + C^{-1/2}\psi|_\Gamma)\ . \tag{10}$$

We introduce the operators A_{ij} defined by

$$\beta g\ P_0(v_2\vec{x_2}) = A_{11}\vec{v}\ ; \qquad \beta g P_0(\frac{\partial\Phi}{\partial x_2}\vec{x_2}) = A_{12}\eta\ ;$$

$$\beta g\ C^{-1/2}\psi|_\Gamma = A_{21}\vec{v}\ ; \quad \beta g\ C^{-1/2}\psi|_\Gamma = A_{22}\eta\ .$$

Then, the equations (6) and (10) can be replaced by the operatorial equation

$$\frac{d^2\, y}{dt^2} + (A+B)\; y = 0\;, \tag{11}$$

with

$$y = \begin{pmatrix} \vec{v} \\ \eta \end{pmatrix} \in \mathcal{H} = \mathcal{J}_0(\Omega) \oplus \widetilde{L}^2(\Gamma)\;,$$

$$A = \begin{pmatrix} A_{11} & A_{12} \\ A_{21} & A_{22} \end{pmatrix}\;; \qquad B = \begin{pmatrix} 0 & 0 \\ 0 & g\; C^{-1} \end{pmatrix}\;.$$

The operator A_{11} was studied in [1,4] where it is denoted by K. It is bounded with $||A_{11}|| = \beta g$, symmetrical, positive definite; its spectrum $\sigma(A_{11})$ coincides with its essential spectrum $\sigma_e(A_{11})$ and it is the closed interval $[0, \beta g]$.

It is easy to prove that A_{12} and A_{21} are mutually adjoint, that $||A_{22}|| \leq \beta g, ||A_{12}|| = ||A_{21}|| \leq \beta g$, and that A is symmetrical and bounded, with $||A|| = \beta g$. On the other hand, B is an unbounded operator of $\mathcal{H}$, self-adjoint and non-negative. Finally, the spectrum of $A+B$ lies on the real positive halfaxis.

4. The point spectrum

We are going to consider the solutions

$$\vec{v}(x_1, x_2, t) = e^{i\omega t}\; \vec{v}(x_1, x_2),\;\; \eta(x_1, t) = e^{i\omega t}\; \eta(x_1),$$

ω real. First, we seek the eigenvalues $\omega^2 > \beta g$. Setting $\mu = \omega^{-2}$, the equation (11) gives

$$(I - \mu A_{11})\,\vec{v} = \mu A_{12}\eta\;; \tag{12}$$

$$(\mu\mathcal{D} - \mathrm{I})\,\eta = -\mu \mathrm{A}_{21}\vec{v}\;, \tag{13}$$

where $\mathcal{D}$ is the unbounded operator $A_{22} + gC^{-1}$. $I - \mu A_{11}$ has a bounded inverse, holomorphic for $|\mu| < (\beta g)^{-1}$. Then, we can eliminate $\vec{v}$ between (12) and (13). Setting

$$\eta' = \mathcal{D}^{1/2}\eta\;,$$
$$\Phi(\mu) = \mathcal{D}^{-1/2}\mathrm{A}_{21}(\mathrm{I} - \mu\mathrm{A}_{11})^{-1}\mathrm{A}_{12}\mathcal{D}^{1/2}\;,$$
$$E(\mu) = \mu I - \mathcal{D}^{-1} + \mu^2\Phi(\mu)\;,$$

we obtain the equation

$$E(\mu)\eta' = 0 \ , \qquad \eta' \in \widetilde{L}^2(\Gamma) \ . \tag{14}$$

$\mathcal{D}^{-1}$ is self-adjoint, compact in $\widetilde{L}^2(\Gamma)$; $\Phi(\mu)$ is an operatorial function which is self-adjoint and holomorphic in $|\mu| < (\beta g)^{-1}$, $E(0) = -\mathcal{D}^{-1}$ is compact, $E'(0) = I$ is strongly positive. Using a theorem of the theory of the operator pencils [2], we obtain the following result:

there is a countable infinity of eigenvalues μ_k in the interval $[0, \beta g]$ which tend to zero, as $k \to +\infty$, so that the corresponding eigenvalues ω_k^2 of the problem tend to infinity as $k \to +\infty$. The corresponding eigenelements η_k' have not associated elements and form a Riesz basis in a subspace of $\widetilde{L}^2(\Gamma)$, which has a finite defect.

5. The essential spectrum

Now, we suppose, $\omega^2 \leq \beta g$.

If $\lambda(\mathcal{D}) > 0$ is the smallest eigenvalue of $\mathcal{D}$, we have

$$\Big((\mu\mathcal{D} - \mathrm{I})\eta, \eta)\Big)_{\widetilde{\mathrm{L}}^2(\Gamma)} \geq [\mu\lambda(\mathcal{D}) - 1]||\eta||^2_{\widetilde{\mathrm{L}}^2(\Gamma)} \ .$$

a) If $\lambda(\mathcal{D}) > \beta\mathrm{g}$, $\mu\mathcal{D} - \mathrm{I}$ has a compact inverse and from (12) and (13), we obtain

$$[A_{11} - A_{12}(\mathcal{D} - \omega^2\mathrm{I})^{-1}\mathrm{A}_{21} - \omega^2\mathrm{I}]\vec{\mathrm{v}} = 0 \ , \qquad \vec{\mathrm{v}} \in \mathcal{I}_0(\Omega) \ , \tag{15}$$

where $A_{12}(\mathcal{D} - \omega^2\mathrm{I})^{-1}\mathrm{A}_{21}$ is self-adjoint and compact for each $\omega^2 < \beta g$. Let be ω_1^2 arbitrary in $[0, \beta g]$. Using Weyl theorems [2,3], it is possible to show that the essential spectrum of the operator $A_{12}(\mathcal{D} - \omega^2\mathrm{I})^{-1}\mathrm{A}_{21}$ is $[0, \beta g]$ and ω_1^2 belongs to the spectrum of the problem, which is, consequently, $[0, \beta g]$.

b) If $\lambda(\mathcal{D}) \leq \beta\mathrm{g}$, $\mathcal{D}$ has a countable infinity of positive eigenvalues $\widetilde{\omega}_k^2$, which tends to infinity as $k \to +\infty$. Then, there is at most finite number of $\widetilde{\omega}_k^2$ in $[0, \beta g]$. For the $\omega^2 \in [0, \beta g]$, different from these $\widetilde{\omega}_k^2$, the results of the first case are valid.

The essential spectrum, being closed, is the closed interval $[0, \beta g]$.

6. Conclusions and remarks

For a heavy almost homogeneous liquid which performs oscillations in a fixed container, the spectrum is formed by the essential spectrum $[0, \beta g]$ – this interval is *an area of resonance* – and by a point spectrum which lies outside of this interval.

We can solve analytically the case of a rectangular container. In this case, the essential spectrum is formed by eigenvalues of infinite multiplicity and their accumulation points.

Acknowledgements. The present paper is dedicated to Professor Tommaso Ruggeri in occasion of His 60th birthday.

References

1. P. Capodanno, D. Vivona, in *Proceedings of WASCOM 05* R. Monaco, G. Mulone, S. Rionero, T. Ruggeri eds., pp. 71-76 (World Scintific, Singapore, 2006).
2. N.D. Kopachevskii, S.G. Krein, *Operator approach in linear problems to hydrodynamics, vol. I* (Birkhäuser, Basel, 2002).
3. M. Reed, B. Simon, *Functional Analysis* (Academic Press, New York, 1980).
4. D. Vivona, in *Trend and Applications of Mathematics to Mechanics, STAMM 2002*, S. Rionero, G. Romano eds. (Springer, 2005).

DIFFUSION-DRIVEN STABILITY FOR BEDDINGTON-DEANGELIS PREDATOR-PREY MODEL

F. CAPONE

Department of Mathematics and Applications "R. Caccioppoli", Federico II
Naples 80126, Italy
fcapone@unina.it

The effect of diffusivities on the coexistence problem for predator-prey Beddington-DeAngelis model is studied. The analysis is essentially based on the use of a peculiar Liapunov functional[12,22] for which the sign of the time derivative along the solutions is linked directly to the eigenvalues of the linear problem.

Keywords: Stability; Liapunov Direct Method; Reaction-Diffusion Systems.

1. Introduction

Denoting by U and V prey and predator density, respectively, the predator-prey Beddington-DeAngelis model is given by:[2–8]

$$\begin{cases} \dfrac{\partial U}{\partial t} = \gamma_1 \Delta U + U(1-U) - a\dfrac{UV}{1+bU+cV} \\ \\ \dfrac{\partial V}{\partial t} = \gamma_2 \Delta V - dV + e\dfrac{UV}{1+bU+cV} \end{cases} \qquad (\mathbf{x}, t) \in \Omega \times I\!R^+ \tag{1}$$

with $\Omega \subset I\!R^3$ bounded domain and a, b, c, d, e, γ_1, γ_2 positive constants.

The dynamics of (1), under the boundary conditions

$$\beta_1 U + (1-\beta_1)\frac{dU}{d\mathbf{n}} = 0, \quad \beta_2 V + (1-\beta_2)\frac{dV}{d\mathbf{n}} = 0, \text{ on } \partial\Omega \times I\!R^+ \tag{2}$$

$\mathbf{n}$ being the outward unit normal to $\partial\Omega$ and $\beta_1, \beta_2 \in [0,1]$, has been studied recently.[2,3] In particular, denoting by $S = (U^*,\, V^*)$ the (positive) equilibrium state

$$\begin{cases} U^* = \dfrac{-[a(e-bd)-ce]+\sqrt{[a(e-bd)-ce]^2+4acde}}{2ce} < 1 \\ \\ V^* = -\dfrac{1}{c} + \dfrac{e-bd}{cd}U^* > 0\,, \quad e > (b+1)d \end{cases} \tag{3}$$

the found stability threshold is that one of the kinetic case. Successively, under the boundary conditions

$$\begin{cases} \dfrac{dU}{d\mathbf{n}} = 0 \ \text{ on } \ \Sigma_1 \times I\!R^+\,, \quad U = U^* \ \text{ on } \ \Sigma_1^* \times I\!R^+ \\ \\ \dfrac{dV}{d\mathbf{n}} = 0 \ \text{ on } \ \Sigma_2 \times I\!R^+\,, \quad V = V^* \ \text{ on } \ \Sigma_2^* \times I\!R^+\,, \end{cases} \tag{4}$$

with $\partial\Omega = \Sigma_i \cup \Sigma_i^*$, $\Sigma_i \cap \Sigma_i^* = \emptyset$, the stability of S has been studied and a stabilizing effect of diffusivities on the (local) stability of S has been found.[4]

Here we reconsider the problem, in the case of the boundary conditions

$$\begin{cases} \beta_1 U + (1-\beta_1)\nabla U \cdot \mathbf{n} = \beta_1 U^* \\ & \text{on} \quad \Sigma \times I\!R^+ \\ \beta_2 V + (1-\beta_2)\nabla V \cdot \mathbf{n} = \beta_2 V^* \end{cases} \tag{5}$$

with $\Sigma = \partial\Omega$, $\beta_i \in]0,1]$ $(i = 1,2)$ and present some results obtained recently.[5] We denote by: $\langle\cdot,\cdot\rangle$ the scalar product in $L^2(\Omega)$; $\langle\cdot,\cdot\rangle_\Sigma$ the scalar product in $L^2(\Sigma)$; $\|\cdot\|$ the $L^2(\Omega)$-norm; $\|\cdot\|_\Sigma$ the $L^2(\Sigma)$-norm; $W_i^{1,2}(\Omega,\beta_i)$ $(i=1,2)$ the functional spaces such that

$$\varphi \in W_i^{1,2}(\Omega,\beta_i) \to \{\varphi \in W_i^{1,2}(\Omega), (5)\,\text{hold with}\, U^* = V^* = 0\}\,; \tag{6}$$

$\bar{\alpha}_i$ $(i=1,2)$ the positive constant appearing in the inequality

$$\|\nabla\varphi\|^2 + \frac{\beta_i}{1-\beta_i}\|\varphi\|_\Sigma^2 \geq \bar{\alpha}_i\|\varphi\|^2\,, \tag{7}$$

holding in $W_i^{1,2}(\Omega,\beta_i)$. As it is well known $\bar{\alpha}_i(\Omega,\beta_i)$ is the lowest eigenvalue of $\Delta\varphi + \lambda\varphi = 0$ in $W_i^{1,2}(\Omega,\beta_i)$ (i.e. the principal eigenvalue of $-\Delta$). We refer here to the positive smooth solutions of (1), under the boundary conditions (5) and the smooth positive initial data:

$$U(\mathbf{x},0) = U_0(\mathbf{x})\,, \quad V(\mathbf{x},0) = V_0(\mathbf{x})(>0) \qquad \mathbf{x} \in \Omega\,. \tag{8}$$

The plan of the paper is as follows. In Section 2 the boundedness and the uniqueness of positive solutions of (1), (5), (8) are proved. Then nonlinear stability analysis (Section 3) of the (positive) equilibrium state S, with respect to the $L^2(\Omega)$-norm, is performed. Finally Section 4 is devoted to the diffusion-driven instability (Turing effect).

2. Boundedness of positive solutions

We denote by $C_1^2(\Omega_T)$ the set of the functions belonging to $C(\Omega_T)$ together with their first (spatial and temporal) derivatives and second spatial derivatives.[9] The following theorem holds.

Theorem 2.1. *Let* $\{U, V \in C_1^2(\Omega_T) \cap C(\bar{\Omega}_T)\}$ *be a positive solution of* (1), (5), (8). *Then* U, V *are bounded according to*

$$\begin{cases} U < M_1 = \max\left\{1, \max\limits_{\bar{\Omega}} U_0(\mathbf{x}), \max\limits_{\partial\Omega\times[0,T]} U^*\right\} \\ V < M_2 = \max\left\{\dfrac{eM_1 - d}{cd}, \max\limits_{\bar{\Omega}} V_0(\mathbf{x}), \max\limits_{\partial\Omega\times[0,T]} V^*\right\}, \quad eM_1 > d. \end{cases} \tag{9}$$

Proof. Here we limit ourselves to prove inequality $(9)_1$. Let $\max\limits_{\bar{\Omega}_T} U = U(\mathbf{x}_0, t_0)$, then if $(\mathbf{x}_0, t_0)$ belongs to the interior of Ω_T, then $(1)_1$ implies

$$\left\{\frac{\partial U}{\partial t} - [\gamma_1 \Delta U + U(1-U)]\right\}_{(\mathbf{x}_0, t_0)} < 0\,. \tag{10}$$

In view of $\dfrac{\partial U}{\partial t}(\mathbf{x}_0, t_0) = 0, ., [\Delta U]_{(\mathbf{x}_0,t_0)} < 0$, it follows that (10) can hold only if $[U(1-U)]_{(\mathbf{x}_0,t_0)} > 0$, i.e. $U(\mathbf{x}_0, t_0) < 1$. If $(\mathbf{x}_0, t_0) \in \Gamma_T$, then in view of the $C^{2+\eta}$, $\eta \in (0,1)$, property of $\partial\Omega$, Ω verifies in any point $\mathbf{x}_0 \in \partial\Omega$, the interior ball condition, i.e. there exists an open ball $B^* \subset \Omega$ with $\mathbf{x}_0 \in \partial B^*$. If $U(\mathbf{x}_0, t_0) > 1$, on choosing the radius of B^* sufficiently small it follows that $\gamma_1 \Delta U - \dfrac{\partial U}{\partial t} > 0$ in B^* and hence the Hopf's Lemma[10,11] implies $\left(\dfrac{dU}{d\mathbf{n}}\right)_{(\mathbf{x}_0,t_0)} > 0$. Then, in view of $(5)_1$, $(9)_1$ immediately follows. □

Theorem 2.2. *System* (1) *under conditions* (5), (8) *can admit only one solution belonging to* $\mathrm{I}_1 = \{\varphi : \Omega_T \to I\!R, \varphi \geq 0, \varphi \in L^2[(0,T), L^2(\Omega)]\}$.[5]

3. Nonlinear stability

Let us set $\phi^* = 1 + bU^* + cV^*$ and

$$\begin{cases} a_{11} = 1 - 2U^* - aV^*(1+cV^*)\phi^{*-2}\,, & a_{12} = -aU^*(1+bU^*)\phi^{*-2} \\ a_{21} = eV^*(1+cV^*)\phi^{*-2}\,, & a_{22} = -cdV^*\phi^{*-1} \\ b_{11} = a_{11} - \bar{\alpha}_1\gamma_1\,, & b_{22} = a_{22} - \bar{\alpha}_2\gamma_2\,. \end{cases} \tag{11}$$

Then the following theorem holds true.[5]

Theorem 3.1. *Let $b_{11} < 0$ holds true. Then $S = (U^*, V^*)$ is nonlinearly asymptotically exponentially stable and (locally) attractive with respect to the $L^2(\Omega)$-norm.*

Theorem 3.2. *Let*

$$e^* = (3 + 2\sqrt{2})d\,, \quad b^* = \frac{e - d + \sqrt{e^2 - 6ed + d^2}}{2d}\,(> 0) \tag{12}$$

and let one of the conditions

$$(b+1)d < e < e^*\,, \quad b < 2(1 + \sqrt{2}) \tag{13}$$

$$b > \max\{2(1 + \sqrt{2});\, b^*\} \tag{14}$$

$$e > e^*\,, \quad b < \min\{2(1 + \sqrt{2});\, b^*\}\,, \quad \bar{\alpha}_1 > \frac{d[-db^2 + (e-d)b - e]}{\gamma_1 e(e - bd)} \tag{15}$$

holds. Then $S = (U^*, V^*)$ is nonlinearly asymptotically exponentially stable and (locally) attractive with respect to the $L^2(\Omega)$-norm.

Proof. It is enough to prove that each of the conditions (13)-(15) implies $b_{11} < 0$. To this end, on taking into account $(11)_1$, from $(11)_5$ it follows that $b_{11} = 1 - 2U^* - a\dfrac{V^*}{\phi^*} + ab\dfrac{V^*}{\phi^*}\dfrac{U^*}{\phi^*} - \gamma_1\bar{\alpha}_1$ with

$$a\frac{V^*}{\phi^*} = 1 - U^*\,, \quad \frac{U^*}{\phi^*} = \frac{d}{e}\,, \quad U^* > \frac{d}{e - bd}\,, \quad e > (b+1)d\,. \tag{16}$$

On taking into account $(16)_1 - (16)_3$, one has

$$b_{11} < \frac{d[-db^2 + (e-d)b - e]}{e(e - bd)} - \gamma_1\bar{\alpha}_1 \tag{17}$$

i.e.

$$db^2 - (e-d)b + e > 0 \Longrightarrow b_{11} < 0\,. \tag{18}$$

On setting $\Delta = e^2 - 6ed + d^2$ it follows that condition $(13)_1$, which compatibility is guaranteed by $(13)_2$, implies $\Delta < 0$ and hence (18) is verified.

Passing now to the condition (14), $b > 2(1 + \sqrt{2})d$ implies $e > e^*$ and hence $\Delta > 0$. Therefore $b > b^*$ implies (18). Finally conditions $(15)_1$, $(15)_2$ imply that $db^2 - (e-d)b + e < 0$ and hence, in view of (17), $(15)_3$ implies $b_{11} < 0$. □

4. Diffusion driven instability: Turing effect

Theorem 4.1. *Let $\beta_1 = \beta_2, \gamma_1 \neq \gamma_2$ and*

$$\begin{cases} a > \max\left\{\dfrac{ec}{b}, \dfrac{ec}{e-bd}\right\}, \quad e > \dfrac{bd(b+1)}{b-1}, \quad b > 1 \\ \\ \dfrac{(ba-ec)d}{(ba-ec)d+ea} < U^* < \dfrac{bd}{e+bd} \\ \\ \gamma_2 > \dfrac{A_0 + \gamma_1 \bar{\alpha}|a_{22}|}{\bar{\alpha}(|a_{11}| - \gamma_1 \bar{\alpha})}, \quad \gamma_1 < \dfrac{|a_{11}|}{\bar{\alpha}} \end{cases} \tag{19}$$

hold. Then the onset of Turing instability occurs.[5]

Acknowledgments

To Professor T. Ruggeri in the occasion of his 60th birthday.
This work has been performed under the auspices of the G.N.F.M. of I.N.D.A. M. and M.I.U.R. (P.R.I.N. 2005): "Nonlinear Propagation and Stability in Thermodynamical Processes of Continuous Media". The author thanks gratefully Prof. S. Rionero, for having proposed the present research and for His helpful suggestions.

References

1. Robert A. Adams, *Sobolev Spaces*, (Academic Press Inc., London, 1975).
2. R.S. Cantrell, C. Cosner, On the dynamics of predator-prey models with the Beddington - De Angelis functional response, *Journal of Mathematical Analysis and Application* **257**, 206 (2001).
3. R.S. Cantrell, C. Cosner, *Spatial ecology via reaction-diffusion equations* (John Wiley and Sons Ltd., Chichester, UK, 2004) ISBN 0-471-49301-5.
4. F. Capone, M. Piedisacco, S. Rionero, Nonlinear stability for reaction-diffusion Lotka-Volterra model with Beddington-DeAngelis functional response, *Rend. Acc. Sc. fis. nat. Napoli* **LXXIII**, 85 (2006).
5. F. Capone, Diffusion Driven Stability and Turing effect for a predator-prey model with Beddington-DeAngelis functional response, *Journal of Mathematical Analysis and Application* (submitted), 2007.
6. D. L. DeAngelis, R. A. Goldstein and R. V. O'Neill, A model for trophic interaction, *Ecology* **56**, 881 (1975).
7. I. Kozlova, M. Singh and A. Easton, Predator-prey models with diffusion based on Luckinbill's experiment with Didinium and Paramecium, *Math. Comp. Modelling* **36**, 83 (2002).
8. L.S. Luckinbill, Coexistence in laboratory population of Paramecium aurelia and its predator didinium nasutum, *Ecology* **54**, 1320 (1973).

9. Mora Xavier, Semilinear parabolic problems define semiflows on C^k spaces, *Trans.Amer. Math. Soc.* **278**, no.1, 21 (1983).
10. M. H. Protter, H. F. Weinberger, *Maximum principles in differential equations* (Prentice-Hall, Inc. 1967).
11. K. Rektoris, *Variational methods in Mathematics* (Science and Engineering. D. Reidal Publ.Co., 1980).
12. S. Rionero, On the stability of binary reaction-diffusion systems, *Nuovo Cimento Soc. Ital. Fis. B* **119**, no.7-9, 773 (2004) MR2136906.
13. S. Rionero, A rigorous link between the L^2-stability of the solutions to a binary reaction-diffusion system of P.D.E. and the stability of the solutions to a binary system of O.D.E., *Rend. Accad. Sci. Fis. Mat. Napoli* **71**, 53 (2004).
14. S. Rionero, Asymptotic properties of solutions to nonlinear possibly degenerated parabolic equations in unbounded domains, *Math. Mech. Solids* **10**, n. 5, 541 (2005).
15. S. Rionero, A rigorous reduction of the L^2-stability of the solutions to a nonlinear binary reaction-diffusion system of P.D.Es. to the stability of the solutions to a linear binary system of O.D.Es., *Journal of Mathematical Analysis and Application* **319**, Issue 2, 377 (2006).
16. S. Rionero, A nonlinear stability analysis for two-species population dynamics with dispersal, *Math. Biosc. Eng.* **3**, n.1, 189 (2006).
17. S. Rionero, L^2-stability of the solutions to a nonlinear binary reaction-diffusion system of P.D.Es., *Rend. Accademia dei Lincei* **4**, 227 (2006).
18. S. Rionero, Nonlinear L^2-stability analysis for two-species population dynamics in spatial ecology under Neumann boundary data, *Rend. Circolo Mat. Palermo*, Serie II, Suppl. 78, 273 (2006).
19. S. Rionero, Long time behaviour of three competing species and mutualistic communities. In: *Asymptotic Methods in Nonlinear Wave Phenomena. In honour of the 65th birthday of Antonio Greco. Eds. T. Ruggeri, M. Sammartino.* (World Scientific Publishing, 171, 2007).
20. S. Rionero, A new approach to nonlinear L^2-stability of double diffusive convection in porous media:Necessary and sufficient conditions for global stability via a linearization principle, *Journal of Mathematical Analysis and Application* **333**, Issue 2, 1036 (2007).
21. S. Rionero, Diffusion-Driven Stability and Turing effect under Robin Boundary Data. (To appear).
22. S. Rionero, On the dynamics of the Brusselator system. In *Proceedings of the Meeting "Mathematical Physics Models and Engineering Sciences"* (Naples, 2006) (To appear).

LOCALIZED STRUCTURES IN A DUSTY PLASMA

P. CARBONARO

Dipartimento di Matematica e Informatica,
Città Universitaria, Viale A. Doria 6, 95125 Catania, Italy
E-mail: carbonaro@dmi.unict.it

By using a multi-scale technique we derive from the equations describing a cold dusty plasma a non-linear Schrödinger-Poisson equation. We show that such equation admits a solution in the form of an oblique dark soliton.

1. Intoduction

A dusty plama is a normal electron-ion plasma with an extra component of heavy charged macro particles whose mass, size and concentration can affect the behaviour and dynamics of the phenomena which take place in a plasma. In the last ten years there has been much interest in studying the properties of dusty plasmas because of their numerous applications both in astrophysical and in laboratory context[1].

A large number of works in this area deal with the propagation of ion-acoustic waves, in particular solitary waves. The standard method consists in using a multi-scale technique wich permits one to reduce the original system to a model equation, such as KdV or non-linear Schrödinger equation, which describes the balance between non-linearity and dispersive effects[2–6]. Several authors have limited their interest to one-dimensional approximation, on the other hand, the variety of phenomena which occur in a plasma suggests one to consider also the multi-dimensional case[7–9].

This work is aimed at deriving a non-linear model equation of Schrödinger type which describes the amplitude modulation of ion-acoustic waves in multi-dimension.

2. Basic equations

The plasma consists of three components: ions, electrons and dust grains. The governing equations in dimensionless form read[1]:

$$\frac{\partial n_d}{\partial t} + \nabla.(n_d \mathbf{u}) = 0, \tag{1}$$

$$\frac{\partial \mathbf{u}_d}{\partial t} + \mathbf{u}_d.\nabla \mathbf{u}_d = \nabla \varphi, \tag{2}$$

$$\nabla^2 \varphi = n_d + n_e - n_i, \tag{3}$$

where n_d and $\mathbf{u}_d$ are respectively the number density and the velocity of the dust massive particles, φ the electrical potential. The electron and ion density is assumed to satisfy the Boltzmann distribution

$$n_e = \mu_e \exp(\beta s \varphi) \quad and \quad n_i = \mu_i \exp(s\varphi)$$

where the electron and ion density μ_e *and* μ_i are normalized so as to have $s(\mu_i + \mu_e) = 1$. The neutrality condition at equilibrium implies also $n_{io} = n_{eo} + n_{do}$, where n_{io}, n_{eo}, *and* n_{do} are unperturbed number densities.

To study the amplitude modulation of dust acoustic waves we develope the solution around an unperturbed state as follows:

$$n = 1 + \epsilon \sum_{l=-\infty}^{\infty} n_l^{(1)}(\underline{\xi}) \; \exp(il\theta) \; + \epsilon^2 \sum_{l=-\infty}^{\infty} n_l^{(2)}(\underline{\xi}) \exp(il\theta) + ... \tag{4}$$

$$u_\alpha = \epsilon \sum_{l=-\infty}^{\infty} u_{\alpha l}^{(1)}(\underline{\xi}) \; \exp(il\theta) \; + \epsilon^2 \sum_{l=-\infty}^{\infty} u_{\alpha l}^{(2)}(\underline{\xi}) \exp(il\theta) + ... \tag{5}$$

$$\varphi = \epsilon \sum_{l=-\infty}^{\infty} \varphi_l^{(1)}(\underline{\xi}) \exp(il\theta) + \epsilon^2 \sum_{l=-\infty}^{\infty} \varphi_l^{(1)}(\underline{\xi}) \exp(il\theta) + ... \tag{6}$$

where $\underline{\xi} = (\xi_1, \xi_2, \xi_3)$, with ξ_α and τ slow variables defined as $\xi_\alpha = \epsilon(x_\alpha - \lambda_\alpha t)$, $\tau = \epsilon^2 t$, $\theta = \mathbf{k}.\mathbf{r} - \omega t$, with $\mathbf{k} = (k_1, k_2, k_3)$ and $\mathbf{r} = (x_1, x_2, x_3)$.

Inserting the foregoing expressions for n, u_α and φ into equations (1)-(3) and equating to zero the coefficients of the different powers of ϵ we obtain, at the lowest order of approximation, $\mu_i = 1 + \mu_e$ (as consequence of neutrality condition) and at the next order (ϵ^1) the following set of equations

$$-l\omega n_l^{(1)} + lk_\alpha u_{\alpha l}^{(1)} = 0, \quad -l\omega u_{\alpha l}^{(1)} + lk_\alpha \varphi_l^{(1)}, \quad n_l^{(1)} + (l^2k^2 + 1)\varphi_l^{(1)} = 0. \tag{7}$$

For $l = 1$, eqs.(7) yield the ion-acoustic wave dispersion relation:

$$\omega^2 = \frac{k^2}{k^2 + 1},$$

and the first order perturbations in terms of an arbitrary function $\Phi(\xi_1, \xi_2, \xi_3, \tau)$, namely

$$n_1^{(1)} = -k^2\Phi, \quad u_{\alpha 1}^{(1)} = -\omega k_\alpha \Phi, \quad \varphi_1^{(1)} = \omega^2 \Phi, \tag{8}$$

while for $l \neq 1$ we assume $n_l^{(1)} = u_{\alpha l}^{(1)} = \varphi_l^{(1)} = 0$.

Going to the next order of approximatio (ϵ^2), we get for $l = 1$

$$-l\omega n_1^{(2)} + ilk_\alpha u_{\alpha 1}^{(2)} - \lambda_\alpha \frac{\partial n_1^{(1)}}{\partial \xi_\alpha} + \frac{\partial u_{\alpha 1}^{(1)}}{\partial \xi_\alpha} = 0, \tag{9}$$

$$-l\omega u_{\alpha 1}^{(2)} - ilk_\alpha \varphi_{\alpha 1}^{(2)} - \lambda_\beta \frac{\partial u_{\alpha 1}^{(1)}}{\partial \xi_\beta} - \frac{\partial \varphi_1^{(1)}}{\partial \xi_\alpha} = 0, \tag{10}$$

$$-n_1^{(2)} - (k^2 + 1)\varphi_1^{(2)} + 2ik_\alpha \frac{\partial \varphi_1^{(1)}}{\partial \xi_\alpha} = 0, \tag{11}$$

from which we obtain the compatibility condition

$$\lambda_\alpha = \frac{\partial \omega}{\partial k_\alpha} = \frac{\omega^3}{k^4} k_\alpha, \tag{12}$$

and the second order corrections

$$n_1^{(2)} = -k^2\Phi' + i2k_\alpha \frac{\partial \Phi}{\partial \xi_\alpha} \tag{13}$$

$$u_{\alpha 1}^{(2)} = -k_\alpha \Phi' + i\omega(\frac{k_\alpha k_\beta}{k^4}\omega^2 + \delta_{\alpha\beta})\frac{\partial \Phi}{\partial \xi_\beta}, \tag{14}$$

$$\varphi_1^{(2)} = \omega^2\Phi' - i\frac{2\omega^4}{k^4}k_\alpha \frac{\partial \Phi}{\partial \xi_\alpha}, \tag{15}$$

where Φ' is an arbitrary function of ξ_1, ξ_2, ξ_3 and τ .

For $l = 2$, we have

$$n_2^{(2)} = \frac{1}{6}[2D\omega^2 + 3k^2(4k^2 + 1)]\Phi^2 \tag{16}$$

$$u_{\alpha 2}^{(2)} = \frac{\omega}{6}[2D\frac{\omega^2}{k^2} + 3(2k^2 + 1)]k_\alpha \Phi^2, \tag{17}$$

$$\varphi_2^{(2)} = -\frac{k^2}{6}[\frac{2D}{(k^2 + 19^2} + 3]\Phi^2, \tag{18}$$

where $D = \frac{s^2}{2}(\mu_e\beta^2 - \mu_i)$, while for $l = 0$ we have simply

$$n_0^{(2)} + \varphi_0^{(2)} + 2D\omega^4|\Phi|^2 = 0. \tag{19}$$

At the order (ϵ^3) and for $l = 0$ the expansion of eq. (1)-(3) reads

$$-\frac{\omega^3}{k^4}k_\alpha\frac{\partial n_0^{(2)}}{\partial\xi_\alpha}+\frac{\partial u_{\alpha 0}^{(2)}}{\partial\xi_\alpha}+2\omega k^2 k_\alpha\frac{\partial|\Phi|^2}{\partial\xi_\alpha}=0 \tag{20}$$

$$-\frac{\omega^3}{k^4}k_\beta\frac{\partial u_{\alpha 0}^{(2)}}{\partial\xi_\beta}+\frac{\partial n_0^{(2)}}{\partial\xi_\alpha}+\omega^2(2D\omega^2+k^2)\frac{\partial|\Phi|^2}{\partial\xi_\alpha}=0 \tag{21}$$

which together (19) yields

$$(\delta_{\alpha\beta}-\lambda^2\frac{k_\alpha k_\beta}{k^2})\frac{\partial u_0^{(2)}}{\partial\xi_\alpha}+\omega k^2[2+\frac{\omega^6}{k^6}(2D+k^2+1)]\frac{\partial|\Phi|^2}{\partial\xi_\alpha}=0. \tag{22}$$

If we contract eq. (21) with k_α we get

$$-\frac{\omega^3}{k^4}k_\alpha k_\beta\frac{\partial u_{\beta 0}^{(2)}}{\partial\xi_\alpha}+k_\alpha\frac{\partial n_0^{(2)}}{\partial\xi_\alpha}+\omega^2(2D\omega^2+k^2)k_\alpha\frac{\partial|\Phi|^2}{\partial\xi_\alpha}=0, \tag{23}$$

from which it follows

$$-\frac{\omega^3}{k^4}k_\beta\frac{\partial u_{\beta 0}^{(2)}}{\partial\xi_\alpha}+\frac{\partial n_0^{(2)}}{\partial\xi_\alpha}+\omega^2(2D\omega^2+k^2)\frac{\partial|\Phi|^2}{\partial\xi_\alpha}=0. \tag{24}$$

Comparing (21) wioth (24) one has

$$k_\beta(\frac{\partial u_{\alpha 0}^{(2)}}{\partial\xi_\beta}-\frac{\partial u_{\beta 0}^{(2)}}{\partial\xi_\alpha})=0, \tag{25}$$

therefore one can write $u_{\alpha 0}^{(2)}$ in terms of a potential function, i.e. $u_{\alpha 0}^{(2)}=\frac{\partial\psi}{\partial\xi_\alpha}$. Then equation (22) takes the form

$$(\delta_{\alpha\beta}-\lambda^2\frac{k_\alpha k_\beta}{k^2})\frac{\partial^2\Psi}{\partial\xi_\alpha\partial\xi_\beta}+\omega k^2[2+\frac{\omega^6}{k^6}(2D+k^2+1)]\frac{\partial|\Phi|^2}{\partial\xi_\alpha}=0. \tag{26}$$

Finally, taking $l = 1$ in the third order of the expansion, the compatibility condition for non-trivial solutions yields the following evolution equation for $\Phi(\xi_1,\xi_2,\xi_3)$,

$$i\frac{\partial\Phi}{\partial\tau}+\frac{1}{2}(\frac{\partial^2\omega}{\partial k_\alpha\partial k_\alpha})\frac{\partial^2\omega}{\partial\xi_\alpha\partial\xi_\alpha}+F(k,\Phi)=0, \tag{27}$$

where

$$\frac{\partial^2\omega}{\partial k_\alpha\partial k_\beta}=\frac{\omega^3}{k^4}[\delta_{\alpha\beta}-(1+3\omega^3)\frac{k_\alpha k_\beta}{k^2}], \tag{28}$$

and

$$F(k,\Phi)=-\omega^3[D+\frac{D^2}{(k^2+1)}(2k^2+1)-\frac{3}{2}E\frac{k^2}{(k^2+1)^2}+$$

$$(k^2+1)(2k^2+\frac{3}{4})]|\Phi|^2\Phi - \frac{1}{2}[\omega(1+\frac{2k}{k^2+1})n_0^{(2)} + 2k_\alpha u_{\alpha 0}^{(2)}]\Phi, \qquad (29)$$

with $E = \frac{s^3}{6}(\mu_e\beta^3 + \mu_i)$.

3. Spatial solitons

The matrix $\frac{\partial^2\omega}{\partial k_\alpha \partial k_\beta}$ of the coefficients of the second derivatives in (27) can be diagonalized by choosing a new set of coordinates. For the two-dimensional case, if we introduce the new variables

$$x = \frac{k_2}{k}\xi_1 - \frac{k_1}{k}\xi_2, \qquad y = \frac{k_1}{k}\xi_1 + \frac{k_2}{k}\xi_2 \qquad (30)$$

eqs.(26) and (27) take the canonical form of a Poisson-Schrödinger equation

$$i\frac{\partial\Phi}{\partial\tau} + \frac{\omega^3}{2k^4}(\frac{\partial^2\Phi}{\partial x^2} - 3\omega^2\frac{\partial^2\Phi}{\partial y^2}) - A|\Phi|^2\Phi - B\frac{\partial\Psi}{\partial y}\Phi = 0, \qquad (31)$$

$$\frac{\partial^2\Psi}{\partial^2 x} + (1-\lambda^2)\frac{\partial^2\Psi}{\partial^2 y} + 2\omega k^2 B\frac{\partial|\Phi|^2}{\partial y} = 0, \qquad (32)$$

where

$$A = \omega^3[D\frac{1-k^2}{1+k^2} + \frac{D^2}{3(k^2+1)} - \frac{3}{2}E\frac{k^2}{(k^2+1})^2 + 2k^4 + \frac{9}{4}k^2 + \frac{3}{4}], \qquad (33)$$

$$B = k[1 + \frac{\omega^4}{2k^4}(1 + \frac{2D}{k^2+1})]. \qquad (34)$$

If we look for steady solutions to eqs. (31) and (32) of the form[10]

$$\Phi = \sqrt{\rho}\exp\{i\,[R\chi(x,y) - S\tau]\}, \qquad (35)$$

we obtain a solution in the form of a spatial dark soliton

$$\rho(\eta) = 1 - \frac{1-p}{\cosh^2(\sqrt{1-p}\;\eta)},$$

where

$$\eta = k^2\sqrt{\frac{Q}{\omega^3(1-3\omega^2a^2)}}\,(x-ay) \quad \textit{and} \quad p = \frac{u_0^2}{1-3\omega^2a^2},$$

with a denoting the slope of the soliton center location with the y-axis, u_0^2 a constant and

$$Q = \omega^3[D\frac{1-k^2}{1+k^2} + \frac{D^2}{3(k^2+1)^2} - \frac{3}{2}E\frac{k^2}{(k^2+1)^2} + 2k^4 + \frac{9}{4}k^2 + \frac{3}{4}] -$$

$$\frac{2\omega k^4 a^2}{1+a^2(1-\lambda^2)}[1 + \frac{\omega^4}{2k^4}(1+\frac{2D}{k^2+1})]^2.$$

Acknowledgments

This work is supported by the Italian Ministry for University and Scientific Research, PRIN:*Problemi matematici non lineari di propagazione e stabilià nei modelli del continuo* (Coordinator Prof. T. Ruggeri), by "Gruppo Nazionale della Fisica Matematica" of the "Istituto Nazionale di Alta Matematica," and by the University of Catania.

References

(1) P.K. Shukla and A.A. Mamum, *Introduction to dusty plasma physics*, Institut of Physics Publishing, Bristol and Philadelphia (2002).
(2) J.K. Xue and L.P. Zhang, *chaos, Solitons and Fractals* **32**, 592 (2007).
(3) N. Akhtar, S. Mahamood and H. Saleem, *Physics Letters A* **361**, 126 (2007).
(4) E.K. El-Shewy,*Chaos, Solitons and Fractals* **31**, 1020 (2007).
(5) B. Tian and Y.T. Gao, *Physics Letters A* **362**, 283 (2007).
(6) S.K. El-Labany, E.F. El-Shami, W.F. El-Taibany and W.M. Moslem, *chaos, Solitons and Fractals* **34**, 1393 (2007).
(7) M.R. Amin, G.E. Morfill and P.K. Shukla, *Phys. Rev. E* **58**, 6517 (1998).
(8) W.S. Duan, X.R. Hong, Y.R. Shi and J.A. Sun,*chaos, Solitons and Fractals* **16**, 767 (2003).
(9) I. Kourakis and P.K. Shukla, *J. Phys. A: Math. Gen.* **36**, 11901 (2003)
(10) G.A. El, A. Gammal and A.M. Kamchatnov, *Phys. Rev. Lett.* **97**, 180405 (2006)

FLUID DYNAMICAL FEATURES OF THE WEAK KAM THEORY*

F. CARDIN

Dipartimento di Matematica Pura ed Applicata, Università di Padova, via Trieste 63 - 35121 Padova, Italy

E-mail: cardin@math.unipd.it

The trend to equilibrium is ascertained for potential motion solutions of a class of systems of nonlinear PDE –the inviscid Burgers equations on tori of any dimension– in the framework of the weak KAM theory. For large time these solutions tend to stationary solutions at a well precise energy (Mañé) critical level, which is independent of initial data. An analogous behaviour is foreseen for the Madelung fluid in the semi-classical limit $\hbar \to 0$.

Keywords: Burgers equation, weak KAM theory, viscosity solutions for Hamilton-Jacobi equations

1. Introduction

Simplified models for fluid dynamics on the tori $\mathbb{T}^n$ are not new in literature: from the pioneering book by Boldrighini [2] up to exhaustive treatises like the handbook by Gallavotti [12] where Foias-Prodi (and others) theory has been translated into this agreeable (compact and boundaryless) geometric environment. In such a framework we will find out some new qualitative properties for the inviscid Burgers model of fluid dynamics, more precisely, for potential solutions $\mathbf{u}(t,x) = \nabla S(t,x)$ we will catch a well precise time asymptotic behaviour towards a sort of equilibrum solutions. This result could be useful from a qualitative point of view, since it is well known that, in general, there are many difficulties to detect trend to equilibria for conservative systems of PDE. The main tools here involved, after a heuristic discussion based on the Cole-Hopf transform, are the viscosity solutions theory and the weak KAM theory: they will be utilized in two directions, (i) giving the existence of global stationary solutions of a suitable

*This paper is dedicated to my friend Tommaso Ruggeri in occasion of its 60th birthday

Hamilton-Jacobi equation connected with our inviscid Burgers model, (ii) constructing the expected time asymptotic behaviour.
In the last Section we will recognize an analogous thermodynamic time-asymptotic property for the quantum mechanics Madelung fluid in the semi-classical limit $\hbar \to 0$; the possible interested readers are invited to papers like [16] and [3] where singularity aspects are treated as well.

2. The inviscid Burgers system and viscosity solutions

Let consider the following evolutive non linear PDE system, a sort of inviscid Burgers-type equation; by components:

$$u^i_{,t} + \sum_{j=1}^{n} u^i_{,x_j} u^j = -V_{,x_i}(x) \qquad i = 1, \ldots, n \tag{1}$$

The unknown vector field is

$$\mathbf{u} : [0, +\infty) \times M \ni (t, x) \longmapsto \mathbf{u}(t, x) = (u^i(t, x))_{i=1,\ldots,n} \in \mathbb{R}^n \tag{2}$$

where $M = \mathbb{T}^n = \mathbb{R}^n/\mathbb{Z}^n$ is the[a] n-torus and $V(x)$ is a given potential energy function on M. This system can be thought as a simplified model for gas or fluid dynamics: for $n = 3$, $\mathbf{u} = (u^1, u^2, u^3)$ is an Eulerian velocity field: following e.g. [13], Burgers equation can be obtained from the Navier-Stokes equations by dropping (i) the pressure[b] term $-\nabla p$, (ii) the viscosity term $\varepsilon \Delta \mathbf{u}$ and (iii) the incompressibility condition $\nabla \cdot \mathbf{u} = 0$ is relaxed too; this model is reasonably connected to low density gases, with a conventionally constant mass density.

Here we are interested to investigate about weak 'potential solutions':

$$u^i(t, x) = S_{,x_i}(t, x) \qquad \text{a.e.} \tag{3}$$

with initial data

$$u^i(0, x) = \sigma_{,x_i}(x), \tag{4}$$

more precisely, we are looking for Lipschitz functions S, so the above system (1), at least from a formal[c] point of view, reads

$$\left(S_{,t} + \frac{1}{2}|\nabla S|^2 + V(x)\right)_{,x_i} = 0. \tag{5}$$

[a]Even though some arguments discussed below could be placed, by some minor changes, in a generic compact connected boundaryless manifold M endowed with a Riemannian metric, in the following we will consider always $M = \mathbb{T}^n$.
[b]As usual: $\nabla = \text{Grad}$, $\nabla \cdot = \text{Div}$, $\Delta = \text{Div Grad}$.
[c]The formula (5) rigorously requires S be twice differentiable a.e.

Hence, we are led to a Hamilton-Jacobi equation for S, for some real valued constant,

$$S_{,t} + \frac{1}{2}|\nabla S|^2 + V(x) = \text{const.} \tag{6}$$

The choice for the above constant does not play a crucial role, so, definitively, we will study Cauchy Problems:

$$S_{,t} + \frac{1}{2}|\nabla S|^2 + V(x) = 0, \qquad S(0,x) = \sigma(x), \qquad x \in \mathbb{T}^n \tag{7}$$

which are related to Cauchy Problems for (1) with $\mathbf{u}(0,x) = \nabla\sigma(x)$. Since the involved Hamiltonian function H,

$$H : T^*\mathbb{T}^n \to \mathbb{R}, \quad (x,p) \mapsto H(x,p) = \frac{1}{2}|p|^2 + V(x), \tag{8}$$

is p-convex and superlinear, the above problem (7) admits one and only one global Lipschitz solution in the weak framework of the viscosity solutions theory, see e.g. [7]. In the next section we review some few topics from this analytical theory.

2.1. *Viscosity solutions for H-J*

Let

$$\frac{\partial u}{\partial t} + H(x, \frac{\partial u}{\partial x}) = 0, \qquad t \in (0,T), \quad x \in M \tag{9}$$

A function $u \in C((0,T) \times M)$ is a viscosity subsolution [supersolution] of (9) if, for any $\phi \in C^1((0,T) \times M)$,

$$\frac{\partial \phi}{\partial t}(\bar{t},\bar{x}) + H(\bar{x}, \frac{\partial \phi}{\partial x}(\bar{t},\bar{x})) \leq 0 \qquad [\geq 0] \tag{10}$$

at any local maximum [minimum] point $(\bar{t},\bar{x}) \in (0,T) \times M$ of $u - \phi$. Finally, u is a viscosity solution of (9) if it is simultaneusly a viscosity sub- and supersolution.

The origin of the term *viscosity solution* is going back to the vanishing viscosity method:

$$\frac{\partial u_\varepsilon}{\partial t}(t,x) + H(x, \frac{\partial u_\varepsilon}{\partial x}(t,x)) = \varepsilon \Delta u_\varepsilon(t,x) \tag{11}$$

Giving a solution of (11), if $\varepsilon \to 0$, does u_ε tend to a function u, solution (in some sense) of the limit equation $\frac{\partial u}{\partial t}(t,x) + H(x, \frac{\partial u}{\partial x}(t,x)) = 0$? The question is not so easy, because the regularizing effect of the term $\varepsilon\Delta u_\varepsilon$ vanishes as $\varepsilon \to 0$ and we end up with an equation that has easily non regular solutions. The answer is that if $u_\varepsilon \to u$ uniformly on every compact sets, then u is a viscosity solution. This is actually the motivation for the terminology, used originally by Crandall and Lions [5].

2.2. *The unique potential motion solution for the inviscid Burgers system*

Going back to our H-J problem, we have obtained the unique viscosity solution $S(t,x)$ for (7), which is Lipschitz, hence, by Rademacher's theorem, almost everywhere differentiable; finally, by (3), we gain the expected potential motion, $\mathbf{u}(t,x) = \nabla S(t,x)$ *a.e.*, solving (1) with the initial (potential-like) data: $\mathbf{u}(0,x) = \nabla\sigma(x)$.

Some remarks can arise:

1. (*Discontinuities*) As said right above, $\mathbf{u}(t,x) = \nabla S(t,x)$ is holding a.e., since $S(t,x)$ is continuous Lipschitz, hence from the regularity point of view, $\mathbf{u}(t,x)$ could carry some (finite) shock wave phenomenon. A standard implementation of Hougoniot-Hadamard relations for weak discontinuities of $S(t,x)$ leads us to the wave propagation speed v in the spatial direction $\mathbf{n}$ ($|\mathbf{n}| \equiv 1$):

$$v = \frac{\nabla S(t,x)^{+} + \nabla S(t,x)^{-}}{2} \cdot \mathbf{n}(t,x),$$

in other words, the possible shock wave speed is

$$v = \frac{\mathbf{u}^{+} + \mathbf{u}^{-}}{2} \cdot \mathbf{n} \tag{12}$$

2. (*Physical character*) Even though the system (1) is not dissipative, viscosity solutions should to encode, to record, a sort of 'vanishing memory' of the dissipative original Burgers systems

$$\mathbf{u}_{\varepsilon\,,t} + \mathbf{u}_{\varepsilon} \cdot \nabla \mathbf{u}_{\varepsilon} = -\nabla V + \frac{\varepsilon}{2}\Delta\, \mathbf{u}_{\varepsilon}, \tag{13}$$

just because these solutions can be thought obtained by the limit $\varepsilon \to 0$ of solutions of

$$S_{\varepsilon\,,t} + \frac{1}{2}|\nabla S_{\varepsilon}|^{2} + V(x) = \frac{\varepsilon}{2}\Delta S_{\varepsilon} \tag{14}$$

All this gives us an interesting physical plausibility to these solutions.

3. (*Turbulence*) Obviously, since $\mathbf{u}(t,x) = \nabla S(t,x)$, we are considering motions out –better, 'after'– of possible (full) turbulence regime. Eventual localized turbulences could arise in correspondence of the above discussed C^1-discontinuities, with sheet and weer shape: in [11] the interested reader can find a survey on it, also from a physical point of view.

By considering the two last points **2.** and **3.**, at least from a heuristic point of view, one can guess that our solution could go for $t \to \infty$ towards some stationary equilibrium solution. The conservative character of the system advises us that this a delicate issue, and we will try to investigate about it from two different points of view: by using (i) the (well known) logarithmic transformation and by (ii) the (maybe, less known) weak KAM theory.

2.3. *The exponential transform: a heuristic reconnaissance*

By using the exponential (Cole-Hopf) transformation technique,

$$\varphi = e^{-\frac{S}{\varepsilon}} \qquad\qquad S = -\varepsilon \ln \varphi \tag{15}$$

we propose a breathing space about the present framework. It is well known, see [14], that this map establishes a weak morphism (of semi-rings) between the set of the solutions of our non linear H-J equation with ε-viscosity (14) and the linear realm of the solutions of a related 'real Schrödinger' equation, see (16) below. In some more details, inserting $(15)_2$ in (14) we easily obtain

$$(-\varepsilon \ln \varphi)_{,t} + \frac{1}{2}\Big| - \varepsilon \frac{\nabla \varphi}{\varphi}\Big|^2 + V = \frac{\varepsilon}{2}\nabla \cdot (-\varepsilon \frac{\nabla \varphi}{\varphi})$$

$$-\varepsilon \frac{\varphi_{,t}}{\varphi} + \frac{1}{2}\varepsilon^2 \frac{|\nabla \varphi|^2}{\varphi^2} + V = -\frac{\varepsilon^2}{2}(\frac{\Delta \varphi}{\varphi} - \frac{|\nabla \varphi|^2}{\varphi^2})$$

$$-\varepsilon \varphi_{,t} + V\varphi = -\frac{\varepsilon^2}{2}\Delta \varphi$$

$$\varphi_{,t} = \frac{\varepsilon}{2}\Delta \varphi + \frac{1}{\varepsilon} V \varphi \tag{16}$$

At this point, in order to study the behaviour of the solutions asymptotically in the time, we begin by shifting the energy potential from $V(x)$ to $V(x) - a$, for some $a \geq \max_{x \in \mathbb{T}^n} V(x)$,

$$V \Longrightarrow V - a, \quad a \geq \max_{x \in \mathbb{T}^n} V(x): \qquad\qquad \varphi_{,t} = \frac{\varepsilon}{2}\Delta \varphi + \frac{1}{\varepsilon}(V - a)\varphi \tag{17}$$

We recall some simple details on the time evolution of the L^2-norm along solutions:

$$\frac{d}{dt}\int_{\mathbb{T}^n} \frac{\varphi^2}{2} dx = \int_{\mathbb{T}^n} \varphi \varphi_{,t} dx = \int_{\mathbb{T}^n} \varphi(\frac{\varepsilon}{2}\Delta \varphi + \frac{1}{\varepsilon}(V(x) - a)\varphi) dx, \tag{18}$$

$$\frac{d}{dt}\int_{\mathbb{T}^n} \frac{\varphi^2}{2} dx = -\frac{\varepsilon}{2}\int_{\mathbb{T}^n} |\nabla \varphi|^2 dx + \frac{1}{\varepsilon}\int_{\mathbb{T}^n} \underbrace{(V(x) - a)}_{\leq 0} \varphi^2 dx \leq \tag{19}$$

$$\leq \frac{1}{\varepsilon}\int_{\mathbb{T}^n}(V(x)-a)\varphi^2 dx \leq -\frac{1}{\varepsilon}\Big(a-\max_{x\in\mathbb{T}^n}V(x)\Big)||\varphi||^2_{L^2} \tag{20}$$

Choosing the constant a strictly greater than the maximum of V,

$$\frac{d}{dt}||\varphi||^2_{L^2} \leq -\frac{2}{\varepsilon}\Big(a-\max_{x\in\mathbb{T}^n}V(x)\Big)||\varphi||^2_{L^2}, \qquad a-\max_{x\in\mathbb{T}^n}V(x)>0, \tag{21}$$

the solutions of (17) are (Liapunov-like) running to zero in L^2 topology for $t \to +\infty$. The corresponding Hamilton-Jacobi with (non-vanishing) ε-viscosity reads

$$S_{\varepsilon,t} + \frac{1}{2}|\nabla S_\varepsilon|^2 + (V(x)-a) = \frac{\varepsilon}{2}\Delta S_\varepsilon \tag{22}$$

and, at the end, we will be interested on the behaviour for $\varepsilon \to 0$.
The above standard L^2 analysis for φ becomes a little more tangled whenever the constant a get down exactly to $\max_{x\in\mathbb{T}^n} V(x)$, and troubles arise for $a < \max_{x\in\mathbb{T}^n} V(x)$.
We conclude this Section by pointing out that, if for some choosen $a \geq \max_{x\in\mathbb{T}^n} V(x)$ the ε-limit does work and supposing realized the expected tendence for $t \to +\infty$ to a stationary equilibrium $\hat{S}(x)$, then this solution $\hat{S}(x)$ has to satisfy, in some weak sense, the stationary H-J equation (see (22))

$$\frac{1}{2}|\nabla \hat{S}(x)|^2 + V(x) - a = 0. \tag{23}$$

The weak KAM theory, sketched in the next section, will help us to overcome these difficulties.

3. Some facts from weak KAM theory

3.1. *The Mañé critical level*

Let M be a compact connected boundaryless manifold. The Hamiltonian $H : T^*M \to \mathbb{R}$ satisfies weak regularity assumptions: more precisely, we take H continuous, coercive and convex in the fibers. Under these conditions, we have already recalled that the Cauchy Problem for the *evolutive* Hamilton-Jacobi equation (9) with a initial datum $S(0,x) = \sigma(x)$ admits one and only one weak solution in the realm of viscosity theory.
This above favorable setting cannot be extended to the *stationary* case:

$$H(x, \nabla S(x)) = a \tag{24}$$

It is nowadays well known that a special, *unique*, value a for the Hamiltonian H is crucial whenever we study the equation (24). It is the so-called *Mañé critical value* $c[0]$. It is characterized by the property that the corresponding Hamilton-Jacobi equation $H(x, \nabla S(x)) = c[0]$ can be solved in the viscosity sense on the *whole* manifold M. In other words: $c[0]$ denotes the unique energy level for which some global viscosity solution does exist. From a technical point of view:

$$c[0] = \inf\{a : \ H(x, \nabla S(x)) = a \text{ has a subsolution}\}. \tag{25}$$

This has been showed, in the case where $M = \mathbb{T}^n$, in [17] using an ergodic approximation of $H(x, \nabla S(x)) = c[0]$ and in the general case in [7], [8] and [9] involving the Lax-Oleinik semigroup for the related evolutive equation; another simple and self-contained proof of this result has been presented also in [10].

It is well known (see e.g. [4]) that for mechanical Hamiltonians like $H(x,p) = \frac{1}{2} g^{-1}(x)(p,p) + V(x)$ one has that

$$c[0] = \max_{x \in M} V(x) \tag{26}$$

3.2. *Asymptotic behaviour of the weak solutions for evolutive H-J*

Recently, an exhaustive very general asymptotic theory has been written by Davini and Siconolfi [6]. Generalizing early works by Roquejoffre [18] and Barles and Souganidis [1], they work out a theory for coercive convex Hamiltonian functions. Among other things, they prove that for any viscosity solution $S(t,x)$ of

$$S_{,t}(t,x) + H(x, \nabla S(t,x)) = 0, \qquad x \in \mathbb{T}^n, \tag{27}$$

it holds that

$$\lim_{t \to +\infty} ||S(t, \cdot) + c[0]\, t - \hat{S}(\cdot)||_\infty = 0 \tag{28}$$

where $\hat{S}(x)$ is a global viscosity solution of

$$H(x, \nabla \hat{S}(x)) = c[0] \tag{29}$$

In other words, asymptotically in the time, the viscosity solution $S(t,x)$ of (27) is uniformely blowing up as

$$\text{for large } t: \quad S(t,x) \ \approx \ S_\infty(t,x) := -c[0]t + \hat{S}(x) \tag{30}$$

For more details on $\hat{S}(x)$ the interested reader is invited to see formula (11) in [6]: a complete explanation of it needs the concept of the non symmetric semi-distance, introduced by Fathi and Siconolfi, and the strict use of the so-called (projected) Aubry set $\mathcal{A}$.

3.3. *The application to fluid dynamics*

By resuming, we have obtained the unique (Lipschitz) viscosity solution $S(t,x)$ for the Cauchy problem (7) on $M = \mathbb{T}^n$; at the same time, we have gained the expected potential motion, $\mathbf{u}(t,x) = \nabla S(t,x)$ solving (1) with the initial potential-like datum: $\mathbf{u}(0,x) = \nabla\sigma(x)$.
The remarks worked out by using the exponential transform gave us a signal about the possible 'good' values a for the energy H in the eventual time asymptotic case (that is, if it there exists), see (23). We have seen that dramatically quality changes from $a > \max_{x\in M} V(x)$ to $a < \max_{x\in M} V(x)$.

By the above KAM theory, we can now state that any weak potential motion solution $\mathbf{u}(t,x) = \nabla S(t,x)$ solving

$$S_{,t}(t,x) + \frac{1}{2}|\nabla S(t,x)|^2 + V(x) = 0 \tag{31}$$

for $t \to +\infty$ goes towards an equilibrium stationary solution $\hat{\mathbf{u}}(x)$, which is generated by a global viscosity time independent solution $\hat{S}(x)$ of the stationary Hamilton-Jacobi equation (23)

$$\frac{1}{2}|\nabla\hat{S}(x)|^2 + V(x) = \max_{x\in M} V(x), \tag{32}$$

that is, at the Mañé critical level

$$c[0] = a = \max_{x\in M} V(x) \tag{33}$$

Finally, we have recognize the trend to equilibria for our non turbulent motions and we are able to say also at which energy level these equilibria are placed, in other words, asymptotically in the time, the viscosity solutions S, generating $\mathbf{u}(t,x) = \nabla S(t,x)$, are splitting into two separated time and spatial terms,

$$S(t,x) \to -\max_{x\in M} V(x)t + \hat{S}(x), \tag{34}$$

so that

$$\mathbf{u}(t,x) \to \hat{\mathbf{u}}(x) = \nabla\hat{S}(x) \tag{35}$$

and the energy density $e(t,x) := \frac{1}{2}|\mathbf{u}(t,x)|^2 + V(x)$ of the fluid is going to

$$\frac{1}{2}|\hat{\mathbf{u}}(x)|^2 + V(x) = \max_{x\in M} V(x). \tag{36}$$

4. Quantum dynamics and time asymptotics for the semiclassical Madelung equation

We consider the Schrödinger equation related to the classical Hamiltonian function $H\,(x,p) = \frac{|p|^2}{2m} + V(x)$:

$$i\hbar\,\Psi_{,t}(t,x) = H\left(x, \frac{\hbar}{i}\nabla\right)\Psi(t,x), \tag{37}$$

and we represent the complex wave function in polar form:

$$\Psi(t,x) = R\, e^{\frac{i}{\hbar}S}$$

A simple standard calculation produces a pair of equations, respectively the real and imaginary parts of the original equation,

$$\begin{cases} S_{,t} + H(x,\nabla S) - \frac{\hbar^2}{2m}\frac{\Delta R}{R} = 0, \\ \\ R^2_{,t} + (R^2\, v^j)_{,j} = 0 \qquad \left(v^j := \frac{\partial H}{\partial p_j}\big|_{p=\nabla S}\right) \end{cases} \tag{38}$$

The first is a Hamilton-Jacobi equation "with a viscosity term", the second has the appearance of a continuity equation for $R^2 = |\Psi|^2$ representing the probability density function for the position x at the time t of the particle described by the wave function Ψ.
Looking closer at (38), we see that, for $\hbar \to 0$, we obtain the setting for the search of a *global solution* $S(t,x)$ of the H-J equation

$$S_{,t} + H(x,\nabla S) = 0 \tag{39}$$

and a related global *invariant measure* $R^2(t,x)$ –the second of (38). This is again extremely close to the core of the viscosity solutions and week KAM theory, even though a difference there exists: this concerns with the vanishing term $\frac{\hbar^2}{2m}\frac{\Delta R}{R}$ in the above Madelung framework, which is different from the 'expected' viscosity H-J term, in fact, the last one should be something like $\frac{\hbar^2}{2m}\Delta S$. Whenever the semi-classic limit $\lim_{\hbar\to 0} S_\hbar$ is running to the viscosity limit $\lim_{\varepsilon\to 0}\bar{S}_\varepsilon$, where $\bar{S}_\varepsilon$ is solving (11), then the above time-asymptotic behaviour does work, and we can claim that for $t \to +\infty$ the semi-classic phase function $S(t,x)$ goes uniformely to $-t\,c[0] + \hat{S}(x)$, where $\hat{S}(x)$ is a stationary solution of

$$H(x,\nabla\hat{S}) = c[0], \tag{40}$$

at the Mañé energy level $c[0] = \max_{x\in\mathbb{T}^n} V(x)$. Although these considerations are holding both for $\hbar \to 0$ and $t \to +\infty$, we stress that this order of ideas draws for the Mañé energy level a quantum meaning as well.

References

1. Barles G., Souganidis P. E. *On the large time behaviour of solutions of Hamilton-Jacobi equations*, SIAM J. Math. Anal., 31 (2000), pp. 925-939.
2. Boldrighini C. *Introduzione alla fluidodinamica* Consiglio Nazionale delle Ricerche (CNR), Quaderno dei Gruppi di Ricerca Matematica, Rome, 1979.
3. Caliari M., Inverso G. and Morato L. M. *Dissipation caused by a vorticity field and generation of singularities in Madelung fluid*, New Journal of Physics 6, 69 (2004)
4. Contreras G., Iturriaga R. *Global minimizers of autonomous Lagrangians.* 22° Colóquio Brasileiro de Matemática. Rio de Janeiro: Instituto de Matemática Pura e Aplicada (IMPA), 148 pp. (1999).
5. Crandall M.G., Lions P.L. *Viscosity solutions of Hamilton-Jacobi equations.* Trans. Am. Math. Soc. 277, (1-42), 1983.
6. Davini A., Siconolfi A. *A generalized dynamical approach to the large time behavior of solutions of Hamilton-Jacobi equations* SIAM J. Math. Anal. 38 (2006), no. 2, 478–502.
7. Fathi A. *Solutions KAM faibles et barriéres de Peierls.* C. R. Aca. Sci. Paris 325, 649-652, (1997).
8. Fathi A. *Sur la convergence du semi-groupe de Lax-Oleinik.* C. R. Aca. Sci. Paris 327, 267-270, (1998).
9. Fathi A. *Weak KAM theory in Lagrangian dynamics.* Preliminary version, (2006).
10. Fathi A., Siconolfi A. *PDE aspects of Aubry-Mather theory for quasiconvex Hamiltonians.* Calc. Var. 22, 185-228, (2005).
11. Frisch U., Bec. J. *Burgulence*, Lesieur, M. (ed.) et al., École de Physique des Houches - UJF & INPG - Grenoble, a NATO Advanced Study Institute. Les Houches session LXXIV, New trends in turbulence. Turbulence: nouveaux aspects. Berlin: Springer; Les Ulis: EDP Sciences. 343-383 (2001).
12. Gallavotti G. *Foundations of fluid dynamics.* Texts and Monographs in Physics. Springer-Verlag, Berlin, xviii+513 pp., 2002
13. Iturriaga R., Khanin K. *Burgers Turbulence and Random Lagrangian Systems*, Communications in Mathematical Physics, Volume 232, Issue 3, pp. 377-428 (2003).
14. Kolokoltsov V. N., Maslov V. P. *Idempotent analysis and its applications.* With an appendix by Pierre Del Moral. Mathematics and its Applications, 401. Kluwer Academic Publishers Group, Dordrecht, 1997. xii+305 pp.
15. Lions P.L. *Generalized solutions of Hamilton-Jacobi equations.* Research Notes in Mathematics, 69. Pitman Advanced Publishing Program (1982).
16. Morato L. M. *Formation of singularities in Madelung fluid: a nonconventional application of Ito calculus to foundations of Quantum Mechanics*, The Abel Symposium 2005, Proceedings to 'Stochastic Analysis and Applications', in Honor of Kiyosi Ito.
17. Lions P.L., Papanicolaou G., Varadhan S.R.S. *Homogenizations of Hamilton-Jacobi equations.* Unpublished, (1987).
18. Roquejoffre J. M. *Convergence to steady states or periodic solutions in a class of HamiltonJacobi equations*, J. Math. Pures Appl. (9), 80 (2001), pp. 85-104.

RICCI FLOW DEFORMATION OF COSMOLOGICAL INITIAL DATA SETS

M. CARFORA[1] and T. BUCHERT[2]

[1] *Dipartimento di Fisica Nucleare e Teorica, Università degli Studi di Pavia and I.N.F.N., Sezione di Pavia*
mauro.carfora@pv.infn.it

[2] *Université Lyon, Centre de Recherche Astrophysique, CNRS UMR 5574*
buchert@obs.univ-lyon1.fr

1. INTRODUCTION

The Ricci flow has been introduced by R. Hamilton[19] with the goal of providing an analytic approach to Thurston's geometrization conjecture for three-manifolds.[30,31] Inspired by the theory of harmonic maps, he considered the geometric evolution equation obtained when one evolves a Riemannian metric g_{ab}, on a three-manifold Σ, in the direction of its Ricci tensor[20] $\mathcal{R}_{ab}$, *i.e.*

$$\begin{cases} \frac{\partial}{\partial \beta} g_{ab}(\beta) = -2\ \mathcal{R}_{ab}(\beta)\ , \\ \\ g_{ab}(\beta = 0) = g_{ab}\ , \quad 0 \leq \beta < T_0\ . \end{cases} \tag{1}$$

In recent years, this geometric flow has gained extreme popularity thanks to the revolutionary breakthroughs of G. Perelman,[25–27] who, taking the whole subject by storm, has brought to completion Hamilton's approach to Thurston's conjecture. The prominent themes recurring in Hamilton's and Perelman's works converge to a proof that the Ricci flow, coupled to topological surgery, provides a natural technique for factorizing and uniformizing a three-dimensional Riemannian manifold (Σ, g) into locally homogeneous geometries. This is a result of vast potential use also in theoretical physics, where the Ricci flow often appears in disguise as a natural real-space renormalization group flow. Non-linear σ-model theory, describing quantum strings propagating in a background spacetime, affords the standard case study in such a setting.[1,2,8,16,23,24] Another paradigmatical, perhaps even more direct, application occurs in relativistic cosmology,[9,10] (for a series of recent results see also[6,7] and the references cited therein). This will be related to the main topic of this talk, and to motivate our interest in it, let us recall that homogeneous and isotropic solutions of Einstein's laws of gravity (the Friedman–Lemaitre–Robertson–Walker (FLRW) spacetimes) do not account for inhomogeneities in the Universe. The question whether they do *on average* is an issue[14] that is the subject of considerable debate especially in the recent literature (see[21,28] and follow–up references; comprehensive lists may be found in[15,29] and[5]). In any case, a member of the family of FLRW cosmologies (the so–called concordance model that is characterized by a dominating cosmological constant in a spatially flat universe model) provides a successful *fitting model* to a large number of observational data, and the generally held view is that the spatial sections of FLRW spacetimes indeed describe the *physical Universe* on a sufficiently large averaging scale. This raises an interesting problem in mathematical cosmology: devise a way to explicitly construct a constant–curvature metric out of a *scale–dependent* inhomogeneous distribution of matter and spatial curvature. It is in such a framework that one makes the basic observation that the Ricci flow (1) and its linearization, provide a natural technique[6,10] for deforming, and under suitable conditions smoothing, the geometrical part of scale–dependent cosmological initial data sets. Moreover, by taking advantage of some elementary aspects of Perelman's results, this technique also provides a natural and unique way for deforming, along the Ricci flow, the matter distribution. The expectation is that in this way we can define a deformation of cosmological initial data sets into a one-parameter family of initial data whose time evolution, along the evolutive part of Einstein's equations, describe the Ricci flow deformation of a cosmological spacetime.

2. Initial data set for cosmological spacetimes

To set notation, we emphasize that throughout the paper we shall consider a smooth three-dimensional manifold Σ, which we assume to be closed and without boundary. We let $C^\infty(\Sigma, \mathbb{R})$ and $C^\infty(\Sigma, \otimes^p T^*\Sigma \otimes^q T\Sigma)$ be the space of smooth functions and of smooth (p, q)–tensor fields on Σ, respectively. We shall denote by $\mathcal{D}iff(\Sigma)$ the group of smooth diffeomorphisms of Σ, and by $\mathcal{R}iem(\Sigma)$ the space of all smooth Riemannian metrics over Σ. The tangent space , $\mathcal{T}_{(\Sigma,g)}\mathcal{R}iem(\Sigma)$, to $\mathcal{R}iem(\Sigma)$ at (Σ, g) can be naturally identified with the space of symmetric bilinear forms $C^\infty(\Sigma, \otimes^2 T^*\Sigma)$ over Σ. The hypothesis of smoothness has been made for simplicity. Results similar to those described below, can be obtained for initial data sets with finite

Hölder or Sobolev differentiability. In such a framework, let us recall that a collection of fields $g \in \mathcal{R}iem(\Sigma)$, $K \in \mathcal{T}_{(\Sigma,g)}\mathcal{R}iem(\Sigma)$, $\varrho \in C^\infty(\Sigma, \mathbb{R}^+)$, $\vec{J} \in C^\infty(\Sigma, T\Sigma)$, defined over the three-manifold Σ, characterizes a set $(\,\Sigma\,,\,g_{ab}\,,\,K_{ab}\,,\,\varrho\,,\,J_a\,)$, of physical cosmological initial data for Einstein equations if and only if the *matter fields* $(\varrho, \vec{J})$ verify the weak energy condition $\varrho \geq 0$, the dominant energy condition $\varrho^2 \geq g_{ab}J^a J^b$, and their coupling with the geometric fields (g, K) is such as to satisfy the Hamiltonian and divergence constraints[a]:

$$\mathcal{R} + k^2 - K^a{}_b K^b{}_a = 16\pi G\varrho + 2\Lambda\,, \tag{2}$$

$$\nabla_b K^b{}_a - \nabla_a k = 8\pi G J_a\,. \tag{3}$$

Here Λ is the cosmological constant, $k := g^{ab}K_{ab}$, and $\mathcal{R}$ is the scalar curvature of the Riemannian metric g_{ab}. If such a set of admissible data is propagated according to the evolutive part of Einstein's equations, then the symmetric tensor field K_{ab} can be interpreted as the extrinsic curvature and k as the mean curvature of the embedding $i_t : \Sigma \to M^{(4)}$ of (Σ, g_{ab}) in the spacetime $(M^{(4)} \simeq \Sigma \times \mathbb{R}, g^{(4)})$ resulting from the evolution of $(\Sigma, g_{ab}, K_{ab}, \varrho, J_a)$, whereas ϱ and J_a are, respectively, identified with the mass density and the momentum density of the material self–gravitating sources on (Σ, g_{ab}).

3. The Heuristics of averaging: Deformation of cosmological initial data sets

The averaging procedure described in[6,9,10] is based on a smooth deformation of the physical initial data $(\Sigma, g_{ab}, K_{ab}, \varrho, J_a)$ into a one-parameter family of initial data sets

$$\beta \longmapsto (\Sigma\,,\, g_{ab}(\beta)\,,\, K_{ab}(\beta)\,,\, \varrho(\beta)\,,\, J_a(\beta)), \tag{4}$$

with $0 \leq \beta \leq \infty$ being a parameter characterizing the averaging scale. The general idea is to construct the flow (4) in such a way as to represent, as β increases, a scale-dependent averaging of $(\Sigma, g_{ab}, K_{ab}, \varrho, J_a)$, and – under suitable hypotheses – reducing it to a constant-curvature initial data set

$$(\Sigma\,,\, \overline{g}_{ab}\,,\, \overline{K}_{ab} = \frac{1}{3}\overline{g}_{ab}\overline{k}\,,\, \overline{\varrho}\,,\, \overline{J}_a = 0), \tag{5}$$

[a]Latin indices run through $1, 2, 3$; we adopt the summation convention. The nabla operator denotes covariant derivative with respect to the 3–metric. The units are such that $c = 1$.

where $\overline{g}_{ab}$ is a constant curvature metric on Σ, $\overline{k}$ is the (spatially constant) trace of the extrinsic curvature (related to the Hubble parameter), and $\overline{\varrho}$ is the averaged matter density. Under the heading of such a general strategy it is easy to figure out the reasons for an important role played by the Ricci flow and its linearization. As we shall recall shortly, they are natural geometrical flows always defining a non-trivial deformation of the metric g_{ab} and of the extrinsic curvature K_{ab}. Moreover, when global, they posses remarkable smoothing properties. For instance, if, for $\beta = 0$, the scalar curvature $\mathcal{R}$ of (Σ, g_{ab}) is > 0, and if there exist positive constants α_1, α_2, α_3, not depending on β, such that $\mathcal{R}_{ab}(\beta) - \alpha_1 g_{ab}(\beta)\mathcal{R}(\beta) \geq 0$, and $\widehat{\mathcal{R}}_{ab}(\beta)\widehat{\mathcal{R}}^{ab}(\beta) \leq \alpha_2 \mathcal{R}^{1-\alpha_3}(\beta)$, where $\widehat{\mathcal{R}}_{ab}(\beta) \doteq \mathcal{R}_{ab}(\beta) - \frac{1}{3}g_{ab}(\beta)\mathcal{R}(\beta)$ is the trace–free part of the Ricci tensor, then[19] the solutions $(g_{ab}(\beta), K_{ab}(\beta))$ of the (volume–normalized) Ricci flow and its linearization exist for all $\beta > 0$, and the pair $(g_{ab}(\beta), K_{ab}(\beta))$, uniformly converges, when $\beta \to \infty$, to $(\overline{g}_{ab}, \mathcal{L}_{\vec{v}}\overline{g}_{ab})$ where $\overline{g}_{ab}$ is a metric with constant positive sectional curvature, and $\vec{v}$ is some vector field on Σ, possibly depending on β. The flow $\beta \mapsto (\overline{g}_{ab}, \mathcal{L}_{\vec{v}}\overline{g}_{ab})$, describing a motion by diffeomorphisms over a constant curvature manifold, can be thought of as representing the smoothing of the geometrical part of an initial data set $(\Sigma, g_{ab}, K_{ab}, \varrho, J_a)$.

It is useful to keep in mind what we can expect and what we cannot expect out of such a Ricci–flow deformation of cosmological initial data set. Let us start by remarking that the family of data (4) will correspond to the initial data for physical spacetimes iff the constraints (2) and (3) hold throughout the β–dependent deformation. This is a very strong requirement and, if we have a technically consistent way $\beta \mapsto (\varrho(\beta), g_{ab}(\beta), K_{ab}(\beta))$ of deforming the matter distribution and the geometrical data, then the most natural way of implementing the constraints is to use them to define scale–dependent backreaction fields $\beta \mapsto \phi(\beta)$, $\beta \mapsto \psi_a(\beta)$ describing the non–linear interaction between matter averaging and geometrical averaging, *i.e.*,

$$\phi(\beta) \doteq \varrho(\beta) - (16\pi\, G)^{-1}\left[\mathcal{R}(\beta) + k^2(\beta) - K^a{}_b(\beta)K^b{}_a(\beta) - 2\Lambda\right]\ , \quad (6)$$

$$\psi_a(\beta) \doteq J_a(\beta) - (8\pi\, G)^{-1}\left[\nabla_b K^b{}_a(\beta) - \nabla_a k(\beta)\right]\ . \quad (7)$$

To illustrate how this strategy works, let us concentrate, in this talk, on the characterization of the scalar field $\beta \mapsto \phi(\beta)$, providing the backreaction between matter and geometrical averaging. The covector field $\beta \mapsto \psi_a(\beta)$ can, in principle, be controlled by the action of a β–dependent diffeomorphism. However, its analysis requires a subtle interplay with the kinematics of spacetime foliations,[4] (*i.e.*, how we deal with the lapse function and with the shift vector field in the framework of Perelman's approach), and will be discussed elsewhere, (for a pre–Perelman approach to this issue see[6,10]). Let us start by observing that the matter averaging flow $\beta \mapsto \varrho(\beta)$ must comply with the preservation of the physical matter content

$$\int_{\Sigma} \varrho(\beta)\, d\mu_{g(\beta)} = \int_{\Sigma} \varrho(\beta = 0)\, d\mu_{g(\beta=0)} \doteq M\,, \;\; \forall \beta\,, \tag{8}$$

and must be explicitly coupled to the scale of geometrical averaging. In other words, if, for some fixed $\beta > 0$, we consider that part of the matter distribution $\varrho(\beta)$ which is localized in a given region $B(x, \tau(\beta)) \subset \Sigma_\beta$ of size $\tau(\beta)$, then we should be able to tell from which localized distribution $(\varrho_m, B(x,\tau))$, at $\beta = 0$, the selected matter content $(\varrho(\beta), B(x, \tau(\beta)))$ has evolved. A natural answer to these requirements is provided by Perelman's backward localization[25] of probability measures on Ricci evolving manifolds. The idea is to probe the Ricci flow with a probability measure whose dynamics can localize the regions of the manifold (Σ, g) of geometric interest. This is achieved by considering, along the solution $g_{ab}(\beta)$ of (1), a β-dependent mapping $\beta \longmapsto f(\beta,\ ) \in C^\infty(\Sigma_\beta, \mathbb{R})$, in terms of which one constructs on Σ_β the measure $d\varpi(\beta) \doteq (4\pi\tau(\beta))^{-\frac{3}{2}}\, e^{-f}\, d\mu_{g(\beta)}$, where $\beta \longmapsto \tau(\beta) \in \mathbb{R}^+$ is a scale parameter chosen in such a way as to normalize $d\varpi(\beta)$ according to the so–called *Perelman's coupling*: $\int_{\Sigma_\beta} d\varpi(\beta) = (4\pi\tau(\beta))^{-\frac{3}{2}} \int_{\Sigma_\beta} e^{-f} d\mu_{g(\beta)} = 1$. It is easily verified that this is preserved in form along the Ricci flow (1), if the mapping f and the scale parameter $\tau(\beta)$ are evolved backward in time $\beta \in (\beta^*, 0)$ according to the coupled flows defined by

$$\begin{cases} \frac{\partial}{\partial \beta} f = -\Delta_{g(\beta)} f - R(\beta) + \frac{3}{2}\tau(\beta)^{-1}, \; f(\beta^*) = f_0 \\[2ex] \frac{d}{d\beta}\, \tau(\beta) = -1, \; \tau(\beta^*) = \tau_0, \end{cases} \tag{9}$$

where $\Delta_{g(\beta)}$ is the Laplacian with respect to the metric $g_{ab}(\beta)$, and f_0, τ_0 are given (final) data. In this connection, note that the equation for f is a backward heat equation, and as such the forward evolution $f(\beta = 0) \to f$ is an ill-posed problem. A direct way for circumventing such a difficulty is to interpret (9) according to the following two-steps prescription: (i) Evolve the metric $\beta \longmapsto (\Sigma, g_{ab}(\beta))$, say up to some β^*, according to the Ricci flow, (if the flow is global we may let $\beta^* \to \infty$); (ii) On the Ricci evolved Riemannian manifold $(\Sigma, \overline{g}_{ab}(\beta^*))$ so obtained, select a function $f(\beta^*)$ and the corresponding scale parameter $\tau(\beta^*)$, and evolve them, backward in β, according to (9).

4. Ricci–flow deformation of the initial data $(\Sigma, g_{ab}, K_{ab}, \varrho)$

With these preliminary remarks along the way, let us characterize the various steps involved in constructing the flow (4). (For ease of exposition, we refer to the standard unnormalized flow; volume normalization can be enforced by a reparametrization of the deformation parameter).

Definition 4.1. (*Geometrical Data Deformation*) Given an initial data set $(\Sigma, g_{ab}, K_{ab}, \varrho, J_a)$ for a cosmological spacetime $(M^{(4)} \simeq \Sigma \times \mathbb{R}, g^{(4)})$, the Ricci–flow deformation of its geometrical part (Σ, g_{ab}, K_{ab}) is defined by the flow $\beta \mapsto (g_{ab}(\beta), K_{hj}(\beta))$, $0 \leq \beta < \beta^*$ provided by the (weakly) parabolic initial value problem (the Ricci flow, proper)

$$\begin{cases} \frac{\partial}{\partial\beta} g_{ab}(\beta) = -2\mathcal{R}_{ab}(\beta), \\ g_{ab}(\beta = 0) = g_{ab} \ , \quad 0 \leq \beta < \beta^* \\ \lim_{\beta \nearrow \beta^*} \left[\sup_{x \in \Sigma} |Rm(x, \beta)|\right] < \infty \ , \end{cases} \tag{10}$$

and by its linearization, which, by suitably fixing the action of the diffeomorphism group $\mathcal{D}iff(\Sigma)$, takes the form of the parabolic initial value problem[13]

$$\begin{cases} \frac{\partial}{\partial\beta} K_{ab}(\beta) = \Delta_L \, K_{ab}(\beta) \ , \\ K_{ab}(\beta = 0) = K_{ab} \ , \quad 0 \leq \beta < \beta^* \ , \end{cases} \ , \tag{11}$$

where Δ_L denotes the Lichnerowicz–DeRham Laplacian $\Delta_L K_{ab} \doteq \nabla^i \nabla_i K_{ab} - R_{as} K^s_b - R_{bs} K^s_a + 2R_{asbt} K^{st}$ acting on symmetric bilinear forms.[22]

Definition 4.2. (*Localization of the Deformed Data*) The geometrical deformation $\beta \mapsto (g_{ab}(\beta), K_{hj}(\beta))$ is controlled by the backward localizing flow $\eta \mapsto (E(x, y; \eta), E^{ab}_{i'k'}(x, y; \eta))$, $\eta \doteq \beta^* - \beta$, defined by the backward heat kernels for the conjugate[11,12] parabolic initial value problem associated with the $g(\eta)$–dependent Laplace-Beltrami and Lichnerowicz operators $\eta \mapsto (\Delta, \Delta_L)$. The operator Δ_L, when acting on (bi)scalars, reduces to Δ. Thus, we can characterize both these kernels in a compact form as solutions of

$$\begin{cases} \frac{\partial}{\partial\eta}\begin{pmatrix} E^{ab}_{i'k'}(x,y;\eta) \\ E(x,y;\eta) \end{pmatrix} = \Delta^{(x)}_{L}\begin{pmatrix} E^{ab}_{i'k'}(x,y;\eta) \\ E(x,y;\eta) \end{pmatrix} - \mathcal{R}\begin{pmatrix} E^{ab}_{i'k'}(x,y;\eta) \\ E(x,y;\eta) \end{pmatrix}, \\ \\ \lim_{\eta\searrow 0^+}\begin{pmatrix} E^{ab}_{i'k'}(x,y;\eta) \\ E(x,y;\eta) \end{pmatrix} = \begin{pmatrix} \delta^{ab}_{i'k'}(x,y;\eta) \\ \delta(x,y;\eta) \end{pmatrix}, \end{cases} \tag{12}$$

where $(y,x;\eta) \in (\Sigma \times \Sigma \backslash Diag(\Sigma \times \Sigma)) \times [0,\beta^*]$, $\Delta^{(x)}_{L}$ denotes the Lichnerowicz–DeRham Laplacian with respect to the variable x, the heat kernels $E^{ab}_{i'k'}(y,x;\eta)$ and $E(x,y;\eta)$ are smooth sections of $(\otimes^2 T\Sigma) \boxtimes (\otimes^2 T^*\Sigma)$ and $\Sigma \boxtimes \Sigma$, respectively, and finally, $\delta^{k_1\ldots k_p}_{i'_1\ldots i'_p}(y,x;\eta)$ is the Dirac p–tensorial measure, $(p = 0,1,\ldots)$ on $(\Sigma, g(\eta))$. The Dirac initial condition is understood in the distributional sense, *i.e.*, $\int_{\Sigma_\eta} K^{ab}_{i'k'}(y,x;\eta)\, w^{i'k'}(y)\, d\mu^{(y)}_{g(\eta)} \rightarrow w^{ab}(x)$ *as* $\eta \searrow 0^+$, for any smooth symmetric bilinear form with compact support $w^{i'k'} \in C^\infty_0(\Sigma, \otimes^2 T\Sigma)$, and where the limit is meant in the uniform norm on $C^\infty_0(\Sigma, \otimes^2 T\Sigma)$.

Note that heat kernels for generalized Laplacians, such as Δ_L, (smoothly) depending on a one–parameter family of metrics $\varepsilon \mapsto g_{ab}(\varepsilon)$, $\varepsilon \geq 0$, are dealt with in.[3,13,17,18]
Let us now consider the matter content localized by taking the $d\varpi(\eta)$–expectation of $\varrho(\eta)$,

$$\int_\Sigma \varrho(\eta)\, d\varpi(\eta) \doteq M(d\varpi(\eta))\,. \tag{13}$$

According to (8), we require that such a local mass is preserved along the η–evolution of the measure $d\varpi(\eta)$, *i.e.*, $\frac{d}{d\beta}\int_\Sigma \varrho(\eta)\, d\varpi(\eta) = 0$. This request motivates the following

Definition 4.3. (*Deformation of Matter Data*) The given matter distribution $\varrho(\beta = 0)$ is deformed according to the heat–flow $\beta \mapsto \varrho(\beta)$ given by

$$\begin{cases} \frac{\partial}{\partial\beta}\varrho(\beta) = \Delta_{g(\beta)}\, \varrho(\beta),\ \beta \in [0, T_0) \\ \\ \varrho(\beta = 0) = \varrho. \end{cases} \tag{14}$$

We are now in the position to characterize the Ricci flow deformation of the cosmological data $(\Sigma, g_{ab}, K_{ab}, \varrho, J_a)$. According to the results described in[11] we have

Proposition 4.1. *Let $\beta \mapsto (g_{ab}(\beta), K_{ik}(\beta), \varrho(\beta))$ be the Ricci flow deformation, on $\Sigma_\eta \times [0, \beta^*]$, of the data $(\Sigma, g_{ab}, K_{ab}, \varrho)$ as defined above. Assume that the underlying Ricci flow $\beta \mapsto (\Sigma, g_{ab}(\beta))$ is of bounded geometry, and let $E(y, x; \eta)$ and $E^{ab}_{i'k'}(y, x; \eta)$ be the (backward) heat kernels Ricci-flow conjugated to the Laplace–Beltrami and to the Lichnerowicz–DeRham operator, respectively. Then, for all $0 \leq \eta \leq \beta^*$,*

$$g_{i'k'}(y, \eta = 0) = \int_\Sigma E^{ab}_{i'k'}(y, x; \eta)\, [g_{ab}(x, \eta) - 2\eta\, \mathcal{R}_{ab}(x, \eta)]\, d\mu_{g(x,\eta)}\,, \quad (15)$$

$$\begin{pmatrix} \mathcal{R}_{i'k'}(y, \eta = 0) \\ K_{i'k'}(y, \eta = 0) \end{pmatrix} = \int_\Sigma E^{ab}_{i'k'}(y, x; \eta) \begin{pmatrix} \mathcal{R}_{ab}(x, \eta) \\ K_{ab}(x, \eta) \end{pmatrix} d\mu_{g(x,\eta)}\,, \quad (16)$$

and

$$\varrho(y, \eta = 0) = \int_\Sigma E(y, x; \eta)\, \varrho(x, \eta)\, d\mu_{g(x,\eta)}\,. \quad (17)$$

Moreover, as $\eta \searrow 0^+$, we have the uniform asymptotic expansion

$$\begin{pmatrix} \mathcal{R}_{i'k'}(y, \eta = 0) \\ K_{i'k'}(y, \eta = 0) \end{pmatrix} = \quad (18)$$

$$\frac{1}{(4\pi\,\eta)^{\frac{3}{2}}} \int_\Sigma \exp\left(-\frac{d_0^2(y, x)}{4\eta}\right) \tau^{ab}_{i'k'}(y, x; \eta) \begin{pmatrix} \mathcal{R}_{ab}(x, \eta) \\ K_{ab}(x, \eta) \end{pmatrix} d\mu_{g(x,\eta)}$$

$$+ \sum_{h=1}^{N} \frac{\eta^h}{(4\pi\,\eta)^{\frac{3}{2}}} \int_\Sigma \exp\left(-\frac{d_0^2(y, x)}{4\eta}\right) \Phi[h]^{ab}_{i'k'}(y, x; \eta) \begin{pmatrix} \mathcal{R}_{ab}(x, \eta) \\ K_{ab}(x, \eta) \end{pmatrix} d\mu_{g(x,\eta)}$$

$$+ O\left(\eta^{N - \frac{1}{2}}\right)\,,$$

where $\tau^{ab}_{i'k'}(y, x; \eta) \in T\Sigma_\eta \boxtimes T^*\Sigma_\eta$ is the parallel transport operator associated with $(\Sigma, g(\eta))$, $d_0(y, x)$ is the distance function in $(\Sigma, g(\eta = 0))$, and $\Phi[h]^{ab}_{i'k'}(y, x; \eta)$ are smooth sections $\in C^\infty(\Sigma \times \Sigma', \otimes^2 T\Sigma \boxtimes \otimes^2 T^*\Sigma)$, (depending on the geometry of $(\Sigma, g(\eta))$), characterizing the asymptotics of the heat kernel $E^{ab}_{i'k'}(y, x; \eta)$. With an obvious adaptation, such an asymptotic behavior also extends to $g_{i'k'}(y, \eta = 0)$ and $\varrho(y, \eta = 0)$.

With these results it is rather straightforward to provide useful characterizations of the (intrinsic) backreaction field $\phi(\beta)$. For illustrative purposes, let us assume that $\beta \mapsto K_{ab}(\beta) \equiv 0$ and $\Lambda \equiv 0$, then from (6) and the above proposition we get that

$$\phi(y,\eta=0)=\int_{\Sigma}E(y,x;\eta)\,\varrho(x,\eta)\,d\mu_{g(x,\eta)} \tag{19}$$
$$-(16\pi\,G)^{-1}\left[\int_{\Sigma}g^{i'k'}(y,\eta=0)\,E^{ab}_{i'k'}(y,x;\eta)\,\mathcal{R}_{ab}(x,\eta)\,d\mu_{g(x,\eta)}\right]\,.$$

If we further assume that the Hamiltonian constraint holds at the *fixed* observative scale η and that $\mathcal{R}_{ab}(x,\eta)\approx[\frac{1}{3}\mathcal{R}(\eta)\,g_{ab}(x,\eta=0)+\delta\mathcal{R}_{ab}(x,\eta)]$, (*i.e.*, curvature fluctuates around the *constant curvature background* $\mathcal{R}_{l'm'}(y,\eta=0)=2C\,g_{l'm'}(y,\eta=0)$), then

$$\phi(y,\eta=0)\approx \tag{20}$$
$$(16\pi\,G)^{-1}\int_{\Sigma}g^{i'k'}(y,\eta=0)\,E^{ab}_{i'k'}(y,x;\eta)\,\delta\mathcal{R}_{ab}(x,\eta)\,d\mu_{g(x,\eta)}\;,$$

from which it follows that the intrinsic backreaction field $\phi(\beta)$ is generated by curvature fluctuations around the given background, as expected.

Acknowledgements

M.C. would like to dedicate this paper to T. Ruggeri on occasion of his ...th birthday. T.B. acknowledges hospitality at and support from the University of Pavia. Research supported in part by PRIN Grant #2006017809.

References

1. I. Bakas, J. High Energy Phys. 0308, 013 (2003);
2. I. Bakas, *Geometric flows and (some of) their physical applications*, AvH conference Advances in Physics and Astrophysics of the 21st Century, 6-11 September 2005, Varna, Bulgaria, hep-th/0511057.
3. N. Berline, E. Getzler and M. Vergne *Heat kernels and Dirac operators*, Grundlehren Math. Wiss., vol. 298, Springer-Verlag, New York, 1992.
4. T. Buchert, Gen. Rel. Grav. **33**, 1381 (2001).
5. T. Buchert, *Dark energy from structure: a status report*, Gen. Rel. Grav. (special issue on dark energy), in press; arXiv:0707.2153 (2007).
6. T. Buchert and M. Carfora, Class. Quant. Grav. **19**, (2002) 6109-6145.
7. T. Buchert and M. Carfora, Phys. Rev. Lett. **90**, (2003) 31101-1-4.
8. M. Carfora, A. Marzuoli, Class. Quantum Grav. 5 (1988) 659-693.
9. M. Carfora and A. Marzuoli, Phys. Rev. Lett. **53**, 2445 (1984).
10. M. Carfora and K. Piotrkowska, Phys. Rev. D **52**, 4393 (1995).
11. M. Carfora, *The Conjugate Linearized Ricci Flow on Closed 3-Manifolds*, arXiv:0710.3342
12. B. Chow, S-C. Chu, D. Glickenstein, C. Guenther, J. Isenberg, T. Ivey, D. Knopf, P. Lu, L. Ni, *The Ricci Flow: Techniques and Applications: PartI:*

Geometric Aspects, Math. Surveys and Monographs Vol. **135**, (2007) Am. Math. Soc.
13. B. Chow, P. Lu, L. Ni, *Hamilton's Ricci Flow*, Graduate Studies in Math. Vol. **77**, (2007) Am. Math. Soc.
14. G.F.R. Ellis, *Relativistic cosmology – its nature, aims and problems*, in *General Relativity and Gravitation* (D. Reidel Publishing Co., Dordrecht), pp. 215–288 (1984).
15. G.F.R. Ellis and T. Buchert, Phys. Lett. A (Einstein Special Issue) **347**, 38 (2005).
16. D. H. Friedan, Ann. Physics **163** (1985), no. 2, 318–419.
17. N. Garofalo, E. Lanconelli, Math. Ann. **283**, 211-239 (1989).
18. C. Guenther, J. Geom Anal. **12** (2002), 425–436.
19. R. S. Hamilton, J. Diff. Geom. **17**, 255-306 (1982).
20. G. Huisken, J. Differential Geom. **17** (1985), 47–62.
21. E. W. Kolb, S. Matarrese and A. Riotto, New J. Phys., **8**, 322 (2006).
22. A. Lichnerowicz, Pub. Math. de l'I.E.H.S. **10**, 5 (1961).
23. J. Lott, Comm. Math. Phys. 107, (1986) 165–176.
24. T Oliynyk, V Suneeta, E Woolgar Nucl.Phys. B739 (2006) 441-458, hep-th/0510239.
25. G. Perelman *The entropy formula for the Ricci flow and its geometric applications* math.DG/0211159.
26. G. Perelman *Ricci flow with surgery on Three-Manifolds* math.DG/0303109.
27. G. Perelman *Finite extinction time for the solutions to the Ricci flow on certain three-manifolds* math.DG/0307245.
28. S. Räsänen, JCAP **042**, 003 (2004).
29. S. Räsänen , JCAP **0611**, 003 (2006).
30. W. P. Thurston, Bull. Amer. Math. Soc. (N.S.) **6** (1982), 357-381.
31. W. P. Thurston, *Three-dimensiional geometry and topology* Vol. 1. Edited by S. Levy. Princeton Math. Series, **35** Princeton Univ. Press, Princeton NJ, (1997).

BÄCKLUND CHARTS & APPLICATIONS

SANDRA CARILLO

Dipartimento di Metodi e Modelli Matematici per le Scienze Applicate,
Università di Roma "La Sapienza", I-00161 Rome, Italy
E-mail: carillo@dmmm.uniroma1.it
www.dmmm.uniroma1.it/ carillo/

Bäcklund transformations are here considered as a tool to investigate nonlinear evolution equations as well as linear ones in the case when a memory term, nonlocal in the time variable, is present. Indeed, the important role played by Bäcklund transformations both in looking for solutions to initial boundary value problems, on one side, and in revealing structural properties, on the other one, are well known. Here, Bäcklund Transformations under the latter viewpoint are applied with the aim to generalize *Bäcklund Charts* to connect hierarchies of operator valued equations in a suitable Banach Space related to the equations of interest. Further new perspectives as well as results are also mentioned.

Keywords: Bäcklund Transformations; Evolution Equations; Recursion operator.

1. Introduction

An overwiew of some on applications of Bäcklund and Reciprocal Transformations to nonlinear differential equations is here presented. Specifically, the attention is focussed on nonlinear evolution equations in $1 + 1$-dimensions. However, some results concerning the link between a linear integro-differential problem with memory and a nonlinear differential equation are also given.

Bäcklund and Reciprocal Transformations represent a powerful tool in investigating nonlinear differential equations. Indeed, they play a key role both in establishing structural properties enjoyed by a nonlinear system, such as the Hamiltonian and/or bi-Hamiltonian structure, symmetry properties, as well as in allowing to find solutions of initial value problems for nonlinear evolution equations.

An extensive bibliography concerning applications of such transformations in solving boundary and initial value problems is given in Rogers and Shadwick [1], and, subsequently, in Rogers and Ames [2].

In particular, nonlinear evolution equations, such as the Korteweg-deVries (KdV), the modified Korteweg-deVries (mKdV), the Harry Dym (Dym), on one side, and the Caudrey-Dodd-Gibbon (CDG), the Kaup-Kupershmidt (KK) and the Kawamoto equations, on the other one, are all connected via two analogous Bäcklund Charts which, according to the terminology introduced in [3], depict the links relating them all via Bäcklund and reciprocal transformations. Specifically, two different *Bäcklund Charts* [4,5] are constructed which, in turn, relate 3rd and 5th order nonlinear evolution equations. It is shown how the same Bäcklund Chart can be extended to the whole hierarchies of equations (all of them admitting a hereditary recursion operator). Thus, their symmetry as well as their bi-Hamiltonian structure can be obtained, or, if already known, retrieved [6]. As further consequences it turns out also to be possible to solve initial value problems on applications of Bäcklund and reciprocal transformations [7–10]. Specifically, via the Bäcklund Chart involving the KdV, mKdV and Dym equations [5], solutions for the Dym equation can be constructed [10].

Here, new perspectives concerning a functional analysis approach [11] are briefly summarized.

The material is organized as follows. The opening Section 2 is devoted to briefly summarize some background material needed in the subsequent Section 3 where some *Bäcklund Charts* are shown pointing out the results they allow to state. The closing Section 4 is concerned about some new perspectives as well as recent achievements.

2. Bäcklund Transformations: background notions

In this Section few notions which are required in the following are recalled to help the reader. Accordingly, here throughout an evolution equation is denoted via $u_t = K(u)$, where $u(x,t) \in M$ where M presents a manifold modeled on a linear topological space so that the typical fiber T_uM of M can be identified with M itself. Specifically, M is assumed to be the Schwartz space S of C^∞-functions rapidly vanishing at infinity; $K : M \to TM$ denotes an appropriate C^∞-vector field on the manifold M. Then, according to [12],

Definition 2.1. Given two evolution equations in $1+1$-dimensions, say $u_t = K(u)$ and $s_t = G(s)$, then $B(u,s) = 0$ represents a Bäcklund transformation between them if , whenever, given two solutions, in turn, denoted as $u(x,t)$ and $s(x,t)$, if $B(u(x,t),s(x,t))|_{t=0} = 0$, it follows that $B(u(x,t),s(x,t)) \equiv 0 \ \forall t > 0 \ \forall x \in \mathbb{R}$.

Hence, a Bäcklund transformation establishes a $1-1$ correspondence between solutions of the two evolution equations. Furthermore, when an evolution equations, say $u_t = K(u)$, admits a hereditary recursion operator [12], denoted as $\Phi(u)$, then, also the second one can be proved to admit such an operator. Indeed, the recursion operator which refers to second evolution equations can be obtained from the recursion operator $\Phi(u)$ of the other one, again, via the Bäcklund transformation[a]. Thus, let $\Psi(s)$ denote the recursion operator of the second equation, then the two equations can be written, in turn

$$u_t = \Phi(u)u_x \quad , \qquad s_t = \Psi(s)s_x \ , \tag{1}$$

and the corresponding hierarchies, which read

$$u_t = \Phi(u)^n u_x \quad , \qquad s_t = \Psi(s)^n s_x \ , n \in \mathbb{N}, \tag{2}$$

are connected via the same Bäcklund transformation which relates the base members [12]. Among the many important consequences implied by the link via Bäcklund transformation between the two hierarchies, such as the possibility to find solutions to initial boundary value problems, on one side, and to investigate structural properties of nonlinear evolution equations, on the other one, here the attention is focussed only some aspects related to this second point of view.

3. Bäcklund Charts, Induced Invariances & New Results

This Section is concerned about how invariance properties can be revealed on application of Bäcklund and reciprocal transformations. As an example to start with, the KdV-Dym Bäcklund Chart [5] which comprises also the mKdV and other nonlinear evolution equation hierarchies whose base member is a third order nonlinear evolution equation, is considered. The links, according to [5] can be sketched in the following Bäcklund Chart:

$$\boxed{\text{KdV}\ (u)} \overset{BT_1}{\leftrightarrow} \boxed{\text{mKdV}\ (v)} \overset{BT_2}{\leftrightarrow} \boxed{\text{KdV Sing}\ (\phi)} \overset{BT_3}{\leftrightarrow} \boxed{\text{Int.So.KdV}\ (s)} \overset{BT_4}{\leftrightarrow} \boxed{\text{Dym}\ (\rho)}$$

[a] a detailed explanation concerning how recursion operators and also the Hamiltonian and bi-Hamiltonian structures are transformed on application of Bäcklund as well as reciprocal transformations is comprised in Refs. 5 and 6.

where: Int.So.KdV (s) : $s_t \;=\; s_{xxx} \;-\; 3s^{\ 1}s_x s_{xx} \;+\; \frac{3}{2}s^{-2}{s_x}^3$ and

$$\text{KdV}\,(u): \;\; u_t = u_{xxx} + 6uu_x \qquad\qquad \text{mKdV}\,(v): \;\; v_t = v_{xxx} - 6v^2 v_x$$

$$\text{KdV Sing}\,(\phi): \;\; \phi_t = \phi_x\{\phi; x\} \qquad\qquad \text{Dym}\,(\rho): \;\; \rho_t = \rho^3 \rho_{\xi\xi\xi}$$

are the equations connected via the following Bäcklund and reciprocal transformations:

$$BT_1: \;\; u + v_x + v^2 = 0 \quad , \quad BT_2: \;\; v = \frac{1}{2}\frac{\phi_{xx}}{\phi_x}$$

$$BT_3: \;\; s = \phi_x \quad , \quad BT_4: \;\; \bar{x} = D^{-1}s(x), \;\; \rho(\bar{x}) = s(x)$$

where $D^{-1} := \int_{-\infty}^{x} d\xi$ represents the inverse of the operator of (partial) derivation with respect to the x-variable and BT_1 is the well known *Miura* transformation relating the KdV and mKdV equations. When the invariance under the Möbius group of transformations, enjoyed by the KdVSing (ϕ) equation, is recalled, *i.e.*, $\forall a, b, c, d \in \mathbb{C}|\; ad - bc \neq 0$

$$\boxed{\phi_t = \phi_x\{\phi; x\}} \overset{M}{\longleftrightarrow} \boxed{\tilde{\phi}_t = \tilde{\phi}_x\{\tilde{\phi}; x\}} \qquad M: \tilde{\phi} = \frac{a\phi + b}{c\phi + d} \tag{3}$$

the following KdV-Dym Bäcklund Chart [5] is obtained

$$\begin{array}{ccccccccc}
\boxed{\text{KdV}\,(u)} & \leftrightarrow & \boxed{\text{mKdV}\,(v)} & \leftrightarrow & \boxed{\text{KdV Sing}\,(\phi)} & \leftrightarrow & \boxed{\text{Int.So.KdV}\,(s)} & \leftrightarrow & \boxed{\text{Dym}\,(\rho)} \\
\updownarrow I_1 & & \updownarrow I_2 & & \updownarrow M & & \updownarrow I_3 & & \updownarrow I_4 \\
\boxed{\text{KdV}\,(\tilde{u})} & \leftrightarrow & \boxed{\text{mKdV}\,(\tilde{v})} & \leftrightarrow & \boxed{\text{KdV Sing}\,(\tilde{\phi})} & \leftrightarrow & \boxed{\text{Int.So.KdV}\,(\tilde{s})} & \leftrightarrow & \boxed{\text{Dym}\,(\tilde{\rho})}
\end{array}$$

where I_i, $i = 1, 2, 3$ or 4 denote auto-Bäcklund transformations enjoyed by the whoole hierarchies of evolution equations which appear in the Bäcklund Chart. Notably, in this way some well known auto-Bäcklund transformations can be retried together with new ones, such as the Dym auto-Bäcklund transformation I_4 [5] . Further invariances enjoyed by other members of the Bäcklund Chart allow to extend it to obtain, according to [5,9] , a wider Bäcklund Chart. The latter turns out to be a powerful tool also in looking for solutions to initial boundary value problems; in particular, it produces solutions to the Dym equation [9,10] from solutions of the KdV equation. Most of the obtained results can be generalized in two different ways; namely, a Bäcklund Chart, similar to the KdV-Dym Bäcklund Chart depicts the links relating the Caudrey-Dodd-Gibbon and Kaup-Kupershmidt hierarchies [3,4,6] , whose base members are 5th order evolution equations. Hence, in Ref.6, such a the Bäcklund Chart induces to prove new invariances as well as the Hamiltonian and bi-Hamiltonian structure, and new symmetry properties enjoyed by such nonlinear evolution equation hierarchies. On the other hand, an analog Bäcklund Chart can be constructed when the $2+1$-dimensional versions of the equations which appear in the KdV-Dym Bäcklund Chart, are considered. A reciprocal trasformation between KP-Dym equations is given by Rogers [13] and, then, sistematically studied in Ref.14.

4. New Results & Perspectives

This Section is devoted to give a flavor of new perspectives and results which are obtained on application of Bäcklund transformations. Specifically, the new idea is to combine to different viewpoints: that is, associate to a given evolution equation, an operator valued one in a suitable Banach space E, according to the approach of Carl and Schiebold [15], and, then, construct an *Operator Bäcklund Chart* to relate the operator valued equations in analogy with the Bäcklund Chart established in the case of nonlinear evolution equations. Results of this new approach are comprised in Ref.11 where the analog of well known links are obtained to connect the corresponding operator equations. Indeed, the operator equations which are associated to the linear heat and Burgers equations, on one side, and to the KdV -mKdV equations, on the other one, are constructed. Then, the Cole-Hopf [16,17] links between the linear heat and Burgers equations as well as the Miura link between KdV-mKdV equations are both constructed at the level of the corresponding operator equations. In addition, in all the mentioned cases, the recursion operators related to all the operator equations are also obtained via Bäcklund transformations.

According to [11], the operator KdV equations reads

$$U_t = U_{xxx} + 3\{U, U_x\}\ , \quad \text{where} \quad \{U, U_x\} := UU_x + U_xU \tag{4}$$

where U denotes an operator acting on a Banach space; note that the anticommutator $\{\cdot,\cdot\}$ is introduced since no commutativity requirement is imposed on the operator U and its derivatives. The Bäcklund transformation, analog of the Miura transformation BT_1 between KdV and mKdV equations, now is given by

$$U + V_x + V^2 = 0 \tag{5}$$

which transforms (4) into

$$V_t = V_{xxx} + 3\{V^2, V_x\}\ . \tag{6}$$

The consequences of such a link are exploited in Ref. 11 where also the operator Bäcklund Chart which comprises the Burgers and linear heat operator equations is considered: it represents as the simplest example of application of the proposed approach.

Finally, a result concerning an application of Bäcklund transformations in investigating a linear heat equation which models a *material with memory* is here briefly considered. A wide literature[b] is concerned about heat conduction with memory, and takes its origins in Cattaneo's work [19]. Specifically, in the case when the temperature within an isotropic rigid heat conductor with memory is modeled, the evolution equation involves an integral term. Such a term is introduced to take into account that the temperature depends on time via both, the present, as well as past times through the (thermal) history of the material.

Consider a 1-dimensional isotropic heat conductor, then, when v denotes the temperature within the heat conductor, $k(t)$, the heat flux relaxation function and $\alpha_0 > 0$ the constant specific heat, the evolution equation reads:

$$\alpha_0\, v_t = -\left[\int_0^t \dot{k}(\tau)\overline{\nabla v}^t(x,\tau)\, d\tau\right]_x, \quad \overline{\nabla v}^t(x,\tau) := \int_{t-\tau}^t \nabla v(x,s)\, ds \tag{7}$$

which, on application of the Cole-Hopf transformation $uv - v_x = 0$ gives [20]

$$u_{tt} = k(0)\left[u_{xx} + 2uu_x\right] \tag{8}$$

where $k(0)$ is the initial value of heat flux relaxation function. Note that, the connection between (7) and (8) provides a reason why the introduction of the nonlocal term in (7) is associated to overcome the *paradox* of infinite propagation speed associated to the linear heat equation. Existence and uniqueness of solution to initial boundary value problems given by (7) supplemented with assigned initial and boundary data has been recently proved [21].

Acknowledgments

The partial support of G.N.F.M.-I.N.D.A.M. and of PRIN2005 *Mathematical models and methods in continuum physics* are gratefully acknowledged.

References

1. C. Rogers and W. F. Shadwick: *Bäcklund Transformations and their Applications*, Mathematics in Science and Engineering Vol. 161, Academic Press, New York - London - Paris - Sydney - Tokyo - Toronto, 1982.

[b]see for instance Fabrizio, Gentili and Reynolds [18] and Ref.s therein

2. C. Rogers and W. F. Ames: *Nonlinear Boundary Value Problems in Science and Engineering*, Academic Press, Boston - San Diego - New York - Berkeley - London - Sydney - Tokyo - Toronto, 1989.
3. C. Rogers and S. Carillo: *Bäcklund Charts for the Caudrey-Dodd-Gibbon and Kaup-Kupershmidt Hierarchies*, in:*Nonlinear Evolutions*, World Sci. Publ., Teaneck, NJ, p. 57-73, 1988.
4. C. Rogers and S. Carillo: *On Reciprocal Properties of the Caudrey-Dodd-Gibbon and Kaup-Kupershmidt Hierarchies*, *Physica Scripta*, **36**, 1987.
5. B. Fuchssteiner and S. Carillo: *Soliton structure versus singularity analysis: Third order completely integrable nonlinear equations in 1+1 dimensions*, *Physica*, **152** A, p. 467-510, 1989.
6. S. Carillo and B. Fuchssteiner: *The abundant symmetry structure of hierarchies of nonlinear equations obtained by reciprocal links*, *J. Math. Phys.*, **30**, p. 1606-1613, 1989.
7. Guo Ben-Yu and C. Rogers: *On the Harry Dym equation and its solution*, *Science in China*, **32**, p. 283-295, 1989.
8. Guo Ben-Yu and S. Carillo: *Infiltration in soils with prescribed boundary concentration: a Burgers model*, *Acta Appl. Math. Sinica*, **6**, p. 365-369, 1990.
9. S. Carillo and B. Fuchssteiner: *Non commutative Symmetries and new solutions of the Harry Dym equation*, *in: Proceedings of the Como conference, Manchester University Press*, 1989.
10. B. Fuchssteiner, T. Schulze and S. Carillo: *Explicit Solutions for the Harry Dym Equation*, *J. Phys. A: Math. Gen.*, **25**, p. 223-230, 1992.
11. S. Carillo and C. Schiebold: *The non-commutative KdV-hierarchy by recursion methods*, preprint,, 2007.
12. A.S. Fokas and B. Fuchssteiner: *Bäcklund Transformations for Hereditary Symmetries*, *Nonlinear Analysis TMA*, **5**, p. 423-432, 1981.
13. C. Rogers: *The Harry Dym equation in 2+1 dimensions: A Reciprocal Link with the Kadomtsev Petviashvili equation*, *Phys. Letters*, **120**, p. 15-18, 1987.
14. W. Oevel and S. Carillo: *Squared Eigenfunction Symmetries for Soliton Equations: Part I and Part II*, *J. Math. Anal. and Appl.*, **217** no.1, p. 161-178, 179-199, 1998.
15. B. Carl and C. Schiebold: *Nonlinear equations in soliton physics and operator ideals*, Nonlinearity,, **12** no.2, p. 333-364, 1999.
16. J.D. Cole: *On a quasi-linear parabolic equation occurring in aerodynamics*, Quart. Appl. Math.,, **9** , p. 225-236, 1951.
17. E. Hopf: *The partial differential equation* $u_t + uu_x = u_{xx}$, Comm. Pure Appl. Math.,, **3** , p. 201-230, 1950.
18. M. Fabrizio, G. Gentili, D.W.Reynolds: *On rigid heat conductors with memory*, *Int. J. Eng. Sci.* , **36**, p. 765-782, 1998.
19. C. Cattaneo: *Sulla conduzione del calore*, Atti Sem. Mat. Fis. Universitá Modena,, **3**, p. 83-101, 1948.
20. S. Carillo: *Bäcklund transformations and Evolution Problem in Heat Conduction with Memory*, in progress, 2008.
21. S. Carillo: *Existence, Uniqueness & Exponential Decay: an Evolution Problem in Heat Conduction with Memory*, *J.Math.Anal. and Appl.*, submitted, 2007.

A PARADOX IN LIFE THERMODYNAMICS: THE LONG–TERM SURVIVAL OF BACTERIAL POPULATIONS*

S. CARNAZZA[1], S. GUGLIELMINO[1], M. NICOLÒ[1], F. SANTORO[2], F. OLIVERI[2]

[1] *Department of Microbiological, Genetic and Molecular Sciences, and*
[2] *Department of Mathematics, University of Messina*
Salita Sperone 31, 98166 Messina, Italy
E-mail: scarnazza@unime.it; sguglielm@unime.it; mnicolo@unime.it; fsantoro@unime.it; oliveri@mat520.unime.it

Pseudomonas aeruginosa is an ubiquitous bacterium that, due to its high metabolic versatility, is able to persist for prolonged periods of time. It is the ethiological agent of cystic fibrosis and is involved in urinary infections, conjunctivitis, otitis and pneumonia. We present the results of a batch culture of *P. aeruginosa* inoculated in LB medium and monitored weekly for a period of 24 months during which no more nutrients are added. A mathematical model suitable to describe the experimental viability data is given.

Keywords: Long–term survival of bacteria; stressed bacteria populations.

1. Introduction

Bacteria — single–cell microorganisms widely spread in every kind of habitat — are highly–complex thermodynamic systems, requiring a source of energy for maintaining their structure and functions. Bacterial growth is influenced by several factors: nutrient availability, pH, temperature, oxygen, osmolarity, presence of toxic compounds (heavy metals, hydrogen peroxide, antibiotics, etc.). When one or more environmental conditions become unfavourable, bacteria react in order to survive. Some genera, such as *Bacillus* and *Clostridium*, when under negative stimuli, produce the endospore, *i.e.*, a differentiated cell with no active metabolism, able to survive for an indefinite time in terms of "latent life". When optimal conditions are restored, the endospore germinates, metabolism is activated, and a new vegetative cell appears. Most of bacterial genera, however, are not able to produce endospores; nonetheless, they can trigger efficient resistance mechanisms. The

*Dedicated to T. Ruggeri on the occasion of his 60th birthday.

survival of non–sporulating bacteria attracted in the last 30 years many microbiologists with the aim of characterizing the molecular mechanisms underlying the bacterial stress response. In general, when under stress, several phenotypic variations are observed (reduction of cell size, morphological transitions from rod to spherical shape,[1] decrease in metabolic activity,[2–4] synthesis repression of constitutive proteins, synthesis activation of stress–induced proteins required for survival[5–7]); parallely, increased resistance against external challenges (*e.g.*, H_2O_2, antibiotics, disinfecting solutions) and improved ability to persist in their habitat are displayed.[6,8] Several authors showed that bacterial populations under negative conditions can undergo very complex genomic rearrangements leading to a more resistant phenotype. This is the case, for example, of the adaptive mutations, described as genetic mutations occurring in resting cells.

Pseudomonas aeruginosa is an ubiquitous bacterium which can be found in marine and estuarine environments, in soils, etc. It is able to use more than one hundred of chemical compounds as carbon and energy sources, and, due to its wide metabolic versatility, is able to persist for prolonged periods of time. Moreover, it is the ethiological agent of cystic fibrosis, a fatal chronic disease, and can be involved in urinary infections, conjunctivitis, otitis and pneumonia. Actually, it represents an emergence in nosocomial infections, because of its high degree of resistance against several classes of antibiotics and persistence also in disinfecting solutions. For these characteristics, the aim of this study is the comprehension and modelization of long-term survival strategies of *P. aeruginosa*.

2. Materials and methods

A batch culture of *P. aeruginosa* ATCC 27853 was inoculated in LB medium and monitored weekly for a period of 24 months. No nutrients were added after bacterial inoculum for all the observation period; weekly, ultrapure sterile water was added, to balance the volumetric reduction of the cultures due to evaporation. Specifically, biomass variations were measured by spectrophotometric OD ($\lambda = 540nm$), viability was measured by colony forming unit (CFU) method, cell morphology was evaluated by scanning electron microscopy (SEM), genomic stability was assessed by restriction fragment length polymorphism (RFLP) resolved in pulsed–field gel electrophoresis (PFGE), fitness of isolates from ageing population was estimated by growth curves and compared with wild type strain growth curves. Viability data throughout the entire period of observation indicated that, after two years, 1% of the starting population was still alive

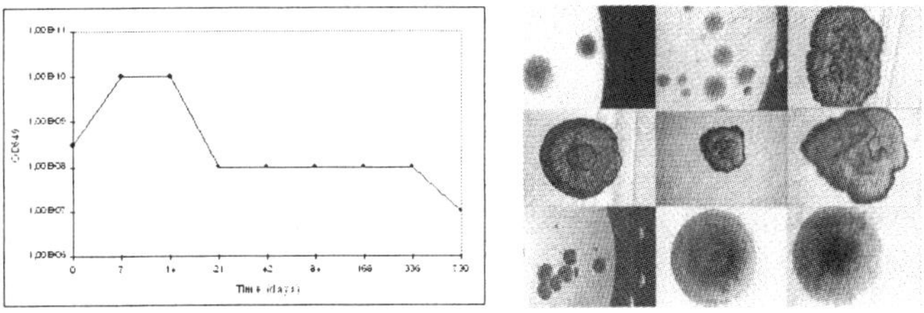

Fig. 1. Viability data (left) and variations in colony morphology after 1 month of incubation (right).

(Fig. 1, left). However, after 1 month of incubation, variations in colony morphology (Fig. 1, right) strongly suggest a high dynamism in the whole population, giving rise to heterogeneous phenotypic variants. Surprisingly, such variants, when regenerated in fresh medium, did not show any increase in growth efficiency with respect to wild type strain, and this is compatible with the observed absence of genomic variations. SEM observations con-

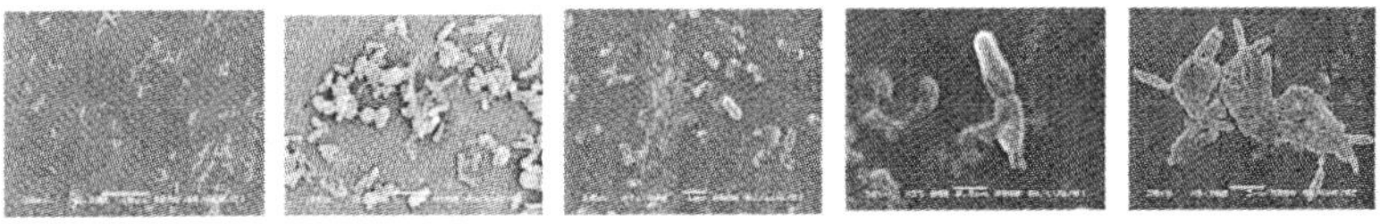

a b c d e

Fig. 2. Micrographs of a population of P. aeruginosa ATCC 27853 after 15 h of incubation (a) and after 7 months of incubation (b-d); structure of bioflocs, showing a lysing cell core, surrounded by whole cells (e).)

firmed data obtained from colony morphology observations, showing that cell morphology was inhomogeneous (Fig. 2, b) inside the population and that aggregations of lysed cells were frequently observed (Fig. 2, c–d). Intriguingly, whole cell aggregates, surrounding lytic cells, were widely present in the tested samples (Fig. 2, e). These observations lead to conjecture that such cell aggregations, called "bioflocs", could be responsible for long-term survival. In fact, lysing cells could be considered as a source of nutrients for living cells; the peculiar aggregation described let nutrients to be released in a confined space, surrounded by alive cells, so becoming readily available. To verify that cell lysis could be a source of nutrients, cell lysates from fresh cultures of *P. aeruginosa* ATCC 27853 were inoculated with both wild type cells and cells from the aged culture. The data show that cytoplasmic fluids from lysing cells could be efficiently used as substrate for survival. On the basis of the experiments, a preliminary mathematical model has been set up in order to describe the viability data.

3. Mathematical model

The mathematical model consists of six coupled ordinary differential equations describing the evolution of active bacteria $A(t)$, inactive bacteria $I(t)$, nutrients $N(t)$, lysed cells $M(t)$, waste material $X(t)$; finally, a function $S(t)$, ruling the transition of bacteria from active to inactive state, is introduced. Active bacteria use the nutrients provided by the LB medium and undergo cellular divisions; on the contrary, inactive bacteria, appearing when fresh nutrients are exhausted, exhibit a reduced metabolism and use lysed cells as a source of nutrients. The whole model can be viewed made by two submodels coupled with a one–way flux. The first submodel describes the interaction (Monod kinetics[11]) between active bacteria and LB medium: essentially, it describes the exponential growth and the stationary phase. In the experiments, the exponential growth phase lasts about 15 hours after which the bacteria reach their maximum density (stationary phase, lasting for about a month) and the fresh nutrients become exhausted. Then a small fraction of active bacteria becomes inactive, and the remaining ones die, so providing a source of nutrients for survivors. The second submodel describes the interaction between inactive bacteria and lysed cells; it involves a dissipative term that contributes to the increase of the compartment of lysed cells and to the amount of *waste* compartment (catabolites not more usable as nutrients). The amount of waste material is assumed to have an inhibitory effect on the metabolism of inactive bacteria. The transition of active bacteria to the inactive state is modelled by a Gompertz–like term[12] involving the function $S(t)$ (a sort of *health function*) defined by $S(t) = \int_{-\infty}^{t} N(u)K_\theta(t-u)du$, with $K_\theta = \frac{\exp(-t/\theta)}{\theta}$, where the delay kernel[13] K_θ accounts for short memory effects on the state of bacteria depending on the availability of fresh nutrients. Thus, the equations of the model are:

$$\dot{A} = \left(\frac{\alpha\nu N(t)}{1+\beta N(t)} - \exp\left(-\frac{S(t)}{\tau}\right)\right)A(t), \qquad \dot{N} = -\frac{\nu N(t)A(t)}{1+\beta A(t)},$$

$$\dot{S} = \frac{N(t)-S(t)}{\theta}, \qquad \dot{I} = \rho\exp\left(-\frac{S(t)}{\tau}\right)A(t) + \eta\mu I(t)M(t) - \gamma I(t),$$

$$\dot{M} = (1-\rho)\exp\left(-\frac{S(t)}{\tau}\right)A(t) - \mu I(t)M(t) + (\gamma-\delta)I(t), \quad \dot{X} = p\delta I(t),$$

where all the parameters involved are positive.

The coefficient ρ is the fraction of active bacteria becoming inactive, α and η measure the efficiency in the transformation of nutrients in biomass for active and inactive bacteria, respectively; γ accounts for the metabolism of inactive bacteria: a part of the catabolites goes to the compartment of lysed cells, a gaseous part is lost and a part goes to the waste compartment ($0 < p < 1$); the meaning of remaining coefficients should be clear. The inhibitory effect of X on metabolism of inactive bacteria is taken into account by assuming the coefficients γ, δ and μ to be not constant but decreasing functions of X: since X increases with time, this captures the idea that the more the environmental conditions become stressful the more the metabolic activity of bacteria is negatively affected. So we assume

$$\gamma(X) = \frac{\gamma_0}{1+\lambda X}, \qquad \delta(X) = \frac{\delta_0}{1+\lambda X}, \qquad \mu(X) = \frac{\mu_0}{1+\lambda X},$$

where γ_0, δ_0, μ_0 and λ are constants. The parameters involved in the model

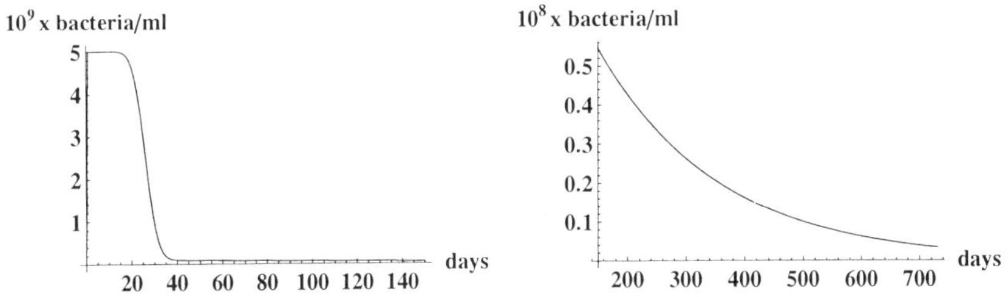

Fig. 3. Plots of total bacteria $(A(t) + I(t))$ *vs.* time (in days); $\alpha = 0.2343$, $\beta = 0.33$, $\gamma_0 = 0.005$, $\delta_0 = 0.003$, $\eta = 0.01$, $\theta = 9$, $\lambda = 1$, $\mu_0 = 5$, $\nu = 7.2$, $\rho = 0.01$, $\tau = 0.1$, $p = 0.6$.

have been determined in order to fit the experimental data, and the system has been numerically integrated. In Fig. 3 the plots of $A(t)+I(t)$ are shown as a function of t expressed in days.

4. Conclusions

The long–term survival of bacteria and the comprehension of their physiology has many implications in several human activities, *e.g.* chronic and recurring pathologies, nosocomial infections, biodegradation of manufacts and artworks, etc. Intriguingly, in recent years it has been shown that similar phenomena occur also in bacterial populations adhering to a solid surface (*biofilms*). This suggests that common metabolism and physiology regulations play a role in phenomena apparently different.[9,10] The experimental data showed that, differently from other bacterial species, after 1 month, neither increased fitness, in terms of growth efficiency, nor genomic variations with respect to wild type population could be observed, even if different colony morphologies emerge, indicating a dynamism in the prolonged stationary phase. Micrographs after 17 months of incubation revealed the presence of bioflocs on which living cells adhere, using them as source of nutrients. This suggests a relationship between recycling of dead cells and topological distribution of living and dead cells; such a relation can be indicated, for the first time, as a factor playing a key role in *P. aeruginosa* survival for a prolonged time in a batch culture. Variations in colony morphology and the appearance of bioflocs, common in stressed/aged bacterial populations, have also been described in biofilms of clinical strains of *P. aeruginosa*, isolated from cystic fibrosis patients.[9,10]

As a next step we aim to build a mathematical model including spatial effects in order to describe the pattern formation in stressed bacterial populations and use it as an advanced tool to investigate the complex phenomena occurring in a biofilm community. In such a way, it could be possible to highlight the main features of such systems and extend our comprehension of mechanisms underlying chronic infections, in which pathogen survival under stressful conditions play a key role.

References

1. C.A. Reeve, P.S. Amy, A. Matin, *J. Bacteriol.* **160**, 1041–1046 (1984).
2. A. G. Chapman, L. Fall, D. E. Atkinson, *J. Bacteriol.* **108**, 1072–1086 (1971).
3. S. Kjelleberg, M. Hermansson, P. Mardén, G.W. Jones, *Ann. Rev. Microbiol.* **41**, 25–49 (1987).
4. P. S. Cabral, *Can. J. Microbiol.* **41**, 372–377 (1995).
5. E. Givskov, L. Eberl, S. Molin, *J. Bacteriol.* **176**, 4816–4824 (1994).
6. L.S. Van Overbeek, L. Eberl, M. Givskov, S. Molin, J.D. Van Elsas, *Appl. Envir. Microbiol.* **61**, 4202–4208 (1995).
7. L. Eberl, M. Givskov, C. Sternberg, S. Møller, G. Christiansen, S. Molin, *Microbiology* **142**, 155–163 (1996).
8. D. Rockabrand, T. Arthur, G. Korinek, K. Livers, P. Blum, *J. Bacteriol.* **177**, 3695–3703 (1995).
9. S.S. Yoon *et al.*, *Dev. Cell.* **3**, 593–603 (2002).
10. D.D. Sriramulu, H. Lünsdorf, J.S. Lam, U. Römling, *J. Med. Microbiol.* **54**, 667–676 (2005).
11. J.R. Lobry, J.P. Flandrois, G. Carret, A. Pave, *Bull. Math. Biol.* **54**, 117–122 (1992).
12. B. Gompertz, *Philos. Trans. R. Soc. London* **115**, 513–585 (1825).
13. J.M. Cushing, *SIAM J. Appl. Math.* **32**, 82–95 (1977).

THE MACROSCOPIC RELATIVISTIC MODEL IN EXTENDED THERMODYNAMICS, PART I: A FIRST TYPE OF ITS SUBSYSTEMS

M.C. CARRISI[1] and S. PENNISI[2]

Dipartimento di Matematica ed Informatica, Università di Cagliari, Via Ospedale 72, 09124 Cagliari, Italy;
E-mail: [1] *cristina.carrisi@tiscali.it,* [2] *spennisi@unica.it*

In paper [1] an interesting and satisfactory model for the case of an arbitrary number of independent variables, satisfying the entropy principle and that of relativity, has been proposed; the closure is obtained except for a numerable family of single-variable functions arising from integration. Our aim is now to show that all these functions are preserved, when we take a subsystem of the previous one, except for the case when the subsystem is constituted only by the conservation laws of mass and of momentum-energy. The only difference is that they will appear at higher orders, with respect to equilibrium, than in the original model. This result is here obtained for a first type of subsystems; the other type will be treated in [2] .

Keywords: Extended Thermodynamics; Subsystems.

1. Introduction

Relativistic Extended thermodynamics with many moments [1] considers the

equations $$\partial_\alpha A^{\alpha\alpha_1\ldots\alpha_M} = I^{\alpha_1\ldots\alpha_M}, \qquad \partial_\alpha B^{\alpha\alpha_1\ldots\alpha_N} = I^{\alpha_1\ldots\alpha_N}. \tag{1}$$

where M is an even number and N is an odd one. Equations involving lower order tensors are already included in (1) because of the following trace

conditions $$\begin{cases} -m^2 A^{\alpha\alpha_1\ldots\alpha_{M-2}} = A^{\alpha\alpha_1\ldots\alpha_M} g_{\alpha_{M-1}\alpha_M}, \\ -m^2 B^{\alpha\alpha_1\ldots\alpha_{N-2}} = B^{\alpha\alpha_1\ldots\alpha_N} g_{\alpha_{N-1}\alpha_N}, \end{cases} \tag{2}$$

where m is the particle mass. The entropy principle for these balance equations, after some standard passages, amounts in assuming the existence of parameters $\lambda_{\alpha_1\cdots\alpha_M}$ and $\mu_{\alpha_1\cdots\alpha_N}$, called Lagrange Multipliers, such that

$$dh^\alpha = \lambda_{\alpha_1\cdots\alpha_M} dA^{\alpha\alpha_1\ldots\alpha_M} + \mu_{\alpha_1\cdots\alpha_N} dB^{\alpha\alpha_1\ldots\alpha_N}\,. \tag{3}$$

Now, if we take $\mu_{\alpha_1\cdots\alpha_N} = -\frac{1}{m^2}\mu_{(\alpha_1\cdots\alpha_{N-2}}g_{\alpha_{N-1}\alpha_N)}$ (4)

and use the trace condition $(2)_2$, we obtain

$$dh^\alpha = \lambda_{\alpha_1\cdots\alpha_M} dA^{\alpha\alpha_1\ldots\alpha_M} + \mu_{\alpha_1\cdots\alpha_{N-2}} dB^{\alpha\alpha_1\ldots\alpha_{N-2}}\,, \tag{5}$$

which is noting else than eq. (3) with N replaced by $N-2$, that is, the entropy principle for a first type of subsystems. In other words, eq. (4) can be called "the passage equation" from the original system to the subsystem. Let us now compare the results of this subsystem with those that can be obtained starting from the beginning with $N-2$ instead of N.

In [1] it has been shown that these results are expressed in terms of the 4-vector $h^{'\alpha}$ related to h^α by

$$h^{'\alpha} = -h^\alpha + \lambda_{\alpha_1\cdots\alpha_M} A^{\alpha\alpha_1\ldots\alpha_M} + \mu_{\alpha_1\cdots\alpha_N} B^{\alpha\alpha_1\ldots\alpha_N}\,. \tag{6}$$

A particular solution found in [1], of the pertinent conditions, is

$$h_1^{'\alpha} = \int F(\lambda_{\alpha_1\ldots\alpha_M} p^{\alpha_1}\ldots p^{\alpha_M}, \mu_{\beta_1\ldots\beta_N} p^{\beta_1}\ldots p^{\beta_N}) p^\alpha dP\,, \tag{7}$$

which, for the passage equation (4), and for $p^\alpha p_\alpha = -m^2$, becomes again eq. (7) but with $N-2$ instead of N. In other words, this particular solution is the same both for the subsystem and for the model obtained by starting from the beginning with $N-2$ instead of N. But what can be said about the general solution? In [1] it was shown that it can be obtained by adding, to the above mentioned particular one, the following 4-vector

$$\Delta h^{'\alpha} = \sum_{h,k=0}^{\infty} \frac{1}{h!k!} C_{h,k}^{\alpha A_1\cdots A_h B_1\cdots B_k} \widetilde{\lambda}_{A_1}\cdots\widetilde{\lambda}_{A_h}\widetilde{\mu}_{B_1}\cdots\widetilde{\mu}_{B_k} \tag{8}$$

where the multi-index notation $A_i = \alpha_{i_1}\cdots\alpha_{i_M}$, $B_j = \alpha_{j_1}\cdots\alpha_{j_N}$ has been used. Let us explain the other particulars of this formula while applying it to our subsystem. Let us start with the following definitions of λ and μ_α

$$\lambda = 2\frac{(M-1)!!}{(M+2)!!}\lambda_{\alpha_1\cdots\alpha_M} g^{\alpha_1\alpha_2}\cdots g^{\alpha_{M-1}\alpha_M}(-m^2)^{\frac{M}{2}}\,,$$

$$\mu_\alpha = 8\frac{N!!}{(N+3)!!}\mu_{\alpha\alpha_1\cdots\alpha_{N-1}} g^{\alpha_1\alpha_2}\cdots g^{\alpha_{N-2}\alpha_{N-1}}(-m^2)^{\frac{N-1}{2}}\,. \tag{9}$$

Obviously, λ remains unchanged for the subsystem, while μ_α, by using the passage equation (4), becomes again eq. $(9)_2$ but with N replaced by $N-2$. Proof of this property is reported in Appendix 2 of [2].

After that, the $\widetilde{\lambda}_{A_h}$ and $\widetilde{\mu}_{B_k}$ appearing in the above expression (8) of $\Delta h^{'\alpha}$ are

$$\begin{cases} \widetilde{\lambda}_{\beta_1\cdots\beta_M} = \lambda_{\beta_1\cdots\beta_M} - \lambda\, g_{(\beta_1\beta_2}\cdots g_{\beta_{M-1}\beta_M)}(-m^2)^{-\frac{M}{2}}\,, \\ \widetilde{\mu}_{\beta_1\cdots\beta_N} = \mu_{\beta_1\cdots\beta_N} - \mu_{(\beta_1}g_{\beta_2\beta_3}\cdots g_{\beta_{N-1}\beta_N)}(-m^2)^{-\frac{N-1}{2}}\,. \end{cases} \tag{10}$$

They are the deviations of the Lagrange Multipliers from their values at equilibrium. What happens now for the subsystem? Obviously, $\widetilde{\lambda}_{A_h}$ remains unchanged, while $\widetilde{\mu}_{B_k}$ becomes

$$\widetilde{\mu}_{\beta_1\cdots\beta_N} = -\frac{1}{m^2}\mu_{(\beta_1\cdots\beta_{N-2}g_{\beta_{N-1}\beta_N)}} - \mu_{(\beta_1 g_{\beta_2\beta_3}\cdots g_{\beta_{N-1}\beta_N)}}(-m^2)^{-\frac{N-1}{2}}$$

$$\text{i.e.,}\quad \widetilde{\mu}_{\beta_1\cdots\beta_N} = -\frac{1}{m^2}\widetilde{\mu}_{(\beta_1\cdots\beta_{N-2}g_{\beta_{N-1}\beta_N)}} \tag{11}$$

where we have used the passage equation (4) and eq. $(10)_2$ with $N-2$ instead of N. In other words, eq. (4) holds also with the $\widetilde{\ }$. This expression we have to substitute in the above expression (8) of $\Delta h'^{\alpha}$. But there it multiplies a symmetric tensor so we can drop the symmetrization. Moreover, we have to convert the multi-index notation of the original system to that of the subsystem, in particular for indexes contracted with $\widetilde{\mu}_{B_k}$. It can be obtained by defining $B_i = \widetilde{B}_i\beta_{i_{N-1}}\beta_{i_N}$. In this way we see that the tensor $C_{h,k}^{\alpha A_1\cdots A_h\tilde{B}_1\cdots\tilde{B}_k}$ of the subsystem is

$$C_{h,k}^{\alpha A_1\cdots A_h B_1\beta_{1_{N-1}}\beta_{1_N}\cdots B_k\beta_{k_{N-1}}\beta_{k_N}} g_{\beta_{1_{N-1}}\beta_{1_N}}\cdots g_{\beta_{k_{N-1}}\beta_{k_N}}(-m^2)^{-k}. \tag{12}$$

So we need the expression of $C_{h,k}^{\alpha A_1\cdots A_h B_1\cdots B_k}$ found in [1]; it is

$$C_{h,k}^{\alpha_1\cdots\alpha_{Mh+kN+1}} = \sum_{s=0}^{\left[\frac{Mh+Nk+1}{2}\right]} C_s^{h,k} g^{(\alpha_1\alpha_2}\cdots g^{\alpha_{2s-1}\alpha_{2s}}\mu^{\alpha_{2s+1}}\cdots\mu^{\alpha_{Mh+Nk+1})}, \tag{13}$$

which determines it except for the scalar functions $C_s^{h,k}$ depending on λ and $\gamma = \sqrt{-\mu^\alpha\mu_\alpha}$. In [1] it has been shown that all the coefficients $C_s^{h,k}$ are determined in terms of that with the highest value of s, which has been called "the leading term" and whose expression, for the case $N>1$, is

$$C^{h,k}_{\left[\frac{Mh+kN+1}{2}\right]} = 2^{\left[\frac{k}{2}\right]}\gamma^{-6-2\left[\frac{k}{2}\right]} \sum_{q=0}^{\frac{Mh+k(N-1)-2}{2}} \frac{(hM+k(N-1)+1+2\left[\frac{k}{2}\right])!!}{[hM+k(N-1)-2q-2]!!}$$

$$(-m^2)^{\frac{N-1}{2}k+\frac{Mh}{2}}\frac{d^h c_{k+Mh,q}}{d\lambda^h}\frac{(q+2+\left[\frac{k}{2}\right])!}{(q+2)!}\left(\frac{1}{\gamma^2}\right)^q. \tag{14}$$

In this expression, the single-variable functions $c_{k,q}$ of λ appear which are

$$\text{restricted by}\; c_{k,q} = c_{k-1,q} \quad \text{for} \quad q = 0,\cdots,\frac{[(k-1)(N-1)-2]}{2}. \tag{15}$$

This equation shows that the functions $c_{k-1,q}$ appearing at the order $k-1$ are present also in the subsequent order k, but $N-1$ new arbitrary of these functions arise so that q now goes from 0 to $\frac{[k(N-1)-2]}{2}$ as in the upper value of the sum in eq. (14) calculated for $h=0$. By substituting

k with $k+Mh$, we see that $c_{k+Mh,q}$ is defined for q that goes from 0 to $\frac{[(k+Mh)(N-1)-2]}{2} > \frac{Mh+k(N-1)-2}{2}$ so that the upper value of q in the sum of eq. (14) can be assumed. From eq. (15) it follows also that

$$c_{k+i,q} = c_{k,q} \quad \text{for} \quad q = 0, \cdots, \frac{[k(N-1)-2]}{2}, \tag{16}$$

which will be used later. But also the tensor $C_{h,k}^{\alpha A_1 \cdots A_h \tilde{B}_1 \cdots \tilde{B}_k}$ is determined except for a leading term; this last one is related to that of $C_{h,k}^{\alpha A_1 \cdots A_h B_1 \cdots B_k}$ for eq. (12), that is to the expression (14). The effective value can be determined by applying the Proposition 6 in [1], with $n = hM + kN + 1$, $r = k$, $p = q + \left[\frac{k}{2}\right]$, finding that the leading term of $C_{h,k}^{\alpha A_1 \cdots A_h \tilde{B}_1 \cdots \tilde{B}_k}$ is

$$2^{\left[\frac{k}{2}\right]}\gamma^{-6-2\left[\frac{k}{2}\right]} \sum_{q=0}^{\frac{Mh+k(N-1)-2}{2}} \frac{(hM+k(N-1)+2\left[\frac{k}{2}\right]+1)!!}{(hM+k(N-1)-2q-2)!!}(-m^2)^{\frac{N-3}{2}k+\frac{M}{2}h}$$

$$\frac{d^h c_{k+Mh,q}}{d\lambda^h} \frac{(q+2+\left[\frac{k}{2}\right])!}{(q+2)!}\left(\frac{1}{\gamma^2}\right)^q \frac{(2\left[\frac{hM+kN+2}{2}\right]-1-2k)!!}{(2\left[\frac{hM+kN+2}{2}\right]-1)!!}\,\eta\Bigg(-2q-2k$$

$$-2+2\left[\frac{hM+kN+2}{2}\right]-2\left[\frac{k}{2}\right], 2\left[\frac{Mh+kN+2}{2}\right]-2q-4-2\left[\frac{k}{2}\right]\Bigg),$$

where $\eta(2a\,,\,2a+2b)$ denotes the product of all even numbers between $2a$ and $2a+2b$. But $\left[\frac{hM+kN+2}{2}\right]-\left[\frac{k}{2}\right] = \frac{1}{2}\left[hM+k(N-1)+2\right]$, so that if $\frac{Mh+k(N-3)-2}{2}+1 \leq q \leq \frac{Mh+k(N-1)-2}{2}$ then $hM+k(N-3)-2q \leq 0$ and $hM+k(N-1)-2-2q \geq 0$ from which it follows that there is a zero factor in $\eta(\ldots,\ldots)$; consequently, we can restrict to values with $q \leq \frac{Mh+k(N-3)-2}{2}$ and, after some simplifications, the above leading term of $C_{h,k}^{\alpha A_1 \cdots A_h \tilde{B}_1 \cdots \tilde{B}_k}$

$$\text{becomes} \quad 2^{\left[\frac{k}{2}\right]}\gamma^{-6-2\left[\frac{k}{2}\right]} \sum_{q=0}^{\frac{Mh+k(N-3)-2}{2}} \frac{(hM+k(N-3)+2\left[\frac{k}{2}\right]+1)!!}{(hM+k(N-3)-2q-2)!!}$$

$$(-m^2)^{\frac{N-3}{2}k+\frac{M}{2}h}\frac{d^h c_{k+Mh,q}}{d\lambda^h}\frac{(q+2+\left[\frac{k}{2}\right])!}{(q+2)!}\left(\frac{1}{\gamma^2}\right)^q. \tag{17}$$

By comparing it with eq. (14), but written with $N-2$ instead of N, we see that the result is the same with the following identification for the single variables functions arising from integration

$$c_{k,q}^{N-2} = c_{k,q}^{N} \quad \text{for } q = 0, \cdots k\frac{N-3}{2} - 1; \tag{18}$$

here we have introduced the apex N and $N-2$ to distinguish the quantities of the general model from those of the subsystem. Two questions now arise;

proof of their answers has to be splitted for the cases $N \geq 5$ and $N = 3$, so let us begin with the first one of these.

2 The case N $\geq$ 5.

$\diamondsuit$ The first of the above mentioned questions is, in certain sense, a test for the theory: The restriction $c_{k,q}^{N-2} = c_{k-1,q}^{N-2}$ for $q = 0, \cdots (k-1)\frac{N-3}{2} - 1$, is respected? (This restriction is eq. (15) with N replaced by $N-2$).
The answer is yes! In fact, we can substitute in the left-hand side from eq. (18) and in the right-hand side the same formula, but with $k-1$ instead of k; so it becomes $c_{k,q}^{N} = c_{k-1,q}^{N}$ for $q = 0, \cdots (k-1)\frac{N-3}{2} - 1$ which is surely verified, also for greater values of q (up to $q = (k-1)\frac{N-1}{2} - 1$) than those requested, because of eq. (15) which holds for the general model.
$\diamondsuit$ The second and more important question is: Let us consider a

$$c_{k,q}^{N} \quad \text{for } q = 0, \cdots k\frac{N-1}{2} - 1; \tag{*}$$

of the general model. Does it occur also in the subsystem? The answer is yes, but at an higher order term, the order $k+i$, with $i \geq \left[\frac{2k}{N-3}\right] + 1$.
Let us prove this trough the following equalities $c_{k,q}^{N} = c_{k+i,q}^{N} = c_{k+i,q}^{N-2}$, where in the first passage we have used eq. (16) connecting functions of the same model N, while in the second passage we have used eq. (18) (with $k+i$ instead of k) connecting functions of the model N to those of the subsystem; but this last equation can be applied only if q belongs to the set $\{0, \cdots, (k+i)\frac{N-3}{2} - 1\}$ and this must be true for all the q in eq. (*); in other words, we must have $k\frac{N-1}{2} - 1 \leq (k+i)\frac{N-3}{2} - 1$, i.e., $i \geq \frac{2k}{N-3}$ which is surely satisfied because we have taken $i \geq \left[\frac{2k}{N-3}\right] + 1$. So, for these values of i we have found the terms $c_{k+i,q}^{N-2}$ of the subsystem which recover $c_{k,q}^{N}$ of the general model.

3 The case N =3.

For the case $N = 3$, we need another proof because the subsystem has $N = 1$. In this case there is no expansion with respect to $\tilde{\mu}$; so only the case $k = 0$ has sense. Moreover, for $N = 1$ the leading term found in [1] is

$$C^{h}_{\frac{M}{2}h} = \gamma^{-6} \sum_{q=0}^{\frac{Mh-2}{2}} (-m^2)^{\frac{M}{2}h} \frac{d^h c_{h,q}(\lambda)}{d\lambda^h} \left(\frac{1}{\gamma^2}\right)^q \frac{(Mh+1)!!}{(Mh-2q-2)!!}$$

$$\text{with} \quad c_{h,q} = c_{h-1,q} \quad \text{for } q = 0, \cdots, \frac{M(h-1)-2}{2}$$

$$\text{and } c_{h,q} \text{ for } q = \frac{M(h-1)}{2}, \cdots, \frac{Mh-2}{2} \text{ are arbitrary functions of } \lambda. \tag{19}$$

$$\text{From these it follows } c_{h,q} = c_{h-j,q} \qquad \text{for } q = 0, \cdots, \frac{M(h-j)-2}{2}. \tag{20}$$

This has to be compared with (17) for $k = 0$ and they are really equal,

$$\text{with} \quad c^1_{h,q} = c^3_{Mh,q} \quad \text{for } q = 0, \cdots, \frac{Mh-2}{2}. \tag{21}$$

$\diamondsuit$ Also in this case two questions arise; the first one is: The restriction $c^1_{h,q} = c^1_{h-1,q}$ for $q = 0, \cdots, \frac{M(h-1)-2}{2}$ is respected? The answer is yes! In fact, we can substitute in the left-hand side from eq. (21) and in the right-hand side the same formula, but with $h-1$ instead of h; so it becomes $c^3_{Mh,q} = c^3_{M(h-1),q}$ for $q = 0, \cdots, \frac{M(h-1)-2}{2}$, which is surely verified, also for greater values of q than those requested, for eq. (16) with $i = M$, $k = M(h-1)$.

$\diamondsuit$ The second and more important question is: Let us consider a

$$c^3_{k,q} \quad \text{for } q = 0, \cdots k-1; \tag{22}$$

of the general model. Does it occur also in the subsystem? The answer is yes, but at the order $h \geq \left[\frac{2k}{M}\right] + 1$. Let us prove this trough the following equalities $c^3_{k,q} = c^3_{Mh,q} = c^1_{h,q}$, where in the first passage we have used eq. (16) with $i = Mh - k$ and $N = 3$, while in the second passage we have used eq. (21); but this last equation can be applied only if q belongs to the set $\{0, \cdots, \frac{Mh-2}{2}\}$ and this must be true for all the q in eq. (22); in other words, we must have $k - 1 \leq \frac{Mh-2}{2}$, which is satisfied because we have taken $h \geq \left[\frac{2k}{M}\right] + 1$. Obviously, this can be said only if $M \neq 0$; on the other hand, when $N = 3$ and $M = 0$ the subsystem is constituted only by the conservation laws of mass and of momentum-energy and, in this case, we have already told that the integration functions are not preserved and our property doesn't hold.

References

1. M.C. Carrisi and S. Pennisi, "An exact fluid model for relativistic electron beams: The many moments case", Journal of Mathematical Physics **48**, 1 (2007).
2. M.C. Carrisi, S. Pennisi and F. Rundo, "The macroscopic relativistic model in E.T. Part II:A second type of its subsystems", presented to Wascom 2007.

THE MACROSCOPIC RELATIVISTIC MODEL IN EXTENDED THERMODYNAMICS, PART II: THE SECOND TYPE OF ITS SUBSYSTEMS

M.C. CARRISI[1], S. PENNISI[2] and F. RUNDO[3]

Dipartimento di Matematica ed Informatica, Università di Cagliari, Via Ospedale 72, 09124 Cagliari, Italy;

E-mail: [1]*cristina.carrisi@tiscali.it,* [2]*spennisi@unica.it,* [3]*rundo@dmi.unict.it*

We consider the problem of finding subsystems of the macroscopic model for Relativistic Extended Thermodynamics with many moments. In part I the subsystem has been obtained by "reducing" one of the fundamental Lagrange Multipliers; here we see what happens when reducing with the other one. The passages are different, although similar, so that a separate treatment is needed. Also in this case, we obtain that all the single-variable functions arising from integration are preserved, except for the case when the subsystem is constituted only by the conservation laws of mass and of momentum-energy.

Keywords: Extended Thermodynamics; Subsystems.

1. Introduction

In this work we do extensive use of [1], where the macroscopic approach to Relativistic Extended Thermodynamics with many moments is treated. Moreover, it has to be intended as a continuation of paper [2], so that we will refer extensively to it, avoiding to report all particulars which can be read in [2]. The second type of subsystems can be obtained by taking

$$\lambda_{\alpha_1\cdots\alpha_M} = -\frac{1}{m^2}\lambda_{(\alpha_1\cdots\alpha_{M-2}g_{\alpha_{M-1}\alpha_M)}; \tag{1}$$

in fact, by using the trace condition $(2)_1$ of paper [2], the eq. (3) of the same paper becomes again eq. (3) of paper [2], but with M replaced by $M-2$, that is, the entropy principle for the second type of subsystems. In other words, eq. (1) can be called "the passage equation" from the original system to the subsystem.

Let us now compare the results of this subsystem with those that can be obtained starting from the beginning with $M-2$ instead of M.

In [1] these results have been found in terms of the 4-vector h'^{α} defined in eq. (6) of [2]. By using the passage equation (1) and the property $p^\alpha p_\alpha = -m^2$, we see that the particular solution (7) of [2], now becomes again eq. (7) of [2], but with $M-2$ instead of M. In other words, this particular solution is the same both for the subsystem and for the model obtained by starting from the beginning with $M-2$ instead of M. But what can be said about the general solution? In [1] it was shown that it can be obtained by adding, to the above mentioned particular one, the 4-vector of eq. (8) in [2]. The tensor $C_{h,k}^{\alpha A_1 \cdots A_h B_1 \cdots B_k}$ depends on λ and μ_α, defined in terms of the Lagrange Multipliers by eq. (9) in [2]. It is obvious that μ_α remains unchanged for the subsystem, while λ, by using the passage equation (1), becomes again eq. $(9)_1$ in ref. [2] but with $M-2$ instead of M. Proof of this property is reported in Appendix 1. In other words, λ is the same both for the subsystem than for the system but with M replaced by $M-2$. After that, the $\widetilde{\lambda}_{A_h}$ and $\widetilde{\mu}_{B_k}$ appearing in the expression of $\Delta h'^{\alpha}$ are expressed by eq. (10) in [2]. What happens now for the subsystem? Obviously, $\widetilde{\mu}_{B_k}$ remains unchanged, while $\widetilde{\lambda}_{A_h}$ becomes

$$\widetilde{\lambda}_{\beta_1 \cdots \beta_M} = -\frac{1}{m^2}\lambda_{(\beta_1 \cdots \beta_{M-2} g_{\beta_{M-1}\beta_M)}} - \lambda g_{(\beta_1\beta_2} \cdots g_{\beta_{M-1}\beta_M)}(-m^2)^{-\frac{M}{2}}$$

$$\text{i.e.,} \quad \widetilde{\lambda}_{\beta_1 \cdots \beta_M} = -\frac{1}{m^2}\widetilde{\lambda}_{(\beta_1 \cdots \beta_{M-2} g_{\beta_{M-1}\beta_M)}} \tag{2}$$

where we have used equation $(10)_1$ in [2], with $M-2$ instead of M; the result shows that eq. (1) holds also with the $\tilde{}$. This expression we have to substitute in the expression (8) in [2] of $\Delta h'^{\alpha}$. But there it multiplies a symmetric tensor so we can drop the symmetrization. Moreover, we have to convert the multi-index notation of the original system to that of the subsystem, in particular for indexes contracted with $\widetilde{\lambda}_{A_h}$. It can be obtained by defining $\widetilde{A}_i$ from $A_i = \widetilde{A}_i \beta_{i_{M-1}} \beta_{i_M}$. In this way we see that the tensor $C_{h,k}^{\alpha \tilde{A}_1 \cdots \tilde{A}_h B_1 \cdots B_k}$ of the subsystem is

$$C_{h,k}^{\alpha \tilde{A}_1 \cdots \tilde{A}_h B_1 \cdots B_k \beta_{1_{M-1}} \beta_{1_M} \cdots \beta_{h_{M-1}} \beta_{h_M}} g_{\beta_{1_{M-1}} \beta_{1_M}} \cdots g_{\beta_{h_{M-1}} \beta_{h_M}} (-m^2)^{-h}. \tag{3}$$

So we need the expression of $C_{h,k}^{\alpha A_1 \cdots A_h B_1 \cdots B_k}$ reported in eq. (13) of [2] and its "leading term" reported in eq. (14) of [2]. But from now on, it will be convenient to split the treatment in the two cases $N > 1$ and $N = 1$, so we can start with the first of these.

2 The case N > 1.

In this case the leading term, as reported also in eq. (14) and (15) of [2] is

$$C^{h,k}_{\left[\frac{Mh+kN+1}{2}\right]} = 2^{\left[\frac{k}{2}\right]}\gamma^{-6-2\left[\frac{k}{2}\right]} \sum_{q=0}^{\frac{Mh+k(N-1)-2}{2}} \frac{\left(hM+k(N-1)+1+2\left[\frac{k}{2}\right]\right)!!}{\left[hM+k(N-1)-2q-2\right]!!}$$
$$(-m^2)^{\frac{N-1}{2}k+\frac{Mh}{2}} \frac{d^h c_{k+Mh,q}}{d\lambda^h} \frac{\left(q+2+\left[\frac{k}{2}\right]\right)!}{(q+2)!} \left(\frac{1}{\gamma^2}\right)^q . \tag{4}$$

In this expression, the single-variable functions $c_{k,q}$ of λ appear which are restricted by

$$c_{k,q} = c_{k-1,q} \quad \text{for} \quad q = 0, \cdots, \frac{[(k-1)(N-1)-2]}{2} \quad . \tag{5}$$

From eq. (5) it follows also that

$$c_{k+i,q} = c_{k,q} \quad \text{for} \quad q = 0, \cdots, \frac{[k(N-1)-2]}{2} \quad , \tag{6}$$

which will be used later. But also the tensor $C^{\alpha\tilde{A}_1\cdots\tilde{A}_h B_1\cdots B_k}_{h,k}$ is determined except for a leading term; this last one is related to that of $C^{\alpha A_1\cdots A_h B_1\cdots B_k}_{h,k}$ for eq. (3). The effective value can be determined by applying the Proposition 6 in [1], with $n = hM + kN + 1$, $r = h$, $p = q + \left[\frac{k}{2}\right]$, finding that the leading term of $C^{\alpha\tilde{A}_1\cdots\tilde{A}_h B_1\cdots B_k}_{h,k}$ is

$$2^{\left[\frac{k}{2}\right]}\gamma^{-6-2\left[\frac{k}{2}\right]} \sum_{q=0}^{\frac{Mh+k(N-1)-2}{2}} \frac{\left(\overline{hM+k(N-1)+1+2\left[\frac{k}{2}\right]}\right)!!}{(hM+k(N-1)-2q-2)!!} (-m^2)^{\frac{N-1}{2}k+\frac{M}{2}h}$$
$$\frac{d^h c_{k+Mh,q}}{d\lambda^h} \frac{\left(q+2+\left[\frac{k}{2}\right]\right)!}{(q+2)!} \left(\frac{1}{\gamma^2}\right)^q \frac{\left(2\left[\frac{hM+kN}{2}\right]-2h+1\right)!!}{\left(2\left[\frac{hM+kN}{2}\right]+1\right)!!}\eta$$
$$\left(2\left[\frac{hM+kN}{2}\right]-2h-2q-2\left[\frac{k}{2}\right], 2\left[\frac{hM+kN}{2}\right]-2-2q-\left[\frac{k}{2}\right]\right).$$

where $\eta(2a\,,\,2a+2b)$ denotes the product of all even numbers between $2a$ and $2a+2b$. But $\left[\frac{hM+kN}{2}\right] = \left[\frac{k}{2}\right] + \frac{[hM+k(N-1)]}{2}$, and, for $\frac{(M-2)h+k(N-1)-2}{2} + 1 \leq q \leq \frac{Mh+k(N-1)-2}{2}$ we have $h(M-2)+k(N-1)-2q \leq 0$ and $hM+k(N-1)-2-2q \geq 0$ from which it follows that there is a zero factor in $\eta(...,...)$; consequently, we can restrict to values with $q \leq \frac{(M-2)h+k(N-1)-2}{2}$. After some simplifications, it becomes eq. (4) with M-2 instead of M provided that

$$c^{M-2}_{k+(M-2)h,q} = c^{M}_{k+Mh,q} \quad \text{for } q = 0, \cdots \frac{(M-2)h+k(N-1)}{2} - 1. \tag{7}$$

Here we have introduced the apex M and $M-2$ to distinguish the quantities of the general model from those of the subsystem. The equation (7) with h=0, becomes

$$c_{k,q}^{M-2} = c_{k,q}^{M} \quad \text{for } q = 0, \cdots \frac{k(N-1)}{2} - 1. \tag{8}$$

Viceversa, this relation with k+(M-2)h instead of k is

$$c_{k+(M-2)h,q}^{M-2} = c_{k+(M-2)h,q}^{M} \quad \text{for } q = 0, \cdots \frac{(M-2)h(N-1) + k(N-1)}{2} - 1$$

which, by using eq. (6) with k+(M-2)h instead of k, and with i=2h, gives again eq. (7) also for greater values of q than those requested. So we can use eq. (8) instead of (7).

Two questions now arise, the first of which is

◇ The restriction $c_{k,q}^{M-2} = c_{k-1,q}^{M-2}$ for $q = 0, \cdots (k-1)\frac{N-1}{2} - 1$, is respected? (This restriction is eq. (5) with M replaced by $M-2$).

The answer is yes! In fact the values of q for which eq. (5) holds, don't depend on M and so are the same, both for M and M-2.

◇ The second question is: Let us consider a $c_{k,q}^{M}$ of the general model. Does it occur also in the subsystem? The answer is yes, because of eq. (8); moreover, it occurs also in the terms with h=0, i.e., in the zero order terms in $\widetilde{\lambda}_{A_h}$.

3 The case N =1.

In the case $N = 1$ the leading term found in [1] is

$$C_{\frac{M}{2}h}^{h} = \gamma^{-6} \sum_{q=0}^{\frac{Mh-2}{2}} (-m^2)^{\frac{M}{2}h} \frac{d^h c_{h,q}(\lambda)}{d\lambda^h} \left(\frac{1}{\gamma^2}\right)^q \frac{(Mh+1)!!}{(Mh-2q-2)!!}$$

$$\text{with} \quad c_{h,q} = c_{h-1,q} \quad \text{for } q = 0, \cdots, \frac{M(h-1)-2}{2}$$

$$\text{and } c_{h,q} \text{ for } q = \frac{M(h-1)}{2}, \cdots, \frac{Mh-2}{2} \text{ are arbitrary functions of } \lambda. \tag{9}$$

$$\text{From these it follows } c_{h,q} = c_{h-j,q} \qquad \text{for } q = 0, \cdots, \frac{M(h-j)-2}{2}. \tag{10}$$

Let's apply proposition 6 of [1], with n=hM+1, r=h, p=q, and with the use of eq. (9), and we find that the leading term of $C_{h,0}^{\alpha \tilde{A}_1 \cdots \tilde{A}_h}$ is

$$\gamma^{-6} \sum_{q=0}^{\frac{Mh-2}{2}} (-m^2)^{\frac{M}{2}h-h} \frac{d^h c_{h,q}}{d\lambda^h} \left(\frac{1}{\gamma^2}\right)^q \frac{(Mh+1)!!}{(Mh-2q-2)!!}$$

$$\frac{(Mh+1-2h)!!}{(Mh+1)!!} \eta(hM-2h-2q, hM-2-2q).$$

But, for $\frac{(M-2)h-2}{2} + 1 \leq q \leq \frac{Mh-2}{2}$ we have $hM - 2h - 2q \leq 0$ and $hM - 2 - 2q \geq 0$, so that $\eta(..., ...) = 0$, so we can limit to the values $q = 0, \cdots, \frac{(M-2)h-2}{2}$ the summation in the equation above that becomes eq. (9) with M-2 instead of M provided that

$$c_{h,q}^{M-2} = c_{h,q}^{M} \qquad\qquad q = 0, \cdots, \frac{(M-2)h-2}{2}. \tag{11}$$

◇ Also in this case two questions arise; the first one is: The restriction $c_{h,q}^{M-2} = c_{h-1,q}^{M-2}$ for $q = 0, \cdots, \frac{(M-2)(h-1)-2}{2}$ is respected? The answer is yes! In fact, we can substitute in the left-hand side from eq. (11) and in the right-hand side the same formula, but with $h-1$ instead of h; so it becomes $c_{h,q}^{M} = c_{h-1,q}^{M}$ for $q = 0, \cdots, \frac{(M-2)(h-1)-2}{2}$, which is surely verified, also for greater values of q than those requested, for eq. $(9)_2$.

◇ The second is: Let us consider a

$$c_{k,q}^{M} \quad \text{for } q = 0, \cdots \frac{Mh-2}{2}; \tag{12}$$

of the general model. Does it occur also in the subsystem? The answer is yes, but at an higher order term, the order h+j, with $j \geq \left[\frac{2h}{M-2}\right] + 1$. Let us prove this trough the following equalities $c_{h,q}^{M} = c_{h+j,q}^{M} = c_{h+j,q}^{M-2}$, where in the first passage we have used eq. (10) with h+j instead of h, while in the second passage we have used eq. (11); but this last equation can be applied only if q belongs to the set $\{0, \cdots, \frac{(M-2)(h+j)-2}{2}\}$ and this must be true for all the q in eq. (12); in other words, we must have $\frac{Mh-2}{2} \leq \frac{(M-2)(h+j)-2}{2}$, which is satisfied because we have taken $j \geq \left[\frac{2h}{M-2}\right] + 1$. Obviously, this can be said only if $M \neq 2$; on the other hand, when $N = 1$ and $M = 2$ the subsystem is constituted only by the conservation laws of mass and of momentum-energy and, in this case, we have already told that the integration functions are not preserved and our property doesn't hold.

4 Conclusions.

We consider the present result very satisfactory, because we have proved that, similarly to the classical approach, even in the relativistic case the single variable functions arising from integration appear in the subsystems but at a different order with respect to the main system. This will be useful, for example in searching the physical meaning of these functions: to this end the many moments case can be considered, but at a first order with respect to equilibrium, or simply the 14 moments case but up to whatever order with respect to equilibrium.

5 Appendix 1.

In section 1 we have said that λ of eq. $(9)_1$ in ref. [2], by using the passage equation (1), becomes again eq. $(9)_1$ in ref. [2] but with $M-2$ instead of M. Let us prove this. It is a consequence of the following identity $\frac{M-1}{M+2}\lambda_{(\alpha_1\cdots\alpha_{M-2}}g_{\alpha_{M-1}\alpha_M)}g^{\alpha_1\alpha_2}\cdots g^{\alpha_{M-1}\alpha_M} = \lambda_{\alpha_1\cdots\alpha_{M-2}}g^{\alpha_1\alpha_2}\cdots g^{\alpha_{M-3}\alpha_{M-2}}$ which is $\lambda_{\alpha_1\cdots\alpha_{M-2}}$ contracted with the following

$$\frac{M-1}{M+2}g_{\alpha_{M-1}\alpha_M}g^{(\alpha_1\alpha_2}\cdots g^{\alpha_{M-1}\alpha_M)} = g^{(\alpha_1\alpha_2}\cdots g^{\alpha_{M-3}\alpha_{M-2})}. \tag{13}$$

We prove now this by observing that its left-hand side is equal to $\frac{M-1}{M+2}g_{\alpha_{M-1}\alpha_M}\ g^{(\alpha_1\alpha_2}\cdots g^{\alpha_{M-1})\alpha_M} = \frac{1}{M+2}g_{\alpha_{M-1}\alpha_M}[g^{(\alpha_1\alpha_2}\cdots g^{\alpha_{M-3}\alpha_{M-2})}$ $g^{\alpha_{M-1}\alpha_M} + (M-2)g^{\alpha_{M-1}(\alpha_1}\cdots g^{\alpha_{M-2})\alpha_M}]$ from which eq. (13) follows.

6 Appendix 2.

In section 1 of [2] we have said that μ_α of eq. $(9)_2$ in ref. [2], by using the passage equation (4) in ref. [2], becomes again eq. $(9)_2$ in ref. [2],but with N replaced by $N-2$. Let us now prove this. It is a consequence of the following identity $\frac{N}{N+3}\mu_{(\alpha\alpha_1\cdots\alpha_{N-3}}g_{\alpha_{N-2}\alpha_{N-1})}g^{\alpha_1\alpha_2}\cdots g^{\alpha_{N-2}\alpha_{N-1}} = \mu_{\alpha\alpha_1\cdots\alpha_{N-3}}g^{\alpha_1\alpha_2}\cdots g^{\alpha_{N-4}\alpha_{N-3}}$ which is $\mu_{\beta\alpha_1\cdots\alpha_{N-3}}$ contracted with the following

$$\frac{N}{N+3}g_{\alpha_{N-2}\alpha_{N-1}}g^{(\alpha_1\alpha_2}\cdots g^{\alpha_{N-2}\alpha_{N-1}}g_\alpha^{\beta)} = g^{(\alpha_1\alpha_2}\cdots g^{\alpha_{N-4}\alpha_{N-3}}g_\alpha^{\beta)}. \tag{14}$$

We prove now this by observing that its left-hand side is equal to

$$\frac{1}{N+3}g_{\alpha_{N-2}\alpha_{N-1}}[g^{(\alpha_1\alpha_2}\cdots g^{\alpha_{N-2}\beta)}g_\alpha^{\alpha_{N-1}}$$
$$+(N-1)g^{\alpha_{N-1}(\alpha_1}\cdots g^{\alpha_{N-3}\alpha_{N-2}}g_\alpha^{\beta)}]$$
$$= \frac{1}{N+3}g_{\alpha_{N-2}\alpha_{N-1}}[g^{(\alpha_1\alpha_2}\cdots g^{\alpha_{N-4}\alpha_{N-3}}g^{\beta)\alpha_{N-2}}g_\alpha^{\alpha_{N-1}}$$
$$+g^{\alpha_{N-1}\alpha_{N-2}}g^{(\alpha_1\alpha_2}\cdots g^{\alpha_{N-4}\alpha_{N-3}}g_\alpha^{\beta)} \qquad +g^{\alpha_{N-1}(\alpha_1}\cdots g^{\alpha_{N-3}\beta)}g_\alpha^{\alpha_{N-2}}$$

$+(N-3)g^{\alpha_{N-1}(\alpha_1}\cdots g^{\alpha_{N-5}\alpha_{N-4}}g_\alpha^\beta g^{\alpha_{N-3})\alpha_{N-2}}]$ from which eq. (14) easily follows.

References

1. M.C. Carrisi and S. Pennisi, "An exact fluid model for relativistic electron beams: The many moments case", Journal of Mathematical Physics **48**, 1 (2007).
2. M.C. Carrisi, S. Pennisi, "The macroscopic relativistic model in E.T. Part I:A first type of its subsystems", presented to Wascom 2007.

FUCHSIAN PARTIAL DIFFERENTIAL EQUATIONS

YVONNE CHOQUET-BRUHAT

Académie des Sciences, 23 Quai de Conti 75006 Paris France

It is generally believed that our cosmos started with an initial singularity, the big bang. In fact all solutions of Einstein equations with compact spacelike sections present a singularity, at least in one time direction, except if the space is the flat three torus. The behavior near the singularity, chaotic or not, is a subject of active research. A case amenable to rigorous mathematical treatment is when the difference of the solution with a given spacetime metric satisfies a Fuchsian system of partial differential equations. In this article we give a more general definition than Kichenassamy and Rendall of a Fuchsian system and give a simpler proof of existence of solutions tending to zero at the singularity. Our generalization is interesting for Hamiltonian systems, as pointed out by Thibault Damour and Sophie de Buyl [10].

1. Introduction

The study of the solutions of the Einstein equations in a neighborhood of an initial singularity, the big bang, is a subject of active research. A case amenable to rigorous mathematical treatment is when the equations can be put in Fuchsian form [1–7]. In this article we simplify a proof of Kichenassamy - Rendall [8] of a Fuchs theorem for partial differential first order systems on a manifold $V = M \times R$, $x \in M$, $t \in R$ of the type

$$t\partial_t u + A(x)u = t\{f(t, x, u, \partial_x u)\}, \tag{1}$$

with f linear* in the derivative $\partial_x u$ and A a linear operator independent of t and such that $|\sigma^A|$ is uniformly bounded for $0 \leq \sigma \leq 1$. We prove a theorem which holds under the weaker hypothesis that $|\sigma|^A \sigma^\alpha$ is bounded in the interval $0 \leq \sigma \leq 1$ for some $\alpha < 1$. The theorem applies then if the eigenvalues of the matrix A have a real part greater than -1. It applies in particular if A is a bounded nilpotent matrix, since in this case $\sigma^A \equiv$

*This property can be attained for more general systems by computing derivatives of the equations and introducing derivatives as new unknowns.

$I + A \log \sigma$, which is not bounded. This case is interesting for Hamiltonian systems, as pointed out by Thibault Damour and Sophie de Buyl [10].

2. Definitions and theorem

We consider the equation (1), with

$$f(t, x, u, \partial_x u) \equiv f_0(t, x, u) + f_1(t, x, u)\partial_x u$$

on a manifold $V \equiv M \times R$, M a paracompact, real analytic $n-$manifold, with u a tensor field, $u : (x, t) \mapsto u(x, t), \ \ x \in M, t \in R$. The linear mapping $A(x)$ is analytic in x, independent of t, while f_0 and f_1 are non linear maps, analytic in $x \in M$, analytic in u if $|u| < c$. continuous in $t \in [0, T]$.

We suppose that M admits a complex analytic extension $\hat{M}$, the linear operator $x \mapsto A(x)$ admits an holomorphic extension to $\hat{M}$ and the mappings $(x, X) \mapsto f_i(t, x, X)$, $i = 0, 1$, admit holomorphic extensions, continuous in $t \in [0, T]$, to $\hat{M} \times U$, where $U := \{|X| < c\}$.

Definition 2.1. The equation is called **Fuchsian** if there exists a neighborhood V of $M \times \{0\}$ in $M \times R$ such that this equation admits one and only one solution u, tensor field on V analytic in $x \in M$, C^1 in t, and such that $u = 0$ for $t = 0$.

Theorem 2.1. *The equation is Fuchsian if there exist a number* $\alpha < 1$ *and a number* $\Sigma > 0$ *such that the linear operator* $\sigma^{A(z)} := e^{A(z) log \sigma}$ *satisfies the inequality*

$$Sup_{z \in \hat{M}} |\sigma^{A(z)}| \sigma^{\alpha} \leq \Sigma, \qquad 0 \leq \sigma \leq 1. \tag{2}$$

We first prove the theorem in a local chart, where u is C^m valued. Standard methods permit the extension to a neighborhood of the manifold M.

3. Equivalence with an integral equation

We consider a chart of $\hat{M}$ whose image contains a ball B_{s_0}, $|z| \leq s_0$. We keep the same notations for the fields and their representatives in the chart, a set of m complex functions taking real values for real z.

To solve the extended complex values equation (1)

$$t\partial_t u + A(z)u = t\{f(t, z, u, \partial_z u)\}, \tag{3}$$

we set $u = Bv$, with B an $m \times m$ matrix such that $\quad t\partial_t B + AB = 0$. We choose $\quad B = t^{-A} \equiv e^{-A \log t}$. $\quad$ The equation (3) reads then

$$\partial_t v = t^A f(t, x, t^{-A} v, \partial_z (t^{-A} v)).$$

Hence the equation (3) together with the condition $u_{|t=0} = 0$ are equivalent to the integral equation

$$u(t,z) = t^{-A} \int_0^t \tau^A f(\tau, z, u(\tau, z), \partial_z u(\tau, z)) d\tau,$$

i.e., setting $\tau = \sigma t$,

$$u(t,z) = t \int_0^1 \sigma^A f(\sigma t, z, u(\sigma t, z), \partial_z u(\sigma t, z)) d\sigma. \tag{4}$$

We define, as [9] and [7], a Banach space $\mathcal{B}_a$ of C^m valued functions $v : (t,z) \mapsto v(t,z)$ holomorphic in z, real valued for z real, and continuous in t in the domain D of $B_{s_0} \times R$, defined as follows.

$$D := \{t, z; \quad 0 \leq t < a(s_0 - |z|)\}, \quad \text{with } |z| < s_0, \quad a > 0. \tag{5}$$

The Banach norm of these functions, called $a-$norm, is given by

$$||v||_a := Sup_D \{t^{-1} |v(t,.)|_s (s_0 - s)\{1 - \frac{t}{a(s_0 - s)}\}^{\frac{1}{2}}\}, \tag{6}$$

with

$$|v(t,.)|_s := Sup_{|z| \leq s} |v(t,z)|. \tag{7}$$

The definition of the $a-$ norm implies for this $s-$ norm

$$|v(t,.)|_s \leq \frac{t}{s_0 - s}\{1 - \frac{t}{a(s_0 - s)}\}^{-\frac{1}{2}} ||v||_a. \tag{8}$$

4. Equivalence with another mapping

The $a-$ norm cannot be used directly to solve the equation (1) by iteration, because the boundedness of $||v||_a$ does not imply the bounds of $|v|_s$ necessary for the holomorphy and estimate of the non linear maps f_i. Kichenassamy and Rendall use the following artefact. They set

$$H(v)(t,z) := \int_0^1 \sigma^{-1} \sigma^A v(\sigma t, z) d\sigma \tag{9}$$

and remark that if v solves the equation

$$v(t,z) = G(v) := tf(t, H(v)(t,z), \partial_z H(v)(t,z)), \tag{10}$$

then $H(v)$ satisfies the integral equation (4). They continue the proof of the Fuchs theorem with the hypothesis that $|\sigma^A|$ is bounded. We relax this hypothesis to the following one

Hypothesis. For $|z| < s_0$, and some α with $0 \le \alpha < 1$ there is a positive number Σ such that

$$|\sigma^{A(z)}|\sigma^{\alpha} \le \Sigma, \quad \text{for } 0 \le \sigma \le 1. \tag{11}$$

Lemma 4.1. *The condition is satisfied if in the ball B_{s_0} the eigenvalues of the matrix $A(z)$ have a real part greater then -1.*

Proof. Elementary calculus using a Jordan decomposition of the matrix A and the definition of the exponential of a matrix show that $\sigma^{\alpha}\sigma^{A}$ is bounded if the real parts of $\lambda + \alpha$, with λ any eigenvalue of A, is positive, and $\alpha > 0$; hence if these real parts are greater than -1 there exists an $\alpha < 1$ such that $\sigma^{\alpha}\sigma^{A}$ is bounded. □

Lemma 4.2. *The mapping $G : v \mapsto G(v) = w$ maps a ball $||v||_a \le R$ into itself if A satisfies the condition (2) and a is small enough.*

Proof. For $t < a(s_0 - s)$, the definition of H and the inequality between the a and s norms give

$$|H(v)(t,.)|_s \le ||v||_a \Sigma \int_0^1 \sigma^{-\alpha}\sigma^{-1}\frac{\sigma t}{s_0 - s}\{1 - \frac{\sigma t}{a(s_0 - s)}\}^{-\frac{1}{2}} d\sigma. \tag{12}$$

One makes in the integral the change of variable from σ to ρ:

$$\sigma t = \rho a(s_0 - s), \quad \text{hence } t d\sigma = a(s_0 - s) d\rho. \tag{13}$$

Then the inequality (12) reads

$$|H(v)(t,.)|_s \le \Sigma ||v||_a a \varepsilon^{\alpha} \int_0^{\varepsilon} \rho^{-\alpha}\{1 - \rho\}^{-\frac{1}{2}} d\rho, \tag{14}$$

with

$$\varepsilon := \frac{t}{a(s_0 - s)}, \quad \sigma \equiv \frac{\rho}{\varepsilon}, \quad 0 \le \varepsilon < 1. \tag{15}$$

To compute the integral in (14), we split it into two parts which we bound, using $0 \le \frac{\varepsilon}{2} < \frac{1}{2}$,

$$\int_0^{\varepsilon} \rho^{-\alpha}\{1-\rho\}^{-\frac{1}{2}} d\rho \le \int_0^{\frac{\varepsilon}{2}} \rho^{-\alpha}\{1 - \frac{\varepsilon}{2}\}^{-\frac{1}{2}} d\rho + \int_{\frac{\varepsilon}{2}}^{\varepsilon} (\frac{\varepsilon}{2})^{-\alpha}\{1-\rho\}^{-\frac{1}{2}} d\rho$$

which, to simplify the writing of computations, we simply bound by

$$\int_0^{\varepsilon} \rho^{-\alpha}\{1 - \frac{\varepsilon}{2}\}^{-\frac{1}{2}} d\rho + \int_0^{\varepsilon} (\frac{\varepsilon}{2})^{-\alpha}\{1-\rho\}^{-\frac{1}{2}} d\rho. \tag{16}$$

We find then, using in the first integral $\frac{\varepsilon}{2} < \frac{1}{2}$ and $\alpha < 1$

$$\int_0^\varepsilon \rho^{-\alpha}\{1-\rho\}^{-\frac{1}{2}} d\rho \leq \frac{\sqrt{2}}{1-\alpha}\varepsilon^{1-\alpha} + 2^\alpha \varepsilon^{-\alpha} 2\{1-\sqrt{1-\varepsilon}).$$

We denote by C any positive number independent of data and unknowns, the above inequality reads

$$\int_0^\varepsilon \rho^{-\alpha}\{1-\rho\}^{-\frac{1}{2}} d\rho \leq C\varepsilon^{1-\alpha}. \tag{17}$$

The inequality (14) gives then

$$|H(v)(t,.)|_s \leq C\Sigma||v||_a a\varepsilon, \quad \varepsilon = \frac{t}{a(s_0-s)} < 1. \tag{18}$$

Choose the ball $||v||_a \leq R$ such that $|H(v)|_s < c$ when $s < s_0$ and $t < a(s_0 - s)$, i.e. $C\Sigma aR < c$. The mappings f_0 and f_1 are then holomorphic and uniformly bounded in absolute value by numbers M_0 and M_1. Therefore if $w := G(v)$

$$|w(t,.)|_s \leq t\{M_0 + M_1|\partial_z H(v)|\}. \tag{19}$$

By the definition of $H(v)$ we have

$$|\partial_z H(v)| \leq \int_0^1 \sigma^{-1}|\sigma^A|\ |\partial_z v(\sigma t, z)|_s d\sigma \tag{20}$$

To estimate $|\partial_z v(\sigma t, z)|$, use a Cauchy inequality which says that, for any s_σ such that $s < s_\sigma < s_0$, one has

$$|\partial_z v(\sigma t,.)|_s \leq C\frac{|v(\sigma t,.)|_{s_\sigma}}{s_\sigma - s}. \tag{21}$$

We deduce from (8) that, if t is such that

$$\sigma t < a(s_0 - s_\sigma), \tag{22}$$

we have

$$|v(\sigma t,.)|_{s_\sigma} \leq ||v||_a \frac{\sigma t}{s_0 - s_\sigma}\{1 - \frac{\sigma t}{a(s_0 - s_\sigma)}\}^{-\frac{1}{2}}, \tag{23}$$

We take as number s_σ, for given s and σ,

$$s_\sigma := \frac{1}{2}(s + s_0 - \frac{\sigma t}{a}), \text{ i.e. } s_0 - s_\sigma = \frac{1}{2}(s_0 - s + \frac{\sigma t}{a}). \tag{24}$$

Then the inequality (23) for t becomes

$$\sigma t < \frac{1}{2}a(s_0 - s) + \frac{\sigma t}{2}, \text{ i.e. } \sigma t < a(s_0 - s),$$

hence the inequality (22) on v holds for all $s < s_0$, $t < a(s_0 - s)$ and $0 \leq \sigma \leq 1$.

Using the the expression of s_σ the inequality (21) becomes

$$|\partial_z v(\sigma t, .)|_s \leq ||v||_a \frac{\sigma t}{(s_0 - s)^2 - (\sigma t/a)^2} \{1 - \frac{\sigma t}{a(s_0 - s_\sigma)}\}^{-\frac{1}{2}}. \tag{25}$$

Elementary calculus gives

$$(s_0 - s)^2 - (\sigma t/a)^2 = (s_0 - s)^2(1 - \rho^2)$$

and, using (24)

$$1 - \frac{\sigma t}{a(s_0 - s_\sigma)} = 1 - \frac{2\sigma t}{a(s_0 - s + \frac{\sigma t}{a})} = \frac{1 - \rho}{1 + \rho} > \frac{1 - \rho}{2}. \tag{26}$$

Therefore

$$(1 - \frac{\sigma t}{a(s_0 - s_\sigma)})^{-\frac{1}{2}} < \sqrt{2}(1 - \rho)^{-\frac{1}{2}}. \tag{27}$$

Elementary calculus gives if $0 \leq \rho \leq 1$

$$\frac{1}{(1 - \rho^2)\sqrt{1 - \rho}} \leq \frac{1}{(1 - \rho)^{\frac{3}{2}}}$$

We then deduce from (20), using the hypothesis on σ^A

$$|\partial_z H(v)| \leq \Sigma ||v||_a \frac{a}{s_0 - s} \varepsilon^\alpha \int_0^\varepsilon \rho^{-\alpha} \frac{d\rho}{(1 - \rho)^{\frac{3}{2}}} \tag{28}$$

We estimate the integral as we estimated (17). We find

$$\int_0^\varepsilon \rho^{-\alpha} \frac{d\rho}{(1 - \rho)^{\frac{3}{2}}} \leq 2^{\frac{3}{2}} \int_0^\varepsilon \rho^{-\alpha} d\rho + (\frac{\varepsilon}{2})^{-\alpha} \int_0^\varepsilon \frac{d\rho}{(1 - \rho)^{\frac{3}{2}}} \tag{29}$$

with

$$\int_0^\varepsilon \frac{d\rho}{(1 - \rho)^{\frac{3}{2}}} = 2\{\frac{1}{\sqrt{1 - \varepsilon}} - 1\} = 2\frac{1 - \sqrt{1 - \varepsilon}}{\sqrt{1 - \varepsilon}} \leq \frac{2\varepsilon}{\sqrt{1 - \varepsilon}}. \tag{30}$$

Therefore

$$|\partial_z H(v)|_s < C\Sigma ||v||_a \frac{a}{s_0 - s} \frac{\varepsilon}{\sqrt{1 - \varepsilon}}. \tag{31}$$

The definition of the $a-$norm gives

$$||t\partial_z H(v)||_a < Ca\Sigma ||v||_a. \tag{32}$$

Hence

$$||w||_a \leq a\{s_0 M_0 + C\Sigma M_1 ||v_a||\}. \tag{33}$$

We have proved that w belongs to the ball $||w||_a < R$ of the Banach space B_a, if a is small enough. □

5. Convergence of iterations

To prove the convergence of the iterations it is sufficient to prove that the mapping is contracting.

Lemma 5.1. *If a is small enough the mapping G is a contracting mapping on B_a.*

Proof. The proof follows the same lines as in the previous section. By the hypothesis on the mappings f_i we have, if $V, v \in B_a$,

$$|f_0(H(V)) - f_0(H(v))| \leq M_0'|H(V-v)| \tag{34}$$

and

$$|f_1(H(V)\partial_z H(V)) - f_1(H(v)\partial_z H(v)| \leq \tag{35}$$

$$M_1'|H(V-v)|\ |\partial_z H(V)| + M_1|\partial_z H(V-v)|.$$

with M_0' and M_1' the bounds of the derivatives with respect to u of f_0 and f_1 in the domain $|u| \leq c$.

We deduce from (34) and (35) the inequality, with $W = G(V)$ and $w = G(v)$,

$$||W - w||_a \leq M_0' I_a' + M_1 II_a' + M_1' \tilde{II}_a',$$

where we have set, the inequalities being consequence of results (17) and (32),

$$I_a' := ||tH(V-v)||_a \leq C s_0 \Sigma ||V-v||_a a,$$

$$II_a' := ||t\partial_z H(V-v)||_a \leq C\Sigma a ||V-v||_a.$$

We consider the last term

$$\tilde{II}_a = ||t|H(V-v)|\ |\partial_z H(V)|\ ||_a$$

By the definition of the $a-$ norm

$$\tilde{II}_a = \sup_{s<s_0, t<a(s_0-s)} \{(s_0-s)|H(V-v)(t,.)|_s\ |\partial_z H(V)(t,.)|_s (1 - \frac{t}{a(s_0-s)})^{\frac{1}{2}}\}$$

$$\leq \{\sup_{s<s_0, t<a(s_0-s)} |H(V-v)(t,.)|_s\} \times$$

$$\{\sup_{s<s_0, t<a(s_0-s)} (s_0-s)|\partial_z H(V)|_s (1 - \frac{t}{a(s_0-s)})^{\frac{1}{2}}\}$$

Hence by these inequalities

$$\tilde{II}_a \leq C\Sigma^2 ||V||_a ||V-v||_a a$$

Assembling these results shows that the mapping G is contracting for small enough a. □

We have proven that the mapping G admits a fixed point $\bar{v} \in B_a$, if a is small enough, which can be constructed by iteration and is unique. The field $H(\bar{v})$ satisfies the integral equation (4) in the domain D, is real for real z, and vanishes for $t = 0$. It is a solution of the Fuchsian problem in the domain $|x| < s_0$, $0 < t < a(s_0 - |x|)$ of the considered chart of M.

6. Global in space theorem

To complete the proof of theorem 2.2, one glues together local solutions, by standard methods using the uniqueness of the solutions we have constructed in neighborhoods $t < a(s - s_0)$. Such neighborhoods cover a neighborhood of $M \times \{0\}$ in $M \times R$, because M has been assumed paracompact.

7. Corollaries

Theorem 7.1. *If all the eigenvalues of A are such that*

$$\mathcal{R}e\lambda > -\mu, \tag{36}$$

then

1. The existence theorem holds for the equation

$$t\partial_t u + Au = t^\mu f(t, x, u, \partial_x u), \quad \mu > 0. \tag{37}$$

2. An analogous theorem holds, but with $u \to 0$ when $t \to +\infty$, for the equation

$$\partial_t u - Au = e^{-\mu t} f(t, x, u, \partial_x u) \quad \mu > 0 \tag{38}$$

Proof. 1. The change of variable $\tau = t^\mu$ transforms the given equation into

$$\mu\tau \frac{d\tau}{dt} \partial_\tau u + Au = \tau f(\tau^{\mu^{-1}}, x, u, \partial_x u);$$

that is,

$$\tau\partial_\tau u + \mu^{-1} Au = \tau\mu^{-1} f(\tau^{\mu^{-1}}, x, u, \partial_x u).$$

The Fuchs theorem applies to this system, but its linear operator is $\mu^{-1}A$ whose eigenvalues are $\mu^{-1}\lambda$ if λ are the eigenvalues of A. The condition is therefore $\mu^{-1}\mathcal{R}e\lambda > -1$.

2. The change of variable $\tau = e^{-\mu t}$ transforms the given equation into

$$\tau\partial_\tau u + \mu^{-1}Au = -\mu^{-1}\tau f(\mu^{-1}\log\tau, x, u, \partial_x u). \tag{39}$$

□

Remark 7.1. The corollary shows that the hypothesis on the boundedness of f in t when t tends to infinity can be replaced by an hypothesis that $e^{-\nu t}f$ is bounded for some $\nu > 0$ such that the eigenvalues of A satisfy the hypothesis, always implied by the original one

$$\mathcal{R}e\lambda > -\mu + \nu. \tag{40}$$

This remark permits the application of the corollary to functions f which have a polynomial growth as t tends to infinity. Corresponding statements hold in the case of the singularity for $t = 0$.

References

1. L. Andersson and A. Rendall, "Quiescent cosmological singularities", Comm. Math. Phys. 218, 479-511 (2001).
2. J. Isenberg and V. Moncrief, "Asymptotic behavior in polarized and half-polarized $U(1)$ symmetric spacetimes", Class. Qtm. Grav. 19, 5361-5386 (2002).
3. Y. Choquet-Bruhat, J. Isenberg and V.Moncrief, "Topologically general $U(1)$ symmetric Einsteinian spacetimes with AVTD behaviour ", Nuovo Cimento 119 B N. 7-9 625-638 (2006).
4. Y. Choquet-Bruhat and J. Isenberg, Half polarized U(1) spacetimes with AVTD behaviour, J. Geom. and Phys. 56 n. 8 1199-2914 (2006).
5. T. Damour, M Henneaux, A. Rendall, and M. Weaver, "Kasner-like behavior for subcritical Einstein-matter systems", Ann. H. Poin. 3, 1049-1111 (2002).
6. J. Isenberg and V, Moncrief, "Asymptotic behavior of the gravitational field and the nature of singularities in Gowdy spacetimes", Ann. Phys. 99, 84-122 (1992).
7. S. Kichenassamy, "Nonlinear Wave Equations" (Dekker, NY) (1996).
8. S. Kichenassamy and A.D. Rendall, "Analytical description of singularities in Gowdy spacetimes", Class. Qtm.Grav. 15, 1339-1355 (1998).
9. S. Baounchi and C. Goulaonic, Comm. PDE 2, 1151-1162 (1977).
10. T. Damour and S. de Buyl "Describing Cosmological Singularities in Iwasawa variables" to be published.

ON THE GENERALIZED RIEMANN PROBLEM FOR A 2 × 2 SYSTEM OF BALANCE LAWS*

F. CONFORTO, S. IACONO, F. OLIVERI

Department of Mathematics, University of Messina
Salita Sperone 31, 98166 Messina, Italy
fiamma@unime.it; iacono@dipmat.unime.it; oliveri@mat520.unime.it

The invariance with respect to a suitable Lie group of point symmetries allows to build invertible point transformations transforming a 2×2 quasilinear system of balance laws into a system of conservation laws. Therefore, the study of the Riemann problem (classical or generalized) for a system of balance laws can be reduced to the investigation of an associated Riemann problem (which can be classical or generalized) for a system of conservation laws. An example of a 2 × 2 system of balance equations for which the procedure can be applied is considered. Numerical simulations are also provided.

Keywords: Lie group invariance; Riemann problem; Balance laws.

1. Introduction

As well known,[1,2] there is deep knowledge of the Riemann problem for a system of conservation laws; on the contrary, not so many results are known on the Riemann problem for a system of balance laws.

If a system of balance laws admits a suitable infinite–dimensional Lie group of point symmetries, then it is possible to build an invertible point transformation allowing to rewrite it under the form of a system of conservation laws.[3,4] Once the solution of the transformed system is known, thanks to the inverse transformation, it is possible to obtain the corresponding solution of the original system.

In the present paper, we study different Riemann problems for a particular system of balance laws that can be mapped to a system of conservation laws; the latter system has the advantage of being more workable both from the numerical and, when possible, the analytical point of view.

*Dedicated to T. Ruggeri on the occasion of his 60th birthday.

Let us consider the system[5]

$$u_t - v_x = 0, \qquad \frac{d_0}{(1+ke_0v)^2}v_t - u_x = -\frac{e_0 v}{1+ke_0v}, \tag{1}$$

with k, d_0 and e_0 constant; if (x,t,u,v) denote distance, time, material velocity and stress, when $d_0 > 0$, (1) can be used to study the dynamical behavior of materials whose responses are both nonlinear and rate dependent. System (1) is also equivalent to a nonlinear telegraph equation[5] suitable to describe electromagnetic waves in a lossy, nonlinear, transmission line,[6] or pressure waves in a relaxing gas.

By means of the following variable rescaling

$$x \longrightarrow \frac{\sqrt{d_0}}{e_0}x, \qquad t \longrightarrow \frac{d_0}{e_0}t, \qquad u \longrightarrow \frac{\sqrt{d_0}}{ke_0}u, \qquad v \longrightarrow \frac{1}{ke_0}v,$$

we obtain the hyperbolic system of balance laws

$$u_t - \left(\frac{1}{w}\right)_x = 0, \qquad w_t + u_x = 1 - w, \tag{2}$$

where $w = 1/(1+v)$. System (2) admits the constant solution $u = u_0$, $w = 1$, but also the nonhomogeneous steady solution

$$u(x) = (1 - w_0)x + u_0, \qquad w = w_0, \tag{3}$$

where u_0 and w_0 ($\neq 0$) are arbitrary constants.

Due to invariance[4] with respect to an infinite–parameter Lie group of point symmetries, in terms of the new independent and dependent variables

$$X = x, \qquad T = \exp(t) - 1, \qquad U = x - u, \qquad W = \exp(t)w, \tag{4}$$

system (2) writes under the form of a system of two conservation laws:

$$U_T + \left(\frac{1}{W}\right)_X = 0, \qquad W_T - U_X = 0. \tag{5}$$

The hyperbolicity condition $w > 0$ ($w < 0$) implies that $W > 0$ ($W < 0$).

2. Riemann problems

A classical piecewise constant initial datum assigned to the system (2), say

$$u(x,0) = \begin{cases} u_\ell & \text{for } x < 0 \\ u_r & \text{for } x > 0 \end{cases}, \qquad w(x,0) = 1, \tag{6}$$

transforms into a piecewise linear initial datum for system (5):

$$U(X,0) = \begin{cases} X - u_\ell & \text{for } X < 0 \\ X - u_r & \text{for } X > 0 \end{cases}, \qquad W(X,0) = 1. \tag{7}$$

In addition, since the original system (2) admits steady solutions of the form (3), a piecewise linear initial value problem of the form

$$u(x,0) = \begin{cases} (1-w_0)\,x + u_\ell \text{ for } x < 0 \\ (1-w_0)\,x + u_r \text{ for } x > 0 \end{cases}, \qquad w(x,0) = w_0, \tag{8}$$

can be assigned to system (2). For system (5), this transforms again into a piecewise linear initial value problem:

$$U(X,0) = \begin{cases} w_0 X - u_\ell \text{ for } X < 0 \\ w_0 X - u_r \text{ for } X > 0 \end{cases}, \qquad W(X,0) = w_0. \tag{9}$$

Let us observe that the first kind of initial data (6) is recovered from the second one (8) by choosing $w_0 = 1$.

Of course, we may consider a classical Riemann problem for the transformed system (5), say

$$U(X,0) = \begin{cases} U_\ell \text{ for } X < 0 \\ U_r \text{ for } X > 0 \end{cases}, \qquad W(X,0) = w_0, \tag{10}$$

U_ℓ, U_r and w_0 ($\neq 0$) being constant, to which corresponds, for system (2):

$$u(x,0) = \begin{cases} x - U_\ell \text{ for } x < 0 \\ x - U_r \text{ for } x > 0 \end{cases}, \qquad w(x,0) = w_0. \tag{11}$$

It is worth to be noticed that initial data (11) and (10) can be viewed as limiting cases, when $w_0 \to 0$ (nevertheless w_0 can not be vanishing), of initial data (8) and (9), respectively. Roughly speaking, if w_0 is sufficiently close to zero, the initial value problem for the transformed system with initial data (9) tends to a piecewise constant initial value problem of the classical Riemann problem.

As well known,[1,2] we are able to solve exactly the Riemann problem for the transformed system (5), that is a set of conservation laws, with classical piecewise constant initial data (10), under the hypothesis of a not too large initial jump. Two possible solutions arise.

The first one is obtained fixing $\mathbf{U}_r = (U_r, w_0)$ ($w_0 > 0$), sufficiently close to $\mathbf{U}_\ell = (U_\ell, w_0)$, with $U_r > U_\ell$. The states $\mathbf{U}_\ell$, $\mathbf{U}_r$ and the unknown intermediate one $\overline{\mathbf{U}}$ are connected by rarefaction curves, *i.e.*, $\overline{\mathbf{U}}$ belongs to the 1–rarefaction curve through $\mathbf{U}_\ell$ and $\mathbf{U}_r$ belongs to the 2–rarefaction curve through $\overline{\mathbf{U}}$; in the original variables the solution reads as

$$u(x,t) = \begin{cases} x - U_\ell, & \frac{x}{\exp(t)-1} \leq -\frac{1}{w_0} \\ x - U_\ell + \ln \frac{w_0|x|}{\exp(t)-1}, & -\frac{1}{w_0} < \frac{x}{\exp(t)-1} < -\frac{1}{w_0} \exp\left(\frac{U_\ell - U_r}{2}\right) \\ x - \frac{U_\ell + U_r}{2}, & \left|\frac{x}{\exp(t)-1}\right| \leq \frac{1}{w_0} \exp\left(\frac{U_\ell - U_r}{2}\right) \\ x - U_r - \ln \frac{w_0 x}{\exp(t)-1}, & \frac{1}{w_0} \exp\left(\frac{U_\ell - U_r}{2}\right) < \frac{x}{\exp(t)-1} < \frac{1}{w_0} \\ x - U_r, & \frac{x}{\exp(t)-1} \geq \frac{1}{w_0} \end{cases}$$

$$w(x,t) = \begin{cases} w_0 \exp(-t), & \frac{x}{\exp(t)-1} \leq -\frac{1}{w_0} \\ -\frac{1-\exp(-t)}{x}, & -\frac{1}{w_0} < \frac{x}{\exp(t)-1} < -\frac{1}{w_0} \exp\left(\frac{U_\ell - U_r}{2}\right) \\ w_0 \exp\left(\frac{U_r - U_\ell}{2} - t\right), & \left|\frac{x}{\exp(t)-1}\right| \leq \frac{1}{w_0} \exp\left(\frac{U_\ell - U_r}{2}\right) \\ \frac{1-\exp(-t)}{x}, & \frac{1}{w_0} \exp\left(\frac{U_\ell - U_r}{2}\right) < \frac{x}{\exp(t)-1} < \frac{1}{w_0} \\ w_0 \exp(-t), & \frac{x}{\exp(t)-1} \geq \frac{1}{w_0} \end{cases}$$

The second solution is obtained fixing $\mathbf{U}_r = (U_r, w_0)$ $(w_0 > 0)$ sufficiently close to $\mathbf{U}_\ell = (U_\ell, w_0)$, with $U_r < U_\ell$. The states $\mathbf{U}_\ell$, $\mathbf{U}_r$ and $\overline{\mathbf{U}}$ are now connected by shock curves, more precisely, $\overline{\mathbf{U}}$ belongs to the 1–shock curve through $\mathbf{U}_\ell$ and $\mathbf{U}_r$ belongs to the 2–shock curve through $\overline{\mathbf{U}}$; in the original variables the solution is

$$u(x,t) = \begin{cases} x - U_\ell, & \frac{x}{\exp(t)-1} < -\frac{1}{\sqrt{w_0 \overline{W}}} \\ x - U_\ell - \frac{\overline{W} - w_0}{\sqrt{w_0 \overline{W}}}, & \left|\frac{x}{\exp(t)-1}\right| < \frac{1}{\sqrt{w_0 \overline{W}}} \\ x - U_r, & \frac{x}{\exp(t)-1} > \frac{1}{\sqrt{w_0 \overline{W}}} \end{cases}$$

$$w(x,t) = \begin{cases} w_0 \exp(-t), & \frac{x}{\exp(t)-1} < -\frac{1}{\sqrt{w_0 \overline{W}}} \\ \overline{W} \exp(-t), & \left|\frac{x}{\exp(t)-1}\right| < \frac{1}{\sqrt{w_0 \overline{W}}} \\ w_0 \exp(-t), & \frac{x}{\exp(t)-1} > \frac{1}{\sqrt{w_0 \overline{W}}} \end{cases}$$

where

$$\overline{W} = \left(\frac{\sqrt{w_0}\,(U_r - U_\ell) + \sqrt{w_0\,(U_r - U_\ell)^2 + 16 w_0}}{4} \right)^2 .$$

Roughly speaking, these solutions can be viewed as the asymptotic limit, when $w_0 \to 0$, of the two possible solutions of the Riemann problem for the original system with piecewise linear initial data (8), in the two cases $u_r < u_\ell$ and $u_r > u_\ell$, respectively. Moreover, let us observe that as $t \to +\infty$ w tends to zero in the whole space domain. This feature, related to the dissipative character of the system, as well as the comparison between the exact solutions of the classical Riemann problem and the solutions of the generalized Riemann problem[7] in the case of w_0 sufficiently close to zero will be inspected in a forthcoming paper.

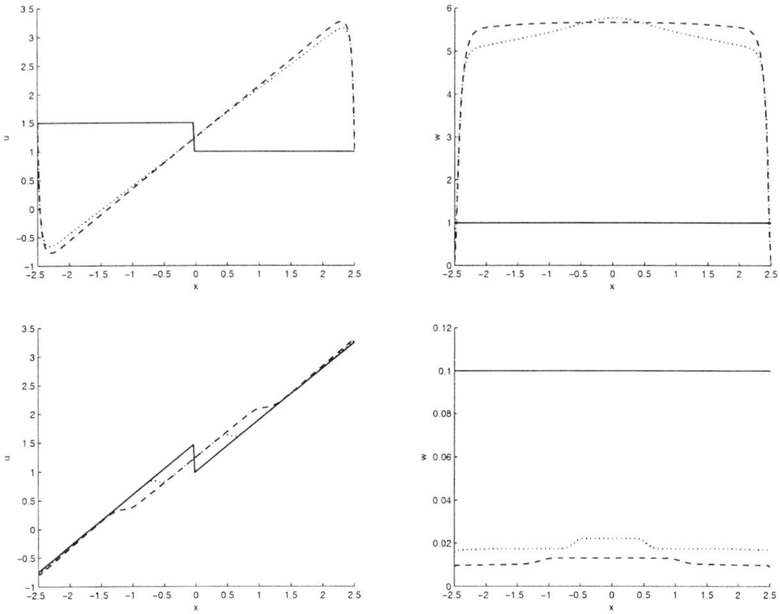

Fig. 1. Numerical solution (u on the left and w on the right) of system (2), (8) corresponding to the solution of (5), (9) with $u_\ell = 1.5$, $u_r = 1$, $w_0 = 1$ (top) and $w_0 = 0.1$ (bottom). The continuous line represents the initial data; dotted line and dashed line represent the solution at different increasing times.

3. Numerical solutions

In this section we present some numerical solutions of system (5) with the initial data presented in the previous section, for different values of w_0. The numerical scheme[8] used, based on the Lax–Friedrichs solver, is a non–oscillatory, second order accurate, central difference method. The integration is performed is a space domain $[-L, L]$, with boundary conditions $u(\pm L, t) = u(\pm L, 0)$, $w(\pm L, t) = w(\pm L, 0)$, $\forall t \in [0, +\infty[$. In Figs. 1 and 2 the plots of the solution of system (2), (8), corresponding to the solution of (5), (9), are shown in order to point out the perturbation in the profiles due to the source term. Moreover, the right plots in both pictures show that w

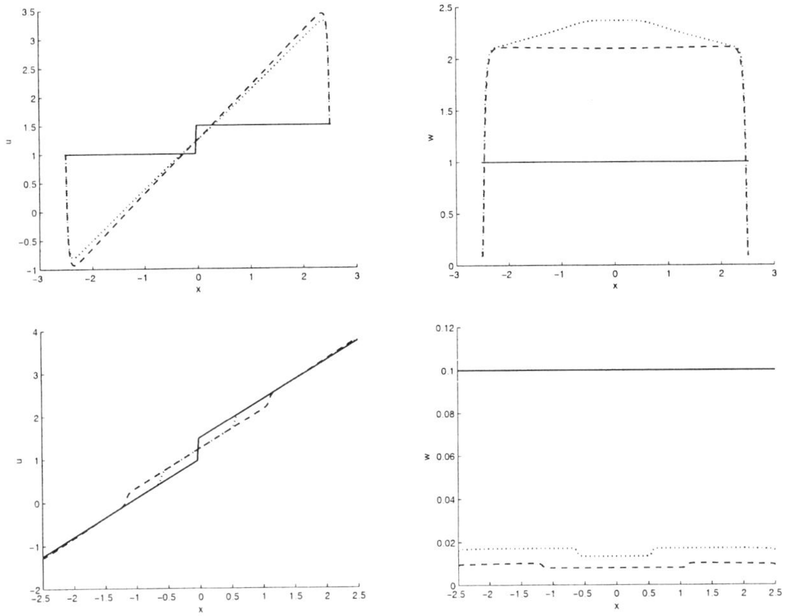

Fig. 2. Numerical solution (u on the left and w on the right) of system (2), (8) corresponding to the solution of (5), (9) with $u_\ell = 1$, $u_r = 1.5$, $w_0 = 1$ (top) and $w_0 = 0.1$ (bottom). The continuous line represents the initial data; dotted line and dashed line represent the solution at different increasing times.

tends to 0 for increasing values of t (this behavior is more evident for small values of w_0 and is related to the dissipative character of the system).

References

1. J. Smoller, *Shock Waves and Reaction–Diffusion Equations* (Springer-Verlag, Berlin, 1983).
2. C. M. Dafermos, *Hyperbolic conservation laws in continuum physics* (Springer-Verlag, Berlin, 2000).
3. G. W. Bluman, S. Kumei, *Symmetries and differential equations* (Springer, New York, 1989).
4. A. Donato, F. Oliveri, *J. Math. Anal. Appl.* **188**, 552–568 (1994).
5. E. Varley, B. Seymour, *Studies Appl. Math.* **72**, 241–262 (1985).
6. I. G. Katayev, *Electromagnetic shock waves* (Iliffe, London, 1966).
7. M. Ben-Artzi, J. Falcovitz, *Generalized Riemann problems in computational fluid dynamics* (Cambridge Univ. Press, New York, 2003).
8. H. Nessyahu, E. Tadmor, *J. Comput. Phys.* **87**, 408–463, 1990.

BOLTZMANN-TYPE EQUATIONS FOR CHAIN REACTIONS MODELING

F. CONFORTO

Dipartimento di Matematica
Università di Messina
E-mail: fiamma@unime.it

R. MONACO, M. PANDOLFI BIANCHI

Dipartimento di Matematica
Politecnico di Torino
E-mail: roberto.monaco@polito.it, miriam.pandolfi@polito.it

The main objective of the present paper consists in showing how a Boltzmann-type model, recently proposed, can be adapted to describe, at the kinetic level, chain reactions in a gaseous mixture of several species where more than one reversible reaction takes place. Comparisons between the results obtained by such a model and those recovered by both classical hydrodynamic equations and quasi-steady-state approximation are then performed.

1. Introduction

In the last years great interest has been given to BGK [1] kinetic models extended to reactive gas mixtures [6,8,9]. The advantages of these models consist mainly in

- simulation of time-space evolution of the system at the kinetic level, that is knowledge, at every time-step and space-point, of the distribution functions of each species involved in the reaction, without heavy numerical computations;
- capability of taking into account different flow regimes [5], where either slow reactions, for which the relaxation of the chemical process is much slower than the elastic one, either fast reactions, for which the order of magnitude of chemical and elastic relaxation time is the same, occur in the mixture.

As already mentioned, a BGK-type model suitable to represent, at the kinetic level, a reactive gas mixture undergoing slow and fast chemical reactions has been proposed in paper [9]. Such a model presents a relatively simple mathematical structure of the collision terms with respect to the other BGK-type models recently derived [6,8], and it has been tested with several simulations at the kinetic scale in a paper [10] presented at the last Wascom conference in 2005. In both papers [9,10] the model equations and numerical results refer to a simple reversible bi-molecular reaction, i.e. $A_1 + A_2 \rightleftharpoons A_3 + A_4$. Let us underline that this model includes in the collision term an equilibrium distribution which depends only on the number density of each species and on the bulk velocity and temperature of the whole mixture. Such a peculiarity allows the model to be more handy for simulations and more convenient for the extensions here proposed.

In fact, the principal aim of the present paper consists in showing that the model [9] can be rather easily applied to the case of a chemical process where chain reactions occur with an arbitrary number of chemical species. In particular, the following two types of chains of elementary reversible reactions will be considered

$$A + B \rightleftharpoons C + D, \qquad E + B \rightleftharpoons A + D, \tag{1}$$

$$A + B \rightleftharpoons C + D, \qquad E + B \rightleftharpoons A + C, \qquad B + F \rightleftharpoons D + G, \tag{2}$$

which, for instance, take place in the oxygen-nitrogen and in the chlorine-iodine systems, respectively. In order to test the reliability of the the proposed model, some comparisons will be proposed, at the macroscopic scale, with the results performed by hydrodynamic equations (see [2]). In particular, in this last paper a procedure to obtain the quasi-steady-state approximation (QSSA) hydrodynamic equations has been proposed for one reversible bi-molecular reaction. Here, this procedure is developed to include also the chain reaction (1) and simulations are then provided in the spatial homogeneous case. Finally, other simulations and comparisons are performed for the chain reaction (2).

2. The basic model

With reference to paper [9], in the spatial homogeneous case and for one reversible chemical reaction $A_1 + A_2 \rightleftharpoons A_3 + A_4$, the BGK-type model has the following general form

$$\partial_t f_i = Q_i[f_i; \widetilde{f_i}] + R_i[f_i; \underline{\widetilde{f}}], \quad i = 1, \ldots, 4, \tag{3}$$

where Q_i and R_i are, respectively, the elastic scattering and reactive interaction operators, and $\underline{\widetilde{f}}(t,\mathbf{v}) = \{\widetilde{f}_1, \ldots, \widetilde{f}_4\}(t,\mathbf{v})$, $t \in \mathbb{R}_+$, $\mathbf{v} \in \mathbb{R}^3$, is the equilibrium distribution given by

$$\widetilde{f}_i = n_i \left(\frac{m_i}{2\pi k_B T}\right)^{3/2} \exp\left(-\frac{m_i(\mathbf{v}-\mathbf{u})^2}{2k_B T}\right). \tag{4}$$

Let us recall that model (3) admits thermodynamical equilibrium when the r.h.s. vanishes and that this occurs when

$$f_i = \widetilde{f}_i, \ \forall i = 1, \ldots, 4, \qquad \frac{n_1 n_2}{n_3 n_4} = M^{\frac{3}{2}} \exp\left(\frac{\Delta E}{k_B T}\right), \tag{5}$$

where the n_i are the species number densities, m_i the molecular masses, T is the temperature, $M = m_1 m_2/(m_3 m_4)$ the reduced mass and $\Delta E = E_3 + E_4 - E_1 - E_2 > 0$ the binding energies difference. Moreover, the model satisfies the equalities

$$\sum_{i=1}^{4} \int_{\mathbb{R}^3} \psi_i^{(k)}(\mathbf{v}) \left(Q_i[f_i; \widetilde{f}_i] + R_i[f_i; \underline{\widetilde{f}}]\right) d\mathbf{v} = 0, \ k = 1, \ldots, 7$$

where

$$\begin{aligned}
\psi^{(1)} &= \{1,0,1,0\}, \quad \psi^{(2)} = \{1,0,0,1\}, \quad \psi^{(3)} = \{0,1,0,1\} \\
\psi^{(4)} &= v_x\{m_1, m_2, m_3, m_4\}, \quad \psi^{(5)} = v_y\{m_1, m_2, m_3, m_4\} \\
\psi^{(6)} &= v_z\{m_1, m_2, m_3, m_4\} \\
\psi^{(7)} &= \left\{\frac{1}{2}m_1 v^2 + E_1, \ \frac{1}{2}m_2 v^2 + E_2, \ \frac{1}{2}m_3 v^2 + E_3, \ \frac{1}{2}m_4 v^2 + E_4\right\}
\end{aligned}$$

are the physical collision invariants, so that conservation of mass, momentum and total energy is assured. Note that the first three invariants are those due to conservation of partial number densities $n_1 + n_3$, $n_1 + n_4$, $n_2 + n_4$.

According to [9], choosing a suitable form [3] of the elastic and reactive collision frequencies, the model equation (3) is detailed in

$$\partial_t f_i = 4\pi \sum_{j=1}^{4} \alpha_{ij} n_j \left[\widetilde{f}_i(\mathbf{v}) - f_i(\mathbf{v})\right] + \gamma_i \widetilde{f}_i(\mathbf{v}) - \sigma_i f_i(\mathbf{v}), \tag{6}$$

where

$$\begin{aligned}
&\gamma_1 = A n_3 n_4/n_1, \quad \gamma_2 = A n_3 n_4/n_2, \quad \gamma_3 = B n_1 n_2/n_3, \quad \gamma_4 = B n_1 n_2/n_4 \\
&\sigma_1 = B n_2, \qquad \sigma_2 = B n_1, \qquad \sigma_3 = A n_4, \qquad \sigma_4 = A n_3 \\
&A(T) = 4\pi\beta, \qquad B(T) = A(T)\left(\frac{1}{M}\right)^{3/2} \exp\left(-\frac{\Delta E}{k_B T}\right),
\end{aligned}$$

α_{ij} being the elastic cross-sections and β a scalar factor characterizing the reaction velocity. Hereinafter, for simplicity, t will be dropped from the argument of f_i, as already done in (6).

3. Comparison with the QSSA equations

Consider the elementary reactions of the oxygen-nitrogen system

$$NO + O \rightleftharpoons O_2 + N \qquad (A + B \rightleftharpoons C + D) \tag{7}$$

$$N_2 + O \rightleftharpoons NO + N \qquad (E + B \rightleftharpoons A + D) \tag{8}$$

and introduce the following quantities

$$A_1 = 4\pi\beta_1, \qquad B_1(T) = A_1 \left(\frac{1}{M_1}\right)^{\frac{3}{2}} \exp\left(-\frac{\Delta E_1}{k_B T}\right)$$

$$A_2 = 4\pi\beta_2, \qquad B_2(T) = A_2 \left(\frac{1}{M_2}\right)^{\frac{3}{2}} \exp\left(-\frac{\Delta E_2}{k_B T}\right),$$

where $M_1 = m_A m_B/(m_C m_D)$, $M_2 = m_E m_B/(m_A m_D)$, $\Delta E_1 = E_C + E_D - E_A - E_B$ and $\Delta E_2 = E_A + E_D - E_E - E_B$.
Then, following the same procedure of paper [9] applied, this time, to reactions (7)-(8), one obtains the following BGK-type equations

$$\partial_t f_i = 4\pi \sum_{j=A}^{E} \alpha_{ij} n_j (\widetilde{f}_i - f_i) + (\gamma_1^i + \gamma_2^i)\widetilde{f}_i - (\sigma_1^i + \sigma_2^i) f_i, \; i = A, \ldots, E \tag{9}$$

where

$$\gamma_1^A = A_1 n_C n_D / n_A, \; \gamma_2^A = B_2 n_B n_E / n_A, \; \sigma_1^A = B_1 n_B, \; \sigma_2^A = A_2 n_D$$
$$\gamma_1^B = A_1 n_C n_D / n_B, \; \gamma_2^B = A_2 n_A n_D / n_B, \; \sigma_1^B = B_1 n_A, \; \sigma_2^B = B_2 n_E$$
$$\gamma_1^C = B_1 n_A n_B / n_C, \; \gamma_2^C = 0, \; \sigma_1^C = A_1 n_D, \; \sigma_2^C = 0$$
$$\gamma_1^D = B_1 n_A n_B / n_D, \; \gamma_2^D = B_2 n_B n_E / n_D, \; \sigma_1^D = A_1 n_C, \; \sigma_2^D = A_2 n_A$$
$$\gamma_1^E = 0, \; \gamma_2^E = A_2 n_A n_D / n_E, \; \sigma_1^E = 0, \; \sigma_2^E = B_2 n_B.$$

Now, according to the procedure outlined in [2], the QSSA equations for the system undergoing reactions (7)-(8) will be derived.
Re-write these reactions in the form

$$A + B \rightleftharpoons X, \qquad X \rightleftharpoons C + D$$
$$E + B \rightleftharpoons Y, \qquad Y \rightleftharpoons A + D$$

where X and Y are much more unstable species than $A, \ldots, E$.

For such a chain we can write the hydrodynamical equations in the spatial homogeneous case as

$$\begin{aligned}
\dot{n}_A &= \frac{1}{\epsilon_1} n_X + \frac{1}{\epsilon_2} n_Y - k_1 n_A n_B - k_2 n_A n_D \\
\dot{n}_B &= \frac{1}{\epsilon_1} n_X + \frac{1}{\epsilon_2} n_Y - k_1 n_A n_B - k_3 n_E n_B \\
\dot{n}_C &= \frac{1}{\epsilon_1} n_X - k_4 n_C n_D \\
\dot{n}_D &= \frac{1}{\epsilon_1} n_X + \frac{1}{\epsilon_2} n_Y - k_2 n_A n_D - k_4 n_C n_D \\
\dot{n}_E &= \frac{1}{\epsilon_1} n_X - k_3 n_E n_B \\
\dot{n}_X &= k_1 n_A n_B + k_4 n_C n_D - \frac{2}{\epsilon_1} n_X \\
\dot{n}_Y &= k_2 n_A n_D + k_3 n_E n_B - \frac{2}{\epsilon_2} n_Y,
\end{aligned}$$

where the reaction rates are such that $\frac{1}{\epsilon_1}$, $\frac{1}{\epsilon_2} >> k_1,\ k_2,\ k_3,\ k_4$, because of the fast disappearance of the species X and Y. Therefore, it follows $\dot{n}_X = \dot{n}_Y = 0$, so that from the above hydrodynamic equations one can derive

$$\epsilon_1 = \frac{2n_X}{k_1 n_A n_B + k_4 n_C n_D}, \qquad \epsilon_2 = \frac{2n_Y}{k_2 n_A n_D + k_3 n_E n_B},$$

and obtain the QSSA hydrodynamic equations of the system in the form

$$\dot{n}_A = \frac{1}{2}\,[k_3 n_E n_B + k_4 n_C n_D - k_1 n_A n_B - k_2 n_A n_D] \tag{10}$$

$$\dot{n}_B = \frac{1}{2}\,[k_2 n_A n_D + k_4 n_C n_D - k_1 n_A n_B - k_3 n_E n_B] \tag{11}$$

$$\dot{n}_C = \frac{1}{2}\,[k_1 n_A n_B - k_4 n_C n_D] \tag{12}$$

$$\dot{n}_D = \frac{1}{2}\,[k_1 n_A n_B + k_3 n_E n_B - k_2 n_A n_D - k_4 n_C n_D] \tag{13}$$

$$\dot{n}_E = \frac{1}{2}\,[k_2 n_A n_D - k_3 n_E n_B]. \tag{14}$$

In order to show a comparison between the behavior, at a macroscopic level, of the chemical process (7)–(8), as provided by the BGK and QSSA models, it is necessary to identify the reaction rates $k_1, \dots, k_4$ with those obtained by the moments $n_A, \dots, n_E$ and T computed by means of the distributions in the kinetic equations (9).

Number densities versus time for BGK and QSSA equations

	t=0	t_{fin}(BGK)	t_{fin}(QSSA)
NO:	0.8837	0.1617	0.1558
O:	0.5293	1.2057	1.2035
O_2:	0.3536	0.3764	0.3805
N:	0.7055	0.0291	0.0314
N_2:	1.0602	1.7594	1.7612

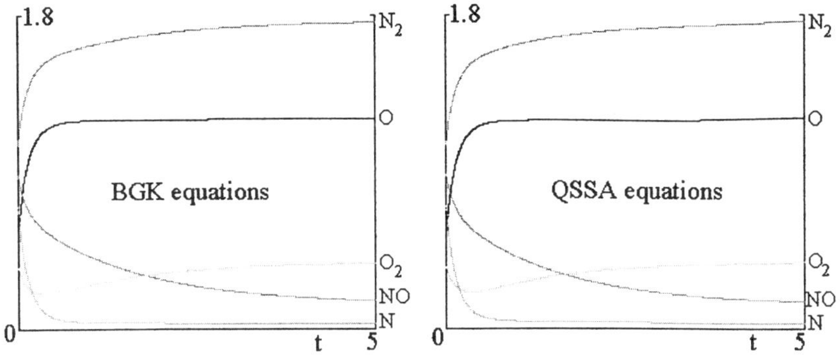

Fig. 1. Comparison between BGK and QSSA equations

After some simple computations such an identification leads to

$$k_1 = 2B_1, \quad k_2 = 2A_2, \quad k_3 = 2B_2, \quad k_4 = 2A_1. \tag{15}$$

In Fig.1 the time evolution of the number densities obtained from the numerical integration of (9), and the consequent computation of the moments (left picture), is compared to that performed by (10)-(14) (right picture).

As shown in the figure, the agreement of results, when the mixture has reached at a time t_{fin} an acceptable chemical equilibrium, is quite satisfactory, except for the atoms of nitrogen which presents a relative error of the order of 7%.

4. Comparison with the hydrodynamic equations

Consider now the elementary reactions of the $H - I - C\ell$ chain

$$HI + HC\ell \rightleftharpoons IC\ell + H_2 \qquad (A + B \rightleftharpoons C + D) \tag{16}$$

$$I_2 + HC\ell \rightleftharpoons IC\ell + HI \qquad (E + B \rightleftharpoons C + A) \tag{17}$$

$$HC\ell + H \rightleftharpoons C\ell + H_2 \qquad (B + F \rightleftharpoons G + D). \tag{18}$$

Following the same line as in the previous section, let us introduce the quantities A_k and $B_k(T)$, $k = 1, 2, 3$, where now $M_1 = m_A m_B/(m_C m_D)$,

$M_2 = m_E m_B/(m_C m_A)$, $M_3 = m_B m_F/(m_G m_D)$, $\Delta E_1 = E_C + E_D - E_A - E_B$, $\Delta E_2 = E_C + E_A - E_B - E_E$ and $\Delta E_3 = E_G + E_D - E_B - E_F$. Then, the kinetic BGK model equations, for $i = A, \ldots, G$, this time, assume the form

$$\partial_t f_i = 4\pi \sum_{j=A}^{E} \alpha_{ij} n_j (\widetilde{f_i} - f_i) + (\gamma_1^i + \gamma_2^i + \gamma_3^i)\widetilde{f_i} - (\sigma_1^i + \sigma_2^i + \sigma_3^i) f_i \qquad (19)$$

where the coefficients are defined by

$$\begin{aligned}
\gamma_1^A &= A_1 n_C n_D / n_A, \quad \gamma_2^A = B_2 n_B n_E / n_A, \quad \gamma_3^A = 0 \\
\sigma_1^A &= B_1 n_B, \quad \sigma_2^A = A_2 n_C, \quad \sigma_3^A = 0 \\
\gamma_1^B &= A_1 n_C n_D / n_B, \quad \gamma_2^B = A_2 n_A n_C / n_B, \quad \gamma_3^B = A_3 n_D n_G / n_B \\
\sigma_1^B &= B_1 n_A, \quad \sigma_2^B = B_2 n_E, \quad \sigma_3^B = B_3 n_F \\
\gamma_1^C &= B_1 n_A n_B / n_C, \quad \gamma_2^C = B_2 n_B n_E / n_C, \quad \gamma_3^C = 0 \\
\sigma_1^C &= A_1 n_D, \quad \sigma_2^C = A_2 n_A, \quad \sigma_3^C = 0 \\
\gamma_1^D &= B_1 n_A n_B / n_D, \quad \gamma_2^D = 0, \quad \gamma_3^D = B_3 n_B n_F / n_D \\
\sigma_1^D &= A_1 n_C, \quad \sigma_2^D = 0, \quad \sigma_3^D = A_3 n_G \\
\gamma_1^E &= 0, \quad \gamma_2^E = A_2 n_A n_C / n_E, \quad \gamma_3^E = 0 \\
\sigma_1^E &= 0, \quad \sigma_2^E = B_2 n_B, \quad \sigma_3^E = 0 \\
\gamma_1^F &= 0, \quad \gamma_2^F = 0, \quad \gamma_3^F = A_3 n_D n_G / n_F \\
\sigma_1^F &= 0, \quad \sigma_2^F = 0, \quad \sigma_3^F = B_3 n_B \\
\gamma_1^G &= 0, \quad \gamma_2^G = 0, \quad \gamma_3^G = B_3 n_B n_F / n_G \\
\sigma_1^G &= 0, \quad \sigma_2^G = 0, \quad \sigma_3^G = A_3 n_D.
\end{aligned}$$

A numerical comparison with the results provided by macroscopic equations is now proposed in order to test the model (19). For this purpose, the hydrodynamic time evolution equations for a chain reaction of type (16)-(18) can be easily derived from well known literature, see book [5]. Introducing the chemical source/sink terms

$$\begin{aligned}
s_1 &= s_1(n_A, n_B, n_C, n_D, T) = A_1 n_C n_D - B_1(T) n_A n_B \\
s_2 &= s_2(n_A, n_B, n_C, n_E, T) = A_2 n_A n_C - B_2(T) n_B n_E \\
s_3 &= s_3(n_B, n_D, n_F, n_G, T) = A_3 n_D n_G - B_3(T) n_B n_F,
\end{aligned}$$

the hydrodynamic equations for the number densities are given by

Number densities versus time for BGK and hydrodynamic equations

	t=0	t_{fin} (BGK)	t_{fin} (Hyd.)
HI:	1.2654	0.8472	0.8426
HCl:	0.6408	1.0743	1.0546
ICl:	0.6959	0.0182	0.0180
H_2:	0.1765	0.3021	0.3131
I_2:	0.6309	1.1889	1.1812
H:	0.4405	0.1619	0.1764
Cl:	0.7430	1.0003	1.0071

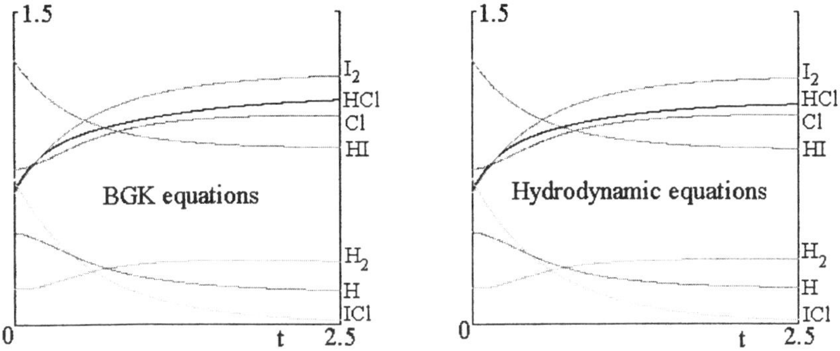

Fig. 2. Comparison between BGK and hydrodynamic equations

$$\begin{aligned} \dot{n}_A &= s_1 - s_2, & \dot{n}_B &= s_1 + s_2 + s_3, & & \\ \dot{n}_C &= -s_1 - s_2, & \dot{n}_D &= -s_1 - s_3, & & \qquad (20) \\ \dot{n}_E &= s_2, & \dot{n}_F &= s_3, & \dot{n}_G &= -s_3. \end{aligned}$$

Since the source/sink terms depend on temperature, an evolution equation of this quantity is needed. In the spatial homogeneous case [4] the equation for T can be easily derived and, in this case, assumes the form

$$\dot{T} = \frac{2}{3nk}(s_1 \Delta E_1 + s_2 \Delta E_2 + s_3 \Delta E_3), \qquad n = \sum_{i=A}^{G} n_i,$$

n being the number density of the whole mixture.

In the same way as in the previous section, the number densities of the species, appearing in the chain (16)-(18), are computed from the moments of the distributions satisfying equations (19), as well as from the system (20). In Fig.2 the time evolution of the species densities are compared (BGK approximation in the left picture and hydrodynamic equations in the right one). Again the agreement between the results of the BGK-type and hydrodynamic equations is quite acceptable. Nevertheless a relative error at

t_{fin} of the order of 8% and 3.5% for the atoms and molecules of hydrogen, respectively, can be noticed.

As a final comment, let us underline that the proposed BGK model provides less accurate results for the evolution of the lightest species of the mixture. This drawback, which has been pointed out in several papers, is confirmed also in book [7] (p. 169-170) and seems to be peculiar of BGK-type models for mixtures formed by chemical components of disparate masses.

Acknowledgements. This piece of research has been supported by the Project PRIN 2005 "Nonlinear Propagation and Stability in Thermodynamical Processes of Continuous Media", coordinator T. Ruggeri, to whom this paper is dedicated.

References

1. P.L. Bhatnagar, E.P. Gross, K. Krook, A model for collision processes in gases, *Phys. Rev.*, **94**, 511–524, 1954.
2. M. Bisi, F. Conforto, L. Desvillettes, Quasi-steady-state approximation for reaction-diffusion equation, *Bull. Inst. Math., Academia Sinica*, to appear in 2007.
3. F. Conforto, R. Monaco, F. Schürrer, I. Ziegler, Steady detonation waves via the Boltzmann equation for a reacting mixture, *J. Phys. A: Math. Gen.*, **36**, 5381-5398, 2003.
4. F. Conforto, A. Jannelli, R. Monaco, T. Ruggeri, On the Riemann problem for a system of balance laws modelling a reactive gas mixture, *Physica A*, **373**, 67-87, 2007.
5. V. Giovangigli, *Multicomponent Flow Modeling*, Boston, USA, Birkhäuser, 1999.
6. M. Groppi, G. Spiga, A BGK-type type approach for chemically reacting gas mixture, *Phys. Fluids*, **16**, 4273–4284, 2004.
7. S. Harris, *An Introduction to the Theory of the Boltzmann Equation*, Dover, New York, 2004.
8. G. Kremer, M. Pandolfi Bianchi, A.J. Soares, A relaxation kinetic model for transport phenomena in a reactive flow, *Phys. Fluids*, **18**, 1-15, 2006.
9. R. Monaco, M. Pandolfi Bianchi, A.J. Soares, BGK-type models in strong reaction and kinetic chemical equilibrium regimes, *J. Phys. A: Math. Gen.*, **38**, 10413–10431, 2005.
10. R. Monaco, M. Pandolfi Bianchi, A.J. Soares, Simulations at a kinetic scale of relaxation models for slow and fast chemical reactions, in *Waves and Stability in Continuous Media*, Eds. R. Monaco et al., World Scientific, Singapore, 378–387, 2006.

ANALYTIC STRUCTURE OF THE FOUR-WAVE MIXING MODEL IN PHOTOREACTIVE MATERIAL

ROBERT CONTE

Service de physique de l'état condensé (CNRS URA 2464)
CEA–Saclay, F–91191 Gif-sur-Yvette Cedex, France
E-mail: Robert.Conte@cea.fr

SVETLANA BUGAYCHUK

Institute of Physics of the National Academy of Sciences of Ukraine
46 Prospect Nauki, Kiev-39, UA 03039, Ukraine
E-mail: bugaich@iop.kiev.ua

In order to later find explicit analytic solutions, we investigate the singularity structure of a fundamental model of nonlinear optics, the four-wave mixing model in one space variable z. This structure is quite similar, and this is not a surprise, to that of the cubic complex Ginzburg-Landau equation. The main result is that, in order to be single valued, time-dependent solutions should depend on the space-time coordinates through the reduced variable $\xi = \sqrt{z}e^{-t/\tau}$, in which τ is the relaxation time.

Keywords: four-wave mixing model; Painlevé test;

1. The four-wave mixing model

Dynamic holography relies on the ability of the refractive index of a nonlinear media to locally change under the action of light, i.e. under the so-called photorefractive response. The most conventional scheme for dynamic holographic applications consists in the formation of a fringe interference pattern between the interacting waves (the light grating), which creates the refractive index grating. During the wave interaction, these are the same waves which create the grating pattern and diffract on this grating. There exist many applications based on dynamic holography with photorefractive crystals [1].

In the four-wave mixing model, the grating, i.e. the modulation of the refractive index induced by the light interference pattern, is created by two pairs of co-propagating waves, 1–2 and 3–4. Let us denote A_1, A_2, A_3, A_4

the complex amplitudes of the four waves, $\Delta\varepsilon$ the complex amplitude of the index grating, τ the characteristic time of the evolution, I_0 the total intensity of light (a real positive constant), γ the photorefractive coupling (a complex constant). The four-wave mixing model is the system of five complex partial differential equations obeyed by these five complex amplitudes [2] (bar denotes complex conjugation)

$$\begin{cases} \partial_z A_1 = -i\Delta\varepsilon A_2,\ \partial_z \overline{A}_2 = i\Delta\varepsilon \overline{A}_1,\ \partial_z \overline{A}_3 = -i\Delta\varepsilon \overline{A}_4,\ \partial_z A_4 = i\Delta\varepsilon A_3, \\ \left(\partial_t + \dfrac{1}{\tau}\right)\Delta\varepsilon = \dfrac{\gamma}{I_0}\left(A_1\overline{A}_2 + \overline{A}_3 A_4\right). \end{cases} \tag{1}$$

The correspondence of notation with Ref. [3] is

$$(A_1, A_2, A_3, A_4, Q, \gamma) \text{ of Ref.} = (A_1, \overline{A}_4, \overline{A}_3, A_2, -i\Delta\varepsilon, i\gamma)_{\text{here}}. \tag{2}$$

At present time, no analytic solution is known to this system, except a stationary solution in which the grating amplitude has a pulse profile [1],

$$\Delta\varepsilon = K \operatorname{sech} k(z - z_0),\ (K, k) = \text{real constants}. \tag{3}$$

The purpose of this short article is to investigate the structure of the singularities of the time-dependent solutions ($\partial_t \neq 0$) in order to find, in a forthcoming paper, closed form analytic solutions by making suitable assumptions dictated by the singularity structure.

The first four equations (1), which do not depend on ∂_t, admit six quadratic first integrals. Two of them,

$$|A_1|^2 + |A_2|^2 = f_1^2(t),\ |A_3|^2 + |A_4|^2 = f_2^2(t), \tag{4}$$

allow us to represent the four complex amplitudes $A_j(z,t)$ with two real functions f_1, f_2 of t and six real functions $\theta_1, \theta_2, \varphi_1, \varphi_2, \varphi_3, \varphi_4$ of (z,t),

$$A_1 = f_1 \sin\theta_1 e^{i\varphi_1}, A_2 = f_1 \cos\theta_1 e^{i\varphi_2}, A_3 = f_2 \cos\theta_2 e^{i\varphi_3}, A_4 = f_2 \sin\theta_2 e^{i\varphi_4} \tag{5}$$

The four other quadratic first integrals are pairwise complex conjugate

$$\begin{cases} A_1 A_3 + A_2 A_4 = f_3(t),\ \overline{A}_1\overline{A}_3 + \overline{A}_2\overline{A}_4 = \overline{f_3(t)}, \\ \overline{A}_1 A_4 - \overline{A}_2 A_3 = f_4(t),\ A_1\overline{A}_4 - A_2\overline{A}_3 = \overline{f_4(t)}, \end{cases} \tag{6}$$

and constrained by one relation,

$$-f_1^2 f_2^2 + |f_3|^2 + |f_4|^2 = 0. \tag{7}$$

2. Local singularity analysis of the FWM model

The dependent variables $A_j, \Delta\varepsilon$ generically present movable singularities, i.e. singularities whose location in the complex domain depends on the initial conditions. The study of the behaviour of $A_j, \Delta\varepsilon$ near these movable singularities is a prerequisite to the possible obtention of closed form analytic solutions to the nonlinear system (1), so let us first perform it. Such a study is made by performing the successive steps of the so-called *Painlevé test*, a procedure explained in detail in Ref. [4,5].

The movable singularities of a PDE lay on a manifold represented by the equation [6]

$$\varphi(z,t) - \varphi_0 = 0, \tag{8}$$

in which φ is an arbitrary function of the independent variables, and φ_0 an arbitrary movable constant. In order to express the behaviour of the ten dependent variables $A_j, \overline{A}_j, \Delta\varepsilon, \Delta\bar{\varepsilon}$, it is convenient to introduce an expansion variable $\chi(z,t)$ [4] which vanishes as $\varphi - \varphi_0$ and which is defined by its gradient

$$\chi_z = 1 + \frac{S}{2}\chi^2, \; \chi_t = -C + C_z\chi - \frac{1}{2}(CS + C_{zz})\chi^2. \tag{9}$$

with the cross-derivative condition

$$X \equiv S_t + C_{zzz} + 2C_z S + CS_z = 0. \tag{10}$$

Since they will be needed later, we give the dependence of χ, S, C on φ,

$$\chi = \frac{\varphi - \varphi_0}{\varphi_z - \dfrac{\varphi_{zz}}{2\varphi_z}(\varphi - \varphi_0)} = \left[\frac{\varphi_z}{\varphi - \varphi_0} - \frac{\varphi_{zz}}{2\varphi_z}\right]^{-1}, \; \varphi_z \neq 0, \tag{11}$$

$$S = \{\varphi; z\} = \frac{\varphi_{zzz}}{\varphi_z} - \frac{3}{2}\left(\frac{\varphi_{zz}}{\varphi_z}\right)^2, \; C = -\varphi_t/\varphi_z. \tag{12}$$

2.1. *Leading behaviour*

The first step is to determine all possible families of noncharacteristic (i.e. $\varphi_x\varphi_t \neq 0$) movable singularities, i.e. all the leading behaviours

$$\chi \to 0: \; \begin{cases} A_k \sim a_k\chi^{p_k}, \; \overline{A}_k \sim b_k\chi^{P_k}, \; a_k b_k \neq 0, \; k = 1,2,3,4, \\ \Delta\varepsilon \sim q_0\chi^{\alpha}, \; \Delta\bar{\varepsilon} \sim r_0\chi^{\beta}, \; q_0 r_0 \neq 0, \end{cases} \tag{13}$$

in which the exponents p_k, P_k, α, β are not all positive integers. If one assumes that the two terms $A_1\overline{A}_4$ and $A_3\overline{A}_2$ have the same singularity order, the ten exponents p_k, P_k, α, β obey the twelve linear equations,

$$\begin{cases} p_1 - 1 = p_2 + \alpha, \quad P_2 - 1 = P_1 + \alpha, \\ p_4 - 1 = p_3 + \alpha, \quad P_3 - 1 = P_4 + \alpha, \\ P_1 - 1 = P_2 + \beta, \quad p_2 - 1 = p_1 + \beta, \\ P_4 - 1 = P_3 + \beta, \quad p_3 - 1 = p_4 + \beta, \\ \alpha - 1 = p_1 + P_2 = p_4 + P_3, \\ \beta - 1 = P_1 + p_2 = P_4 + p_3, \end{cases} \tag{14}$$

and the ten coefficients a_k, b_k, q_0, r_0 obey the ten nonlinear equations

$$\begin{cases} p_1 a_1 = -q_0 a_2, \ p_2 a_2 = -r_0 a_1, \ p_4 a_4 = -q_0 a_3, \ p_3 a_3 = -r_0 a_4, \\ P_1 b_1 = r_0 b_2, \ P_2 b_2 = q_0 b_1, \ P_4 b_4 = r_0 b_3, \ P_3 b_3 = q_0 b_4, \\ C\alpha q_0 = -(\gamma/I_0)(a_1 b_2 + a_4 b_3), \\ C\beta r_0 = -(\bar{\gamma}/I_0)(b_1 a_2 + b_4 a_3). \end{cases} \tag{15}$$

Therefore the five squared moduli $|A_k|^2, |\Delta\varepsilon|^2$ behave like double poles,

$$p_k + P_k = -2, \ k = 1, 2, 3, 4, \ \alpha + \beta = -2, \tag{16}$$

and the linear system (14) is solved as

$$\begin{cases} p_1 = -1 + s_1 + \delta, \ P_1 = -1 - s_1 - \delta, \\ p_2 = -1 + s_1 - \delta, \ P_2 = -1 - s_1 + \delta, \\ p_4 = -1 + s_2 + \delta, \ P_4 = -1 - s_2 - \delta, \\ p_3 = -1 + s_2 - \delta, \ P_3 = -1 - s_2 + \delta, \\ \alpha = -1 + 2\delta, \ \beta = -1 - 2\delta, \end{cases} \tag{17}$$

in which δ, s_1, s_2 are to be determined by the nonlinear system (15).

The first set of relations

$$-q_0 r_0 = p_1 p_2 = p_4 p_3 = P_1 P_2 = P_4 P_3, \tag{18}$$

imply

$$s_1 = 0, \ s_2 = 0, \ -q_0 r_0 = 1 - \delta^2. \tag{19}$$

The second set of relations

$$\frac{a_1}{a_2} = -\frac{a_4}{a_3} = -\frac{b_2}{b_1} = \frac{b_3}{b_4} = \frac{i q_0}{1 - \delta} = \frac{1 + \delta}{i r_0} \tag{20}$$

is solved as

$$\begin{cases} a_1 = a_{12}\lambda, \ b_2 = -b_{12}\lambda, \ a_4 = a_{34}\lambda, \ b_3 = b_{34}\lambda, \ i q_0 = (1 - \delta)\lambda^2, \\ a_2 = a_{12}\lambda^{-1}, \ b_1 = b_{12}\lambda^{-1}, \ a_3 = -a_{34}\lambda^{-1}, \ b_4 = b_{34}\lambda^{-1}, \ i r_0 = (1 + \delta)\lambda^{-2}, \end{cases} \tag{21}$$

in which $a_{12}, a_{34}, b_{12}, b_{34}, \lambda$ are nonzero functions of (z, t) to be determined.

Finally, the two remaining equations

$$\begin{cases} a_{12}b_{12} - a_{34}b_{34} + \dfrac{I_0 C}{i\gamma}(1-\delta)(1-2\delta) = 0, \\ a_{12}b_{12} - a_{34}b_{34} - \dfrac{I_0 C}{i\bar{\gamma}}(1+\delta)(1+2\delta) = 0, \end{cases} \tag{22}$$

first define δ as a root of the second degree equation

$$\gamma(1+\delta)(1+2\delta) + \bar{\gamma}(1-\delta)(1-2\delta) = 0, \tag{23}$$

then put one constraint among $a_{12}, b_{12}, a_{34}, b_{34}$.

The equation (23) which defines δ only depends on the argument of γ,

$$(2\cos g)\delta^2 + (3i \sin g)\delta + \cos g = 0, \quad g = \arg\gamma. \tag{24}$$

When the photorefractive coupling constant γ is purely imaginary, the exponent δ vanishes, otherwise it can take two purely imaginary values

$$\delta = i\frac{-3\sin g \pm \sqrt{9\sin^2 g + 8\cos^2 g}}{4\cos g}. \tag{25}$$

To conclude, there generically exist two families (13) of movable singularities, defined by the equations (17), (19), (21), (22), i.e.

$$\begin{cases} \delta = \text{ one the two roots of Eq. (24)}, \\ p_1 = p_4 = P_2 = P_3 = -1 + \delta, \\ P_1 = P_4 = p_2 = p_3 = -1 - \delta, \\ a_1 = \ \ N\lambda \ \ p_{12} \ \cosh\mu, \ b_2 = -N\lambda \ \ p_{12}^{-1} \cosh\mu, \\ a_4 = \ \ N\lambda \ \ p_{34} \ \sinh\mu, \ b_3 = \ \ N\lambda \ \ p_{34}^{-1} \sinh\mu, \\ a_2 = \ \ N\lambda^{-1} p_{12} \ \cosh\mu, \ b_1 = \ \ N\lambda^{-1} p_{12}^{-1} \cosh\mu, \\ a_3 = -N\lambda^{-1} p_{34} \ \sinh\mu, \ b_4 = \ \ N\lambda^{-1} p_{34}^{-1} \sinh\mu, \\ iq_0 = (1-\delta)\lambda^2, \ ir_0 = (1+\delta)\lambda^{-2}, \\ N^2 = -\dfrac{I_0 C}{iG}(1-\delta^2)(1-4\delta^2), \\ G = (1+\delta)(1+2\delta)\gamma = -(1-\delta)(1-2\delta)\bar{\gamma}. \end{cases} \tag{26}$$

and they depend on four arbitrary complex functions $\lambda, \mu, p_{12}, p_{34}$ of (z,t).

2.2. *Fuchs indices*

The leading behaviour (13) is the first term of a Laurent series

$$\chi \to 0: \begin{cases} A_k = \chi^{p_k} \displaystyle\sum_{j=0}^{+\infty} a_{k,j}\chi^j, \ \Delta\varepsilon = \chi^\alpha \sum_{j=0}^{+\infty} q_j \chi^j, \\ \overline{A}_k = \chi^{P_k} \displaystyle\sum_{j=0}^{+\infty} b_{k,j}\chi^j, \ \Delta\bar{\varepsilon} = \chi^\beta \sum_{j=0}^{+\infty} r_j \chi^j, \end{cases} \tag{27}$$

and the indices j at which arbitrary coefficients enter this expansion are computed as follows. If one symbolically denotes $E(A_k, \overline{A}_k, \Delta\varepsilon, \Delta\bar{\varepsilon}) = 0$ the ten equations (1), one builds the linearized system of (1),

$$\lim_{\zeta \to 0} \frac{E(A_k + \zeta C_k, \overline{A}_k + \zeta \overline{C}_k, \Delta\varepsilon + \zeta D, \Delta\bar{\varepsilon} + \zeta \overline{D}) - E(A_k, \overline{A}_k, \Delta\varepsilon, \Delta\bar{\varepsilon})}{\zeta} = 0,$$

i.e.

$$\begin{cases} \partial_z C_1 + i\Delta\varepsilon C_2 + iA_2 D = 0, \ \dots \\ \left(\partial_t + \frac{1}{\tau}\right) D - \frac{\gamma}{I_0}\left(\overline{A}_2 C_1 + A_1 \overline{C}_2 + \overline{A}_3 C_4 + A_4 \overline{C}_3\right) = 0. \end{cases} \tag{28}$$

At the point (13), this ten-dimensional linear system displays near $\chi = 0$ a singularity which has the Fuchsian type [7, Chap. 15], therefore it admits a solution

$$\chi \to 0: \ \begin{cases} C_k \sim c_k \chi^{p_k + j}, \ \overline{C}_k \sim d_k \chi^{P_k + j}, \ k = 1, 2, 3, 4, \\ D \sim s_0 \chi^{\alpha + j}, \ \overline{D} \sim t_0 \chi^{\beta + j}. \end{cases} \tag{29}$$

The condition that this solution be nonidentically zero results in the vanishing of a tenth order determinant whose roots j, called Fuchs indices, take the ten values

$$j = -1, 0, 0, 0, 0, 2, 2, 2, \frac{5 \pm \sqrt{1 + 48\delta^2}}{2}. \tag{30}$$

In the two-wave mixing case, the four roots $0, 0, 2, 2$ are to be removed from this list. The necessary condition, required by the Painlevé test, that all indices be integer is violated when δ is nonzero, therefore the FWM model fails the test in this case. This failure is quite similar to what occurs in the complex cubic Ginzburg-Landau equation [8], and conclusions similar to those of Ref. [9] can probably be drawn.

Let us remark that the Fuchs index 0 has multiplicity four, in agreement with the number of arbitrary functions involved in the leading behaviour, see Eq. (26).

2.3. *Conditions at the Fuchs indices*

The coefficients $a_{k,j}, b_{k,j}, q_j, r_j$ of the Laurent series (27) are computed for $j \geq 1$ by solving a linear system. Whenever j reaches an integer Fuchs index, an obstruction may occur, resulting in the introduction of a movable logarithmic branching, and some necessary conditions need to be satisfied in order to avoid it. In the generic case $\delta \neq 0$, these obstructions may only arise at $j = 2$, value of a triple Fuchs index. In the nongeneric case $\delta = 0$, in addition to the quadruple index $j = 2$, one must also check the simple index $j = 3$. The results are as follows.

At $j = 2$, whatever be δ, a movable logarithm exists, unless the following necessary condition is satisfied,

$$Q_2 \equiv a\,(C_t + CC_z - 2aC) = 0,\ a = \frac{1}{\tau}. \tag{31}$$

Since $a = 0$ is excluded, the second factor defines a precise dependence of C on (z,t), hence a similar dependence for S, χ, φ and ultimately for the five complex amplitudes *via* their Laurent expansion (27). Let us find this dependence explicitly.

The general solution of (31) is provided by the method of characteristics and is defined implicitly by the relation

$$2az = C + F(e^{-2at}C), \tag{32}$$

in which F is an arbitrary function of one variable.

The invariant S is then defined by the first order linear PDE (10), which reads

$$S(x,t) = \Sigma(C,t),\ D = 1 + e^{-2at}F',$$
$$e^{2at}\left(\Sigma_t + 2aC\Sigma_C + 4a\frac{\Sigma}{D}\right) + 24a^3e^{-6at}\frac{F''^2}{D^5} - 8a^3e^{-4at}\frac{F'''}{D^4} = 0, \tag{33}$$

and admits the general solution

$$S = c_0e^{-4at}D^{-2} - 4a^2e^{-6at}D^{-3}F''' + 6a^2e^{-8at}D^{-4}F''^2, \tag{34}$$

in which c_0 is an arbitrary constant.

We have not been able to prove whether the function F is a gauge which can be arbitrarily chosen (e.g. $F = 0$) or whether it is essential. If it is essential, there could exist much more intricate solutions than those which we now outline.

When F is arbitrary, we have not yet succeeded to integrate the system (9) for $\chi(z,t)$, hence to compute the induced dependence of $A_j, \Delta\varepsilon$ on (z,t). However, if one restricts to a constant value for F, which can then be chosen equal to zero, the equation (32) for φ integrates as

$$F = 0,\ \varphi = \Phi(\xi),\ \xi = \sqrt{z}e^{-at}, \tag{35}$$

in which Φ is an arbitrary function, and it is straightforward to check that the Laurent expansion (27) defines the reduction $(z,t) \to \xi = \sqrt{z}e^{-at}$. As opposed to the stationary reduction $\partial_t = 0$, this reduction is noncharacteristic (i.e. it does not lower the differential order) it is defined with a reduced variable of the factorized type $\xi = f(z)g(t)$,

$$\begin{cases} \xi = z^{1/2}e^{-at}, \ a = 1/\tau, \\ \Delta\varepsilon(z,t) = e^{-at-i\omega t}z^{-1/2}\Delta\varepsilon_r(\xi), \ \Delta\bar{\varepsilon}(z,t) = e^{-at+i\omega t}z^{-1/2}\Delta\bar{\varepsilon}_r(\xi), \\ A_j(z,t) = e^{-at-i\omega t/2}A_{j,r}(\xi), \ \overline{A}_j(z,t) = e^{-at+i\omega t/2}\overline{A}_{j,r}(\xi), \ j = 1,4, \\ A_j(z,t) = e^{-at+i\omega t/2}A_{j,r}(\xi), \ \overline{A}_j(z,t) = e^{-at-i\omega t/2}\overline{A}_{j,r}(\xi), \ j = 2,3, \\ \dfrac{\mathrm{d}}{\mathrm{d}\xi}A_{1,r} = -2i\Delta\varepsilon_r A_{2,r}, \ \dfrac{\mathrm{d}}{\mathrm{d}\xi}\overline{A}_{2,r} = 2i\Delta\varepsilon_r \overline{A}_{1,r}, \\ \dfrac{\mathrm{d}}{\mathrm{d}\xi}\overline{A}_{3,r} = -2i\Delta\varepsilon_r \overline{A}_{4,r}, \ \dfrac{\mathrm{d}}{\mathrm{d}\xi}A_{4,r} = 2i\Delta\varepsilon_r A_{3,r}, \\ \dfrac{\mathrm{d}}{\mathrm{d}\xi}\Delta\varepsilon_r + i\omega a\xi^{-1}\Delta\varepsilon_r = -\dfrac{\gamma}{I_0}\left(A_{1,r}\overline{A}_{2,r} + \overline{A}_{3,r}A_{4,r}\right), \end{cases} \tag{36}$$

and it depends on the additional real parameter ω.

To summarize the information obtained from the Fuchs index 2, provided F can be arbitrarily chosen, the four-wave mixing model (1) (as well as the two-wave mixing model) admits no single valued dynamical solution other than the possible solutions of the reduction $(z,t) \to \xi = \sqrt{z}e^{-at}$.

When $\delta \neq 0$, this ends the Painlevé test. When $\delta = 0$, the Fuchs index 3 is found to be free of movable logarithms.

Finally, the information provided by the Painlevé test is the following.

(1) Whatever be γ, and under the mild restriction that F can be arbitrarily chosen, no single valued dynamical solution of (1) exists other than the possible solutions of the reduction $(z,t) \to \xi = \sqrt{z}e^{-at}$.
(2) For this reduction, two cases must be distinguished. When $\Re(\gamma) \neq 0$, the ten-dimensional ODE system (36) possesses at most an eight-parameter single valued solution. When $\Re(\gamma) = 0$, the system (36) passes the Painlevé test therefore it may admit a ten-parameter single valued solution, which then would be its general solution. Finding a Lax pair in this case would considerably help to perform the explicit integration.

Similar conclusions apply to the six-dimensional two-wave mixing: the only possibility for a single valued solution is the reduction (36) (with $A_3 = A_4 = 0$), this solution depending on four movable constants for $\Re(\gamma) \neq 0$, and six movable constants for $\Re(\gamma) = 0$.

3. Conclusion

The present study proves, like for the complex Ginzburg-Landau equation, that physically relevant analytic solutions quite certainly exist for the four-wave mixing model. The present counting of the possible arbitrary constants in the solutions displays, as expected, the crucial role of the photorefractive complex constant. Explicit solutions based on the present study will be presented elsewhere.

The results can be used for predicting new phenomena in optical self-diffraction of waves in photorefractive media which use the non-local response as well as for optimization of optical dynamic holographic settings. Among these applications let us quote: (i) the formation of a localized grating to increase optical information density; (ii) the methods of all-optical control of output wave characteristics versus input beam intensities and phases; (iii) the optimization of the parameters of optical phase-conjugation; (iv) the use of the new holographic topographic technique to material parameter characterization. Just like similar processes of nonlinear self-action of waves arise in models of optical networks, optical information processing, quantum information processing, interacted neural chains, then various other problems of nonlinear wave interaction can become the subject of further independent research.

Acknowledgments

We warmly acknowledge the financial support of the Max Planck Institut für Physik komplexer Systeme, and RC thanks the WASCOM organizers for invitation.

References

1. S. Bugaychuk, R.A. Rupp, G. Mandula and L. Kóvacs, Soliton profile of the dynamic grating amplitude and its alteration by photorefractive wave mixing, 404–409, *Photorefractive effects, materials and devices*, eds. P. Delayer, C. Denz, L. Mager and G. Montemezzani, *Trends in optics and photonics series* **87** (Optical society of America, Washington DC, 2003).
2. S. Bugaychuk, L. Kóvacs, G. Mandula, K. Polgár and R.A. Rupp, Nonuniform dynamic gratings in photorefractive media with nonlocal response, Phys. Rev. E **67** 046603-1 to -8 (2003).
3. A. Błędowski, W. Królikowski and A. Kujawski, Temporal instabilities in single-grating photorefractive four-wave mixing, J. Opt. Soc. Am. B **6** 1544–1547 (1989).
4. R. Conte, The Painlevé approach to nonlinear ordinary differential equations, *The Painlevé property, one century later*, 77–180, ed. R. Conte, CRM series in mathematical physics (Springer, New York, 1999). http://arXiv.org/abs/solv-int/9710020
5. R. Conte, Exact solutions of nonlinear partial differential equations by singularity analysis, *Direct and inverse methods in nonlinear evolution equations*, 1–83, ed. A. Greco, Lecture notes in physics **632** (Springer Verlag, Berlin,

2003). http://arXiv.org/abs/nlin.SI/0009024. CIME school, Cetraro, 5–12 September 1999.
6. J. Weiss, M. Tabor and G. Carnevale, The Painlevé property for partial differential equations, J. Math. Phys. **24** 522–526 (1983).
7. E. L. Ince, *Ordinary differential equations* (Longmans, Green, and co., London and New York, 1926). Reprinted (Dover, New York, 1956). Russian translation (GTIU, Khar'kov, 1939).
8. F. Cariello and M. Tabor, Painlevé expansions for nonintegrable evolution equations, Physica D **39** 77–94 (1989).
9. O. Thual and U. Frisch, Natural boundary in the Kuramoto model, *Combustion and nonlinear phenomena*, 327–336, eds. P. Clavin, B. Larrouturou, and P. Pelcé (Éditions de physique, Les Ulis, 1986).

ON NON–HOMOGENEOUS QUASILINEAR 2×2 SYSTEMS EQUIVALENT TO HOMOGENEOUS AND AUTONOMOUS ONES*

C. CURRÒ, F. OLIVERI

Department of Mathematics, University of Messina
Salita Sperone 31, 98166 Messina, Italy
E-mail: curro@mat520.unime.it; oliveri@mat520.unime.it

Within the framework of Lie group analysis, necessary and sufficient conditions are determined in order to map through a one–to–one point transformation non–homogeneous and non–autonomous quasilinear 2×2 systems to homogeneous and autonomous form. The procedure is constructive and the new independent and dependent variables are the canonical variables associated to the admitted Lie symmetries. Various examples of physical interest arising from different contexts are considered.

Keywords: Lie group analysis; Non–homogeneous and homogeneous systems.

1. Introduction

In mathematical modelling of one–dimensional nonlinear wave propagation through non–dissipative and homogeneous media, a relevant role is played by 2×2 quasilinear homogeneous and autonomous systems

$$A^0(\mathbf{U})\,\mathbf{U}_t + A^1(\mathbf{U})\,\mathbf{U}_x = \mathbf{0}, \qquad \mathbf{U} = [u_1, u_2]^T, \tag{1}$$

where A^0, A^1 are 2×2 matrices depending of the vector field $\mathbf{U}$, while x and t are space and time coordinates. In the case where (1) is strictly hyperbolic, by introducing the Riemann invariants, several wave problems have been investigated in order to get an insight into the occurrence of nonlinear wave breaking[1,2] under smooth initial data, or to characterize special classes of nonlinear homogeneous models for which the associated hodograph system is explicitly integrable.[1–3] A physically interesting problem related to (1) is the Riemann problem where the solution connecting two different constant states is expressed in terms of centered waves.

*Dedicated to T. Ruggeri on the occasion of his 60th birthday.

For non–homogeneous and dissipative media, we have systems like

$$A^0(\mathbf{U},x,t)\,\mathbf{U}_t + A^1(\mathbf{U},x,t)\,\mathbf{U}_x = \mathbf{B}(\mathbf{U},x,t)\,, \tag{2}$$

where $\mathbf{B}(\mathbf{U},x,t)$ is a column vector. Models belonging to the class (2) arise in different physical contexts such as viscoelastic materials,[4] nonlinear elastic rod with variable cross-section[5] or nonlinear heat conduction,[6] as well as problems in cylindrical or spherical symmetry.

The structure of the system (2) does not allow the use of the linearizing hodograph transformation; furthermore, as far as the hyperbolic wave propagation is concerned, system (2) does not admit in general Riemann invariants along characteristics so that, usually, asymptotic methods are used in order to get approximate simple wave solutions and to point out the main features of the wave interaction process.

In the last years much effort has been devoted to develop reduction procedures, based on hodograph–like or Backlünd transformations or direct methods, allowing to map models belonging to the class (2) in the form (1);[7–9] in Ref. 10 it has also considered the case of two opposite travelling *quasi–simple waves* exhibiting a soliton–like behaviour when interacting.

The aim of this paper is to show that the possibility of reducing systems belonging to class (2) to autonomous and homogeneous form is strictly related to the symmetry properties of the model under investigation. The underlying idea relies on the fact that any 2×2 homogeneous and autonomous first order quasilinear system admits an infinite–parameter Lie group of transformations to which it must correspond, as the mapping is invertible, an infinite parameter Lie group admitted by the original system.

2. The main result

The use of Lie symmetries to construct a mapping transforming a given set of differential equations into a different but equivalent set of differential equations is a well known procedure and it is widely applied in literature; the procedure is of relevant interest when the source system is nonlinear and the target system is linear.[11,12]

In this section we give necessary and sufficient conditions for the existence of an invertible mapping linking (2) to homogeneous and autonomous form; remarkably, the theorem is constructive.

Theorem 2.1. *The non–homogeneous and non–autonomous* 2×2 *quasilinear system*

$$A^0(\mathbf{U},x,t)\,\mathbf{U}_t + A^1(\mathbf{U},x,t)\,\mathbf{U}_x = \mathbf{B}(\mathbf{U},x,t) \tag{3}$$

transforms, under the action of the invertible point transformation

$$z = X(x,t), \quad \tau = T(x,t), \quad \mathbf{W} = \mathbf{V}(x,t,\mathbf{U}), \tag{4}$$

to the homogeneous and autonomous form, say

$$\widehat{A}^0(\mathbf{W})\mathbf{W}_\tau + \widehat{A}^1(\mathbf{W})\mathbf{W}_z = 0, \tag{5}$$

if and only if it is left invariant by the Lie group of point symmetries

$$\Xi = F_1(\mathbf{W})\Xi_1 + F_2(\mathbf{W})\Xi_2, \tag{6}$$

with Ξ_i $(i=1,2)$,

$$\Xi_i = \xi_i(x,t,\mathbf{U})\frac{\partial}{\partial x} + \gamma_i(x,t,\mathbf{U})\frac{\partial}{\partial t} + \eta_i^{(1)}(x,t,\mathbf{U})\frac{\partial}{\partial u_1} + \eta_i^{(2)}(x,t,\mathbf{U})\frac{\partial}{\partial u_2},$$

commuting infinitesimal operators, i.e., $[\Xi_1, \Xi_2] = 0$*, whereas* F_1 *and* F_2 *are solutions of the linear system of partial differential equations*

$$\widehat{A}^1(\mathbf{W})\begin{bmatrix} F_{1,w_2} \\ -F_{1,w_1}\end{bmatrix} - \widehat{A}^0(\mathbf{W})\begin{bmatrix} F_{2,w_2} \\ -F_{2,w_1}\end{bmatrix} = 0, \tag{7}$$

and it is

$$\Xi_1 X = 0, \quad \Xi_1 T = 1, \quad \Xi_2 T = 0, \quad \Xi_2 X = 1, \quad \Xi_1\mathbf{W} = \Xi_2\mathbf{W} = 0. \tag{8}$$

Proof. Let us assume that there exists a one–to–one point variable transformation like (4) mapping (3) to (5). Since the homogeneous and autonomous 2×2 quasilinear system (5) admits[11] the infinite–parameter Lie group of point symmetries generated by

$$\Xi = F_1(\mathbf{W})\frac{\partial}{\partial \tau} + F_2(\mathbf{W})\frac{\partial}{\partial z}, \tag{9}$$

with F_1 and F_2 solutions of a linear system, and the mapping (4) is invertible, to (9) it must correspond an infinite–parameter Lie group of point symmetries admitted by the non–autonomous and non–homogeneous source system. Due to (8), the new variables z, τ and $\mathbf{W}$ are canonical variables for the operators Ξ_1 and Ξ_2.

Conversely, let us assume that (3) is left invariant by (6); by introducing the canonical variables associated to both Ξ_1 and Ξ_2 (this is possible because the two operators commute), we are able to write (6) in the form (9). In terms of these canonical variables, we get the quasilinear system

$$\widehat{A}^0\mathbf{W}_\tau + \widehat{A}^1\mathbf{W}_z = \mathbf{N}$$

with $\widehat{A}^0 = \left(\frac{\partial T}{\partial t}A^0 + A^1\frac{\partial T}{\partial x}\right)(\nabla_\mathbf{U}\mathbf{V})^{-1}$, $\widehat{A}^1 = \left(\frac{\partial X}{\partial t}A^0 + A^1\frac{\partial X}{\partial x}\right)(\nabla_\mathbf{U}\mathbf{V})^{-1}$, $\mathbf{N} = \mathbf{B} + A^0(\nabla_\mathbf{U}\mathbf{V})^{-1}\mathbf{V}_t + A^1(\nabla_\mathbf{U}\mathbf{V})^{-1}\mathbf{V}_x$; since this system is invariant with respect to the infinite–parameter Lie group of point transformations generated by (9), it must be autonomous and homogeneous. □

3. Applications

3.1. *Nonlinear Hyperbolic Heat Equation*

Let us consider the nonlinear hyperbolic heat equation proposed in Ref. 6

$$\frac{\partial u}{\partial t}+\frac{\partial v}{\partial x}=0, \qquad \frac{\partial v}{\partial t}+\frac{\overline{\chi}^2}{u^2}\frac{\partial u}{\partial x}=-\frac{v}{\overline{\overline{T}}}, \tag{10}$$

where $\overline{\chi}$, $\overline{T}$ are positive constants. By eliminating v from (10) we have

$$\frac{\partial}{\partial t}\left(\frac{\partial u}{\partial t}+\frac{u}{\overline{\overline{T}}}\right)-\frac{\partial}{\partial x}\left(\frac{\overline{\chi}^2}{u^2}\frac{\partial u}{\partial x}\right)=0, \tag{11}$$

which has been reduced[6] to the standard linear telegraph equation by using a reciprocal Backlünd transformation. Furthermore, by setting $u=1/\overline{u}$, $\overline{T}=\overline{\chi}=1$, the equation (11) specializes to the nonlinear telegraph equation that is linearizable by using its potential symmetries.[11] Actually, by introducing a potential function $r\,(x,t)$, we are led to the system

$$\frac{\partial u}{\partial t}-\frac{\partial r}{\partial x}=-\frac{u}{\overline{\overline{T}}}, \qquad \frac{\partial r}{\partial t}-\frac{\overline{\chi}^2}{u^2}\frac{\partial u}{\partial x}=0, \tag{12}$$

which is left invariant by the Lie group whose infinitesimal operator is

$$\Xi=F_1\,(w_1,w_2)\exp\left(-\frac{t}{\overline{\overline{T}}}\right)\left(\overline{T}\frac{\partial}{\partial t}-u\frac{\partial}{\partial u}\right)+F_2\,(w_1,w_2)\,\frac{\partial}{\partial x},$$

where $w_1=\exp\left(\frac{t}{\overline{\overline{T}}}\right)u$, $w_2=r$, and F_1, F_2 satisfy the linear system

$$\frac{\partial F_2}{\partial w_1}-\frac{\overline{\chi}^2\overline{T}}{w_1^2}\frac{\partial F_1}{\partial w_2}=0, \qquad \frac{\partial F_2}{\partial w_2}-\overline{T}\frac{\partial F_1}{\partial w_1}=0.$$

The hypotheses of theorem 2.1 are satisfied and the variable transformation

$$w_1=\exp\left(\frac{t}{\overline{\overline{T}}}\right)u, \qquad w_2=r, \qquad z=x, \qquad \tau=\exp\left(\frac{t}{\overline{\overline{T}}}\right)$$

reduces the system (12) to the autonomous and homogeneous form

$$\frac{\partial w_1}{\partial \tau}-\overline{T}\frac{\partial w_2}{\partial z}=0, \qquad \frac{\partial w_2}{\partial \tau}-\frac{\overline{\chi}^2\overline{T}}{w_1^2}\frac{\partial w_1}{\partial z}=0.$$

3.2. *Rate–Type Materials*

Within the framework of viscoelasticity, in order to describe the experimental behaviour of special classes of materials under loading–unloading 1D processes, the following quasilinear first order system has been proposed:[4]

$$\frac{\partial u}{\partial t}-\frac{\partial v}{\partial x}=0, \qquad \frac{\partial v}{\partial t}-\phi\,(t,v)\,\frac{\partial u}{\partial x}=\psi\,(t,v)\,, \tag{13}$$

where v and u denote the stress and the Lagrangian velocity, whereas the functions $\phi(t,v)$ and $\psi(t,v)$ measure the non–instantaneous and the instantaneous response of the material to an increment of the stress, respectively. The functional forms of $\phi(t,v)$ and $\psi(t,v)$ are usually suggested by experimental investigations. In the particular case where $\phi = v^2$, $\psi = v(1-v)$, system (13) has been linearized by a hodograph–like transformation.[7,8]

By carrying on a full group analysis for the system (13), owing to the arbitrariness of the constitutive functions $\phi(t,v)$ and $\psi(t,v)$, several possibilities arise. In particular, we find that (13) is left invariant by the Lie group whose infinitesimal operator is

$$\Xi = F_1(w_1,w_2)\,T(t)\left(\frac{\partial}{\partial t} - (\eta'(t) + vT'(t))\frac{\partial}{\partial v}\right) + F_2(w_1,w_2)\left(\frac{\partial}{\partial x} + \frac{\partial}{\partial u}\right),$$

with $T(t)$ and $\eta(t)$ arbitrary functions, $w_1 = u - x$, $w_2 = vT(t) + \eta(t)$, and F_1, F_2 solutions of the linear system of partial differential equations

$$\frac{\partial F_2}{\partial w_2} - \frac{\partial F_1}{\partial w_1} = 0 \qquad \frac{\partial F_2}{\partial w_1} - \varphi(w_2)\frac{\partial F_1}{\partial w_2} = 0,$$

with the constitutive functions $\phi(t,v)$ and $\psi(t,v)$ such that

$$\phi(t,v) = \frac{\varphi(vT(t) + \eta(t))}{T^2(t)}, \qquad \psi(t,v) = -\phi(t,v) - \frac{vT'(t) + \eta'(t)}{T(t)}. \tag{14}$$

Use of theorem 2.1 permits to build the variable transformation

$$w_1 = u - x, \qquad w_2 = vT(t) + \eta(t), \qquad z = x, \qquad \tau = \int \frac{dt}{T(t)}, \tag{15}$$

whereupon the system (13) reduces to autonomous and homogeneous form:

$$\frac{\partial w_1}{\partial \tau} - \frac{\partial w_2}{\partial z} = 0, \qquad \frac{\partial w_2}{\partial \tau} - \varphi(w_2)\frac{\partial w_1}{\partial z} = 0.$$

We remark that the choice

$$T(t) = \exp(-t), \qquad \eta(t) = 0, \qquad \varphi(w_2) = w_2^2$$

in (14) allows one to obtain the constitutive relations used in Ref. 7.

3.3. *Nonlinear Elastic Rod*

Let us consider a nonlinear elastic incompressible rod with a variable cross–section which is assumed everywhere symmetrical with respect to the axis of the rod and two fixed mutually orthogonal axes that are normal to it.

Under the assumptions that the cross–sectional area is a (slowly) varying function of the distance measured along the rod, we have the system[5]

$$\frac{\partial p}{\partial t} - \frac{\partial v}{\partial x} = 0, \qquad \frac{\partial v}{\partial t} - \frac{1}{\rho} T_p \frac{\partial p}{\partial x} = \frac{T}{\rho} \frac{S'(x)}{S(x)}, \tag{16}$$

where $v = \frac{\partial u}{\partial t}$, $p = \frac{\partial u}{\partial x}$, u being the displacement of the particle along the rod with respect to the axis; moreover, ρ is the constant density, $T = T(p)$ denotes the stress, and $S = S(x)$ is the cross–sectional area.

By carrying on a full group analysis for the system (16), owing to the arbitrariness of the functions $T(p)$ and $S(x)$, several possibilities arise. In particular, (16) is invariant with respect to the Lie group generated by

$$\Xi = F_1(w_1, w_2) \frac{\partial}{\partial t} + \frac{1}{S(x)} F_2(w_1, w_2) \left(\frac{\partial}{\partial x} + (p + \beta) \frac{\partial}{\partial p} \right),$$

with $w_1 = \frac{p+\beta}{S(x)}$, $w_2 = v$, and F_1, F_2 solutions of the linear system of partial differential equations

$$\frac{\partial F_2}{\partial w_2} - \frac{\partial F_1}{\partial w_1} = 0, \qquad \rho w_1^2 \frac{\partial F_2}{\partial w_1} + \gamma \frac{\partial F_1}{\partial w_2} = 0,$$

whereas β, γ are arbitrary constants, $S(x)$ is arbitrary and $T(p) = \frac{\gamma}{\beta + p}$.

By using theorem 2.1, the variable transformation

$$w_1 = \frac{p + \beta}{S(x)}, \qquad w_2 = v, \qquad z = \int S(x)\, dx, \qquad \tau = t, \tag{17}$$

reduces system (16) to the following autonomous and homogeneous form:

$$\frac{\partial w_1}{\partial \tau} - \frac{\partial w_2}{\partial z} = 0, \qquad \rho w_1^2 \frac{\partial w_2}{\partial \tau} + \gamma \frac{\partial w_1}{\partial z} = 0.$$

References

1. P. D. Lax, *J. Math. Phys* **5**, 611–620 (1964).
2. A. Donato, A. Jeffrey, *Wave Motion* **1**, 11–20 (1979).
3. C. Currò, D. Fusco, *Int. J. Non-Linear Mech* **23**, 25–35 (1988).
4. N. Cristescu, *Dynamic Plasticity* (Wiley, New York, 1964).
5. A. Jeffrey, *Wave Motion* **4**, 173–180 (1982).
6. C. Rogers, T. Ruggeri, *Lett. Nuovo Cimento* **2**, 289–296 (1985).
7. B. Seymour, E. Varley, *Studies Applied Math.* **72**, 241–262 (1985).
8. D. Fusco, N. Manganaro, *J. Math, Phys.* **32**, 3043–3046 (1991).
9. C. Currò, G. Valenti, *Int. J. Non-Linear Mech.* **31**, 25–35 (1996).
10. C. Currò, D. Fusco, N. Manganaro, *Applicable Analysis* **57**, 47–62 (1995).
11. G. W. Bluman, S. Kumei, *Symmetries and Differential Equations* (Springer–Verlag, Berlin, 1989).
12. A. Donato, F. Oliveri, *J. Math. Anal. Appl.* **188**, 552–568 (1994).

ON THE FITZHUGH - NAGUMO MODEL

M. DE ANGELIS, P. RENNO

Faculty of Engineering. Dept. of Mathematics and Appl. "R. Caccioppoli", Federico II
Naples 80125, Italy
modeange@unina.it

The initial value problem $\mathcal{P}_0$, in all of the space, for the spatio-temporal FitzHugh - Nagumo equations is analyzed. When the reaction kinetics of the model can be outlined by means of piecewise linear approximations, then the solution of $\mathcal{P}_0$ is explicitly obtained. For periodic initial data are possible damped travelling waves and their speed of propagation is evaluated. The results imply applications also to the non linear case.

Keywords: Parabolic systems. Biological models. Laplace transforms

1. Introduction

One of the reaction diffusion systems which models various important biological phenomena is given by

$$\begin{cases} \dfrac{\partial u}{\partial t} = \varepsilon \dfrac{\partial^2 u}{\partial x^2} - v + f(u) \\ \\ \dfrac{\partial v}{\partial t} = b u - \beta v \end{cases} \tag{1}$$

where the appropriate class of functions $f(u)$ depends on the reaction kinetics of the model [1–4]. In the theory of nerve membranes, for example, the system (1) is related to the spatio-temporal FitzHugh - Nagumo equations (FHN) with

$$f(u) = u(a-u)(u-1) \qquad (0 < a < 1) \tag{2}$$

and where $u(x,t)$ models the transmembrane voltage of a nerve axon at distance x and time t, $v(x,t)$ is an auxiliary variable that acts as recovery variable. Further, the diffusion coefficient ε and the parameters b, β are all non negative [5–9].

In addition to the propagation of nerve action potentials, (1) - (2) govern other several biological and biochemical phenomena; the list of references is very long and the variety of analytical aspects examined is wide [10–15]. Further, we must remark travelling pulses and periodic wavetrains obtained by means of piecewise linear approximations of $f(u)$ as

$$f(u) = \eta(u-a) - u \qquad (0 < a < 1) \tag{3}$$

where η denotes the unit- step function [16–21].

Typical boundary value problems related to the linear case (3) can be explicitly solved. Aim of this paper is the analysis of the initial value problem $\mathcal{P}_0$ in all of the space; the fundamental solution and the explicit solution of $\mathcal{P}_0$ are determined when (3) holds. More, in the non linear case of the FHN model given by (1)-(2), the problem $\mathcal{P}_0$ is reduced to appropriate integral equations whose kernels are functions characterized by basic properties. All this implies existence and uniqueness properties, together with a priori estimates.

2. Statement of the problem and results

Both the non linear source (2) and the linear approximation (3) involve a linear term $-k\,u$ with $k = a$ for (2), and $k = 1$ for (3). As consequence, the system (1) becomes

$$\begin{cases} u_t - \varepsilon\, u_{xx} + k\,u + v = \varphi(u) \\ v_t + \beta v - b\,u = 0 \ , \end{cases} \qquad (x,t) \in \Re \tag{1}$$

where $\varphi(u) = u^2(a+1-u)$ when f is given by (2), while, in the linear case, $\varphi(u)$ is equal to the constant $\bar{\eta}$ that holds zero or one.

The initial- value problem $\mathcal{P}_0$ related to (1) with $\varphi(u) = \bar{\eta}$ is analyzed in the set

$$\Omega_T = \{(x,t) : x \in \Re, \ \ 0 < t \leq T\}, \tag{2}$$

with the conditions

$$u(x,0) = u_0(x), \qquad v(x,0) = v_0(x), \qquad x \in \Re. \tag{3}$$

When the linear approximation of f holds, then the explicit solution of the problem $\mathcal{P}_0$ can be obtained by means of functional transforms (Fourier with respect to x and Laplace as for t). If one puts formally

$$\hat{u}(x,s) = \int_0^\infty e^{-st}\, u(x,t)\, dt\ , \quad \hat{v}(x,s) = \int_0^\infty e^{-st}\, v(x,t)\, dt\ , \qquad (4)$$

from (1) (3) one deduces:

$$\hat{u} = \int_{\Re} [\, u_0(\xi) + \frac{\bar{\eta}}{s}\,]\, \hat{K}_0\, [(x-\xi, s)\, d\xi - \int_{\Re} v_0(\xi)\, \hat{K}_1\, (x-\xi, s)\, d\xi \qquad (5)$$

$$\hat{v} = b\int_{\Re} [u_0(\xi) + \frac{\bar{\eta}}{s}]\hat{K}_1(x-\xi,s)d\xi - b\int_{\Re} v_0(\xi)\hat{K}_2(x-\xi,s)d\xi + \frac{v_0(x)}{s+\beta}, \qquad (6)$$

where

$$\hat{K}_n(x,s) = \frac{e^{-\frac{|x|}{\sqrt{\varepsilon}}\,\sigma}}{2\,\sqrt{\varepsilon}\ \sigma\,(s+\beta)^n} \qquad (n=0,1,2) \qquad (7)$$

with $\sigma^2 = s + k + \dfrac{b}{s+\beta}.$

Now, let us consider the fundamental solution

$$g(x,y) = \frac{1}{2\,\sqrt{\pi\varepsilon y}}\; e^{-k\,y - \frac{x^2}{4\varepsilon\, y}} \qquad (8)$$

of the heat equation $\varepsilon\, g_{xx} - g_y - k\, g = 0$ and, for $n = 0,1,2,$ let

$$G_n(x,t) = \int_0^t e^{-\beta(t-y)} g(x,y) \Big(\frac{t-y}{by}\Big)^{\frac{n-1}{2}} J_{n-1}\left(2\sqrt{b(t-y)y}\,\right) dy, \qquad (9)$$

where $J_n(z)$ denotes the Bessel function of first kind. Then, if one puts

$$K_0 = g(x,t) + G_0(x,t), \quad K_i = G_i(x,t)\ \ (i=1,2), \qquad (10)$$

the following theorems hold.

Theorem 2.1. *In the half-plane* $\Re e\, s > max(-k, -\beta)$, *the Laplace integrals of* $K_n(x,t)$ $(n = 0,1,2)$ *converge absolutely for all* $|x| > 0$, *and one has* $\mathcal{L}_t\, K_n(x,t) = \hat{K}_n(x,s)$.

Theorem 2.2. *The functions K_0, K_1, K_2 are $C^\infty(\Omega_T)$ solutions of the integro differential equation:*

$$z_t - \varepsilon\, z_{xx} + k\, z + b \int_0^t e^{-\beta(t-\tau)}\, z(x,\tau)\, d\tau = 0 \tag{11}$$

and have the same basic properties of the fundamental solution (8) of the heat operator. Further, for $i = 0$ and $i = 1$, it results:

$$K_i(x,t) = (\partial_t + \beta)\, K_{i+1}, \qquad \lim_{t \downarrow 0} K_{i+1} = 0 \ \ (i = 0,1) \tag{12}$$

while $\lim\limits_{t\to 0} K_0(x,t) = 0$ only for $|x| > 0$.

These properties assure the convergence of the convolutions

$$K_n * \psi = \int_{\Re} K_n(x-\xi,t)\, \psi(\xi)\, d\xi \quad (n = 0,1,2) \tag{13}$$

for all the functions that satisfy a *growth condition* of the form

$$|\psi(x)| < c_1\, exp\, [\, c_2\, |x|^{\alpha+1}\,], \quad 0 < \alpha < 1 \tag{14}$$

with c_1 and c_2 positive constants. Further, let consider the following functions

$$N_1(t) = \frac{\bar\eta\beta}{b+\beta}\Big[1 - e^{-\frac{1+\beta}{2}t}\cos(\gamma t)\Big] + \frac{\bar\eta(2b+\beta-\beta^2)}{2\gamma(b+\beta)}\, e^{-\frac{1+\beta}{2}t}\sin(\gamma t) \tag{15}$$

$$N_2(t) = \frac{\bar\eta}{b+\beta}\Big[1 - e^{-\frac{1+\beta}{2}t}\cos(\gamma t)\Big] - \frac{\bar\eta(1+\beta)}{2\gamma(b+\beta)}\, e^{-\frac{1+\beta}{2}t}\sin(\gamma t), \tag{16}$$

where $\gamma = \Big[b - (\beta-1)^2/4\Big]^{1/2}$. Then, by (5)-(6) and the foregoing statements, the explicit solution of the linear problem $\mathcal{P}_0$ is given by

$$\begin{cases} u = N_1(t) + u_0 * K_0 - v_0 * K_1 \\[2ex] v = N_2(t) + u_0 * K_1 - v_0 * K_2 + v_0(x)\, e^{-\beta t}, \end{cases} \tag{17}$$

and the following conclusion is deduced.

Theorem 2.3. *When the data (u_0, v_0) are continuous functions that satisfy the growth condition (14), then the formulae (17) represents the unique solution of the problem $\mathcal{P}_0$ in the class of solutions compatible with (14).*

3. Travelling waves and a priori estimates

A first example of applications is related to the linear case and concerns the analysis of travelling waves. By the explicit solution (17), for instance, when $\bar{\eta} = 0$, $v_0, = 0$ $u_0 = A\,cos(w\,x)$, then one obtains:

$$u = (\,\partial_t + \beta\,)w, \quad v = b\,w \tag{1}$$

where

$$w = \frac{A}{2\,\alpha}\,e^{-\frac{1+\beta+\varepsilon w^2}{2}\,t}\left[\sin(\,wx + \alpha t) - \sin(\,wx - \alpha t)\right] \tag{2}$$

with $\alpha = \left[b - \left(\frac{1+\varepsilon w^2-\beta}{2}\right)^2\right]^{1/2}$. So, when $b > \left(\frac{1+\varepsilon w^2-\beta}{2}\right)^2$, there exist damped travelling waves with speed equal to α/w.

Moreover, in the non linear case, by (5),(6) one deduces the following integral equations

$$\begin{cases} u = \,u_0 * K_0 - v_0 * K_1 + \int_0^t \varphi(u) * K_0 \;d\tau \\ \\ v = b\,u_0 * K_1 - b\,v_0 * K_2 + v_0\,(x)\,e^{-\beta t} + b\,\int_0^t \varphi(u) * K_1 \;d\tau \end{cases} \tag{3}$$

that imply a priori estimates. In fact, in the class of bounded solutions, if one puts

$$||u_0|| = \sup_{x\in\Re}|u_0(x)|, \;\; ||v_0|| = \sup_{x\in\Re}|v_0(x)|, \;\; ||\varphi|| = \sup_{u\in\Re}|\varphi(u)| \tag{4}$$

by means of the basic properties of the kernels K_0, K_1, K_2, one obtains:

$$\begin{cases} |u| < \;c_1\,||\varphi|| + c_2\,(\;||u_0||\,t + ||v_0||\,)\,E(t) \\ \\ |v| < \;c_1\,||\varphi|| + c_2\,(\;||u_0|| + ||v_0||\,t\;)\,E(t) \end{cases} \tag{5}$$

where

$$E(t) = \frac{e^{-kt} - e^{-\beta t}}{\beta - k} \tag{6}$$

and where the constants c_1, c_2 depend on the parameters b, k, β. Worthy of remark is the fact that the estimates (5) hold for all t, also when T tends to infinity.

References

1. E.M. Izhikevich *Dynamical Systems in Neuroscience: The Geometry of Excitability and Bursting.* (The MIT press. England, 2007)

2. J.D. Murray, *Mathematical Biology. II. Spatial models and biomedical applications* , (Springer-Verlag, N.Y, 2003)
3. J.D. Murray, *Mathematical Biology. I. An Introduction* , (Springer-Verlag, N.Y , 2002)
4. J. P. Keener - J. Sneyd *Mathematical Physiology* (Springer-Verlag, N.Y , 1998)
5. M. Krupa, B. Sandstede, P. Szmolyan, Fast and slow waves in the FitzHugh-Nagumo equation, *J. Differential Equations* **133** , no. 1, 49-97 (1997).
6. L.Glass, M.E. Josephson, Resetting and annihilation of reentrant abnormally rapid heartbeat, *Phys. Rev. Lett* **75,10** 2059-2062 (1995)
7. J. Rinzel *Models in neurobiology* Nonlinear Phenomena in Physics and Biology. Edit by R. H. Enns, B. L. Jones, R. M. Miura, and S.nd S. Rangnekar. D. Reidel Publishing Company, Dordrecht-Holland. NATO Advanced Study Institutes Series. B75, (1981), p.345
8. J. Nagumo, S. Animoto, S. Yoshizawa, An active pulse transmission line simulating nerve axon, *Proc. Inst. Radio Engineers*, **50**, 2061-2070 (1962).
9. R. FitzHugh, Impulses and physiological states in theoretical models of nerve membrane, *Biophysical journal* **1** 445- 466 (1961).
10. Wang, Jiang; Zhang, Ting; Deng, Bin, Synchronization of FitzHugh-Nagumo neurons in external electrical stimulation via nonlinear control, *Chaos Solitons Fractals* **31** , no. 1, 30-38 (2007)
11. J.G. Alford, G. Auchmuty, Rotating wave solutions of the FitzHugh-Nagumo equations, *J. Math. Biol.* **53** 797-819 (2006)
12. D. Bini, c. Cherubini, S. Filippi, Heat transfer in FitzHugh-Nagumo models, *Physical Review E.* **74**, (2006)
13. D. Bini, C. Cherubini, S. Filippi, Viscoelastic FitzHugh-Nagumo models, *Phys. Rev. E* (3) **72** no. 4, (2005)
14. J. Rinzel, D. Terman, Propagation phenomena in a bistable reaction-diffusion system, *SIAM J. Appl. Math.* **42** , no. 5, 1111–1137 (1982).
15. J. Nagumo, S. Yoshizawa, S. Animoto, Bistable Transmission Lines, *IEEE Transactions on Circuits and Systems*, **12**, Issue 3, Page(s): 400 - 412 (1965)
16. McKean, H. P., Jr., Nagumo's equation, *Advances in Math.* **4** 209–223 (1970).
17. J. Rinzel, J.B. Keller, Travelling wave solutions of a nerve conduction equation, *Biophysical Journal,* **13** 1313-1337, (1973).
18. J.Rinzel, Spatial stability of travelling wave solutions of a nerve conduction equation, *Biophys J.* **15** 975-988 (1975).
19. J.A. Feroe, Temporal stability of solitary impulse solutions of a nerve equation, *Biophys. J.* **21** 103-110 (1978)
20. J.A. Feroe, Existence of travelling wave trains in nerve axon equations, *SIAM J. Appl. Math.* **46** , 1079-1097 (1986).
21. A. Tonnelier, The McKean's caricature of the FitzHugh-Nagumo model. I : The space-clamped system, *SIAM Journal on Applied Mathematics* **63**, pp. 459-484 (2002)

HIGH REYNOLDS NUMBER NAVIER–STOKES SOLUTIONS AND BOUNDARY LAYER SEPARATION INDUCED BY A RECTILINEAR VORTEX ARRAY

G. DELLA ROCCA[1], F. GARGANO[2], M. SAMMARTINO[2], and V. SCIACCA[2]

[1] *Mathematics Department, CSULB,*
1250 Bellflower Blvd, Long Beach, CA 90840, USA
[2] *Department of Mathematics and Applications, University of Palermo,*
Palermo, via Archirafi 34, 90123, Italy
gdellaro@csulb.edu *gargano,marco,sciacca@math.unipa.it*

Numerical solutions of Prandtl's equation and Navier Stokes equations are considered for the two dimensional flow induced by an array of periodic rectilinear vortices interacting with an infinite plane. We show how this initial datum develops a separation singularity for Prandtl equation. We investigate the asymptotic validity of boundary layer theory considering numerical solutions for the full Navier Stokes equations at high Reynolds numbers.

Keywords: unsteady separation; zero viscosity limit; Navier Stokes solutions; Prandtl equation, interactive viscous–inviscid equation.

1. Introduction

The study of the behavior of a high Reynolds fluid interacting with a solid boundary is a central problem both in the mathematical analysis of fluid dynamics as well in many technical applications of fluid mechanics. To resolve the behavior of a fluid close to a wall, Prandtl introduced a thin layer of thickness $\mathcal{O}(Re^{-1/2})$. In this layer the Navier Stokes solution experiences a sharp transition from the inviscid regime,ruled by the Euler equations, to the no-slip boundary condition valid at the wall. Using asymptotic expansion[20] , he derived Eqs.(1)–(2), which rule the flow inside the boundary layer and which are at the hearth of the boundary layer theory:

$$\frac{\partial u}{\partial t}+u\frac{\partial u}{\partial x}+V\frac{\partial u}{\partial Y}-(\frac{\partial U_\infty}{\partial t}+U_\infty\frac{\partial U_\infty}{\partial x})=\frac{\partial^2 u}{\partial Y^2},\qquad \frac{\partial u}{\partial x}+\frac{\partial V}{\partial Y}=0, \quad (1)$$

$$u(x,0,t)=v(x,0,t)=0,\quad u(x,Y\to\infty,t)=U_\infty(x,t). \quad (2)$$

These equations are supplemented with the initial condition $u(x, Y, 0) = u_0(x, Y)$. In the above equations $Y = y\sqrt{Re}$ and $V = v\sqrt{Re}$ are respectively the normal variable and velocity written according to the boundary–layer scale, while U_∞ is the outer Euler's solution at the boundary. Equations (1) express the streamwise momentum conservation and the incompressibility condition, while Eqs. (2) are the no–slip and no–flux conditions at the wall and the matching condition with outer Euler flow.

Prandtl's equations have been successfully applied in many situations of practical relevance but, regarding the mathematical theory, several questions remain unsolved[8] ; we mention three of them. The first one is the well–posedness of Prandtl equation: there exist local in time well posedness results for monotone initial data[15] and global results if the streamwise pressure gradient is favourable[22] . A local in time well posedness result exists also if the initial datum is analytic and in this case the convergence of NS solution to Prandtl and Euler's solution is proved[12,17,18] . The second question is the problem of convergence of the solutions of NS to the solutions of Euler or Prandtl equations for non analytic initial data. Finally there is the problem of the finite time blow up for Prandtl's solutions, found numerically by Van Dommelen & Shen[6] and classified as a shock type singularity[5,19] by the complex singularity tracking method. How the singularity is related to the separation phenomenon and how the Prandtl's solutions compare with the NS solutions are the main topic of this paper.

Singularity formation of Prandtl's equation is the first non–interactive stage of unsteady boundary layer separation theory[4,21] . In this stage, the adverse streamwise pressure gradient imposed across the boundary layer induces the formation of a back–flow region that grows in time in the streamwise direction, and ejects farther in the normal direction forming a spike that reaches, at a given time, the outer external flows. It is clear that, as the spike grows, the external flow begins to respond, leading to interaction between the viscous boundary layer and the inviscid outer flow. The viscous–inviscid interaction represents the second stage of unsteady separation, governed by an interactive boundary–layer IBL equation[16] . Unfortunately, the IBL equation develops a singularity in the streamwise pressure gradient earlier then the time of the separation singularity for Prandtl. For this reason, some authors[9] consider a third stage where the normal pressure gradient effects are considered in the IBL equation. In[3,14] , it was raised the doubt that IBL solution doesn't converge effectively to Prandtl's solution, for Reynolds number going to infinity. It was observed the presence of a large–scale interaction that occurs prior to spike formation, influencing the small–scale interaction described by the second interactive stage. In this paper we consider the asymptotic validity of boundary layer theory for a two dimensional incompressible fluid confined in an infinite half plane with an array of rectilinear vortices as initial datum. We find that this initial datum develops a singularity for Prandtl's equation in a finite time. We compute numerically the solutions for NS at high Reynolds numbers, comparing with the asymptotic theory of boundary layer separation. In our investigation we analyze some important numerical quantities such as the pressure gradient at the wall and the skin friction coefficient, by which it is possible to determine when the transition to the various stages of unsteady separation occurs. The plain of the paper is the following: in the next section we introduce the datum and we explain our numerical method. In section 3 we show the respective evolution and the singularity formation for Prandtl's equation and we make a comparison with the NS numerical results.

2. Rectilinear vortices array

2.1. *The initial datum*

Our case study is the two-dimensional inviscid flow due to a row of infinite equidistant vortices in an incompressible fluid which is at rest at infinity. This problem is studied in[11] , and consists in point-vortices having the same strength k with positive rotations interacting with an infinite rectilinear wall. In a cartesian frame we consider the vortices centered in $(ma+\pi, b)_{m\epsilon\mathbb{Z}}$, where b is the distance of the the row from the wall and a is the distance of two consecutive vortices. This datum is obtained by the superposition of the flows induced by two single rows of vortices symmetrically placed at distance b from the axes $y=0$, with positive rotation for the upper vortices and negative for the lower ones. The velocity induced by this vortices configuration is such that each vortex moves with uniform velocity $U=\frac{k}{2a}\coth(\frac{2\pi b}{a})$ parallel to the wall. We study the system in the reference frame comoving with the vortices; the initial streamfunction, is therefore:

$$\Psi(x,y)=Uy-\frac{k}{4\pi}\log\left(\frac{\cosh(\frac{2\pi}{a}(y-b))-\cos(\frac{2\pi}{a}(x-\pi))}{\cosh(\frac{2\pi}{a}(y+b))-\cos(\frac{2\pi}{a}(x-\pi))}\right). \tag{3}$$

This is an a-periodic datum, and the velocity components obtained are such that $u=k/a$, $v=0$ for $y=0$, and $u,v\to 0$ for $y\to\pm\infty$.

The initial vorticity is singular and it is given by $w_0=\sum_{m\in\mathbb{Z}}\delta_{ma,b}$,where $\delta_{x,y}$ is Dirac's mass. In the boundary-layer calculations this is not a problem, as the only quantity needed is the streamwise pressure gradient along the surface. In NS calculations instead, we want to resolve the flow in the entire domain, including the region in which the datum is singular. For the spectral method we will use, this datum would lead to the appearance of the Gibb's phenomenon. We solve this problem by smoothing the initial vorticity, approximating it with a finite sum of vortex blobs. This is implemented by convolving the vorticity with the mollifier $\Phi^\sigma(x,y) = \frac{1}{\sigma^2\pi}e^{-\frac{(x^2+y^2)}{\sigma^2}}$, obtaining the regularized vorticity:

$$\omega^\sigma = \omega * \Phi^\sigma = \sum_{m\in\mathbb{Z}} \Phi^\sigma(x - ma - \pi, y - b). \tag{4}$$

This is a typical procedure used in computational Vortex Methods when the initial data have point singularities or for vortex-sheet motion[10,13] . In the NS calculations we have chosen $\sigma^2 = 5 \times 10^{-3}$

2.2. *The numerical scheme for the Navier–Stokes equations*

We solve the Navier-Stokes equations in the vorticity–streamfunction formulation where the vorticity is $\omega = \frac{\partial v}{\partial x} - \frac{\partial u}{\partial y}$ and the streamfunction is defined to be $\nabla^2\Psi = \omega$. We set $a = 2\pi$ and solve our problem in the domain $[0,2\pi]$x$[0,\infty]$. We pass to non dimensional variables using the distance of the vortex from the wall b, and the the velocity $U_0 = k/2a$. The Reynolds number is therefore defined as $Re = U_0 b/\nu$. We map the semi-infinite domain in a finite one using the transformation for the normal variable $\bar{y}=\frac{2}{\pi}\arctan(\frac{y}{c})$, where c is the parameter determining the degree of focusing of the grid points near the wall. This maps $y = \infty$ to $\bar{y} = 1$, focusing the grid points near the wall to better resolve the region where large gradients appears. The vorticity–transport equations becomes:

$$\frac{\partial\omega}{\partial t} + u\frac{\partial\omega}{\partial x} + (\Gamma_y(\bar{y}) - \frac{1}{Re}\Gamma_y(\bar{y})\Gamma_y'(\bar{y}))v\frac{\partial\omega}{\partial\bar{y}} = \frac{1}{Re}(\frac{\partial^2\omega}{\partial x^2} + \Gamma_y^2(\bar{y})\frac{\partial^2\omega}{\partial\bar{y}^2}), \tag{5}$$

$$\frac{\partial^2\Psi}{\partial x^2} + \Gamma_y^2(\bar{y})\frac{\partial^2\Psi}{\partial\bar{y}^2} + \Gamma_y(\bar{y})\Gamma_y'(\bar{y})\frac{\partial\Psi}{\partial\bar{y}} = \omega, \tag{6}$$

$$\Gamma_y(\bar{y})\frac{\partial\Psi}{\partial\bar{y}} = v, \qquad \frac{\partial\Psi}{\partial x} = -u, \tag{7}$$

with $\Gamma_y(\bar{y}) = \frac{1}{\pi c}[1 + \cos(\pi\bar{y})]$ and a prime denotes the differentiation with respect to the y variable. As $\Psi \to \infty$ for $y \to \infty(\bar{y} \to 1)$ we need to truncate the computational domain at a finite value $y_{\max}$ of the normal variable . Applying the procedure used in[1] for solving the Poisson problem with a correct upper boundary condition, we chose this value requiring that the vorticity remains negligible for $y \geqslant y_{\max}$ for all computational time (let us say $\omega \leqslant 10^{-15}$). We take $y_{\max} = 1.5$ which we find to be large enough to satisfy the required condition for the vorticity. We have also made some tests using a bigger value for $y_{\max}$ which gives no appreciable differences in our numerical results. To solve the equations we have used a mixed spectral-finite differences method for the variables x and y. Denoting by Δy and Δt the normal spatial mesh and the time step we approximate the solution as:

$$\omega(x, j\Delta y, n\Delta t) = \sum w^n_{k,j} e^{ikx}.$$

The first and second order normal derivative are treated in an explicit way using centered finite differences approximation of second order. Regarding the nonlinear terms we have used the pseudo-spectral approximation involving multiplication in the physical space via inverse FFT, and using the 3/2-rule to handle aliasing effects[2] . For the temporal scheme we have used the well known Runge–Kutta scheme of fourth order which prevents our calculation from the formation of numerical instability that we have encountered using the simpler Euler scheme also with small temporal step. At the first step we solve the Poisson equation (6) getting the initial regularized vorticity (4) and then we compute the velocity components by equations (7). To solve the system at the step n we use the following procedure:

- We update the vorticity in the interior grid $[0,2\pi]$x$(0,y_{\max})$ using the spectral-finite differences scheme. The boundary condition for the vorticity at the wall is obtained using Jensen's method[7] which gives a second order accurate estimate for the vorticity at the surface. The boundary condition at $y_{\max}$ is $\omega_{y_{max}} = 0$.
- The value Ψ^{n+1} is computed by solving the Poisson problem (6). The value of the boundary condition at $y_{\max}$ is recovered using the procedure introduced in[1] . The boundary value at the wall is the $no-flux$ condition $\Psi = 0$.
- We compute the velocity components u^{n+1}, v^{n+1} by equations (7).

The numerical solutions are grid independent, and we show the result obtained with the finest grid according to the following table:

Re	GRID	*c*	*dt*
10^3	512x600	0.4	$5 \cdot 10^{-4}$
10^4	1024x800	0.15	$2.5 \cdot 10^{-5}$
$5 \cdot 10^4$	2048x800	0.07	$5 \cdot 10^{-6}$

3. Results

3.1. *Boundary Layer results*

To solve numerically Prandtl's equations we use the mixed spectral-finite differences numerical scheme proposed in[19] with IMEX122 temporal scheme and implicit treatment of the diffusive term. The computational domain is $[0,2\pi]$x$[0,Y]$ where Y is chosen big enough so that the solution $u(x,Y,t)$ satisfy, for all computational time, the matching condition (2). In our calculation a value of $Y = 20$ has satisfied the required condition. We have chosen as finest grid 4096x800 with adaptive temporal step according to CFL stability condition as in[19] .

The first relevant physical process is formation of a recirculation eddy near the wall approximatively at time $t \approx 0.3$ due to the adverse streamwise pressure gradient. This eddy may be seen in Fig.1(a) at a later stage of development at $t = 0.35$, and as time passes thickens rapidly in the streamwise direction, until a kink is formed at about $t \approx 0.67$ (Fig.1(c)). This evolves in a sharp spike at $t \approx 0.74$ revealing singularity formation (Fig.1(d)). To determine the exact instant at which the singularity forms, we apply the singularity tracking method[19] to Prandtl's equation, following in time the evolution of the exponential decay rate δ of the Fourier spectrum. The result confirms the singularity formation at time $t \approx 0.74$ (see Fig.2) at streamwise and normal location respectively $x \approx 3.15$ and $Y \approx 2$.

3.2. *Comparison*

In this section we compare Navier-Stokes solutions at different Re numbers ($10^3, 10^4$ and $5 \cdot 10^4$), with the solution of the boundary layer equations. We continue our calculations up to the singularity time $t_s = 0.74$. In particular we investigate the asymptotic viscous–inviscid interaction of unsteady boundary layer separation as explained in the introduction. In order to evaluate when the viscous–inviscid interaction begins it is useful to analyze the streamwise pressure gradient at the wall, which is defined by $\partial_x p_w = -\frac{1}{Re}\Gamma_y(\bar{y})\frac{\partial \omega}{\partial y}|_{y=0}$. The streamwise pressure gradient does not change for boundary layer calculations contrary to the NS results. In Fig.3 it is shown the temporal evolution of $\partial_x p_w$ starting at $t = 0.2$ until $t = 0.5$, with increments of 0.1 for the three Re numbers considered. The pressure gradient is observed to change at the same streamwise location, close to the maximum, where a first region of recirculation forms. For $Re = 10^3$ the streamwise pressure gradient remains the same essentially up to $t \approx 0.2$

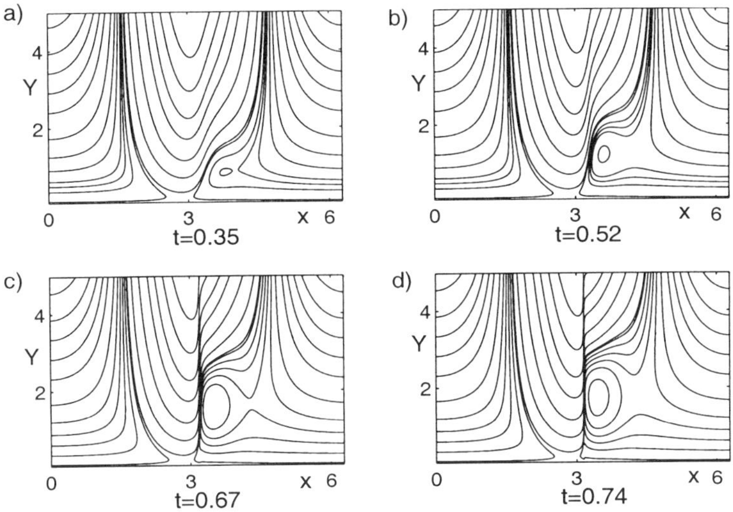

Fig. 1. Streamline from Prandtl's calculations are shown. In Fig.(a) a recirculating eddy is formed and growth in both streamwise and normal lenght (Fig.(b)). At $t = 0.67$ a kink seems to be formed in streamlines (Fig.(c)) and thickens in streamwise direction, evolving in a sharp spike at singularity time (Fig.(d)).

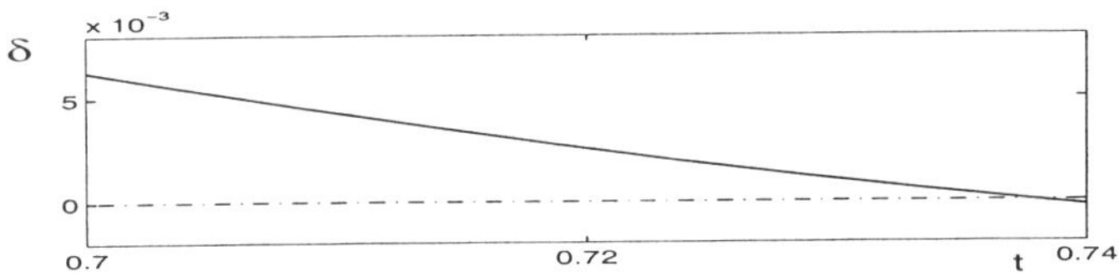

Fig. 2. Time evolution of exponential decay of Fourier modes of Prandtl's solution. At $t \approx 0.74$ the δ vanish and singularity forms.

while for $Re = 10^4 - 5 \cdot 10^5$ the interaction seems to start later, approximatively at $t \approx 0.3$. The time of first interaction is quite early with respect to theoretical prediction, according to which it is the rapidly growth of the spike that initiates viscous–inviscid interaction: in the boundary layer results shown in Fig.1 the spike forms at $t \approx 0.67$. It is clear that the interaction starts later for higher Re. These results are similar to other NS numerical solutions with other vortex induced initial data[14] and support the conclusion[16] that viscous–inviscid interaction accelerates the unsteady separation process. In Fig.3 the formation of a kink is visible for all Re considered at $t = 0.5$, but with increasing streamwise thickness as Re de-

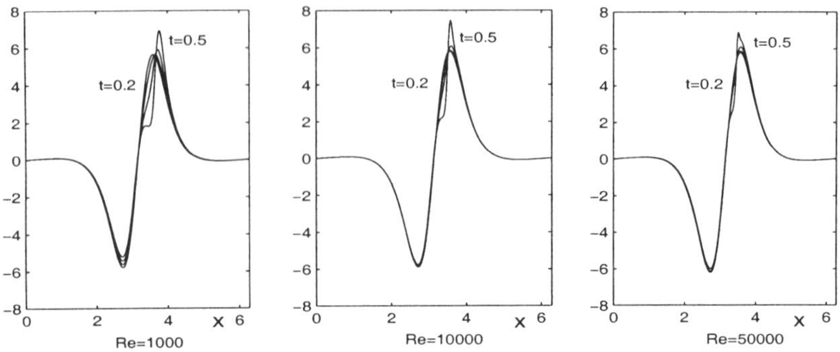

Fig. 3. The evolution in time, starting at $t = 0.2$ with increments of 0.1, of the streamwise pressure gradient at the wall for the Navier Stokes solutions with different Re numbers.

creases. This rapidly evolves in a spike for $Re = 10^4 - 5 \cdot 10^5$ as in the boundary layer results, and the formation of other spikes close to singularity time are evident, as shown in Fig.5(c) and Fig.6(d) (note that for $Re = 5 \cdot 10^5$ we report the result at time $t = 0.67$ as at singularity time the presence of many sharp spike makes the analysis difficult).

In Fig.4 and Fig.5, it is shown the evolution of streamlines at various time for $Re = 10^4$, compared with the streamwise pressure gradient along the surface and the scaled skin friction coefficient $C_f = -2\omega_{y=0}/Re$.

At time $t = 0.35$ the recirculation region is formed according with the boundary layer results. At time $t = 0.52$ the recirculation region has grown in size with respect to the normal direction and the streamwise pressure gradient starts to change locally. At this time it is also visible the formation of a local minimum in the streamwise pressure gradient in correspondence to the formation of a small spike in the recirculation region. The formation of a minimum after the maximum position in the streamwise pressure gradient accelerates the formation of the spike because the flow across the boundary is compressed in the streamwise direction, and this compression leads the recirculation region growth in the normal direction. At time $t = 0.74$, the recirculation region has split into a series of corotating eddies in correspondence to the local maxima of streamwise pressure gradient or the local minima of skin coefficient. The presence of more recirculating regions becomes more evident at time $t = 0.85$. This is quite different from boundary–layer solution behavior, which shows only one recirculation region without splitting. For comparison with the series of results described

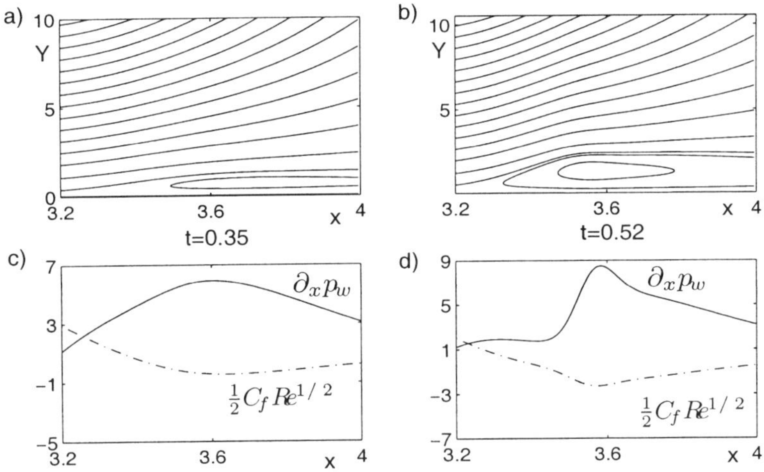

Fig. 4. The streamlines, the streamwise pressure gradient at the wall and the skin friction coefficient (dashed) for Navier Stokes solution with $Re = 10^4$, at times $t = 0.35$ and $t = 0.52$.

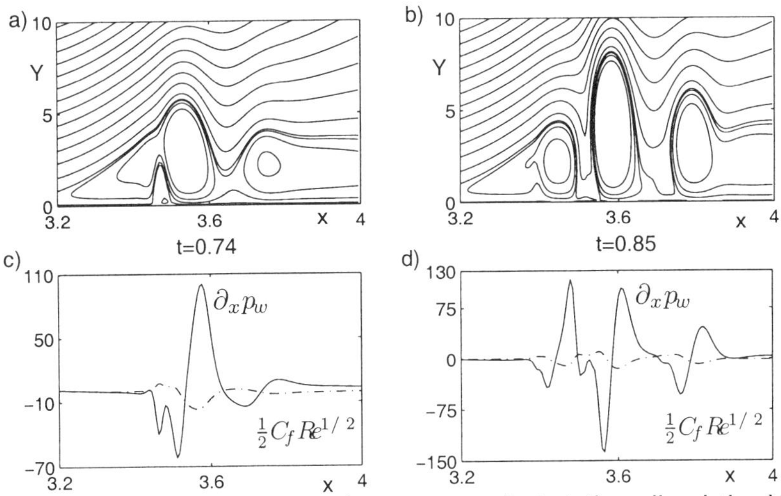

Fig. 5. The streamlines, the streamwise pressure gradient at the wall and the skin friction coefficient (dashed) for Navier Stokes solution with $Re = 10^4$, at times $t = 0.74$ and $t = 0.85$.

for $Re = 10^4$, in Fig.6 we show the results for $Re = 10^3$ and $Re = 5 \cdot 10^4$. The results for $Re = 10^3$ are different from those for moderately higher Re. There is no spike formation on the upstream side of the recirculation region, the recirculation region does not split, and no other recirculation region exists close to the singularity time.

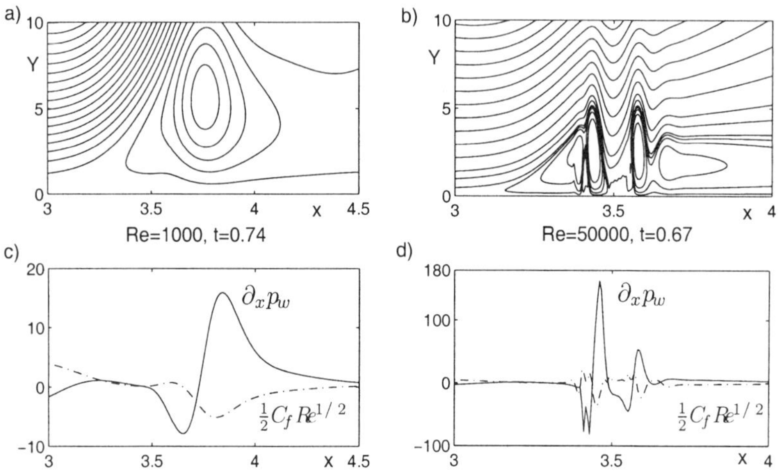

Fig. 6. The streamlines, the streamwise pressure gradient at the wall and the skin friction coefficient (dashed) for Navier Stokes solution with $Re = 10^3$ and $Re = 5 \cdot 10^5$.

4. Conclusions

We have computed the solutions of Prandtl's and Navier-Stokes equations at different Re numbers, for an array of rectilinear vortices interacting with a boundary. These results show that the unsteady separation process evolves differently for the three Re considered. The solutions agree with the BL results until large–scale interaction begins, according to the local change of the streamwise pressure gradient. The subsequent detachment process shows a different near–wall development. No splitting of the recirculation region is present for low Re, while the formation of various corotating eddies and of a spike–like behavior, is visible at moderate–high Re=10^4, $5 \cdot 10^5$, as a consequence of a small-scale interaction. In these Re regime the formation of the first spike in the solution happens at a time earlier then the time of BL spike formation. However we note that the larger the Re, the later the spike forms, supporting the conjecture that for a very large Re the separation time is very close to the BL singularity, without the occurrence of the large–scale interaction[14] .

Acknowledgments

This paper is dedicated to Tommaso Ruggeri with appreciation and friendship. The work of the third (MS) and fourth (VS) author has been supported by the INDAM and by the PRIN grant *Propagazione non lineare e stabilità nei processi termodinamici del continuo.* This work makes use of results produced by the PI2S2 Project managed by the Consorzio COMETA, a project co-funded by the Italian MIUR within the PON "Ricerca Scientifica, Sviluppo Tecnologico, Alta Formazione" (PON 2000-2006).

References

1. C. Anderson, M. Reider, *J. Comp. Phys.* **125** 207–224 (1996).
2. C. Canuto, M.Y. Hussaini, A. Quarteroni, T. Zang, *Spectral Methods in Fluid Dynamics*, 2th ed., (Springer Verlag, Berlin Heidelberg, 1988).
3. K.W. Cassel, *Phil. Trans. R. Soc. Lond. A* **358** 3207–3227 (2000).
4. S.J. Cowley, in *Proceedings of ICTAM 2000*, eds. H. Aref and J.W. Phillips, (Kluwer, 2001).
5. G. Della Rocca, M.C. Lombardo, M. Sammartino, V. Sciacca, *Appl. Numer. Math.* **56** 1108–1122 (2006).
6. L.L. van Dommelen, S.F. Shen, *J. Comp. Phys.* **38** 125–140 (1980).
7. W. E, J.G. Liu, *J. Comp. Phys.* **124** 368–382, (1996).
8. W. E, *Acta Math. Sin.* **16** 207–218 (2000).
9. J.M. Hoyle, F.T. Smith, J.D.A. Walker, *Comput. Phys. Commun.* **65** 151–157 (1991).
10. R. Krasny, *J. Fluid. Mech.* **167** 292–213 (1986).
11. O. Lamb, *Hydrodynamics*, sixth edition (Cambridge University Press, Cambridge, 1932).
12. M.C. Lombardo, M. Cannone, M. Sammartino, *SIAM J. Math. Anal.* **35** 987–1004 (2003).
13. A. Majda, A.L. Bertozzi, *Vorticity and Incompressible Flow*, Cambridge University Press, Cambridge, 2002.
14. A.V. Obabko, K.W. Cassel, *J. Fluid Mech.* **465** 99–130 (2002).
15. O.A. Oleinik, V.N. Samokhin, *Mathematical Models in Boundary Layer Theory* (Chapman & Hall/CRC, Boca Raton, FL, 1999).
16. V.J. Peridier, F.T. Smith, J.D.A. Walker, *J. Fluid Mech.* **232** 99–165 (1991).
17. M. Sammartino, R.E. Caflisch, *Comm. Math. Phys.*, **192** 433-461 (1998).
18. M. Sammartino, R.E. Caflisch, *Comm. Math. Phys.*, **192** 463-491 (1998).
19. M. Sammartino , V. Sciacca, *submitted* (2007).
20. H. Schlichting, *Boundary Layer Theory*, 4th ed., (McGraw-Hill, Karlsruhe, 1960).
21. V.V. Sychev, A.I. Ruban, V.V. Sychev, G.L. Korolev, *Asymptotic theory of separated flows*, (Cambridge University Press, Cambridge, 1998).
22. Z. Xin, L. Zhang, *Adv. Math.*, **181** 88–133 (2004).

A NOTE ABOUT WAVES IN DISSIPATIVE AND DISPERSIVE SOLIDS

M. DESTRADE

Institut Jean Le Rond d'Alembert (UMR7190),
CNRS / Université Pierre et Marie Curie,
4 place Jussieu, case 162,
75252 Paris Cedex 05, France
E-mail: destrade@lmm.jussieu.fr

G. SACCOMANDI

Dipartimento di Ingegneria Industriale,
Università degli Studi di Perugia,
06125 Perugia, Italy
E-mail: saccomandi@mec.dii.unipg.it

We study shear waves propagating in a special viscoelastic model proposed first by Fosdick and Yu in 1996. We deduce an asymptotic approximation which reduces the full balance equations to a system of evolution equations which are a vectorial generalization of the Modified KDV-Burger equation. In such a way we show that the model takes into account not only dissipative effects but also dispersive effects.

1. Introduction

We consider a transverse wave, polarized in the (XY) plane, and propagating in the Z direction, of a Cartesian coordinate system associated with an unbounded solid,

$$x = X + u(z,t), \qquad y = Y + v(z,t), \qquad z = Z. \tag{1}$$

Here u and v are the unknown scalar functions describing the motion. We compute the usual geometrical quantities of interest, here the left Cauchy-Green strain tensor and its inverse,

$$\mathbf{B} = \begin{bmatrix} 1+u_z^2 & u_z v_z & u_z \\ u_z v_z & 1+v_z^2 & v_z \\ u_z & v_z & 1 \end{bmatrix}, \qquad \mathbf{B}^{-1} = \begin{bmatrix} 1 & 0 & -u_z \\ 0 & 1 & -v_z \\ -u_z & -v_z & 1+(u_z^2+v_z^2) \end{bmatrix}. \tag{2}$$

For a general hyperelastic, incompressible, isotropic material, we have the following *constitutive equation* for the Cauchy stress tensor $\mathbf{T}$ (see Chadwick [1] for instance),

$$\mathbf{T} = -p\mathbf{I} + 2(\partial\Sigma/\partial I_1)\mathbf{B} - 2(\partial\Sigma/\partial I_2)\mathbf{B}^{-1}, \tag{3}$$

where $\mathbf{I}$ is the identity tensor and $p = p(\mathbf{x}, t)$ is the arbitrary Lagrange multiplier associated with the constraint of incompressibility ($\sqrt{\det \mathbf{B}} = 1$ at all times). In Eq. (3), Σ is the strain energy function, which for an incompressible material depends only on I_1 and I_2, the first two principal invariants of $\mathbf{B}$: $I_1 \equiv \text{tr } \mathbf{B}$, $I_2 \equiv [I_1^2 - \text{tr } (\mathbf{B}^2)]/2$. For the motion of Eq. (1), the first two invariants are

$$I_1 = I_2 = 3 + u_z^2 + v_z^2. \tag{4}$$

The *equations of motion*, in the absence of body forces, are given in current form as $\text{div } \mathbf{T} = \rho\ddot{\mathbf{x}}$, where ρ is the mass density; here they read

$$-\frac{\partial p}{\partial x} + \frac{\partial T_{13}}{\partial z} = \rho u_{tt}, \qquad -\frac{\partial p}{\partial y} + \frac{\partial T_{23}}{\partial z} = \rho v_{tt}, \qquad \frac{\partial T_{33}}{\partial z} = 0. \tag{5}$$

Differentiating these equations with respect to x, we find $p_{xx} = p_{yx} = p_{zx} = 0$, so that $p_x = p_1(t)$, say. Similarly, by differentiating the equations with respect to y, we find $p_y = p_2(t)$, say. The first two equations in (5) reduce to

$$-p_1(t) + (Qu_z)_z = \rho u_{tt}, \qquad -p_2(t) + (Qv_z)_z = \rho v_{tt}, \tag{6}$$

and the third equation determines p. Here, $Q = Q(u_z^2 + v_z^2)$ is the *generalized shear modulus* of nonlinear elasticity, defined by

$$Q = 2(\partial\Sigma/\partial I_1 + \partial\Sigma/\partial I_2). \tag{7}$$

Following Destrade and Saccomandi [2], we take the derivative of Eqs. (6) with respect to z, we introduce the notations $U = u_z$, $V = v_z$, and the complex function $W = U + \mathrm{i}V$, and we recast Eq. (6) as the single complex equation

$$(QW)_{zz} = \rho W_{tt}, \tag{8}$$

where Q is now a function of $U^2 + V^2$ alone, $Q = Q(U^2 + V^2)$. In nonlinear elasticity, it is common to require that the acoustic tensor be positive, in which case (8) is an hyperbolic system.

Now we decompose the complex function W into modulus Ω and argument θ as

$$W(z, t) = \Omega(z, t)\exp(\mathrm{i}\theta(z, t)), \tag{9}$$

and we focus on *travelling wave solutions* in the form

$$W = W(s) = \Omega(s)\mathrm{e}^{\mathrm{i}\theta(s)}, \quad \text{where} \quad s = z - ct, \tag{10}$$

c being the speed. This ansatz reduces Eq. (8) to $(QW)'' = \rho c^2 W''$. We integrate it twice taking each integration constant to be zero in order to eliminate the rigid and the homogeneous motions. We end up with the simple equation,

$$\left(Q - \rho c^2\right) W = 0. \tag{11}$$

Eq.(11) says, in accordance with the theory of hyperbolic second order nonlinear equations, that waves of permanent form are impossible unless $Q - \rho c^2 = 0$. This last opportunity is possible only if Q is independent of z, a situation that may happen only (a) for special classes of constitutive equations or (b) for a special classes of initial data.

For the first possibility (a), we determine that the most general constitutive class for which the generalized modulus is independent of z is the *Mooney-Rivlin model*, for which $2\Sigma = C(I_1 - 3) + E(I_2 - 3)$, where C and E are positive material constants; then $Q = C + E = \text{const}$. This is a special case of a general result by Ruggeri [3] about the existence of a double exceptional wave in unconstrained isotropic elastic materials.

For the second possibility (b), Carroll [4] determined the special solutions known as *circularly polarized harmonic waves*,

$$u(z,t) = A\cos k(z - ct), \quad v(z,t) = \pm A\sin k(z - ct), \tag{12}$$

where A and k are arbitrary constants. For these motions, $U^2 + V^2 = A^2k^2$ and therefore Q is independent of z; then the equation of motion Eq. (11) leads to the following dispersion equation,

$$Q(A^2k^2) = \rho c^2, \tag{13}$$

which may be solved for any *reasonable* constitutive equation.

Smooth solutions of initial-value problems for nonlinear hyperbolic systems are rare. Usually singularities will develop after a finite time, even when the initial data are smooth. To the best of our knowledge, theorems of global-in-time well-posedness to the initial-value problem for quasi-linear wave equations may be achieved only under the assumption of small initial data and the additional *null condition* [5]. Since our solutions (12) are smooth also for arbitrary large initial-data, it is clear that our knowledge of the mathematics of hyperbolic systems is still incomplete.

Kolsky [6] and Mason [7] showed that it is possible to produce experimentally tensile waves in stretched natural rubber bands, where the front becomes sharper as they progress. This is because natural vulcanized rubber becomes increasingly stiff with increasing tensile stress and because the stress-strain curve changes from concave to convex when very large deformations are involved. We argue that the hyperbolic system (8) may be a good mathematical approximation of such shock-like phenomena (For a more recent account of similar experiments, we refer to Vermorel et al. [8].)

On the other hand, rubber-like materials exhibit strong attenuation in the usual range of applications; it is for this reason that rubber is often used to damp out vibrations and to absorb shocks. Moreover, the underlying microstructure of polymeric materials introduces a characteristic nonlocal scale which is nearly 30 times the characteristic scale of face centered cubic materials such as copper [9]. This means that in many interesting applications there is an important range of wave-lengths where wave phenomena in elastomers and soft-tissues must be dispersive.

Thus it is necessary to improve our mathematical models of dynamic phenomena in rubber-like materials and to take into account both *dissipation* and *dispersion*. The aim of the present Note is to introduce a simple model accounting for these two effects in the framework of nonlinear solid mechanics.

2. Dispersion and dissipation

Guided by preliminary work [2], we now augment the constitutive equation Eq. (3) to $\mathbf{T} + \mathbf{T}^{\mathrm{D}}$, with

$$\mathbf{T}^{\mathrm{D}} = \nu \mathbf{A}_1 + \alpha[\mathbf{A}_2 - \mathbf{A}_1^2], \tag{14}$$

where $\mathbf{D}$ is the stretching tensor, and $\mathbf{A}_1$ and $\mathbf{A}_2$ are the first two Rivlin-Ericksen tensors,

$$\mathbf{D} \equiv (\mathbf{L} + \mathbf{L}^{\mathrm{T}})/2, \qquad \mathbf{A}_1 \equiv 2\mathbf{D}, \qquad \mathbf{A}_2 \equiv \dot{\mathbf{A}}_1 + \mathbf{A}_1\mathbf{L} + \mathbf{L}^{\mathrm{T}}\mathbf{A}_1. \tag{15}$$

The *viscosity function* $\nu = \nu(\mathbf{D} \cdot \mathbf{D})$, and the *dispersion material function* $\alpha = \alpha(\mathbf{D} \cdot \mathbf{D})$, must be positive due to thermodynamics restrictions.

Destrade and Saccomandi [10], show that at $\nu \equiv 0$, Eq. (14) coincides exactly with the dispersion function proposed by Rubin et al. [11], and that $\mathbf{T}^{\mathrm{D}}$ is a straightforward generalization of the extra Cauchy stress tensor associated with a non-Newtonian fluid of second grade [12], which is

$$\nu \mathbf{A}_1 + \alpha_1 \mathbf{A}_2 + \alpha_2 \mathbf{A}_1^2, \tag{16}$$

where ν is the classical viscosity and α_1, α_2 are the microstructural coefficients.

In the case of the motion (1), the kinematical quantities of interest are $\mathbf{A}_1$, $\mathbf{A}_1^2$, and $\mathbf{A}_2$, given by

$$\begin{bmatrix} 0 & 0 & u_{zt} \\ 0 & 0 & v_{zt} \\ u_{zt} & v_{zt} & 0 \end{bmatrix}, \quad \begin{bmatrix} u_{zt}^2 & u_{zt}v_{zt} & 0 \\ u_{zt}v_{zt} & v_{zt}^2 & 0 \\ 0 & 0 & u_{zt}^2+v_{zt}^2 \end{bmatrix}, \quad \begin{bmatrix} 0 & 0 & u_{ztt} \\ 0 & 0 & v_{ztt} \\ u_{ztt} & v_{ztt} & 2(u_{zt}^2+v_{zt}^2) \end{bmatrix}, \tag{17}$$

respectively. Hence we find that

$$\mathbf{D}\cdot\mathbf{D} = \tfrac{1}{2}(u_{zt}^2+v_{zt}^2). \tag{18}$$

Following the process conducted in Section 2 for hyperelastic materials, and introducing the complex function $W \equiv U + \mathrm{i}V$, we recast the determining equations for the transverse wave motions as the following single complex equation,

$$(QW)_{zz} + (\nu W_t + \alpha W_{tt})_{zz} = \rho W_{tt}, \tag{19}$$

where Q is again a function of $U^2+V^2 = \Omega^2$ alone: $Q = Q(\Omega^2)$, and ν and α are now functions of $U_t^2+V_t^2$ alone: $\nu = \nu(U_t^2+V_t^2)$, $\alpha = \alpha(U_t^2+V_t^2)$.

3. A vector evolution equation

As is usual, we now perform a *moving frame expansion* for equation Eq. (19), with the new scales $s = z - ct$, $\tau = \epsilon t$ (here ϵ is a small parameter).

We assume that W is of the form

$$W = \epsilon^{1/2} w, \quad \text{where} \quad w = O(1). \tag{20}$$

Then $\Omega = |W| = \epsilon^{1/2}|w|$ and we expand the terms in (8) as

$$\begin{aligned}
(QW)_{zz} &= \epsilon^{1/2}Q(0)w_{ss} + \epsilon^{3/2}Q'(0)\left(|w|^2 w\right)_{ss} + \dots, \\
\rho W_{tt} &= \epsilon^{1/2}\rho c^2 w_{ss} - 2\epsilon^{3/2}\rho c w_{s\tau} + \dots, \\
(\nu W_t)_{zz} &= -\epsilon^{1/2}c\nu(0)w_{sss} + \epsilon^{3/2}\nu(0)w_{ss\tau} + \dots, \\
(\alpha W_{tt})_{zz} &= \epsilon^{1/2}c^2\alpha(0)w_{ssss} - 2\epsilon^{3/2}c\alpha(0)w_{sss\tau} + \dots
\end{aligned} \tag{21}$$

In order to recover the linear wave speed at the lowest order (here, $\epsilon^{1/2}$) given by $Q(0) = \rho c^2$, we must assume that

$$Q(0) = O(1), \quad \nu(0) = O(1), \quad \text{and} \quad \alpha(0) = O(\epsilon) = \epsilon\alpha_0 \text{ (say)}, \tag{22}$$

where α_0 is a constant of order $O(1)$.

Then we find that at the next order,

$$Q'(0)\left(|w|^2 w\right)_{ss} - c\nu(0)w_{sss} + c^2\alpha_0 w_{ssss} = -2\rho c w_{s\tau}, \tag{23}$$

which we integrate once with respect to s to get the *vectorial MKdV-Burgers equation*,

$$w_\tau + q\left(|w|^2 w\right)_s - n w_{ss} + p w_{sss} = 0, \tag{24}$$

where $q \equiv Q'(0)/(2\rho c)$, $n \equiv \nu(0)/(2\rho)$, and $p \equiv c\alpha_0/(2\rho)$.

In this equation, the third derivative term is associated with dispersive phenomena whereas the second derivative term is associated with dissipative phenomena. To derive (24), we assumed that the nonlinear elastic effects are of the same order as the dissipative effects whilst the dispersive effects are of smaller order than elastic and dissipative effects. This assumption is quite realistic in biological applications at the length scales of interest in the framework of elastography.

4. Travelling waves solutions

We search for *travelling wave solutions* to equation (24). Introducing the variable $\xi = s - v\tau$, where v is the speed in the moving frame, we reduce (24) to the ordinary differential equation

$$-vw' + q\left(|w|^2 w\right)' - nw'' + pw''' = 0, \tag{25}$$

where a prime denotes the derivative with respect to ξ. With the usual asymptotic boundary conditions, we integrate once to obtain

$$\left(q|w|^2 - v\right) w - nw' + pw'' = d, \tag{26}$$

(here d is a real integration constant, to be considered null if we are interested in drop boundary conditions). Then, separating the real part of this equation from the imaginary part, by using the notation $w(\xi) = \omega(\xi)\mathrm{e}^{\mathrm{i}\vartheta(\xi)}$ say, gives

$$p\left(\omega'' - \omega\vartheta'^2\right) - n\omega' + \left(q\omega^2 - v\right)\omega = d, \qquad \omega^2\vartheta' = I\mathrm{e}^{\frac{n}{p}\xi}, \tag{27}$$

where the imaginary part has been integrated directly, with I as the integration constant. If we set $n = 0$ (no dissipation) in (27), then we recover a classical result by Gorbatcheva and Ostrosky [13].

The system (27) may be reduced to a single non-autonomous second order equation in the general case ($I \neq 0$.) When we consider linearly-polarized waves ($I = 0$) we get an autonomous Duffing-like equation with damping

$$\omega'' - \frac{n}{p}\omega' + \left(\frac{q}{p}\omega^2 - \frac{v}{p}\right)\omega = d. \tag{28}$$

In fact, if we focus on linearly polarized waves from the outset, then $|w| = w$, and Eq. (24) is the scalar MKdV-Burgers equation,

$$w_\tau + \left(w^3\right)_x = \alpha w_{xx} + \beta w_{xxx}, \tag{29}$$

where $x \equiv s/q$, $\alpha = n/q^2$, and $\beta = -p/q^3$. The possibility of travelling wave solutions to the modified Korteweg-deVries-Burgers equation has been studied in great details over the years, in particular by Jacobs et al. [14], Wang [15], Feng [16], or Vladimirov et al. [17]. However it seems that the combination: $\alpha > 0$, $\beta < 0$, — as found here — precludes the possibility of exact solutions.

We recall that the governing equations derived above have been obtained for any type of nonlinear incompressible solid, for which dispersion and dissipation can be modelled by Eq. (14); however, the equations were the result of a small parameter expansion, see Section 3. We now remark that it is also possible to study travelling waves for the *exact equation* (19), both in the case of linearly-polarized waves and in the non-linearly polarized case, when a given constitutive behaviour is chosen. For instance, we make the constitutive assumptions that $\alpha =$ const., $\nu =$ const., which are the simplest assumptions we can make for the modelling of dispersion and dissipation. For the shear modulus, the choice $Q =$ const. corresponds to a Mooney-Rivlin type of elastic behaviour and it has been treated elsewhere [18]: it leads to linear differential equations. Then the assumption $Q = \mu_0 + \mu_1 \Omega^2$ (where μ_0, μ_1 are positive constants) is the simplest one we can make to uncover nonlinear governing equations; it corresponds to fourth-order elasticity theory [10]. With these assumptions, equation (19) reduces to

$$\left[(\mu_0 + \mu_1 \Omega^2) W\right]_{zz} + \nu W_{tzz} + \alpha W_{ttzz} = \rho W_{tt}, \tag{30}$$

a vectorial version of the damped good Boussinesq equation. A thorough review on several mechanical aspects and applications of this equation may be found in the recent paper by Christov et al. [19]. When we study travelling waves of (30) in the linearly polarized case, we obtain

$$\alpha c^2 \Omega'' - \nu c \Omega' + \left(\mu_1 \Omega^2 + \mu_0 - \rho c^2\right) \Omega = D, \tag{31}$$

which is equivalent to (28).

By standard phase plane analysis it is possible to obtain the conditions on the coefficients of (31) for which *kink*-like solutions are possible. For example in the case $D = 0$ (or $d = 0$ if we are considering (28)) a detailed discussion of these solutions is provided by Feng [20] where it is shown that monotonous kinks (like in dissipative systems) are possible when the classical viscosity is sufficiently strong. Otherwise, when dispersive and/or nonlinear elastic effects are more important than dissipative effects, we observe kink-like solutions with an oscillatory character.

References

1. P. Chadwick, *Continuum Mechanics* (Dover, New York, 1999).
2. M. Destrade and G. Saccomandi, *Phys. Rev. E* **72**, 016620 (2005).
3. T. Ruggeri,, *Annali di Matematica pura ed applicata (IV)* **CXII**, 315 (1977).
4. M. M. Carroll, *Q. J. Mech. appl. Math.* **30**, 223 (1977).
5. W. Domański and R.W. Ogden *Arch. Mech* **58**, 339 (2006).
6. H. Kolsky, *Proc. Phys. Soc. B* **62**, 676 (1949).
7. P. Mason, *Proc. R. Soc. A* **272**, 315 (1963).
8. R. Vermorel, N Vandeberghe, and E. Villermaux *Proc. Roy. Soc. A* **463**, 641-658 (2007).
9. R. Marangani and P. Sharma, *Phys. Rev. Lett.* **98**, 195504 (2007).
10. M. Destrade and G. Saccomandi, *Wave Motion* (to appear 2007).
11. M. B. Rubin, P. Rosenau and O. Gottlieb, *J. Appl. Phys.*, **77**, 4054 (1995).
12. R. L. Fosdick and J. H. Yu, *Int. J. Nonlinear Mech.* **31**, 495 (1996).
13. O.B. Gorbacheva, L.A. Ostrovsky, *Physica D*, **8**, 223 (1983).
14. D. Jacobs, B. McKineey, and M. Shearer, *J. Diff. Eq.* **116**, 448 (1995).
15. M. Wang, *Phys. Lett. A* **213**, 279 (1996).
16. Z. Feng, *Phys. Lett. A* **318** 522 (2003).
17. V.A. Vladimirov, E.V. Kutafina, and A. Pudelko, *SIGMA* **2**, Paper 061 (2006).
18. M. Destrade and G. Saccomandi, *Proc. 13th Conf. Waves Stab. Continuous Media WASCOM 2005* (World Scientific, 2006).
19. C.C. Christov, G.A. Maugin, and A.V. Porubov, *C. R. Mécanique* doi:10.1016/j.crme.2007.08.006 (2007).
20. Z. Feng, *Chaos, Solitons and Fractals* **28** 463 (2006).

NON LOCAL THERMODYNAMICAL EQUILIBRIUM LINE RADIATIVE TRANSFER QUASI-STATIONARY APPROXIMATION

LAURENT DESVILLETTES[1] AND CHUNJIN LIN[2]

[1] *CMLA, ENS Cachan, CNRS, PRES UniverSud*
61, Av. du Pdt. Wilson, 94235 Cachan Cedex, FRANCE
E-mail: desville@cmla.ens-cachan.fr

[2] *College of Science, Hohai University, Nanjing 210098, CHINA,*
E-mail: chun-jin.lin@163.com

In this paper, we prove existence and uniqueness of solutions to the coupling between the radiative transfer equation and equations for the population of atoms in a certain state. We also prove the validity of the quasi-static approximation in this context.

Keywords: Radiative transfer, quasi-static approximation

1. Introduction

We consider a coupling between a radiating field and a plasma. The radiation is described by the following transport equation

$$\frac{1}{c}\partial_t f + v \cdot \nabla_x f = \eta - \chi f, \tag{1}$$

where the constant c is the speed of light, $v \in \mathbb{S}^2$ is the direction of propagation of photons, η is the emission coefficient (or emissivity) of matter, χ is the absorption coefficient (or extinction coefficient) of matter, and the unknown is the specific intensity $f := f(t, x, v, \nu)$ which is a function of time $t \in \mathbb{R}^+$, space position $x \in X \subset \mathbb{R}^3$, velocity direction $v \in \mathbb{S}^2$, and frequency $\nu \in \mathbb{R}^+$.

If we consider the coefficients η and χ as given, the transfer equation (1) is linear and its solution can be written explicitly by integrating along the characteristics. These coefficients depend however in reality upon the internal excitation and ionization states of the plasma. These states are fixed in part by radiative processes that populate and depopulate atomic levels. For the line radiative transfer (bound-bound transitions without ionization), they depend on the Einstein coefficients, the spontaneous emission probability A_{ji} (with $i, j \in \{1, .., K\}$, $i < j$), the absorption probability B_{ij} and the induced (stimulated) emission probability B_{ji}, and can be written as

$$\eta = \sum_i \sum_{j>i} n_j A_{ji} h\nu \phi_{ij}(\nu), \tag{2}$$

$$\chi = \sum_i \sum_{j>i} (n_i B_{ij} - n_j B_{ji})\, h\nu \phi_{ij}(\nu), \tag{3}$$

where n_i (respectively n_j) denotes the population density at the atomic level $i = 1, .., K$ (respectively j), and $\phi_{ij}(\nu)$ represents the line profile for these transitions (it can for example be approximated by a Gaussian function of ν centered around the frequency ν_{ij} of the transition). Finally, h is the Planck constant.

The population density n_i at level i satisfies the following rate equation, in a static medium,

$$\frac{\partial n_i}{\partial t} = \sum_{j \neq i} n_j P_{ji} - n_i \sum_{j \neq i} P_{ij}, \tag{4}$$

where P_{ij} denotes the total (radiative plus collisional) transition rate from level i to level j. Note that the total population of atoms $n := \sum_{i=1}^{K} n_i$ is clearly conserved along time.

Bound-bound transitions (line transitions) between the lower energy level i and the upper energy level j may occur as radiative excitation, spontaneous radiative de-excitation, induced radiative de-excitation, collisional excitation and collisional de-excitation. Let us denote C_{ij} (respectively C_{ji}) the rate of collisional excitation (respectively the rate of collisional de-excitation). In (4), the total excitation rate P_{ij} and the total de-excitation rate P_{ji} can be written as

$$P_{ij} = B_{ij}\rho_{ij} + C_{ij}, \qquad P_{ji} = A_{ji} + B_{ji}\rho_{ij} + C_{ji}, \tag{5}$$

where ρ_{ij} is the integrated mean intensity over the line profile $\phi_{ij}(\nu)$:

$$\rho_{ij}(t, x) = \int_{\mathbb{R}^+} \int_{\mathbb{S}^2} f(t, x, v, \nu)\, \phi_{ij}(\nu)\, \mathrm{d}v \mathrm{d}\nu, \tag{6}$$

with $\mathrm{d}v$ denoting the normalized Lebesgue measure on $\mathbb{S}^2$. For the physical background underlying eq. (1) – (6), we refer to to [9] §85, [10] §2.6.

In general, the radiation field and the internal state of the matter must be determined simultaneously and self-consistently. In many situtions, the characteristic time of the excitation and de-excitation processes of the matter is much smaller than the characteristic time of the evolution of the radiative field. After adimensionalizing the time variable in eq. (1) and (4), it is therefore possible to introduce a parameter $\epsilon > 0$ such that our coupled system becomes

$$\begin{cases} \frac{1}{c}\partial_t f^\epsilon + v \cdot \nabla_x f^\epsilon \\ \qquad = \sum_i \sum_{j>i} n_j^\epsilon A_{ji} h\nu \phi_{ij}(\nu) - \sum_i \sum_{j>i} \left(n_i^\epsilon B_{ij} - n_j^\epsilon B_{ji} \right) h\nu \phi_{ij}(\nu) f^\epsilon, \\ \epsilon \frac{\partial n_i^\epsilon}{\partial t} = \sum_{j \neq i} n_j^\epsilon P_{ji} - \sum_{j \neq i} n_i^\epsilon P_{ij}. \end{cases} \tag{7}$$

We consider the system (7) in the case when the position variable x varies in a bounded (regular, open) domain $X \subset \mathbb{R}^3$. We add therefore the initial condition

$$f^\epsilon(0, x, v, \nu) = f_0(x, v, \nu), \tag{8}$$

and the incoming boundary condition

$$f^\epsilon|_{\mathbb{R}^+ \times (\partial X \times \mathbb{S}^2)_- \times \mathbb{R}^+} = g(t, x, v, \nu), \tag{9}$$

where $(\partial X \times \mathbb{S}^2)_- := \{(x, v) \in \partial X \times \mathbb{S}^2 : \Gamma_x \cdot v < 0\}$, with Γ_x denoting the outward normal to X at the point $x \in \partial X$. Finally, the initial population densities $n_i^\epsilon(0, x)$ are given by

$$\forall i = 1, .., K, \qquad n_i^\epsilon(0, x) = n_{i0}(x) \geq 0. \tag{10}$$

We are interested in the existence of solutions f^ϵ, $(n_i^\epsilon)_{i=1,..,K}$, to (5) – (10) (when $\epsilon > 0$ is fixed), and in the behavior of the solutions f^ϵ, n_i^ϵ, as $\epsilon \to 0$ (quasi-stationary approximation).

In the sequel, we shall consider the following assumption on the data:

Assumption A: The initial condition f_0 and the boundary condition g satisfy

$$0 \leq f_0 \in L^\infty(X \times \mathbb{S}^2 \times \mathbb{R}^+), \qquad 0 \leq g \in L^\infty(\mathbb{R}^+ \times (\partial X \times \mathbb{S}^2)_- \times \mathbb{R}^+), \tag{11}$$

and the initial occupation numbers n_{i0} are such that $n(x) = \sum_{i=1}^K n_{i0}(x) \in L^\infty(X)$.

The Einstein Coefficients A_{ji}, B_{ij} and B_{ji} are (strictly positive) constants, and the collisional coefficients C_{ij} and C_{ji} are (nonnegative) functions of the position $x \in X$ verifying

$$\delta_* \leq C_{ij}(x),\ C_{ji}(x) \leq \delta^*, \tag{12}$$

for some δ_*, $\delta^* > 0$.

Finally, the line profile ϕ_{ij} is integrable on $\mathbb{R}^+$ and satisfies, for some $\delta > 0$,

$$\forall \nu \in \mathbb{R}^+, \qquad 0 \leq \phi_{ij}(\nu) h\nu \leq \delta. \tag{13}$$

Our main result is stated as

Theorem 1.1. *Let assumption A on the data be satisfied. Then, for any given $T > 0$, there exists a unique nonnegative solution f^ϵ, $(n_i^\epsilon)_{i=1,..,K}$, to (5) – (10), which belongs to $L^\infty([0,T] \times X \times \mathbb{S}^2 \times \mathbb{R}^+) \times (L^\infty([0,T] \times X))^K$. Furthermore, as $\epsilon \to 0$, this solution converges in $L^\infty([0,T] \times X \times \mathbb{S}^2 \times \mathbb{R}^+) \times (L^\infty([0,T] \times X))^K$ weak * to f, $(n_i)_{i=1,..,K}$, unique nonnegative solution in $L^\infty([0,T] \times X \times \mathbb{S}^2 \times \mathbb{R}^+) \times (L^\infty([0,T] \times X))^K$ to the system*

$$\begin{cases} \frac{1}{c}\partial_t f + v \cdot \nabla_x f \\ \qquad = \sum_i \sum_{j>i} n_j A_{ji} h\nu \phi_{ij}(\nu) - \sum_i \sum_{j>i} (n_i B_{ij} - n_j B_{ji}) h\nu \phi_{ij}(\nu) f, \\ 0 = \sum_{j \neq i} n_j P_{ji} - \sum_{j \neq i} n_i P_{ij}, \\ f(0,x,v,\nu) = f_0(x,v,\nu), \qquad f|_{\mathbb{R}^+ \times (\partial X \times \mathbb{S}^2)_- \times \mathbb{R}^+}(t,x,v,\nu) = g(t,x,v,\nu), \end{cases} \tag{14}$$

where P_{ji}, P_{ij} are given by formulas (5), (6).

Most of the rest of the paper is devoted to the proof of Theorem 1.1. Existence and uniqueness of a solution to (5) – (10) (for a given ϵ) are proven in section 2. At the end of this section, we also show a result of existence and uniqueness for the limiting system (5), (6), (14).

Then, in section 3, we prove the validity of the quasi-stationary approximation, that is the convergence of solutions of (5) – (10) when $\epsilon \to 0$ toward solutions of (5), (6), (14).

Finally, we present a numerical test in order to illustrate this convergence in section 4.

In all the sequel, we shall restrict ourselves in the proof, for the sake of simplicity, to a two-level molecular model (that is, $K = 2$). The proof in the general case is identical.

In this paper we limit our discussion to the bound-bound transitions, we refer for details on the bound-free transitions or the free-free transitions to [9,10], or the papers [2,3,5].

We refer to [1] for the existence theory of the radiative transfer equation for a 'grey' model, by using the compactness result introduced in [6,7], that is, the averaging lemma.

In [4,11], the authors studied some numerical methods for the line radiative transfer, and the comparison was given between a number of independent computer programs for radiative transfer in molecular rotational lines. Our numerical tests are inspired from the data introduced in [4,11].

2. Proof of existence and uniqueness to system (5) – (10) for a given ϵ

We begin with a classical explicit resolution of the linear kinetic equation.

Lemma 2.1. *Let X be a bounded regular open set in $\mathbb{R}^3$. We consider the following system:*

$$\begin{cases} \dfrac{1}{c}\partial_t f + v\cdot\nabla_x f = \eta - \chi f, \\ f(0,x,v,\nu) = f_0(x,v,\nu) \geq 0, \\ f|_{\mathbb{R}^+\times(\partial X\times\mathbb{S}^2)_-\times\mathbb{R}^+}(t,x,v,\nu) = g(t,x,v,\nu) \geq 0, \end{cases} \tag{15}$$

where the initial data f_0, the boundary data g, and the coefficients η, χ are bounded.

Then, for any given $T>0$, there exists a constant $\delta(T)>0$ (depending only on T and the L^∞ norms of η, χ, f_0 and g) such that

$$\forall (t,x,v,\nu)\in[0,T]\times X\times\mathbb{S}^2\times\mathbb{R}^+, \qquad 0\leq f(t,x,v,\nu)\leq\delta(T). \tag{16}$$

Proof of Lemma 2.1.

Let us denote $Q=\{(t,x)|t\in\mathbb{R}^+, x\in X\}$, and denote by Σ the boundary of Q. The boundary Σ has thus two parts:

$$\Sigma=\Sigma_1\bigcup\Sigma_2=\{(0,x)|x\in X\}\bigcup\{(t,x)|t\in\mathbb{R}^+, x\in\partial X\}.$$

Let us fix a point $M^*=(t^*,x^*)$ in Q, and introduce a characteristic line through M^* as

$$t\longmapsto x(t)=x^*-c\,v(t^*-t). \tag{17}$$

We look for the intersection of this characteristic line with Σ, the boundary of Q. There are two cases: either the line remains in Q and intersects Σ_1, (that is, the plane $t=0$) at the point $x(0)=x_0=x^*-c\ v\ t^*$, or the line intersects $\Sigma_2=\{(t,x)|x\in\partial X, t>0\}$ at some point $(t_0,x(t_0))$ with $0\leq t_0<t^*$.

In both cases, it is possible to write f explicitly in terms of f_0, g, χ and η using the characteristic lines (17) and Duhamel's formula. The estimate is obtained by taking L^∞ norms in this explicit formulation of f. We refer to [8] for details. □

Proof of Theorem 1.1: We begin by proving the existence of solutions to system (5) – (10) (when ϵ is fixed) thanks to an iterative procedure. In order to keep notations tractable, we denote f instead of f^ϵ and n_i instead of n_i^ϵ.

This procedure is defined in this way:

For $t \geq 0$, we set

$$f^0(t,x,v,\nu) = f_0(x,v,\nu), \quad n_i^0(t,x) = n_{i0}(x), \;\; i = 1,2;$$

For $k = 0,1,2,...,$ we assume that (f^k, n_1^k, n_2^k) are defined. We define $f^{k+1}, n_1^{k+1}, n_2^{k+1}$ by

$$\begin{cases} \frac{1}{c}\partial_t f^{k+1} + v\cdot\nabla_x f^{k+1} = \left(n_2^k A_{21} - (n_1^k B_{12} - n_2^k B_{21}) f^{k+1}\right)\phi_{12}(\nu) h\nu, \\ f^{k+1}(0,x,v,\nu) = f_0(x,v,\nu), \\ f^{k+1}|_{\mathbb{R}^+\times(\partial X\times\mathbb{S}^2)_-\times\mathbb{R}^+}(t,x,v,\nu) = g(t,x,v,\nu), \end{cases} \tag{18}$$

and

$$\begin{cases} \epsilon\partial_t n_1^{k+1} = n_2^{k+1}A_{21} + (n_2^{k+1}B_{21} - n_1^{k+1}B_{12})\rho^{k+1} + (n_2^{k+1}C_{21} - n_1^{k+1}C_{12}) \\ \epsilon\partial_t n_2^{k+1} = -\big[n_2^{k+1}A_{21} + (n_2^{k+1}B_{21} - n_1^{k+1}B_{12})\rho^{k+1} \\ \qquad\qquad +(n_2^{k+1}C_{21} - n_1^{k+1}C_{12})\big] \\ n_i^{k+1}(0,x) = n_{i0}(x), \;\; i = 1,2. \end{cases} \tag{19}$$

Note that n_1^{k+1} and n_2^{k+1} can be written explicitly in terms of f^{k+1} in eq. (19) thanks to Duhamel's formula.

Using Lemma 2.1 and the nonnegativity of the initial population densities n_{i0} (and of f_0), we see that $(f^{k+1}, n_1^{k+1}, n_2^{k+1})$ are well-defined, and that

$$\forall k\in\mathbb{N},\; i=1,2, \qquad n_i^k \geq 0, \qquad n_1^k + n_2^k = n_{10} + n_{20} = n.$$

Moreover, still thanks to lemma 2.1, we see that f^k satisfies $0 \leq f^k(t,x,v,\nu) \leq \delta(T)$, for all $(t,x,v,\nu) \in [0,T]\times X\times\mathbb{S}^2\times\mathbb{R}^+$.

Using the equation satisfied by $f^{k+1} - f^k$ and the characteristics, it is possible to show that (when $t\in[0,T]$)

$$\left\|f^{k+1} - f^k\right\|_{L^\infty_{x,v,\nu}}(t) \leq \xi_1(T)\int_0^t \left\|\left(n_1^k - n_1^{k-1}\right)(s)\right\|_{L^\infty_x} \mathrm{d}s, \tag{20}$$

for some constant $\xi_1(T) \geq 0$.

Using then the equation satisfied by $n_1^{k+1} - n_1^k$, it is possible to show that (when $t \in [0,T]$)

$$\|n_1^{k+1} - n_1^k\|_{L_x^\infty}(t) \leq \frac{\xi_2(T)}{\epsilon} \sup_{s\in[0,t]} \left\| \left(f^{k+1} - f^k\right)(s) \right\|_{L^\infty_{x,v,\nu}}, \tag{21}$$

for some constant $\xi_2(T) \geq 0$. The proof of estimates (20) and (21) is detailed in [8].

Using (20) and (21), a classical induction argument shows that for all $k \in \mathbb{N}$, $p \in \mathbb{N}^*$,

$$\|n_1^{k+p} - n_1^k\|_{L^\infty([0,T]\times X)} \leq Cst \sum_{l=k}^{k+p-1} \frac{(\xi_1(T)\, \xi_2(T)/\epsilon)^l}{l!}.$$

Thus we obtain that $(n_1^k)_k$ is a Cauchy sequence in $L^\infty([0,T] \times X)$. The same holds of course for $(n_2^k)_k$. Then, using estimate (20), we see that $(f^k)_k$ is also a Cauchy sequence in $L^\infty([0,T] \times X \times \mathbb{S}^2 \times \mathbb{R}^+)$. We can therefore pass to the limit in (the Duhamel formulations of) equations (18) and (19), and obtain a bounded solution f, n_1, n_2 to the coupled system (5) – (10).

Uniqueness is obtained by simply considering two solutions (f, n_1, n_2) and $(\overline{f}, \overline{n}_1, \overline{n}_2)$ to (5) – (10) with the same initial and boundary conditions, and by using estimates (20), (21) (with f, $\overline{f}$ instead of f^{k+1}, f^k, and the same for the populations). This ends the proof of the first part of theorem 1.1.

We conclude this section by observing that when we replace (19) by

$$\begin{cases} n_1^{k+1} = \dfrac{A_{21} + B_{21}\rho^{k+1} + C_{21}}{A_{21} + (B_{21} + B_{12})\rho^{k+1} + C_{12} + C_{21}}\, n, \\ n_2^{k+1} = \dfrac{B_{12}\rho^{k+1} + C_{12}}{A_{21} + (B_{21} + B_{12})\rho^{k+1} + C_{12} + C_{21}}\, n, \end{cases} \tag{22}$$

the inductive procedure (18), (22) together with estimate (20) enables to build a solution to system (5), (6), (14). Uniqueness for this system is also a consequence of estimate (20). We refer to [8] for details.

3. Quasi-stationary approximation, convergence

In this section, we prove the second part of Theorem 1.1, that is the convergence of the solution f^ϵ, $(n_i^\epsilon)_{i=1,2}$ toward the solution f, $(n_i)_{i=1,2}$ of the limiting system (5), (6), (14).

We already know that for $i = 1, 2$, $0 \le n_i^\epsilon(t, x) \le ||n||_{L^\infty}$. As a consequence of lemma 2.1 and this estimate, we obtain that $(f^\epsilon)_\epsilon$ is bounded in $L^\infty([0, T] \times X \times \mathbb{S}^2 \times \mathbb{R}^+)$, so that $\int_{\mathbb{R}^+} f^\epsilon(t, x, v, \nu)\, \phi_{12}(\nu)\, \mathrm{d}\nu$ is bounded in $L^\infty([0, T] \times X \times \mathbb{S}^2)$. Furthermore, this quantity solves the following system:

$$\begin{cases} \frac{1}{c}\partial_t \int_{\mathbb{R}^+} f^\epsilon \phi_{12}(\nu)\mathrm{d}\nu + v \cdot \nabla_x \int_{\mathbb{R}^+} f^\epsilon \phi_{12}(\nu)\mathrm{d}\nu \\ \qquad = n_2^\epsilon A_{21} \int_{\mathbb{R}^+} \phi_{12}^2(\nu) h\nu \mathrm{d}\nu - (n_1^\epsilon B_{12} - n_2^\epsilon B_{21}) \int_{\mathbb{R}^+} f^\epsilon \phi_{12}^2(\nu) h\nu \mathrm{d}\nu, \\ \left(\int_{\mathbb{R}^+} f^\epsilon \phi_{12}(\nu)\mathrm{d}\nu \right)(0, x, v) = \int_{\mathbb{R}^+} f_0(x, v, \nu)\phi_{12}(\nu)\mathrm{d}\nu, \\ \left(\int_{\mathbb{R}^+} f^\epsilon \phi_{12}(\nu)\mathrm{d}\nu \right)\Big|_{\mathbb{R}^+ \times (\partial X \times \mathbb{S}^2)_-} = \int_{\mathbb{R}^+} g(t, x, v, \nu)\phi_{12}(\nu)\mathrm{d}\nu. \end{cases}$$

Using the L^∞ bounds of f^ϵ, n_i^ϵ and the properties of ϕ_{12}, we see that

$$(\frac{1}{c}\partial_t + v \cdot \nabla_x) \int_{\mathbb{R}^+} f^\epsilon \phi_{12}(\nu)\mathrm{d}\nu$$

is bounded in $L^\infty([0, T] \times X \times \mathbb{S}^2)$. Thanks to an averaging lemma ([6, 7]), we obtain that the family $\int_{\mathbb{R}^+}\int_{\mathbb{S}^2} f^\epsilon(t, x, v, \nu)\phi(\nu)\mathrm{d}v\mathrm{d}\nu = \rho^\epsilon(t, x)$ is strongly compact in $L^2([0, T] \times X)$. This ensures that ρ^ϵ converges (up to a subsequence) a.e.

Thus (still up to a subsequence), we can assume that

$$\begin{aligned} &n_i^\epsilon \rightharpoonup n_i \text{ weakly* in } L^\infty([0, T] \times X), \qquad i = 1, 2; \\ &f^\epsilon \rightharpoonup f \text{ weakly* in } L^\infty([0, T] \times X \times \mathbb{S}^2 \times \mathbb{R}^+); \\ &\rho^\epsilon \to \rho \text{ strongly in } L^1([0, T] \times X), \end{aligned}$$

where

$$\rho = \int_{\mathbb{R}^+} \int_{\mathbb{S}^2} f \phi_{12}(\nu)\mathrm{d}v\mathrm{d}\nu.$$

The sequence $n_i^\epsilon \rho^\epsilon$ converges therefore to $n_i \rho$ weakly in $L^1([0, T] \times X)$.

It remains also to pass to the limit in the quantity $n_i^\epsilon\, f^\epsilon\, \phi_{12}(\nu)\, h\nu$. This is done by observing that for any test function $\psi_1(v)\, \psi_2(\nu)$ (with $\psi_1, \psi_2 \in \mathcal{D}$), the quantity

$$\int_{\mathbb{R}^+} \int_{\mathbb{S}^2} f^\epsilon(t, x, v, \nu)\, \phi_{12}(\nu)\, h\nu\, \psi_1(v)\, \psi_2(\nu)\, \mathrm{d}v\mathrm{d}\nu$$

converges for a.e. t, x. This is due to the fact that the quantity $\int_{\mathbb{R}^+} f^\epsilon(t, x, v, \nu)\, \phi_{12}(\nu)\, h\nu\, \psi_2(\nu)\, \mathrm{d}\nu$ satisfies a kinetic equation (like $\int_{\mathbb{R}^+} f^\epsilon(t, x, v, \nu)\, \phi_{12}(\nu)\, \mathrm{d}\nu$), so that it is possible to use an averaging lemma.

Finally, when ϵ tend to 0, the solution to (5) – (10) converges up to extraction to a solution of (5), (6), (14).

Thanks to the result of uniqueness for (5), (6), (14) obtained in the previous section, the convergence is in fact not restricted to a subsequence. This ends the proof of Theorem 1.1.

We notice that in the limiting equation, no initial data are needed for the populations n_1, n_2. As a consequence, an initial layer appears if the initial data of the problem for a given $\epsilon > 0$ are not compatible with the limiting equation.

4. Numerical simulation

We introduce a numerical test in order to see how the quasi-static approximation is valid in practice. This test is inspired from the problem that was introduced in [4,11]. It consists in a 3D computation with two populations of atoms ($K = 2$), and no initial layer. For a detailed description of the data, we refer to [8].

The rate equations of the atomic populations are discretized with usual methods for the ODEs, while for solving the kinetic equation, we use a particle method.

In order to verify the convergence of solutions, we compute the (relative) difference between n_1^ϵ and n_1, (solution of the limiting system) i.e

$$\frac{|n_1^\epsilon(t,x) - n_1(t,x)|}{n_1(t,x)}, \tag{23}$$

obtained at a given time for different values of ϵ. This quantity is presented as a function of $|x|$, for a given direction of the space variable.

The validity of the quasi-static approximation is observed on our simulation, see fig. 1. In practice, the value of ϵ is usually extremely small (smaller than in the simulations presented here).

References

1. Bardos, C.; Golse, F.; Perthame, B. and Sentis, R.; *The nonaccretive radiative transfer equations: existence of solutions and Rosseland approximation*, J. Funct. Anal., **77**, n.2, (1988), 434–460.
2. Belkov, S.; Gasparyan, P.; Kochubei, Yu. and Mitrofanov, E.; *Average-ion model for calculating the state of a multicomponent transient nonequilibrium highly charged ion plasma*, J. Exp. Theor. Phys., **84**, (1997), 272-280.
3. Djaoui A. and Rose S.J.; *Calculation of the time-dependent excitation and ionization in a laser-produced plasma*, J. Phys. B: Atomic, Molecular and Optical Physics, **25**, (1992), 2745-2762,

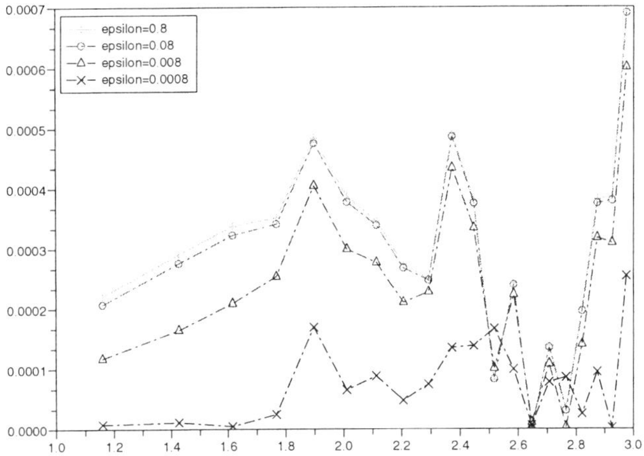

Fig. 1. $|n_1^\epsilon - n_1|/n_1$ in terms of $|x|$

4. Dullemond, C. P.; Radiative transfer in compact circumstellar nebulae, University of Leiden, 1999.
5. Faussurier, G.; Blancard, C. and Berthier, E. ; *Nonlocal thermodynamic equilibrium self-consistent average-atom model for plasma physics*, Phys. Rev. E, **63**, n. 2, (2001).
6. Golse, F.; Perthame, B.; Sentis, R.; *Un résultat de compacité pour les équations de transport et application au calcul de la limite de la valeur propre principale d'un opérateur de transport*, C. R. Acad. Sci. Paris Sér. I Math., **301**, n.7, (1985), 341–344.
7. Golse, F., Lions; P.-L.; Perthame, B. and Sentis, R.; *Regularity of the moments of the solution of a transport equation*, J. Funct. Anal., **76**, n.1 (1988), 110–125.
8. Lin, C.; Thèse de l'Université Lille 1, 2007.
9. Mihalas, D. and Mihalas. B.; Foundations of radiation hydrodynamics, Oxford University Press, New York, 1984.
10. Rutten, R.J.; Radiative Transfer in Stellar Atmospheres, Utrecht University Lecture notes, 2003, 8th Edition.
11. Van Zadelhoff, G. -J.; Dullemond C. P.; Van der Tak F. F. S.; Yates J. A.; Doty S. D.; Ossenkopf V.; Hogerheijde M. R.; Juvela M.; Wiesemeyer H. and Schoeier F. L.; *Numerical methods for non-LTE line radiative transfer: Performance and convergence characteristics*, Astron. Astrophys., **395**, (2002), 373.

EXPONENTIAL AND ALGEBRAIC RELAXATION IN KINETIC MODELS FOR WEALTH DISTRIBUTION

B. DÜRING

Institut für Analysis und Scientific Computing, Technische Universität Wien, 1040 Wien, Austria.

D. MATTHES AND G. TOSCANI

Department of Mathematics at the University of Pavia, via Ferrata 1, 27100 Pavia, Italy.

Two classes of kinetic models for wealth distribution in simple market economies are compared in view of their speed of relaxation towards stationarity in a Wasserstein metric. We prove fast (exponential) convergence for a model with *risky investments* introduced by Cordier, Pareschi and Toscani,[7] and slow (algebraic) convergence for the model with *quenched saving propensities* of Chakrabarti, Chatterjee and Manna.[3] Numerical experiments confirm the analytic results.

Keywords: Econophysics, Maxwell molecules, relaxation, Wasserstein metric.

1. Kinetic Models in Econophysics

One of the founding ideas in the rapidly growing field of econophysics is that the laws of statistical mechanics for particle systems also govern the trade interactions between agents in a simple market. Just as a classical kinetic model is defined by prescribing the collision kernel for the microscopic particle interactions, the econophysical model is defined by prescribing the exchange rules for wealth in trades. In dependence on these "microscopic" rules, the system develops "macroscopic" features in the long-time limit. Such macroscopic correlations are visible for instance in the form of a non-trivial stationary wealth distribution curve.

Many different (and somewhat justifiable) approaches to create a good model exist. Nevertheless, up to now little is known about which factors *should* enter into the exchange rules (in order to make the model realistic), and which *should not* (in order to keep it simple). Typically, the value of a model is estimated *a posteriori* by comparing its predictions with real-world data. For instance, it is widely accepted that the stationary wealth distribution $f_\infty(v)$ (denoting the density of agents with wealth $v > 0$) should possess a *Pareto tail*,

$$F_\infty(v) = \int_v^\infty f_\infty(w)\,dw \propto v^{-\alpha}. \tag{1}$$

The exponent α is referred to as *Pareto index*, named after the economist Vilfredo Pareto,[13] who proposed formula (1) more than a hundred years ago. According to recent empirical data, the wealth distribution among the population in a western country follows the Pareto law, with an index α ranging between 1.5 and 2.5. We refer e.g. to Ref. 1 and the references therein.

Below, we compare two types of econophysical models, which are able to produce Pareto tails.

1.1. *The Cordier-Pareschi-Toscani model*

The *Cordier-Pareschi-Toscani model* (CPT model) has been introduced in Ref. 7, and was intensively studied only recently.[10] When two agents with pre-trade wealths v and w interact, then their post-trade wealth v^* and w^*, respectively, is given by

$$v^* = (\lambda + \eta_1)v + (1-\lambda)w, \qquad w^* = (1-\lambda)v + (\lambda + \eta_2)w. \tag{2}$$

Here $\lambda \in (0,1)$ is the *saving propensity*, which models the fact that agents never exchange their entire wealth in a trade, but always retain a certain fraction λ of it. The quantities η_1 and η_2 are random variables with mean zero, satisfying $\eta_i \geq -\lambda$. They model *risky investments* that each agent performs in addition to trading. A crucial feature of the CPT model is that it preserves the total wealth in the statistical mean,

$$\langle v^* + w^* \rangle = \big(1 + \langle \eta_1 \rangle\big)\, v + \big(1 + \langle \eta_1 \rangle\big)\, w = v + w, \tag{3}$$

where $\langle \cdot \rangle$ denotes the statistical expectation value. The behavior of the homogeneous Boltzmann equation corresponding to (3) is to a large extend determined[10] by the convex function

$$\mathfrak{S}(s) := (1-\lambda)^s - 1 + \frac{1}{2}\big\langle (\lambda + \eta_1)^s + (\lambda + \eta_2)^s \big\rangle. \tag{4}$$

Clearly, $\mathfrak{S}(1) = 0$ by (3). Provided $\mathfrak{S}'(0) < 0$, the model possesses a unique steady state f_∞. If $\mathfrak{S}(s) < 0$ for all $s > 1$, then f_∞ has an exponentially small tail. On the contrary, if there exits a *non-trivial* root $\bar{s} \in (1,\infty)$ of $\mathfrak{S}$, then f_∞ possesses a Pareto tail (1) of index $\alpha = \bar{s}$.

Theorem 2.1 below states that (in both cases) any solution $f(t)$ converges to f_∞ exponentially fast in suitable Fourier and Wasserstein metrics.

1.2. *The Chakrabarti-Chatterjee-Manna model*

The *Chakrabarti-Chatterjee-Manna model* (CCM model) was introduced in Ref. 3, and heavily investigated in the last decade. While the saving propensity λ is a *global quantity* in the CPT model, in the CCM model it is a characteristics of the individual agents. The "state" of an agent is now described by his wealth *and* his personal saving propensity. The latter does not change with time. In a trade between two agents with wealth v, w and saving propensities λ, μ, respectively, wealth is exchanged according to

$$v^* = \lambda v + \epsilon\Delta, \quad w^* = \mu w + (1-\epsilon)\Delta \quad \text{with} \quad \Delta = (1-\lambda)v + (1-\mu)w. \tag{5}$$

Here ϵ is a random variable in $(0,1)$. The key ingredient for this model is the (time-independent) density $g(\lambda)$ of saving propensities among the agents. The homogeneous Boltzmann equation associated to the rules (5) has been heavily investigated numerically in terms of Monte Carlo simulations;[2–6] we present further simulation results here. Also, some theoretical investigations exist.[6,11,14] At least in the deterministic case $\epsilon \equiv 1/2$, the wealth distribution of the steady state is explicitly known,[12]

$$f_\infty(w) = \frac{C}{w^2}\, g\Big(1 - \frac{C}{w}\Big). \tag{6}$$

In the non-deterministic case, the choice of the random quantity ϵ has seemingly little influence[2] on the shape of f_∞. Thus, by prescribing g suitably, cf. section 3, steady states with a Pareto tail of arbitrary index α can be generated. Furthermore, we prove below (see Theorem 2.2) that for the majority of initial conditions the Wasserstein distance between the solution and the corresponding steady state with a Pareto tail can at best decay algebraically in time. For the proof, we shall need no properties of the CCM model other than the pointwise conservation of wealth,

$$v^* + w^* = v + w, \tag{7}$$

which is much stricter than conservation in the mean (3). The argument is based on the result from Ref. 10, that in pointwise conservative models initially finite moments of the solution diverge at most at algebraic rate.

1.3. *Other approaches*

Many econophysical models for wealth distribution — including the ones mentioned above — are still very basic as the agents just trade randomly with each other, do not adapt their saving strategy, and so on. Slightly more realistic economic models have been proposed, which admit the agents to have a little bit of intelligence, or at least trading preferences, see e.g. the collected works in Ref. 5 for an overview on recent developments. Finally, we mention that one may consider mean-field equations or hydrodynamic limits[9] instead of the full kinetic model.

2. Analytical Estimation of Convergence Rates

2.1. *Preliminaries*

We consider weak solutions f to the homogeneous Boltzmann equation,

$$\frac{d}{dt}\int_0^\infty \Phi(v)df(v)$$
$$= \frac{1}{2}\left\langle \int_0^\infty \int_0^\infty \big(\Phi(v^*)+\Phi(w^*)-\Phi(v)-\Phi(w)\big)df(v)df(w)\right\rangle, \quad (8)$$

where Φ is a regular test function, v, w denote the pre-collisional, and v^*, w^* the post-collisional wealths, according to the rules (2) and (5), respectively. Further, we assume that f is a probability density with mean wealth equal to one. (Notice that both models preserve mass and mean wealth.) In order to measure the convergence to equilibrium, $f(t,v) \to f_\infty(v)$ for $t \to \infty$, we introduce the following distances.

Definition 2.1. Let two probability densities f and g on $\mathbb{R}_+$ be given, both with first moment equal to one, and finite moments of some order $\bar{s} \in (1,2]$.

- For $s \in [1,\bar{s}]$, the *Fourier distance* d_s is defined by

$$d_s(f,g) := \sup_{\xi\in\mathbb{R}\setminus\{0\}} \left(|\xi|^{-s}\left|\widehat{f}(\xi)-\widehat{g}(\xi)\right|\right), \quad (9)$$

where $\widehat{f}$ and $\widehat{g}$ denote the Fourier transforms of f and g.
- The *Wasserstein-one-distance* is defined by

$$W(f,g) := \int_{\mathbb{R}_+} \left|F(v)-G(v)\right| dv, \quad (10)$$

where F and G denote the distribution functions of f and g,

$$F(v) = \int_v^\infty f(w)\,dw, \qquad G(v) = \int_v^\infty g(w)\,dw.$$

Equivalently, the Wasserstein distance between f and g can be defined as the infimum of the costs for transportation,

$$W(f,g) := \inf_{\pi\in\Pi} \int_{\mathbb{R}_+\times\mathbb{R}_+} |v-w|\, d\pi(v,w). \tag{11}$$

Here Π is the collection of all probability measures on $\mathbb{R}_+ \times \mathbb{R}_+$ with marginals $f(x)\,dx$ and $g(x)\,dx$, respectively. The infimum in (11) is in fact a minimum, and is realized by some *optimal transport plan* π_{opt}. For details, see Ref. 8 and references therein.

2.2. *Exponential convergence for the CPT model*

Theorem 2.1. *Assume η_1 and η_2 in the CPT model (2) are such that $\mathfrak{S}'(1) < 0$ with $\mathfrak{S}$ defined in (4). Then there exists a unique steady state f_∞ for (8), which is of mean wealth one. Further, there is some $\bar{s} \in (1,2]$ for which $\lambda := \mathfrak{S}(\bar{s}) < 0$, and any solution $f(t)$ to (8) — with initially bounded $\bar{s}$th moment — is exponentially attracted by f_∞:*

$$d_{\bar{s}}\big(f(t), f_\infty\big) \le d_{\bar{s}}\big(f(0), f_\infty\big)\exp(\lambda t), \tag{12}$$

$$W\big(f(t), f_\infty\big) \le C\exp\left(\frac{(\bar{s}-1)\lambda}{\bar{s}(2\bar{s}-1)}t\right), \tag{13}$$

with some finite, time-independent constant $C > 0$.

Estimate (12) is a consequence of Theorem 3.3 in Ref. 10. Estimate (13) is new, and follows from (12) and estimate (14) below. We remark that (12) is relatively easy to obtain, working on the Fourier representation of the Boltzmann equation (8), and using the homogeneity properties of d_s. On the contrary, a direct proof of (13) seems difficult. Notice that we cannot resort to the more convenient Wasserstein-*two*-metric here since the second moment of f_∞ might be infinite.

Lemma 2.1. *Assume that two probability densities f and g have first moment equal to one, and some moment of order $s \in (1,2]$ bounded. Then there exists a constant $C > 0$, depending only on s and the values of the sth moments of f and g, such that*

$$W(f,g) \le C d_s(f,g)^{\frac{s-1}{s(2s-1)}}. \tag{14}$$

Conversely, one has

$$d_1(f,g) \le W(f,g), \tag{15}$$

even if no moments of f and g above the first are bounded.

Proof. To prove (14), we extend the proof of Theorem 2.21 in Ref. 8, corresponding to $s = 2$ in the theorem above. Define

$$M = \max\left\{ \int_{\mathbb{R}} |v|^s f(v)\, dv,\ \int_{\mathbb{R}} |v|^s g(v)\, dv \right\}.$$

Starting from the definition of the Wasserstein distance in (10), we estimate

$$\begin{aligned} W(f,g) &= \int_{\mathbb{R}_+} \big|F(v) - G(v)\big|\, dv \\ &\le \int_0^R \big|F(v) - G(v)\big|\, dv + R^{1-s} \int_R^\infty v^{s-1} \big|F(v) - G(v)\big|\, dv \qquad (16) \\ &\le R^{1/2} \left(\int_{\mathbb{R}_+} \big|F(v) - G(v)\big|^2 dv \right)^{1/2} + R^{1-s} \int_R^\infty v^{s-1} \big|F(v) - G(v)\big|\, dv, \end{aligned}$$

where the parameter $R = R(t) > 0$ is specified later. By Parseval's identity,

$$\begin{aligned} &\int_{\mathbb{R}_+} \big|(F-G)(v)\big|^2 dv = \int_{\mathbb{R}} \big|\widehat{(F-G)}(\xi)\big|^2 d\xi \\ &= \int_{\mathbb{R}} \big|(i\xi)^{-1} \big(\widehat{f}(\xi) - \widehat{g}(\xi)\big)\big|^2 dx \le d_s(f,g)^2 \int_{|\xi|<r} |\xi|^{2(s-1)}\, d\xi + \\ &+4 \int_{|\xi|\ge r} \xi^{-2}\, d\xi = (2s-1)^{-1} r^{2s-1} d_s(f,g)^2 + 8r^{-1} \le C_1 d_s(f,g)^{1/s}. \end{aligned}$$

The last estimate follows by optimizing in the previous line with respect to $r > 0$. The constant C_1 depends only on $s > 1$. This gives a bound on the first term in (16) above. We estimate the second term, integrating by parts:

$$\begin{aligned} &\int_R^\infty v^{s-1} \big|F(v) - G(v)\big|\, dv \le \int_R^\infty v^{s-1} \big(F(v) + G(v)\big)\, dv \\ &= \frac{1}{s} \int_R^\infty v^s \big(f(v) + g(v)\big)\, dv + \Big(v^s \big(F(v) + G(v)\big)\Big)\Big|_R^\infty \\ &\le \frac{2M}{s} + \lim_{r\to+\infty} \Big(r^s \big(F(r) + G(r)\big)\Big). \end{aligned}$$

The last expression is easily estimated by Chebyshev's inequality, i.e.,

$$\lim_{r\to\infty} \big(r^s F(r)\big) \le \lim_{r\to\infty} \big(r^s\, \mathrm{P}_f\, [v > r]\big) \le \lim_{r\to\infty} \int_r^\infty v^s f(v)\, dv = 0,$$

since the sth moment of f is finite. In summary, (16) yields

$$W(f,g) \le C_1^{1/2} R^{1/2} d_s(f,g)^{1/(2s)} + 2s^{-1} M R^{1-s}.$$

Optimizing this over R yields the desired inequality (14).

The other inequality (15) is derived from the alternative definition (11) of $W(f,g)$. With π_{opt} being the optimal transport in (11),

$$d_1(f,g) = \sup_{\xi\neq 0}\Big(|\xi|^{-1}\Big|\int_{\mathbb{R}_+} e^{-iv\xi} f(v)\,dv - \int_{\mathbb{R}_+} e^{-iw\xi} g(w)\,dw\Big|\Big)$$

$$\leq \sup_{\xi\neq 0}\Big(|\xi|^{-1}\int_{\mathbb{R}_+}\big|e^{-iv\xi} - e^{-iw\xi}\big|\,d\pi_{opt}(v,w)\Big)$$

$$\leq \int_{\mathbb{R}_+}\sup_{\xi\neq 0}\Big(\frac{|1-e^{i(v-w)\xi}|}{|v-w||\xi|}\Big)|v-w|\,d\pi_{opt}(v,w) = \sup_{x\in\mathbb{R}}\Big(\frac{|1-e^{ix}|}{|x|}\Big)\,W(f,g).$$

In view of the elementary inequality $|1-\exp(ix)| \leq |x|$ for $x \in \mathbb{R}$, this yields the claim (15). □

2.3. *Algebraic convergence for the CCM model*

Theorem 2.2. *Assume the steady state f_∞ for the CCM model possesses a Pareto tail of index $\alpha > 1$. Let $f(t)$ be a solution of the associated Boltzmann equation* (8), *whose initial condition $f(0)$ has a finite moment of some order $n > \alpha$. Then*

$$W\big(f(t), f_\infty\big) \geq ct^{-\frac{n(\alpha-1)}{n-\alpha}} \tag{17}$$

with some time-independent constant $c > 0$.

Proof. By definition of the Pareto tail (1), one has $F_\infty(v) \geq 2\varepsilon v^{-\alpha}$ for $v \gg 0$ with some $\varepsilon > 0$. On the other hand, Theorem 3.2 in Ref. 10 (we refer also to the discussion of Example 8 in section 4.2 in Ref. 10) yields that the n-th moment $M_n(t)$ of $f(t)$ satisfies $M_n(t) \leq Ct^n$ for $t \gg 0$ with a finite constant $C > 0$. Consequently,

$$F(t;v) = \int_v^\infty f(t;w)\,dw \leq v^{-n}\int_0^\infty w^n f(t;w)\,dw \leq Cv^{-n}t^n.$$

Hence $F_\infty(v) - F(t;v) \geq \varepsilon v^{-\alpha}$ for all $v \geq V(t) := (Ct^n/\varepsilon)^{1/(n-\alpha)}$. By definition of the Wasserstein distance,

$$W\big(f(t), f_\infty\big) \geq \varepsilon\int_{V(t)}^\infty v^{-\alpha}dv \geq c\cdot t^{-\frac{n(\alpha-1)}{n-\alpha}},$$

where $c > 0$ depends on ε, n, α and C, but *not* on t. □

3. Numerical Experiments

In order to verify the analytically estimated bounds on the relaxation behavior, we have performed a series of kinetic Monte Carlo simulations for both the CPT and the CCM models. We compare numerical results for systems consisting of $N = 200$, $N = 1000$ and $N = 5000$ agents, respectively. In these rather basic simulations, pairs of agents are randomly selected for binary collisions, and exchange wealth according to the trade rules (2) and (5), respectively. One *time step* corresponds to N such interactions.

As parameters for the CPT model, we have chosen a saving propensity of $\lambda \equiv 0.7$ and independent random variables η_1, η_2 attaining the values ± 0.5 with probability 1/2 each. The non-trivial root of $\mathfrak{S}(s)$ in (4) is $\bar{s} \approx 3.7$. In the CCM model, we assign the saving propensities by means of $\lambda = (1-\omega)^{2.5}$, with $\omega \in (0,1)$ being a uniformly distributed random variable. We restricted simulations to the deterministic situation $\epsilon \equiv 1/2$. In all our experiments, every agent possesses unit wealth initially.

In order to compute a good approximation of the steady state, the simulation is carried out for about 10^5 time steps, and then the wealth distribution is averaged over another 10^4 time steps. The thus obtained reference state is used in place of the (unknown) steady wealth distribution when calculating the decay of the Wasserstein distance in Fig. 1 and Fig. 2, respectively. The evolution of the wealth distributions over time for $N = 1000$ agents is illustrated in Fig. 3 and Fig. 4. The agents are sorted by wealth, and the wealth distribution at different time steps is compared to the approximate steady state.

Some words are in order to explain the results. The first remark concerns the seemingly poor approximation of the steady state in the CPT model, with a residual Wasserstein distance of the order $10^{-1} \ldots 10^{-2}$. The reason for this behavior lies in the essentially statistical nature of this model, which *never* reaches equilibrium in finite-size systems, due to persistent thermal fluctuations. Strictly speaking, a comparison with the CMM model is misleading here, since simulations for the latter are performed in the purely deterministic setting $\epsilon \equiv 1/2$. Second, the almost perfect exponential (instead of algebraic) decay displayed in Fig. 2 is due to finite-size effects of the system. The decrease of the exponential rates when the system size N increases strongly indicates that in the theoretical limit $N \to \infty$ relaxation is indeed sub-exponential as expected. We stress that — in contrast — the decay rate in Fig. 1 for the CPT model is independent of the system size.

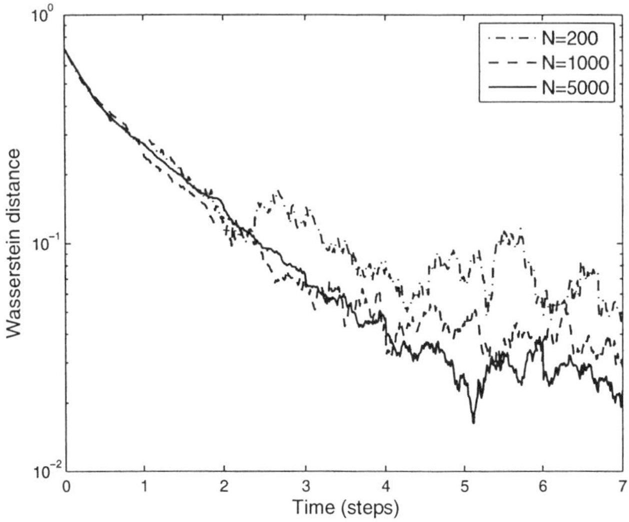

Fig. 1. Decay of the Wasserstein distance to the steady state in the CPT model.

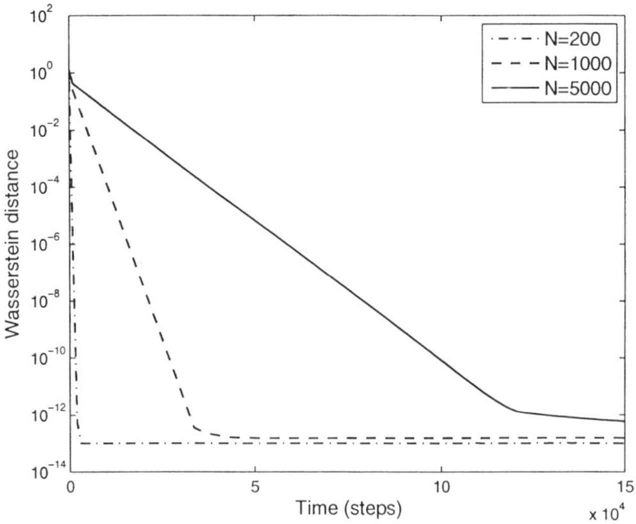

Fig. 2. Decay of the Wasserstein distance to the steady state in the CCM model.

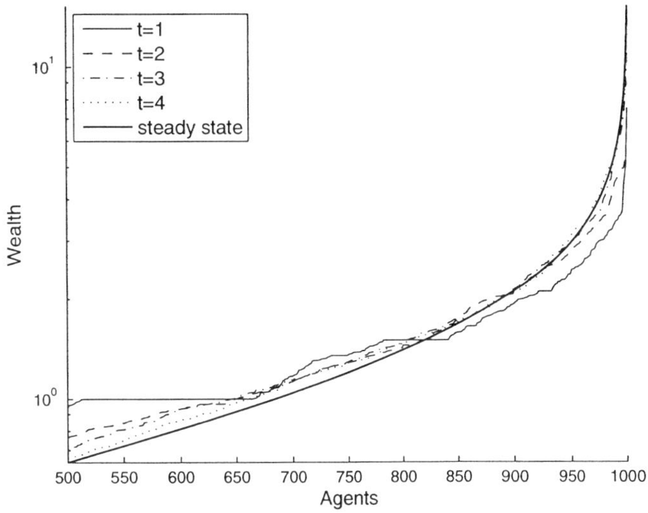

Fig. 3. Evolution of the wealth distribution towards the steady state for the upper half of the population in the CPT model ($N = 1000$).

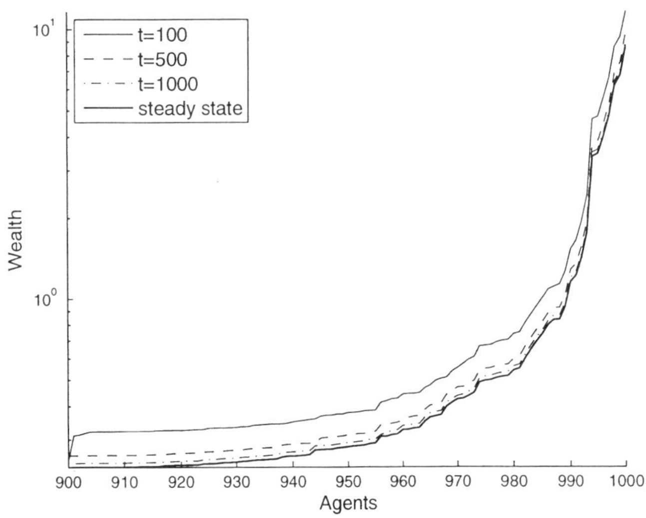

Fig. 4. Evolution of the wealth distribution towards the steady state for the upper tenth part of the population in the CCM model ($N = 1000$).

References

1. Y. Fujiwara, W. Souma, H. Aoyama, T. Kaizoji and M. Aoki, *Physica A* **321** 598–604 (2003).
2. B. K. Chakrabarti and A. Chatterjee, `arXiv: 0709.1543`
3. B. K. Chakrabarti, A. Chatterjee and S. S. Manna, *Physica A* **335** 155–163 (2004).
4. B. K. Chakrabarti, A. Chatterjee and R. B. Stinchcombe, *Phys. Rev. E* **72** (2005).
5. B. K. Chakrabarti, A. Chatterjee and Y. Sudhakar (Eds.), *Economics of Wealth Distribution*, Milan (2005).
6. A. Chakraborti, G. Germano, E. Heinsalu and M. Patriarca, *Europ. Phys. J. B* **57** 219–224 (2007).
7. S. Cordier, L. Pareschi and G. Toscani, *J. Stat. Phys.* **120** 253–277 (2005).
8. J. A. Carrillo and G. Toscani, *Riv. Mat. Univ. Parma* (7) **6** 75–198 (2007).
9. B. Düring and G. Toscani, *Physica A* **384** 493–506 (2007).
10. D. Matthes and G. Toscani, *J. Stat. Phys.*, to appear (2007).
11. D. Matthes and G. Toscani, *Kinetic Rel. Mod.*, to appear (2008).
12. P. K. Mohanty, *Phys. Rev. E* **74** (1) 011117 (2006).
13. V. Pareto, *Cours d Economie Politique*, Lausanne and Paris, 1897.
14. P. Repetowicz, S. Hutzler and P. Richmond *Physica A* **356** 641–654 (2005).

SOLITARY WAVES IN DISPERSIVE MATERIALS

J. ENGELBRECHT, A. BEREZOVSKI, A. SALUPERE

Centre for Nonlinear Studies, Institute of Cybernetics at Tallinn UT, Akadeemia 21, 12618 Tallinn, Estonia
E-mail: je@ioc.ee, arkadi.berezovski@cs.ioc.ee, salupere@ioc.ee

The Mindlin-type models are able to describe microstructured solids with a good accuracy. In this paper, nonlinear deformation waves in such solids are analysed. Similarly to classical models, nonlinear and dispersive effects can be balanced although the emergence and propagation of solitary waves is more complicated due to more complicated character of effects. We show also that the solitary waves can be used also in the Nondestructive Testing.

1. Introduction

The propagation of deformation waves in microstructured solids is characterised by strong dispersion effects. This is clearly evident in Mindlin-type materials [13] which involve an internal scale to distinguish between macro- and microstructure. It has been shown that not only materials like alloys, crystallites, ceramics, functionally graded materials, *etc.* can be described by such a model [4, 14] but also layered composites bear a certain resemblance to a Mindlin material [19]. Leaving aside the general description (see [4–6] and references therein), we concentrate here on possible balance of dispersive and nonlinear effects in microstructured Mindlin-type materials. It is well-known that such a balance may give rise to solitary waves or solitons which retain their shape in a conservative case. Here we refer only to a recent Handbook [18] where the long history of soliton studies is explicitly presented. However, the classical case of the Korteweg–de Vries equation [9] describes the balance of the cubic dispersion and quadratic nonlinearity. Although it has been shown that for certain crystal structures this is the case [12], in most of the materials the situation is much more complicated. One such a case is martensitic-austenitic alloys [17] with quadratic-quartic nonlinearity and cubic-quintic dispersion. The existence of internal scales refers also to possible different nonlinearities [3]. For the full Mindlin-type model, the derivation of the 1D nonlinear mathematical model with nonlinearities on both the macro- and microscale is given in [6]. Here we analyse this case in more detail.

2. Mathematical models

The linear models and corresponding hierarchies are presented in [4–6]. Two important parameters are $\delta = l^2/L^2$ and $\epsilon = U_0/L$ where l is the characteristic scale of the microstructure, L is the wavelength of the excitation and U_0 its amplitude. Here we enlarge the potential energy function W by cubic terms:

$$W = \frac{1}{2}\left(\alpha u_x^2 + 2A\varphi u_x + B\varphi^2 + C\varphi_x^2\right) + \frac{1}{6}\left(\beta u_x^3 + M\varphi^3 + N\varphi_x^3\right), \quad (1)$$

where u is the displacement and φ – the microdeformation. The coefficients $\alpha, \beta, A, B, C, M, N$ are the material parameters. Here and further, the indices denote differentiation. ¿From Euler–Lagrange equations (see [6]) the governing equations for 1D longitudinal wave yield

$$\rho u_{tt} = \alpha u_{xx} + N u_x u_{xx} + D\varphi_x\,, \quad (2)$$

$$I\varphi_{tt} = C\varphi_{xx} + M\varphi_x\varphi_{xx} - Du_x - B\varphi\,, \quad (3)$$

where ρ is the density of the macrostructure and I is the microinertia. In order to reduce system (2), (3) into one equation, we use the scaling with δ and ε, series representation and the slaving principle (for details, see [4, 6]). The result is a hierarchical equation in terms of $U = u/U_0, X = x/L$, $T = tc_0/L, c_0^2 = \alpha/\rho$:

$$U_{TT} = b\,U_{XX} + \frac{\mu}{2}\left(U_X^2\right)_X + \delta\left(\beta U_{TT} - \gamma U_{XX}\right)_{XX} - \delta^{3/2}\frac{\lambda}{2}\left(U_{XX}^2\right)_{XX}, \quad (4)$$

where $b, \mu, \beta, \gamma, \lambda$ are constants. Equation (4) involves hierarchically two nonlinear wave operators

$$L_{ma} = U_{TT} - bU_{XX} - \frac{1}{2}\mu\left(U_X^2\right)_X, \quad (5)$$

$$L_{mi} = \delta\left(\beta U_{TT} - \gamma U_{XX} - \delta^{1/2}\lambda U_{XX}^2\right)_{XX}, \quad (6)$$

characteristic of macro- and microstructure, respectively. In [4], the corresponding linear case is studied with a special attention to dispersion effects. Here dispersion and nonlinearities are both taken into account and the problem of their possible balance is studied. The standard approach [18] calls for the derivation of the evolution equation, i.e. of the one-wave equation like in the KdV case. Here, however, with wave operators of the second order, the governing equation is of the two-wave type, i.e. involving both right- and left-going waves. Although the reduction to an evolution equation is possible [15], we retain the model equation (4) that permits to model also the collision of solitary waves. For the further analysis, however, we rewrite Eq. (4) in terms of $v = U_x$ and use further x for the spatial coordinate. Then Eq. (4) yields

$$u_{tt} = b\, v_{xx} + \frac{\mu}{2}\left(v^2\right)_{xx} + \delta\left(\beta\, v_{tt} - \gamma\, v_{xx}\right)_{xx} - \delta^{3/2}\,\frac{\lambda}{2}\left(v_x^2\right)_{xxx}, \tag{7}$$

that must be solved with proper initial and boundary conditions.

3. Solitary waves

The model equation (7) involves complicated dispersion (terms v_{ttxx} and v_{xxxx}) and complicated nonlinearities (terms $\left(v^2\right)_{xx}$ and $\left(v_x^2\right)_{xxx}$). The first question to be answered is: could nonlinear and dispersive terms be balanced so that a solitary wave with a constant shape can exist? It has been proved by Janno and Engelbrecht [8] that a solitary wave solution to Eq. (7) exists provided

$$\left(\frac{\beta c_1^2 - \gamma}{c_1^2 - b}\right)^3 > \frac{4\lambda^2}{\mu^2}\,, \qquad \frac{c_1^2 - b}{\beta c_1^2 - \gamma} > 0\,, \tag{8}$$

$$\beta c_1^2 - \gamma \neq 0\,,\;\; c_1^2 - b \neq 0\,,\;\; \mu \neq 0\,. \tag{9}$$

Here c_1 is the velocity of the solitary wave $v(x,t) = v(x - c_1 t)$. If $\lambda = 0$, i.e. the nonlinearity exists only on the macroscale ($\mu \neq 0$), then the classical sech^2-type solitary wave exists. This case is similar to wave motion in rods [16], where geometrical dispersion is also described by the fourth-order derivatives u_{ttxx} and u_{xxxx}. If $\lambda \neq 0, \mu \neq 0$, i.e. both nonlinearities are taken into account, then the situation is different — a solitary wave will demonstrate asymmetry. Here we demonstrate this by the numerical solution of Eq. (7) using the pseudo-spectral method. The initial condition is chosen as an analytical solution for Eq. (7) with $\lambda = 0$: $v(x,0) = A\ \mathrm{sech}^2[(x - 6\pi)/\Delta]$, $A = 0.6413$, $\Delta = 1.638$. In Fig. 1 the initial wave-profile, wave-profile at $t = 136$ and corresponding $v_x - v$ phase curves are presented.

Once solitary waves exist in materials modelled by Eq. (7), the next intriguing problem is the emergence of solitary waves from arbitrary initial conditions, either localised or periodic in space. The classical KdV

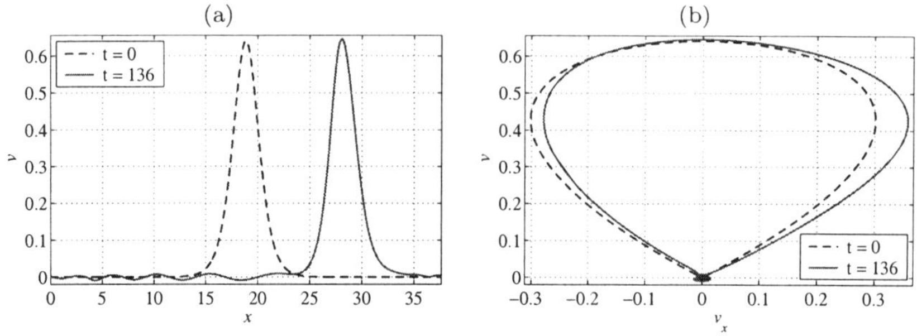

Fig. 1. Propagation of a solitary wave – (a) and the $v_x - v$ phase-plane of a solitary wave – (b); $0 <= x < 12\pi$, $b = 0.5$, $\mu = 1.45$, $\delta = 0.2$, $\beta = 1.65$, $\gamma = 0.297$, $\lambda = 2.3597$.

equation as an one-wave equation permits the formation of a soliton train. Equation (7) has a main wave operator of the second order and therefore describes both right- and left-going waves. Figure 2 demonstrates clearly such a situation for an initial condition $v(0, x) = A\,\mathrm{sech}^2[(x - 600\pi)/\Delta]$, $A = 1$, $\Delta = 20$.

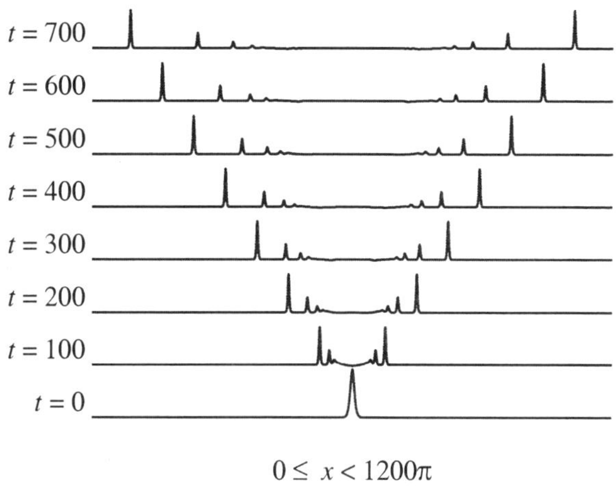

Fig. 2. Emergence of two trains of solitons; $b = 0.775$, $\mu = 16.67$, $\delta = 0.25$, $\beta = 40.04$, $\gamma = 7.5$, $\lambda = 0.8333$.

According to the definition of a soliton [18], its shape should not be changed after interaction with another soliton, only the phase shift is possible. The interaction test is needed in order to determine this quality. The KdV-equation permits to model interaction of solitons propagating in one direction and interaction is possible due to different speeds of solitons with different amplitudes. Equation (7), however, permits to analyse also the head-on collision. In Fig. 3 such a collision is presented in case when two

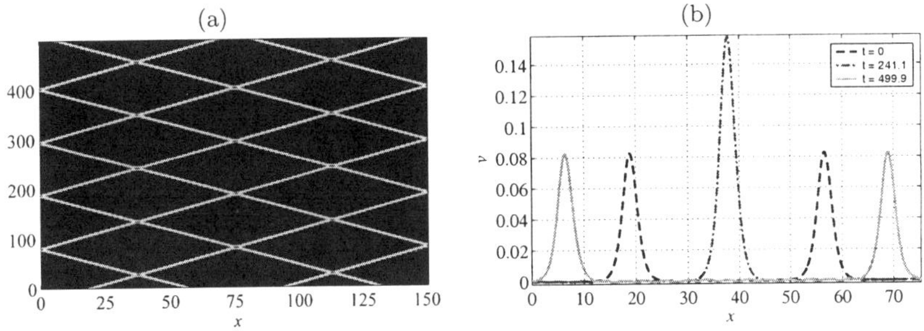

Fig. 3. Head-on collision of solitons for $b = 0.45$, $\mu = 1.45$, $\delta = 0.25$, $\beta = 0.44$, $\gamma = 0.0792$, $\lambda = 0.1305$. Pseudocolor plot over two space periods for $0 \leq t < 500$ – (a), single wave-profiles – (b).

pulses that have equal amplitude $A = 0.0828$ and equal width $\Delta = 3.6932$ propagate at equal, but opposite velocities $c_1 = -c_2 = 0.7$; corresponding initial condition $v(0, x) = A\,\mathrm{sech}^2[(x - 6\pi)]/\Delta + [A\,\mathrm{sech}^2(x - 18\pi)]/\Delta$.

In the 1D study, the propagation of longitudinal waves across the laminated composites and in the Mindlin material is characterized by a similar dispersion [19]. Bearing in mind a clear discrete structure of laminates, the straightforward numerical simulations can be performed for the wave motion. As an example, wave propagation in a laminated composite with alternating layers of copper (density $\rho = 8960$ kg/m^3 and wave velocity 4361 m/s) and aluminium (density $\rho = 2703$ kg/m^3 and wave velocity 6149 m/s) was studied by means of wave-propagation algorithm [10]. The effects of nonlinearity introduce the dependence of velocity on the deformation and the real velocity of propagation c_1 is

$$c_1 = c\,(1 + 2\,Au_x)\,, \tag{10}$$

where A is constant equal to 0.8 for both materials.

Physically such a wave process in nonlinear laminated composites should be similar to processes analysed above. Indeed, Figs. 4 and 5 demonstrate, how a single pulse generated initially is transformed into a train of soliton-like pulses. The periodic variation of density (normalized by maximal value) is schematically shown by its dashed lines. In linear case, such a train cannot emerge [1]. We are careful here calling these pulses solitons because in this case the solitonic properties of formed pulses is not clear (cf. [11]).

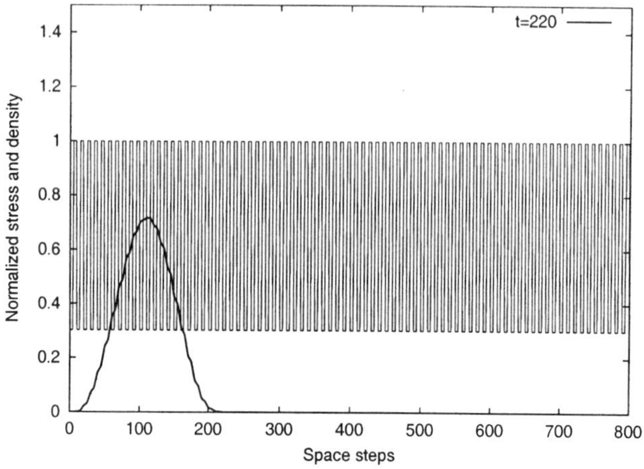

Fig. 4. Initial pulse shape in a laminated composite

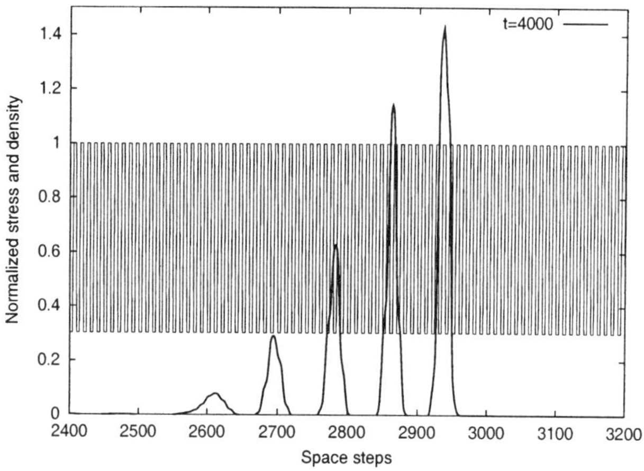

Fig. 5. Train of pulses in a laminated composite at $t = 4000$ time steps

4. Ideas for Nondestructive Testing

The aim of the Nondestructive Testing (NDT) is to determine the physical and/or geometrical properties of materials (specimens) by measuring the wave fields at given excitations. Waves carry information about the material, therefore on the basis of given excitations and the measured quantities, the NDT is possible. Usually the harmonic waves are used for the NDT in the ultrasound range and there are many applications known in engineering. In the case of microstructured materials, the number of material coefficients to be determined is certainly larger than in the case of homogeneous materials. However, there are more physical effects to be accounted for and consequently, one can use these effects for advanced NDT. If nonlinearity is weak then dispersive effects are governing. In this case, the dependence of velocity on frequency can be used for determining the coefficients of the model equation. Corresponding algorithms can be constructed for harmonic waves and for localized harmonic waves [7]. These algorithms, although specific for microstructured solids, actually use the well-known concept of harmonic excitations. As shown above, in nonlinear microstructured solids the balance of dispersive and nonlinear effects is possible which results in solitary waves. The novel idea is to use solitary waves in the NDT [8]. The remarkable effect due to nonlinearity of the microstructure is the asymmetry of a solitary wave. Asymmetry can be measured at the level of $A/2$ where A is the amplitude of a wave. In the model equation (7) there are 5 material parameters: b, μ, β, γ, λ which need to be determined. Here we assume the scale parameter δ known. Measuring just a single solitary wave , one can recover maximally 3 parameters (cf. a standard soliton solution). That is why for solving the full problem, one should use the measurements of two independent solitary waves with different amplitudes and consequently with two different velocities. The details of such an algorithm with the proof of the uniqueness and a stability estimate are given in [8]. In practical realizations the ultrasound transducers generate wave beams which are not one-dimensional but the diffractional expansion in the transverse direction is rather weak [2]. On the axis of a wave beam, the one-dimensional approximation is possible.

Acknowledgements

The support from the Estonian Science Foundation (grants 7035 and 7037) is gratefully acknowledged. J.E would like to thank the LOC of the WASCOM 2007 for the support to attend the conference.

References

1. A. Berezovski, M. Berezovski, J. Engelbrecht and G. A. Maugin, Numerical simulation of waves and fronts in inhomogeneous solids. In: W. K. Nowacki and Han Zhao (eds), *Multi-Phase and Multi-Component Materials under Dynamic Loading*. Inst. Fundam. Technol. Research, Warsaw, 71 (2007).
2. J. Engelbrecht, *Nonlinear Wave Dynamics. Complexity and Simplicity*. Kluwer, Dordrecht (1997).

3. J. Engelbrecht and F. Pastrone, *Proc. Estonian Acad. Sci. Phys. Math.*, **52**, 12 (2003).
4. J. Engelbrecht, A. Berezovski, F. Pastrone and M. Braun, *Phil. Mag.*, **85**, 4127 (2005).
5. J. Engelbrecht, A. Berezovski, F. Pastrone, and M. Braun, Deformation waves in microstructured solids and dispersion. In: R. Monaco et al. (eds), *Proc. WASCOM 2005*, World Scientific, New Jersey et al., 204 (2006).
6. J. Engelbrecht, F. Pastrone, M. Braun and A. Berezovski, Hierarchies of waves in nonclassical materials. In: P.-P. Delsanto (ed), *The Universality of Nonclassical Nonlinearity. Applications to NDE and Ultrasonics*, Springer, New York, 29 (2007).
7. J. Engelbrecht and J. Janno, *Rend. Sem. Mat. Univ. Pol. Torino*, **65**, 159 (2007).
8. J. Janno and J. Engelbrecht, An inverse solitary wave problem related to microstructured materials. *Inverse Problems*, **21**, 2019 (2005).
9. D. J. Korteweg and G. de Vries, *Phil. Mag.*, **39**, 422 (1895).
10. R. J. LeVeque, *Finite Volume Methods for Hyperbolic Problems.* Cambridge University Press, Cambridge (2002).
11. R. J. LeVeque and D. H. Yong, Solitary waves in layered nonlinear media. *SIAM J. Appl. Math.*, **63**, 1539–1560 (2003).
12. G. A. Maugin, *Nonlinear Waves in Elastic Crystals*, Oxford University Press, Oxford (1999).
13. R. D. Mindlin, *Arch. Rat. Mech. Anal.*, **16**, 51 (1964).
14. F. Pastrone, P. Cermelli and A. V. Porubov, *Math. Phys. Mech.*, **7**, 9 (2004).
15. T. Peets, M. Randrüüt and J. Engelbrecht, *Wave Motion.*, (in press).
16. A. V. Porubov, *Amplification of Nonlinear Strain Waves in Solids.* World Scientific, Singapore, (2003).
17. A. Salupere, J. Engelbrecht and G. A. Maugin, *Wave Motion*, **34**, 51 (2001).
18. A. Scott (ed), *Encyclopedia of Nonlinear Science.* Routledge, New York and London (2005).
19. C.-T. Sun, J. D. Achenbach and G. Hermann, *J. of Appl. Mech. Trans. ASME*, **35E**, 467 (1968).

A GINZBURG-LANDAU MODEL FOR THE ICE-WATER AND LIQUID-VAPOR PHASE TRANSITIONS

MAURO FABRIZIO

Department of Mathematics, University of Bologna
Piazza Porta S. Donato 5., 40127 Bologna. Italy

We suggest the study of the ice-water and liquid-vapor transformations as a first order phase transition, by means of a suitable system realized by the Ginzburg-Landau and Navier-Stokes equations.

1. Introduction

The first order phase transition, which occurs in the ice-water and water-vapor transitions is studied by a mathematical model, which describes the phase by means of a Ginzburg-Landau equation in the order parameter φ [7], [6], [4], [3], such that $0 \leq \varphi \leq 1$ and of a decomposition of the density ρ in two components ρ_0, ρ_1 so as to have

$$\frac{1}{\rho} = \frac{1}{\rho_0} + \frac{1}{\rho_1} \tag{1}$$

The density ρ_0 is connected with the pressure p, and hence any variation of ρ_0 involves a variation of the pressure. On the contrary, the component ρ_1 is independent of the pressure and its variation is due to the phase transition; in other words, ρ_1 is a function of the order parameter φ.

When we deal with incompressible materials it is not possible to have a dependence of the pressure on ρ_0. Anyway, we always have the formula (1), where ρ_1 is a function of the phase φ and ρ_0 will be constant.

Moreover, by the Second Law of Thermodynamics, we obtain the restrictions which bring to the relation between the density ρ_1 and the phase φ. In particular, in the study of the liquid-vapor transition, the model provides a good representation of the Andrews diagram; thus, it allows us to get over the incoherences of the van der Waals model, which cannot describe adequately the isothermal-lines during the phase transition, the form of which is stated by the experiments.

2. Ice-water transition

In this section we provide a model for the study of a general first order phase transition in a domain $\Omega \subset \mathbb{R}^3$ by means of the Ginzburg-Landau equation

$$\rho\gamma\dot{\varphi} = \varkappa \bigtriangleup \varphi - \rho F'(\varphi) - \rho(\theta + \lambda p)G'(\varphi) \tag{2}$$

where ρ denotes the *density*, p the *pressure*, θ the *absolute temperature* and $\gamma, \varkappa, \lambda$ are positive constants. Moreover, the functions F and G are polynomials of the type [4], [1]

$$F(\varphi) = \frac{\varphi^4}{4} - \frac{\varphi^3}{3}, \quad G(\varphi) = \frac{\varphi^4}{4} - \frac{2\varphi^3}{3} + \frac{\varphi^2}{2} \tag{3}$$

Finally, the points that locate the transition in the (θ, p) plane are such that in the equation (2) we have

$$\theta + \lambda p = 1 \tag{4}$$

When we want to describe the ice-water transition, we have to regard the density ρ as a variable quantity and also to introduce the *velocity* $\mathbf{v}$ and the motion equation

$$\rho\dot{\mathbf{v}} = -\nabla p + \nabla \cdot \mathbf{T}_E + \rho\mathbf{b} \tag{5}$$

where $\mathbf{T}_E$ is the *extra-stress* and $\mathbf{b}$ the *body forces*. If in this framework, we also consider the ice and the water as incompressible viscous materials, then the extra-stress is expressed by

$$\mathbf{T}_E = \nu(\varphi)\nabla\mathbf{v} \tag{6}$$

where $\nu(\varphi) > 0$ is the viscosity. Besides the equation (5), we have the continuity equation

$$\dot{\rho} + \rho\nabla \cdot \mathbf{v} = 0 \tag{7}$$

Moreover, since for the ice as well as the water we have assumed the hypothesis of incompressibility, it is correct to suppose that $\dot{\rho}$ and $\nabla \cdot \mathbf{v}$ are different from zero only during the phase transition. Then, the pressure will be undetermined in the ice and water phase, but also during the transition, because in such a case p does not change even if we have a change of the density.

Multiplying (2) by $\dot{\varphi}$ we obtain

$$\rho\kappa\dot{\varphi} + \mathbf{p} \cdot \nabla\dot{\varphi} = \nabla \cdot (\mathbf{p}\dot{\varphi}). \tag{8}$$

where

$$\kappa = \gamma\dot{\varphi} + F'(\varphi) + (\theta + \lambda p)G'(\varphi). \tag{9}$$

$$\mathbf{p} = \varkappa\nabla\varphi, \ \varkappa > 0 \tag{10}$$

Then, the left-hand side of (8) provides the *internal structural power* $\mathcal{P}_i^\varphi$ and the right-hand side $\mathcal{P}_e^\varphi$, the *external structural power*,

$$\mathcal{P}_i^\varphi = \rho\kappa\dot{\varphi} + \mathbf{p}\cdot\nabla\dot{\varphi}, \tag{11}$$

$$\mathcal{P}_e^\varphi = \nabla\cdot(\mathbf{p}\dot{\varphi}). \tag{12}$$

3. Thermodynamic restrictions

The thermo-mechanical properties of a material system are defined by means of the concepts of *state* σ and *process* P .

Let are consider the principles of thermodynamics [5], [2].

The first law of thermodynamics

There exists a state function $e(\sigma)$, called *internal energy*, such that

$$\rho\dot{e}(\sigma) = \rho h(\sigma, P) + \mathcal{P}_i^\varphi(\sigma, P) + \mathcal{P}_i^v(\sigma, P) \tag{13}$$

where h is the *internal heat power*, $\mathcal{P}_i^\varphi$ is given by (11) and the *internal mechanical* powers $\mathcal{P}_i^v$ is defined by

$$\mathcal{P}_i^v = p\frac{\dot{\rho}}{\rho} + \nu(\varphi)\nabla\mathbf{v}\cdot\nabla\mathbf{v} \tag{14}$$

The second law of thermodynamics

There exists a state function η, called the *entropy* function, such that

$$\rho\dot{\eta}(\sigma) \geq \rho\frac{h(\sigma, P)}{\theta} - \frac{1}{\theta^2}\mathbf{q}\cdot\nabla\theta \tag{15}$$

In the following we suppose that the heat flux obeys the Fourier law

$$\mathbf{q} = -k_0\nabla\theta$$

k_0 being a positive constant. From the balance heat

$$\rho h = -\nabla\cdot\mathbf{q} + \rho r \tag{16}$$

where r is the heat supply. Then from (13), (11), (12) and (16), it follows that

$$\rho(\dot{e} - \kappa\dot{\varphi}) - \mathbf{p}\cdot\nabla\dot{\varphi} - p\frac{\dot{\rho}}{\rho} - \nu(\varphi)\nabla\mathbf{v}\cdot\nabla\mathbf{v} = k_0\Delta\theta + \rho r. \tag{17}$$

where κ is defined in (9) and $\mathbf{p}$ is given by (10).

Let F and G be the integrals of F' and G'. Upon the substitution of κ from (17) we obtain

$$\rho(\dot{e}-\dot{F}-\theta\dot{G}-\gamma\dot{\varphi}^2)-p(\frac{\dot{\rho}}{\rho}+\lambda\rho\dot{G})-\nu(\varphi)\nabla\mathbf{v}\cdot\nabla\mathbf{v}-(\frac{\varkappa(\nabla\varphi)^2}{2})^{\cdot}= \quad (18)$$

$$=k_0\Delta\theta+\rho r.$$

From the Second law (15) we have using (13)

$$\rho\theta\dot{\eta}\geq\rho\dot{e}-\mathcal{P}^{\varphi}-\mathcal{P}^{v}-\frac{1}{\theta}\mathbf{q}\cdot\nabla\theta \quad (19)$$

By introducing the free energy $\psi=e-\theta\eta$ we have

$$\rho\dot{\psi}+\rho\eta\dot{\theta}-p\frac{\dot{\rho}}{\rho}-\gamma\rho\dot{\varphi}^2-\rho\dot{F}(\varphi)-\rho(\theta+\lambda p)\dot{G}(\varphi)$$

$$-\nu(\varphi)\nabla\mathbf{v}\cdot\nabla\mathbf{v}-\varkappa\nabla\varphi\cdot\nabla\dot{\varphi}-\frac{1}{\theta}\mathbf{q}\cdot\nabla\theta\leq 0.$$

The free energy ψ and the entropy η are assumed to depend on $\theta,\varphi,\nabla\varphi$. Hence upon substitution of $\mathcal{P}^{\varphi}$ and $\mathcal{P}^{v}$ we obtain

$$\rho(\psi_\theta+\eta)\dot{\theta}-\rho\gamma\dot{\varphi}^2+\rho(\psi_\varphi-F'-\theta G')\dot{\varphi}-p\left[\frac{\dot{\rho}}{\rho}+\lambda\rho\dot{G}(\varphi)\right] \quad (20)$$

$$-\nu(\varphi)\nabla\mathbf{v}\cdot\nabla\mathbf{v}+(\rho\psi_{\nabla\varphi}-\varkappa\nabla\varphi)\cdot\nabla\dot{\varphi}-\frac{k_0}{\theta}|\nabla\theta|^2\leq 0.$$

We now evaluate the restrictions placed by the second law through (20). The arbitrariness and the linearity in $\dot{\theta}$ imply that

$$\eta=-\psi_\theta. \quad (21)$$

Moreover, the arbitrariness of $p>0$ requires that

$$\frac{\dot{\rho}}{\rho}+\lambda\rho\dot{G}(\varphi)=0 \quad (22)$$

from which

$$\frac{1}{\rho}-\lambda G(\varphi)=const. \quad (23)$$

We suppose that $\varphi=0$ represents the water phase with density ρ_0, then we have

$$\frac{1}{\rho}-\frac{1}{\rho_0}=\lambda G(\varphi) \quad (24)$$

or equivalently

$$\rho = \frac{\rho_0}{1 + \lambda\rho_0 G(\varphi)} \tag{25}$$

The equation (24) and (25) are in agreement with the physical condition which sees the water density greater of the ice one, because $0 = G(0) < G(1) = \frac{1}{12}$,

Then, the density ρ is composed of two terms ρ_0 and ρ_1, such that

$$\frac{1}{\rho} = \frac{1}{\rho_0} + \frac{1}{\rho_1} \tag{26}$$

where

$$\frac{1}{\rho_1} = \lambda G(\varphi) \tag{27}$$

Otherwise, if we denote by $v = \frac{1}{\rho}$ the *specific volume*, we have

$$v = v_0 + v_1\ , \quad v_0 = \frac{1}{\rho_0}\ , \quad v_1 = \frac{1}{\rho_1} \tag{28}$$

Moreover, from the inequality (20) and the arbitrariness of $\nabla\dot{\varphi}$ we obtain

$$\rho\psi_{\nabla\varphi} = \varkappa\nabla\varphi. \tag{29}$$

Now, $\dot{\varphi}$ is subject to (2) and hence may be chosen arbitrarily because of non locality of $\Delta\varphi$. Then, if we put $\nabla\theta = 0$ and $\nabla\mathbf{v} = 0$ we have from (20)

$$-\rho\gamma\dot{\varphi}^2 + \rho(\psi_\varphi - F' - \theta G')\dot{\varphi} \leq 0. \tag{30}$$

By considering $|\dot{\varphi}|$ sufficiently small in (30), from the arbitrariness of $\dot{\varphi}$, we conclude that

$$F' + \theta G' = \psi_\varphi. \tag{31}$$

Hence, the inequality (20) is satisfied and reduced to the form

$$-\rho\gamma\dot{\varphi}^2 - \frac{k_0}{\theta}|\nabla\theta|^2 \leq 0 \tag{32}$$

The free energy is a state function depending on $(\theta, \varphi, \rho, \nabla\varphi)$

$$\psi = \int e_\theta(\theta)d\theta - \theta\int\frac{e_\theta}{\theta}d\theta + F(\varphi) + \theta G(\varphi) + \frac{\varkappa}{2\rho}(\nabla\varphi)^2 \tag{33}$$

Then, the entropy is given by

$$\eta = \int\frac{e_\theta}{\theta}d\theta - G(\varphi) \tag{34}$$

Now, we are in a position to study the differential system connected with the ice-water phase transition problem. In this frame-work let us disregard the infinitesimal terms of second order. Then from the equation (18) on the temperature we have

$$e_\theta\dot{\theta} - \theta\dot{G}(\varphi) = \frac{k_0}{\rho_0} \triangle \theta + r \tag{35}$$

Using the same arguments, the equation (5) yields

$$\dot{\mathbf{v}} = -\frac{1}{\rho_0}(\nabla p - \nabla \cdot \mathbf{T}_E) + \mathbf{b} \tag{36}$$

Moreover, the differential problem is completed by the equation (2) and the continuity equation (7), which under these hypothesis assume the forms

$$\gamma\dot{\varphi} = \frac{\varkappa}{\rho_0} \triangle \varphi - F'(\varphi) - (\theta + \lambda p)G'(\varphi) \tag{37}$$

$$-\nabla \cdot \mathbf{v} + \lambda\rho_0\dot{G}(\varphi) = 0 \tag{38}$$

Finally, ρ and ρ_0 are related by the equation (27), viz

$$\frac{1}{\rho} - \frac{1}{\rho_0} = \lambda G(\varphi) \tag{39}$$

4. Water-vapor phase transition

For the water-vapor phase transition, the order parameter equation is given by

$$\rho\gamma\dot{\varphi} = \varkappa \triangle \varphi - \rho\theta_c F'(\varphi) - \rho\left[\theta + \alpha(\theta,\theta_c)(p - p_0)\right] G'(\varphi) \tag{40}$$

where θ_c is the *critical temperature*, while p_0 is a suitable constant.

In this paper we suggest that

$$\alpha(\theta,\theta_c) = \begin{cases} \alpha_0 \frac{\theta_c - \theta}{\theta} , & \theta < \theta_c \\ 0 , & \theta \geq \theta_c \end{cases} \tag{41}$$

the coefficient $\alpha_0 > 0$ is related with the angular coefficient of the line between the water ($\varphi = 0$) and vapor ($\varphi = 1$). In (40) we have the same function F and G given in (3). Now the material will be considered compressible. Hence the pressure will not be a indeterminate quantity, but a function of the density and temperature.

The first law is always represented by

$$\rho\dot{e} = \rho h + \mathcal{P}_i^\varphi + \mathcal{P}_i^v \tag{42}$$

where

$$\mathcal{P}_i^\varphi = \rho\left\{\gamma\dot{\varphi}^2 + \theta_c\dot{F}(\varphi) + [\theta + \alpha(\theta,\theta_c)(p-p_0)]\,\dot{G}(\varphi)\right\} + \varkappa\nabla\varphi\cdot\nabla\dot{\varphi}$$

$$\mathcal{P}_i^v = (p-p_0)\frac{\dot{\rho}}{\rho} + \nu(\varphi)\nabla\mathbf{v}\cdot\nabla\mathbf{v} \tag{43}$$

In this framework the second law (19) takes the form

$$\rho(\psi_\theta+\eta)\dot{\theta} - \gamma\rho\dot{\varphi}^2 + \rho(\psi_\varphi - \theta_c F' - \theta G')\dot{\varphi} + \rho\psi_{\rho_0}\dot{\rho}_0 - \nu\nabla\mathbf{v}\cdot\nabla\mathbf{v} \tag{44}$$

$$-\rho(p-p_0)(\frac{\dot{\rho}}{\rho^2} + \alpha(\theta,\theta_c)\dot{G}(\varphi)) + \rho(\psi_{\nabla\varphi} - \varkappa\nabla\varphi)\cdot\nabla\dot{\varphi} - \frac{k_0}{\theta}|\nabla\theta|^2 \le 0.$$

As for the ice-water transition, we suppose the same composition given by (26) or (28) for the density and the following constitutive equation

$$\frac{\dot{\rho}}{\rho^2} + \alpha(\theta,\theta_c)\dot{G}(\varphi) = \frac{\dot{\rho}_0}{\rho_0^2} \tag{45}$$

where ρ_0 denotes the water density.

Moreover, because during the transition the temperature is constant, then on any isothermal process, the equation (45) can be written

$$\frac{d}{dt}\left[\frac{1}{\rho} - \alpha(\theta,\theta_c)G(\varphi)\right] = \frac{d}{dt}\frac{1}{\rho_0} \tag{46}$$

hence on any isothermal process we have the following constraint between ρ, ρ_0 and φ

$$\frac{1}{\rho} - \alpha(\theta,\theta_c)G(\varphi) = \frac{1}{\rho_0} \tag{47}$$

Thus, also during the water-vapor transition, the density separates in two elements ρ_0 and ρ_1 such that

$$\frac{1}{\rho_0} + \frac{1}{\rho_1} = \frac{1}{\rho} \tag{48}$$

where

$$\frac{1}{\rho_1} = \alpha(\theta,\theta_c)G(\varphi) \tag{49}$$

In this framework, the pressure is related with the density ρ_0 and with the temperature by the constitutive equation

$$p = p(\rho_0,\theta) \tag{50}$$

while, the density ρ_1 does not enter in (50), but it is a function of the phase and temperature only.

Therefore, the length $(\frac{1}{\rho_1})_{\max}$ of the horizontal line of the Andrews diagram will be a function of the temperature. Of course, when $\theta = \theta_c$ the horizontal line is equal to zero. Moreover, $(\frac{1}{\rho_1})_{\max}$ increases with the temperature by

$$(\frac{1}{\rho_1})_{\max} = \alpha_0 \frac{(\theta_c - \theta)}{12\theta}$$

In the following the free energy ψ and the entropy η will be assumed as state functions, where the state is defined by

$$\sigma = (\rho_0, \varphi, \theta, \nabla\varphi).$$

Thus, taking account of (45, the inequality (44) takes the new form

$$\begin{aligned} &\rho(\psi_\theta + \eta)\dot{\theta} - \gamma\rho\dot{\varphi}^2 + \rho(\psi_\varphi - \theta_c F' - \theta G')\dot{\varphi} \\ &-\nu\nabla\mathbf{v}\cdot\nabla\mathbf{v} + \rho(\psi_{\rho_0} - \frac{p(\rho_0, \theta) - p_0}{\rho_0^2})\dot{\rho}_0 \\ &+(\rho\psi_{\nabla\varphi} - \varkappa\nabla\varphi)\cdot\nabla\dot{\varphi} - \frac{k_0}{\theta}|\nabla\theta|^2 \leq 0. \end{aligned} \tag{51}$$

Therefore, from (51) because of the arbitrarity of $(\dot{\varphi}, \dot{\theta}, \dot{\rho}_0, \nabla\dot{\varphi})$,we obtain

$$\eta = -\psi_\theta, \quad \psi_\varphi - \theta_c F' - \theta G' = 0 \ , \tag{52}$$

$$\rho\psi_{\nabla\varphi} - \varkappa\nabla\varphi = 0, \tag{53}$$

$$\psi_{\rho_0} - \frac{p(\rho_0, \theta) - p_0}{\rho_0^2} = 0 \tag{54}$$

In order to obtain the Andrews diagram it is convenient to choose for ψ a function such that

$$\psi_{\rho_0} = \frac{\beta(\theta)}{\rho_0} \ , \quad \beta(\theta) > 0 \tag{55}$$

in particular, if $\beta(\theta) = nR\theta$, from (54) we have the classical equation of a perfect gas

$$p(\rho_0, \theta) - p_0 = \rho_0^2\psi_{\rho_0} = \beta(\theta)\rho_0 = nR\rho_0\theta \tag{56}$$

where R is a universal constant and n the number of moles of the gas.

Thus, for the free energy we obtain the function

$$\psi = \int e_\theta(\theta)d\theta - \theta \int \frac{e_\theta}{\theta} d\theta \qquad (57)$$
$$+\theta_c F(\varphi) + \theta G(\varphi) + \frac{\varkappa}{2\rho}(\nabla\varphi)^2 + nR\theta \log \rho_0$$

while the entropy is given by

$$\eta = \int \frac{e_\theta}{\theta} d\theta - G(\varphi) - nR \log \rho_0 \qquad (58)$$

from (57-58), we obtain the internal energy

$$e = \theta_c F(\varphi) + \frac{\varkappa}{2\rho}(\nabla\varphi)^2 + \int e_\theta d\theta$$

Finally, the inequality (51) reduces to

$$-\gamma\rho\dot{\varphi}^2 - \frac{k_0}{\theta}|\nabla\theta|^2 \leq 0. \qquad (59)$$

In this framework, the differential problem is defined by the system

$$\gamma\dot{\varphi} = \frac{\varkappa}{\rho} \triangle \varphi - \theta_c F'(\varphi) - (\theta + \alpha(\theta, \theta_c)(p - p_0))G'(\varphi) \qquad (60)$$

$$\rho\dot{\mathbf{v}} = -\nabla p(\rho_0, \theta) + \rho\mathbf{b} \qquad (61)$$

$$\dot{\rho} + \rho\nabla \cdot \mathbf{v} = 0 \qquad (62)$$

$$\frac{\dot{\rho}}{\rho^2} = \frac{\dot{\rho}_0}{\rho_0^2} - \alpha(\theta, \theta_c)\dot{G}(\varphi) \qquad (63)$$

$$\rho(\frac{e_\theta}{\theta}\dot{\theta} - \dot{G}(\varphi) - nR(\log \rho_0)^{\cdot}) = k_0 \nabla^2 \log\theta + \frac{r}{\theta} \qquad (64)$$

Multiplying (60) for $\dot{\varphi}$, (62) for $\mathbf{v}$, and (64) for θ we have the energy balance

$$\rho(\theta_c \dot{F}(\varphi) + (\frac{\mathbf{v}^2}{2})^{\cdot} + e_\theta\dot{\theta}) + (\frac{(\nabla\varphi)^2}{2})^{\cdot} \quad +\rho\gamma\dot{\varphi}^2 + k_0\frac{1}{\theta}(\nabla\theta)^2$$
$$= \nabla \cdot (\dot{\varphi}\nabla\varphi) + \nabla \cdot (k_0\theta\nabla \log\theta) - nR\rho_0\nabla \cdot (\theta\mathbf{v}) + \rho\mathbf{b} \cdot \mathbf{v} + \rho r$$

Hence, by an integration on the domain Ω we obtain the global form of energy theorem

$$\frac{d}{dt}\int_\Omega \left[\rho(\theta_c F(\varphi) + (\frac{\mathbf{v}^2}{2}) + \int e_\theta d\theta) + (\frac{(\nabla\varphi)^2}{2})\right] dx \leq$$

$$\leq \int_{\partial\Omega} (\varphi\nabla\varphi - k_0\theta\nabla \log\theta + nR\rho_0\theta\mathbf{v}) \cdot \mathbf{n} dS \qquad (65)$$

$$+ \int_{\Omega} \rho \left(\mathbf{b} \cdot \mathbf{v} + r\right) dx$$

Finally, in order to obtain the Andrews diagram, we study the $p - \rho$ function, which we obtain by (56) and by the equation

$$\theta - \alpha(\theta, \theta_c)(p - p_0) = \theta_c \tag{66}$$

that describes the line which demarcates the liquid- vapor phase transition. Then, for $\theta < \theta_c$, we have

$$\theta - \theta_c = \alpha_0 \frac{(\theta - \theta_c)}{\theta}(p - p_0) \tag{67}$$

from which

$$p = \frac{\theta}{\alpha_0} + p_0 \tag{68}$$

or by means of (56)

$$p(1 - \frac{1}{\alpha_0 n R \rho_0}) = p_0 \tag{69}$$

or

$$p(1 - \frac{v_0}{\alpha_0 n R}) = p_0$$

hence

$$p = \frac{n p_0 \alpha_0 R \rho_0}{n \alpha_0 R \rho_0 - 1} = \frac{n p_0 \alpha_0 R}{n \alpha_0 R - \frac{1}{\rho_0}} \tag{70}$$

Moreover, the critical value p_c and θ_c are jointed by

$$p_c = \frac{\theta_c}{\alpha_0} + p_0 \tag{71}$$

Therefore, from (70) and (71) we obtain

$$\frac{\theta_c}{\alpha_0} + p_0 = \frac{n p_0 \alpha_0 R \rho_c}{n \alpha_0 R \rho_c - 1}$$

from which $\frac{1}{\rho_c} < \alpha_0 n R$. Then, the function $p = p(v, \theta, \varphi)$ $(v = \frac{1}{\rho})$ is able to describe the Andrews diagram, because if we consider the constitutive equation of a perfect gas

$$p - p_0 = n R v_0^{-1} \theta = n R \rho_0 \theta$$

we have by (47)

$$p - p_0 = n R \left(v - \alpha(\theta, \theta_c) G(\varphi)\right)^{-1} \theta \tag{72}$$

with

$$v - \alpha(\theta, \theta_c)G(\varphi) = v_0 \tag{73}$$

Hence,

$$p - p_0 = \begin{cases} nRv^{-1}\theta \,, & if \quad \theta > \theta_c \\ nR\left(v - \alpha(\theta, \theta_c)G(\varphi)\right)^{-1}\theta = nRv_0^{-1}\theta \,, & if \quad \theta < \theta_c \end{cases}$$

then

$$if \;\; \theta > \theta_c, \qquad (p - p_0)v = nR\theta$$

$$if \quad \theta < \theta_c \;\; and \;\; \varphi \in [0,1] \qquad pv_0 = R\theta$$

References

1. Berti V, Fabrizio M and Giorgi C., Well-posedness for solid-liquid phase transitions with a fourth-order nonlinearity. Physica D: Nonlinear Phenomena 236 (2007) 13-21.
2. Bonetti E., Colli P. and Fremond M., A phase field model with thermal memory governed by the entropy balance. Mathematical Models and Methods in Applied Sciences 13 (2003) 1565-1588.
3. Fabrizio M, Giorgi C and Morro A., A continuum theory for first-order phase transitions through the balance of structure order. Mathematical Models in Applied Science, to apper.
4. Fabrizio M., Ginzburg-Landau equations and first and second order phase transitions. Int. J. Engng Sci. 44 (2006), 529-539.
5. Frémond M., Non-smooth Thermodynamics. Springer. Berlin, 2002.
6. Fried E and Gurtin ME., Continuum theory of thermally induced phase transitions based on an order parameter. Physica D 68 (1993), 326–343.
7. Ginzburg V. L. and Landau L. D., On the theory of superconductivity, Zh.Eksp. Teor. Fiz. 20 (1950) 1064-72.

LOCAL ERROR REDUCTION FOR FIRST ORDER IMPLICIT PSEUDO-SPECTRAL METHODS APPLIED TO LINEAR ADVECTION MODELS

RICCARDO FAZIO and SALVATORE IACONO

Department of Mathematics, University of Messina
Salita Sperone 31, 98166 Messina, Italy
E-mail: rfazio@dipmat.unime.it and iacono@dipmat.unime.it

By using any second or higher order pseudo-spectral method applied to linear non dispersive wave equations, one can experience the appearance in the numerical solution of the celebrated Gibbs phenomenon, located at the discontinuity points. On the contrary, by working out the numerical solution through a first order spectral method such a drawback is avoided. In particular, for stability reasons, we chose to consider the implicit Euler method and experienced that, even through a Richardson's extrapolation of such a method, the numerical solution obtained is always affected by the same phenomenon. In this paper we propose a way to improve the accuracy of the implicit Euler first order pseudo-spectral method by reducing the coefficient of its truncation error leading term via a time one-step extrapolation-like technique.

1. Introduction

Spectral methods are considered to be a valid or at least equivalent alternative to other numerical approaches in working out the solution to a partial differential equation. A very broad and deep treatment of spectral methods is done in various monographies and other books like the ones by Gottlieb and Orszag [1], Vichnevetsky and Bowles [2], Canuto et al. [3], Fonberg [4], Gottlieb and Hestaven [5] beyond the plenty of references quoted therein. For instance, as long as problems in acoustics or optics are concerned, in the book [6, pag.7], LeVeque states that the primary computational difficulty arises from the fact that the domain of interest is many orders of magnitude larger than the wavelength of interest and as a consequence methods with higher order of accuracy are typically used, for example, fourth order finite difference methods or spectral methods.

Dealing with an analytic function, it is proved that, by means of a spectral decomposition, it is possible to achieve a uniform convergence, exponentially increasing with the number of harmonics taken. On the contrary, if we consider functions that are piecewise smooth, even the point-wise convergence is lost and it does arise the celebrated Gibbs phenomenon consisting in a few overshoots located at function discontinuities. As in many practical applications the solution has to be strictly included within a predefined interval, it is obvious that the appearance of the Gibbs phenomenon might make the numerical solution lacking of physical meaning.

Our attention in this paper is devoted to numerically solving the linear advection equation

$$\frac{\partial u}{\partial t} + \nabla \cdot (\mathbf{v}\, u) = 0\ , \tag{1}$$

where $u = u(\mathbf{x}, t)$ with $u \in \mathbb{R}$ and the space variables $\mathbf{x} \in \Omega \subset \mathbb{R}^3$, whereas the time variable is t. As long as the advection velocity field $\mathbf{v}$ is concerned, we assume that it is constant. Furthermore, we have a prescribed initial condition and consider only periodic boundary conditions, so that it is straightforward to apply the Fourier decomposition approach.

2. Derivation of pseudo-spectral methods.

By defining the differential spacial operators

$$D = \frac{\partial}{\partial x}, \qquad E = \frac{\partial}{\partial y}, \qquad F = \frac{\partial}{\partial z},$$

we can rewrite the equation (1) as follows

$$\frac{\partial u}{\partial t} + v_1 \mathrm{D}(u) + v_2 E(u) + v_3 F(u) = 0,$$

where v_1, v_2, v_3 are the three constant components of the velocity vector $\mathbf{v}$. The same equation can be rewritten in integral form on the time interval $[t, t + \Delta t]$, as

$$u(\mathbf{x}, t+\Delta t) = u(\mathbf{x}, t) - \int_t^{t+\Delta t} v_1 \mathrm{D}(u)dt - \int_t^{t+\Delta t} v_2 E(u)dt - \int_t^{t+\Delta t} v_3 F(u)dt.$$

In order to obtain a first order implicit scheme by applying a quadrature rule, we can use the end-point rectangle one, so as to get the in time implicit discrete numerical scheme

$$\begin{aligned} u(\mathbf{x}, t+\Delta t) - u(\mathbf{x}, t) = -\Delta t\, \{ & v_1 \mathrm{D}\left[u(\mathbf{x}, t+\Delta t)\right] \\ & + v_2 E\left[u(\mathbf{x}, t+\Delta t)\right] + v_3 F\left[u(\mathbf{x}, t+\Delta t)\right] \} . \end{aligned}$$

It is interesting to notice that the resulting integration method have the same discretization error of the quadrature rule considered, because no errors were introduced so far and that this approach is equivalent to integrate directly in time by means of the implicit Euler scheme.

By introducing the symbolic operator $\mathrm{R}(\Delta t)$, this scheme can be rewritten more compactly in the form

$$u(\mathbf{x}, t+\Delta t) = \mathrm{R}(\Delta t) u(\mathbf{x}, t) \ , \quad \mathrm{R}(\Delta t) = \frac{1}{\mathrm{I} + \Delta t(v_1 \mathrm{D} + v_2 E + v_3 F)} \ . \quad (2)$$

By indicating with $v(\xi, \eta, \zeta, t) = \mathtt{fft}\{u(\mathbf{x}, t)\}$ the spacial FFT of the unknown function $u(\mathbf{x}, t)$, the equation (2) is rewritten in time-spectral domain like this

$$v(\xi, \eta, \zeta, t+\Delta t) = \hat{R}(\Delta t) v(\xi, \eta, \zeta, t) \ , \quad \hat{R}(\Delta t) = \frac{1}{I + i\Delta t(v_1 \xi + v_2 \eta + v_3 \zeta)}$$

where i is the imaginary unit.

More details on the pseudo-spectral algorithm can be found in [7].

3. Numerical approach

The first order discrete numerical scheme at a given time provides the vector $T(\Delta t)$, by indicating with $T(0)$ its value obtained as the discretization parameter Δt tends to zero, we can always write

$$T(0) = T(\Delta t) + \alpha_1 \Delta t + \alpha_2 \Delta t^2 + \alpha_3 \Delta t^3 + \dots + \alpha_m \Delta t^m + \dots,$$

where it is assumed that the coefficients α_i are independent from Δt. If we refer to the same scheme, but for the reduced time steps $q^k \Delta t$, with $0 < q < 1$, and the exponent $k \geq 0$ and integer, then for each k we can also write

$$T(0) = T(q^k \Delta t) + q^k \alpha_1 \Delta t + q^{2k} \alpha_2 \Delta t^2 + q^{3k} \alpha_3 \Delta t^3 + \dots + q^{mk} \alpha_m \Delta t^m + \dots,$$

By indicating with $T(k, 0)$ the discrete value $T(q^k \Delta t)$, we can introduce an extrapolation scheme

$$T(k+1, n+1) = T(k, n) + \frac{T(k, n) - T(k+1, n)}{q^w - 1}. \quad (3)$$

As long as the parameter w is equal to the accuracy order owned by the values $T(k, 0)$, such a scheme is just the one by Richardson and the value $T(k, n)$ would have actually an accuracy order increased by the integer n. Otherwise we get an extrapolation-like scheme. Let us now consider the implicit Euler method that is first order accurate, if we extrapolate properly, i.e. by using the value $w = 1$, then for any extrapolated order higher than one, the solution will go on showing the Gibbs phenomenon. On the contrary, by extrapolating once using the value $w = 2$, i.e. we are doing an extrapolation-like technique, we get discrete values that are still first order accurate, but the leading term of the truncation error becomes $q(q-1)\alpha_1 \Delta t$, whereas the second one is canceled. If we want to involve the values $T(0,0)$ and a generic $T(k,0)$, then we must modify (3) by using $w = 2k$. In such a case it is easy to prove that for the resulting extrapolated value the leading term becomes $q^k(q^k - 1)\alpha_1 \Delta t$. As a consequence, being $0 < q < 1$, in any case we get a reduction of the local error. Finally, obtaining always a first order method, there is no point in going on extrapolating beyond the first extrapolation step.

4. Numerical tests

In this section we consider the advection equation in 2D

$$u_t + v_1 \mathrm{D}u + v_2 \mathrm{E}u = 0 \ , \qquad x \in [0, L], \quad y \in [0, L]$$

with the Heavyside function as initial condition jointly to periodic boundary conditions

$$u(x,y,0) = \begin{cases} 0, & \text{for} \quad 0 \le x, y \le L/2 \ , \\ 1, & \text{for} \quad L/2 < x, y \le L \ , \end{cases} \qquad \begin{aligned} u(x,0,t) &= u(x,L,t) \ , \\ u(0,y,t) &= u(L,y,t) \ . \end{aligned}$$

For our numerical calculation we decided to fix a square domain with $L = 10$ and as constant vector field $\mathbf{v} = (1, 1)$. As a consequence, at the final time $t_{\max} = 10$, the exact solution will reproduce the initial condition. We carried out two numerical experiments, for the above $2D$ case and the analogous $1D$ case, using $q = 1/2$ and $\Delta t = 0.025$. They consisted in calculating the solution by applying one step of the extrapolation-like technique involving $T(0,0)$ and $T(k,0)$ for $1 \le k \le 5$, so that the obtained solution, $T(k,1)$, can be compared with the one obtained simply by refining the mesh by the same step reduction factor used for each $T(k,0)$. Besides, the FFT was computed involving up to $N = 1024$ harmonics for the $1D$ case, whereas for the $2D$ case it involved $N = 128$ harmonics.

In both cases, there is an improvement in slope recovery at the discontinuities. This remark can be appreciated graphically by looking at Fig. 1, for the $1D$ case, and Fig. 2, for the $2D$ case. In order to prove numerically such a result, we can define err_{extra} as the difference between the initial condition and the final time solution obtained by one step of extrapolation-like technique, i.e. $err_{\text{extra}} = |T(0) - T(k,1)|$, whereas analogously err_{ref} is referred to the solution obtained by mesh refining, i.e.

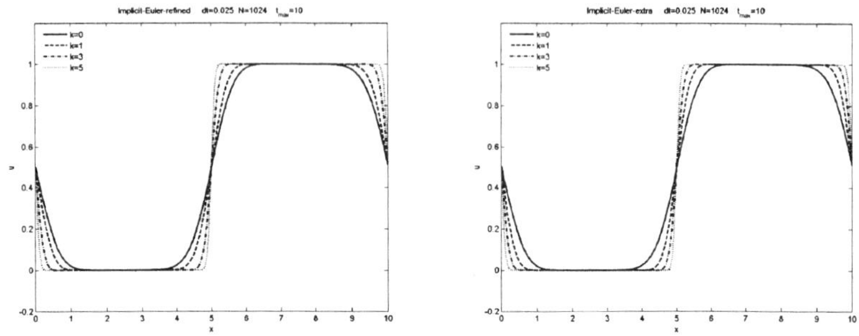

Fig. 1. $1D$ solutions for $k = 0, 1, 3, 5$: on the left, $T(k, 0)$ obtained for mesh refining, whereas on the right, $T(k, 1)$ obtained for extrapolation-like.

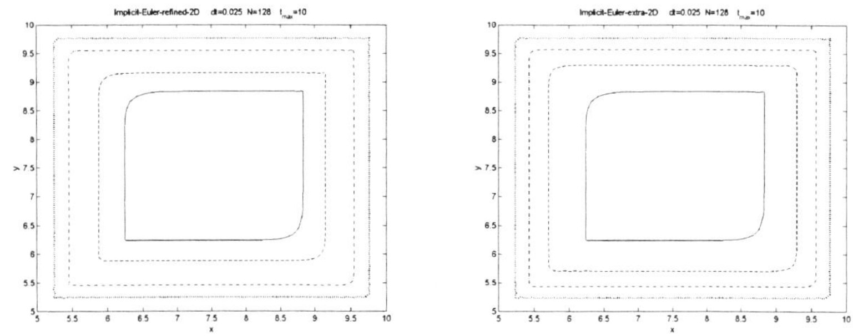

Fig. 2. $2D$ contour plot solution at $u = 0.99$ for $k = 0, 1, 3, 5$: on the left, $T(k, 0)$ obtained for mesh refining, whereas on the right, $T(k, 1)$ obtained for extrapolation-like.

$err_{\text{ref}} = |T(0) - T(k, 0)|$. In Table 1, for $1D$ case, and in Table 2, for the $2D$ case, there are reported the err_{extra} and err_{ref} 2-norm values for increasing values of k and it can be seen that also numerically it is proved a real improvement.

It is clear that the higher is the value taken for k the less is the numerical profit in applying the present technique and, as a consequence the maximum profit is achieved by choosing $k = 1$.

5. Conclusions

This work was motivated by a preliminary study concerning the implicit Euler and the second order Adams-Moulton methods, in the ordinary differential context, for more details see Fazio [8]. We have found that, when dealing with discontinuous functions, the Gibbs phenomenon can be avoided

Table 1. $1D$ numerical comparison for increasing k.

k	$\|err_{\text{extra}}\|_2$	$\|err_{\text{ref}}\|_2$	$\|err_{\text{ref}}\|_2 - \|err_{\text{extra}}\|_2$
0	0.047818	0.047818	
1	0.036378	0.040208	0.003830
2	0.032009	0.033808	0.001799
3	0.027628	0.028425	0.000798
4	0.023558	0.023896	0.000338
5	0.019945	0.020083	0.000138

Note: $\Delta t = .025$, N=1024, $t_{\max} = 10$.

Table 2. $2D$ numerical comparison for increasing k.

k	$\|err_{\text{extra}}\|_2$	$\|err_{\text{ref}}\|_2$	$\|err_{\text{ref}}\|_2 - \|err_{\text{extra}}\|_2$
0	0.065722	0.065722	
1	0.048061	0.053928	0.05867
2	0.041443	0.044170	0.02727
3	0.034838	0.036046	0.01208
4	0.028638	0.029163	0.00525
5	0.022861	0.023092	0.00231

Note: $\Delta t = .025$, N=128, $t_{\max} = 10$.

by using a pseudo-spectral method of order one. As a final remark it is clear that the proposed extrapolation-like error reduction technique can be easily extended to $3D$ problems.

References

1. G. B. Gottlieb and S. A. Orszag, *Numerical Analysis of Spectral Methods: Theory and Applications* (SIAM, 1977).
2. R. Vichnevetsky and J. B. Bowles, *Fourier Analysis of Numerical Approximations of Hyperbolic Equations* (SIAM, 1982).
3. C. Canuto, M. Y. Hussaini, A. Quarteroni and T. A. Zang, *Spectral Methods in Fluid Dynamics* (Springer-Verlang, New York, 1988).
4. B. Fornberg, *A Practical Guide to Pseudospectral Methods* (Cambridge University Press, Cambridge, 1996).
5. D. Gottlieb and J. S. Hesthaven, *J. Comput. Appl. Math.* **128**, 83 (2001).
6. R. J. LeVeque, *Finite Volume Methods for Hyperbolic Problems* (Cambridge Univesity Press, Cambridge, 2002).
7. R. Fazio and A. Jannelli, Implicit pseudo-spectral methods for dispersive and wave propagation problems, *Communications to SIMAI Congress*, ISSN 1827-9015, DOI: 10.1685/CSC06078.
8. R. Fazio, *Int. J. Math. Educ. Sci. Technol.* **32**, 752 (2001).

POSITIVE SCHEMES FOR THE ADVECTION EQUATION

RICCARDO FAZIO and ALESSANDRA JANNELLI

Department of Mathematics, University of Messina
Salita Sperone 31, 98166 Messina, Italy
E-mail: rfazio@dipmat.unime.it and jannelli@dipmat.unime.it

In this paper, we present some positive numerical methods for the advection equation. In particular, we consider two different classes of schemes for linear advection equation, the first one based on direct discretization and one based on method of lines. By theoretical point of view, the accuracy and positivity property of numerical methods are investigated. Some numerical results follow.

1. Introduction

Th aim of this study is to present positive numerical methods for the advection equation

$$\frac{\partial c}{\partial t} + \nabla \cdot (\mathbf{u}c) = 0 \tag{1}$$

where $c = c(\mathbf{x}, t)$ with $c \in \mathbb{R}$, $\mathbf{x} \in \Omega \subset \mathbb{R}^3$ and $t \in \mathbb{R}^+$; $\mathbf{u}(\mathbf{x}) \in \mathbb{R}^3$ is supposed to be given. This is a time-dependent partial differential equation in three spatial dimensions used in the applied sciences to describe several problems of great interest. Different methods have been developed in order to solve (1) numerically. Together with computational efficiency, another important property that should be possessed by an advection scheme is preservation of positivity, to avoid instabilities in the numerical solution. As point out by Liu and Lax [1], in the modern numerical treatment of conservation laws, the positivity is a key requirement. According to the second Godunov's barrier, second order numerical methods are not positive. We consider two different classes of schemes for linear advection equation, the first one based on direct discretization and another based on method of lines (MOL). The MOL approach represents an efficient tool to solve numerically more complex models. An example is represented by time dependent advection–diffusion–reaction models in 3D, used in many applications. Among others we can quote the application to the pollutant transport in the atmosphere [4], the mucilage dynamics [5], the ash-fall from vulcano [6], and groundwater and surface water [7]. Significant applications solved numerically by MOL scheme can be found in the work by Verwer et al. [4].

The considered methods satisfy the positive property, at first order accuracy, but not at second order accuracy. First order upwind methods have the advantage of keeping the solution monotonically varying in regions where the solution should be monotone, even though the accuracy is not very good. Second order accurate methods give much better accuracy on smooth solutions than the first order upwind method, but fail near discontinuities, where oscillations may appear due to their dispersive nature. On the other side, the methods implemented with flux limiters, used in order to suppress spurious oscillations, perform much better. The idea of flux limiter is to combine the best features of both methods. A reasonable large class of flux-limiter has been studied by Sweby [2], who derived algebraic conditions on limiter function which guarantee second order accuracy and positivity. For a more recent discussion on this topics see the paper [3] by LeVeque.

In this context, we present numerical methods implemented with flux limiter function. These methods are at least second order accurate on smooth solutions and yet give well resolved, non-oscillatory discontinuities. Two specific numerical tests are reported in the last section in order to show the behaviour of the considered direct discretization methods.

2. Numerical methods

In this section we consider numerical methods for solving scalar advection equation in three space dimensions

$$c_t + \sum_{s=1}^{3} (u_s c)_{x_s} = 0 \tag{2}$$

with given initial condition and appropriate boundary conditions (for instance: Dirichlet conditions at the inflow and no conditions at the outflow boundaries, or periodic boundary conditions, etc.). Here and in the sequel the subscripts indicate partial derivatives, and the velocity field (u_1, u_2, u_3) may depend on the independent variables x_s. We notice that in many applications the velocity field can be taken as divergence-free, that is $\sum_{s=1}^{3} (u_s)_{x_s} = 0$.

Set a uniform Cartesian grid $\Omega_J \subseteq \Omega \subseteq \mathbb{R}^3$, where $J = (j_1, j_2, j_3)^T$ is a lattice of points in which all j_s are integers. The grid points are $x_J = (j_1 \Delta x_1, j_2 \Delta x_2, j_3 \Delta x_3)^T$, where Δx_s are fixed step-sizes. Let C_J be an approximation to the value of the solution $c(x_J, t)$ at current time t, and $C_J^{\Delta t}$ an approximation to the value of the solution $c(x_J, t + \Delta t)$ at time $t + \Delta t$. The velocity j_s-components, for $s = 1, 2, 3$, are centered at the right, back and top face of the Cartesian cell, respectively, whereas the approximation C_J are located at its center. This is the so-called MAC (marker-and-cell) method, see Harlow and Welsh [8].

Being the governing equation linear, it would be appropriate to consider numerical schemes in the linear form

$$C_J^{\Delta t} = \sum_K \gamma_{JK} C_{J+K} \ . \tag{3}$$

It was proposed by Friedrichs [9] to require that the coefficients in equation (3) should verify the following conditions:

(a) $\gamma_{JK} \geq 0$ for each coefficient;
(b) $\sum_K \gamma_{JK} = 1$;
(c) $\gamma_{JK} = 0$ except for a finite set of K in a neighbourhood of J;
(d) γ_{JK} depends Lipschitz continuously on $\mathbf{x}$.

The first two properties imply that the solution at future $C_J^{\Delta t}$ is a convex combination of the values of solution at current step C_J, and this leads to a local maximum-minimum principle. That is, the solution $C_J^{\Delta t}$ is bounded above and below by solution C_J locally. The third property is a discrete consequence of the finite propagation speed of waves for the advection equation. Under the reported conditions, Friedrichs has shown that numerical approximations to solution that depend Lipschitz continuously on the space variables and that have positive coefficients are bounded under the discrete l^2 norm of the numerical solution, that is, they verify a bounded growth property:

$$\|C^{\Delta t}\| \leq (1 + M\ \Delta)\|C\| \tag{4}$$

where $\|\cdot\|$ is the discrete l^2 norm defined by

$$\|C\|^2 = \sum_J (C_J, C_J) \ ,$$

with $\Delta = \min\{\Delta t, \Delta x_1, \Delta x_2, \Delta x_3\}$, the value of the constant M in the above inequality depends on the Lipschitz constant. This suggests positivity as a design principle for solving system of conservation laws in more than one space variables. The schemes (3) verifying the property (a) are called *positive* in the sense of Friedrichs.

2.1. *1D Advection*

We start by considering the 1D advection equation

$$c_t + u c_x = 0 \ , \tag{5}$$

where u is constant. A numerical approximation can be obtained by considering a direct discretization. In the case $u \geq 0$, for instance, within the finite difference approach and a five-point stencil, we can consider the α-scheme defined by

$$C_j^{\Delta t} = \gamma_{-2} C_{j-2} + \gamma_{-1} C_{j-1} + \gamma_0 C_j + \gamma_1 C_{j+1} + \gamma_2 C_{j+2} \ , \tag{6}$$

with coefficients γ, depending on the Courant number $\nu = u\Delta t/\Delta x$, given as follows

$$\gamma_{-2} = -\frac{1}{2}\nu\,(1-\nu)\,\alpha \ , \qquad \gamma_{-1} = \nu\left[1 + \frac{1}{2}(1-\nu)(3\alpha - 1)\right] \ ,$$

$$\gamma_0 = 1 - \nu\left[1 + \frac{1}{2}(1-\nu)(3\alpha - 2)\right]$$

$$\gamma_1 = -\frac{1}{2}\nu(1-\nu)(1-\alpha) \ , \qquad \gamma_2 = 0 \ .$$

From this scheme we recover some classical second order schemes for particular choices of α: central (Lax–Wendroff $\alpha = 0$), Fromm ($\alpha = 1/2$), and upwind (Beam–Warming for $\alpha = 1$). The α-scheme has a local truncation error given by

$$\frac{1}{6}\ u\ \Delta x^2 \left[3\alpha - 1 - 3\alpha\ \nu + \nu^2\right] c_{xxx} + O\left(\Delta t^3\right) \ . \tag{7}$$

Moreover, the α-scheme is stable for $\nu \leq 1$ and computes the advection (exact) solution by setting $\nu = 1$. Note that, the α-scheme verifies all the conditions listed before, but the (a), i.e., the positivity condition. This is in agreement to the second Godunov's barrier: a linear positive numerical scheme is at most first order accurate.

The α-scheme can be written in conservation form

$$C_j^{\Delta t} = C_j - \frac{\Delta t}{\Delta x}\left[F_{j+1/2} - F_{j-1/2}\right]$$

where F is a numerical flux function, by setting

$$F_{j+1/2} = u\left\{C_j + \frac{1}{2}(1-\nu)\left[\alpha\,(C_j - C_{j-1}) + (1-\alpha)\,(C_{j+1} - C_j)\right]\right\} \ . \tag{8}$$

This is important because by the Law-Wendroff theorem we know that the shock wave velocity is computed correctly if and only if the scheme is in conservation form.

Now, we describe a different numerical approach: the method of lines or MOL. We introduce a spatial discretization, a given five-point discretization formula defines the semi-discrete system

$$\frac{d}{dt}c_j(t) = \frac{u}{\Delta x}\left[\delta_{-2}c_{j-2}(t) + \delta_{-1}c_{j-1}(t) + \delta_0 c_j + \delta_1 c_{j+1}(t) + \delta_2 c_{j+2}(t)\right] \tag{9}$$

where $c_j(t) \approx c(t, x_j)$ with $x_j = j\Delta x$, and the δ coefficients are given constants. The system (9) can be solved by an ODE scheme to get a numerical approximation $c_j^n \approx c(t^n, x_j)$, with $t^n = n\Delta t$. If we consider the limit as Δt goes to zero, for fixed Δx and t^n, then the approximate solution C_j^n of (6) converges to the solution $c_j(t^n)$ of (9). Therefore, for ν going to zero, and fixed Δx, the direct scheme becomes the MOL scheme with an exact time integration. In fact, as an example, when $u \geq 0$, from the α-scheme we can derive its associated MOL semi-discrete system (9) with parameters

$$\delta_{-2} = -\frac{1}{2}\alpha\ , \quad \delta_{-1} = 1 + \frac{1}{2}(3\alpha - 1)\ , \quad \delta_0 = -1 - \frac{1}{2}(3\alpha - 2)$$
$$\delta_1 = -\frac{1}{2}(1 - \alpha)\ , \qquad \delta_2 = 0\ .$$

2.2. *Multi-D peculiarities.*

One of the simplest way to deal with advection problems, in 2D or 3D, is to apply a MOL approach. In this approach the 1D formula can be used as they are in all the Cartesian directions, with or without limiting. To the people coming from an ODE background, this kind of semi-discretization seems to give simple and effective schemes, which in the case of advection-diffusion-reaction models, can be easily complemented by diffusion and reaction terms. Moreover, when the conversion from the PDE to the ODE system is done, we have at our fingertips a good deal of numerical methods of different accuracy and stability properties. However, this kind of approach has its drawbacks. As a first step, it is necessary to write the 2D (worse in 3D) problem as a single system of ODE and the multi-D structure with initial and boundary conditions has to be rendered by a single vector. Then, we have to take into account the stability of the ODE system. Accurate solution can be computed by using higher order spatial discretization formulas with limiting, the interested reader can find full details on this topic in the research paper by Hundsdorfer et al. [10] where the positivity of the so called k-schemes and a fourth order central scheme are investigated or Hundsdorfer and Trompert [11] where the comparison of two related third order MOL and dimensional splitting (from a fully discrete one-dimensional) methods are described. However, these authors point out that the full discrete method is more efficient and reliable than the MOL method, so that in the following we consider a full discrete method (without dimensional splitting).

In order to derive the high order scheme in 2D, we start considering the first order upwind method obtained by a direct discretization. In the case $u_1, u_2 \geq 0$, for instance, within the finite difference approach, we can consider the scheme defined by

$$C_{jh}^{\Delta t} = \gamma_{-1,0} C_{j-1h} + \gamma_{0,0} C_{jh} + \gamma_{0,-1} C_{jh-1} \tag{10}$$

with coefficients γ, depending on Courant numbers $\nu_1 = u_1 \Delta t / \Delta x_1$ and $\nu_2 = u_2 \Delta t / \Delta x_2$, given as follows

$$\gamma_{-1,0} = \nu_1 \ , \qquad \gamma_{0,0} = 1 - \nu_1 - \nu_2 \ , \qquad \gamma_{0,-1} = \nu_2 \ . \tag{11}$$

This method is often called donor-cell upwind (DCU) method. In this method we assume that the only contribution to each flux is coming from the adjacent cell on the upwind side (the donor cell) and the numerical flux approximates the amount of concentration that is flowing normal to the corresponding cell edge. It is clear that the new value $C_{jh}^{\Delta t}$ is computed by using only the values C_{jh}, C_{j-1h} and C_{jh-1}. This is correct only in the special cases when $u_1 = 0$ or $u_2 = 0$. In the general case, when the velocity field, at some times, may be at an angle to the grid, it is clear that also the value of C_{j-1h-1} should be involved to define $C_{jh}^{\Delta t}$. Then, in order to evaluate the contribution to fluxes by corner transport, we have to consider the transverse propagation due to mixed derivative terms in the Taylor formula

$$\begin{aligned} c\left(x_1, x_2, t^{n+1}\right) &= c\left(x_1, x_2, t^n\right) + \Delta t c_t + \frac{1}{2}(\Delta t)^2 c_{tt} + \ldots \\ &= c\left(x_1, x_2, t^n\right) - u_1 \Delta t c_{x_1} - u_2 \Delta t c_{x_2} + \\ &+ \frac{1}{2}(\Delta t)^2 \left[u_1^2 c_{x_1 x_1} + u_1 u_2 c_{x_1 x_2} + u_2 u_1 c_{x_2 x_1} + u_2^2 c_{x_2 x_2}\right] + \ldots \end{aligned} \tag{12}$$

where all derivatives are evaluated at (x_1, x_2, t^n). In this way we obtain the following method of first order accuracy in 2D, the so called corner transport upwind (CTU),

$$\begin{aligned} C_{jh}^{\Delta t} = C_{jh} &- \frac{u_1 \Delta t}{\Delta x_1}\left(C_{jh} - C_{j-1h}\right) - \frac{u_2 \Delta t}{\Delta x_2}\left(C_{jh} - C_{jh-1}\right) + \\ &+ \frac{1}{2}(\Delta t)^2 \left\{ \frac{u_1}{\Delta x_1}\left[\frac{u_2}{\Delta x_2}\left(C_{jh} - C_{jh-1}\right) - \frac{u_2}{\Delta x_2}\left(C_{j-1h} - C_{j-1h-1}\right)\right] + \right. \\ &\left. + \frac{u_2}{\Delta x_2}\left[\frac{u_1}{\Delta x_1}\left(C_{jh} - C_{j-1h}\right) - \frac{u_1}{\Delta x_1}\left(C_{jh-1} - C_{j-1h-1}\right)\right]\right\} \end{aligned} \tag{13}$$

that can be written also in the following form

$$C_{jh}^{\Delta t} = \gamma_{-1,-1} C_{j-1h-1} + \gamma_{-1,0} C_{j-1h} + \gamma_{0,0} C_{jh} + \gamma_{0,-1} C_{jh-1} \tag{14}$$

with coefficients γ given as follows

$$\begin{aligned} \gamma_{-1,-1} &= \nu_1 \nu_2 , & \gamma_{-1,0} &= \nu_1 - \nu_1 \nu_2 , \\ \gamma_{0,0} &= 1 - \nu_1 - \nu_2 + \nu_1 \nu_2 , & \gamma_{0,-1} &= \nu_2 - \nu_1 \nu_2 . \end{aligned} \tag{15}$$

By the von Neumann analysis we can determine step-size restrictions in order to guarantee the stability property of numerical methods. For the DCU method (10-11) we have

$$\frac{\Delta t}{\Delta x_1} |u_1| + \frac{\Delta t}{\Delta x_2} |u_2| \leq 1 , \tag{16}$$

in contrast with the less restrictive stability condition obtained for the CTU method (13)

$$\max \left(\frac{\Delta t}{\Delta x_1} |u_1| , \frac{\Delta t}{\Delta x_2} |u_2| \right) \leq 1 . \tag{17}$$

As far as the derivation of higher order scheme is concerned, we have to consider also the high order derivative terms in (12). By using second order central finite difference approximation, we obtain the following second order Lax–Wendroff method

$$\begin{aligned} C_{jh}^{\Delta t} = C_{jh} &- \frac{u_1 \Delta t}{\Delta x_1} (C_{jh} - C_{j-1h}) - \frac{u_2 \Delta t}{\Delta x_2} (C_{jh} - C_{jh-1}) + \\ &+ \frac{1}{2} (\Delta t)^2 \left\{ \frac{u_1}{\Delta x_1} \left[\frac{u_2}{\Delta x_2} (C_{jh} - C_{jh-1}) - \frac{u_2}{\Delta x_2} (C_{j-1h} - C_{j-1h-1}) \right] + \right. \\ &\qquad + \frac{u_2}{\Delta x_2} \left[\frac{u_1}{\Delta x_1} (C_{jh} - C_{j-1h}) - \frac{u_1}{\Delta x_1} (C_{jh-1} - C_{j-1h-1}) \right] + \\ &+ \left. \left[\frac{u_1^2}{\Delta x_1^2} (C_{j-2h} - 2C_{j-1h} + C_{jh}) + \frac{u_2^2}{\Delta x_2^2} (C_{jh-2} - 2C_{jh-1} + C_{jh}) \right] \right\} , \end{aligned} \tag{18}$$

that can also be written in the form (3). By the von Neumann analysis, the stability condition for (18) is given again by (17).

2.3. *Test Problems*

In this section we consider a first test problem:

$$\begin{aligned} & c_t + u_1 c_{x_1} + u_2 c_{x_2} = 0 , \\ & c(x_1, x_2, 0) = \begin{cases} 1 & \text{if } r = \sqrt{(x_1 + 0.2)^2 + (x_2 + 0.2)^2} \leq 0.1 \\ 0 & \text{otherwise} \end{cases} \\ & c(0, x_2, t) = c(1, x_2, t) , \quad c(x_1, 0, t) = c(x_1, 1, t) , \end{aligned} \tag{19}$$

where $0 \leq x_1, x_2 \leq 1$, $u_1 = u_2 = 1$ and $t \geq 0$. The initial datum is a unitary cylinder with radius 0.1, centered at $(0.2, 0.2)$; at the boundaries we impose periodic boundary conditions. We report the numerical results

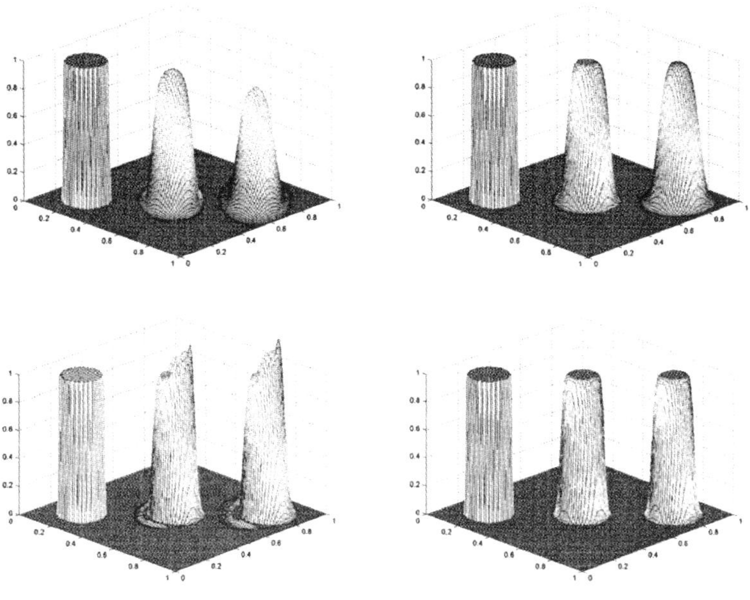

Fig. 1. Advection equation with 100×100 spatial grid, Courant number 0.95 and final time $t_{\max} = 0.6$. Top left: first order DTU method; top right: first order CTU method; bottom Left: second order Lax-Wendroff method without limiter; bottom right: second order Lax-Wendroff method with van Leer limiter.

obtained with direct discretization numerical methods of first and second order accuracy, with and without flux limiter in Fig. 1. By considering the different behaviour of the methods and limiters used, we have found that the best results are obtained by the second order method with van Leer limiter.

The second order Lax-Wendroff method, with van Leer limiter, has been used for solving a second test problem

$$\begin{aligned} &c_t + (u_1\, c)_{x_1} + (u_2\, c)_{x_2} = 0 \\ &c(x_1, x_2, 0) = \tanh\left(-\frac{1}{2}x_2\right) \\ &c(-4, x_2, t) = c(4, x_2, t)\ , \quad c(x_1, -4, t) = c(x_1, 4, t)\ , \end{aligned} \tag{20}$$

with $-4 \le x_1, x_2 \le 4$ and $t \ge 0$ and where the velocity field is given by

$$u_1(x_1,x_2) = -\omega(r)\, x_2\ , \quad u_2(x_1,x_2) = \omega(r)\, x_1\ , \quad r = \sqrt{x_1^2 + x_2^2}\ ,$$
$$\omega(r) = \frac{V(r)}{r\, V_{\max}}\ , \quad V(r) = \frac{\tanh(r)}{\cosh^2(r)}\ . \tag{21}$$

Here $V(r)$ represents the tangential velocity around the center of the do-

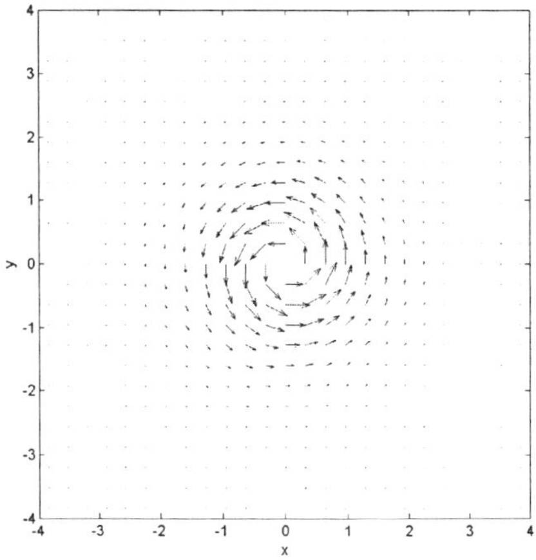

Fig. 2. Velocity field for problem (20)-(21) with $V_{\max} = 0.385$.

main, and $V_{\max}$ is its maximum value. Fig. 2 shows the velocity on a 25×25 grid – with this number of cells it is possible to grasp the structure of the velocity field – far from the center the velocity is smaller compared to its vicinity, and at the center the velocity is zero. This problem provides a simple model describing the mixing of cold and hot air, due to the rotational velocity field, which is similar to the cyclonic air motion at low pressure systems observed on the weather maps.

Problem (20)-(21) was used by Tamamidis and Assanis [12] and by Hundsdorfer and Trompert [11] to evaluate the performance of different numerical methods. In fact, for this problem it is possible to carry on convergence tests since its exact solution is known:

$$c(x_1,x_2,t) = \tanh\left(\frac{1}{2}x_1 \sin(\omega(r)t) - \frac{1}{2}x_2 \cos(\omega(r)t)\right)\ . \tag{22}$$

Fig. 3 shows a side view of the numerical solution, at final time $t_{\max} = 4$

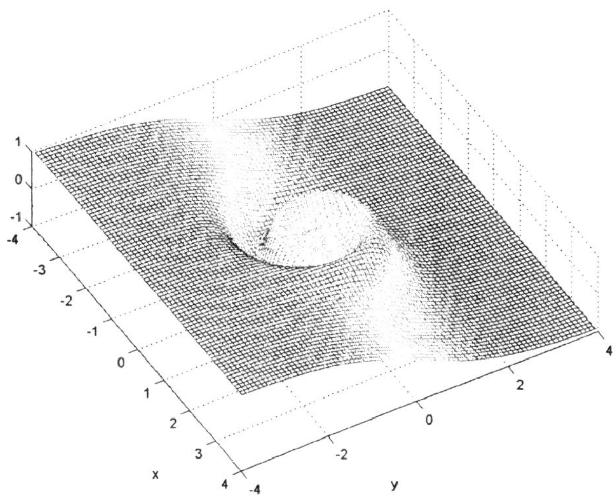

Fig. 3. Sample numerical solution for problem (20)-(21) with $V_{\max} = 0.385$: side view.

on a 80×80 spatial grid, for $V_{\max} = 0.385$. For this test we used a Courant number equal to 0.9. Two sample animations are available from the web page `http://mat520.unime.it/fazio/Mixing.html`

References

1. X.-D. Liu and P. D. Lax, *J. Comput. Phys.* **187**, 428 (2003).
2. P. K. Sweby, *SIAM J. Num. Anal.* **21**, 995 (1984).
3. R. J. LeVeque, *SIAM J. Numer. Anal.* **33**, 627 (1996).
4. J. Verwer, W. Hundsdorfer and J. Blom, *Sur. Math. Ind.* **2**, 107 (2002).
5. A. Jannelli, R. Fazio and D. Ambrosi, *Comput. & Fluids* **32**, 47 (2003).
6. R. McKibbin, L. L. Lim, T. A. Smith and W. L. Sweatman, A model for dispersal of eruption ejecta, in *Proceedings World Geothermal Congress*, 24-29 April 2005, Antalya, Turkey.
7. M. Toro, L. van Rijn and K. Meijer, Three-dimensional modelling of sand and mud transport in current and waves Technical Report No. H461, Delft Hydraulics, Delft, The Netherlands, 1989.
8. F. Harlow and J. Welsh, *Phys. Fluids* **8**, 2182 (1965).
9. K. O. Friedrichs, *Commun. Pure Appl. Math.* **7**, 345 (1954).
10. W. Hundsdorfer, B. Koren, M. van Loon and J. G. Verwer, *J. Comput. Phys.* **117**, 35 (1995).
11. W. Hundsdorfer and R. A. Trompert, *Appl. Num. Math.* **13**, 469 (1994).
12. P. Tamamidis and D. N. Assanis, *Internat. J. Numer. Methods Fluids.* **16**, 931 (1993).

TRAVELLING-WAVE SOLUTIONS OF A MODIFIED SINE-GORDON EQUATION USED IN SUPERCONDUCTIVITY

GAETANO FIORE

Dip. di Matematica e Applicazioni, Università "Federico II",
V. Claudio 21, 80125 Napoli, Italy
E-mail: gaetano.fiore@na.infn.it

We briefly illustrate a non-perturbative analysis [4] of the modified sine-Gordon equation

$$\varphi_{tt} - \varphi_{xx} + \sin\varphi = \delta - \alpha\varphi_t \tag{1}$$

($\alpha \geq 0, \delta \in \mathbb{R}$) on the real axis ($x \in \mathbb{R}$) and on the circle ($x \in S^1$), and its application to the Josephson effect in the theory of superconductors.

The Josephson effect (for an introduction see e.g. Ch. 1 in [1]) is importantant both for fundamental research (in macroscopic effects in quantum physics, quantum computation, etc.) and for applications to electronic devices [e.g. highly sensitive magnetometers (SQUIDS), radiation detectors, etc.]. There φ is the phase difference of the macroscopic quantum wavefunctions describing the Bose-Einstein condensates of Cooper pairs in two superconductors separated by a very thin, narrow and long dielectric (a "Josephson Junction", JJ). In normalized units the constant forcing term δ (providing energy to the system) is the "bias current" density, whereas $-\alpha\varphi_t$ is a dissipative term due to Joule effect of the residual single-electron current across the junction; their energetic balance may enable persistent solutions. The Josephson equations $j = \sin\varphi$, $\mathcal{V} = \frac{1}{2\pi}\varphi_t$ give the tension $\mathcal{V}(x,t)$ and the supercurrent density $j(x,t)$ in terms of $\varphi(x,t)$. Treating as usual (see e.g. [2,3]) the rhs(1) as a small perturbation to the pure sine-Gordon eq. (sGe) [i.e. (1) with $\alpha = \delta = 0$] is unjustified unless $\alpha, \delta \ll 1$, hence is not sufficient for predicting important observables like the whole $I-V$ characteristic (the relation between the total current I and the average tension V across a finite JJ). In [4] we stick to travelling-wave solutions of (1) and point out that phase space analysis of the corresponding ordinary differential equations (o.d.e.'s) allows their complete non-perturbative classification within the *whole* range of parameters α, δ. We shall call physically *relevant* those stable and with bounded derivatives. Constant ones exist only if $|\delta| \leq 1$ and are $\varphi^s(x,t) \equiv \theta := \sin^{-1}\delta$ and $\varphi^u(x,t) \equiv \pi - \theta$ (mod 2π). They coincide and are unstable if $|\delta| = 1$. If $|\delta| < 1$ φ^s (resp. φ^u) is stable (resp. unstable), as at each x it minimizes (resp. maximizes) the potential energy density. Beside φ^s, the relevant φ turn out to be of the following types: (anti)kink (i.e. solitons), array of (anti)kinks, or half-array of (anti)kinks. They belong to the finite-dimensional global attractor of (1) predicted in [5]. The first two are deformations of solutions of the sGe, the third interpolate between them and have no sGe analog.

The sGe describes also the dynamics of the continuum limit of a sequence of pendula constrained to rotate around an horizontal axis ($\mathbb{R}$) or circle (S^1) and mutually coupled by a torque spring [6]; $\varphi(x,t)$ is the deviation angle from the lower vertical position at time t of the pendulum located at x. One can model also the terms $-\alpha\varphi_t, \delta$ of (1) by immersing the pendula in a linearly viscous fluid and by assuming that a constant torque distribution is applied to them. This mechanical analog allows a qualitative comprehension of many features of the solutions, e.g. φ^s, φ^u respectively describe pendula hanging all down or standing all up. The problem on S^1 can be reduced to that on $\mathbb{R}$, as any solution of (1) on a circle of length L can be represented as one defined on all $\mathbb{R}$ and fulfilling

$$\varphi(x+L,t) = \varphi(x,t) + 2\pi n \tag{2}$$

with some (topologically invariant) $n \in \mathbb{Z}$: the pendula sequence twists around the circle $|n|$ times, (anti)clockwise according to the sign of n.

The space of solutions of (1) is invariant under x, t translations; we shall call $[\varphi]$ the orbit containing φ and exhibit one φ for each orbit. Without loss of generality we can assume $\gamma := -\delta \geq 0$ (if originally $\gamma < 0$, this is obtained by replacing $\varphi \to -\varphi$). We refine the travelling-wave Ansatz $\varphi(x,t) = f(x - vt)$ (v is the constant phase velocity of the wave) as follows:

$$\begin{aligned}
&\varphi(x,t) = -g(\xi) && \xi := -\text{sign}(v)\tfrac{x-vt}{\sqrt{v^2-1}} && \text{if } v^2 > 1,\\
&\varphi(x,t) = \tilde{g}(\tilde{\xi}) - \pi && \tilde{\xi} := \pm x - t && \text{if } v = \pm 1,\\
&\varphi(x,t) = g(\xi) - \pi && \xi := \text{sign}(v)\tfrac{x-vt}{\sqrt{1-v^2}} && \text{if } 0 < v^2 < 1,\\
&\varphi(x,t) = g(\xi) - \pi && \xi := x && \text{if } v = 0.
\end{aligned} \tag{3}$$

We replace it in (1). If $v = \pm 1$ the latter reduces to the 1st order o.d.e.

$$\alpha\tilde{g}' = \gamma - \sin\tilde{g} \tag{4}$$

(integrable even in closed form [7]). Integrating it by quadrature and replacing in (3) one finds, beside φ^s, φ^u: if $\gamma \le 1$, $\alpha > 0$ manifestly unstable solutions φ, since $\varphi \to \varphi^u$ as either $x \to \infty$ or $x \to -\infty$; if $\gamma > 1$, $\alpha > 0$ virtually stable solutions fulfilling

$$\varphi(x+X,t) = \varphi(x,t) \pm 2\pi \tag{5}$$

with $X = 2\pi\alpha/\sqrt{\gamma^2-1}$, and (2) with $L = |n|X$. They describe arrays of (anti)kinks. If $|v| \ne 1$ (1) reduces to the 2nd order o.d.e. at the lhs of

$$g'' + \mu g' + \frac{dU}{dg} = 0 \qquad \Leftrightarrow \qquad \begin{cases} g' = u, \\ u' = -\mu u - \sin g + \gamma \end{cases} \tag{6}$$

or the equivalent 1st order system at the rhs, where

$$U(g) = -(\cos g + \gamma g), \qquad \mu = \alpha/\sqrt{|v^{-2}-1|} \tag{7}$$

[note that in (6) α, v appear only through μ]. This is the equation of motion w.r.t. 'time' ξ of a damped pendulum with a constant torque γ, or equivalently of a particle with unit mass, position g, subject to a 'washboard' potential energy $U(g)$ and a viscous force. The solutions of (6) can be classified [8–10] by phase space analysis and some useful monotonicity properties. By the Peano-Picard theorem all $g(\xi)$ are defined on all $\mathbb{R}$ (existence), the paths in phase space $p = (g,u)$ do not intersect (uniqueness), so each is uniquely singled out by any point $p_0 = (g_0, u_0)$ of it. They are continuous and monotonic functions of μ, γ, p_0 (away from singular p_0). Singular p_0 exist only if $\gamma \le 1$ and lie all on the $u = 0$ axis:

$$\begin{aligned} A_k &= \big((2k+1)\pi - \theta, 0\big), \qquad & B_k &= (2k\pi - \theta, 0), \ \gamma < 1 \\ C_k &= \big((2k+1/2)\pi, 0\big), & & \gamma = 1, \end{aligned} \tag{8}$$

where $k \in \mathbb{Z}$. A_k are saddles; B_k are nodes, foci or centers (depending on μ, γ); C_k are saddle-nodes. The constant solutions $p(\xi) \equiv A_k, B_k, C_k$ lead to φ^s, φ^u. We assume $p(\xi)$ non-constant henceforth. Then each path is cut by the $u = 0$ axis into pieces lying either in the upper or in the lower half-plane; each piece can be parametrized as $u = u(g)$, where $u(g)$ fulfills the 1st order problem (invariant under $g \to g + 2\pi$)

$$uu_g(g) + \mu u(g) + \sin g - \gamma = 0, \qquad u(g_0) = u_0. \tag{9}$$

Once this is solved, one needs only integrating by quadrature

$$g'(\xi) = u\big(g(\xi)\big), \qquad g(\xi_0) = g_0, \tag{10}$$

to solve the problem (6) & $p(\xi_0) = p_0$. If $\mu = 0$ the system is Hamiltonian, and (9) is solved by the energy integral. If $\gamma \ge 1$ all non-constant $p(\xi)$ are unbounded with diverging momentum u as ξ goes to either ∞ or $-\infty$; if $0<\gamma<1$ there are in addition only homoclinic paths (with both ends on some A_k) and periodic ones (cycles around some B_k). None of them yields relevant φ [12]. If also $\gamma=0$ (sGe), (6) reduces to the pendulum equation: the heteroclinic paths [going from some A_k to $A_{k\pm1}$] replace the homoclinic ones and yield [11] unstable φ for $|v|>1$ (since $\varphi\to\varphi^u$ as $x\to\pm\infty$) and stable φ for $|v|<1$, the celebrated (anti)kink solutions

$$\hat{\varphi}^{\pm}_{(0)}(x,t;v)=4\tan^{-1}\left\{\exp\left[\pm(x-vt)/\sqrt{1-v^2}\right]\right\}; \tag{11}$$

the unbounded ones have periodic positive- (or negative-) definite $u(\xi)$ and yield solutions φ fulfilling (2) with L depending on p_0: those with $|v|>1$ are unstable, those $\check{\varphi}^{\pm}_{(0)}$ with $|v|<1$ [array of (anti)kinks] are stable [11]. Note: v is a *free* parameter $v\in]-1,1[$ in both $\hat{\varphi}^{\pm}_{(0)}$ and $\check{\varphi}^{\pm}_{(0)}$. Whereas the periodic paths yield unstable [11] φ oscillating around φ^u or φ^s (plasma waves).

If $\mu>0$, $\gamma=0$, the paths are: unbounded with diverging u as $\xi\to-\infty$; or going from some A_k to B_k or B_{k+1} (implying $\varphi\to\varphi^u$ as either $x\to\infty$ or $x\to-\infty$). All the corresponding φ are not relevant.

Finally, the region $\mu>0$, $\gamma>0$. Using continuity and monotonicity w.r.t. p_0 one can show [8] the existence of paths $\check{p}(\xi)$ with periodic $\check{u}(\xi)>0$; these exist for all $\mu>0$ if $\gamma>1$, for $\mu\in]0,\hat{\mu}(\gamma)[$ if $\gamma\leq1$, where $\hat{\mu}(\gamma)$ is [9] a strictly increasing function such that $\hat{\mu}(0)=0$, $\hat{\mu}(1)\approx1.193$. Replacing in (3), with v one of the four solutions of $(7)_2$, one finds φ fulfilling (5), (2) with

$$\frac{L}{|n|}=X=\sqrt{|1-v^2|}\int_0^{2\pi}\frac{dg}{\check{u}(g;\gamma,\mu)}; \tag{12}$$

the two with $|v|>1$ are unstable, the two $\check{\varphi}^{\pm}$ with $|v|<1$ are virtually stable and describe arrays of (anti)kinks (see Thm), going to the ones mentioned before (5) as $v\to\pm1$. If $\gamma>1$, in all the remaining paths u diverges as $\xi\to-\infty$, yielding irrelevant φ. If $\gamma\leq1$ and $\mu\in]0,\hat{\mu}(\gamma)[$ there are in addition paths $\bar{p}(\xi)$ with $\bar{u}>0$, starting from A_k (resp. C_k if $\gamma=1$) and asymptotically approaching a corresponding $\check{p}(\xi)$ as $\xi\to\infty$. Replacing $\bar{g}(\xi)$ in (3) with v one of the two solutions of $(7)_2$ s.t. $|v|<1$ (resp. $|v|>1$) one finds virtually stable (resp. unstable) solutions; the stables ones $\bar{\varphi}^{\pm}$ describe "half-arrays" of (anti)kinks (see Thm). Moreover, if $\mu=\hat{\mu}(\gamma)$ the stable and unstable manifolds of two nearby saddle points are matched, and there are heteroclinic paths $\hat{p}(\xi)$ going from A_k to A_{k+1} (C_k to C_{k+1} if $\gamma=1$). Replacing $\hat{g}(\xi)$ in (3), with v one of the two solutions of $(7)_2$ s.t. $|v|<1$ (resp. $|v|>1$) one finds virtually stable (resp. unstable) solutions; the stables ones $\hat{\varphi}^{\pm}$ describe (anti)kinks (see theorem), going to sGe ones as $\gamma\to0$.

If $\gamma = 1$ there are no other paths. If $\gamma < 1$, the remaining paths are not relevant, as they are: either unbounded with $u < 0$ diverging as $\xi \to -\infty$; or going from A_k to either B_k (for any $\mu > 0$) or B_{k+1} (for $\mu \in]\hat{\mu}, \infty[$), implying $\varphi \to \varphi^u$ as either $x \to \infty$ or $x \to -\infty$). See also [12].

We summarize the main results in the figure and quoting

Theorem 1 in [4]. *Assume* $-\delta = \gamma > 0$, $\alpha > 0$ *and let*

$$\xi^{\pm}(\mu) := \frac{\pm x - \check{v}(\mu) t}{\sqrt{1 - v^2(\mu)}}, \qquad \check{v}(\mu) := \frac{\mu}{\sqrt{\alpha^2 + \mu^2}} \leq 1. \tag{13}$$

Mod. 2π, *relevant*[a] *travelling-wave solutions of (1) are only of the types:*

(1) Static, uniform $\varphi^s(x,t) \equiv \theta := \sin^{-1}\delta$, *if* $\gamma < 1$.

(2) Kink $\hat{\varphi}^+$ *or antikink* $\hat{\varphi}^-$, *where* $\hat{\varphi}^{\pm}(x,t) := \hat{g}\left[\xi^{\pm}\left(\hat{\mu}(\gamma)\right); \gamma\right] - \pi$, *only if* $\gamma < 1$. *The function* $\hat{\mu}(\gamma)$ *is [9] continuous and strictly increasing in* $[0,1]$, *with* $\hat{\mu}(0) = 0$, $\hat{\mu}(1) \approx 1,193$, $\hat{\mu}(\gamma) = \pi\gamma/4 + O(\gamma^2)$ *[4].* $\hat{\varphi}^{\pm}$ *respectively has velocity* $v = \pm\check{v}\left(\hat{\mu}(\gamma)\right)$ *and fulfill*

$$\lim_{x \to -\infty} \hat{\varphi}^{\pm}(x,t) = \theta, \qquad \lim_{x \to \infty} \hat{\varphi}^{\pm}(x,t) = \theta \pm 2\pi, \tag{14}$$

(3) Arrays of kinks $\check{\varphi}^+$ *or antikinks* $\check{\varphi}^-$*:* $\check{\varphi}^{\pm}(x,t) := \check{g}\left[\xi^{\pm}(\mu); \gamma, \mu\right] - \pi$, *for any* $\mu \in]0, \hat{\mu}(\gamma)[$ *if* $\gamma \leq 1$ *and for any* $\mu \in]0, \infty[$ *if* $\gamma > 1$, *whereas* $\check{\varphi}^{\pm}(x,t) = \tilde{g}(\pm x - t; \gamma) - \pi$ *if* $\gamma > 1$, $\mu = \infty$. $\check{\varphi}^{\pm}$ *has resp. velocity* $v = \pm\check{v}(\mu)$ *and fulfills (5), (2) with X,L given by (12) or* $X = 2\pi\alpha/\sqrt{\gamma^2 - 1}$ *if* $\mu = \infty$.

(4) Half-array of kinks $\bar{\varphi}^+$ *or antikinks* $\bar{\varphi}^-$, *with* $\bar{\varphi}^{\pm}(x,t) := \bar{g}\left[\xi^{\pm}(\mu); \gamma, \mu\right] - \pi$, *only if* $\gamma < 1$ *and for any* $\mu \in]0, \hat{\mu}(\gamma)[$. $\bar{\varphi}^{\pm}$ *respectively have velocity* $v = \pm\check{v}(\mu)$ *and fulfill*

$$\lim_{x \to \mp\infty} \bar{\varphi}^{\pm}(x,t) = \theta, \qquad \lim_{x \to \pm\infty} \left[\bar{\varphi}^{\pm}(x,t;\mu) - \check{\varphi}'^{\pm}(x,t;\mu)\right] = 0^+ \tag{15}$$

for a suitable $\check{\varphi}'^{\pm} \in [\check{\varphi}^{\pm}]$. *Both limits are approached exponentially fast.*

W.r.t. the sGe $|v|$ is no longer a free parameter, but is determined by α, γ for the $\hat{\varphi}^{\pm}$, α, μ for the $\check{\varphi}^{\pm}$. In a circular JJ of length L the Josephson eq.'s imply for the total current I and the average tension V associated to $\check{\varphi}^{\pm}$

$$I := \int_0^L dx\, j(x,t) = L\delta \pm 2\pi\alpha n v \qquad V := \int_0^L dx\, \frac{\mathcal{V}(x,t)}{L} = \mp \frac{nv}{L}. \tag{16}$$

Eq.'s (16), $(13)_2$, (12) implicitly determine the branches of the $I - V$ characteristic [1], which are parametrized by the number n of magnetic fluxons trapped in the JJ ($V = 0$, $I \in]-L, L[$ is the well-known [1] $n = 0$ branch).

[a]We have tested the stability numerically, but note that the key property used in the stability proof of [11], $g'(\xi) > 0\ \forall \xi \in \mathbb{R}$, is fulfilled by the families of solutions 2,3,4.

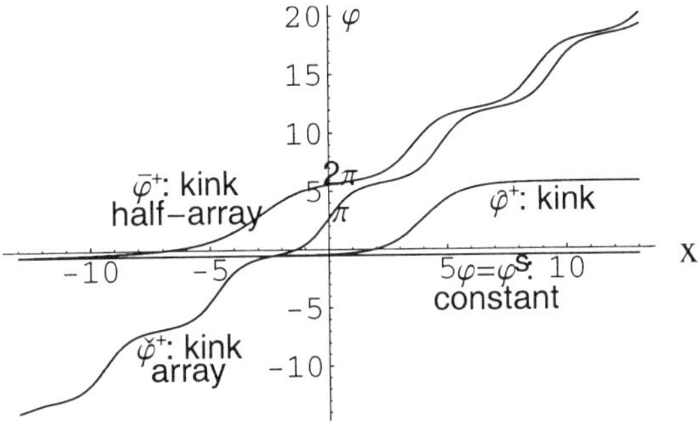

References

1. A. Barone, G. Paternó *Physics and Applications of the Josephson Effect* (Wiley-Interscience, New-York, 1982); and references therein.
2. J. P. Keener, D. W. McLaughlin, *Solitons under perturbations*, Phys. Rev. **A16** (1977), 777-790; *A Green's function for a linear equation associated with solitons*, J. Math. Phys. **18**, 2008-2013 (1977).
3. D. W. McLaughlin, A. C. Scott, *Fluxon interactions*, Appl. Phys. Lett. **30** (1977), 545-547; *Perturbation analysis in fluxon dynamics*, Phys. Rev. **A 18**, 1652-1680 (1978).
4. G. Fiore, *Soliton and other travelling-wave solutions for a perturbed sine-Gordon equation*, math-ph/0512002.
5. J. M. Ghidaglia, R. Temam, *Attractors for damped nonlinear hyperbolic equations*, J. Math. pures et appl. **66**, 273-319 (1987).
6. A. C. Scott, *Active and Nonlinear Wave Propagation in Electronics*, Chapters 2,5 (Wiley-Interscience, New-York, 1970).
7. G. Fiore, *Some explicit travelling-wave solutions of a perturbed sine-Gordon equation*. Proceedings of "Mathematical Physics Models and Engineering Sciences" (Napoli, June 2006), celebrating P. Renno's 70th birthday (in press).
8. F. Tricomi, *Sur un'equation differentielle de l'electrotechnique*, C.-R. Acad. Sci. Paris, **198** (1931), 635; *Integrazione di un'equation differenziale presentatasi in elettrotecnica*, Ann. Sc. Norm. Sup. Pisa **2**, 1-20 (1933).
9. M. Urabe, *The least upper bound of a damping coefficient ensuring the existence of a periodic motion of a pendulum under a constant torque*, J. Sci. Hiroshima Univ, Ser. A, **18**, 379-389 (1954).
10. G. Sansone, R. Conti, *Equazioni differenziali nonlineari* (CNR - Monografie Matematiche 3. Ed. Cremonese, Roma, 1956).
11. A. C. Scott, *Waveform stability of a nonlinear Klein-Gordon Equation*. Proc. IEEE **57** (1969), 1338.
12. A. D'Anna, M. De Angelis, G. Fiore, *Towards soliton solutions of a perturbed sine-Gordon equation*, Rend. Acc. Sc. Fis. Mat. Napoli **LXXII** (2005), 95-110.

SOME REMARKS ON ACCELERATION WAVES IN POROUS SOLIDS

L. FIORINO AND P. GIOVINE*

Dipartimento di Meccanica e Materiali, Università "Mediterranea", Via Graziella 1, Località Feo di Vito, 89122 Reggio Calabria, Italy
E-mail: giovine@unirc.it

A linear theory of thermoelastic continua with big irregular pores, which includes inelastic surface effects associated with changes in the deformation of the holes in the vicinity of the void boundaries, is here used to study the propagation of homothermal macro–acceleration waves.

1. Introduction

The general mechanical balance equations for a continuum with ellipsoidal microstructure was presented in Ref. 1 to describe bodies whose material elements contain a big pore filled by an inviscid fluid, or an elastic inclusion, both of negligible mass, which could have a microstretch different from (and independent of) the local affine deformation ensuing from the macromotion and so could allow distinct microstrains along the principal axes of microdeformation, in absence of microrotations (*e.g.*, composite materials reinforced with chopped elastic fibers, porous media with elastic granular inclusions, real ceramics, *etc.*). When the microstretch is constrained to be spherical, the voids theory of Nunziato and Cowin,[2] which describes continua with "small" spherical pores that may only contract and expand homogeneously, is obtained.[1]

Cowin[3] itself pointed out the necessity of a more general theory for the importance of the shape of the pores in his description of bone canaliculi and of lacunae containing osteocytes: in fact in the human bone, *e.g.*, the lacunae are roughly ellipsoidal with mean values along the axes of about 9 μm, 22μm and 4 μm. Grioli[4] also observed that the refinement of the

*Work supported by the Italian M.I.U.R. through the project PRIN2005 "Modelli Matematici per la Scienza dei Materiali".

Cauchy theory is necessary to characterize these more complex structures and that the new mathematical models have more physical concreteness than the classical ones, even if some problems of mathematical hardness could arose[5] (see, furthermore, Ref. 6).

In this work we use the linear theory developed in Ref. 7 for a porous material to derive the propagation conditions and the growth equations governing the motion of the homothermal macro–acceleration waves and to discuss the eventual couplings between the discontinuities. In particular, we generalize the studies made in Ref. {8–10} by including a rate effect in the pores' response, which results in internal dissipation from experimental evidence,[11] and by considering singularities of order 2 only for the macro–displacement vector, because the microstructural kinematic variable is here directly related to the micro–displacement vector.

2. Linear Theory

The linear theory of continua with ellipsoidal microstructure deals with small changes from a reference placement $\mathcal{B}_*$ of a porous body, that is homogeneous, free of macro- and micro–stress and with zero heat flux rate. The independent kinematic variables in the linear theory are the displacement field $\mathbf{u}$, the temperature change ϑ and the microstrain tensor $\mathbf{V}$:

$$\mathbf{u} := \mathbf{x}(\mathbf{x}_*, \tau) - \mathbf{x}_*, \quad \vartheta := \theta(\mathbf{x}_*, \tau) - \theta_* \quad \text{and} \quad \mathbf{V} := \mathbf{U}(\mathbf{x}_*, \tau) - \mathbf{U}_*, \tag{1}$$

where $\mathbf{x}$ is the spatial position at time τ of the material point which occupied the position $\mathbf{x}_*$ in the reference placement $\mathcal{B}_*$ (and, now and in the course, the subscript $(\cdot)_*$ refers to the referential value of the quoted quantity); θ is the positive temperature; $\mathbf{U}$ is the second–order symmetric and positive definite tensor field which describes the changes in the pore microstructure, namely a left micro Cauchy–Green tensor: in particular, it is supposed that the big pores may contract and expand, but they have not rotary inertia.

The general system of referential equations which rules thermo–kinetic processes for a linear isotropic homogeneous thermoelastic solid, with large irregular pores and which possesses a center of symmetry, is obtained by the local ones stated in Ref. 7 in terms of fields of displacement, temperature variation and microstrain, and of their spatial and/or time derivatives; they are the conservation of mass, the energy balance, the Cauchy equation and the micromomentum balance in the Lagrangian description, respectively:

$$\rho = \rho_* \left(1 - \mathrm{Div}\, \mathbf{u}\right), \quad \tilde{\xi}\Delta\vartheta + \gamma_1\dot{\vartheta} + \gamma_2 \mathrm{Div}\, \dot{\mathbf{u}} + \kappa_*\gamma_3 \mathrm{tr}\dot{\mathbf{V}} + \theta_*^{-1}\lambda = 0, \tag{2}$$

$$\ddot{\mathbf{u}} = v_t^2\Delta\mathbf{u} + \nabla\left[(v_l^2 - v_t^2)\mathrm{Div}\, \mathbf{u} + \kappa_*\lambda_5 \mathrm{tr}\mathbf{V} + \gamma_2\vartheta\right] + \kappa_*\lambda_6 \mathrm{Div}\, \mathbf{V} + \mathbf{f}, \tag{3}$$

$$\ddot{\mathbf{V}} = v_{sm}^2\, \Delta\mathbf{V} + 2(v_{tm}^2 - v_{sm}^2)\,\mathrm{sym}\,[\nabla(\mathrm{Div}\,\mathbf{V})] + \lambda_1\nabla^2(\mathrm{tr}\,\mathbf{V}) +$$
$$+ \Big[\lambda_1\mathrm{Div}\,(\mathrm{Div}\,\mathbf{V}) + \lambda_8\Delta(\mathrm{tr}\mathbf{V}) - \mathrm{tr}(\lambda_3\mathbf{V} + \omega\dot{\mathbf{V}}) - \lambda_5\mathrm{Div}\,\mathbf{u} - \gamma_3\vartheta\Big]\,\mathbf{I} -$$
$$-\lambda_6\,\mathrm{sym}\,(\nabla\mathbf{u}) - 2\lambda_4\mathbf{V} - 2\sigma\dot{\mathbf{V}} + \kappa_*^{-1}\mathbf{C}, \tag{4}$$

where ρ is the mass density of the body; $\mathbf{I}$ is the identity tensor; $\nabla(\cdot) := (\partial(\cdot))/(\partial\mathbf{x}_*)$; $\mathrm{tr}(\cdot) := \mathbf{I}\cdot(\cdot)$; $\mathrm{Div}\,(\cdot) := \mathrm{tr}\,[\nabla(\cdot)]$; $\Delta(\cdot) := \mathrm{Div}\,[\nabla(\cdot)]$; a superposed dot $(\dot{\cdot})$ denotes the time derivative; λ is the rate of heat generation due to irradiation; the vector $\mathbf{f}$ and the symmetric tensor $\mathbf{C}$ are the referential external macro– and micro–actions, respectively; $\mathrm{sym}\,(\cdot) := \frac{1}{2}\left[(\cdot) + (\cdot)^T\right]$ is the symmetric part of a second–order tensor; $\tilde{\xi}$ is the thermal conductivity and κ_* is a non-negative microinertia coefficient. In each equation the contributions due to microstructural quantities $\mathbf{V}$ are easily recognizable: in particular, the terms on the right hand side of Eq. (4) are owed to controlled pore pressures, to interactive forces between the gross and fine structures as well as to internal dissipative contributions due to the stir of the pores' surface; instead, the higher order spatial derivatives are related to boundary microtractions, even if, in some cases, they could express weakly non–local internal effects. The thermoelastic and inelastic constants have to satisfy the following inequalities:[7]

$$\begin{gathered}
3v_l^2 > 4v_t^2 > 0, (3v_l^2 - 4v_t^2)(3\,\lambda_3 + 2\,\lambda_4) > \kappa_*(3\,\lambda_5 + \lambda_6)^2, 4\,\lambda_4 v_t^2 > \kappa_*\lambda_6^2,\\
v_{tm}^2 > v_{sm}^2 > 0,\ v_{sm}^2\left(6\lambda_1 + 9\lambda_8 + 2v_{tm}^2 + v_{sm}^2\right) > 4(v_{tm}^2 - v_{sm}^2)^2,\\
\gamma_1[(3v_l^2 - 4v_t^2)\tfrac{\rho_*}{\kappa_*}(3\,\lambda_3 + 2\,\lambda_4) - (3\,\lambda_5 + \lambda_6)^2] + 6\gamma_2\gamma_3(3\,\lambda_5 + \lambda_6) >\\
> 3\tfrac{\gamma_2^2}{\kappa_*}(3\,\lambda_3 + 2\,\lambda_4) + 3\gamma_3^2(3v_l^2 - 4v_t^2),\ \gamma_1(3v_l^2 - 4v_t^2) > 3\gamma_2^2,\\
\tilde{\xi} \geq 0,\quad \gamma \geq 0,\quad 3\alpha + 2\gamma \geq 0,
\end{gathered} \tag{5}$$

where the last three derive from the dissipation inequality. If the pores were absent, v_l and v_t can be identified with the propagation speeds of dilatational and distortional waves in the linear isothermal elasticity.

3. Homothermal Macro–Acceleration Waves

In a three–dimensional body a wave is represented by a smoothly propagating surface Σ across which certain kinematical and thermal fields suffer jump discontinuities. Σ has a non-zero speed of propagation v_n in the direction of its unit normal $\mathbf{n}$ (see Ref. 12 for the general wave theory).

The propagating surface Σ is called a *macro–acceleration wave* if the fields $\mathbf{u}$, $\dot{\mathbf{u}}$, $\nabla\mathbf{u}$, $\mathbf{V}$ and ϑ are continuous everywhere in the body, while the derivatives of $\mathbf{u}$ of order 2 and of $\mathbf{V}$ and ϑ of order 1 are continuous everywhere except at Σ, where they may suffer jump discontinuities; moreover, all external forces and supplies, *i.e.*, $\mathbf{f}$, $\mathbf{C}$ and λ, are supposed continuous across it with all their derivatives.

Remark: We wish to observe that in our context the microstrain tensor $\mathbf{V}$ is directly related to a micro–displacement vector, hence jumps of the first derivatives of $\mathbf{V}$ correspond to jumps of order 2 for the micro–displacement and so we will study singular surfaces for macro– and micro–displacement of order 2. Instead in Refs. {8–10} weak singularities of order 2 for both macro– and micro–structural kinematic variables $\mathbf{u}$ and $\mathbf{V}$ are examined.

By employing the usual notation $[\![\,\cdot\,]\!]$ for jumps, the following Hugoniot–Hadamard compatibility condition for the jump across Σ of the derivatives of an arbitrary field $\mathbf{\Psi}$ in $\mathcal{B}_*$ holds:

$$[\![\nabla\mathbf{\Psi}]\!] = \nabla[\![\mathbf{\Psi}]\!] - v_n^{-1}[\![\dot{\mathbf{\Psi}}]\!]\otimes\mathbf{n}. \tag{6}$$

In the linear theory of porous thermoelasticity, the Fourier condition also holds,[10] hence the first derivatives of ϑ are continuous and so *every macro–acceleration wave is homothermal.* Moreover, by using the constitutive equation (28) of Ref. 7 into the jump condition (7) for the micromomentum of Ref. 10, we deduce the following relation:

$$\begin{aligned} & v_n[\![\dot{\mathbf{V}}]\!] + \{v_{sm}^2[\![\nabla\mathbf{V}]\!] + 2(v_{tm}^2 - v_{sm}^2)\,\mathrm{syml}\,([\![\mathrm{Div}\,\mathbf{V}]\!]\otimes\mathbf{I}) + \\ & \quad +\lambda_1\left[\mathbf{I}\otimes[\![\mathrm{Div}\,\mathbf{V}]\!] + \mathrm{syml}\,([\![\nabla(\mathrm{tr}\mathbf{V})]\!]\otimes\mathbf{I})\right] + \lambda_8\,\mathbf{I}\otimes[\![\nabla(\mathrm{tr}\mathbf{V})]\!]\}\,\mathbf{n} = \mathbf{0}, \end{aligned} \tag{7}$$

where the symbol "syml" means $(\mathrm{syml}\,\mathbf{\Omega})_{ijl} := \frac{1}{2}(\Omega_{ijl} + \Omega_{jil})$, $\forall$ third–order tensor $\mathbf{\Omega}$. Instead the form of the jumps across Σ of higher order time derivatives is obtained from the balance laws (2) and (3):

$$\begin{aligned} & [\![\dot{\rho}]\!] = -\rho_*[\![\mathrm{Div}\,\dot{\mathbf{u}}]\!], \quad \tilde{\xi}\,[\![\Delta\vartheta]\!] + \gamma_2[\![\mathrm{Div}\,\dot{\mathbf{u}}]\!] + \kappa_*\gamma_3\mathrm{tr}[\![\dot{\mathbf{V}}]\!] = 0, \\ & [\![\ddot{\mathbf{u}}]\!] = v_t^2[\![\Delta\mathbf{u}]\!] + (v_l^2 - v_t^2)[\![\nabla(\mathrm{Div}\,\mathbf{u})]\!] + \kappa_*\left(\lambda_5[\![\nabla(\mathrm{tr}\mathbf{V})]\!] + \lambda_6[\![\mathrm{Div}\,\mathbf{V}]\!]\right). \end{aligned} \tag{8}$$

Now we are able to apply the compatibility condition (6), to get a system of algebraic equations where the amplitudes $[\![\dot{\rho}]\!]$, $[\![\ddot{\mathbf{u}}]\!]$, $[\![\dot{\mathbf{V}}]\!]$ and $[\![\ddot{\vartheta}]\!]$ of the discontinuities are the unknown quantities:

$$\begin{aligned} & v_n[\![\dot{\rho}]\!] = \rho_*[\![\ddot{\mathbf{u}}]\!]\cdot\mathbf{n}, \quad \tilde{\xi}\,[\![\ddot{\vartheta}]\!] = \gamma_2 v_n[\![\ddot{\mathbf{u}}]\!]\cdot\mathbf{n} - \kappa_*\gamma_3 v_n^2\mathrm{tr}[\![\dot{\mathbf{V}}]\!], \\ & \left[v_n^2\mathbf{I} - \mathcal{U}(\mathbf{n}\otimes\mathbf{n})\right][\![\ddot{\mathbf{u}}]\!] = -\kappa_* v_n(\lambda_5\,\mathbf{n}\otimes\mathbf{I} + \lambda_6\,\mathbf{I}\otimes\mathbf{n})[\![\dot{\mathbf{V}}]\!], \\ & \left[v_n^2\mathcal{I} - \mathcal{C}(\mathbf{n}\otimes\mathbf{n})\right][\![\dot{\mathbf{V}}]\!] = \mathbf{0}; \end{aligned} \tag{9}$$

here we employed the instantaneous *homothermal acoustic macro–tensor* $\mathcal{U}(\mathbf{n}\otimes\mathbf{n})$ and *micro–tensor* $\mathcal{C}(\mathbf{n}\otimes\mathbf{n})$ per unit mass defined by

$$\mathcal{U}(\mathbf{n}\otimes\mathbf{n}) := v_t^2\,\mathbf{I} + (v_l^2 - v_t^2)\mathbf{n}\otimes\mathbf{n}, \tag{10}$$

$$\begin{aligned} \mathcal{C}(\mathbf{n}\otimes\mathbf{n}) := {} & v_{sm}^2\,\mathcal{I} + (v_{tm}^2 - v_{sm}^2)(\mathbf{\Phi}\otimes\mathbf{n} + \mathbf{n}\otimes\mathbf{I}\otimes\mathbf{n}) + \\ & +\lambda_8\,\mathbf{I}\otimes\mathbf{I} + \lambda_1\,(\mathbf{I}\otimes\mathbf{n}\otimes\mathbf{n} + \mathbf{n}\otimes\mathbf{n}\otimes\mathbf{I}), \end{aligned} \tag{11}$$

where we introduced the tensors $\mathcal{I}$ and $\mathbf{\Phi}$ of components: $\mathcal{I}_{ijkl} := \delta_{ik}\delta_{jl}$ and $\Phi_{ijk} := \delta_{ik}n_j$ (δ_{ik} is the delta of Kronecker).

Remark: We observe that the jumps of macro–acceleration are in general coupled to the discontinuities in the microstructural variable unlike the purely acceleration waves studied in Refs. 8 and 10.

We are also interested in the growth or the decay of the wave Σ which travels through a heat–conducting porous elastic material, thus we restrict ourselves to plane waves which are of uniform scalar amplitude with assigned initial value, uniform in the sense that the scalar amplitude does not vary with position on Σ. To this task, we differentiate each term of Eqs. (2) and (3) with respect to time, take into account the balance of micromomentum (4) and form jumps across the wave by introducing the jumps $\mathbf{d}$ and Θ in the third time-derivative of the displacement field $\mathbf{u}$ and of the temperature field ϑ, respectively ($d_n := \mathbf{d}\cdot\mathbf{n}$). Without report all the algebraic computation that is similar to the one in Ref. 10, we obtain the following evolution equations for the propagating wave Σ:

$$
\begin{aligned}
&[\![\ddot\rho]\!] = \rho_*(v_n^{-1}d_n - \mathrm{Div}\,[\![\ddot{\mathbf{u}}]\!]), \quad \left[v_n^2\mathbf{I} - \mathcal{U}(\mathbf{n}\otimes\mathbf{n})\right]\mathbf{d} = \\
&\quad = v_n\left\{(v_t^2 - v_l^2)\left[(\mathrm{Div}\,[\![\ddot{\mathbf{u}}]\!])\mathbf{n} + \nabla([\![\ddot{\mathbf{u}}]\!]\cdot\mathbf{n})\right] - 2v_t^2(\nabla[\![\ddot{\mathbf{u}}]\!])\mathbf{n} - \gamma_2[\![\ddot\vartheta]\!]\mathbf{n}\right\} + \\
&\quad +\kappa_* v_n\left\{v_n\left[\lambda_5\nabla(\mathrm{tr}[\![\dot{\mathbf{V}}]\!]) + \lambda_6\mathrm{Div}\,[\![\dot{\mathbf{V}}]\!]\right] - \left[\lambda_5(\mathrm{tr}[\![\ddot{\mathbf{V}}]\!])\mathbf{n} + \lambda_6[\![\ddot{\mathbf{V}}]\!]\mathbf{n}\right]\right\}, \\
&\left[v_n^2\mathcal{I} - \mathcal{C}(\mathbf{n}\otimes\mathbf{n})\right][\![\ddot{\mathbf{V}}]\!] = 2v_n\left\{-\left\{\lambda_8[\mathbf{n}\cdot\nabla(\mathrm{tr}[\![\dot{\mathbf{V}}]\!])] + \omega\,\mathrm{tr}[\![\dot{\mathbf{V}}]\!]\right\}\mathbf{I} + \right. \\
&\quad +(v_{sm}^2 - v_{tm}^2)\,\mathrm{sym}\left[(\mathrm{Div}\,[\![\dot{\mathbf{V}}]\!])\otimes\mathbf{n} + \nabla([\![\dot{\mathbf{V}}]\!]\mathbf{n})\right] - 2\sigma[\![\dot{\mathbf{V}}]\!] - \\
&\quad \left. - v_{sm}^2(\nabla[\![\dot{\mathbf{V}}]\!])\mathbf{n} - \lambda_1\left\{\mathrm{sym}\left[\nabla(\mathrm{tr}[\![\dot{\mathbf{V}}]\!])\otimes\mathbf{n}\right] + (\mathbf{n}\cdot\mathrm{Div}\,[\![\dot{\mathbf{V}}]\!])\mathbf{I}\right\}\right\}, \\
&\tilde\xi\Theta = v_n\left[2\,\tilde\xi\,(\mathbf{n}\cdot\nabla[\![\ddot\vartheta]\!]) + \gamma_2 d_n\right] - v_n^2\left(\gamma_1[\![\ddot\vartheta]\!] + \gamma_2\mathrm{Div}\,[\![\ddot{\mathbf{u}}]\!] + \kappa_*\gamma_3\,\mathrm{tr}[\![\ddot{\mathbf{V}}]\!]\right).
\end{aligned}
\tag{12}
$$

Therefore, in this linearized case, the transport equations will give standard evolution laws of the type $f' = -\mu f$ and hence $f = f_0\,\mathrm{e}^{-\mu\phi}$, where ϕ is the distance, measured along the normal to the wave, from the wave front at $\tau = \tau_0$ and f_0 the strength of the wave at the same time.

4. Solutions

The analysis of Eq. $(9)_4$ gives three possible speed of propagation v_n for the surface Σ related to:

i) *shear optical micro-waves*, which propagate at a constant speed $v_n = v_{sm}$ and where the amplitude of discontinuity for the micro–velocity is

$$
[\![\dot{\mathbf{V}}]\!]_{sm} = \alpha(\mathbf{e}\otimes\mathbf{e} - \mathbf{f}\otimes\mathbf{f}) + \beta(\mathbf{e}\otimes\mathbf{f} + \mathbf{f}\otimes\mathbf{e}); \tag{13}
$$

α and β are the scalar components of the wave amplitude and $\mathbf{e}$ and $\mathbf{f}$ are the unit vectors in the plane orthogonal to $\mathbf{n}$, such that $\mathbf{e} \cdot \mathbf{f} = 0$.

By inserting this solution in Eqs.$(9)_{1,2,3}$, we obtain that $\ddot{\mathbf{u}}$, $\dot{\rho}$ and $\ddot{\vartheta}$ are continuous through the wave, namely, in essence, the waves carry predominantly a change in the pore microstructure without altering the thermoelastic features of the matrix material. Moreover, by analysing Eqs. (12), we obtain the following results: $[\![\ddot{V}]\!]_{11} = [\![\ddot{V}]\!]_{12} = [\![\ddot{V}]\!]_{13} = 0$, $[\![\ddot{V}]\!]_{33} = -[\![\ddot{V}]\!]_{22}$ and they remains undefined together with $[\![\ddot{V}]\!]_{23}$, while

$$\alpha(\tau) = \alpha_0 \exp\left(-\frac{\sigma}{v_{sm}^3}\phi\right) \quad \text{and} \quad \beta(\tau) = \beta_0 \exp\left(-\frac{\sigma}{v_{sm}^3}\phi\right).$$

Finally, we have that $\mathbf{d} = \mathbf{0}$, $[\![\ddot{\rho}]\!] = 0$ and $\Theta = 0$.

This kind of micro-wave does not cause any disturbance in the mechanical and thermal fields and the scalar amplitudes decay to zero as the time interval $(\tau - \tau_0)$ increases indefinitely;

ii) *transverse micro-waves*, which propagate at a constant speed $v_n = v_{tm}$ of amplitude:

$$[\![\dot{\mathbf{V}}]\!]_{tm} = \chi_e(\mathbf{n} \otimes \mathbf{e} + \mathbf{e} \otimes \mathbf{n}) + \chi_f(\mathbf{n} \otimes \mathbf{f} + \mathbf{f} \otimes \mathbf{n}), \tag{14}$$

with χ_i the components of the amplitude itself. In this case, the study of equation $(9)_{1,2,3}$ gives the following coupled *transverse macro–wave*:

$$[\![\ddot{\mathbf{u}}]\!]_{tm} = \frac{k_* v_{tm} \lambda_6}{v_t^2 - v_{tm}^2}(\chi_e\, \mathbf{e} + \chi_f\, \mathbf{f}). \tag{15}$$

while the first order derivative of the mass density and the second order derivative of the temperature are continuous along Σ, as in the classical transverse–case, in fact $[\![\dot{\rho}]\!]_{tm} = 0$ and $[\![\ddot{\vartheta}]\!]_{tm} = 0$.

Now the system of evolution Eqs. (12) gives the following results:

$$\chi_i(\tau) = \chi_{i0} \exp\left(-\frac{\sigma}{v_{tm}^3}\phi\right), \quad \text{for } i = e, f,$$

and all the components of $[\![\ddot{\mathbf{V}}]\!]$ are equal to zero except for the undetermined $[\![\ddot{V}]\!]_{12}$ and $[\![\ddot{V}]\!]_{13}$; instead, the other equations give $d_n = [\![\ddot{\rho}]\!] = \Theta = 0$ and

$$d_i = \frac{\kappa_* v_{tm} \lambda_6}{v_t^2 - v_{tm}^2}\left([\![\ddot{V}]\!]_{1i} + \frac{\sigma(v_{tm}^2 + v_t^2)}{v_{tm}^2(v_{tm}^2 - v_t^2)}\chi_i\right), \quad \text{for } i = e, f.$$

Also in this case the scalar amplitudes decay to zero as the time interval $(\tau - \tau_0)$ increases indefinitely, but, unlike shear optical waves, we have here a macro–acceleration jump with a third order discontinuity related to the elastic properties of pores and to a part that decays to zero.

iii) *extensional micro-waves*, that propagate at a constant speed $v_n = v_{em}$ with $v_{em}^2 := \mathrm{A} + \frac{1}{2}\sqrt{\mathrm{D}}$; for inequalities (5), the discriminant $\mathrm{D}\,(:=$

$12\lambda_1^2+12\lambda_1\lambda_8+9\,\lambda_8^2+4(v_{tm}^2-v_{sm}^2)(v_{tm}^2-v_{sm}^2+2\lambda_1-\lambda_8))$ and the constant $\mathrm{A}\,(:=v_{tm}^2+\lambda_1+\frac{3}{2}\lambda_8)$ are positive by inspection.

The scalar and vector amplitudes are, respectively, δ and

$$[\![\,\dot{\mathbf{V}}]\!]_{em}=\delta(\zeta\,\mathbf{n}\otimes\mathbf{n}+\mathbf{e}\otimes\mathbf{e}+\mathbf{f}\otimes\mathbf{f}), \tag{16}$$

where $\zeta:=(\lambda_1+\lambda_8)^{-1}\left(v_{tm}^2-v_{sm}^2+\lambda_1-\frac{1}{2}\lambda_8+\frac{1}{2}\sqrt{\mathrm{D}}\right)$ and $(\lambda_1+\lambda_8)\neq 0$. The coupled *macro-waves* are now *longitudinal*, that is,

$$[\![\,\ddot{\mathbf{u}}]\!]_{em}=\delta\,\nu\,\mathbf{n},\quad \text{with}\quad \nu:=\frac{k_* v_{em}}{v_l^2-v_{em}^2}[2\lambda_5+(\lambda_5+\lambda_6)\zeta], \tag{17}$$

and the discontinuity amplitudes in the mass density and the temperature are, respectively:

$$[\![\,\dot{\rho}]\!]_{em}=\rho_*\delta\,\nu\,v_{em}^{-1}\quad\text{and}\quad[\![\,\ddot{\vartheta}]\!]_{em}=\delta\tilde{\xi}^{-1}[\gamma_3 v_{em}^2(2+\zeta)-\gamma_2 v_{em}\nu]. \tag{18}$$

In this case, the solutions of evolution Eqs. (12) are $[\![\,\ddot{V}]\!]_{ij}=0$, if $i\neq j$, and $[\![\,\ddot{V}]\!]_{33}=[\![\,\ddot{V}]\!]_{22}$, while, to determine the amplitude δ, it is necessary to have prior knowledge of one of the jumps $[\![\,\ddot{V}]\!]_{11}$ or $[\![\,\ddot{V}]\!]_{22}$. Moreover, we have that $d_e=d_f=0$, while $[\![\,\ddot{\rho}]\!]$, d_n and Θ also are related to the previous undeterminacies. The micro–wave is accompanied by second and third order discontinuities in macro–mechanical and thermal fields.

References

1. P. Giovine, Porous Solids as Materials with Ellipsoidal Structure, in *Contemporary Research in the Mechanics and Mathematics of Materials*, eds. R.C. Batra and M.F. Beatty (CIMNE, Barcelona, 1996) pp.335–342.
2. J.W. Nunziato, S.C. Cowin, *Arch. Rat. Mech. Analysis* **72**, 175–201 (1979).
3. S.C. Cowin, *Bone* **22**, Supplement, 119S–125S (1998).
4. G. Grioli, *Cont. Mech. Thermodyn.* **15**, 441–450 (2003).
5. M. Cieszko, Extended Description of Pore-Space Structure. Application of Minkowski Space, in *Proc. XIVth Int. Symp. Trends Applic. Math. Mech. (STAMM'04)*, (Shaker Verlag, Aachen, 2005) pp.93–102.
6. G. Capriz, *Continua with Microstructure* (Springer–Verlag, New York, 1989).
7. P. Giovine, *Transport in Porous Media* **34**, 305–318 (1999).
8. D. Ieşan, *Acta Mechanica* **60**, 67–89 (1986).
9. P.M. Mariano, L. Sabatini, *Int. J. Non-Linear Mech.* **35**, 963–977 (2000).
10. P. Giovine, On Acceleration Waves in Continua with Large Pores, in *Proc. XIVth Int. Symp. Trends Applic. Math. Mech. (STAMM'04)*, (Shaker Verlag, Aachen, 2005) pp.113–124.
11. D.P.H.Hasselman, J.P.Singh, Criteria for Thermal Stress Failure of Brittle Structural Ceramics, in *Thermal Stresses* **I**,(Northolland, Amsterdam, 1982).
12. T.Y. Thomas, *Plastic Flow and Fracture in Solids*, (Academic Press, New York, 1961).

STABILITY CONSIDERATIONS FOR REACTION-DIFFUSION SYSTEMS

J.N. FLAVIN

Department of Mathematical Physics, National University of Ireland, Galway, Ireland
E-mail: james.flavin@nuigalway.ie

The main purpose of the article is to outline how novel Liapunov functionals may be used to obtain pointwise stability estimates for some reaction-diffusion systems.

1. Introduction

The main purpose of the article is to outline how novel Liapunov functionals may be used to obtain pointwise stability estimates for some reaction diffusion systems.

Section 2 considers a pair of reaction-diffusion p.d.e.s subject to zero boundary conditions, discusses an analogue of a functional due to Rionero (see[1−3]) appropriate to the context, and uses the functional to obtain sufficient conditions for simple stability of the system.

Section 3 considers a Lotka-Volterra reaction-diffusion system in one dimension, and uses a functional of the aforesaid type to obtain a conditional pointwise stability estimate for an equilibrium state of the system.

Section 4 outlines how the considerations of Section 3 may be extended to a region with prescribed moving boundaries.

Section 5 outlines how pointwise stability estimates, analogous to those of Section 3, may be obtained for a three-dimensional (fixed) region.

2. Reaction-Diffusion equations and a new Liapunov functional

Consider smooth solutions of the reaction-diffusion system, in the fixed spatial domain Ω,

$$\left.\begin{aligned} \partial u_1/\partial t &= a_{1j}u_j + \gamma_1 \nabla^2 u_1 + F_1(u_1, u_2), \\ \partial u_2/\partial t &= a_{2j}u_j + \gamma_2 \nabla^2 u_2 + F_2(u_1, u_2), \end{aligned}\right\} \tag{1}$$

subject to

$$u_i = 0 \quad \text{on} \quad \partial\Omega, \tag{2}$$

where $\partial\Omega$ is the smooth boundary of Ω. The summation convention is used over repeated dummy indices; $a_{ij}, \gamma_\bullet\ (> 0)$ are given constants; and F_i are given smooth functions of u_i such that

$$F_i(0,0) = 0. \tag{3}$$

The latter condition ensures that $u_i = 0$ is a solution to the system. It proves convenient to introduce the positive scaling constants α, β, which may be chosen subsequently, and new dependent variables u, v such that

$$u_1 = \alpha u, u_2 = \beta v. \tag{4}$$

Further write

$$f(u,v) = \alpha^{-1}F_1(u,v), g(u,v) = \beta^{-1}F_2(u,v), \tag{5}$$

where the foregoing $F_1(u,v), F_2(u,v)$ mean $F_1(u_1,u_2), F_2(u_1,u_2)$ expressed in terms of the new variables u and v. Define the constants

$$\begin{aligned} &b_1 = a_{11} - \bar{\alpha}\gamma_1,\ b_4 = a_{22} - \bar{\alpha}\gamma_{22}, \\ &b_2 = (\beta/\alpha)a_{12},\ b_3 = (\alpha/\beta)a_{21}, \end{aligned} \tag{6}$$

where $\bar{\alpha}$ is a positive constant, yet to be chosen.

We shall denote by $< \bullet, \bullet >$ the L^2 scalar product, and by $\|\bullet\|$ the L^2 norm, for scalar and vector functions as appropriate.

Defining

$$\left.\begin{aligned} &A = b_1b_4 - b_2b_3 = b_1b_4 - a_{12}a_{21}, \\ &I = b_1 + b_4, \\ &\alpha_1 = A + b_3^2 + b_4^2,\ \alpha_2 = A + b_1^2 + b_2^2,\ \alpha_3 = b_1b_3 + b_2b_4, \end{aligned}\right\} \tag{7}$$

one has the following Theorem, encapsulating the fundamental property of an analogue of a Liapunov functional, which had earlier been introduced by Rionero in this context (see[3]).

Theorem 2.1. *Defining*

$$\begin{aligned} V(t) = (1/2)\Big[&A(\|\nabla u\|^2 + \|\nabla v\|^2) + \|b_1 \nabla v - b_3 \nabla u\|^2 \\ &+ \|b_2 \nabla v - b_4 \nabla u\|^2\Big] \end{aligned} \tag{8}$$

one has

$$dV/dt = AI(\|\nabla u\|^2 + \|\nabla v\|^2) + \Psi^* + \Psi, \tag{9}$$

where

$$\Psi^* = \gamma_1\alpha_1\left[-\left\|\nabla^2 u\right\| + \bar{\alpha}\left\|\nabla u\right\|^2\right] + \gamma_2\alpha_2\left[-\left\|\nabla^2 u\right\| + \bar{\alpha}\left\|\nabla u\right\|^2\right]$$

$$+(\gamma_1+\gamma_2)\alpha_3[<\nabla^2 u, \nabla^2 v> - \bar{\alpha} < \nabla u, \nabla v>], \tag{10}$$

$$\begin{aligned}\Psi &= (\alpha_1 \nabla u - \alpha_3 \nabla v, \nabla f) + (\alpha_2 \nabla v - \alpha_3 \nabla u, \nabla g) \\ &= (\alpha_1 \nabla u - \alpha_3 \nabla v, f_u \nabla u + f_v \nabla v) + \\ &\quad (\alpha_2 \nabla v - \alpha_3 \nabla u, g_u \nabla u + g_v \nabla v),\end{aligned} \tag{11}$$

where the subscripts u, v denote partial differentiation with respect to these variables.

One may, of course, prove this directly (albeit with some difficulty) but it can be deduced from the original result of Rionero, by using a simple lemma (see[3]). Rionero's original functional is formally identical to that of (8) but the gradients are absent. Other analogues involving higher derivatives may also be derived from Rionero's result.

The remainder of this section discusses some sufficient conditions for the simple stability of the system (1),(2) in the absence of nonlinear/forcing terms. Stability in the presence of such terms is discussed in subsequent sections.

The functional V (defined by (8)) is required to be positive-definite in u, v. A sufficient condition for this is

$$A > 0 \text{ (i.e. } b_1b_4 - a_{12}a_{21} > 0). \tag{12}$$

This is *assumed throughout.*

For stability in the measure V, one requires

$$dV/dt \leq 0. \tag{13}$$

In the absence of nonlinear/forcing terms, sufficient conditions for (13) are given by

$$\left.\begin{array}{c} I < 0, \quad i.e. \quad b_1 + b_4 < 0, \\ \Psi^* \leq 0, \end{array}\right\} \tag{14}$$

in addition to (12).

Conditions sufficient for (14_2) are most easily realised when

$$b_1b_2b_3b_4 < 0, \tag{15}$$

in which case the scaling constants α, β, defined by (4), are chosen so that

$$\alpha/\beta = \left|b_2b_4/b_1b_3\right|^{1/2}, \tag{16}$$

ensuring that

$$\alpha_3 = 0. \tag{17}$$

With a view to realising (14_2), let us recall that

$$\left\|\nabla^2\Phi\right\|^2 \geq \left\|\nabla\Phi\right\|^2 \tag{18}$$

for arbitrary smooth functions Φ such that

$$\Phi = 0 \text{ on } \partial\Omega, \tag{19}$$

where λ_1 is the lowest positive eigenvalue of

$$\nabla^2\Phi + \lambda\Phi = 0 \text{ in } \Omega, \;\; \Phi = 0 \text{ on } \partial\Omega \tag{20}$$

etc. The *constant* $\bar{\alpha}$, arising in the functional (8),is assumed to be λ_1 *henceforward.* We encapsulate the foregoing results in the following:

Theorem 2.2. *Sufficient conditions for simple stability (i.e. (13)) of the system (1),(2), in the absence of nonlinear/forcing terms, are given by (12),(14), (15)-(17),where it is assumed that the constant $\bar{\alpha}$ arising in (6) is given by λ_1, the lowest eigenvalue of the eigenvalue problem (20)*

3. Lotka-Volterra system: stability estimates for the solution gradient

Here we consider a Lotka-Volterra system of reaction-diffusion equations in one spatial dimension, with Dirichlet boundary conditions: Theorem 1 is used to obtain a stability estimate in the measure V, for an equilibrium configuration, from which a *pointwise* stability estimate, may be deduced.

We discuss the Lotka-Volterra system (discussed in,[4] for example).

$$\begin{aligned} \partial S_1/\partial t &= \gamma_1 S_{1,xx} + a_1 S_1 - c_1 S_1 S_2\,, \\ \partial S_2/\partial t &= \gamma_2 S_{2,xx} - a_2 S_2 + c_2 S_1 S_2, \end{aligned} \tag{21}$$

where $\gamma_\bullet$, $a_\bullet$, $c_\bullet$ are all positive constants, in the interval $0 < x < 1$ (the symbol x is used, instead of x_1, as the spatial variable, and subscripts in x denote partial differentiation with respect to x). We consider the equilibrium configuration of (21), in the presence of constant boundary conditions:

$$S_1 = a_2/c_2, \quad S_2 = a_1/c_1\,. \tag{22}$$

We write

$$S_1 = (a_2/c_2) + u_1, \quad S_2 = (a_1/c_1) + u_2\,, \tag{23}$$

where the perturbations $u_1(x,t)$, $u_2(x,t)$ satisfy (in the interval $0 < x < 1$)

$$\begin{aligned} \partial u_1/\partial t &= \gamma_1 u_{1,xx} - c_1(a_2/c_2)u_2 - c_1 u_1 u_2 \,, \\ \partial u_2/\partial t &= \gamma_2 u_{2,xx} + c_2(a_1/c_1)u_1 + c_2 u_1 u_2, \end{aligned} \tag{24}$$

subject to

$$u_i = 0 \quad \text{on} \ \ x = 0, 1. \tag{25}$$

The equations (24) are of the type (1) with

$$F_1 = -c_1 u_1 u_2 \,, \quad F_2 = c_2 u_1 u_2 \,. \tag{26}$$

We use scaled variables (see (4)) and in the context of these we have

$$f = -c_1 \beta uv \,, \quad g = c_2 \alpha uv \,, \tag{27}$$

and

$$\begin{gathered} b_1 = -\pi^2\gamma_1, \quad b_4 = -\pi^2\gamma_2, \quad b_2 = (\beta/\alpha)(-c_1 a_2/c_2) \\ b_3 = (\alpha/\beta)(a_1 c_2/c_1), \end{gathered} \tag{28}$$

on noting that the relevant eigenvalue in this case is π^2.

Prior to using Theorem 1 we note that the condition $(14)_1$ is automatically satisfied here. Moreover, since (15) is automatically satisfied here, we choose α/β in accordance with (16) in which case $\alpha_3 = 0$. Using Theorem 1 in the context described above, we obtain the following: the measure of the perturbation u,v

$$\overline{V}(t) = (1/2)\left[A\left(\|u_x\|^2 + \|v_x\|^2\right) + \|b_1 v_x - b_3 u_x\|^2 + \|b_2 v_x - b_4 u_x\|^2\right], \tag{29}$$

satisfies

$$dV/dt \leq AI\left(\|u_x\|^2 + \|v_x\|^2\right) + \Psi, \tag{30}$$

where Ψ is given by

$$\begin{aligned} \Psi = \textstyle\int_0^1 & \left[-\alpha_1 c_1 \beta v u_x^2 + \alpha_2 c_2 \alpha u v_x^2 + (-\alpha_1 c_1 \beta u + \alpha_2 c_2 \alpha v)\, u_x v_x\right] dx \\ & \leq M \textstyle\int_0^1 (|u| + |v|)\left(u_x^2 + v_x^2 + |u_x v_x|\right) dx \\ & \leq (3/2) M \textstyle\int_0^1 (|u| + |v|)\left(u_x^2 + v_x^2\right) dx \end{aligned} \tag{31}$$

where

$$M = \max\left[c_1\alpha_1\beta, c_2\alpha_2\alpha\right] \,. \tag{32}$$

The following fundamental inequality [e.g.[5]] is used hereunder:

$$|\Phi(x)|^2 \leq x(1-x)\,\|\Phi_x\|^2 \,, \tag{33}$$

or the weaker version thereof,

$$|\Phi(x)| \leq (1/2) \|\Phi_x\| \,, \tag{34}$$

where Φ is any smooth function of x vanishing at $x = 0, 1$.

Using (34) we obtain

$$\int_0^1 (|u| + |v|) \left(u_x^2 + v_x^2\right) dx \leq 2^{-(1/2)} \left(\|u_x\|^2 + \|v_x\|^2\right)^{(3/2)} . \tag{35}$$

This together with (31) gives

$$\Psi \leq \delta_1 \left(\|u_x\|^2 + \|v_x\|^2\right)^{(3/2)} \tag{36}$$

where

$$\delta_1 = 3 \cdot 2^{-(3/2)} M \,. \tag{37}$$

It follows from (29) etc. that

$$k_1 \left(\|u_x\|^2 + \|v_x\|^2\right) \leq V \leq k_2 \left(\|u_x\|^2 + \|v_x\|^2\right) \tag{38}$$

where

$$k_1 = A/2; \quad k_2 = A/2 + \sum_{i=1}^{4} b_i^2 \,. \tag{39}$$

Thus, using (29)-(30),(36),(38)-(39), we obtain the differential inequality

$$dV/dt \leq -dV + d_1 V^{(3/2)} \tag{40}$$

where

$$d = A|I|/k_2; \quad d_1 = \delta_1 / k_1^{(3/2)} \,. \tag{41}$$

It may be noted, *en passant*, that

$$A|I| = (\pi^4 \gamma_1 \gamma_2 + a_1 a_2) \pi^4 \gamma_1 \gamma_2 \tag{42}$$

using (7) etc.

From (40) we obtain the following. Supposing that the initial perturbation (assumed known) is such that

$$\{V(0)\}^{1/2} \leq d/d_1 \,, \tag{43}$$

then

$$dV/dt \leq -\eta V \tag{44}$$

where

$$\eta = d \left[1 - d_1 d^{-1} \{V(0)\}^{3/2}\right] , \tag{45}$$

whence

$$V(t) \leq V(0)e^{-\eta t} \tag{46}$$

A pointwise estimate follows from this on using (33), (46):

$$\{u(x,t)\}^2 + \{v(x,t)\}^2 \leq x(1-x)k_1^{-1}\overline{V}(0)e^{-\eta t}. \tag{47}$$

Thus we have

Theorem 3.1. *The equilibrium configuration (22) of the Lotka-Volterra system (21) etc. is conditionally exponentially stable*

(a) *in the measure V, as conveyed by (43), (45), (46);*
(b) *pointwise, as conveyed by (47) etc.*

4. Stability estimates in the context of moving boundaries

The estimates derived for (24), (25) can be extended to cater for a one-dimensional region with prescribed moving boundaries, in certain circumstances. We confine the discussion to the essentially novel issues.
Suppose the boundaries $x = \{0,1\}$ are replaced by prescribed moving boundaries

$$x = x_+(t), x = x_-(t)$$

where x_+, x_- are continuously differentiable functions such that

$$x_+(t) > x_-(t).$$

We adopt the notation

$$\phi(x,t)\dot{x}]_-^+ = \phi(x_+(t),t)\dot{x}_+(t) - \phi(x(t),t)\dot{x}_{--}(t) \tag{48}$$

where a superposed dot denotes ordinary differentiation with respect to t.

One may deduce from Theorem 1, by elementary means, an analogue thereof, appropriate to the current context.

Define

$$V(t) = \int_{x_-(t)}^{x_+(t)} P(u_x, v_x)dx \tag{49}$$

where

$$P(u_x, v_x) = \frac{1}{2}[A(u_x^2 + v_x^2) + (b_1 v_x - b_3 u_x)^2 + (b_2 v_x - b_4 u_x)^2] \tag{50}$$

wherein b_i, A, I (used here) are as defined in (6),(7) but over the relevant region with moving boundaries.

One has

Theorem 4.1.

$$\frac{dV}{dt} = AI(\| u_x \|^2 + \| v_x \|^2) + \Psi^* + \Psi + P(u_x, v_x)x^\bullet]_-^+ \qquad (51)$$

where Ψ^*, Ψ *as defined previously, but over the time-varying interval.*

This may be used in the following way: noting that $A > 0$ implies that $P(u_x, v_x) \geq 0$ and that $V(t)$ is positive definite in u, v we have

Theorem 4.2. *If* $A > 0$ *and*

$$x_+^\bullet(t) \geq 0, x_-^\bullet(t) \leq 0,$$

then

$$\frac{dV}{dt} \leq AI(\| u_x \|^2 + \| v_x \|^2) + \Psi^* + \Psi$$

where A, I, Ψ, Ψ^x *are as defined previously, but over the region with moving boundaries.*

It is clear that the foregoing can be used to derive stability estimates for (24) etc. in the context of a one dimensional region with prescribed moving boundaries, with zero boundary conditions, on using analogous techniques.

5. Pointwise stability in a three dimensional region

One may also derive pointwise stability estimates for a Lotka-Volterra system of the type (24),(25) in a fixed, three dimensional region Ω, but at the expense of considerably more complicated computations. We content ourselves with an outline of the main points of novelty. The previous discussion, as contained in (21)-(28), continues to be valid in Ω with the obvious exceptions:

(a) The operator ∇^2 replaces $\frac{\partial^2}{\partial x^2}$;
(b) The eigenvalue λ_1 (defined by (20)) replaces the eigenvalue π^2 arising in the one dimensional context.

One further notes that in the current context

$$\nabla^2 u = \nabla^2 v = 0 \text{ on } \partial\Omega. \qquad (52)$$

Further, the relevant Liapunov functional and the inequality which it satisfies are as follows:

$$\overline{V}(t) = \left(\frac{1}{2}\right)\Big[A\left(\| \nabla^2 u \|^2 + \| \nabla^2 v \|\right)^2 + \| b_1 \nabla^2 v - b_3 \nabla^2 u \|^2 \\ + \| b_2 \nabla^2 v - b_4 \nabla^2 u \|^2\Big] \tag{53}$$

satisfies

$$\frac{d\overline{V}}{dt} \leq AI(\| \nabla^2 u \|^2 + \| \nabla^2 v \|^2) + \bar{\Psi}, \tag{54}$$

where

$$\bar{\Psi} = \alpha_1(\nabla^2 u, \nabla^2 f) + \alpha_2(\nabla^2 v, \nabla^2 g) \tag{55}$$

where $A, I, b_i, \alpha_1, \alpha_2, f, g$ are as defined in Section 2.

The essential novelty in the present context is that one must obtain an inequality of the type

$$\overline{\Psi} \leq C(\| \nabla^2 u \|^2 + \| \nabla^2 v \|^2)^{1+f} \tag{56}$$

where C, f are positive constants. This is required in order to establish an inequality of the type

$$\overline{\Psi} \leq D\bar{V}^{1+f} \tag{57}$$

D being a positive constant.

Two fundamental inequalities are needed to realise this programme:

For smooth functions $\Phi(x)$ vanishing on the boundary Ω, one has

(a)

$$\sup \Omega |\Phi(\mathbf{x})| \leq m_1 \| \nabla^2 \Phi \|, \tag{58}$$

(b)

$$\| \nabla \Phi \|_4 \leq m_2 \left\| \nabla^2 \Phi \right\|, \tag{59}$$

where $m_\bullet$ are constants (depending on the region Ω), both here and subsequently.

Whereas (58) is well known (e.g.[5]) (59) does not appear to be readily available in standard recipes. The inequality (59) may be obtained by combining a particular case of the Ehrling-Browder inequality (valid for smooth functions Φ),

$$\| \nabla \Phi \|_4 \leq m_3 \| \nabla^2 \Phi \|^{7/8} \| \Phi \|^{1/8}, \tag{60}$$

[e.g.[6]] with the standard inequality (for smooth functions vanishing on $\partial\Omega$)

$$\| \Phi \| \leq m_4 \| \nabla^2 \Phi \| . \tag{61}$$

The essentially novel step in proving (58) is the establishment of the following:

$$\begin{aligned} J &:= \int (|\nabla^2 u| + |\nabla^2 v|)|\nabla u \cdot \nabla v| d\Omega \\ &\leq m_5[\| \nabla^2 u \|^2 + \| \nabla^2 v \|^2]^{3/2}. \end{aligned} \tag{62}$$

One proceeds as follows: by elementary means

$$J \leq m_6 \sqrt{(\| \nabla^2 u \|^2 + \| \nabla^2 v \|^2) \int [(\nabla u)^4 + (\nabla u)^4] d\Omega}. \tag{63}$$

On combining this with (59) one obtains the inequality (56).
One may obtain an analogue of Theorem 3 using analogous techniques used in connection therewith, incorporating therein an inequality of the type (58).

The required estimates are embodied in this following theorem.

Theorem 5.1. *The functional* $\overline{V}(t)$ *satisfies*

$$\overline{V}(t) \leq \overline{V}(0) exp[-\{d - d_1 \overline{V}^{\frac{1}{2}}(0)\} t]$$

provided that

$$\overline{V}^{\frac{1}{2}}(0) \leq d/d_1,$$

where d, d_1 *are the computable constants; and one has the associated pointwise estimate for a computable constant* N

$$\sup \Omega\{|u(\mathbf{x},t)| + |v(\mathbf{x},t)|\} \leq N[\overline{V}(t)]^{1/2}.$$

Acknowledgments. Discussions with Professor S. Rionero are gratefully acknowledged. The support of the Mathematics Applications Consortium for Science and Industry (MACSI , *www.macsi.ul.ie*), supported by Science Foundation Ireland Mathematics Initiative grant 06/M1 05, is also gratefully acknowledged.

References

1. S. Rionero, *Rend. Mat. Acc. Lincei* s. 9 **16**, 227 (2005).
2. S. Rionero, *J. Math. Anal. Appl.* **319**, 372 (2006).
3. J.N. Flavin, S. Rionero, *Note di Matematica* **27** (2), 97 (2007).
4. A. Okubo, S.A. Levin, *Diffusion and Ecological Problems: Modern Perspectives*, vol. 14 (Springer-Verlag, Berlin, Heidelberg, New York, 2001).
5. J.N.Flavin, S. Rionero, *Qualitative Estimates for P.D.E.s. An Introduction* (CRC Press, Boca Raton, FL, 1995).
6. C. Bandle, M. Flucher, in *Recent Progress in Inequalities*, G. V. Milovanovic ed., 97, (Kluwer Academic Publishers, 1998).

CROSS-DIFFUSION DRIVEN INSTABILITY FOR A LOTKA-VOLTERRA COMPETITIVE REACTION–DIFFUSION SYSTEM *

G. GAMBINO, M. C. LOMBARDO, M. SAMMARTINO

Dept. of Mathematics, University of Palermo
Via Archirafi 34, 90123 Palermo, Italy
E-mail: {gaetana},{lombardo},{marco}@math.unipa.it

In this work we investigate the possibility of the pattern formation for a reaction-diffusion system with nonlinear diffusion terms. Through a linear stability analysis we find the conditions which allow a homogeneous steady state (stable for the kinetics) to become unstable through a Turing mechanism. In particular, we show how cross-diffusion effects are responsible for the initiation of spatial patterns. Finally, we find a Fisher amplitude equation which describes the weakly nonlinear dynamics of the system near the marginal stability.

Keywords: nonlinear diffusion, Turing instability, Fisher equation

1. Introduction

Since the pioneering work of Turing[6] it is well known that, in a reaction-diffusion system describing the interaction between two species (or reactants), different diffusion rates can lead to the destabilization of a constant steady state followed by the transition to a nonhomogeneous steady state (pattern).

A steady state is Turing unstable if it is stable as a solution to the reaction system (without diffusion terms), but unstable as a solution of the full reaction-diffusion system.[2,4] This mechanism, known as diffusion driven instability, is used to model the pattern appearance.

In certain models, as the classical Lotka-Volterra competition-diffusion system, classical diffusion is not sufficient to describe the pattern formation since no Turing unstable steady state arise no matter what the diffusion rates are. Thus, to model segregation, Shigesada, Kawasaki and Teramoto[5]

*This paper is supported by the INDAM and by the MURST under the PRIN grant:"Nonlinear Propagation and Stability in Thermodynamical Processes of Continuous Media, 2005-2007"

proposed the nonlinear evolution system (1). Here u, v are the population densities of two competing species. The system tries to describe the tendency of the species to diffuse (faster than predicted by the usual linear diffusion) toward lower density areas:

$$\frac{\partial u}{\partial t} = \nabla \cdot \nabla(u(c_1 + a_1 u + b_1 v)) + \Gamma u(\mu_1 - \gamma_{11}u - \gamma_{12}v),$$
$$\frac{\partial v}{\partial t} = \nabla \cdot \nabla(v(c_2 + a_2 v + b_2 u)) + \Gamma v(\mu_2 - \gamma_{21}u - \gamma_{22}v). \qquad (1)$$

The non–negative parameters a_i, b_i and c_i are respectively the self diffusion, the cross diffusion and the diffusion coefficients. The constants $\gamma_{ij} > 0$ represent the competitive interaction and the parameter Γ describes alternatively the relative strength of reaction terms or the size of the spatial domain. In this work we are interested to describe the mechanism of pattern formation for the system (1) with homogeneous Neumann boundary conditions. In section 2 we perform a linear stability analysis of the system (1) and we recover how the cross-diffusion is the key mechanism of pattern formation. In section 3, through a weakly nonlinear analysis, we derive the Fisher equation which describes the dynamics near marginal stability; as a special case we find travelling wave solutions.

2. Linear analysis

The kinetic part of the system (1) has four fixed points:

$$O(0,0);\ A\left(0, \frac{\mu_2}{\gamma_{22}}\right);\ B\left(\frac{\mu_1}{\gamma_{11}}, 0\right);\ C\left(\frac{\mu_1\gamma_{22} - \mu_2\gamma_{12}}{\gamma_{11}\gamma_{22} - \gamma_{12}\gamma_{21}}, \frac{\mu_2\gamma_{11} - \mu_1\gamma_{21}}{\gamma_{11}\gamma_{22} - \gamma_{12}\gamma_{21}}\right).$$

Let (u_0, v_0) denote one of the fixed point. The linearized system in the neighborhood of (u_0, v_0) is:

$$\dot{\mathbf{w}} = \Gamma J \mathbf{w} + D\nabla^2 \mathbf{w}, \quad \mathbf{w} = \begin{pmatrix} u - u_0 \\ v - v_0 \end{pmatrix}; \qquad (2)$$

where:

$$J = \begin{pmatrix} \mu_1 - 2\gamma_{11}u_0 - \gamma_{12}v_0 & -\gamma_{12}u_0 \\ -\gamma_{21}v_0 & \mu_2 - \gamma_{21}u_0 - 2\gamma_{22}v_0 \end{pmatrix}; \qquad (3)$$

$$D = \begin{pmatrix} c_1 + 2a_1u_0 + b_1v_0 & b_1u_0 \\ b_2v_0 & c_2 + 2a_2v_0 + b_2u_0 \end{pmatrix}. \qquad (4)$$

We look for conditions on the parameters such that (u_0, v_0) is stable to spatially uniform perturbations (stable for the kinetics) but unstable to non-homogeneous perturbations (diffusion driven instability). It is well known that (u_0, v_0) is stable for the kinetics if $\mathrm{tr}(J(u_0, v_0)) < 0$ and $\det(J(u_0, v_0)) > 0$. One can therefore derive the following classification of the equilibria in absence of the diffusion terms:

(i) the trivial equilibrium is always unstable for the kinetics;
(ii) A is a stable equilibrium point if $\mu_1 - \mu_2 \frac{\gamma_{12}}{\gamma_{22}} < 0$;
(iii) B is a stable equilibrium point if $\mu_1 \frac{\gamma_{21}}{\gamma_{11}} - \mu_2 > 0$;
(iv) C is a stable equilibrium point if $\gamma_{11}\gamma_{22} - \gamma_{12}\gamma_{21} > 0$ (weak inter-specific competition). Moreover, the conditions $\mu_1\gamma_{22} - \mu_2\gamma_{12} > 0$ and $\mu_2\gamma_{11} - \mu_1\gamma_{21} > 0$ must hold for the existence of the point C. This means that C exists and is a stable equilibrium point when both the points A and B are unstable.

Standard linear analysis leads to the following dispersion relation, which gives the eigenvalue λ as a function of the wavenumber k:

$$\det(\lambda I - \Gamma J + Dk^2) = 0. \tag{5}$$

Spatial patterns arise in correspondence of those modes k for which $Re(\lambda) > 0$. The analysis of the dispersion relation for the A and B fixed points shows that the diffusion terms have a stabilizing effect. The dispersion relation of the C point reads:

$$\lambda^2 - (k^2 \mathrm{tr}(D) - \Gamma \mathrm{tr}(J))\lambda + h(k^2) = 0, \tag{6}$$

where $h(k^2) = \det(D)k^4 + \Gamma q k^2 + \Gamma^2 \det(J)$ and

$$\begin{aligned} q = {} & \gamma_{11}u_0(2a_2v_0 + c_2) + \gamma_{22}v_0(2a_1u_0 + c_1) + b_1v_0(\gamma_{22}v_0 - \gamma_{21}u_0) \\ & + b_2u_0(\gamma_{11}u_0 - \gamma_{12}u_0). \end{aligned} \tag{7}$$

Being $\mathrm{tr}(D) > 0$ and $\mathrm{tr}(J) < 0$, the only way one can have $Re(\lambda) > 0$ for some $k \neq 0$ is when $h(k^2) < 0$. This implies that, for Turing instability, the following two conditions must hold:

$$\begin{cases} q < 0\,, \\ \min(h(k^2)) < 0 \Leftrightarrow q^2 - 4\det(D)\det(J) > 0\,. \end{cases} \tag{8}$$

By the expression (7) of q it follows that the only potential destabilizing mechanism is the presence of the cross–diffusion terms. The conditions on the existence and stability of the point C imply that only one of the two expressions $\gamma_{22}v_0 - \gamma_{21}u_0$ or $\gamma_{11}u_0 - \gamma_{12}v_0$ can be less than zero. Therefore when b_1 has a destabilizing effect then b_2 has a stabilizing one and vice versa. In what follows we choose the case $\gamma_{22}v_0 - \gamma_{21}u_0 < 0$ without loss of generality.

Let us now define the quantities α and β as:

$$\begin{aligned} \alpha &= v_0(\gamma_{21}u_0 - \gamma_{22}v_0) \\ \beta &= \gamma_{11}u_0(2a_2v_0 + c_2) + \gamma_{22}v_0(2a_1u_0 + c_1) + b_2u_0(\gamma_{11}u_0 - \gamma_{12}u_0) \end{aligned}$$

and the quantity ξ^+ as the positive root of:

$$\frac{\alpha^2}{4(\gamma_{11}\gamma_{22}-\gamma_{12}\gamma_{21})u_0v_0}\xi^2 - v_0(2a_2v_0+c_2)\xi - [v_0\beta/\alpha(2a_2v_0+c_2) + (2a_1u_0+c_1)(2a_2v_0+b_2u_0+c_2)] = 0. \quad (9)$$

One can easily see that the critical value $b_1^* = \beta/\alpha + \xi^+$ is such that, when $b_1 > b_1^*$, the conditions (8) are satisfied and the system has a finite k pattern-forming stationary instability.

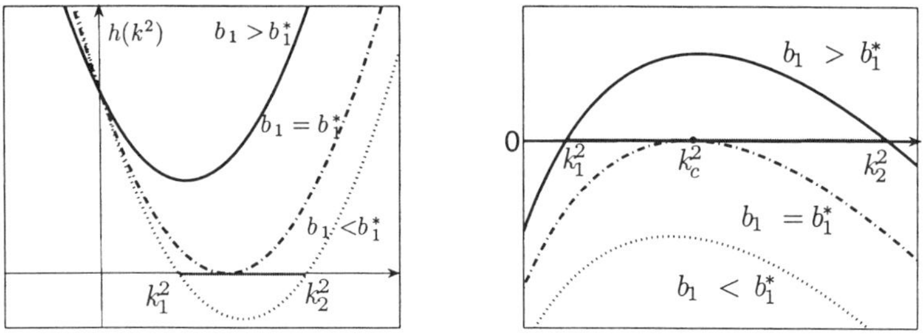

Fig.1 *Left* Plot of $h(k^2)$. *Right* Growth rate the k-th mode. A band of positively growing modes is present.

The unstable wavenumbers stay in between the roots of $h(k^2)$, denoted by k_1^2 and k_2^2. They are proportional to Γ.[3] Hence, for Γ small enough, the pattern does not form, because no allowable modes would be in $[k_1^2, k_2^2]$.

In Fig 2 we show a pattern, computed using a spectral methods. The initial datum is a random periodic perturbation of the equilibrium C.

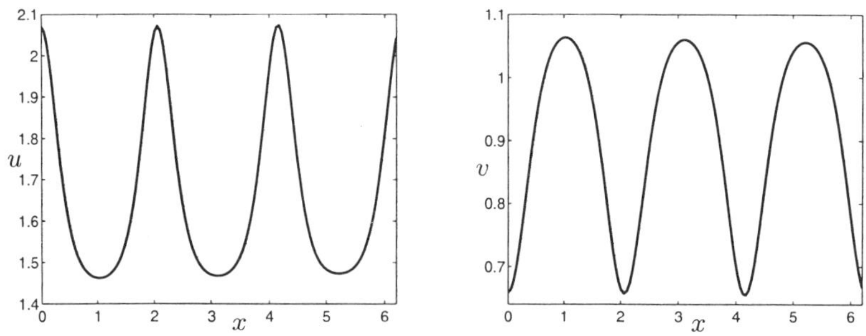

Fig.2 The parameters are $\mu_1 = 1.2, \mu_2 = 1, \gamma_{11} = 0.5, \gamma_{12} = 0.4, \gamma_{21} = 0.38, \gamma_{22} = 0.41, a_1 = 10^{-4}, a_2 = 0.1, c_i = 0.2, \Gamma = 49.75$, $b_2 = 0.3$, $b_1 = 6.65 > b_1^* = 6.15$.

3. Amplitude equation near marginal stability

The stability diagram in the (μ_2, μ_1) plane of the system (1) is:

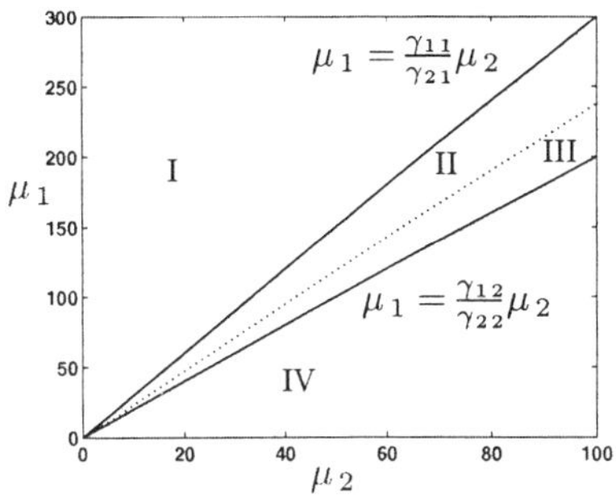

The dotted curve separating regions II and III is the C stability boundary: C is stable in region II, unstable in region III and does not exist in regions I and IV. The point A is stable in region IV and unstable otherwise; the pooint B is stable in region I and unstable otherwise.

We want to perform a weakly nonlinear analysis near the A state stability boundary. The growth rate of the A fixed point is $\lambda = \Gamma\left(\mu_1 - \mu_2\frac{\gamma_{12}}{\gamma_{22}}\right)$ which is zero along the stability boundary. Close to the boundary we define:

$$\delta^2 = \left(\mu_1 - \mu_2\frac{\gamma_{12}}{\gamma_{22}}\right); \quad \tau = \delta^2 t; \quad \eta = \delta x$$

Substituting the perturbation expansion:

$$u = u_1\delta^2 + u_2\delta^3 + \dots$$
$$v = \frac{\mu_2}{\gamma_{22}} + v_1\delta^2 + v_2\delta^3 + \dots$$

into (1) written in terms of the rescaled variables, it follows:

$$\frac{\partial u_1}{\partial\tau} = \left(c_1 + b_1\frac{\mu_2}{\gamma_{22}}\right)\frac{\partial^2 u_1}{\partial\eta^2} + \Gamma\left(1 - \left(\gamma_{11} - \frac{\gamma_{12}\gamma_{21}}{\gamma_{22}}\right)u_1\right)u_1 \tag{10}$$

$$v_1 = -\frac{\gamma_{21}}{\gamma_{22}}u_1 \tag{11}$$

We have obtained a slave condition for v_1 and the Fisher equation which describes the weakly nonlinear dynamics of u_1.
Travelling wave solutions to the Fisher equation (10) are obtained substituting $z = \eta - \omega t$ into (10):

$$h\frac{du_1}{dz^2} + \omega\frac{du_1}{dz} + \Gamma(1 - gu_1)u_1 = 0, \tag{12}$$

where $h = c_1 + b_1\frac{\mu_2}{\gamma_{22}}$ and $g = \gamma_{11} - \frac{\gamma_{12}\gamma_{21}}{\gamma_{22}}$. Looking for solutions $u_1 \sim e^{-\nu z}$ we obtain the eigenvalue relation:

$$\nu = \frac{1}{2}\left(-\frac{\omega}{h} \pm \left(\frac{\omega^2}{h^2} - 4\right)^{\frac{1}{2}}\right). \tag{13}$$

The critical damping $\omega = \omega^* = 2\sqrt{h}$ defines the minimum front speed. There exists a non-negative trajectory which connects the states C and A when $\omega > \omega^*$. Moreover, the Fisher equation has the following exact analytical solution:

$$u_1^1(\xi) = -\frac{1}{4g}\left(\mathrm{sech}^2(\xi) - 2\tanh(\xi) - 2\right), \tag{14}$$

where $\xi = \pm\sqrt{\frac{\Gamma}{24h}}\eta + \frac{5\Gamma}{12}\tau$, which is a travelling wave with front speed $5/\sqrt{6}\omega^*$. Numerical simulations show that, near marginal stability, the front propagates following (with very good approximation) the profile (14), see Fig 3.

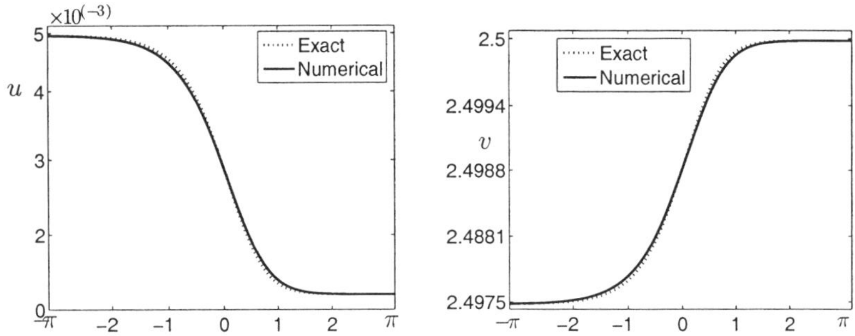

Fig. 3 Uniform travelling front profile connecting the equilibria A and C, for $a_1 = a_2 = 10^{-4}, b_1 = b_2 = c_1 = c_2 = 10^{-2}, \mu_1 = 0.5, \mu_2 = 0.501, \gamma_{11} = \gamma_{21} = 0.2, \gamma_{12} = 0.1, \gamma_{22} = 0.3, \Gamma = 50$.

Extensive numerical simulations show that the solution assumes the travelling front profile independently of the initial conditions.

Analogous results are recovered if one performs the same analysis for the point B. Finally, the weakly nonlinear analysis near the marginal stability of the point C is much more complicated and it will be the subject of a forthcoming paper.

References

1. W. S. Duan, H. J. Yang, Y. R. Shi, *An exact solution of Fisher equation and its stability*, Chin. Phys. Soc., **18**,(7) (2006), 1414–1417.
2. B. I. Henry, S. L. Wearne, *Existence of Turing instabilities in a two-species reaction-diffusion system*, SIAM J. Appl. Math., **62**,(3) (2002), 870–887.
3. J. D. Murray, *Mathematical Biology*, Springer, New York, (1989).
4. R. A. Satnoianu, M. Menzinger, P. K. Miani, *Turing instability in general systems*, J. Math. Biol., **41**, (6) (2000), 493–512.
5. N. Shigesada, K. Kawasaki, E. Teramoto, *Spatial segregation of interacting species*, J. Theo. Biology, **79**, (1979), 83–99.
6. A. M. Turing, *The chemical basis of morphogenesis*, Phil. Trans. Roy. Soc. London, B, **237**, (1952), 37–72.

ON A PROCEDURE FOR FINDING HIDDEN POTENTIAL SYMMETRIES

M. L. GANDARIAS

Department of Mathematics, University of Cadiz,
Puerto Real, 11510, Spain
E-mail:marialuz.gandarias@uca.es

In this paper we introduce a new procedure for deriving new nonlocal symmetries that seem not to be recorded in the literature. These hidden potential symmetries are determined by considering a generalized potential system, rather than the natural potential system or a general integral variable. An inhomogeneous diffusion equation and the Fokker-Planck equation have been considered as application of this procedure.

Keywords: Non local symmetries; potential symmetries.

1. Introduction

An obvious limitation of group-theoretic methods based in local symmetries, in their utility for particular partial differential equations (PDE's), is that many of these equations do not have local symmetries.

In [2] Bluman introduced a method to find a new class of symmetries for a PDE. Suppose a given scalar PDE of second order

$$F(x, t, u, u_x, u_t, u_{xx}, u_{xt}, u_{tt}) = 0, \tag{1}$$

where the subscripts denote the partial derivatives of u, can be written as a conservation law

$$\frac{D}{Dt} f(x, t, u, u_x, u_t) - \frac{D}{Dx} g(x, t, u, u_x, u_t) = 0, \tag{2}$$

for some functions f and g of the indicated arguments. Here $\frac{D}{Dx}$ and $\frac{D}{Dt}$ are total derivative operators defined by

$$\begin{aligned} \frac{D}{Dx} &= \frac{\partial}{\partial x} + u_x \frac{\partial}{\partial u} + u_{xx} \frac{\partial}{\partial u_x} + u_{xt} \frac{\partial}{\partial u_t} + \ldots, \\ \frac{D}{Dt} &= \frac{\partial}{\partial t} + u_t \frac{\partial}{\partial u} + u_{xt} \frac{\partial}{\partial u_x} + u_{tt} \frac{\partial}{\partial u_t} + \ldots. \end{aligned}$$

Through the conservation law (2) one can introduce an auxiliary potential variable v and form an auxiliary potential system (system approach)

$$\begin{aligned} v_x &= f(x,t,u,u_x,u_t), \\ v_t &= g(x,t,u,u_x,u_t). \end{aligned} \tag{3}$$

Any Lie group of point transformations

$$\mathbf{v} = \xi(x,t,u,v)\frac{\partial}{\partial x} + \tau(x,t,u,v)\frac{\partial}{\partial t} + \phi(x,t,u,v)\frac{\partial}{\partial u} + \psi(x,t,u,v)\frac{\partial}{\partial v}, \tag{4}$$

admitted by (3) yields a nonlocal symmetry *potential symmetry* of the given PDE (2) if and only if the following condition is satisfied

$$\xi_v^2 + \tau_v^2 + \phi_v^2 \neq 0. \tag{5}$$

The method introduced by Bluman in [1] only requires the PDE to be written in a conserved form (2). Nevertheless, in most papers concerning evolution equations the auxiliary system considered is the so called *natural* potential system

$$\begin{aligned} v_x &= u, \\ v_t &= g(x,t,u,u_x), \end{aligned} \tag{6}$$

and a different system is searched only when the given equation is not in a conserved form.

In [3] the authors developed a method in order to find potential symmetries which are missed by standard potential symmetry analysis. These symmetries were determined by using a general integrable variable,

The aim of this paper is to find new nonlocal symmetries by considering conservations laws more general than the ones studied before. Some of these potential symmetries are also missed by introducing a general weighted integral variable [3].

In this paper we consider a more general potential system

$$\begin{aligned} v_x &= f_1(x)h_1(u), \\ v_t &= f_2(x)h_2(u)u_x + f_3(x)h_3(u). \end{aligned} \tag{7}$$

By requiring the governing PDE (1) to be equivalent to the potential system (7) we get that functions $f_i(x)$ and $h_i(u)$ $i = 1,2,3$ must satisfy several conditions that lead to a *hidden potential system*. A point symmetry (4) admitted by this system yields a *hidden potential symmetry* of the given PDE (2) if this symmetry is not a potential symmetry derived from the *natural* potential system and condition (5) is satisfied. We present some non-trivial examples of nonlocal symmetries. This is interesting, since in a nonlocal setting one cannot use algorithmic methods as in the local symmetry case . These examples are: an inhomogeneous diffusion equation, and a Fokker-Planck equation. These two equations have been respectively considered in [5], [3], and [4]. by standard potential symmetry analysis. We show that this procedure yields new hidden potential symmetries for these equations, which as far as we know do not appear in the literature.

2. Inhomogeneous diffusion equation

One of the mathematical models for diffusion processes is the generalised inhomogeneous nonlinear diffusion equation

$$f(x)u_t = [g(x)u^n u_x]_x. \tag{8}$$

The diffusion processes appear in many physics processes such as plasma physics, kinetic theory of gases, solid state, metallurgy and transport in porous medium. In (8) $u(x,t)$ is a function of position x and time t and may represent the temperature, $f(x)$ and $g(x)$ are arbitrary smooth functions of position and may denote the density and the density-dependent part of thermal diffusion, respectively. In [5], C. Sophocleous has classified the potential symmetries of (8). He proved that potential symmetries exist only if the parameter n takes the values -2 or $-\frac{2}{3}$ and $f(x)g(x) = constant$ or $g(x) = \frac{1}{f(x)}[\int f(x)dx]^4$. It was pointed out in [5] that equation (8) is equivalent by means of a point transformation to

$$u_t = [g(x)u^n u_x]_x. \tag{9}$$

Consequently, by considering the natural potential system C. Sophocleous has derived potential symmetries, for

$$u_t = [g(x)u^{-2} u_x]_x, \tag{10}$$

only when $g(x) = constant$ or $g(x) = x^4$. In [3] hidden nonlocal symmetries were determined by considering an integrated equation, obtained, using a general integrable variable

$$\phi = \int k(x)u(x,t)dx + J(t).$$

In [3] the authors found extra potential symmetries of (10) for $g(x) = x^2$ that corresponds to the case in which (10) is linearizable.

We are considering the more general auxiliary system (7) and we require (10) to be expressed in a conserved form as this system, then system (7) becomes

$$v_x = k(x)u,$$
$$v_t = k(x)g(x)\frac{u_x}{u^2} - \frac{a}{u}, \tag{11}$$

with $a = constant$ and $k'(x)g(x) + a = 0$.

If system (11) is invariant under a Lie group of point transformations with infinitesimal generator (4), then

$$\xi = \xi(x,t,v),\ \tau = \tau(t),\ \psi = \psi(t,v)$$

$$\phi = -k\xi_v u^2 + (\psi_v - \xi\tfrac{k'}{k} - \xi_x)u, \tag{12}$$

where ξ, τ, ψ and k must satisfy three equations We can distinguish the following two cases:

Case 1 $2gk' + g'k = 0$. Setting $a = 1$ we get that $k(x) = \frac{1}{x}$, and $g(x) = x^2$. This function $g(x) = x^2$ was already derived in [3] by deriving the classical symmetries of an integrated equation with a general weighted variable.

Case 2 $2gk' + g'k \neq 0$ then the corresponding determining equations give rise to

$$\xi = \frac{gk(\psi_v - \tau_t)}{2gk' + g'k},\ \tau = k_3 t + k_4,\ \psi = k_1 v^2 + \frac{k_3}{2}v + k_6. \tag{13}$$

After setting $\frac{k'}{k} = f$, the compatibility of the remaining determining equations leads to

$$f'' - \frac{5}{4}\frac{(f')^2}{f} - \frac{ff'}{2} - \frac{f^3}{4} = 0. \tag{14}$$

A particular solution of (14) is $f = \frac{d}{x}$ with $d = -1$ or $d = 3$. By setting $c = a = 1$ we get $g(x) = x^2$ already considered in Case 1 and $g(x) = -\frac{1}{x^2}$. For $g(x) = -\frac{1}{x^2}$ the generators of the Lie algebra are

$$\begin{aligned} &\mathbf{w_1} = \partial_t, && \mathbf{w_2} = \partial_v, \\ &\mathbf{w_3} = 2t\partial_t + u\partial_u + v\partial_v, && \mathbf{w_4} = xv\partial_x + v^2\partial_v - (x^4u^2 + 2uv)\partial_u, \\ &\mathbf{w_5} = x\partial_x + 2v\partial_v - 2u\partial_u. && \end{aligned} \tag{15}$$

We point out that for $g(x) = -\dfrac{1}{x^2}$ equation (10) admits a new hidden potential symmetry $\mathbf{w_4}$ which, as far as we know, has not been derived by considering the natural potential system [5] nor has been derived by considering the general weighted integral variable [3]. We have derived this generator by considering the hidden potential system

$$\begin{aligned} v_x &= x^3 u, \\ v_t &= -\frac{ax}{3}\frac{u_x}{u^2} - \frac{a}{u}. \end{aligned} \tag{16}$$

3. The Fokker-Planck equation

The classical potential symmetries for the Fokker-Planck equation with drift

$$u_t = u_{xx} + [f(x)u]_x, \tag{17}$$

were derived by Pucci and Saccomandi in [4] by using the *natural* potential system They found [4] that, besides the infinite dimensional generator, when $f(x)$ satisfies any of these Riccati equations

$$\frac{f'}{2} + \frac{f^2}{4} = ax^2 + bx + c, \quad \frac{f'}{2} + \frac{f^2}{4} = a(x+\lambda)^2 + c + \frac{d}{(x+\lambda)^2}, \tag{18}$$

then (17) is invariant under a Lie group with four or two parameters respectively.

We are now considering the general auxiliary system (7) and we require the governing PDE (17) to be expressed in conserved form as this system, and system (7) becomes

$$\begin{aligned} v_x &= a(x)u, \\ v_t &= a(x)u_x + (\frac{a(x)a(x)''}{a(x)'} - a(x)')u, \end{aligned} \tag{19}$$

with $f(x) = \dfrac{a''(x)}{a'(x)}$. If system (19) is invariant under a Lie group of point transformations with infinitesimal generator (4) then we get

$$\begin{aligned} \xi &= \frac{\tau'(t)x}{2} + \varphi(t), \; \psi = \alpha(x,t)v + \beta(x,t), \\ \tau &= \tau(t), \qquad\qquad \phi = (\psi_v - \xi_x - \frac{a'}{a}\xi)u + \frac{\xi_x}{a}, \end{aligned} \tag{20}$$

where $\alpha(x,t)$, $\beta(x,t)$, $a(x)$, $\tau(t)$ and $\varphi(t)$ must satisfy four equations In general it is difficult to find all solutions of the overdetermined nonlinear system of PDEs. Next we consider the following solution

$$\xi = 1, \; \tau = c, \; \psi = \tan(x)v$$

$$\phi = (\tan(x) + \frac{2(1+\cos(2x))}{2x+\sin(2x)})u + \frac{k_1}{4}\sec^2(x)(2x+\sin(2x))v, \tag{21}$$

with

$$a = -\frac{4}{k_1(\sin(2x)+2x)}, \; f = -\frac{2(\cos(4x)+4x\sin(2x)+8\cos(2x)+7)}{\sin(4x)+2\sin(2x)+4x\cos(2x)+4x}. \tag{22}$$

From generator (21) we obtain the similarity variable and the similarity solution and the reduced ODE

$$z = x - ct, \; v = \frac{h(z)}{cos(x)}, \; h'' + ch' + h = 0. \tag{23}$$

Some solutions of the ODE, for $a(x)$ given by (22) are:

$$v = \frac{(k_2(x-2t)+k_3)e^{-(x-2t)}}{\cos(x)}. \tag{24}$$

$$v = \frac{e^{-\frac{c(x-ct)}{2}}}{cos(x)}[k_2\sinh(k(x-ct)) + k_3\cosh(k(x-ct))], \tag{25}$$

for $(c-2)(c+2) > 0$, $k = \frac{1}{2}\sqrt{c^2-4}$.

$$v = \frac{e^{-\frac{c(x-ct)}{2}}}{cos(x)}[k_2\sin(k(x-ct)) + k_3\cos(k(x-ct))], \tag{26}$$

for $(c-2)(c+2) < 0$, $k = \frac{1}{2}\sqrt{4-c^2}$.

The corresponding solutions for (17) are given by $u = \dfrac{v_x}{a}$. It can be checked that f *does not* satisfy any of the Riccati equations (18), consequently generator (21) can not be derived as a potential symmetry by considering the natural potential system [4]. Generator (21) is a *new* potential symmetry derived from the *hidden* potential system (19).

References

1. Bluman G.W., Kumei S. *Symmetries and Differential Equations.* Berlin: Springer, 1989.
2. Bluman G.W., Reid G.J., Kumei S. New classes of symmetries for partial differential equations. *J. Math. Phys.* 1988; **29**: 806-811.
3. Moitsheki R J, Broadbridge P and Edwards M P *J. Phys A: Math and General.* Sistematic construction of hidden nonlocal symmetries for the inhomogeneous nonlinear diffuesion equation. 2004 **37** 8279-8286.
4. Pucci E, Saccomandi G. *Modern group analysis : advanced analytical and computational methods in mathematical physics.* Nordfjordeid: ed. N.H. Ibragimov et. all, 1997: 291-298. Kluwer, Dordrecht, 1993.
5. Sophocleous C. Classification of potential symmetries of generalised inhomogeneous nonlinear diffusion equations. *Phys. A.* 2003; **320**:169-183.

TURING INSTABILITY FOR THE SCHNACKENBERG SYSTEM

M. GENTILE and A. TATARANNI

Dipartimento di Matematica e Applicazioni "R. Caccioppoli", Universitá Federico II, Complesso Universitario Monte S'Angelo, via Cintia, I-80126 Napoli, ITALY
m.gentile@unina.it, assunta.tataranni@dma.unina.it

The linear stability-instability of the equilibrium state of the Schnackenberg system is studied. The onset of Turing instability is obtained.

Keywords: Linear stability, Reaction-diffusion systems, Turing instability

1. Introduction

The reaction-diffusion system of P.D.Es, often are used for modelling the chemicals reactions (see [1–6] and the references quoted therein). Here we consider the system

$$\begin{cases} U_t = \gamma(a - U + U^2 V) + \Delta U \\ \\ V_t = \gamma(b - U^2 V) + d\Delta V \end{cases} \tag{1}$$

in a smooth domain $\Omega \subset \mathbb{R}^3$ under Dirichlet boundary conditions

$$\begin{cases} U = a + b \\ \\ V = \dfrac{b}{(a+b)^2} \end{cases} \qquad \forall (x,t) \in \partial\Omega \times \dot{\mathbb{R}}^+ \tag{2}$$

with a, b, γ and d constants such that

$$\begin{cases} a + b > 0 \\ b > 0, d > 0 \end{cases} \tag{3}$$

System (1) contains as particular case the system introduced by Schnackenberg and is contained in the Segel-Jackson ([6]). Our aim is to study the longtime behaviour of the solution of (1)-(2). Setting

$$U = u^* + C_1 \tag{4}$$

$$V = v^* + C_2 \tag{5}$$

with

$$\begin{cases} u^* = a + b \\ v^* = \dfrac{b}{(a+b)^2} \end{cases} \tag{6}$$

critical point of (1)-(2), the longtime behaviour of the solution of (1)-(2) is then reduced to the stability of (6).

2. Preliminaries

The perturbed system is given by

$$\begin{cases} \dfrac{\partial C_1}{\partial t} = a_1 C_1 + a_2 C_2 + \Delta C_1 + f(C_1, C_2) \\ \\ \dfrac{\partial C_2}{\partial t} = a_3 C_1 + a_4 C_2 + d\Delta C_2 + g(C_1, C_2) \end{cases} \tag{7}$$

under Dirichlet homogeneous boundary conditions

$$C_1 \equiv C_2 \equiv 0, \qquad (x,t) \in \partial\Omega \times \mathbb{R}^+; \tag{8}$$

with

$$\begin{cases} a_1 = \gamma \dfrac{b-a}{a+b}, \qquad a_2 = \gamma(a+b)^2 \\ \\ a_3 = -2\gamma \dfrac{b}{a+b}, \; a_4 = -\gamma(a+b)^2 \end{cases} \tag{9}$$

and

$$\begin{cases} f(C_1, C_2) = \gamma(C_1^2 v^* + C_1^2 C_2 + 2u^* C_1 C_2) \\ g(C_1 C_2) = -f(C_1, C_2) \end{cases} \tag{10}$$

Following the methodology introduced by Rionero ([7–10]) we use the scalings

$$C_1 = \alpha u, \qquad C_2 = \beta v \tag{11}$$

with α and β constants and set $\mu = \alpha/\beta$. Then it turns out that

$$\begin{cases} \dfrac{\partial u}{\partial t} = a_1 u + \dfrac{a_2}{\mu} v + \Delta u + f^* \\ \dfrac{\partial v}{\partial t} = \mu a_3 u + a_4 v + d\Delta v + g^* \end{cases} \tag{12}$$

under the boundary conditions

$$u = v = 0, \qquad (x,t) \in \partial\Omega \times \mathbb{R}^+ \tag{13}$$

with

$$\begin{cases} f^* = \dfrac{1}{\alpha}[f]_{\substack{C_1 = \alpha u \\ C_2 = \beta v}} \\ g^* = \dfrac{1}{\beta}[g]_{\substack{C_1 = \alpha u \\ C_2 = \beta v}} \end{cases} \tag{14}$$

Denoting by
$< \cdot,\cdot >$ the $L^2(\Omega)$ scalar product, $\|\cdot\|$ the $L^2(\Omega)$-norm, we study the problem in the variable functional space $W_0^{1,2}[\Omega(t)]$. In this space the Poincaré inequality

$$\|\nabla\varphi\|^2 \geq \bar{\alpha}\|\varphi\|^2 \tag{15}$$

holds, where $\bar{\alpha} = \bar{\alpha}(\Omega)$ is the lowest eigenvalue of

$$\Delta\varphi + \lambda\varphi = 0, \quad \varphi \in W_0^{1,2}(\Omega). \tag{16}$$

Setting

$$\begin{cases} b_1 = a_1 - \bar{\alpha} \\ b_4 = a_4 - d\bar{\alpha} \end{cases} \tag{17}$$

(12) becomes

$$\begin{cases} \dfrac{\partial u}{\partial t} = b_1 u + \dfrac{a_2}{\mu} v + f^* + f_1^* \\ \dfrac{\partial v}{\partial t} = \mu a_3 u + b_4 v + g^* + g_1^* \end{cases} \tag{18}$$

with

$$\begin{cases} f_1^* = \Delta u + \bar{\alpha} u \\ g_1^* = d(\Delta v + \bar{\alpha} v) \end{cases} \tag{19}$$

3. Linear stability

Theorem 3.1. *Let*

$$\begin{cases} I = b_1 + b_4 < 0 \\ A = b_1 b_4 - a_2 a_3 > 0 \end{cases} \tag{20}$$

then (u^*, v^*) is linearly asymptotically stable with respect to the $L^2(\Omega)$-norm.

Proof. Let $\{\bar{\alpha_n}\}, \{\varphi_n\}$ be respectively the sequence of the eigenvalues of (16) and the sequence of the associate eigenfunction in $W_0^{1,2}(\Omega)$. Assuming that:
(i) $\{\varphi_n\}$ is complete and orthogonal in $W^{1,2}(\Omega)$;
(ii) u, v and their first and second (spatial) derivatives can be expanded in a (Fourier) series absolutely and uniformly converging in Ω according to

$$\begin{cases} u = \sum_{n=1}^{+\infty} X_n(t)\varphi_n \\ \\ v = \sum_{n=1}^{+\infty} Y_n(t)\varphi_n \end{cases} \tag{21}$$

and differentiable term by term, from the linearized system (18) (for $\mu = 1$) we obtain

$$\begin{cases} \dfrac{dX_n}{dt} = b_{1n}X_n + a_2 Y_n \\ \\ \dfrac{dY_n}{dt} = a_3 X_n + b_{4n} Y_n \end{cases} \tag{22}$$

with

$$\begin{cases} b_{1n} = a_1 - \bar{\alpha}_n \\ b_{4n} = a_4 - d\bar{\alpha}_n \end{cases} \tag{23}$$

Then

$$\begin{cases} I_n < 0 \\ A_n > 0 \end{cases} \tag{24}$$

imply the stability of the zero solution with respect the perturbation (u_n, v_n). In fact, setting

$$W_n = \frac{1}{2}[A_n(\|u_n\|^2 + \|v_n\|^2) + \|b_{1n}v_n - a_3 u_n\|^2 + \|a_2 v_n - b_{4n}u_n\|^2]$$

it follows that

$$W_n \leq W_n(0)e^{-\delta_n t} \tag{25}$$

where

$$\delta_n = \frac{2A_n|I_n|}{k_2^{(n)}} \tag{26}$$

with

$$k_2{}^{(n)} = A_n + 2[b_{1n}{}^2 + b_{4n}{}^2 + a_2{}^2 + a_3{}^2]. \tag{27}$$

But

$$\begin{cases} I < 0 \Rightarrow I_n < 0 \\ \\ A > 0 \Rightarrow A_n > 0 \end{cases} \tag{28}$$

hence, setting

$$W = \sum_{n=1}^{+\infty} W_n \tag{29}$$

it follows that ([10])

$$W \leq W_0 e^{-\delta t} \tag{30}$$

with δ positive constant independent of n. □

4. Instability

Theorem 4.1. *Let either $I > 0$ or $A < 0$, then (u^*, v^*) is unstable.*

Proof. We refer for the sake of concreteness to the case $I > 0$. Then for $n = 1$, the the eigenvalues of (22) are such that at least one either has positive real part or is zero. □

5. Diffusion driven instability

Let us consider the system inequalities

$$\begin{cases} I_0 = a_1 + a_4 = \gamma \dfrac{b-a}{a+b} - \gamma(a+b)^2 < 0 \\ \\ A_0 = \gamma^2(a+b)^2 > 0 \\ \\ A = A_0 - \bar{\alpha}(a_1 d + a_4) + d\bar{\alpha}^2 < 0 \end{cases} \tag{31}$$

Condition (32) are the conditions for the onset of driven instability (Turing effect) i.e. (u^*, v^*) *is stable in the absence of diffusion, but is destabilized by diffusion*. It is easily seen that (32) hold if and only if ([10])

$$\begin{cases} b > a \\ \\ b - a < (a+b)^3 \\ \\ d > \dfrac{A_0 + \bar{\alpha}|a_4|}{\bar{\alpha}(|a_1| - \bar{\alpha})} \end{cases} \tag{32}$$

In view of (33), in particular it follows that the onset of Turing instability is guaranteed by $\{l = 1, a < 1, b > 1, \gamma < \pi^2\}$.

6. Acknowledgements

To Professor T.Ruggeri in the occasion of his 60th birthday. This work has been performed under the auspicies of the G.N.F.M. of I.N.D.A.M. and M.I.U.R. (P.R.I.N. 2005): "Nonlinear Propagation and Stability in Thermodinamical Processes of Continuous Media." The authors thank gratefully Prof. S.Rionero, for having proposed the present research and for his helpful suggestions.

References

1. Schnackenberg, J. (1979) *Simple chemical reaction systems with limit cycle behaviour*, J. Theor. Biol., **81**, 389-400.
2. Turing, A.M. (1952) *The chemical basis of morphogenesis*, Phil. Trans. R. Soc. B, **237**, 37-72
3. Dillon, R. Maini, P.K. and Othmer, H.G. (1994)*Pattern formation in generalized Turing systems*, J. Math. Biol., **32**, 345-393.
4. Madzvamuse, A. Thomas, R.D.K. Maini, P.K. and Wathen, A.J.(2002) *A numerical approach to the study of spatial pattern formation in the ligaments of arcoid bivalves*, Bull. Math. Biol., **64**, 501-530.
5. Murray, J.D. (2003) *Mathematical Biology. I-II*, Third edition, Interdisciplinary Applied Mathematics **17-18**, Springer Verlag, New-York.
6. Segel, L. and Jackson, J. (1972) *Dissipative structure: an explanation and an ecological example*, J. Theor. Biol., **37**, 545-559.
7. Rionero, S. (2006) *A rigorous reduction of the $L^2(\Omega)$-stability of the solutions to a nonlinear binary reaction-diffusion system of P.D.Es. to the stability of the solutions to a linear binary system of O.D.E's*, Journal of Math. Anal. Appl., **319**, 377-397.
8. Rionero, S. (2006) *A nonlinear L^2-stability analysis for two-species population dynamics with dispersal*, Math. Biosci. Eng., **3**, 189-204.
9. Rionero, S. (2005) *L^2 stability of the solutions to a nonlinear binary reaction-diffusion system of P.D.ES.*, Rend. Accad. Naz. Lincei, Serie IX, **XVI**, 227-238.
10. Rionero, S. *Diffusion Driven Stability and Turing Effect Under Robin Boundary data*, (to appear).
11. Flavin, J.N. Rionero, S. (1996) *Qualitative estimates for partial differential equations: an introduction*, CRC Press, Boca Raton, Florida.

ON NONLINEAR STABILITY FOR A REACTION-DIFFUSION SYSTEM CONCERNING CHEMICAL REACTIONS

M. GENTILE and A. TATARANNI

Dipartimento di Matematica e Applicazioni "R. Caccioppoli", Universitá Federico II, Complesso Universitario Monte S'Angelo, Via Cinthia, I-80126 Napoli, ITALY
m.gentile@unina.it, assunta.tataranni@dma.unina.it

The nonlinear stability of the equilibrium state of a reaction-diffusion system of P.D.Es is studied by using a peculiar Liapunov functional and a maximum principle.

Keywords: Nonlinear stability, Liapunov direct method, Reaction-diffusion systems

1. Introduction

The reaction-diffusion system of P.D.Es, are often used for modeling the chemical reactions.[1-4] Here we consider the initial-boundary value problem[11]

$$\begin{cases} \dfrac{\partial U}{\partial t} = \gamma(a - U + U^2V) + \Delta U, \\ \\ \dfrac{\partial V}{\partial t} = \gamma(b - U^2V) + d\Delta V, \end{cases} \tag{1}$$

in a fixed open bounded domain $\Omega \subset I\!R^3$, with boundary at least C^2, with a, b, γ and d constants such that: $a + b > 0,\ b > 0,\ d > 0$.
System (1) contains as particular case the system introduced by Schnackenberg for trimolecular autocatalytic reactions.[4]
The aim of this paper is to study the nonlinear stability of the steady solution of (1):

$$u^* = a + b \qquad v^* = \frac{b}{(a+b)^2} \tag{2}$$

The plan of the paper is as follows. In section 2, we state a maximum principle for regular solutions of system (1). Then, section 3, we introduce a peculiar Liapunov functional. Finally, a nonlinear locally stability result is proven in section 4.

2. General properties of regular solutions

In this section we consider system (1)with the following general initial boundary conditions:

$$U(\mathbf{x},0) = U_0(\mathbf{x}), \quad V(\mathbf{x},0) = V_0(\mathbf{x}), \qquad \forall \mathbf{x} \in \Omega \tag{3}$$

$$U(\mathbf{x},t) = U^*(\mathbf{x},t), \quad V(\mathbf{x},t) = V^*(\mathbf{x},t), \qquad \forall (\mathbf{x},t) \in \partial\Omega \times I\!R^+ \tag{4}$$

where U^*, V^*, U_0, V_0 are regular functions. For any $T > 0$, setting $\Omega_T = \Omega \times (0,T]$, the following theorem holds:[12]

Theorem 2.1. *Let $\{U, V \in C_1^2(\Omega_T) \cap C(\bar{\Omega}_T\}$ be a positive solution of (1),(3),(4) where U^*, V^*, U_0, V_0 positive continuous functions. Then*

$$\begin{aligned} U(\mathbf{x},t) &\geq m = \inf\left\{a, \min_{\bar{\Omega}} U_0, \min_{\partial\Omega\times[0,T]} U^*\right\}, \\ V(\mathbf{x},t) &\leq M = \sup\left\{\frac{b}{m^2}, \max_{\bar{\Omega}} V_0, \max_{\partial\Omega\times[0,T]} V^*\right\}. \end{aligned} \tag{5}$$

3. Preliminaries to longtime behaviour of solutions

In order to study the non linear stability of the steady state (2) of system (1), we introduce the perturbations $(u^* + C_1, v^* + C_2)$. It can be proved easily that the equations governing the perturbations (C_1, C_2) are:

$$\begin{cases} \dfrac{\partial C_1}{\partial t} = a_1 C_1 + a_2 C_2 + \Delta C_1 + f(C_1, C_2), \\[2ex] \dfrac{\partial C_2}{\partial t} = a_3 C_1 + a_4 C_2 + d\Delta C_2 + g(C_1, C_2), \end{cases} \tag{6}$$

under the homogeneous Dirichlet boundary conditions

$$C_1 = C_2 = 0, \qquad \forall (\mathbf{x},t) \in \partial\Omega \times I\!R^+, \tag{7}$$

with:

$$a_1 = \gamma\frac{b-a}{a+b}, \; a_2 = \gamma(a+b)^2, \; a_3 = -2\gamma\frac{b}{a+b}, \; a_4 = -\gamma(a+b)^2, \tag{8}$$

and

$$f(C_1, C_2) = \gamma(v^* C_1^2 + C_1^2 C_2 + 2u^* C_1 C_2), \; g(C_1, C_2) = -f(C_1, C_2). \tag{9}$$

Following the Rionero's method[6,7] we introduce the scalings:

$$C_1 = \alpha u, \qquad C_2 = \beta v, \tag{10}$$

with α and β constants, to be chosen suitably later, and $\mu = \alpha/\beta$. By using the above scalings in (6), denoting by $\bar{\alpha}$ the lowest eigenvalue λ of

$$\Delta\phi + \lambda\phi = 0 \tag{11}$$

in $H_0^1(\Omega)$, we get:

$$\begin{cases} \dfrac{\partial u}{\partial t} = b_1 u + \dfrac{a_2}{\mu} v + f^* + f_1^*, \\ \\ \dfrac{\partial v}{\partial t} = \mu a_3 u + b_4 v + g^* + g_1^*, \end{cases} \tag{12}$$

where

$$\begin{cases} b_1 = a_1 - \bar{\alpha}, \quad b_4 = a_4 - d\bar{\alpha}, \quad f^* = \alpha^{-1} f(\alpha u, \beta v), \\ g^* = \beta^{-1} g(\alpha u, \beta v), \quad f_1^* = \Delta u + \bar{\alpha} u, \quad g_1^* = d(\Delta v + \bar{\alpha} v). \end{cases} \tag{13}$$

Setting $I = b_1 + b_4$, $A = b_1 b_4 - a_2 a_3$, and introducing the Rionero-Liapunov functional:

$$W = \frac{1}{2}\left[A(\| u \|^2 + \| v \|^2) + \| b_1 v - \mu a_3 u \|^2 + \| \frac{a_2}{\mu} v - b_4 u \|^2\right], \tag{14}$$

along the solutions of (12) it turns out:

$$\frac{dW}{dt} = AI(\| u \|^2 + \| v \|^2) + \Psi^* + \Psi_1^*, \tag{15}$$

$$\begin{cases} \Psi^* = < \alpha_1 u - \alpha_3 v, f^* > + < \alpha_2 v - \alpha_3 u, g^* >, \\ \Psi_1^* = < \alpha_1 u - \alpha_3 v, f_1^* > + < \alpha_2 v - \alpha_3 u, g_1^* >, \\ \alpha_1 = A + b_4^2 + \mu^2 a_3^2, \ \alpha_2 = A + b_1^2 + \dfrac{a_2^2}{\mu^2}, \ \alpha_3 = \mu a_3 b_1 + \mu^{-1} a_2 b_4. \end{cases} \tag{16}$$

Remark 1
It can be shown that[7] W is equivalent to the usual L^2-norm. In other words, there exist two positive constants k_1 and k_2 such that:

$$k_1(\| u \|^2 + \| v \|^2) \leq W \leq k_2(\| u \|^2 + \| v \|^2). \tag{17}$$

4. Nonlinear stability analysis via the boundedness of V

Theorem 4.1. *Let* $\dfrac{b-a}{a+b} < \dfrac{\bar{\alpha}}{\gamma}$, *then* (u^*, v^*) *is nonlinearly asymptotically stable with respect the* $L^2(\Omega)$*-norm.*

Proof. Let us observe that:

$$\frac{b-a}{a+b} < \frac{\bar{\alpha}}{\gamma} \Longleftrightarrow b_1 < 0. \tag{18}$$

This implies that:

$$I < 0, \qquad A > 0, \qquad b_1 b_4 a_2 a_3 < 0. \tag{19}$$

Following,[7] for any constant $\bar{\varepsilon}$ such that

$$0 < \bar{\varepsilon} < \inf\left\{\frac{|b_1|}{\bar{\alpha}}, \frac{|b_4|}{\bar{\alpha}}, \frac{|I|}{2\bar{\alpha}}, \frac{A}{\bar{\alpha}|I|}, 1, d\right\}, \tag{20}$$

setting

$$\bar{b}_i = b_i + \bar{\alpha}\bar{\varepsilon}, \ (i = 1, 4) \quad \bar{\gamma}_1 = 1 - \bar{\varepsilon}, \quad \bar{\gamma}_2 = d - \bar{\varepsilon}, \tag{21}$$

we can write (12) as follows

$$\begin{cases} \dfrac{\partial u}{\partial t} = \bar{b}_1 u + \dfrac{a_2}{\mu} v + f^* + \bar{f}_1^*, \\[2ex] \dfrac{\partial v}{\partial t} = \mu a_3 u + \bar{b}_4 v + g^* + \bar{g}_1^*, \end{cases} \tag{22}$$

where

$$\bar{f}_1^* = \bar{\gamma}_1(\Delta u + \bar{\alpha} u) + \bar{\varepsilon}\Delta u, \quad \bar{g}_1^* = \bar{\gamma}_2(\Delta v + \bar{\alpha} v) + \bar{\varepsilon}\Delta v, \tag{23}$$

and we observe that from conditions (19) the following inequalities hold:

$$\bar{I} = \bar{b}_1 + \bar{b}_4 < 0, \qquad \bar{A} = \bar{b}_1 \bar{b}_4 - a_2 a_3 > 0. \tag{24}$$

Along the solutions of (22), it turns out:

$$\frac{d\bar{W}}{dt} = \bar{A}\bar{I}(\| u \|^2 + \| v \|^2) + \bar{\Psi}^* + \bar{\Psi}_1^*, \tag{25}$$

$$\begin{cases} \bar{\Psi}^* = < \bar{\alpha}_1 u - \bar{\alpha}_3 v, f^* > + < \bar{\alpha}_2 v - \bar{\alpha}_3 u, g^* >, \\[2ex] \bar{\Psi}_1^* = < \bar{\alpha}_1 u - \bar{\alpha}_3 v, \bar{f}_1^* > + < \bar{\alpha}_2 v - \bar{\alpha}_3 u, \bar{g}_1^* >, \\[2ex] \bar{\alpha}_1 = \bar{A} + \bar{b}_4^2 + \mu^2 a_3^2, \ \bar{\alpha}_2 = \bar{A} + \bar{b}_1^2 + \dfrac{a_2^2}{\mu^2}, \ \bar{\alpha}_3 = \mu a_3 \bar{b}_1 + \mu^{-1} a_2 \bar{b}_4. \end{cases} \tag{26}$$

Now choosing $\mu^2 = |a_2\bar{b}_4|/|\bar{b}_1 a_3|$, it follows that $\bar{\alpha}_3 = 0$ and therefore

$$\begin{aligned}\bar{\Psi}^* &= \bar{\alpha}_1 < u, f^* > + \bar{\alpha}_2 < v, g^* >, \\ \bar{\Psi}_1^* &= \bar{\alpha}_1 < u, \bar{f}_1^* > + \bar{\alpha}_2 < v, \bar{g}_1^* > .\end{aligned} \tag{27}$$

But it can be proved:

$$\bar{\Psi}_1^* \le -k^* \left(\| \nabla u \|^2 + \| \nabla v \|^2 \right) \tag{28}$$

with $k^* = \bar{\varepsilon} \inf(\bar{\alpha}_1, \bar{\alpha}_2)$. In order to prove the decay of $\bar{W}$, and then the stability of (2), we must control suitably the nonlinear term $\bar{\Psi}^*$.
Coming back to (27), choosing $\beta = 1$, recall that:

a) from theorem (2.1), there exists a positive constant Γ_2 such that

$$|C_2(\mathbf{x}, t)| \le \Gamma_2, \qquad \forall (\mathbf{x}, t) \in \Omega \times [0, T]; \tag{29}$$

b) from Sobolev embedding theorem, there exists a positive constant $k(\Omega)$ such that

$$(< \phi^4 >)^{1/2} \le k(\Omega) \| \nabla \phi \|^2 . \tag{30}$$

By virtue of the above inequalities and the Cauchy-Schwarz inequality it turns out that:

$$\Psi^* \le \sqrt{2}\Gamma \left(\| C_1 \|^2 + \| C_2 \|^2 \right)^{\frac{1}{2}} \left(\| \nabla C_1 \|^2 + \| \nabla C_2 \|^2 \right), \tag{31}$$

where

$$\Gamma = \gamma k \left[\frac{\bar{\alpha}_1}{\alpha^2} (v^* + 2u^* + \Gamma_2) + \bar{\alpha}_2 (v^* + 2u^*) \right]. \tag{32}$$

Finally, from (25),(28), (31) it turns out that:

$$\frac{d\bar{W}}{dt} \le -\frac{\bar{A}|\bar{I}|}{k_2} \bar{W} + \left(\Gamma^* \sqrt{\frac{2}{k_1}} \bar{W}^{\frac{1}{2}} - k^* \right) (\| \nabla u \|^2 + \| \nabla v \|^2), \tag{33}$$

where $\Gamma^* = \Gamma \left(\sup\{\alpha^2, 1\} \right)^{\frac{3}{2}}$. So, provided that

$$\bar{W}_0^{\frac{1}{2}} < \frac{k^*}{\Gamma^*} \sqrt{\frac{k_1}{2}} \Longrightarrow \bar{W} \le \bar{W}_0 \exp(-\delta t), \tag{34}$$

with

$$\delta = \frac{1}{k_2} \left[\bar{A}|\bar{I}| - \bar{\alpha} \left(k^* - \Gamma^* \sqrt{\frac{2}{k_1}} \bar{W}_0^{\frac{1}{2}} \right) \right]. \tag{35}$$

Acknowledgements
To Professor T.Ruggeri on the occasion of his 60th birthday. This work has been performed under the auspicies of the G.N.F.M. of I.N.D.A.M. and

M.I.U.R. (P.R.I.N. 2005): "Nonlinear Propagation and Stability in Thermodinamical Processes of Continuous Media." The authors thank gratefully Prof. S.Rionero, for having proposed the present research and for his helpful suggestions.

References

1. J. Schnackenberg, *J. Theor. Biol.* **81**, 389-400. (1979)
2. Murray, J.D. (2003) *Mathematical Biology.I. An introduction*, Third edition,Interdisciplinary Applied Mathematics **17**, Springer Verlag, New York.
3. Murray, J.D. (2003) *Mathematical Biology.II. Spatial model and Biomedical Applications*, Third edition,Interdisciplinary Applied Mathematics **18**, Springer Verlag, New York.
4. L. Segel and J. Jackson, *J. Theor. Biol.* **37**, 545-559. (1972)
5. S. Rionero *Asymptotic Methods in Nonlinear Wave Phenomena*, In honor of the 65th birthday of A.Greco. (World Scientific, 2007), , 171-185.
6. S. Rionero , *Journal of Math. Anal. Appl.* **319** , 377-397(2006) .
7. S. Rionero, *Math. Biosci. Eng.* **3**, 189-204(2006).
8. S. Rionero , *Rend. Accad. Naz. Lincei*, Serie IX, **XVI**, 227-238 (2005).
9. S. Rionero *Diffusion Driven Stability andTuring Effect Under Robin Boundary data*, (to appear).
10. J.N. Flavin and S. Rionero, *Qualitative estimates for partial differential equations: an introduction*, (CRC Press, Boca Raton, Florida, 1996).
11. M.Gentile, S.Rionero, A.Tataranni, *Rend. Accad. Sc. fis. mat. Napoli*, **LXXIII**, 481-490 (2006).
12. M.Gentile and A.Tataranni, On L^2-stability of the Schnackemberg trimolecular autocatalytic chemical reaction via the Liapunov direct method (2007),(To appear).

RADIATIVE TRANSFER EQUATIONS AND ROSSELAND APPROXIMATION IN GRAY MATTER

F. GOLSE

Centre de Mathématiques Laurent Schwartz, Ecole Polytechnique
91128 Palaiseau Cedex, France
E-mail: golse@math.polytechnique.fr

F. SALVARANI

Dipartimento di Matematica, Università degli Studi di Pavia
Via Ferrata 1 - 27100 Pavia, Italy
E-mail: francesco.salvarani@unipv.it

In this paper we consider the radiative transfer equations in a bounded domain with non-homogeneous boundary conditions, when the opacity does not depend on the frequency of the photons. We discuss the existence of weak solutions of the radiative transfer system and show that the corresponding Rosseland approximation is robust even when the target equation is parabolic degenerate and the flux on the boundary is non vanishing.

Keywords: Radiative transfer; Rosseland approximation; Diffusive limit.

1. Introduction

Radiative transfer equations describe the process of transmission of the electromagnetic radiation through an host medium. The interactions between the photons and the particles which compose the medium result in phenomena of absorption, emission and scattering.

Even if the microphysics of the collisions between the single photon and the components of the host medium is essentially of quantum type, the description of the interaction between radiation and matter can be fruitfully discussed, at a larger scale, by means of kinetic-type models in a classical setting.

As established in all textbooks on Radiation Hydrodynamics,[1] if we suppose that the host medium can be viewed as a background, this physical process can be described by the following set of partial differential equations (with independent variables given by the time $t \in [0, \tau]$, $\tau > 0$, the position

$x \in \Omega \subseteq \mathbf{R}^3$, the spherical angle with respect to a suitable reference frame $\omega \in S^2$ and the frequency $\nu \in \mathbf{R}^+$) for the *spectral intensity* $I = I(t, x, \omega, \nu)$ and the *temperature* $T = T(t, x)$:

$$\begin{cases} \dfrac{1}{c}\dfrac{\partial I}{\partial t} + \omega \cdot \nabla_x I + \Sigma(T,\nu)(I - B_\nu(T)) = 0 \\ \\ \dfrac{\partial E}{\partial t}(T) = \displaystyle\int_0^{+\infty} \Sigma(T,\nu)(\bar{I} - B_\nu(T))\, d\nu, \end{cases} \tag{1}$$

where

$$B_\nu = \frac{\alpha \nu^3}{e^{h\nu/kT} - 1} \qquad \bar{I} = \frac{1}{4\pi}\int_{S^2} I(t,x,\omega,\nu)\, d\omega.$$

The parameter $c > 0$ is the speed of light and B_ν is Planck's law for the black-body radiation; the quantities h and k are the Planck and the Boltzmann constants respectively.

We have introduced the normalization constant $\alpha = 15\,[h/(k\pi)]^4$, so that Stefan's law assumes the simple form: $\int_0^{+\infty} B_\nu(T)\, d\nu = T^4$.

The function $\Sigma = \Sigma(T,\nu)$ is the *opacity* of the background. It describes the details of the interaction between the photons and the medium. Physical experiments[2] indicate that the opacity depends on the temperature and on the frequency of the radiation in a very complicated – and highly nonlinear – way. In particular, it tends to infinity as $T \to 0$, and tends to zero as $T \to +\infty$. Between these two extremes, the opacity oscillates very strongly, according to the structure of the energy levels of the atoms which compose the host medium and to the effects of the temperature on the particles which compose the background.

Another important nonlinear term in System (1) is the *internal energy* $E = E(T)$, an increasing monotone function such that $E'(T) \sim T^\gamma$ for $T \to 0$, where $\gamma \in [0, 2]$. This prescription is sufficiently general to allow the most common physical situations; for example the perfect gases law, $E(T) = c_V T$, is allowed by this hypothesis. The internal energy having an arbitrary degree of freedom, it is customary to impose that $E(0) = 0$ (even if this condition is irrelevant, since E appears in System (1) only through its time derivative).

In some relevant physical situations the opacity does not depend on the frequency of the electromagnetic radiation.[1] It is therefore convenient to integrate System (1) with respect to the frequency variable and obtain the following system of equations:

$$\begin{cases} \dfrac{1}{c}\dfrac{\partial u}{\partial t} + \omega \cdot \nabla_x u + \Sigma(T)(u - T^4) = 0 \\ \\ \dfrac{\partial E}{\partial t}(T) = \Sigma(T)(\bar{u} - T^4), \qquad \bar{u} = \dfrac{1}{4\pi}\displaystyle\int_{S^2} u(t,x,\omega)\, d\omega. \end{cases} \tag{2}$$

Since the dependence on the frequency of the radiation has been dropped out, this system is usually called the *gray model*. In (2) the unknowns are now the *photon density* $u = u(t,x,\omega) = \int_0^{+\infty} I(t,x,\omega,\nu)\, d\nu$ and, again, the temperature $T = T(t,x)$.

Through multiple interactions, the photons experience a sort of random walk in the host medium. Since the equilibrium has mean zero velocity, it is therefore natural to look at the possibility of deducing a diffusive behavior of the photon gas.

From a mathematical point of view, this would mean that it is possible to obtain diffusion-type equations through the usual parabolic scaling:

$$\begin{cases} \dfrac{\varepsilon}{c}\dfrac{\partial u_\varepsilon}{\partial t} + \omega \cdot \nabla_x u_\varepsilon + \dfrac{1}{\varepsilon}\Sigma(T_\varepsilon)(u_\varepsilon - T_\varepsilon^4) = 0 \\ \\ \varepsilon^2 \dfrac{\partial E}{\partial t}(T_\varepsilon) = \Sigma(T_\varepsilon)(\bar{u}_\varepsilon - T_\varepsilon^4), \qquad \bar{u}_\varepsilon = \dfrac{1}{4\pi}\displaystyle\int_{S^2} u_\varepsilon(t,x,\omega) d\omega. \end{cases} \tag{3}$$

Here $\varepsilon > 0$ is a non-dimensional quantity which represents the ratio of the mean free path to a characteristic length of the host medium.

We assume that the opacity and the internal energy in System (3) satisfy the following properties:

Definition 1.1. Let $\Sigma = \Sigma(T)$ and $E = E(T)$ be the opacity and the internal energy of System (3). We say that Σ and E are *admissible* if and only if

(1) $\Sigma(y) > 0$ for any $y > 0$ and is of class $C^1((0,+\infty))$;
(2) $\lim_{y \to +\infty} \Sigma(y) = 0$;
(3) $\lim_{y \to 0} \Sigma(y) = +\infty$ and $\Sigma(y) \sim y^{-\beta}$ as $y \to 0$, with $\beta \in (0,1]$.
(4) E is continuous in $[0,+\infty)$ and belongs to the class $C^1((0,+\infty))$. Moreover $E(0) = 0$ and $E'(y) \geq 0$ for all $y > 0$.
(5) $E'(y) \sim y^\alpha$ for $y \to 0$, where $\alpha \in [0,2]$.

In next section we will describe our results on the initial-boundary value problem for System (3), with $(t,x,\omega) \in (0,\tau) \times \Omega \times S^2$, where $\Omega \subset \mathbf{R}^3$ is a bounded domain with boundary of class C^1, supplemented with appropriate initial and boundary conditions. Finally, in Section 3, we will give some directions in order to generalize our results.

2. Results

Two questions arise in a natural way: first, the proof of the existence of a solution for System (3), then the limiting behavior of System (3) as $\varepsilon \to 0$, which will result in a (degenerate) nonlinear equation of parabolic type for the *macroscopic density* $\rho_\varepsilon(t,x) = \bar{u}_\varepsilon$. This diffusive approximation is known under the name of *Rosseland limit*.

The authors have proven the following result:[3]

Theorem 2.1. *Let us consider System* (3) *with initial conditions* $u_\varepsilon(0,x,\omega) = u^{\text{in}}(x,\omega) \in L^\infty(\Omega \times S^2)$, $T_\varepsilon(0,x) = T^{\text{in}}(x) \in L^\infty(\Omega)$ *and boundary data* $u_\varepsilon(t,x,\omega)|_{\Gamma_-} = u^{\text{b}}(t,x,\omega) \in L^\infty((0,\tau)\times\Gamma_-)$, *where* $\Omega \subset \mathbf{R}^3$ *is a bounded domain with boundary of class* C^1, $\Gamma_- = \{x \in \partial\Omega \,:\, \omega \cdot n_x < 0\}$ *and* $\varepsilon > 0$.

If Σ *and* E *satisfy the prescriptions of Definition* 1.1, *then System* (3) *admits a weak solution*

$$(u_\varepsilon(t,x,\omega), T_\varepsilon(t,x)) \in L^\infty((0,\tau) \times \Omega \times S^2) \times L^\infty((0,\tau) \times \Omega).$$

The hypotheses of this theorem are noticeably weaker with respect to the existing literature. Previous works study indeed either the accretive case[4–6] (with a monotone opacity, which is a non-physical assumption) or the non-accretive case when $E'(T) \ll \Sigma(T)$.[7] The latter simplification reduces System (3) to the single equation

$$\frac{\varepsilon}{c}\frac{\partial u_\varepsilon}{\partial t} + \omega \cdot \nabla_x u_\varepsilon + \frac{1}{\varepsilon}\Sigma(\bar{u}_\varepsilon)(u_\varepsilon - \bar{u}_\varepsilon) = 0.$$

In the proof of Theorem 2.1 we cannot use the theory of accretive operators because of the nonlinear oscillating behavior of the opacity allowed by Definition 1.1. Our method for proving existence uses a compactness argument based on the maximum principle and a velocity averaging lemma.

The second result proved by the authors is a rigorous justification of the Rosseland limit of the radiative transfer equations. Here, in order to avoid additional difficulties caused by the presence of a boundary layer, we have supposed that the boundary data do not depend on ω. The result is summarized in the following theorem:[3]

Theorem 2.2. *Let us consider System* (3) *in* $\Omega \subset \mathbf{R}^3$, *with admissible opacity* Σ *and internal energy* E, *posed for* $(t,x,\omega) \in (0,\tau) \times \Omega \times S^2$, *with* *initial conditions* $u_\varepsilon(0,x,\omega) = u^{\rm in}(x,\omega) \in L^\infty(\Omega \times S^2)$, $T_\varepsilon(0,x) = T^{\rm in}(x) \in L^\infty(\Omega)$ *and boundary data* $u_\varepsilon(t,x,\omega)|_{\Gamma_-} = u^{\rm b}(t,x) \in W^{1,\infty}((0,\tau) \times \Gamma_-)$, *where* $\Omega \subset \mathbf{R}^3$ *is a bounded domain with boundary of class* C^1, $\Gamma_- = \{x \in \partial\Omega \,:\, \omega \cdot n_x < 0\}$ *and* $\varepsilon > 0$.

Let $\{u_\varepsilon\}$ *be a family of solutions of System* (3), *obtained by varying the parameter* ε. *Then, there exists a function* $\rho \in L^\infty((0,\tau) \times \Omega)$ *such that* $\rho = \lim_{\varepsilon\to 0} \bar{u}_\varepsilon$ *strongly in* $L^2((0,\tau) \times \Omega)$. *The limit function* ρ *moreover solves the following initial-boundary value problem in* $(0,\tau) \times \Omega$, *for* $\tau > 0$:

$$(\rho + E(\rho^{1/4}))_t - \frac{1}{3}\nabla_x \cdot \left[(\Sigma(\rho^{1/4}))^{-1}\nabla_x \rho\right] = 0$$

with initial conditions $\rho(0,x) = \bar{u}^{\rm in}$ *and boundary data* $\rho(t,x)|_{\partial\Omega} = u^{\rm b}(t,x)$.

For technical reasons, the regularity condition on the boundary data for the family $\{u_\varepsilon\}$ in Theorem 2.2 is slightly more stringent than the requirements on the boundary conditions in Theorem 2.1; moreover, we suppose that $u^{\rm b}$ does not depend on the variable ω. In any case, we allow the possibility to have temperature gradients on the boundary of Ω.

It is interesting to note that, even if only the incoming density on the boundary is prescribed for the family $\{u_\varepsilon\}$, we obtain that the boundary condition for the limit is fully defined.

Because of the conditions on the opacity given in Definition 1.1, the Rosseland equation is degenerate parabolic, so that the temperature may display some singularity in the form of a propagating front when the radiation penetrates into cold matter. Hence the result of Theorem 2.2 proves that the Rosseland approximation is robust in the sense that it holds even in situations in which such singularities may occur — something that is not at all clear with classical asymptotic expansions.

The method for proving the Rosseland limit is based on the control of the radiation flux by means of a relative entropy functional. Normally, relative entropy methods are used when the solutions of the target equation are known to be smooth. This is not the case here, as the temperature may fail to be differentiable at the interfaces with cold matter. Here, as in the case of Carleman-like kinetic models,[8] we use both relative entropy and a compactness argument that allow for possibly singular solutions of the target equation. Relative entropy is not used all the way through the proof to control the distance between the radiative intensity and the (fourth power of) the temperature that solves the Rosseland limiting equation, but only to control the radiation flux. This control is in turn one essential step in the following compactness argument.

3. Perspectives

Besides the questions on the uniqueness of the solution of System (3), which require new techniques of proof since the involved operators are not accretive, the next step would be to study System (1) in its fully generality, with hypotheses on Σ and E which reduces to the conditions listed in Definition 1.1 when the opacity does not depend on the frequency of the radiation.

Both existence and diffusive limit of (1), with admissible (in the sense of Definition 1.1) opacity Σ and internal energy E, are open problems: existence of weak solutions[9] has been indeed obtained only in the setting of nonlinear accretive techniques, which are not adapted to our case; on the other hand, also the Rosseland limit studied by Dogbe[10] requires to deal with accretive operators.

We finally note that the strategy based on relative entropy coupled with a compactness argument to establish the validity of a macroscopic (or hydrodynamic) limit seems to be useful only in the case of limits leading to a pure diffusion equation (degenerate or not), but without streaming term. In particular, the method presented here does not seem to apply to the derivation of the incompressible Navier-Stokes equations from the Boltzmann equation.

Acknowledgments

This work has been supported by the Italian National Group for Mathematical Physics (GNFM) and the Franco-Italian group GREFI-MEFI.

References

1. D. Mihalas, B. Weibel Mihalas, *Foundations of Radiation Hydrodynamics* (Oxford University Press, New York, 1984).
2. D. R. Alexander, J. W. Ferguson, *Astrophysical Journal* **437**, 879-891 (1994).
3. F. Golse, F. Salvarani, The Rosseland limit for radiative transfer in gray matter Preprint (2007).
4. B. Mercier, *SIAM J. Math. Anal.* **18**, 393-408 (1987).
5. C. Bardos, F. Golse, B. Perthame, *Comm. Pure Appl. Math.* **40**, 691-721 (1987).
6. C. Bardos, F. Golse, B. Perthame, *Comm. Pure Appl. Math.* **42**, 891-894 (1989).
7. C. Bardos, F. Golse, B. Perthame, R. Sentis, *J. Funct. Anal.* **77**, 434-460 (1988).
8. F. Golse, F. Salvarani, *Nonlinearity* **20**, 927-942 (2007).
9. F. Golse, B. Perthame, *Comm. Math. Phys.* **106**, 211-239 (1986).
10. C. Dogbe, *Comput. Math. Appl.* **42**, 783-791 (2001).

A MECHANICAL MODEL FOR LIQUID NANOLAYERS

H. GOUIN

University of Aix-Marseille & U.M.R. C.N.R.S. 6181,
Case 322, Av. Escadrille Normandie-Niemen, 13397 Marseille Cedex 20 France
E-mail: henri.gouin@univ-cezanne.fr

Liquids in contact with solids are submitted to intermolecular forces making liquids heterogeneous and stress tensors are not any more spherical as in homogeneous bulks. The aim of this article is to show that a square-gradient functional representing liquid-vapor interface free energy corrected with a liquid density functional at solid surfaces is a well adapted model to study structures of very thin nanofilms near solid walls. This result makes it possible to study the motions of liquids in nanolayers and to generalize the approximation of lubrication in long wave hypothesis.

Keywords: Nanolayers, disjoining pressure, thin flows, approximation of lubrication.

1. Introduction

At the end of the nineteenth century, the fluid inhomogeneity in liquid-vapor interfaces was taken into account by considering a volume energy depending on space density derivative.[1] This van der Waals square-gradient functional is unable to model repulsive force contributions and misses the dominant damped oscillatory packing structure of liquid interlayers near a substrate wall.[2] Furthermore, the decay lengths are correct only close to the liquid-vapor critical point where the damped oscillatory structure is subdominant.[3] In mean field theory, weighted density-functional has been used to explicitly demonstrate the dominance of this structural contribution in van der Waals thin films and to take into account long-wavelength capillary-wave fluctuations as in papers that renormalize the square-gradient functional to include capillary wave fluctuations.[4] In contrast, fluctuations strongly damp oscillatory structure and it is mainly for this reason that van der Waals' original prediction of a *hyperbolic tangent* is so close to simulations and experiments.[5] The recent development of experimental technics allows us to observe physical phenomena at length scales

of a few nanometers.[6] To get an analytic expression in density-functional theory for liquid film of a few nanometer thickness near a solid wall, we add a liquid density-functional at the solid surface to the square-gradient functional representing closely liquid-vapor interface free energy. This kind of functional is well-known in the literature.[7] It was used by Cahn in a phenomenological form, in a well-known paper studying wetting near a critical point.[8] An asymptotic expression is obtained in[9] with an approximation of hard sphere molecules and London potentials for liquid-liquid and solid-liquid interactions: we took into account the power-law behavior which is dominant in a thin liquid film in contact with a solid.
For fluids submitted to this density-functional, we recall the equation of motion and boundary conditions. We point out the definition of disjoining pressure and analyze the consequences of the model. Finally, we study the motions in liquid nanolayers; these motions are always object of many debates. Within lubrication and long wave approximations, a relation between disjoining pressure, viscosity of liquid, nanolayer thickness variations along the layer and tangential velocity of the liquid is deduced.

2. The density-functional

The free energy density-functional of an inhomogeneous fluid in a domain O of boundary ∂O is taken in the form

$$F = \int\int\int_O \varepsilon \, dv + \int\int_{\partial O} \varphi \, ds. \tag{1}$$

The first integral is associated with square-gradient approximation when we introduce a specific free energy of the fluid at a given temperature θ, $\varepsilon = \varepsilon(\rho, \beta)$ as a function of density ρ and $\beta = (\mathrm{grad}\,\rho)^2$. Specific free energy ε characterizes together fluid properties of *compressibility* and *molecular capillarity* of liquid-vapor interfaces. In accordance with gas kinetic theory, $\lambda = 2\rho\,\varepsilon'_\beta(\rho, \beta)$ is assumed to be constant at given temperature[10] and

$$\rho\,\varepsilon = \rho\,\alpha(\rho) + \frac{\lambda}{2}\,(\mathrm{grad}\ \rho)^2, \tag{2}$$

where term $(\lambda/2)\,(\mathrm{grad}\ \rho)^2$ is added to the volume free energy $\rho\,\alpha(\rho)$ of a compressible fluid. Specific free energy α enables to connect continuously liquid and vapor bulks and pressure $P(\rho) = \rho^2\alpha'_\rho(\rho)$ is similar to van der Waals one. Near a solid wall, London potentials of liquid-liquid and liquid-solid interactions are

$$\begin{cases} \varphi_{ll} = -\dfrac{c_{ll}}{r^6}\,, & \text{when } r > \sigma_l \text{ and } \varphi_{ll} = \infty \text{ when } r \leq \sigma_l\,, \\ \varphi_{ls} = -\dfrac{c_{ls}}{r^6}\,, & \text{when } r > \delta \text{ and } \varphi_{ls} = \infty \text{ when } r \leq \delta\,, \end{cases}$$

where c_{ll} et c_{ls} are two positive constants associated with Hamaker constants, σ_l and σ_s denote fluid and solid molecular diameters, $\delta = \frac{1}{2}(\sigma_l + \sigma_s)$ is the minimal distance between centers of fluid and solid molecules.[11] Forces between liquid and solid have short range and can be described simply by adding a special energy at the surface. This is not the entire interfacial energy: another contribution comes from the distortions in the density profile near the wall.[9,12] For a plane solid wall (at a molecular scale), this surface free energy is

$$\phi(\rho) = -\gamma_1 \rho + \frac{1}{2}\,\gamma_2\,\rho^2. \tag{3}$$

Here ρ denotes the fluid density value at the wall; constants γ_1, γ_2 are positive and given by relations $\gamma_1 = \dfrac{\pi c_{ls}}{12\delta^2 m_l m_s}\,\rho_{sol}$, $\gamma_2 = \dfrac{\pi c_{ll}}{12\delta^2 m_l^2}$, where m_l et m_s denote respectively masses of fluid and solid molecules, ρ_{sol} is the solid density.[9] Moreover, we have $\lambda = \dfrac{2\pi c_{ll}}{3\sigma_l\, m_l^2}$.

We consider a horizontal plane liquid interlayer contiguous to its vapor bulk and in contact with a plane solid wall (S); the z-axis is perpendicular to the solid surface. The liquid film thickness is denoted by h. Conditions in vapor bulk yield $\operatorname{grad}\rho = 0$ and $\Delta\rho = 0$. Another way to take into account the vapor bulk contiguous to the liquid interlayer is to compute a density-functional of the complete *liquid-vapor interlayer* by adding a supplementary surface energy ψ on a geometrical surface (Σ) at $z = h$ to volume energy (2) in liquid interlayer (L) and surface energy (3) on solid wall (S).[13] This assumption corresponds to a liquid interlayer included between $z = 0$ and $z = h$, a liquid-vapor interface of a few Angström thickness assimilated to surface $z = h$ and a vapor layer included between $z = h$ and $z = \infty$. Due to small vapor density, let us denote by ψ the surface free energy of a liquid in contact with a vacuum,

$$\psi(\rho) = \frac{\gamma_4}{2}\,\rho^2 \tag{4}$$

where $\gamma_4 \simeq \gamma_2$ and ρ is the liquid density in a convenient point inside the liquid-vapor interface.[13] Density-functional (1) of the liquid-vapor layer gets the final form

$$F = \iiint_{(L)} \varepsilon\, dv + \iint_{(S)} \phi\, ds + \iint_{(\Sigma)} \psi\, ds$$

3. Equation of motion and boundary conditions

In case of equilibrium, functional F is minimal and yields the equation of equilibrium and boundary conditions. In case of motions we simply add the inertial forces $\rho\,\boldsymbol{\Gamma}$ and the dissipative stresses to the results.[14–16]

3.1. *Equation of motion*

The equation of motion is[14,15]

$$\rho\,\boldsymbol{\Gamma} = \mathrm{div}\,(\sigma + \sigma_v) - \rho\,\mathrm{grad}\,\Omega\;, \tag{5}$$

where $\boldsymbol{\Gamma}$ is the acceleration, Ω the body force potential and σ the stress tensor generalization

$$\sigma = -\,p\,\mathbf{1} - \lambda\,\mathrm{grad}\,\rho\,\otimes\,\mathrm{grad}\,\rho,$$

with $p = \rho^2\varepsilon'_\rho - \rho\,\mathrm{div}\,(\lambda\,\mathrm{grad}\,\rho)$. The viscous stress tensor is $\sigma_v = \kappa_1(\,\mathrm{tr}\,D)\,\mathbf{1} + 2\,\kappa_2\,D$ where D denotes the velocity strain tensor; κ_1 and κ_2 are the coefficients of viscosity.

For a horizontal layer, in an orthogonal system of coordinates such that the third coordinate is the vertical direction, the stress tensor σ of the thin film takes the form :

$$\sigma = \begin{bmatrix} a_1, & 0, & 0 \\ 0, & a_2, & 0 \\ 0, & 0, & a_3 \end{bmatrix}, \quad \text{with} \quad \begin{cases} a_1 = a_2 = -P + \dfrac{\lambda}{2}\left(\dfrac{d\rho}{dz}\right)^2 + \lambda\,\rho\,\dfrac{d^2\rho}{dz^2} \\ a_3 = -P - \dfrac{\lambda}{2}\left(\dfrac{d\rho}{dz}\right)^2 + \lambda\,\rho\,\dfrac{d^2\rho}{dz^2} \end{cases}$$

Let us consider a thin film of liquid at equilibrium (*gravity forces are neglected* but the variable of position is the ascendant vertical). The equation of equilibrium is :

$$\mathrm{div}\,\sigma = \mathbf{0} \tag{6}$$

Eq. (6) yields a constant value for the eigenvalue a_3,

$$P + \frac{\lambda}{2}\left(\frac{d\rho}{dz}\right)^2 - \lambda\rho\,\frac{d^2\rho}{dz^2} = P_{v_b}\,.$$

where P_{v_b} denotes pressure $P(\rho_{v_b})$ in the vapor bulk of density ρ_{v_b} bounding the liquid layer. Eigenvalues a_1, a_2 are not constant but depend on the distance z to the solid wall.[17] At equilibrium, Eq. (5) yields:[14]

$$\mathrm{grad}\;[\,\mu\,(\rho) - \lambda\Delta\rho\,] = 0, \tag{7}$$

where μ is the chemical potential at temperature θ defined to an unknown additive constant. The chemical potential is a function of P (and θ) but it can be also expressed as a function of ρ (and θ). We choose as *reference chemical potential* $\mu_o = \mu_o(\rho)$ null for bulks of densities ρ_l and ρ_v of phase equilibrium. Due to Maxwell rule, the volume free energy associated with μ_o is $g_o(\rho) - P_o$ where $P_o = P(\rho_l) = P(\rho_v)$ is the bulk pressure and $g_o(\rho) = \int_{\rho_v}^{\rho} \mu_o(\rho)\, d\rho$ is null for liquid and vapor bulks of phase equilibrium. The pressure P is

$$P(\rho) = \rho\, \mu_o(\rho) - g_o(\rho) \ + P_o. \tag{8}$$

Thanks to Eq. (7), we obtain in the fluid *and not only in the fluid interlayer,*

$$\mu_o(\rho) - \lambda\Delta\rho = \mu_o(\rho_b),$$

where $\mu_o(\rho_b)$ is the chemical potential value of a liquid *mother* bulk of density ρ_b such that $\mu_o(\rho_b) = \mu_o(\rho_{v_b})$, where ρ_{v_b} is the density of the vapor *mother* bulk bounding the layer. We must emphasis that $P(\rho_b)$ and $P(\rho_{v_b})$ are unequal as for drop or bubble bulk pressures. Density ρ_b is not a fluid density in the interlayer but density in the liquid bulk from which the interface layer can extend (this is the reason why Derjaguin used the term *mother liquid,*[17] *page 32*). In the interlayer

$$\lambda \frac{d^2\rho}{dz^2} = \mu_b(\rho), \quad \text{with} \quad \mu_b(\rho) = \mu_o(\rho) - \mu_o(\rho_b) \tag{9}$$

3.2. *Boundary conditions*

Condition at the solid wall (S) is associated with Eq. (3)[15] :

$$\lambda \left(\frac{d\rho}{dn}\right)_{|S} + \phi'(\rho)_{|S} = 0, \tag{10}$$

where n is the external normal direction to the fluid; Eq. (10) yields

$$\lambda \left(\frac{d\rho}{dz}\right)_{|z=0} = -\gamma_1 + \gamma_2\ \rho_{|z=0}.$$

Condition at the liquid-vapor interface (Σ) is associated with Eq. (4):

$$\lambda \left(\frac{d\rho}{dz}\right)_{|z=h} = -\gamma_4\ \rho_{|z=h}\,. \tag{11}$$

Eq. (11) defines the film thickness by introducing a reference point inside the liquid-vapor interface bordering the liquid interlayer with a convenient density at $z = h$.[13]

We must also add the classical surface conditions on the stress vector associated with the total stress tensor $\sigma + \sigma_v$ to these conditions on density.

4. The disjoining pressure for horizontal liquid films

We consider fluids and solids *at a given temperature* θ. The hydrostatic pressure in a thin liquid interlayer included between a solid wall and a vapor bulk differs from the pressure in the contiguous liquid phase. At equilibrium, the additional pressure interlayer is called the *disjoining pressure*.[17] The measure of a disjoining pressure is either the additional pressure on the surface or the drop in the pressure within the *mother bulks* that produce the interlayer. The disjoining pressure is equal to the difference between the pressure P_{v_b} on the interfacial surface (pressure of the vapor mother bulk of density ρ_{v_b}) and the pressure P_b in the liquid mother bulk (density ρ_b) from which the interlayer extends :

$$\Pi(h) = P_{v_b} - P_b.$$

If $g_b(\rho) = g_o(\rho) - g_o(\rho_b) - \mu_o(\rho_b)(\rho - \rho_b)$ denotes the primitive of $\mu_b(\rho)$ null for ρ_b, we get from Eq. (8)

$$\Pi(\rho_b) = -g_b(\rho_{v_b}), \tag{12}$$

and an integration of Eq. (9) yields

$$\frac{\lambda}{2}\left(\frac{d\rho}{dz}\right)^2 = g_b(\rho) + \Pi(\rho_b). \tag{13}$$

The reference chemical potential linearized near ρ_l (*respectively* ρ_v) is $\mu_o(\rho) = \dfrac{c_l^2}{\rho_l}(\rho - \rho_l)$ (*respectively* $\mu_o(\rho) = \dfrac{c_v^2}{\rho_v}(\rho - \rho_v)$) where c_l (*respectively* c_v) is the isothermal sound velocity in liquid bulk ρ_l (*respectively vapor bulk* ρ_v) at temperature θ.[18] In the liquid and vapor parts of the liquid-vapor film, Eq. (9) yields

$$\lambda\frac{d^2\rho}{dz^2} = \frac{c_l^2}{\rho_l}(\rho - \rho_b) \quad \text{(liquid)} \quad \text{and} \quad \lambda\frac{d^2\rho}{dz^2} = \frac{c_v^2}{\rho_v}(\rho - \rho_{v_b}) \quad \text{(vapor)}.$$

The values of $\mu_o(\rho)$ are equal for the mother densities ρ_{v_b} and ρ_b,

$$\frac{c_l^2}{\rho_l}(\rho_b - \rho_l) = \mu_o(\rho_b) = \mu_o(\rho_{v_b}) = \frac{c_v^2}{\rho_v}(\rho_{v_b} - \rho_v), \quad \text{and consequently,}$$

$$\rho_{v_b} = \rho_v\left(1 + \frac{c_l^2}{c_v^2}\frac{(\rho_b - \rho_l)}{\rho_l}\right).$$

In liquid and vapor parts of the liquid-vapor interlayer we have,

$$g_o(\rho) = \frac{c_l^2}{2\rho_l}(\rho - \rho_l)^2 \quad \text{(liquid)} \quad \text{and} \quad g_o(\rho) = \frac{c_v^2}{2\rho_v}(\rho - \rho_v)^2 \quad \text{(vapor)}.$$

From definition of $g_b(\rho)$ and Eq. (12) we deduce the disjoining pressure

$$\Pi(\rho_b) = \frac{c_l^2}{2\rho_l}(\rho_l - \rho_b)\left[\rho_l + \rho_b - \rho_v\left(2 + \frac{c_l^2}{c_v^2}\frac{(\rho_b - \rho_l)}{\rho_l}\right)\right]. \tag{14}$$

Due to $\rho_v\left(2 + \frac{c_l^2}{c_v^2}\frac{(\rho_b - \rho_l)}{\rho_l}\right) \ll \rho_l + \rho_b$, we get $\Pi(\rho_b) \approx \frac{c_l^2}{2\rho_l}(\rho_l^2 - \rho_b^2)$.
Now, we consider a film of thickness h; the density profile in the liquid part of the liquid-vapor film is solution of system :

$$\begin{cases} \lambda \dfrac{d^2\rho}{dz^2} = \dfrac{c_l^2}{\rho_l}(\rho - \rho_b) & (S1) \\ \text{with} \quad \lambda \dfrac{d\rho}{dz}_{|z=0} = -\gamma_1 + \gamma_2\, \rho_{|z=0} \quad \text{and} \quad \lambda \dfrac{d\rho}{dz}_{|z=h} = -\gamma_4\, \rho_{|z=h}\,. \end{cases}$$

Quantity τ is defined such that $\tau = c_l/\sqrt{\lambda\rho_l}$, where $1/\tau$ is a reference length and $\gamma_3 = \lambda\tau$. Solution of system $(S1)$ is

$$\rho = \rho_b + \rho_1\, e^{-\tau z} + \rho_2\, e^{\tau z} \tag{15}$$

where boundary conditions at $z = 0$ and h yield the values of ρ_1 and ρ_2 :

$$\begin{cases} (\gamma_2 + \gamma_3)\rho_1 + (\gamma_2 - \gamma_3)\rho_2 = \gamma_1 - \gamma_2\rho_b & (S2) \\ -e^{-h\tau}(\gamma_3 - \gamma_4)\rho_1 + e^{h\tau}(\gamma_3 + \gamma_4)\rho_2 = -\gamma_4\rho_b. \end{cases}$$

The liquid density profile is a consequence of Eq. (15) when $z \in [0, h]$. Taking Eq. (15) into account in Eq. (13) and $g_b(\rho) = (c_l^2/2\,\rho_l)(\rho - \rho_b)^2$ in linearized form for the liquid part of the interlayer, we get

$$\Pi(\rho_b) = -\frac{2\,c_l^2}{\rho_l}\,\rho_1\,\rho_2. \tag{16}$$

By identification of expressions (14), (16) and using $(S2)$, we get a relation between h and ρ_b. We denote finally the disjoining pressure by $\Pi(h)$.
Due to the fact that $\rho_b \simeq \rho_l$,[17] the disjoining pressure reduces to

$$\Pi(h) = \frac{2\,c_l^2}{\rho_l}\left[(\gamma_1 - \gamma_2\rho_l)(\gamma_3 + \gamma_4)e^{h\tau} + (\gamma_2 - \gamma_3)\gamma_4\rho_l\right] \times \frac{\left[(\gamma_2 + \gamma_3)\gamma_4\rho_l - (\gamma_1 - \gamma_2\rho_l)(\gamma_3 - \gamma_4)e^{-h\tau}\right]}{\left[(\gamma_2 + \gamma_3)(\gamma_3 + \gamma_4)e^{h\tau} + (\gamma_3 - \gamma_4)(\gamma_2 - \gamma_3)e^{-h\tau}\right]^2}.$$

Let us notice an important property of mixture of van der Waals fluid and perfect gas where the total pressure is the sum of partial pressures of components:[18] at equilibrium, the partial pressure of the perfect gas is constant through the liquid-vapor-gas interlayer -where the perfect gas is dissolved in the liquid. The disjoining pressure of the mixture is the same than for a single van der Waals fluid and calculations and results are identical to those previously obtained.

5. Motions along a liquid nanolayer

When the liquid layer thickness is small with respect to transverse dimensions of the wall, it is possible to simplify the Navier-Stokes equation which governs the flow of a classical viscous fluid in the approximation of lubrication.[19] When $h \ll L$, where L is the wall transversal characteristic size, *i)* the velocity component along the wall is large with respect to the normal velocity component which can be neglected ; *ii)* the velocity vector varies mainly along the direction orthogonal to the wall and it is possible to neglect velocity derivatives with respect to coordinates along the wall compared to the normal derivative ; *iii)* the pressure is constant in the direction normal to the wall. It is possible to neglect the inertial term when $Re \ll L/h$ (Re is the Reynolds number of the flow). Equation of Navier-Stokes is not valid in a liquid nanolayer because the fluid is strongly inhomogeneous and the elastic stress tensor is not scalar. However, it is possible to adapt the approximation of lubrication for viscous flows in a liquid nanolayer. We are in the case of long wave approximation: $\epsilon = h/L \ll 1$. We denote the velocity by $\mathbf{V} = (u, v, w)$ where (u, v) are the tangential components. In the approximation of lubrication we have : $e = \sup\left(|w/u|\,, |w/v|\right) \ll 1$. The main parts of terms associated with second derivatives of liquid velocity components correspond to $\partial^2 u/\partial z^2$ and $\partial^2 v/\partial z^2$. The density is constant along each stream line ($\dot{\rho} = 0 \Longleftrightarrow div\mathbf{V} = 0$) and iso-density surfaces contain the trajectories. Then, $\partial u/\partial x, \partial v/\partial y$ and $\partial w/\partial z$ have the same order of magnitude and $\epsilon \sim e$. As in Rocard model, we assume that the kinematic viscosity coefficient $\nu = \kappa_2/\rho$ depends only on the temperature.[10] In motion equation, the viscosity term is $(1/\rho)\ \mathrm{div}\ \sigma_v = 2\nu\ [\ \mathrm{div}\ D + D\ \mathrm{grad}\ \{\mathrm{Ln}\,(2\,\kappa_2)\}\ ]$; D grad$\{$Ln $(2\,\kappa_2)\}$ is negligible with respect to div D. In both lubrication and long wave approximations the liquid nanolayer motion verifies

$$\mathbf{\Gamma} + \mathrm{grad}[\,\mu_o(\rho) - \lambda\,\Delta\rho\,] = \nu\,\Delta\mathbf{V} \quad \text{with} \quad \Delta\mathbf{V} \simeq \left[\frac{\partial^2 u}{\partial z^2}, \frac{\partial^2 v}{\partial z^2}, 0\right] \tag{17}$$

In approximation of lubrication, the inertial term is neglected and Eq. (17) separates into tangential and normal components to the solid wall. As in equilibrium, the normal component of Eq. (17) is

$$\frac{\partial}{\partial z}\left[\,\mu_o(\rho) - \lambda\,\Delta\rho\,\right] = 0 \quad \Rightarrow \quad \mu_o(\rho) - \lambda\,\Delta\rho = \mu_o(\rho_b).$$

To each value ρ_b (different of liquid bulk density value ρ_l of the plane interface at equilibrium) is associated a liquid nanolayer thickness h. We can write $\mu_o(\rho) - \lambda\,\Delta\rho = \eta(h)$, where η is such that $\eta(h) = \mu_o(\rho_b)$. For one-dimensional motions colinear to the solid wall (direction $\mathbf{i_o}$ and velocity $u\,\mathbf{i_o}$), the tangential component of Eq. (17) yields :

$$\text{grad}\ \mu_o(\rho_b) = \nu \frac{\partial^2 u}{\partial z^2}\ \mathbf{i_o} \quad \Longleftrightarrow \quad \frac{\partial \mu_o}{\partial \rho_b}\,\frac{\partial \rho_b}{\partial x} = \nu \frac{\partial^2 u}{\partial z^2}. \tag{18}$$

A liquid can slip on a solid wall only at a molecular level.[20] The sizes of solid walls are several orders of magnitude higher than slipping distances which are negligible and kinematic condition at solid walls is the adherence condition ($z = 0 \ \Rightarrow\ u = 0$). From the continuity of fluid tangential stresses through a liquid-vapor interface of molecular size and assuming that vapor viscosity stresses are negligible, we obtain ($z = h, \ \Rightarrow \ \dfrac{\partial u}{\partial z} = 0$). Consequently Eq. (18) implies $\nu\, u = \dfrac{\partial \mu_o}{\partial \rho_b}\,\dfrac{\partial \rho_b}{\partial x}\left(\dfrac{1}{2}\,z^2 - h\,z\right)$. The mean spatial velocity $\overline{u}$ of the liquid in the nanolayer is $\overline{u} = \frac{1}{h}\int_o^h u\; dz$; previous computations yield $\nu\, \overline{\mathbf{u}} = -\dfrac{h^2}{3}\ \text{grad}\ \mu_o(\rho_b)$ with $\overline{\mathbf{u}} = \overline{u}\ \mathbf{i_o}$. Let us remark that

$$\frac{\partial \mu_o(\rho_b)}{\partial x} = \frac{\partial \mu_o}{\partial \rho_b}\,\frac{\partial \rho_b}{\partial h}\,\frac{\partial h}{\partial x} \equiv \frac{1}{\rho_b}\,\frac{\partial P(\rho_b)}{\partial \rho}\,\frac{\partial \rho_b}{\partial h}\,\frac{\partial h}{\partial x}.$$

The pressure P_{v_b} in the vapor bulk is constant along flow motions and $\Pi(h) = P_{v_b} - P(\rho_b)$; consequently, we get $\dfrac{\partial \mu_o(\rho_b)}{\partial x} = -\dfrac{1}{\rho_b}\,\dfrac{\partial \Pi(h)}{\partial h}\,\dfrac{\partial h}{\partial x}$ and

$$\chi_b\ \overline{\mathbf{u}} = \frac{h^2}{3}\ \text{grad}\ \Pi(h). \tag{19}$$

where $\chi_b = \rho_b \nu$ is the liquid kinetic viscosity.

Eq. (19) yields the mean spatial velocity of the isothermal liquid nanolayer as a function of the disjoining pressure gradient. Like as the disjoining pressure depends on the nanolayer thickness, the mean flow velocity is a function of thickness variations along the flow.

Taking into account that in the liquid nanolayer $\rho \simeq \rho_b$, then

$$\left(\int_o^h \rho\ dz\right)\,\overline{\mathbf{u}} \simeq \int_o^h \rho\,\mathbf{u}\ dz$$

and the mean spatial velocity corresponds also to the mean velocity with respect to the mass density.

In shallow water approximation, the equation of continuity yields

$$\frac{\partial}{\partial t}\left(\int_o^h \rho\ dz\right) + \text{div}\left\{\left(\int_o^h \rho\ dz\right)\overline{\mathbf{u}}\right\} = 0$$

and we obtain Eq. (14) in ref. ([21]) associated with $h-$perturbations :

$$\frac{\partial h}{\partial t} + h\ \mathrm{div}\overline{\mathbf{u}} = 0.$$

Thanks to Eq. (19) an equation for $h-$perturbations is :

$$\frac{\partial h}{\partial t} + \frac{h}{3\,\chi_b}\frac{\partial}{\partial x}\left(h^2\frac{\partial}{\partial x}\Pi(h)\right) = 0. \tag{20}$$

Eq. (20) is an *equation of diffusion* in parabolic structure with a good sign of diffusion coefficient associated with stability when $\dfrac{\partial\Pi(h)}{\partial h} < 0.$[17]

Acknowledgments

This paper supported by PRIN 2005 (Nonlinear Propagation and Stability in Thermodynamical Processes of Continuous Media) is dedicated to Prof. Tommaso Ruggeri.

References

1. B. Widom, *Physica A,* **263**, 500 (1999).
2. A.A. Chernov, L.V. Mikheev, *Phys. Rev. Lett,* **60**, 2488 (1988).
3. R. Evans, *Adv. Phys.* **28**, 143 (1979).
4. M.E. Fisher, A.J. Jin, *Phys. Rev. B,* **44**, 1430 (1991).
5. J.S. Rowlinson, B. Widom, *Molecular theory of capillarity,* (Clarendon Press, Oxford, 1984).
6. *Springer handbook of nanotechnology,* Ed. B. Bhushan (Berlin, 2004).
7. H. Nakanishi, M.E. Fisher, *Phys. Rev. Lett.* **49**, 1565 (1982).
8. J.W. Cahn, *J. Chem. Phys.* **66**, 3667 (1977).
9. H. Gouin, *J. Phys. Chem. B* **102**, 1212 (1998).
10. Y. Rocard, *Thermodynamique,* (Masson, Paris, 1967).
11. J. Israelachvili, *Intermolecular and surface forces,* (Academic Press, New York, 1992).
12. P.G. de Gennes, *Rev. Mod. Phys.* **57**, 827 (1985).
13. S. Gavrilyuk, I. Akhatov, *Phys. Rev. E.,* **73**, 021604 (2006).
14. H. Gouin, *Physicochemical Hydrodynamics, B Physics* **174**, 667 (1987).
15. H. Gouin, W. Kosiński, *Arch. Mech.* **50**, 907 (1998).
16. H. Gouin, T. Ruggeri, *Eur. J. Mech. B/fluids,* **24**, 596 (2005).
17. B.V. Derjaguin, N.V. Churaev, V.M. Muller, *Surfaces forces,* (Plenum Press, New York, 1987).
18. H. Gouin, L. Espanet, *C. R. Acad. Sci. Paris* **328**, IIb, 151 (2000).
19. G. K. Batchelor, *An Introduction to Fluid Dynamics,* (Cambridge University Press, 1967).
20. N.V. Churaev, *Colloid. J.* **58**, 681 (1996).
21. V.S. Nikolayev, S.L. Gavrilyuk, H. Gouin, *J. Coll. Interf. Sci.* **302**, 605 (2006).

A PARTICLE METHOD FOR A LOTKA–VOLTERRA SYSTEM WITH NONLINEAR CROSS AND SELF–DIFFUSION

A.M. GRECO AND M. SAMMARTINO

Department of Mathematics and Applications
University of Palermo
Via Archirafi 34, 90123 Palermo, Italy
greco@math.unipa.it marco@math.unipa.it

In this paper we apply a particle method to a system of two non linearly coupled reaction–diffusion equations. Time discretization is based on the Peaceman-Rachford operator splitting scheme. Numerical simulations show pattern formation and the possibility of survival of the dominated species due to segregation.

Keywords: Particle Methods; Nonlinear Diffusion; Reaction–Diffusion; ADI Schemes; Pattern Formation

1. Background

We consider the following system:

$$\partial_t u + \nabla \cdot \boldsymbol{J}_u = f(u,v) \ , \tag{1}$$

$$\partial_t v + \nabla \cdot \boldsymbol{J}_v = g(u,v) \ , \tag{2}$$

where $(\boldsymbol{x},t) \in \Omega \times [0,T]$ with $\Omega \subset \mathbb{R}^n$ and u and v are the densities of the two competing species, whose fluxes are respectively:

$$\boldsymbol{J}_u = -\nabla(c_1 u + a_1 u^2 + b_1 uv) + d_1 u \nabla\phi \ , \tag{3}$$

$$\boldsymbol{J}_v = -\nabla(c_2 v + a_2 v^2 + b_2 uv) + d_2 v \nabla\phi \ . \tag{4}$$

The competition is described by the Lotka-Volterra terms

$$f(u,v) = K\,(\mu_1 - \gamma_{11} u - \gamma_{12} v)\, u \ , \tag{5}$$

$$g(u,v) = K\,(\mu_2 - \gamma_{21} u - \gamma_{22} v)\, v \ . \tag{6}$$

The quantities a_i, b_i, c_i and d_i are non negative constants, $\phi(\boldsymbol{x},t)$ is the ecological potential, and K, μ_i and γ_{ij} are positive constants.

The system (1)-(2) must be supplemented with the initial conditions:

$$u(\boldsymbol{x},0)=u_0(\boldsymbol{x})\,,\qquad v(\boldsymbol{x},0)=v_0(\boldsymbol{x})\,,\tag{7}$$

and with boundary conditions. We shall impose Neumann boundary conditions assuming a zero net flux through the boundary:

$$\boldsymbol{J}_u\cdot\boldsymbol{n}=\boldsymbol{J}_v\cdot\boldsymbol{n}=0\qquad\text{for}\quad \boldsymbol{x}\in\partial\Omega\,,\tag{8}$$

where $\boldsymbol{n}$ is the outgoing normal to the boundary.

The numerical solution of Eqs.(1)-(2) in the 1D case has been tackled by other authors by using a finite–differences[1,2] and finite–elements[3] numerical schemes. Here the numerical method we use is different, in that the spatial discretization is achieved through a ***particle approximation*** of the solution. Time evolution is treated through operator splitting (convection–diffusion terms and reaction terms are treated separately) and an Alternating Direction Implicit (ADI) scheme.

We now give an outline of the diffusion–velocity particle method as introduced by Degond and Mustieles.[4] Consider the equation:

$$\partial_t u+\boldsymbol{\nabla}\cdot(\boldsymbol{V}(\boldsymbol{x},t)u)=0\,.\tag{9}$$

for the scalar quantity $u(\boldsymbol{x},t)$. We seek a particle approximation of the solution in the form:

$$u_N(\boldsymbol{x},t)=\sum_{i=1}^{N}w_i\delta\left(\boldsymbol{x}-\boldsymbol{x}_i(t)\right)\,,\tag{10}$$

where w_i is the mass of the i-th particle, δ is the Dirac distribution. Using the weak formulation of (9) one derives:

$$\dot{\boldsymbol{x}}_i=\boldsymbol{V}(\boldsymbol{x}_i,t)\,.\tag{11}$$

The above characteristic equation is initialized by a particle approximation of the initial datum. To use the method to deal with the diffusion equation $\partial_t u=\Delta u$, one has to recast the equation in the form:

$$\partial_t u-\boldsymbol{\nabla}\cdot\left(u\left(\frac{\boldsymbol{\nabla}u}{u}\right)\right)=0\,.$$

By approximating the δ distribution with the kernel $\zeta_\epsilon(\boldsymbol{x})$ defined as:

$$\zeta_\epsilon(\boldsymbol{x})=\epsilon^{-n}\zeta(\boldsymbol{x}/\epsilon)\qquad\text{with}\qquad\int_{\mathbb{R}^n}\zeta(\boldsymbol{x})d\boldsymbol{x}=1\,,$$

One can write the following approximation of (10):

$$u_N^\epsilon(\boldsymbol{x},t)=\sum_{i=1}^{N}w_i\,\zeta_\epsilon\left(\boldsymbol{x}-\boldsymbol{x}_i(t)\right)\,.\tag{12}$$

Approximating the convective velocity as $\boldsymbol{V} \approx \boldsymbol{V}^{\epsilon} = -\boldsymbol{\nabla} u_N^{\epsilon}/u_N^{\epsilon}$ we are led to solve the characteristic equation in the form:

$$\dot{\boldsymbol{x}}_i = -\frac{\boldsymbol{\nabla} u_N^{\epsilon}(\boldsymbol{x}_i,t)}{u_N^{\epsilon}(\boldsymbol{x}_i,t)} = -\frac{\sum_{j=1}^{N} w_j \, \boldsymbol{\nabla}\zeta_{\epsilon}\,(\boldsymbol{x}_i - \boldsymbol{x}_j)}{\sum_{j=1}^{N} w_j \, \zeta_{\epsilon}\,(\boldsymbol{x}_i - \boldsymbol{x}_j)} \,. \tag{13}$$

Next we shall apply these ideas to our system.

2. Application of the particle method to the Lotka-Volterra system

Let us write the system in the symbolic form

$$\frac{d\,\mathrm{U}(t)}{dt} = F(\mathrm{U}(t)) + G(\mathrm{U}(t))\,, \tag{14}$$

where $F(\mathrm{U})$ is the convection-diffusion part and $G(\mathrm{U})$ is the reactive part.

We solve the above equation using an operator splitting in the form of a Peaceman–Rachford ADI scheme (see e.g. Ref.[5–8] and references therein). The solution at $(m+1)$-th time step, U_{m+1}, is given in terms of the solution at m-th time step, U_m, by:

$$\mathrm{U}_{m+1} = \left(\boldsymbol{I} - \frac{\Delta t}{2}G\right)^{-1}\left(\boldsymbol{I} + \frac{\Delta t}{2}F\right)\left(\boldsymbol{I} - \frac{\Delta t}{2}F\right)^{-1}\left(\boldsymbol{I} + \frac{\Delta t}{2}G\right)\mathrm{U}_m\,.$$

The implicit steps (both reactive and diffusive) are done through iteration using a suitable convergence criterion.[9]

Reaction terms change the mass of the particles. We kept into account this effect through the following formula:

$$W_i^{u,new}(\boldsymbol{x}_i) = W_i^{u,old}(\boldsymbol{x}_i)\left(1 + \frac{u^{new}(\boldsymbol{x}_i) - u^{old}(\boldsymbol{x}_i)}{u^{old}(\boldsymbol{x}_i)}\right)\,.$$

Therefore the mass of the particle relative to a species is changed proportionally to the change of the species at the particle location.

2.1. *The creation of a spatial niche*

An interesting effect due to space dependence and nonlinear diffusion is the survivability of the dominated species in an ecological niche.[9]

We have imposed the following coefficients for the kinetics: the scale factor is $K = 2$, the reproduction coefficients are $\mu_1 = 0.7$ and $\mu_2 = 0.75$, while the competition matrix is:

$$\Gamma = \begin{pmatrix} 0.05 & 0.2 \\ 0.05 & 0.05 \end{pmatrix}\,.$$

With these coefficients the species u is dominated by the species v and without diffusion it would go extinct. However, in some instances, spatial dependence and diffusions effects lead to segregation and the survival of the species u.

I: Repulsive potential.
We have supposed, in the spatial domain $x \in [-\pi, \pi]$, the ecological potential to be of the form:

$$\phi = -\exp\left(-2x^2\right),$$

modelling the tendency, for both species, to escape from places nearby the origin. Moreover the diffusion and transport coefficients are $a_i = 10^{-4}$, $b_1 = 4.1$, $b_2 = 0.3$, $c_i = 0.2$ and $d_i = 2$. The meaning of the above coefficients is that cross–diffusion is much stronger than self–diffusion, while the effect of linear diffusion is moderate. Starting from the initial conditions $u_0(x) = v_0(x) = 3$, we have seen that the solution stabilizes in the configuration shown to the left of Fig.1.
The simulation was performed, with spatial resolution $N = 256$, $M = 8 \cdot 10^4$ time steps, and particle width of $\epsilon = 0.5\sqrt{\Delta x}$. Notice how, in places where v dominates, the v density gets closer to the value 15, which is the value μ_2/γ_{22} predicted by the logistic equation (in absence of species u).

II: Attractive potential.
In the same spatial domain as before we have supposed the ecological potential to be of the form:

$$\phi = \exp\left(-0.15x^4\right).$$

Both species are attracted by a spot close to the origin, and the constants in the above potential regulate the size of the spot. The diffusion and transport coefficients are: $a_1 = c_1 = 0.1$, $b_1 = 2$, $d_1 = 10$, $a_2 = c_2 = 0.04$, $b_2 = 10$, $d_2 = 0.4$. In this case the high transport coefficient d_1 allows species u to build up enough density close to the origin; the high cross–diffusion b_2 prevents species v from invading the niche created by u.

2.2. *Pattern formation driven by cross–diffusion*

Another effect shown by the model is pattern formation. The spatial domain we have considered is $[-\pi, \pi]$, with no ecological potential. We chose $a_1 =$

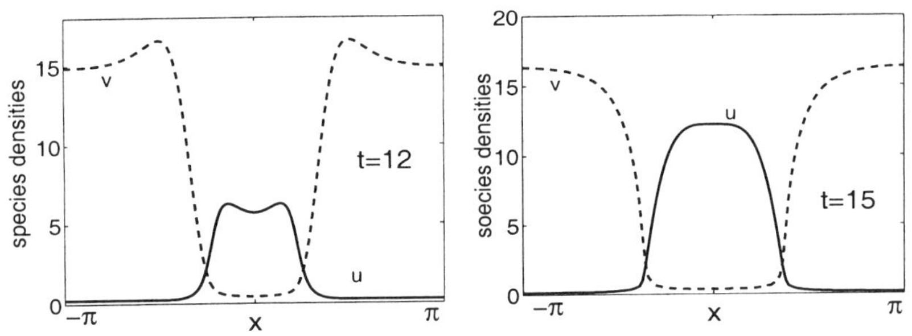

Fig. 1. A striking effect of the cross-diffusion. The dominated species u (solid line) is able to create a spatial niche where it can avoid extinction. The species v (dashed line) is not able to invade the ecological niche. *Left:* Here both species have the tendency to escape from places nearby the origin. This creates an ecological niche for u where it confines itself due to the strong cross–diffusion coefficient $b_1 = 4.1$. We show the final equilibrium configuration. *Right:* Both species are attracted toward the origin, but species v does not invade the niche occupied by u because the cross–diffusion dominates all other effects, $b_2 >> a_2, c_2, d_2$. We show the final equilibrium configuration.

10^{-4}, $a_2 = 0.1$, $c_i = 0.2$, $b_1 = 7.15$, $b_2 = 0.3$, $\mu_1 = 1.2$, $\mu_2 = 1.0$; the competition matrix is:

$$\Gamma = \begin{pmatrix} 0.5 & 0.4 \\ 0.38 & 0.41 \end{pmatrix} .$$

while the scale factor is $K = 100$. As initial data we have chosen a periodic perturbation of the stable equilibrium $(u^{eq}, v^{eq}) \approx (1.73, 0.83)$ predicted by the kinetics:

$$u_0(x) = u^{eq} + 0.02\cos(3x) + 0.01\cos(11x/2),$$
$$v_0(x) = v^{eq} + 0.03\cos(2x) + 0.01\cos(4x) .$$

In Fig.2 one can see how, after a transient that seems to lead the solution toward the constant equilibrium, instabilities develop which eventually evolve into a stationary periodic pattern. This is due to a Turing instability driven by cross–diffusion, see Ref.[10] and references therein.

In this case we have tested the convergence of the numerical scheme at time $t = 2$, using a particle size of $\epsilon = 0.5 \cdot (\Delta x)^{3/4}$, where Δx is the average inter-particle distance. The results show a convergence of $O((\Delta x)^{3/2})$, which is consistent with the chosen particle size and with the fact that the Gaussian kernel gives second order accuracy.

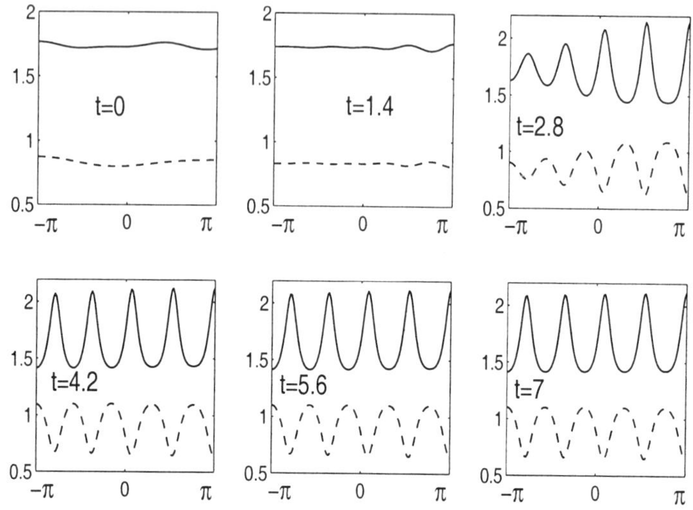

Fig. 2. Pattern formation starting from a perturbation of the equilibrium predicted by the kinetics. The u (solid line) and the v (dashed line) represented are computed using a resolution of $N = 256$ particles.

Acknowledgments

This paper is dedicated to our distinguished colleague and dear friend Prof. Tommaso Ruggeri on the occasion of his sixtieth birthday .

This work has been partially supported by the INDAM and by the PRIN grant *Propagazione non lineare e stabilità nei processi termodinamici del continuo*.

References

1. G. Galiano, M. L. Garzon and A. Jüngel, *RACSAM Rev. R. Acad. Cienc. Exactas Fs. Nat. Ser. A Mat.* **95**, 281 (2001).
2. G. Galiano, M. L. Garzon and A. Jüngel, *Numer. Math.* **93**, 655 (2003).
3. J. Barrett and J. Blowey, *Numer. Math.* **98**, 195 (2004).
4. P. Degond and F.-J. Mustieles, *SIAM J. Sci. Stat. Comput.* **11**, 293 (1990).
5. S. Descombes and M. Massot, *Numer. Math.* **97**, 667 (2004).
6. S. Descombes and M. Ribot, *Numer. Math.* **95**, 503 (2003).
7. K. J. in 't Hout and B. D. Welfert, *Appl. Numer. Math.* **57**, 19 (2007).
8. T. P. Witelski and M. Bowen, *Appl. Numer. Math.* **45**, 331 (2003).
9. G. Gambino, M.C. Lombardo and M. Sammartino, *submitted* (2007).
10. J. N. Flavin and S. Rionero, *IMA J. Appl. Math.* **72**, 540 (2007).

ON SHOCK STRUCTURE IN REACTIVE MIXTURES

M. GROPPI* and G. SPIGA

Dipartimento di Matematica Università di Parma,
Viale G. P. Usberti 53/A - 43100 Parma, Italy
**E-mail: maria.groppi@unipr.it*

The steady shock–wave problem in a chemically reacting gas mixture is addressed at the kinetic level. A consistent BGK model is employed for constructing the shock profiles, and preliminary numerical results are presented.

Keywords: Shock wave structure, Reactive kinetic equations, BGK models

1. Introduction

Wave propagation in a gaseous mixture is a classical challenging problem which has attracted considerable attention from several points of view in recent years.[1,2] The analysis is quite harder for chemical reactions, also due to the relevant exchanges of masses and energies of chemical link. Reactive kinetic theory is a growing subject in the literature, but, to the authors' knowledge, its application to space dependent problems is confined up to now to the study of detonation waves at the Euler level,[3] and to the numerical investigation of the classical Riemann problem.[4] This paper is aimed at moving a step in such direction, and will apply to the steady shock problem a simple kinetic model[5] for a bimolecular reaction $A^1 + A^2 \rightleftharpoons A^3 + A^4$, whose essential features are briefly summarized below. The four species mixture is governed by the integro–differential Boltzmann–like equations

$$\frac{\partial f^i}{\partial t} + \mathbf{v} \cdot \frac{\partial f^i}{\partial \mathbf{x}} = Q^i[\underline{f}] \qquad i = 1, \ldots, 4 \tag{1}$$

where $\underline{f}$ is the vector of the four distribution functions $f^i(\mathbf{x}, \mathbf{v}, t)$. The collision term may be split into its mechanical and chemical parts, accounting for the proper conservations of mass, momentum, and energy, where the latter involves also the bond energies E^i, with energy gap $\Delta E = E^3 + E^4 - E^1 - E^2$ assumed conventionally to be positive. Collision invariants constitute a seven dimensional linear subspace of the continuous functions of $\mathbf{v}$, and represent conservation of momentum, of total (kinetic plus chemical) energy, and of particles in the independent pairs of species $(1,3)(1,4)(2,4)$. The global collision term $\underline{Q}$ obeys an extended version of Boltzmann's lemma in terms of a suitable Lyapunov functional, and collision equilibria are identified as the seven parameter family of local Maxwellians

$$f_M^i(\mathbf{v}) = n^i \left(\frac{m^i}{2\pi KT}\right)^{3/2} \exp\left[-\frac{m^i}{2KT}(\mathbf{v}-\mathbf{u})^2\right] \qquad i = 1,\cdots,4 \tag{2}$$

with $\mathbf{u}$ and T standing for mass velocity and temperature of the mixture, and where number densities n^i must be related by the mass action law

$$\frac{n^1 n^2}{n^3 n^4} = \left(\frac{m^1 m^2}{m^3 m^4}\right)^{3/2} \exp\left(\frac{\Delta E}{KT}\right). \tag{3}$$

The main features of the one–dimensional shock problem have been investigated[6] with reference to Rankine–Hugoniot conditions and entropy inequality. Physical solutions exist only for upstream Mach number greater than unity (supersonic wave front, like for the inert gas), and chemical reaction always proceeds in a well defined direction, namely products (A^3,A^4) increase their concentration fractions across the shock at the expenses of reactants (A^1,A^2). In this work we shall proceed to the determination of the shock profiles by solving numerically the kinetic equations in a suitable relaxation time approximation,[7] whose main features are recalled in the next Section. Then, the BGK model equations are implemented and the stationary shock profile is achieved as time asymptotic limit of actual space-time computations. The Chu decomposition is used to reduce three–dimensional calculations versus the velocity vector to only one scalar dimension, and a suitable second order splitting scheme is used.[4] Preliminary numerical results for illustrative input data are finally presented and briefly commented on in Section 3.

2. Reactive BGK equations

The reactive relaxation model used here originated from an idea introduced for (inert) gas mixtures,[8] in which all typical drawbacks encountered by BGK models for more than one species were overcome. The strategy is very simple, and consists in introducing only one BGK operator for each species, accounting for any type of interaction with whatever other species. The model is constructed in such a way that the i-th BGK operator drifts the distribution function f^i towards a suitable local Maxwellian $\mathcal{M}_i$, with fictitious macroscopic parameters $n_i, \mathbf{u}_i, T_i$ (different from the actual ones, $n^i, \mathbf{u}^i, T^i$) , and such parameters are chosen in such a way that the exact exchange rates for mass, momentum and energy of each species are correctly reproduced. Using such a philosophy, it has been possible to prove that BGK equations retain most of the significant features of the original kinetic equations, in particular, correct collision invariants and conservation equations, local equilibria (2) and mass action law (3). The key point for the practical realization of this BGK approximation, that looks promising a priori and has proved satisfactory in several homogeneous and space-dependent applications, is the availability of the exact Boltzmann exchange rates for mass, momentum, and energy. They are known in closed analytical form for the elastic scattering operator in the collision model of Maxwellian molecules, so that we will stick for simplicity also here to this common option. The situation is even more complicated for the chemical collision integrals. To our knowledge, analytical exchange rates have been evaluated only under an assumption equivalent to Maxwellian molecules,[9] namely microscopic collision frequency for endothermic reaction constant versus the relative speed on its support, and in a physical regime in which mechanical relaxation is significantly faster than the chemical one. Though these restrictions are quite strong (negligible activation energy and slow chemical reactions only), we will adopt them here for this first non–macroscopic approach to the reactive shock wave problem. Model equations for our reactive mixture in one space dimension take the form

$$v_1 \frac{\partial f^i}{\partial x} = \bar{Q}^i[\underline{f}] = \nu_i(\mathcal{M}_i - f^i) \qquad i = 1, \ldots, 4 \tag{4}$$

where

$$\mathcal{M}_i = n_i \left(\frac{m^i}{2\pi K T_i} \right)^{\frac{3}{2}} \exp\left[-\frac{m^i}{2KT_i} (\mathbf{v} - \mathbf{u}_i)^2 \right] \qquad i = 1, \ldots, 4. \tag{5}$$

and ν_i are suitable $\mathbf{v}$-independent (macroscopic) collision frequencies, defined later. The exchange rates underlying the ansatz (4) and relevant to the test functions $(1, m^i\mathbf{v}, m^i v^2/2)$ can be easily evaluated analytically in terms of the considered auxiliary and actual moments. Under the hypotheses above, the same exchange rates for the true[5] mechanical and chemical collision operators can also be evaluated[9] in terms of the actual moments of the distribution functions and of the quantity

$$\mathcal{S} = \nu_{12}^{34} \frac{2}{\sqrt{\pi}} \Gamma\left(\frac{3}{2}, \frac{\Delta E}{KT} \right) \left[n^3 n^4 \left(\frac{m^1 m^2}{m^3 m^4} \right)^{\frac{3}{2}} \exp\left(\frac{\Delta E}{KT} \right) - n^1 n^2 \right], \tag{6}$$

which represents the reaction rate for reactants at thermal equilibrium, and vanishes only when the mass action law is satisfied. Here ν_{12}^{34} is the chemical endothermic microscopic collision frequency and Γ denotes incomplete gamma function. The quite long expressions[9] are skipped here for brevity. Equating rates, one can determine exactly the 20 auxiliary parameters in terms of the actual fields, and close the highly nonlinear, but numerically treatable, equations (4). The choice of the inverse collision times ν_i is suggested by an estimate of the number of collisions.[7] They depend on the constant collision parameters ν_k^{ij} and ν_{12}^{34}, which represent suitable angular integrations of the microscopic collision frequencies.[7] Then, the model kinetic equations have been numerically processed as stated in the Introduction in their time–dependent form by a suitable splitting method,[4] and steady profiles have been obtained as asymptotic time limits. First results are shown below.

3. Results and comments

We present here one of the simplest test case in order to illustrate preliminarily the essential features of the reactive shock structure. We refer to the case of all equal masses (only one type of molecule, but in different energy states, no mass exchange) and of all equal elements of the scattering collision matrices. All such quantities, as well as the energy gap ΔE, may be then scaled to unity, and we let the strength of the chemical collision frequency vary, focusing the attention on the effects of the reactive term. Upstream conditions are determined as $n_- = 1, u_- = 1.5791$, with chemical composition given by the string $(0.25, 0.3655, 0.25, 0.1345)$ of the concentration fractions $\chi_-^i = n_-^i/n_-$. This implies a temperature $T_- = 1.0003$, and a Mach number $Ma_- = 1.2316$. Rankine-Hugoniot conditions define uniquely the downstream equilibrium[6] as $n_+ = 1.3446, u_+ = 1.1742$, while concentrations χ_+^i are given by the string $(0.24, 0.3555, 0.26, 0.1445)$; we have consequently $T_+ = 1.2192$. Figures 1 to 3 show the results for a reaction characterized by $\nu_{12}^{34} = 0.2$. The initial condition for the numerical solution corresponds to a smoothed step at $x = 0$ between upstream and downstream distributions, and is plotted as a dashed line. Profiles correspond to the asymptotic steady states reached from such initial data, but of course may be shifted along the x-axis due to invariance under translation of the stationary problem. We report in particular the steady shock profiles for the various densities n^i, and for velocity u and temperature T. In order to illustrate non-equilibrium conditions in the shock region, we show also the deviations of the species temperatures from the global temperature. In

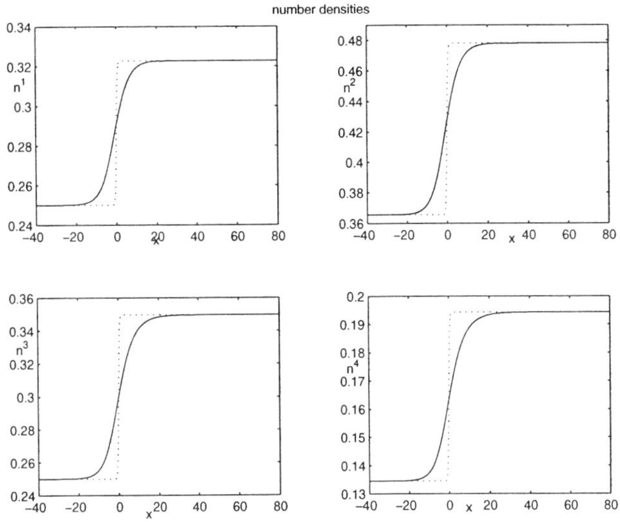

Fig. 1. Spatial distribution of number densities $n^i, i = 1, \ldots, 4$ in the case $\nu_{12}^{34} = 0.2$ (dotted line: initial data).

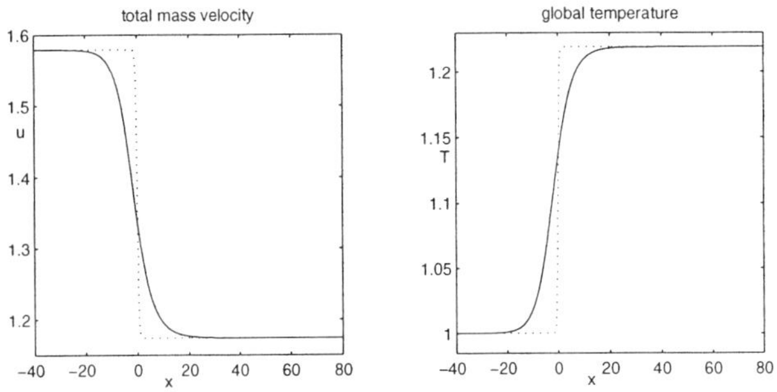

Fig. 2. Spatial distribution of total mass velocity u (left) and global temperature T (right) in the case $\nu_{12}^{34} = 0.2$ (dotted line: initial data).

order to see the impact on the described structure of the strength of the chemical reaction, we make it slower by decreasing the reactive collision frequency to $\nu_{12}^{34} = 0.0333$. The most significant variations of the previous scenario can be seen in figures 4 and 5, and will be matter of future investigation and comparison to hydrodynamic results.[1] In figure 4, one can observe presence of overshooting in the profile of some densities, while figure 5 shows, as a measure of chemical non–equilibrium, the scaled reaction rate for products (the negative of the square bracket in (6)) in the two cases. It

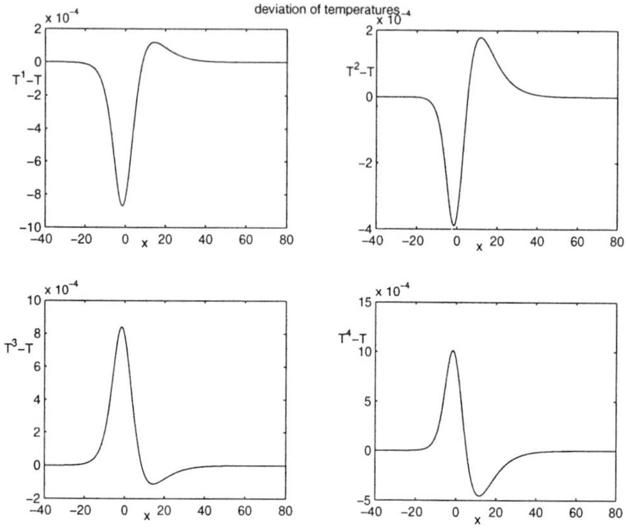

Fig. 3. Difference between single-species and global temperatures in the case $\nu_{12}^{34} = 0.2$.

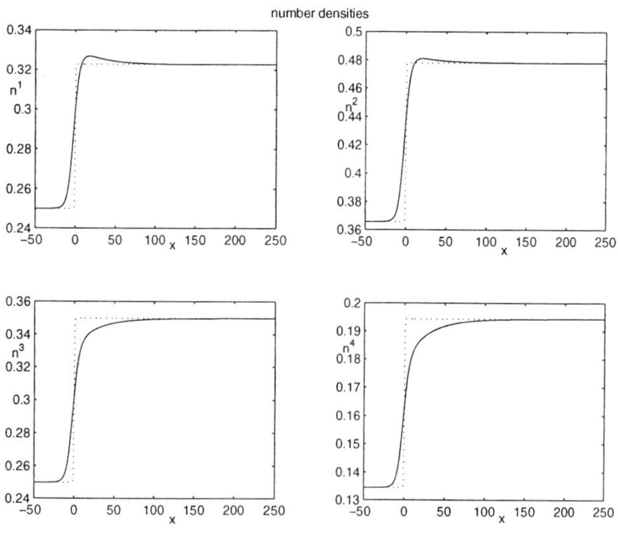

Fig. 4. Spatial distribution of number densities $n^i, i = 1, \ldots, 4$ in the case $\nu_{12}^{34} = 0.0333$ (dotted line: initial data).

is evident for the slower reaction a sensible increase of the shock thickness, and the occurrence of a fat and slowly–damped tail, clear indication of the

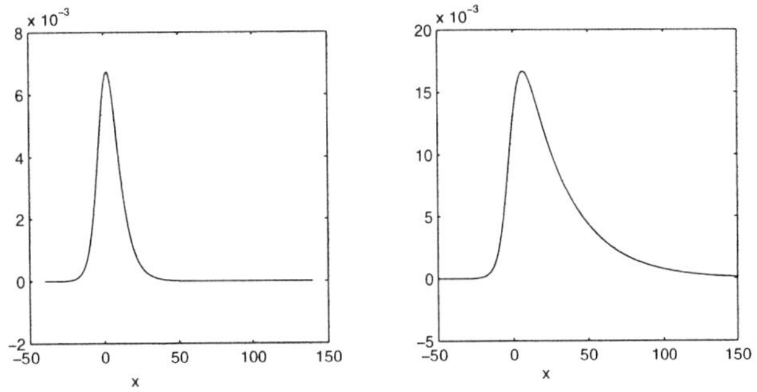

Fig. 5. Opposite of the (scaled) chemical source term versus x in the cases $\nu_{12}^{34} = 0.2$ (left) and $\nu_{12}^{34} = 0.0333$ (right).

singular limit expected for vanishing ν_{12}^{34}. In fact, for given upstream conditions, the reactive downstream state is fixed and independent from the strength of the reaction, whereas in the inert limit it changes abruptly, due to the different nature of Rankine-Hugoniot conditions.

Acknowledgments

This work was performed in the frame of the activities sponsored by MIUR (Project "Nonconservative binary interactions in various types of kinetic models"), by INdAM, by GNFM, and by the University of Parma. Authors are gratefully indebted to prof. K. Aoki for extensive and enlightening discussions on the subject of the present paper.

References

1. Ya. B. Zel'dovich and Yu. P.Raizer, *Physics of Shock Waves and High-Temperature Hydrodynamic Phenomena*, Dover, Mineola, NY, 2002.
2. T. Ruggeri, in *New Trends in Mathematical Physics*, World Sci. Publ., Singapore, 205 (2004).
3. F. Conforto, R. Monaco, F. Schürrer, and I. Ziegler, *J. Phys. A* **36**, 5381 (2003).
4. A. Aimi, M. Diligenti, M. Groppi, and C. Guardasoni, *Europ. J. Mech./B Fluids* **26**, 455 (2007).
5. A. Rossani and G. Spiga, *Physica A* **272**, 563 (1999).
6. M. Groppi, G. Spiga, S. Takata, The steady shock problem in reactive gas mixtures, *Bull. Inst. Math. Acad. Sin. (N.S.)*, to appear (2007).
7. M. Groppi and G. Spiga, *Phys. of Fluids* **16**, 4273 (2004).
8. P. Andries, K. Aoki, and B. Perthame, *J. Stat. Phys.* **106**, 993 (2002).
9. M. Bisi, M. Groppi, and G. Spiga, *Continuum Mech. Thermodyn.* **14**, 207 (2002).

TRANSPORT PROPERTIES OF CHEMICALLY REACTING GAS MIXTURES*

GILBERTO M. KREMER

Departamento de Física, Universidade Federal do Paraná, Brazil
E-mail: kremer@fisica.ufpr.br

The non-equilibrium effects arising on reactive systems described by Boltzmann equations are examined for different chemical processes. First, the influence of a bimolecular reaction $A_1 + A_2 \rightleftharpoons A_3 + A_4$ on transport coefficients and its trend to equilibrium in the hydrogen-chlorine system are investigated. Then, the effects of a single symmetric reaction $A + A \rightleftharpoons B + B$ on both sound speed propagation and light scattering are analyzed. Finally, relativistic corrections to Arrhenius law are also determined.

Keywords: Boltzmann equation; Chemical reactions; Trend to equilibrium; Light scattering; Sound propagation; Relativistic reaction rate.

1. Introduction

Chemical reactions play an important role in either physical chemistry or their applications to engineering problems, chemical technologies and other processes in both physics and chemistry. Since the pioneer works of Prigogine and co-workers,[1,2] in the early fifties, on kinetic theory of chemically reacting gas mixtures, several researchers have investigated these kind of mixtures by using the Boltzmann equation.

In this work chemically reacting systems are analyzed within the framework of the Boltzmann equation. For a simple reversible bimolecular reaction of type $A_1+A_2 \rightleftharpoons A_3+A_4$ it is shown how the chemical reactions have influence on the transport coefficients and how such gas mixture evolves with time from a non-equilibrium to an equilibrium state. For a simple symmetric reaction of the type $A + A \rightleftharpoons B + B$ an analysis of the influence of chemical reactions on both wave propagation and light scattering is investigated. Furthermore, for a single bimolecular reaction the relativistic

*Dedicated to Professor Tommaso Ruggeri on the occasion of his sixtieth birthday.

corrections to Arrhenius law are determined.

2. Transport coefficients for a single bimolecular reaction

For a gaseous mixture undergoing a simple reversible bimolecular reaction of the type $A_1 + A_2 \rightleftharpoons A_3 + A_4$, it is considered that a state – in the phase space characterized by the positions and velocities of the molecules – is defined by the set of one-particle distribution functions, $f_\alpha \equiv f(\mathbf{x}, \mathbf{c}_\alpha, t)$ for each $\alpha = 1$ to 4, such that $f_\alpha d\mathbf{x}d\mathbf{c}_\alpha$ gives at time t the number of α particles in the volume element $d\mathbf{x}d\mathbf{c}_\alpha$ around the position $\mathbf{x}$ and the velocity $\mathbf{c}_\alpha$. Furthermore, each distribution function f_α is assumed to satisfy a Boltzmann equation of the form

$$\frac{\partial f_\alpha}{\partial t} + c_i^\alpha \frac{\partial f_\alpha}{\partial x_i} = \sum_{\beta=1}^{4} \int \left(f'_\alpha f'_\beta - f_\alpha f_\beta\right) g_{\beta\alpha} \sigma_{\beta\alpha} d\Omega_{\beta\alpha} d\mathbf{c}_\beta + \mathcal{Q}_\alpha^R, \qquad (1)$$

where external body forces are not taken into account. The first term on the right-hand side of Eq. (1) is related to elastic collisions where the primes denote post-collision velocities, $\sigma_{\alpha\beta}$ is the differential elastic cross section, $g_{\beta\alpha} = |\mathbf{c}_\beta - \mathbf{c}_\alpha|$ a relative velocity and $d\Omega_{\beta\alpha}$ an element of solid angle. The second one, $\mathcal{Q}_\alpha^R$, refers to the inelastic collisions due to chemical reactions and is given by

$$\mathcal{Q}_{1(2)}^R = \int \left[f_3 f_4 \left(\frac{m_1 m_2}{m_3 m_4}\right)^3 - f_1 f_2 \right] \sigma_{12}^\star g_{21} d\Omega d\mathbf{c}_{2(1)}, \qquad (2)$$

with a similar expression for $\mathcal{Q}_{3(4)}^R$ by changing (1,2) with (3,4). The quantities $\sigma_{12}^\star$ and $\sigma_{34}^\star$ are differential reactive cross sections for forward and backward reaction, respectively, and m_α ($\alpha = 1$ to 4) denotes the mass of a molecule of constituent α. The relationship between $\sigma_{12}^\star$ and $\sigma_{34}^\star$ is given by micro-reversibility principle: $\sigma_{34}^\star/\sigma_{12}^\star = (m_{12}g_{21})^2/(m_{34}g_{43})^2$, where $m_{\alpha\beta} = m_\alpha m_\beta/(m_\alpha + m_\beta)$ is a reduced mass. Moreover, the conservation law of total energy can be written in terms of the relative velocities as $m_{12}g_{21}^2 = m_{34}g_{43}^2 + 2E$, where $E = \varepsilon_3 + \varepsilon_4 - \varepsilon_1 - \varepsilon_2$ is the binding energy difference between the products and reactants with ε_α denoting the formation energy of constituent α.

The deviation of the system from the chemical equilibrium is characterized by the affinity $\mathcal{A}$, which is written in terms of the equilibrium n_α^{eq} and non-equilibrium n_α particle number densities as

$$\mathcal{A} = kT \ln\left(\frac{n_1 n_2 n_3^{\text{eq}} n_4^{\text{eq}}}{n_3 n_4 n_1^{\text{eq}} n_2^{\text{eq}}}\right), \qquad \text{with} \qquad \mathcal{A}^{\text{eq}} = 0. \qquad (3)$$

In this work it is assumed elastic cross-sections of rigid spheres, namely, $\sigma_{\alpha\beta} = d_{\alpha\beta}^2/4$, where $d_{\alpha\beta} = (d_\alpha + d_\beta)/2$ refers to a mean diameter. For the reactive cross-sections the line-of-centers energy model is adopted, i.e.,

$$\sigma_{\alpha\beta}^\star = 0, \text{ for } \gamma_{\alpha\beta} \leq \epsilon_\sigma^\star, \text{ and } \sigma_{\alpha\beta}^\star = d_\sigma^2\left(1 - \epsilon_\sigma^\star/\gamma_{\alpha\beta}\right)/4, \text{ for } \gamma_{\alpha\beta} > \epsilon_\sigma^\star. \quad (4)$$

Above, $\gamma_{\alpha\beta} = m_{\alpha\beta} g_{\beta\alpha}^2/2kT$ denotes the relative translational energy, $\epsilon_\sigma^\star = \epsilon_\sigma/kT$ the activation energy in units of kT, and the index σ assumes either the value $+1$ for the reactants ($\alpha = 1, 2$), or -1 for the products ($\alpha = 3, 4$) of the reaction. The relationship between the forward and backward activation energies reads $\epsilon_{-1}^\star = \epsilon_1^\star - E/kT$. Moreover, d_σ is a reactive collision diameter which is connected with the elastic one by $d_\sigma = s_\sigma\, d_{\alpha\beta}$, where s_σ is the steric factor. The forward and backward steric factors are related by $s_{-1} = \sqrt{m_{12}/m_{34}}(d_{12}/d_{34})s_1$ with $0 \leq s_1 \leq 1$.

The constitutive quantities – that appear in the balance equations of a eight field theory characterized by partial particle number densities n_α, velocity v_i and temperature T of the mixture – are: reaction rate densities τ_α, diffusion velocities u_i^α, pressure tensor p_{ij} and heat flux q_i of the mixture. They are defined as follows

$$\tau_\alpha = \int \mathcal{Q}_\alpha^R d\mathbf{c}_\alpha, \qquad q_i = \sum_{\alpha=1}^{4}\left(\int \frac{1}{2} m_\alpha \xi_\alpha^2 \xi_i^\alpha f_\alpha d\mathbf{c}_\alpha + n_\alpha \varepsilon_\alpha u_i^\alpha\right), \quad (5)$$

$$u_i^\alpha = \frac{1}{n_\alpha}\int \xi_i^\alpha f_\alpha d\mathbf{c}_\alpha, \qquad p_{ij} = \sum_{\alpha=1}^{4}\int m_\alpha \xi_i^\alpha \xi_j^\alpha f_\alpha d\mathbf{c}_\alpha. \quad (6)$$

In the above equations $\xi_i^\alpha = c_i^\alpha - v_i$ is the peculiar velocity and the second term on the right-hand side of the heat flux refers to the transport of formation energy due to diffusion.

In order to have the constitutive equations, one has to determine the non-equilibrium distribution functions from the four coupled Boltzmann equations (1). Here it is analyzed the case where the system is close to chemical equilibrium so that the elastic and inelastic frequencies are of the same order. This condition refers to the last stage of a chemical reaction where the affinity is a small quantity, i.e., $|\mathcal{A}| < 1$. For that end the Chapman-Enskog method is applied with the distributions functions written as

$$f_\alpha = f_\alpha^M\left(1 + \Phi_\alpha\right), \qquad f_\alpha^M = n_\alpha\left(\frac{m_\alpha}{2\pi kT}\right)^{3/2}\exp\left(\frac{-m_\alpha \xi_\alpha^2}{2kT}\right), \quad (7)$$

where f_α^M refers to a Maxwellian distribution and Φ_α is a small deviation from equilibrium. It turns out that the deviation Φ_α is a function of the following thermodynamic forces:

$$\mathcal{A}, \quad \left(\frac{\partial v_i}{\partial x_j}+\frac{\partial v_j}{\partial x_i}-\frac{2}{3}\frac{\partial v_r}{\partial x_r}\delta_{ij}\right), \quad \frac{\partial T}{\partial x_i}, \quad d_i^\alpha=\frac{1}{p}\frac{\partial p_\alpha}{\partial x_i}-\frac{\varrho_\alpha}{\varrho p}\frac{\partial p}{\partial x_i}, \tag{8}$$

which are identified as affinity, deviator of the gradient of velocity, gradient of temperature and gradient of particle number densities, respectively.

From a coupled system of four linear integral equations for each thermodynamic force one can determine the deviation Φ_α and obtain the following laws for the constitutive quantities:

REACTION RATE LAW

$$\tau_1 = n_1^{eq} n_2^{eq}\left(\kappa_1+\kappa_{-1}\right)\frac{\mathcal{A}}{kT}, \tag{9}$$

where κ_1 and κ_{-1} are the forward and backward rate constants, respectively. The first approximation of κ_1 and κ_{-1} correspond to Arrhenius law

$$\kappa_1^{(0)}=s_1^2 d_{12}^2\sqrt{\frac{8\pi kT}{m_{12}}}\exp\left(-\frac{\epsilon_1}{kT}\right), \quad \kappa_{-1}^{(0)}=s_{-1}^2 d_{34}^2\sqrt{\frac{8\pi kT}{m_{34}}}\exp\left(-\frac{\epsilon_{-1}}{kT}\right). \tag{10}$$

FICK LAW

$$d_i^\alpha+\kappa_T^\alpha\frac{\partial \ln T}{\partial x_i}=-\sum_{\beta=1}^{4}\frac{x_\alpha x_\beta}{D_{\alpha\beta}}\left(u_i^\alpha-u_i^\beta\right), \tag{11}$$

where $D_{\alpha\beta}$ denote the diffusion coefficients, κ_T^α the thermal-diffusion ratios and $x_\alpha=n_\alpha/n$ – with $n=\sum_{\alpha=1}^4 n_\alpha$ – the molar fractions.

FOURIER LAW

$$q_i=-\tilde{\lambda}\frac{\partial T}{\partial x_i}-\sum_{\beta=1}^{4}\tilde{D}_\beta d_i^\beta, \qquad \lambda=\tilde{\lambda}-\sum_{\alpha=1}^{4}\tilde{D}_\alpha\kappa_T^\alpha, \tag{12}$$

where λ is the coefficient of thermal conductivity and $\tilde{D}_\alpha$ refers to a coefficient of cross-effect, the so-called diffusion-thermal coefficient.

NAVIER-STOKES LAW

$$p_{ij}=p\delta_{ij}-\eta\left(\frac{\partial v_i}{\partial x_j}+\frac{\partial v_j}{\partial x_i}-\frac{2}{3}\frac{\partial v_r}{\partial x_r}\delta_{ij}\right), \tag{13}$$

where η denotes the coefficient of shear viscosity and $p=\sum_{\alpha=1}^4 n_\alpha kT$ the pressure of the mixture.

To observe the effect of the chemical reactions on the transport coefficients the reversible reaction $H_2+Cl \rightleftharpoons HCl+H$ is analyzed. For this reaction it is known the experimental values for the Arrhenius coefficient, forward and backward activation energies and reaction heat, so that the

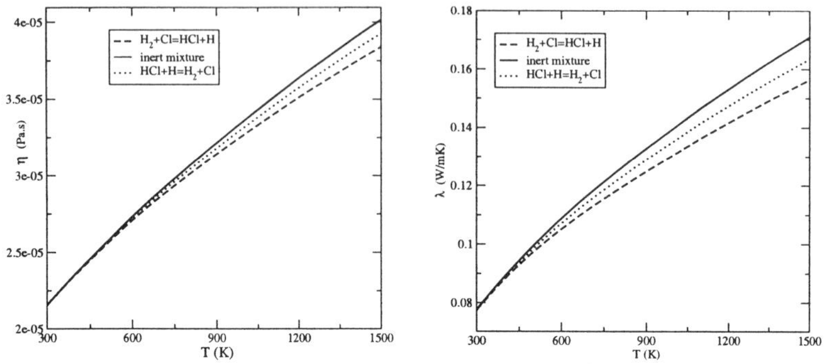

Fig. 1. Shear viscosity η (left frame) and thermal conductivity λ (right frame) as functions of the temperature T for the reaction $H_2 + Cl \rightleftharpoons HCl + H$.

transport coefficients do depend only on the temperature of the mixture as well as on the molar fraction of the constituents and for more details one is referred to the works.[3,4] In figure 1 the coefficients of shear viscosity and thermal conductivity are plotted as function of the temperature when $x_1 = x_2$ and $x_3 = x_4$. One can infer that: (i) for both coefficients the effects of chemical reactions are negligible at lower temperatures; (ii) each coefficient becomes smaller than the corresponding one in the inert case as the temperature increases and (iii) the coefficients are more affected at high temperatures for endothermic reactions, which for the reaction $H_2 + Cl \rightleftharpoons HCl + H$ is the one which proceeds from left to right.

3. Trend to equilibrium of a single bimolecular reaction

The trend to equilibrium of the reaction $H_2 + Cl \rightleftharpoons HCl + H$ can be determined by considering that the constituents are displaced from an equilibrium state – characterized by constant temperature and particle number densities – by small deviations which do depend only on time, i.e.

$$n_\alpha(t) = n_\alpha^{\rm eq}[1 + \overline{n}_\alpha(t)], \qquad |\overline{n}_\alpha(t)| < 1, \qquad (\alpha = 1, 2, 3, 4) \tag{14}$$

$$T(t) = T_{\rm eq}[1 + \overline{T}(t)], \qquad |\overline{T}(t)| < 1. \tag{15}$$

The search for the spatially homogeneous solutions is based on the coupled linearized system of differential equations for $\overline{n}_\alpha$ and $\overline{T}$, which reads

$$x_\alpha^{\rm eq}\frac{d\overline{n}_\alpha}{dt} = \nu_\alpha x_\alpha^{\rm eq}\zeta_{\alpha\gamma}^R(\overline{n}_1 + \overline{n}_2 - \overline{n}_3 - \overline{n}_4) \equiv \tau_\alpha, \qquad \alpha = 1, 2, 3, 4, \tag{16}$$

$$\frac{d\overline{T}}{dt} = -\frac{2}{3}x_1^{\rm eq}\zeta_{12}^R E^\star(\overline{n}_1 + \overline{n}_2 - \overline{n}_3 - \overline{n}_4), \tag{17}$$

where it was introduced the reactive collision frequencies:

$$\zeta_{12}^{R} = n_2^{\rm eq}\sqrt{\frac{8\pi kT}{m_{12}}}\,{\rm e}^{-\epsilon_1^\star}(s_1 d_{12})^2, \qquad \zeta_{34}^{R} = \frac{n_1^{\rm eq}}{n_3^{\rm eq}}\zeta_{12}^{R}. \tag{18}$$

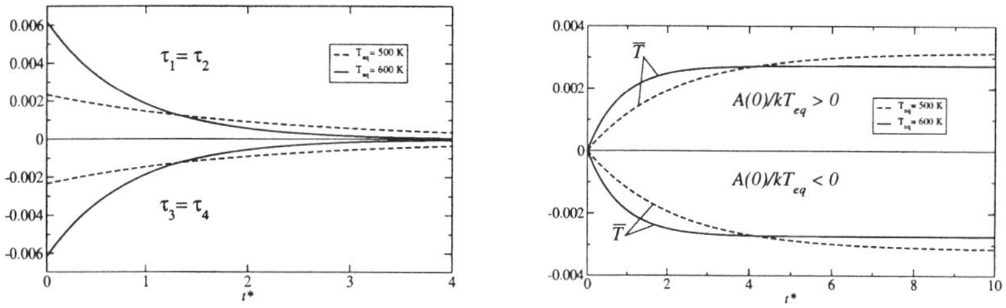

Fig. 2. Reaction rate densities (left frame) and temperature perturbations (right frame) versus time for $T_{\rm eq} = 500$ K (dashed lines) and $T_{\rm eq} = 600$ K (solid lines).

The system of field equations (16) and (17) was solved for given initial conditions and for more details one is referred to the work.[5] In figure 2 the time evolution of the reaction rate densities τ_α and temperature perturbations $\overline{T}$ are represented for the reaction $H_2 + Cl \rightleftharpoons HCl + H$. For this reaction the conditions are those of the previous section. One can infer from the left frame of figure 2 that the reaction rate densities become larger by increasing the temperature but decrease more rapidly with time and, as was expected, tend to zero with time. Furthermore, from the right frame of figure 2 one can conclude that for positive (negative) affinities the temperature of the mixture increases (decreases) and an exothermic (endothermic) reaction happens.

4. Light scattering and sound propagation in single symmetric reactions

Symmetric reactions of type $A + A \rightleftharpoons B + B$ were also analyzed within the framework of Boltzmann equation in the works.[6,7] This kind of binary reaction is also called isomerization, e.g., the reactions $CH_3NC \rightleftharpoons CH_3CN$ and cyclopropane $\rightleftharpoons$ propene. For theses reactions the molecules of both constituents have equal masses $m_A = m_B = m$ and the law of mass action implies that the binding energy difference is connected to the molar concentration through $E/kT = 2\ln\left[x_A/(1 - x_A)\right]$.

Here the search for solutions $n_\alpha(\mathbf{x},t) = n_0 x_\alpha + \bar{n}_\alpha(\mathbf{x},t)$, $(\alpha = A, B)$ $v_i(\mathbf{x},t) = \bar{v}_i(\mathbf{x},t)$ and $T(\mathbf{x},t) = T_0 + \bar{T}(\mathbf{x},t)$ – that represent small perturbations about an equilibrium state of constant partial particle densities and temperature, and vanishing velocity – is analyzed.

The linearized system of field equations for $\bar{n}_A$, $\bar{n}_B$, $\bar{v}_i$ and $\bar{T}$ reads

$$\frac{\partial \bar{n}_\alpha}{\partial t} + n_0 x_\alpha \frac{\partial \bar{v}_i}{\partial x_i} \mp D\left[x_B \frac{\partial^2 \bar{n}_A}{\partial x_i \partial x_i} - x_A \frac{\partial^2 \bar{n}_B}{\partial x_i \partial x_i} + \frac{n_0}{T_0} k_T \frac{\partial^2 \bar{T}}{\partial x_i \partial x_i}\right]$$
$$\mp 2\frac{\kappa}{n_0}\left(\frac{\bar{n}_B}{x_B} - \frac{\bar{n}_A}{x_A}\right) = 0, \qquad \alpha = A(-),\ B(+)$$
$$\frac{\partial \bar{v}_i}{\partial t} + \frac{kT_0}{\varrho_0}\frac{\partial \bar{n}_A}{\partial x_i} + \frac{kT_0}{\varrho_0}\frac{\partial \bar{n}_B}{\partial x_i} + \frac{k}{m}\frac{\partial \bar{T}}{\partial x_i} = \frac{\eta}{\varrho_0}\left[\frac{\partial^2 \bar{v}_i}{\partial x_j \partial x_j} + \frac{1}{3}\frac{\partial}{\partial x_i}\left(\frac{\partial \bar{v}_j}{\partial x_j}\right)\right],$$
$$\frac{\partial \bar{T}}{\partial t} - \frac{2}{3}\frac{\lambda}{n_0 k}\frac{\partial^2 \bar{T}}{\partial x_i \partial x_i} - \frac{2}{3}\frac{T_0}{n_0}\frac{k_T}{x_A x_B} D\left[x_B \frac{\partial^2 \bar{n}_A}{\partial x_i \partial x_i} - x_A \frac{\partial^2 \bar{n}_B}{\partial x_i \partial x_i}\right.$$
$$\left. + \frac{n_0}{T_0} k_T \frac{\partial^2 \bar{T}}{\partial x_i \partial x_i}\right] + \frac{2}{3} T_0 \frac{\partial \bar{v}_i}{\partial x_i} = \frac{2}{3}\frac{\kappa E}{k n_0^2}\left(\frac{\bar{n}_A}{x_A} - \frac{\bar{n}_B}{x_B}\right). \tag{19}$$

In Eq. $(19)_1$ the minus sign refers to the constituent A whereas the plus sign to the constituent B. All transport coefficients (D, k_T, κ, η and λ) in the above equations were determined from a coupled system of Boltzmann equations for the constituents.[6]

From the linearized system of partial differential equations (19) one can analyze two problems, namely, the sound propagation and the light scattering. For the problem of sound propagation one can obtain the phase velocity v and the absorption coefficient α as functions of the angular frequency of the wave ω. For the problem of light scattering one can determine the so-called dynamic structure factor $S(\mathbf{q},\omega)$ which is a function of both the scattering vector $\mathbf{q}$ and of the shift in the angular frequency ω in the scattering process.

In the left frame of figure 3 the reduced reciprocal phase velocity c/v and the reduced absorption coefficient $\alpha c/\omega$ are plotted as functions of the reduced frequency $\omega\tau$ for an activation energy $\epsilon_1 = 2.0$. The quantities $c = \sqrt{5kT_0/3m}$ and τ denote the adiabatic sound speed and the mean free time, respectively. One can infer from this figure that chemical reactions have influence on the phase velocity and absorption coefficient. The effect is more pronounced for endothermic reactions ($x_A = 0.7$) than for exothermic reactions ($x_A = 0.3$) indicating a decrease of the phase velocity. The dynamic structure factor $S(\mathbf{q},\omega)$ is plotted in the right frame of figure 3 as function of the reduced frequency ω/cq for a uniformity parameter $y = 1/\tau cq = 7.0$. Here the effect of endothermic reactions is also more pro-

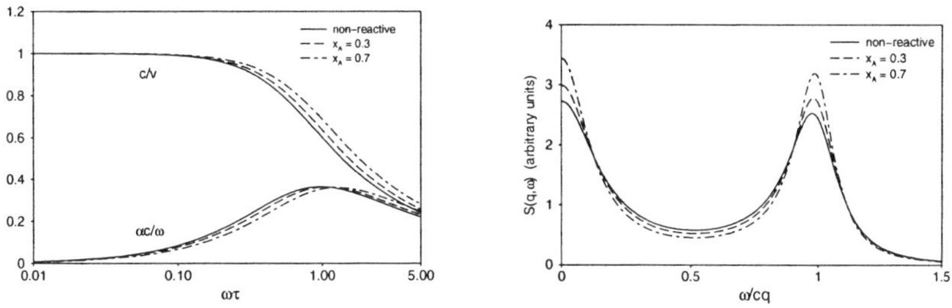

Fig. 3. Left frame: sound propagation; right frame: light scattering.

nounced than the one of exothermic reactions, and one can conclude that chemical reactions reduce the width of the Rayleigh and Brillouin lines and shift the Brillouin lines outwards the center of the spectrum.

5. Relativistic reaction rate coefficient

In this section the relativistic corrections on the reaction rate coefficient for a single bimolecular chemical reaction $A_1 + A_2 \rightleftharpoons A_3 + A_4$ is investigated. The relativistic reaction rate coefficient is given by[8]

$$\kappa = \frac{1}{n_1 n_2} \int f_1 f_2 \sqrt{(\mathbf{v}_1 - \mathbf{v}_2)^2 - \frac{1}{c^2}(\mathbf{v}_1 \times \mathbf{v}_2)^2} \sigma_{12}^{\star} d\Omega' d^3 p_1 d^3 p_2. \qquad (20)$$

Here only the relativistic corrections to the Arrhenius law are determined. The Arrhenius law is characterized by the equilibrium Maxwell-Jüttner distribution function

$$f_a^{(0)} = \frac{n_a}{4\pi k T m_a^2 c K_2(\zeta_a)} \exp\left(\frac{-U_\alpha p_a^\alpha}{kT}\right), \qquad a = 1 \ldots 4. \qquad (21)$$

Above, U^α – such that $U^\alpha U_\alpha = c^2$ – is the four-velocity and $K_n(\zeta_a)$ are Bessel functions of second kind with $\zeta_a = m_a c^2/kT$ denoting the ratio between the rest energy of a particle $m_\alpha c^2$ and the thermal energy of the gas kT.

To obtain the Arrhenius law the following differential cross-section was proposed[9]

$$\begin{aligned} \sigma_{12}^{\star} &= 0, && \text{for} \quad \gamma = Q^2/(8m_{12}) < \varepsilon, \\ \sigma_{12}^{\star} &= \sigma_{12}^{\mathrm{R}}(T)\,(1 - \varepsilon/\gamma)\,W_{12}, && \text{for} \quad \gamma = Q^2/(8m_{12}) > \varepsilon, \end{aligned} \qquad (22)$$

where ε denotes the activation energy in units of kT, Q is the modulus of the relative momentum four-vector and W_{12} is a factor for mathematical convenience so that the integrals of the Bessel functions can be performed analytically, and for more details one is referred to the work.[9] In the non-relativistic limit where $\zeta_1 = m_1c^2/kT \gg 1$, and $\zeta_2 = m_2c^2/kT \gg 1$ the differential cross-section (22) reduces to the non-relativistic line-of-centers model (4).

With the above differential cross-section the reaction rate coefficient (20) becomes

$$\kappa = \frac{\pi c\sigma_{12}^{\rm R}(T)}{\zeta_1^2\zeta_2^2K_2(\zeta_1)K_2(\zeta_2)(\zeta_1+\zeta_2)}\Bigg\{x_0^5K_5(x_0) - 6x_0^4K_4(x_0) - \Bigg[2(\zeta_1^2+\zeta_2^2) + \frac{8\varepsilon m_1m_2(\zeta_1+\zeta_2)}{(m_1+m_2)^2}\Bigg]x_0^3K_3(x_0) + (\zeta_1^2-\zeta_2^2)^2\Bigg[x_0K_1(x_0)+2K_0(x_0)\Bigg]\Bigg\}. \quad (23)$$

Above, it was introduced the following abbreviations

$$x_0=\sqrt{\zeta_1^2+\zeta_2^2+Z_{12}+\sqrt{(Z_{12}+2\zeta_1^2)\,(Z_{12}+2\zeta_2^2)}},\quad Z_{12}=\frac{4\varepsilon(\zeta_1+\zeta_2)m_1m_2}{(m_1+m_2)^2}. \quad (24)$$

In the non-relativistic limiting case, characterized by $\zeta = \zeta_1 + \zeta_2 = (m_1 + m_2)c^2/(kT) \gg 1$, it follows

$$\kappa=4\sigma_{12}^{\rm R}\sqrt{\frac{8\pi kT}{m_{12}}}e^{-\varepsilon}\Bigg\{1-\underline{\Bigg[\frac{\varepsilon^2}{2}-\frac{3\varepsilon}{2}-\frac{17}{8}-\frac{m_1^2+m_2^2}{m_1m_2}\left(\frac{\varepsilon^2}{2}+\varepsilon-\frac{7}{8}\right)\Bigg]\frac{1}{\zeta}}+\dots\Bigg\}, \quad (25)$$

where the underlined term refers to the first relativistic correction to Arrhenius law.

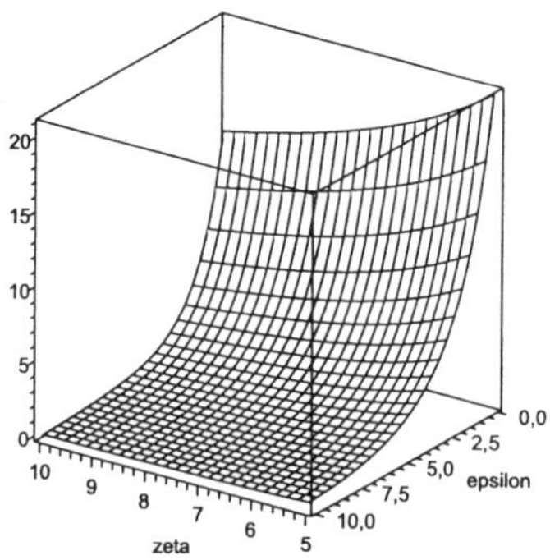

Fig. 4. Reaction rate coefficient vs. ε and ζ for $m_1 = m_2$.

In figure 4 the reaction rate coefficient $\kappa/(c\sigma_{12}^{\mathrm{R}}(T))$ is plotted as function of both the activation energy ε and of the parameter ζ for $m_1 = m_2$. The behavior of the reaction rate coefficient in this figure shows that it decreases by increasing the activation energy and increases when the relativistic effects are more predominant for the mixture.

6. Conclusions

To sum up [a] **transport coefficients:** (i) become smaller than the corresponding ones in the inert case as the temperature increases; (ii) are more affected at high temperatures for endothermic reactions; [b] **trend to equilibrium:** (i) reaction rate densities become larger by increasing the temperature but decrease more rapidly with time; (ii) for positive (negative) affinities the temperature of the mixture increases (decreases) and an exothermic (endothermic) reaction happens; [c] **sound propagation and light scattering:** (i) chemical reactions decrease the phase velocity with a larger effect for endothermic reactions; (ii) chemical reactions reduce the width of the Rayleigh and Brillouin lines and shift the Brillouin lines outwards the center of the spectrum; [d] **relativistic theory:** relativistic effects tend to increase the reaction rate coefficient.

References

1. I. Prigogine and E. Xhrouet, "On the perturbation of Maxwell distribution function by chemical reaction in gases", *Physica*, **15**, 913-932 (1949).
2. I. Prigogine and M. Mahieu, "Sur la perturbation de la distribution de Maxwell par des réactions chimiques en phase gazeuse", *Physica*, **16**, 51-64 (1950).
3. G. M. Kremer, M. P. Bianchi and A. J. Soares, "A relaxation kinetic model for transport phenomena in a reactive flow", *Phys. Fluids*, **18**, 037104 (2006).
4. A. W. Silva, G. M. Alves and G. M, Kremer, "Transport phenomena in a reactive quaternary gas mixture", *Physica A*, **374**, 533–548 (2007).
5. G. M. Kremer, M. P. Bianchi and A. J. Soares, "Analysis of the trend to equilibrium of a chemically reacting system", *J. Phys. A*, **40**, 2553–2571 (2007).
6. G. M. Alves and G. M. Kremer, "Effect of chemical reactions on the transport coefficients of binary mixtures", *J. Chem. Phys.*, **117**, 2205–2215 (2002).
7. W. Marques Jr., G. M. Alves and G. M. Kremer, "Light scattering and sound propagation in a chemically reacting binary gas mixture", *Physica A*, **323**, 401–412 (2003).
8. C. Cercignani and G. M. Kremer, *The relativistic Boltzmann equation: theory and applications* (Basel, Birkhäuser, 2002).
9. G. M. Kremer, "Note on the relativistic reaction rate coefficient", *Physica A*, **380**, 61–65 (2007).

THE CONTRIBUTION OF THE REACTION HEAT TO NON-EQUILIBRIUM EFFECTS OF CHEMICALLY REACTING GASES

GILBERTO M. KREMER

Departamento de Física, Universidade Federal do Paraná,
Curitiba, Brazil
E-mail: kremer@fisica.ufpr.br

ANA JACINTA SOARES

Departamento de Matemática, Universidade do Minho,
Braga, Portugal
E-mail: ajsoares@math.uminho.pt

The effects of the reaction heat induced by a reversible reaction of type $A + A \rightleftharpoons B + B$ on the Maxwellian distribution functions as well as on the reaction rate and temperature exchange rate are investigated. Assuming hard sphere elastic cross sections and reactive cross sections with activation energy, the Chapman-Enskog method and Sonine polynomial representation of the distribution functions are used to obtain the non-equilibrium solution of the Boltzmann equation in the early stage of the reaction. The departure from the equilibrium state of the reaction rate is also evaluated for both exothermic and endothermic reactions. The calculations performed in this paper show that the reaction heat changes the trend to equilibrium, namely the concentrations and temperature of the reactants.

Keywords: Boltzmann equation; chemical reactions; reaction heat

1. Introduction

In the kinetic theory extended to chemically reacting gases the non-equilibrium effects induced by chemical reactions on the distribution function have attracted attention of many researchers. Prigogine and collaborators[1,2] were the first to investigate this topic and, in particular, the effects of the reaction heat induced on a gaseous system undergoing a chemical reaction of type $A_0 + B \to A_1 + B$. Some years later, Baras and Mansour[3] have analyzed some stationary properties of an exothermic gas-phase chemical system considering also the effects of the reaction heat. Further contributions to the evaluation of on the non-equilibrium effects have been directed to other features, namely transport processes,[4,5] rate of reaction[6–9] relaxation models,[10] combustion processes,[11] sound propagation and light scattering,[12] among others.

In this paper, the influence of the reaction heat is re-examined for a binary mixture undergoing a reversible reaction of type $A + A \rightleftharpoons B + B$ and the extent of the disturbances induced on the velocity distributions as well as on the reaction rate is explicitly evaluated. The theoretical treatment is based on the Chapman-Enskog method and Sonine polynomial representation of the distribution functions, assuming that the chemical reaction is in the early stage, and adopting hard sphere cross sections for elastic collisions and two different models with activation energy for reactive cross sections.

2. Reacting gas system

The Boltzmann equation for a reacting binary mixture in the unknown one-particle distribution function $f_\alpha \equiv f(\boldsymbol{x}, \boldsymbol{c}_\alpha, t)$, $\alpha = A, B$, reads

$$\frac{\partial f_\alpha}{\partial t} + c_i^\alpha \frac{\partial f_\alpha}{\partial x_i} = \mathcal{Q}_\alpha^E + \mathcal{Q}_\alpha^R, \qquad \alpha = A, B, \tag{1}$$

where $\mathcal{Q}_\alpha^E$ and $\mathcal{Q}_\alpha^R$ describe elastic and reactive collisions, respectively,

$$\mathcal{Q}_\alpha^E = \sum_{\alpha^\star = A}^{B} \int [f'_\alpha f'_{\alpha^\star} - f_\alpha f_{\alpha^\star}]\, \mathsf{d}^2 (\boldsymbol{g}_{\alpha^\star\alpha} \cdot \boldsymbol{k}_{\alpha^\star\alpha}) d\boldsymbol{k}_{\alpha^\star\alpha}\, d\boldsymbol{c}_{\alpha^\star}, \tag{2}$$

$$\mathcal{Q}_\alpha^R = \int [f_\delta f_{\delta_1} - f_\alpha f_{\alpha_1}]\, \sigma_\alpha^\star (\boldsymbol{g}_\alpha \cdot \boldsymbol{k}_\alpha) d\boldsymbol{k}_\alpha\, d\boldsymbol{c}_{\alpha_1}. \tag{3}$$

Above, the primes refer to outgoing molecules and the indices α, δ $(\alpha \neq \delta)$ represent one of the constituents A, B; d is the molecular diameter, $\boldsymbol{k}_{\alpha^\star\alpha}$ and $d\boldsymbol{k}_{\alpha^\star\alpha}$ are the unit collision vector and the element of solid angle for elastic collisions, $\boldsymbol{k}_\alpha$ and $d\boldsymbol{k}_\alpha$ are the corresponding quantities for reactive collisions, $\sigma_\alpha^\star$ is the reactive cross section, $\boldsymbol{g}_A = \boldsymbol{c}_{A_1} - \boldsymbol{c}_A$ and $\boldsymbol{g}_B = \boldsymbol{c}_{B_1} - \boldsymbol{c}_B$ denote the relative velocities and the sub-index 1 is used in order to distinguish two identical molecules. Elastic and reactive collisions preserve momentum and total energy, i.e.

$$\boldsymbol{c}_A + \boldsymbol{c}_{A_1} = \boldsymbol{c}_B + \boldsymbol{c}_{B_1}, \quad 2\epsilon_A + \frac{m}{2}(\boldsymbol{c}_A^2 + \boldsymbol{c}_{A_1}^2) = 2\epsilon_B + \frac{m}{2}(\boldsymbol{c}_B^2 + \boldsymbol{c}_{B_1}^2), \tag{4}$$

where ϵ_A and ϵ_B denote the formation energies of the constituents. The heat of the reaction Q_R is given in terms of the formation energies by $Q_R \equiv -E = 2(\epsilon_A - \epsilon_B)$.

• *Macroscopic framework.* The number densities n_α, velocity v_i^α and temperature T_α, as well as the corresponding mixture quantities, are given by

$$n_\alpha = \int f_\alpha d\mathbf{c}_\alpha, \quad v_i^\alpha = \frac{1}{n_\alpha}\int c_i^\alpha f_\alpha d\mathbf{c}_\alpha, \quad T_\alpha = \frac{m}{3n_\alpha k}\int \xi_\alpha^2 f_\alpha d\mathbf{c}_\alpha, \qquad (5)$$

$$n = \sum_{\alpha=A}^{B} n_\alpha, \quad v_i = \frac{1}{n}\sum_{\alpha=A}^{B} n_\alpha v_i^\alpha, \quad T = \frac{1}{n}\sum_{\alpha=A}^{B} n_\alpha T_\alpha, \qquad (6)$$

where $\xi_i^\alpha = c_i^\alpha - v_i$. The constituents will be considered at the same temperature, $T_\alpha = T$, since the constituent temperature effect disappears when the reactants are identical[13] . The evolution equations for the fields (5) are obtained from the Boltzmann equation (1) in the usual way[14] , and read

$$\frac{\partial n_\alpha}{\partial t} + \frac{\partial}{\partial x_i}(n_\alpha v_i + n_\alpha u_i^\alpha) = \int (\mathcal{Q}_\alpha^E + \mathcal{Q}_\alpha^R) d\mathbf{c}_\alpha \equiv \tau_\alpha, \qquad (7)$$

$$\frac{\partial m n_\alpha v_i^\alpha}{\partial t} + \frac{\partial}{\partial x_j}\left[p_{ij}^\alpha + m n_\alpha (u_i^\alpha v_j + u_j^\alpha v_i + v_i v_j)\right] = \int m c_i^\alpha (\mathcal{Q}_\alpha^E + \mathcal{Q}_\alpha^R) d\mathbf{c}_\alpha, \quad (8)$$

$$\begin{aligned} &\frac{\partial}{\partial t}\left[\frac{3}{2}n_\alpha kT + n_\alpha \epsilon_\alpha + m n_\alpha\left(u_i^\alpha v_i + \frac{1}{2}v^2\right)\right] + \frac{\partial}{\partial x_i}\Big\{q_i^\alpha + p_{ij}^\alpha v_j + n_\alpha \epsilon_\alpha u_i^\alpha \\ &\qquad + \frac{1}{2} m n_\alpha v^2 u_i^\alpha + \left[\frac{3}{2}n_\alpha kT + n_\alpha \epsilon_\alpha + m n_\alpha\left(u_i^\alpha v_i + \frac{1}{2}v^2\right)\right] v_i\Big\} \qquad (9) \\ &\qquad = \int \left(\frac{m}{2}c_\alpha^2 + \epsilon_\alpha\right)(\mathcal{Q}_\alpha^E + \mathcal{Q}_\alpha^R) d\mathbf{c}_\alpha \equiv \chi_\alpha, \end{aligned}$$

where τ_α is the reaction rate, χ_α and the term on the *r.h.s.* of Eq. (8) are the α-production rates of total energy and momentum, u_i^α, p_{ij}^α and q_i^α denote the α-constituent diffusion velocity, pressure tensor and heat flux,

$$u_i^\alpha = \frac{m}{\varrho_\alpha}\int \xi_i^\alpha f_\alpha d\mathbf{c}_\alpha, \quad p_{ij}^\alpha = m\int \xi_i^\alpha \xi_j^\alpha f_\alpha d\mathbf{c}_\alpha, \quad q_i^\alpha = \frac{m}{2}\int \xi_\alpha^2 \xi_i^\alpha f_\alpha d\mathbf{c}_\alpha. \quad (10)$$

• *Chemical kinetics.* The equilibrium of a reacting mixture with constituents at the same temperature and negligible vibrational and rotational degrees of freedom is defined by Maxwellian distributions

$$f_\alpha^M = n_\alpha \left(\frac{m}{2\pi kT}\right)^{\frac{3}{2}} e^{-m\xi_\alpha^2/2kT}, \qquad (11)$$

where k is the Boltzmann constant. The chemical potential of each constituent – with h being the Planck constant – is given by

$$\mu_\alpha = \varepsilon_\alpha - kT\left[\frac{3}{2}\ln T - \ln n_\alpha + \frac{3}{2}\ln\left(\frac{2\pi m_\alpha k}{h^2}\right)\right]. \qquad (12)$$

The deviation of the mixture from the chemical equilibrium can be specified by the affinity of the forward reaction,

$$\mathcal{A} = \mu_A + \mu_{A_1} - \mu_B - \mu_{B_1} \qquad \text{or} \qquad \mathcal{A} = kT \ln\left(\frac{n_A}{n_B}\right)^2 - E.$$

The chemical equilibrium condition is then

$$\mu_A^{\rm eq} + \mu_{A_1}^{\rm eq} = \mu_B^{\rm eq} + \mu_{B_1}^{\rm eq}, \qquad \text{or} \qquad E^\star \equiv \frac{E}{kT} = \ln\left(\frac{n_A^{\rm eq}}{n_B^{\rm eq}}\right)^2.$$

3. The distribution function and the production terms

The velocity distribution functions contain all information about the non-equilibrium effects induced by chemical reactions, and can be determined once the solution of the Boltzmann equations (1) is obtained. The Chapman-Enskog method can be applied and the integral equation for the Maxwellian deviation $f_\alpha^{(0)}$ reads (for more details see[14])

$$\begin{aligned}\mathcal{D} f_\alpha^M - \int_B \left[f_\gamma^M f_{\gamma_1}^M - f_\alpha^M f_{\alpha_1}^M\right] \sigma_\alpha^\star (\boldsymbol{g}_\alpha \cdot \boldsymbol{k}_\alpha) d\boldsymbol{k}_\alpha\, d\boldsymbol{c}_{\alpha_1} \\ = \sum_{\alpha^\star = A}^{B} \int \left[f_\alpha^{(0)\prime} f_{\alpha^\star}^{(0)\prime} - f_\alpha^{(0)} f_{\alpha^\star}^{(0)}\right] \mathsf{d}^2 (\boldsymbol{g}_{\alpha^\star\alpha} \cdot \boldsymbol{k}_{\alpha^\star\alpha}) d\boldsymbol{k}_{\alpha^\star\alpha}\, d\boldsymbol{c}_{\alpha^\star}.\end{aligned} \tag{13}$$

The balance equations (7-9) give further conditions to be used to eliminate the material time derivative $\mathcal{D} = (\partial/\partial t + v_i \partial/\partial x_i)$ in Eq. (13). When one assumes that the reaction heat affects the distribution function, one could expect that the solution of Eqs. (13) is no longer Maxwellian and write $f_\alpha^{(0)}$ in terms of second order Sonine polynomials (with $\beta = m/2kT$)

$$f_\alpha^{(0)} = f_\alpha^M \left[1 + a_1^\alpha \left(\frac{3}{2} - \beta\xi_\alpha^2\right) + a_2^\alpha \left(\frac{15}{8} - \frac{5\beta\xi_\alpha^2}{2} + \frac{\beta^2\xi_\alpha^4}{2}\right)\right], \tag{14}$$

where a_1^α, a_2^α are scalar coefficients to be determined, for a particular choice of reactive cross sections. In this paper, two models with activation energy are considered, namely the line-of-centers and the step cross section models,

$$\sigma_\alpha^{\star\,(1)} = \begin{cases} 0 & \text{if } \gamma_\alpha < \varepsilon_\alpha^\star \\ \mathsf{d}_\mathsf{r}^2 \left(1 - \varepsilon_\alpha^\star/\gamma_\alpha\right) & \text{if } \gamma_\alpha > \varepsilon_\alpha^\star \end{cases} \quad \text{and} \quad \sigma_\alpha^{\star\,(2)} = \begin{cases} 0 & \text{if } \gamma_\alpha < \varepsilon_\alpha^\star \\ \mathsf{d}_\mathsf{r}^2 & \text{if } \gamma_\alpha > \varepsilon_\alpha^\star, \end{cases} \tag{15}$$

where $\gamma_\alpha = mg_\alpha^2/4kT$, $\varepsilon_A^\star$, $\varepsilon_B^\star$ are the forward and backward activation energies in units of kT and d_r is a reactive diameter. Some rather cumbersome calculations lead to $a_1^\alpha = 0$ and $a_2^A = a_2^B \equiv a_2$, with

$$a_2^{(1)} = S\left(1 - 2\varepsilon^\star - Q_R^\star\right) \quad \text{and} \quad a_2^{(2)} = S\left(1 - Q_R^\star - Q_R^\star\varepsilon^\star + \varepsilon^\star - 2\varepsilon^{\star 2}\right), \tag{16}$$

for the models $(15)_1$ and $(15)_2$, respectively. Above, $Q_R^\star$ is the reaction heat in units of kT, and $S = -\left(\frac{n_A \mathsf{d}}{n \mathsf{d}_\mathsf{r}}\right)^2 \frac{Q_R^\star}{8}\, e^{-\varepsilon^\star}$. The contributions which involve the affinity, namely $e^{-\mathcal{A}/kT}$, have been neglected since in the early stage of the reaction the concentrations of the products are negligible and the affinity is very large ($\mathcal{A} \to \infty$). Expressions (14) with coefficients $a_1^\alpha = 0$ and coefficients a_2 alternatively given by expressions $(16)_1$ and $(16)_2$ completely determine the distributions $f_\alpha^{(0)}$ showing an explicit dependence on the reaction heat $Q_R^\star$. In the left frame of Fig.1, the dimensionless distribution $\frac{1}{n_A} f_A^{(0)} \left(\frac{\pi kT}{m}\right)^{3/2}$ is plotted versus $Q_R^\star$ and $x = \frac{m\xi_\alpha^2}{2kT}$, for an exothermic reaction, in the case of the step cross section, with data $x_A = 1$, $\mathsf{d_r} = \mathsf{d}$ and $\varepsilon^\star = 5$. From this figure, one can infer that the reaction heat modifies the profile of the Maxwellian when the reaction heat increases. In the right frame of Fig.1, the dimensionless reaction rate coefficient $\kappa_A^{(0)}/(\mathsf{d_r}^2\sqrt{\pi kT/m})$ is plotted versus the activation energy, in the case of the step cross-section, with the same data as before, in the cases $Q_R^\star = -2$ (endothermic), $Q_R^\star = 2$ (exothermic) and $Q_R^\star = 0$. From this figure, one can conclude that the effects on the reaction rate are more evident for an exothermic reaction (lower curve) and that the reaction rate coefficient becomes smaller by increasing the activation energy.

4. Application

The time decay of the particle number density and temperature of the reactants, when the chemical reaction advances, can be consistently studied by considering spatially homogeneous balance equations for particle number density and temperature, see Eqs. (7,9). The corresponding dimensionless system, when reactive line-of-centers model is adopted, is given by[14]

$$\frac{d\tilde{n}}{d\tilde{t}} = -\left(\frac{\mathsf{d}}{\mathsf{d_r}}\right)^2 \tilde{n}^2 \sqrt{\tilde{T}} e^{-\tilde{\varepsilon}/\tilde{T}} \left[1 - \frac{\tilde{n}^2}{128}\left(\frac{\mathsf{d}}{\mathsf{d_r}}\right)^2 \frac{\tilde{Q}}{\tilde{T}}\left(1 - \frac{\tilde{Q}}{\tilde{T}} - \frac{2\tilde{\varepsilon}}{\tilde{T}}\right)\left(\frac{4\tilde{\varepsilon}^2}{\tilde{T}^2} - \frac{4\tilde{\varepsilon}}{\tilde{T}} - 1\right) e^{-\tilde{\varepsilon}/\tilde{T}}\right]$$

$$\frac{d\tilde{T}}{d\tilde{t}} = -\frac{1}{96}\left(\frac{\mathsf{d}}{\mathsf{d_r}}\right)^2 \tilde{n}\sqrt{\tilde{T}^3} e^{-\tilde{\varepsilon}/\tilde{T}} \left[16 + \frac{32\tilde{\varepsilon}}{\tilde{T}} - \frac{\tilde{n}^2}{8}\left(\frac{\mathsf{d}}{\mathsf{d_r}}\right)^2 \frac{\tilde{Q}}{\tilde{T}}\right.$$
$$\left. \times \left(1 - \frac{\tilde{Q}}{\tilde{T}} - 2\frac{\tilde{\varepsilon}}{\tilde{T}}\right)\left(\frac{8\tilde{\varepsilon}^3}{\tilde{T}^3} - \frac{4\tilde{\varepsilon}^2}{\tilde{T}^2} + \frac{66\tilde{\varepsilon}}{\tilde{T}} + 35\right) e^{-\tilde{\varepsilon}/\tilde{T}}\right].$$

This system was numerically solved with initial data $\tilde{n}(0) = 1$, $\tilde{T}(0) = 1$, $\mathsf{d_r} = \mathsf{d}$ and $\tilde{\varepsilon} = 5$. The time decay of $\tilde{n}$ and $\tilde{T}$ is plotted in Fig.2 for $\tilde{Q} = -2$, $\tilde{Q} = 2$ and $\tilde{Q} = 0$. From this figure, one can infer that the time decay is more striking for an exothermic chemical reaction. This behavior is justified by the fact that the reaction velocity of exothermic reactions is larger than the one of endothermic reactions as well as the one of chemical processes for which the heat of reaction is not taken into account.

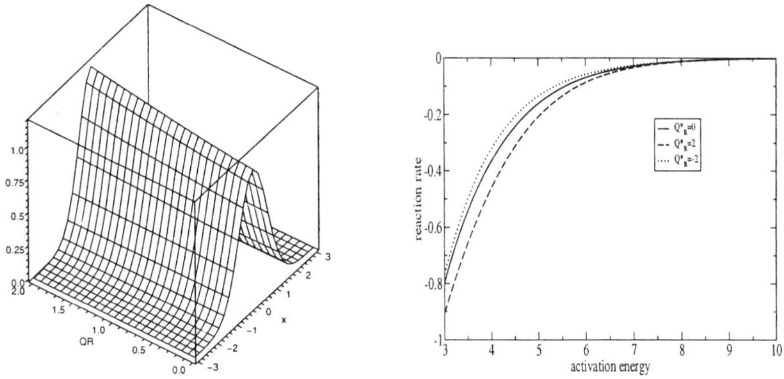

Fig. 1. Left figure: distribution function of the constituent A for an exothermic reaction. Right figure: reaction rate coefficient for the reactive step cross section and exothermic (dashed line) or endothermic (dotted line) reactions as well as for $Q_R^\star = 0$ (solid line).

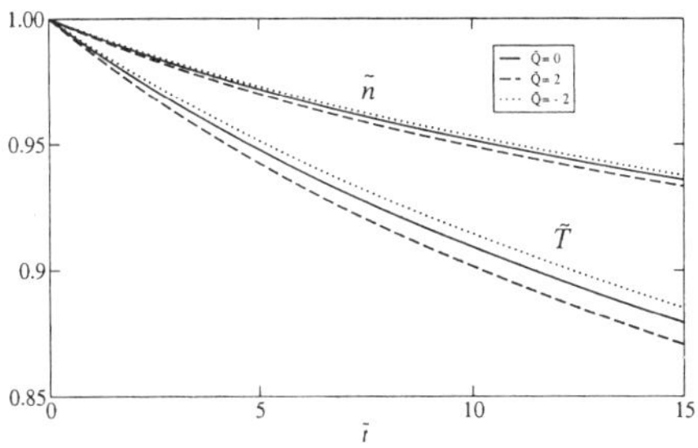

Fig.2 Time decay of particle number density and temperature for exothermic (dashed lines) and endothermic (dotted lines) reactions, as well as for $\tilde{Q} = 0$ (solid lines).

References

1. Prigogine I, Xhrouet E, 1949 *Physica* **15** 913
2. Prigogine I, Mahieu M, 1950 *Physica* **16** 51
3. Baras F, Malek Mansour M, 1989 *Phys Rev Lett* **63** 2429
4. Ross J, Mazur P, 1961 *J Chem Phys.* **35** 19
5. Alves GM, Kremer GM, 2002 *J Chem Phys* **117** 2205
6. Shizgal BD, Karplus M, 1970 *J Chem Phys* **52** 4262; 1971 *J Chem Phys* **55** 76
7. Shizgal BD and Chikhaoui A, 2006 *Physica A* **365** 317
8. Cukrowski AS, Popielawski J, 1986 *Chem Phys* **109** 215
9. Cukrowski A S, 2006 *Acta Phys Pol B* **37** 1715
10. Kremer GM, Pandolfi Bianchi M, Soares AJ, 2006 *Phys Fluids* **18** 037104
11. Ern A, Giovangigli V, 2003 *Transp Theory Stat Phys* **32** 657
12. Marques W Jr., Alves GM and Kremer GM, 2003 *Physica A* **323** 401
13. Shizgal B D, Napier D G, 1996 *Physica A* **223** 50
14. Kremer GM, Soares AJ, "Effect of reaction heat on Maxwellian distribution functions and rate of reactions" *J Stat Mech*, in press.

NONLINEAR CLOSURE RELATIONS: A CASE FROM SEMICONDUCTORS

S. LA ROSA AND V. ROMANO

Dipartimento di Matematica e Informatica,
Università di Catania, viale A. Doria, 6, 95125 Catania, Italy
larosa@dmi.unict.it, romano@dmi.unict.it

G. MASCALI

Dipartimento di Matematica, Universitá della Calabria and INFN-Gruppo c. Cosenza,
87036 Cosenza, Italy, g.mascali@unical.it

Assuming the energy band described by the Kane dispersion relation, the exact closure for the 8-moment model based on the maximum entropy principle is analyzed, investigating in particular the region of realizability by means of the simulation of some test cases.

1. The mathematical model

We consider the hydrodynamical model for semiconductors given by the 8-moment system closed with the maximum entropy principle (hereafter MEP), obtained in [1]. We skip the details of the derivation, but we remark that the exact closure has been obtained explicitly expressing the fluxes and the production terms as functions of the Lagrangian multipliers. The model consists of the balance equations for the electron density n, the average velocity V, the energy W and the energy flux S. In the one dimensional case it reads

$$\begin{aligned}
&\frac{\partial n}{\partial t} + \frac{\partial (nV)}{\partial x} = 0, \\
&\frac{\partial (nV)}{\partial t} + \frac{\partial (nU)}{\partial x} + neE\,H = nC_V, \\
&\frac{\partial (nW)}{\partial t} + \frac{\partial (nS)}{\partial x} + neVE = nC_W, \\
&\frac{\partial (nS)}{\partial t} + \frac{\partial (nF)}{\partial x} + neEG = nC_S.
\end{aligned}$$

All the quantities entering into the previous system are expressed as function of the relevant components of the Lagrangian multipliers λ, λ^W, $\boldsymbol{\lambda}^V$ and $\boldsymbol{\lambda}^S$ relative to the density, the energy, the velocity and the energy-flux. U, H, F and G are the fluxes, C_V, C_W and C_S are the production terms and e is the absolute value of the electron charge. E represents the significant component of the electric field which satisfies the Poisson equation

$$E = -\frac{\partial \phi}{\partial x}, \quad \epsilon \Delta \phi = -e(N_D - N_A - n),$$

where ϕ is the electric potential, N_D and N_A are the donor and acceptor densities, and ϵ is the semiconductor dielectric constant. The fields are given in terms of the Lagrangian multipliers by

$$n = \frac{2\pi (2m^*)^{3/2}}{\hbar^3} e^{-\frac{\lambda}{k_B}} d_0,$$

$$V = \frac{1}{d_0}\sqrt{\frac{2}{m^*}} \int_0^\infty \sigma(\varepsilon)\, \varepsilon (1 + \alpha\varepsilon)\, e^{-\lambda^W \varepsilon} \left[\frac{\sinh A(\varepsilon)}{A(\varepsilon)^2} - \frac{\cosh A(\varepsilon)}{A(\varepsilon)} \right] d\varepsilon,$$

$$W = \frac{1}{d_0} \int_0^\infty \varepsilon\, e^{-\lambda^W \varepsilon} \frac{\sinh A(\varepsilon)}{A(\varepsilon)} \sqrt{\varepsilon(1+\alpha\varepsilon)}(1 + 2\alpha\varepsilon)\, d\varepsilon,$$

$$S = \frac{1}{d_0}\sqrt{\frac{2}{m^*}} \int_0^\infty \sigma(\varepsilon)\, \varepsilon^2 (1 + \alpha\varepsilon)\, e^{-\lambda^W \varepsilon} \left[\frac{\sinh A(\varepsilon)}{A(\varepsilon)^2} - \frac{\cosh A(\varepsilon)}{A(\varepsilon)} \right] d\varepsilon,$$

with $\hbar$ and k_B respectively the Planck and the Boltzmann constants, ε the microscopic electron energy and

$$\sigma(\varepsilon) = \text{sign}(\lambda^V + \varepsilon\lambda^S), \; d_0 = \int_0^\infty e^{-\lambda^W \varepsilon} \frac{\sinh A(\varepsilon)}{A(\varepsilon)} \sqrt{\varepsilon(1+\alpha\varepsilon)}(1 + 2\alpha\varepsilon)\, d\varepsilon,$$

where $A(\varepsilon) = |\lambda^V + \lambda^S \varepsilon|\, v$,

$$\text{sign}(x) = \begin{cases} 1 \text{ if } x \geq 0 \\ -1 \text{ if } x < 0 \end{cases} \quad \text{and} \quad v = \sqrt{\frac{2\varepsilon\,(1+\alpha\varepsilon)}{m^*\,(1+2\alpha\varepsilon)^2}}$$

is the modulus of the electron velocity. m^* is the effective electron mass and α the non parabolicity factor. Similarly, the fluxes are given by

$$m^* U = \frac{2}{d_0} \int_0^\infty e^{-\lambda^W \varepsilon} \frac{[\varepsilon(1+\alpha\varepsilon)]^{3/2}}{1 + 2\alpha\varepsilon} B(\varepsilon)\, d\,\varepsilon,$$

$$m^* F = \frac{2}{d_0} \int_0^\infty \varepsilon\, e^{-\lambda^W \varepsilon} \frac{[\varepsilon(1+\alpha\varepsilon)]^{3/2}}{1 + 2\alpha\varepsilon} B(\varepsilon)\, d\,\varepsilon,$$

$$m^* H = \frac{1}{d_0} \int_0^\infty e^{-\lambda^W \varepsilon} \left[\sqrt{\varepsilon(1+\alpha\varepsilon)} \frac{\sinh A(\varepsilon)}{A(\varepsilon)} - 4\alpha \frac{[\varepsilon(1+\alpha\varepsilon)]^{3/2}}{(1+2\alpha\varepsilon)^2} B(\varepsilon) \right] d\varepsilon,$$

$$m^* G = \frac{1}{2\alpha}\left(1 - m^* H\right),$$

with

$$B(\varepsilon) = \left[\frac{\sinh A(\varepsilon)}{A(\varepsilon)} - \frac{2}{A(\varepsilon)^2}\left(\cosh A(\varepsilon) - \frac{\sinh A(\varepsilon)}{A(\varepsilon)}\right)\right].$$

For the significant components of the production terms, in the case of elastic electron-phonon scattering, one has

$$\begin{pmatrix} C_V^{(el)} \\ C_S^{(el)} \end{pmatrix} = \frac{8\pi K_{ac} m^*}{\hbar^3\, d_0} \int_0^\infty \sigma(\varepsilon) \begin{pmatrix} 1 \\ \varepsilon \end{pmatrix} e^{-\lambda^W \varepsilon} [\varepsilon(1+\alpha\varepsilon)]^{3/2} (1+2\alpha\varepsilon) B_1(\varepsilon) d\varepsilon$$

and $C_W^{(el)} = 0$,

with

$$B_1(\varepsilon) = \left(\frac{\cosh A(\varepsilon)}{A(\varepsilon)} - \frac{\sinh A(\varepsilon)}{A(\varepsilon)^2}\right), \quad \text{and} \quad K_{ac} \quad \text{a physical constant,}$$

while in the case of inelastic electron-phonon scattering, we find

$$\begin{pmatrix} C_V^{(in)} \\ C_S^{(in)} \end{pmatrix} = \frac{8\pi K_\alpha m^*}{\hbar^3 d_0} \int_0^\infty \Bigg[e^{-\lambda^W \varepsilon} \begin{pmatrix} \varepsilon \\ \varepsilon^2 \end{pmatrix} (1+\alpha\varepsilon) N_B^{(\alpha)}\, \sigma(\varepsilon)\, N_+(\varepsilon) B_1(\varepsilon)$$

$$+ (N_B^{(\alpha)} + 1)\, \sigma(\varepsilon_+) \begin{pmatrix} \varepsilon_+ \\ \varepsilon_+^2 \end{pmatrix} (1+\alpha\varepsilon_+) e^{-\lambda^W \varepsilon_+} N(\varepsilon) B_1(\varepsilon_+) \Bigg] d\varepsilon,$$

$$C_W^{(in)} = \frac{8\pi K_\alpha (m^*)^{3/2}}{\sqrt{2}\hbar^3 d_0} \hbar\omega_\alpha \int_0^\infty e^{-\lambda^W \varepsilon} N_+(\varepsilon) N(\varepsilon)$$

$$\left[N_B^{(\alpha)} \frac{\sinh A(\varepsilon)}{A(\varepsilon)} - (N_B^{(\alpha)} + 1)\, e^{-\lambda^W \hbar\omega_\alpha} \frac{\sinh A(\varepsilon_+)}{A(\varepsilon_+)} \right] d\varepsilon,$$

$$N(\varepsilon) = \sqrt{\varepsilon(1+\alpha\varepsilon)}(1+2\alpha\varepsilon), \quad \varepsilon_+ = \varepsilon + \hbar\omega_\alpha, \quad N_+(\varepsilon) = N(\varepsilon_+),$$

$$N_B^{(\alpha)} = \frac{1}{\exp(\hbar\omega_\alpha / k_B T_L) - 1},$$

K_α is a physical constant, T_L the lattice temperature and $\hbar\omega_\alpha$ the phonon energy in the α-th valley.

2. The realizability region

The previous relations have been obtained by using the maximum entropy function

$$f_{ME} = \exp\left\{ -\lambda(x,t) - \lambda^W(x,t)\, \varepsilon(\mathbf{k}) - \left[\boldsymbol{\lambda}^V(x,t) + \boldsymbol{\lambda}^S(x,t)\, \varepsilon(\mathbf{k}) \right] \cdot \mathbf{v} \right\}, \qquad \mathbf{k} \in \mathbb{R}^3.$$

The integrability of f is guaranteed (see [2]) if and only if the condition

$$\lambda^W > 0 \quad \text{and} \quad \lambda^W - |\boldsymbol{\lambda}^S| v_\infty \geq 0 \tag{1}$$

is satisfied, with $v_\infty = \lim_{\varepsilon \to \infty} v(\varepsilon) = 1/\sqrt{2\alpha m^*}$.

It is well known [3],[4] that there exist moments that are not moments of the maximum entropy distribution. In [2], it is proved that, in the case of the Kane dispersion relation, the set of the moments generated by the maximum entropy distribution function is a convex cone $\mathcal{M}$, named realizability region, which is characterized in the 8-moments case by condition (1). Moreover it is shown that the equilibria are interior points of this region, guaranteeing the validity of expansions around equilibrium states.

In this paper, we check if the boundary of $\mathcal{M}$ is reached in the typical benchmark problem constituted by the $n^+ - n - n^+$ silicon diode (see [5],[6] for details) by varying the channel length.

The moment equations closed with the MEP closure relations form a quasilinear hyperbolic system of conservation laws

$$\frac{\partial F^{(0)}(\Lambda)}{\partial t} + \frac{\partial F^{(1)}(\Lambda)}{\partial x} = G(\Lambda, E), \tag{2}$$

with the Lagrangian multipliers $\Lambda = \left(\lambda, \lambda^W, \lambda^V, \lambda^S\right)$ as variables, x denoting the significant spatial variable.

It is convenient to perform the further change of variables

$$\Lambda \to \left(n, \lambda^W, \lambda^V, \lambda^S\right)$$

taking the density and the Lagrangian multipliers of the energy, the velocity and the energy flux as variables. It is simple to verify that the transformation of coordinates is regular and the new system of PDEs is still hyperbolic.

We solve numerically system (2) by using a splitting strategy where a semi-implicit Euler scheme for the relaxation part is alternated with a suitably modified Nessyahu-Tadmor scheme for homogeneous hyperbolic system of conservation laws. The evaluation of the fields, fluxes and production terms is performed with Gaussian quadrature formulas. For details the interested reader is referred to [1].

We consider a silicon diode where the n^+ regions are $0.1\mu m$ long, while various lengths of the channel are taken into account. We study the stationary solution which is reached after about 5 picoseconds. The condition $\lambda^W > 0$ is always satisfied. In figure 1 there is a plot of the function $g(x) = \lambda^W(x) - |\lambda^S(x)| v_\infty$, where $\lambda^W(x)$ and $\lambda^S(x)$ are given by the stationary solution. For channels no shorter than three microns we find $g(x) > 0$, that is the solution is inside the realizability region. For channels of 2 micron or shorter there is a loss of integrability beside the first junction. In

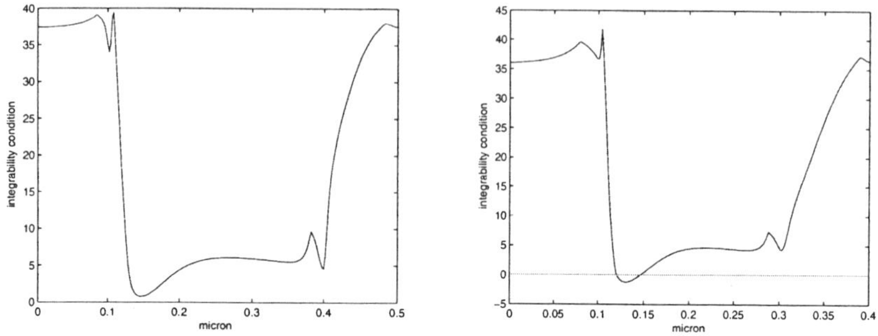

Fig. 1. Plot of $g(x) = \lambda^W(x) - |\lambda^S(x)|v_\infty$ versus the position. On the left the channel is 0.3 μm, on the right 0.2 μ. In both cases the bias voltage is 1V, the doping in the n^+ regions is 10^{18} cm^{-3}, while in the n region is 10^{16} cm^{-3}

figure 2 we report the electron velocity and energy to assess also the importance of the non linearity comparing the results with those obtained by linearizing f_{ME} with respect to a small anisotropy parameter [7] (semilinear MEP model) and with the standard Blotekjaer-Baccarani-Wordeman (BBW) model. The difference between the nonlinear MEP model and the semilinear MEP model is related to the importance of the nonlinearity in the closure relations and increases as the channel length decreases. The classical BBW model is the least accurate with a considerable difference with respect to Monte Carlo simulation and direct Boltzmann integration.

The nonlinearity improves the results but it presents the drawback that the realizability region could be not so large to contain all the physical cases. For sub-micron diodes of moderately small channel length the model is adequate but for very small channels must be improved. One possibility is to increase the number of moments. Alternatively a more sophisticated energy band description could be introduced. These topics are currently under investigation by the authors.

Acknowledgments

The authors acknowledge support by M.I.U.R. (PRIN 2007 *Equazioni cinetiche e idrodinamiche di sistemi collisionali complessi*), the EU Marie Curie RTN project **COMSON** grant n. MRTN-CT-2005-019417, the INdAM project *Mathematical modeling and numerical analysis of quantum systems with applications to nanosciences*, the contribution by P.R.A. University of Catania (ex 60 %).

References

1. S. La Rosa, G. Mascali and V. Romano, *Exact maximum entropy closure for the hydrodynamical model of Si semiconductors: the 8-moment case.* Preprint

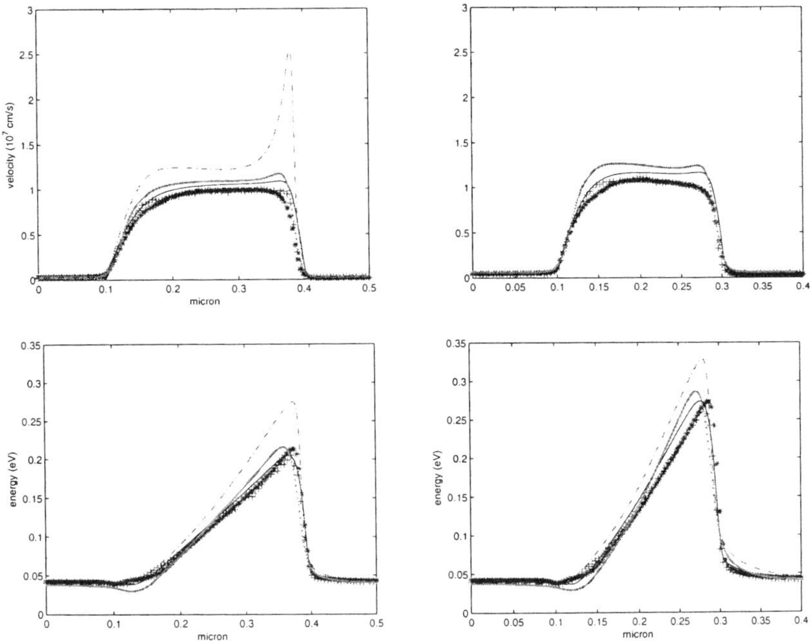

Fig. 2. Stationary solutions for the velocity and energy in the silicon diode. The plots refer to the nonlinear MEP model (continuous line), the semilinear MEP model (dotted line), MC simulation (crossed line), direct Boltzmann integration (stars) and BBW model (dashed-dotted line). Doping and bias voltage are the same as in fig. 1.

(see archive EU Marie Curie COMSON project: www.comson.org).
2. M. Junk and V. Romano, *Cont. Mech. Thermodyn.* **17** 247 (2005).
3. M. Junk, *J. Stat. Phys.* **93**, 1143 (1998).
4. M. Junk, *Math. Models Methods Appl. Sci.* **10**, 1001 (2000).
5. V. Romano, *Cont. Mech. Thermodyn.* **12**, 31 (2000).
6. A.M. Anile, G. Mascali and V. Romano, in *Mathematical Problems in Semiconductor Physics, Lecture Notes in Mathematics* Vol. 1832, (Springer, Berlin, 2003).
7. A. M. Anile and V. Romano , *Cont. Mech. Thermodyn.* **11**, 307 (1999).

STABILITY IN THE BÉNARD PROBLEMS WITH COMPETING EFFECTS VIA THE REDUCTION METHOD*

S.LOMBARDO

Dipartimento di Matematica e Informatica,
Città Universitaria, Viale A. Doria, 6, 95125, Catania, Italy
E-mail: lombardo@dmi.unict.it

The linear and nonlinear stability of the non-convecting motion of a uniformly rotating layer of a binary fluid mixture heated and salted from below, in the Oberbeck-Boussinesq scheme, is studied through the Lyapunov direct method. Necessary and sufficient stability conditions are obtained, for Schmidt P_C and Prandtl P_T numbers equal to 1 and any Taylor number. An optimal Lyapunov function is built by means of the reduction method. A computable radius of attraction for the initial data is also obtained.

Keywords: Lyapunov stability; thermohaline convection; competing effects.

1. Introduction

The Bénard problem in the presence of two or more fields (as rotation, magnetic and concentration fields) is important in many mathematical, physical and technological applications[1-11] . In the case of not rotating homogeneous fluid, the linear stability has been studied in Ref. 1 by means of classical normal modes. Nonlinear energy stability has been shown by Joseph,[5] who found coincidence of the linear and nonlinear critical numbers. For a uniformly rotating homogenous fluid, the *stabilizing effect* of the rotation predicted by the linear theory, has been obtained in the nonlinear context with the use of a suitable Lyapunov function different from the classical energy[2] . The nonlinear stability, in the case $P_T = 1$, has been studied in Ref. 3,4. The coincidence of linear and nonlinear critical Rayleigh parameters, for any Prandtl and Taylor numbers, has been proved in Ref. 4 . The Bénard problem for a mixture salted and heated from below has been studied in[5-7,9] . In Ref. 6, the authors obtained exponential

*Dedicated to Prof. Tommaso Ruggeri on the occasion of his 60th birthday

nonlinear stability and coincidence of the linear and nonlinear critical parameters for Prandtl numbers greater than Schmidt numbers. The case of Prandtl numbers less than Schmidt numbers has been studied in Ref. 8. The rotating effect in a binary fluid mixture is also of importance for the applications. The combined effect of different fields may show unexpected *conflicting tendencies* like in the rotating magnetic Bénard problem (see Chandrasekhar[1]).

Here we consider an infinite horizontal layer of an incompressible newtonian binary fluid mixture heated and salted from below, subject to a vertical gravity field, uniformly rotating around a vertical axis. We study the combined effect of rotation and concentration fields simultaneously acting, in the theoretical case $P_T = P_C = 1$ (for computation's simplicity). By using the reduction method (see Ref. 14,15), we obtain

i) the coincidence of the linear and nonlinear critical parameters,

ii) a radius of attraction for the initial data.

2. Basic equations

The evolution equations of a perturbation $(\mathbf{u}(x,y,z,t),\ \vartheta(x,y,z,t),\ \gamma(x,y,z,t),\ \tilde{p}(x,y,z,t))$ to the basic motionless state with linear temperature and concentration profiles are[8]

$$\begin{cases} \mathbf{u}_t + \mathbf{u}\cdot\nabla\mathbf{u} = -\nabla\tilde{p} + (\mathcal{R}\vartheta - \mathcal{C}\gamma)\mathbf{k} + \Delta\mathbf{u} + \mathcal{T}\,\mathbf{u}\times\mathbf{k} \\ \nabla\cdot\mathbf{u} = 0, \\ P_T(\vartheta_t + \mathbf{u}\cdot\nabla\vartheta) = \mathcal{R}w + \Delta\vartheta \\ P_C(\gamma_t + \mathbf{u}\cdot\nabla\gamma) = \mathcal{C}w + \Delta\gamma \end{cases} \tag{1}$$

in $\Omega_1 \times (0,\infty)$ where $\Omega_1 = \mathbb{R}^2 \times (0,1)$. The parameters $\mathcal{R}^2 = \frac{g\beta_T\alpha_T d^4}{\nu k_T}$, $\mathcal{C}^2 = \frac{g\beta_C\alpha_C d^4}{\nu k_C}$, and $\mathcal{T}^2 = \frac{4\check{\Omega}^2 d^4}{\nu^2}$ are the Rayleigh number for heat and solute and the Taylor number. The Prandtl and Schmidt numbers are respectively $P_T = \frac{\nu}{k_T}$, $P_C = \frac{\nu}{k_C}$. The initial and boundary conditions are $\mathbf{u}(\mathbf{x},0) = \mathbf{u}_0(\mathbf{x})$, $\vartheta(\mathbf{x},0) = \vartheta_0(\mathbf{x})$, $\gamma(\mathbf{x},0) = \gamma_0(\mathbf{x})$, $\vartheta(\hat{\mathbf{x}},t) = \gamma(\hat{\mathbf{x}},t) = w(\hat{\mathbf{x}},t) = u_z(\hat{\mathbf{x}},t) = v_z(\hat{\mathbf{x}},t) = 0$ (stress-free) where $t > 0, \mathbf{x} = (x,y,z) \in \Omega_1$ and $\hat{\mathbf{x}} = (x,y,z)$ with $z = 0$ or $z = 1$. As usual, the perturbations $\mathbf{u}, \vartheta, \gamma, \tilde{p}$ are periodic functions of x and y of periods $2\pi/a_x$, $2\pi/a_y$, respectively, $(a_x > 0,\ a_y > 0)$. Periodicity cell and wave number are denoted by $\Omega = [0, 2\pi/a_x] \times [0, 2\pi/a_y] \times [0,1]$, $a = (a_x^2 + a_y^2)^{1/2}$, respectively. Moreover we require the "average velocity condition" in order that m_0 is unique.

We study the stability of the basic motion, following Chandrasekhar,[1] by taking the third component of the *curl* of $(1)_1$ and the third component of the *double curl* of $(1)_1$. We easily obtain

$$\begin{cases} \Delta w_t = \Delta\Delta w - \mathcal{T}\zeta_z + \mathcal{R}\Delta_1\vartheta - \mathcal{C}\Delta_1\gamma + \mathbf{k}\cdot\nabla\times\nabla\times(\mathbf{u}\cdot\nabla\mathbf{u}) \\ \zeta_t = \mathcal{T}w_z + \Delta\zeta - \mathbf{k}\cdot\nabla\times(\mathbf{u}\cdot\nabla\mathbf{u}) \\ P_T\vartheta_t = \mathcal{R}w + \Delta\vartheta - P_T\mathbf{u}\cdot\nabla\vartheta, \quad P_C\gamma_t = \mathcal{C}w + \Delta\gamma - P_C\mathbf{u}\cdot\nabla\gamma \end{cases} \tag{2}$$

where $\zeta = \mathbf{k}\cdot\nabla\times\mathbf{u}$, $\Delta_1 = \frac{\partial^2}{\partial x^2} + \frac{\partial^2}{\partial y^2}$. In the following, we denote $\zeta_z = \xi$, $(P_T)^{-1} = p_1$, $(P_C)^{-1} = p_2$ and the nonlinear part of i-th equation of (2) by N_i. Since the linear operator is not symmetric (in the usual scalar product $(.,.)$ of $L_2(\Omega)$), by using the classical energy stability analysis, see Ref. 9,,[11] the optimal stability thresholds are not reached. In fact,
$\frac{d}{dt}[\frac{1}{2}(\|\mathbf{u}\|^2 + P_T\|\vartheta\|^2 + P_C\|\gamma\|^2)] = 2\mathcal{R}(w,\vartheta) - [\|\nabla\mathbf{u}\|^2 + \|\nabla\vartheta\|^2 + \|\nabla\gamma\|^2]$, the terms $(\mathcal{T}\mathbf{u}\times\mathbf{k},\mathbf{u})$ and $(\mathcal{C}\gamma, w)$ disappear, and the critical parameter is equal to $\mathcal{R}_c^2(\mathcal{T}=\mathcal{C}=0) = 657.511$.
In order to reach the critical parameters of linearized instability in the nonlinear context, we use the principal eigenvalues of the Laplacian with zero boundary conditions on $z=0$, $z=1$ and introduce a transformation of variables with a reduction method[12] ,[14] ,[15] . For this we take the z-derivative of $(2)_2$ and apply the Laplacian operator to the evolution equations relative to ξ, ϑ and γ. Then we obtain

$$\begin{cases} \Delta w_t = \Delta\Delta w - \mathcal{T}\xi + \mathcal{R}\Delta_1\vartheta - \mathcal{C}\Delta_1\gamma + N_1 \\ \Delta\xi_t = \mathcal{T}\Delta w_{zz} + \Delta\Delta\xi + \Delta N_{2,z} \\ \Delta\vartheta_t = p_1(\mathcal{R}\Delta w + \Delta\Delta\vartheta) + \Delta N_3, \quad \Delta\gamma_t = p_2(\mathcal{C}\Delta w + \Delta\Delta\gamma) + \Delta N_4 \end{cases} \tag{3}$$

and add the further boundary conditions $\xi = \Delta w = \vartheta = \gamma = 0$ on the boundaries $z=0$ and $z=1$. In order to study linear instability, as in Ref. 1, we have to compute the eigenvalues λ which are solutions of equation

$$\lambda^4 + s\mathcal{A}_1\lambda^3 + s^2\mathcal{A}_2\lambda^2 + s^3\mathcal{A}_3\lambda + s^4\mathcal{A}_4 = 0 \tag{4}$$

where $s = n^2\pi^2 + a^2$, $\mathcal{A}_1 = 2 + p_1 + p_2$, $\mathcal{A}_2 = 1 + 2(p_1+p_2) + p_1p_2 + p_2\hat{\mathcal{C}} - p_1\hat{\mathcal{R}} + \hat{\mathcal{T}}$, $\mathcal{A}_3 = p_1 + p_2 + 2p_1p_2 + p_2(1+p_1)\hat{\mathcal{C}} - p_1(1+p_2)\hat{\mathcal{R}} + (p_1+p_2)\hat{\mathcal{T}}$, $\mathcal{A}_4 = p_1p_2(1 + \hat{\mathcal{C}} - \hat{\mathcal{R}} + \hat{\mathcal{T}})$ and $\hat{\mathcal{C}} = \frac{\mathcal{C}^2a^2}{s^3}$, $\hat{\mathcal{R}} = \frac{\mathcal{R}^2a^2}{s^3}$, $\hat{\mathcal{T}} = \frac{\mathcal{R}^2\pi^2}{s^3}$.
By splitting (4) in its real and imaginary parts, we find the marginal region where the instability rises by stationary secondary motions (exchange of instabilities principle, PES, holds): $\mathcal{A}_4 = 0$ and the overstability region where instability rises by oscillatory motions: ${\mathcal{A}_3}^2 - \mathcal{A}_1\mathcal{A}_2\mathcal{A}_3 + {\mathcal{A}_1}^2\mathcal{A}_4 = 0$.
In general, the critical linear number is given by $\min(\mathcal{R}_{L_1}^2, \mathcal{R}_{L_2}^2)$, where $\mathcal{R}_{L_1}^2$ is obtained by solving the equation $\mathcal{A}_4 = 0$ with respect to $\mathcal{R}^2$ and minimizing with respect to n^2 and a^2, while $\mathcal{R}_{L_2}^2$ is obtained, analogously,

by solving ${\mathcal{A}_3}^2 - \mathcal{A}_1\mathcal{A}_2\mathcal{A}_3 + {\mathcal{A}_1}^2\mathcal{A}_4 = 0$ with respect to $\mathcal{R}^2$ and minimizing with respect to n^2 and a^2. For simplicity, here we study analytically the case $P_T = P_C = 1$ for which $\mathcal{R}^2_{L_1} \leq \mathcal{R}^2_{L_2}$ and PES holds. So, the critical linear number is $\mathcal{R}^2_{L_1}$ which is given by

$$\min_{n, a^2} \frac{(n^2\pi^2 + a^2)^3}{a^2} + \mathcal{C}^2 + \frac{(n^2\pi^2\mathcal{T}^2)}{a^2} = \frac{(\pi^2 + a_c^2)^3}{a_c^2} + \mathcal{C}^2 + \frac{(\pi^2\mathcal{T}^2)}{a_c^2} \quad (5)$$

where a_c^2 is the real root of $2x^3 + 3\pi^2x^2 - \pi^2(\pi^4 + \mathcal{T}^2) = 0$. For this case a direct computation of the eigenvalues, by using (4), is easy. We have: $\lambda_1 = \lambda_2 = -s$, $\lambda_3 = -s[1 + \sqrt{\frac{\mathcal{R}^2a^2}{s^3} - \frac{\mathcal{C}^2a^2}{s^3} - \frac{\mathcal{T}^2\pi^2}{s^3}}]$ and $\lambda_4 = -s[1 - \sqrt{\frac{\mathcal{R}^2a^2}{s^3} - \frac{\mathcal{C}^2a^2}{s^3} - \frac{\mathcal{T}^2\pi^2}{s^3}}]$. They all are real numbers (if we consider the value of $\mathcal{R}^2$ not far from the marginality) and the condition $\lambda_4 > 0$ assures linear instability.
In order to prove the stabilizing effects of the solute and rotation and to reach optimal stability parameters, we use the reduction method. For this we fix $n = 1$, characterizing first linear instability mode, and consider the new field variables obtained by means of the (inverse) matrix of the eigenvectors of (3):

$$\begin{aligned}
\phi &= \frac{1}{\alpha^2}[-\mathcal{C}\xi - \frac{\mathcal{RC}a^2}{\mathcal{T}}\vartheta + \frac{\mathcal{R}^2a^2 - \pi^2\mathcal{T}^2}{\mathcal{T}}\gamma] \\
\psi &= \frac{1}{\alpha^2}[\mathcal{R}\xi + \frac{\mathcal{C}^2a^2 + \pi^2\mathcal{T}^2}{\mathcal{T}}\vartheta - \frac{\mathcal{RC}a^2}{\mathcal{T}}\gamma] \\
\chi &= \frac{1}{2}[w - \frac{1}{\alpha\sqrt{s}}(\mathcal{T}\xi + \mathcal{R}a^2\vartheta - \mathcal{C}a^2\gamma)] \\
\Phi &= \frac{1}{2}[w + \frac{1}{\alpha\sqrt{s}}(\mathcal{T}\xi + \mathcal{R}a^2\vartheta - \mathcal{C}a^2\gamma)]
\end{aligned} \quad (6)$$

where $\alpha = \sqrt{a^2\mathcal{R}^2 - a^2\mathcal{C}^2 - \pi^2\mathcal{T}^2}$. We differentiate respect to t and apply the Laplacian. By taking into account of system (3), we obtain the new equivalent PDE system:

$$\Delta\phi_t = \Delta\Delta\phi - \frac{\mathcal{TC}}{\alpha^2}[\Delta\chi_{zz} + \pi^2\Delta\chi + \Delta\Phi_{zz} + \pi^2\Delta\Phi] + \bar{N}_1 \quad (7)$$

$$\Delta\psi_t = \Delta\Delta\psi + \frac{\mathcal{TR}}{\alpha^2}[\Delta\chi_{zz} + \pi^2\Delta\chi + \Delta\Phi_{zz} + \pi^2\Delta\Phi] + \bar{N}_2 \quad (8)$$

$$\begin{aligned}
\Delta\chi_t = \tfrac{1}{2}[&-a^2\mathcal{CT}\phi - a^2\mathcal{RT}\psi - \tfrac{\mathcal{T}^2\pi^2\sqrt{s}}{\alpha}(\chi - \Phi) - \mathcal{CT}\Delta_1\phi - \\
&\mathcal{RT}\Delta_1\psi + \tfrac{\sqrt{s}}{\alpha}(\mathcal{C}^2 - \mathcal{R}^2)\Delta_1(\chi - \Phi) + \tfrac{a^2}{\alpha\sqrt{s}}(\mathcal{C}^2 - \\
&\mathcal{R}^2)\Delta(\chi + \Phi) - \tfrac{\mathcal{T}^2}{\alpha\sqrt{s}}\Delta\chi_{zz} - \tfrac{\mathcal{T}^2}{\alpha\sqrt{s}}\Delta\Phi_{zz}] + \Delta\Delta\chi + \bar{N}_3
\end{aligned} \quad (9)$$

$$\Delta\Phi_t = \tfrac{1}{2}[-a^2\mathcal{C}\mathcal{T}\phi - a^2\mathcal{R}\mathcal{T}\psi - \tfrac{\mathcal{T}^2\pi^2\sqrt{s}}{\alpha}(\chi-\Phi) - \mathcal{C}\mathcal{T}\Delta_1\phi - \mathcal{R}\mathcal{T}\Delta_1\psi + \tfrac{\sqrt{s}}{\alpha}(\mathcal{C}^2-\mathcal{R}^2)\Delta_1(\chi-\Phi) - \tfrac{a^2}{\alpha\sqrt{s}}(\mathcal{C}^2 - \mathcal{R}^2)\Delta(\chi+\Phi) + \tfrac{\mathcal{T}^2}{\alpha\sqrt{s}}\Delta\chi_{zz} + \tfrac{\mathcal{T}^2}{\alpha\sqrt{s}}\Delta\Phi_{zz}] + \Delta\Delta\Phi + \bar{N}_4 \tag{10}$$

where:
$\bar{N}_1 = \frac{1}{\alpha^2}[-\mathcal{C}\Delta N_{2,z} - \frac{a^2\mathcal{R}\mathcal{C}}{\mathcal{T}}\Delta N_3 + \frac{a^2\mathcal{R}^2-\pi^2\mathcal{T}^2}{\mathcal{T}}\Delta N_4]$,
$\bar{N}_2 = \frac{1}{\alpha^2}[\mathcal{R}\Delta N_{2,z} + \frac{a^2\mathcal{C}^2+\pi^2\mathcal{T}^2}{\mathcal{T}}\Delta N_3 - \frac{a^2\mathcal{R}\mathcal{C}}{\mathcal{T}}\Delta N_4]$,
$\bar{N}_3 = \frac{1}{2\alpha\sqrt{s}}[\alpha\sqrt{s}N_1 - \mathcal{T}\Delta N_{2,z} - a^2\mathcal{R}\Delta N_3 + a^2\mathcal{C}\Delta N_4]$,
$\bar{N}_4 = \frac{1}{2\alpha\sqrt{s}}[\alpha\sqrt{s}N_1 + \mathcal{T}\Delta N_{2,z} + a^2\mathcal{R}\Delta N_3 - a^2\mathcal{C}\Delta N_4]$.

As concerns the linear part, the use of the (canonical) Lyapunov function $E_1 = \frac{1}{2}[\|\nabla\phi\|^2 + \|\nabla\psi\|^2 + \|\nabla\chi\|^2 + \|\nabla\Phi\|^2]$ permits to have, along the solution to (7)-(10), $\frac{dE_1}{dt} = \mathcal{I} - \mathcal{D} + \mathcal{N} \le (m-1)\mathcal{D} + \mathcal{N}$ where

$$\begin{aligned}\mathcal{I} = {} & \tfrac{\mathcal{C}\mathcal{T}}{\alpha^2}[(\phi,\Delta\chi_{zz}) + \pi^2(\phi,\Delta\chi) + (\phi,\Delta\Phi_{zz}) + \pi^2(\phi,\Delta\Phi)] - \\ & \tfrac{\mathcal{R}\mathcal{T}}{\alpha^2}[(\psi,\Delta\chi_{zz}) + \pi^2(\psi,\Delta\chi) + (\psi,\Delta\Phi_{zz}) + \pi^2(\psi,\Delta\Phi)] + \\ & \tfrac{\mathcal{C}\mathcal{T}}{2}[a^2(\chi,\phi) + (\chi,\Delta_1\phi) + a^2(\Phi,\phi) + (\Phi,\Delta_1\phi)] + \\ & \tfrac{\mathcal{R}\mathcal{T}}{2}[a^2(\chi,\psi) + (\chi,\Delta_1\psi) + a^2(\Phi,\psi) + (\Phi,\Delta_1\psi)] \\ & + \tfrac{\pi^2\mathcal{T}^2\sqrt{s}}{2\alpha}[\|\chi\|^2 - \|\Phi\|^2] + \tfrac{\sqrt{s}}{2\alpha}(\mathcal{C}^2-\mathcal{R}^2)[(\Phi,\Delta_1\Phi) - (\chi,\Delta_1\chi)] + \\ & \tfrac{a^2}{2\alpha\sqrt{s}}(\mathcal{C}^2-\mathcal{R}^2)[\|\nabla\chi\|^2 + \|\nabla\Phi\|^2] + \tfrac{\mathcal{T}^2}{2\alpha\sqrt{s}}[(\chi,\Delta\chi_{zz}) - (\Phi,\Delta\Phi_{zz})],\end{aligned} \tag{11}$$

$\mathcal{D} = \|\Delta\phi\|^2 + \|\Delta\psi\|^2 + \|\Delta\chi\|^2 + \|\Delta\Phi\|^2$, $\mathcal{N} = (\phi,\bar{N}_1) + (\psi,\bar{N}_2) + (\chi,\bar{N}_3) + (\Phi,\bar{N}_4)$ and

$$m = \max_{\mathcal{H}} \mathcal{I}/\mathcal{D},$$

where $\mathcal{H}$ is the space of the admissible fields: $\mathcal{H} = \{\ \phi,\psi,\chi,\Phi$ regular fields, periodic in x and y of periods $2\pi/a_x, 2\pi/a_y\}$. By solving the Euler-Lagrange equations we have $m = \dfrac{\alpha}{s^{3/2}}$. Then we have linear stability, respect to E_1 norm, if $m < 1$, that is $\alpha < s^{3/2}$. By taking into account the expression of α and by minimizing with respect to a^2, we obtain the same limit of the linear instability theory.

In order to control the nonlinear terms, we add to E_1 the complementary function: $V_1 = \dfrac{1}{2}[\|\Delta\mathbf{u}\|^2 + \|\Delta\vartheta\|^2 + \|\Delta\gamma\|^2]$ and the full Lyapunov norm is $\mathcal{E} = E_1 + bV_1$ with b positive parameter (see Ref. 2). Then

$$\frac{d\mathcal{E}}{dt} = \mathcal{I} - \mathcal{D} + \mathcal{N} + b\mathcal{I}_1 - b\mathcal{D}_1 + b\mathcal{N}_1$$

where $\mathcal{I}_1 = 2\mathcal{R}(\Delta w, \Delta\vartheta)$, $\quad \mathcal{D}_1 = \|\nabla\Delta\mathbf{u}\|^2 + \|\nabla\Delta\vartheta\|^2 + \|\nabla\Delta\gamma\|^2$ and $\mathcal{N}_1 = -(\Delta\Delta\mathbf{u}, \mathbf{u}\cdot\nabla\mathbf{u}) - (\Delta\Delta\vartheta, \mathbf{u}\cdot\nabla\vartheta) - (\Delta\Delta\gamma, \mathbf{u}\cdot\nabla\gamma)$.

We soon observe that $b\mathcal{I}_1 \leq \mathcal{D}_2$ with $b = (\frac{1-m}{2})(\frac{2\sqrt{s}}{\alpha}\mathcal{R}^2 + 3\mathcal{T}\mathcal{R})^{-1}$, where $\mathcal{D}_2 = \frac{1-m}{2}\mathcal{D} + \frac{b}{2}\mathcal{D}_1$. The estimates of the nonlinear terms are obtained by using the Wirtinger-Poincaré inequality $\pi_0^2\|f\|^2 \leq \|\nabla f\|^2$, where $\pi_0^2 = \min(\pi^2, a_x^2, a_y^2)$ (see Ref. 3) and the imbedding result $\sup|F| \leq c_0\|\Delta F\|$, where $F = 0$ or $F_z = 0$ on $z = 0$, $z = 1$ and periodic in x and y, and c_0 is the maximum of the two constants given by (A.7) and (A.8) of Ref. 11. We obtain

$$\mathcal{N} + b\mathcal{N}_1 \leq \mathcal{A}\mathcal{D}_2\mathcal{E}^{1/2}$$

where $\mathcal{A}$ is a computable constant depending on $\pi_0, c_0, (1-m), \mathcal{R}, \mathcal{C}, \mathcal{T}$. Then we can write $\frac{d\mathcal{E}}{dt} \leq -\mathcal{D}_2(1 - \mathcal{A}\mathcal{E}^{1/2})$ which implies the following nonlinear stability theorem:

Theorem 2.1. *Assume that $\mathcal{R}^2 < \mathcal{R}_{L1}^2$ with $\mathcal{R}_{L1}^2$ given by (5), and $\mathcal{E}(0) < \mathcal{A}^{-2}$. Then the basic motionless state is nonlinearly stable with respect to the Lyapunov function $\mathcal{E}(t)$, and*

$$\mathcal{E}(t) \leq \mathcal{E}(0)\exp\left\{-k_0\left[1 - \mathcal{A}\mathcal{E}(0)^{1/2}\right]t\right\}, \tag{12}$$

where k_0 is a positive constant depending on m, c_0, π_0.

References

1. S. Chandrasekhar, *Hydrodynamic and hydromagnetic stability.* Oxford: Clarendon Press (1961).
2. G. P. Galdi and B. Straughan, *Proc. Roy. Soc. London A* **402**, 257-283 (1985).
3. R. Kaiser and L. X. Xu, *Nonlinear Diff. Eqns. Appl.* **5(3)**, 283-307 (1998).
4. G. Mulone and S. Rionero, *Continuum Mech.Thermodyn.***9**, 347-363 (1997).
5. D. D. Joseph, *Stability of Fluid Motions.* Springer Tracts in Natural Philosophy, vols. 27 and 28. Berlin, Springer-Verlag 1976.
6. G. Mulone, *Continuum Mech.Thermodyn.***6**, 161-184 (1994).
7. G. Mulone and S. Rionero, *Rend. Mat. Acc. Lincei* **9**,221-236(1998).
8. S. Lombardo, G. Mulone and S. Rionero, *J. Mat. Anal. Appl.* **262(1)**, 191-207 (2001).
9. S. Rionero and G. Mulone, *ZAMM* **69**, 441-446 (1989).
10. J. Ortiz de Zrate, F. Peluso and J. Sengers, *The Eur. Phys. J. E - Soft Matter* **15,** 319-333 (2004).
11. B. Straughan, *The Energy Method, Stability, and Nonlinear Convection,* Springer-Verlag: New-York, 2nd Ed., 2004.
12. G. Mulone, *Far East J. Appl. Math.* **15** (2004), no. 2, 117–135.
13. D. Laoroze, J. Martinez-Mardones, J. Bragard and P. Vargas, *Phyisca A,* **371** 46–49 (2006).
14. S. Lombardo and G. Mulone, *Nonlinear Anal.* **63/5-7**, e2091-e2110 (2005).
15. S. Lombardo, G. Mulone and M.Trovato, (JMAA, 2008 to appear).

NAVIER-STOKES IN APERTURE DOMAINS: EXISTENCE WITH BOUNDED FLUX AND QUALITATIVE PROPERTIES

P. MAREMONTI

Mathematical Department
via Vivaldi, 43,
Caserta, I-81100, Italy
E-mail: paolo.maremonti@unina2.it

In this note we show a result of existence and some qualitative properties of solution to the Navier-Stokes equations in aperture domains. Roughly speaking, an aperture domain is an open connected consisting of two separated half-spaces connected by a hole. As it is known, in this special but physically interesting geometry the IBVP is a well posed if a flux condition trhough the hole is given as data. Our results correspond to an assumption on the flux that we consider *physically reasonable*. The results of this note will appear in the papers [2,18].

1. Introduction

This note is concerned with the initial boundary value problem in aperture domains for the Navier-Stokes system, which is assumed as model for the motion of an incompressible homogeneous viscous fluid. Roughly speaking, an aperture domain is an open connected consisting of two separated half-spaces connected by a hole. We denote by $\Omega \subset \mathbb{R}^3$ a three-dimensional aperture domain. Since our question are not connected with the regularity of Ω, we assume $\partial\Omega$ smooth. $\mathbb{R}^3 \setminus \overline{\Omega}$ can be a non-connected domain. We set for $d > 0$

$$\mathbb{R}^3_{+d} = \{x \in \mathbb{R}^3 : x_3 > d\},\ \mathbb{R}^3_{-d} = \{x \in \mathbb{R}^3 : x_3 < -d\}.$$

We say that $\Omega \subset \mathbb{R}^3$ is an aperture domain, if Ω is a smooth domain and there exists a ball $B_R(0)$ such that

$$\Omega \cup B_R(0) = \mathbb{R}^3_{+d} \cup \mathbb{R}^3_{-d} \cup B_R(0).$$

In this paper we assume that the following decomposition holds: there exist Ω_+ and Ω_- disjoint domains $\subset \Omega$ and a smooth two dimensional manifold

M such that

$$\Omega_+ \cup B_R = \mathbb{R}^3_{+d} \cup B_R, \ \Omega_- \cup B_R = \mathbb{R}^3_{-d} \cup B_R,$$
$$M = \partial\Omega_+ \cap \partial\Omega_- \subset \overline{B}_R, \ \Omega = \Omega_+ \cup \Omega_- \cup M.$$

Since $\mathbb{R}^3 \backslash B_R$ is like of the "half-space" $\mathbb{R}^3_{+d}$ and $\mathbb{R}^3_{-d}$, following [14], we call Ω_+ (respectively Ω_-), a "perturbation" of $\mathbb{R}^3_{+d} \cup \mathbb{R}^3_{-d} \cup B_R(0)$. We introduce the notion of flux through the aperture. Let us consider a field ψ:

$$\psi : \Omega \to \mathbb{R}^3.$$

We define the functional

$$\Phi[\psi] = \int_M \psi \cdot n d\sigma.$$

$\Phi[\psi]$ is the flux of ψ trough M, with n normal on M which we assume directed into Ω_-. If $\nabla \cdot \psi = 0$, given a smooth two-dimensional manifold $\widetilde{M}$ such that $\widetilde{M} \cup M$ is the boundary of a bounded domain $\widetilde{\Omega} \subset \overline{\Omega}$, then,

$$\int_{\widetilde{M}} \psi \cdot n d\sigma = - \int_M \psi \cdot n d\sigma.$$

Hence for $\nabla \cdot \psi = 0$, $\Phi[\psi]$ is independent of M, hence of the particular partition $\Omega = \Omega_+ \cup \Omega_- \cup M$. We sketch two typical example of aperture domains *:

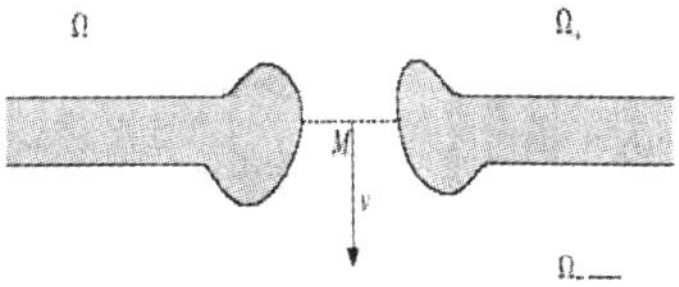

Fig. 1. aperture domain with $\mathbb{R}^3 - \Omega$ connected.

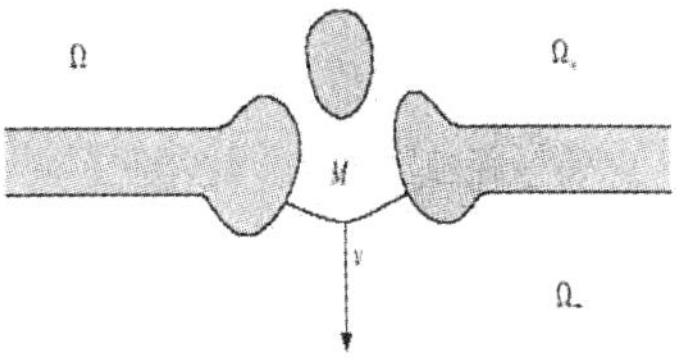

Fig. 2. aperture domain with $\mathbb{R}^3 - \overline{\Omega}$ non-connected.

Although the geometry of the domain and the physical meaning of the dynamic of the fluid are immediately, the analytic problem is very involved already in the most simple kind of domain. Now, we introduce

*From the paper [8].

the analytic difficulties of the problem. We start giving the definition of some spaces of functions. By $L^p(\Omega)$ we denote the usual Lebesgue space; by $W^{1,p}(\Omega)$ the Sobolev spaces, that is function in $L^p(\Omega)$ together with its first generalized derivatives. Fundamental are the completion spaces. We indicate by $W_0^{1,p}(\Omega)$ = completion of $C_0^\infty(\Omega)$ in $W^{1,p}(\Omega)$. Denoted by $\mathscr{C}_0(\Omega) = \{\psi \in C_0^\infty(\Omega) : \nabla \cdot \psi = 0\}$, we consider

- $p > 1$ $J^p(\Omega)$ = completion of $\mathscr{C}_0(\Omega)$ in $L^p(\Omega)$;
- $p > 1$, $J^{1,p}(\Omega)$ = completion of $\mathscr{C}_0(\Omega)$ in $W^{1,p}(\Omega)$.

The above functional spaces as subspaces are respectively contained in

- $p \geq 1$, $J_*^p(\Omega) = \{\psi \in L^p(\Omega) : \nabla \cdot \psi = 0,$ in a generalized sense$\}$;
- $p \geq 1$, $J_*^{1,p}(\Omega) = \{\psi \in W_0^{1,p}(\Omega) : \nabla \cdot \psi = 0\}$.

Of course we have also $J^{1,p}(\Omega) \subseteq J_*^{1,p}(\Omega) \subset W_0^{1,p}(\Omega)$. The coincidence of $J^{1,p}(\Omega)$ and $J_*^{1,p}(\Omega)$ arises our analytic problem. There is a nonempty set of domains, physically meaning, for which the coincidence holds (cf. [9]):

- $\Omega \subset \mathbb{R}^n, n \geq 2$, bounded,
- $\Omega = \mathbb{R}_+^n$, $n \geq 2$, half-space,
- $\Omega \subset \mathbb{R}^n, n \geq 2$, exterior domain,
- $\Omega = \mathbb{R}^n$,
- domains with noncompact boundary provided that suitable conditions are satisfied;

hence for each of them $J^{1,p}(\Omega) = J_*^{1,p}(\Omega)$. For the aperture domain:

$$p > \tfrac{n}{n-1},\ J^{1,p}(\Omega) = \{\psi \in J_*^{1,p}(\Omega) : \Phi[\psi] = 0\} \subset J_*^{1,p}(\Omega);$$
$$p \in [1, \tfrac{n}{n-1}],\ J^{1,p}(\Omega) = J_*^{1,p}(\Omega),$$

in the sense that,

$$p \in [1, \tfrac{n}{n-1}],\ \psi \in J_*^{1,p}(\Omega) \Rightarrow \Phi[\psi] = 0.$$

Usually $J^{1,p}(\Omega)$ is the natural space to establish the existence and uniqueness of solutions to the Navier-Stokes equations, but the above characterization becomes a problem. We start with the problem from the physical point of view. Condition $\psi \in J^{1,p}(\Omega)$, then $\phi[\psi] = 0$, means that the possible fluid motion holds without flux trough M. The motion develops in Ω_+ and Ω_-, this becomes an *a priori* physical constrain. The above considerations lead to consider $J_*^{1,p}(\Omega)$, for some $p > \frac{3}{2}$. Now, we look to the problem from the analytic point of view. It is well known that the usual IBVP for the Navier-Stokes equations is

$$\begin{aligned} & u_t + u \cdot \nabla u + \nabla \pi = \nu \Delta u + F \quad \text{in } \Omega \times (0,T), \\ & \nabla \cdot u = 0, \quad \text{in } \Omega \times (0,T), \\ & u(x,t) = 0 \quad \text{on } \partial\Omega \times (0,T), \ \lim_{|x|\to\infty} u(x,t) = 0, \\ & u(x,0) = u_\circ(x), \quad \text{in } \Omega, \end{aligned} \tag{1}$$

with u kinetic field, π pressure field, ν kinetic viscosity; F body force. J. Heywood in 1976 in [11] has proved that already in the steady linear case:

$$\begin{aligned} & \nabla \pi = \nu \Delta u \,, \ \nabla \cdot u = 0 \quad \text{in } \Omega, \\ & u(x) = 0 \quad \text{on } \partial\Omega, \ \lim_{|x|\to\infty} u(x) = 0, \end{aligned} \tag{2}$$

the problem of the uniqueness is not well posed in $J_*^{1,p}(\Omega)$, $p > \frac{3}{2}$. It is well posed in $J_*^{1,p}(\Omega)$, $p \in (1, \frac{3}{2}]$, but this last space is not one of the physically meaning cases. In other words there exist solutions of the problem (2) with zero data but with flux $\Phi[u] \neq 0$. Of course, to avoid the difficulties, it is sufficient to give the flux condition trough M

$$\Phi[u](t) = \alpha(t), \ \forall t \in [0,T).$$

However, the condition on the flux can be substituted with the pressure drop

$$[\pi](t) = \pi_+(t) - \pi_-(t) = \beta(t)$$

where

$$\pi_+(t) = \lim_{|x|\to\infty} \pi(x,t) \ \text{ in } \Omega_+, \ \pi_-(t) = \lim_{|x|\to\infty} \pi(x,t), \ \text{ in } \Omega_-.$$

Hence the usual IBVP must changed in the following:

$$\begin{aligned} & u_t + u \cdot \nabla u + \nabla \pi = \Delta u \,, \ \nabla \cdot u = 0, \quad \text{in } \Omega \times (0,T), \\ & u(x,t) = 0 \quad \text{on } \partial\Omega \times (0,T), \ \lim_{|x|\to\infty} u(x,t) = 0, \\ & u(x,0) = u_\circ(x), \quad \text{in } \Omega, \ \Phi[u](t) = \alpha(t), \ \forall t \in [0,T) \end{aligned} \tag{3}$$

otherwise, unlike the flux condition, the pressure drop:

$$[\pi](t) = \pi_+(t) - \pi_-(t) = \beta(t), t \in [0,T).$$

I recall the remark on the independence of the flux

$$\phi[u](t) = \int_M u(x,t) \cdot n d\sigma$$

from the manifold M, which makes the new IBVB with flux condition independent of M. Hence the measure of the flux data $\alpha(t)$ makes uniquely resolving the problem. Since 1976, this problem has been studied by several authors. Pioneer papers are [11,19,22,23], more recently on steady problem [1,9]. Results essentially concerning the space of the hydrodynamics and the flux compatibility condition are given in [4–6]. The first results related to the unsteady problem can be found in [12,22,23] and recent contributes for unsteady problem in [3,7,8,13,15].

2. Statement of the results for the INVP (3)

To study the initial boundary value problem (3), which is a well posed problem in an aperture domains, we start translating in a suitable way the problem from $J_*^{1,2}(\Omega)$ to $J^{1,2}(\Omega)$. For this task we introduce an auxiliary function. Indeed, in order to solve the existence problem, we look to an extension in Ω of the flux data on M. In such way we introduce a suitable body force and the flux trough M becomes zero. As a consequence, we can employ $J^{1,p}(\Omega)$ as space of functions for the existence of solutions. Let us consider the Stokes problem:

$$\begin{aligned}
&\Delta b = \nabla \tilde{q}\,,\ \nabla \cdot b = 0, \quad \text{in } \Omega,\\
&b(x) = 0, \quad \text{on } \partial\Omega,\ \lim_{|x|\to\infty} b(x) = 0,\\
&\Phi(b) = 1.
\end{aligned} \tag{4}$$

Lemma 2.1. *There exists a unique solution (b, q) of the problem (8), smooth in $\overline{\Omega}$ and such that*

$$|(1+|x|)\nabla b(x)| + |b(x)| \le c(1+|x|)^{-2}, \forall x \in \Omega.$$

Proof. See, e.g., [9].

The existence of $b(x)$ allows to look to a solution $u(x,t)$ in the form $u(x,t) = U(x,t) + \alpha(t)b(x)$, where $U(x,t)$ is a solution of the system

$$\begin{aligned}
&U_t + U\cdot\nabla U + \nabla P = \nu\Delta U - \alpha U\cdot\nabla b - \alpha b\cdot\nabla U\\
&\qquad\qquad\qquad\qquad\qquad\qquad -\alpha^2 b\cdot\nabla b - \alpha' b,\\
&\nabla\cdot U = 0, \quad \text{in } \Omega\times(0,T),\\
&U(x,t) = 0, \quad \text{su } \partial\Omega\times(0,T),\ \lim_{|x|\to\infty} U(x,t) = 0,\\
&U(x,0) = u_\circ(x) - \alpha(0)b(x), \ \text{in } \Omega,\ \Phi[U](t) = 0,\ \forall t\in[0,T).
\end{aligned} \tag{5}$$

This approach leads to the flux condition

$$\Phi[U](t) = 0, \forall t \in [0, T),$$

which allows to study (5) in $J^{1,p}(\Omega)$, for some $p > 1$, again. In the following we consider $J^{1,2}(\Omega)$. With the position $u(x,t) = U(x,t) + \alpha(t)b(x)$ we come back to the problem in the space $J^{1,2}(\Omega)$, which is congenial to exhibit an existence theorem. However, this simple idea arises questions connected with the mechanics of the fluids. The position

$$u(x,t) = U(x,t) + \alpha(t)b(x),$$

produces, among other terms, a "*body force*" in (5):

$$-\alpha^2(t)b \cdot \nabla b - \alpha'(t)b(x).$$

It is evident that we need of suitable assumptions on the flux $\alpha(t)$ just to obtain an existence theorem, possibly, of regular solutions defined for any $t > 0$. Supposed the regularity of the flux $\alpha(t)$, assumption "physically reasonable" for $\alpha(t)$ is the *small size* in L^∞:

$$|\alpha(t)| << \varepsilon \ \text{ for any } t > 0.$$

To our result we premise the definition of regular solution:

Definition 2.1. A pair of functions (u, π) is said $p-$regular solution of system (3) if, for some $p \in (1, \infty)$ and for any $\eta \in (0, T)$, $u(x,t) \in C(0,T; L^p(\Omega)) \cap L^p(\eta, T; W^{2,p}(\Omega) \cap J^{1,p}_*(\Omega))$, $\nabla \pi(x,t)$, $u_t(x,t) \in L^p(\eta, T; L^p(\Omega))$, (u, π) satisfies system (3) a.e. in $\Omega \times (0,T)$ and $\lim\limits_{t \to 0^+} |u(t) - u_\circ|_p = 0$.

Our result is the following

Theorem 2.1. *Let $u_\circ(x) \in W^{2,2}(\Omega) \cap L^p(\Omega)$, for some $p \in (\frac{3}{2}, 2]$, and $D^k\alpha(t) \in L^\infty(0,T) \cap C([0,T))$, for any $T > 0$ and $k \in \{0,1\}$, $D^2\alpha(t) \in L^1(0,T) \cap C(0,T)$, for any $T > 0$. Assume that $u_\circ(x)$ and $\alpha(t)$ satisfy the compatibility conditions $u_{\circ|\partial\Omega} = 0$, $\nabla \cdot u_\circ = 0$, $\Phi(u_\circ) = \alpha(0)$ and, in particular, that*

$$|u_\circ|_p + |u_\circ|_{2,2} + \sum_{k=0}^{1} |D^k\alpha(t)|_\infty + \sup_{t \geq 0} \int_t^{t+1} |D^2\alpha(\tau)| d\tau < \mu, \tag{6}$$

for a positive constant μ suitably small. Then, system (3) has a unique regular solution (u, π) with

$$|u(t)|_p + |u(t)|_{2,2} + |u_t(t)|_2 \leq c(\mu), \ \ \forall t \geq 0. \tag{7}$$

It is quite natural to inquire:

- what are the differences between our result and results already known in literature;
- what are the mechanical meaning of our result.

Already for Stokes IBVP, our theorem is the first result of *global existence* of regular, and weak, solutions, where, with respect to suitable norms, the solution $u(x,t)$ is *uniformly bounded in time*, without requiring integrability on $(0,+\infty)$ to the flux $|\alpha(t)|$.
Concerning the point 2), we remark that a flux only bounded on $(0,+\infty)$ cannot arise infinity "kinetic energy". In our theorem the kinetic energy is uniformly finite in $t>0$. Moreover, if the flux is constant, then, we can attend to steady solutions as limit of unsteady solutions; in particular if the flux is time periodic, then, we obtain time periodic solutions. *The last two statements are not compatible with $|\alpha(t)|$ integrable on $(0,+\infty)$.*

3. Some qualitative properties of the solutions

In this section we show results concerning with the existence and uniqueness of time periodic solutions and with the asymptotic spatial properties of the solutions of Theorem 2.1. We begin giving the definition of time periodic function

Definition 3.1. A field $f:\Omega\times(0,T)\to\mathbb{R}^3$ is said *time periodic*, with period ω, if $f\in C(0,\omega;X)$, X Banach space, and

- $\max\limits_{[0,\omega]}|f(t)|_X<+\infty$;
- $|f(t+\omega)-f(t)|_X=0,\quad \forall t\geq 0.$

Theorem 3.1. *Let μ_ω be a positive constant. Let $\alpha(t)$ be a time-periodic function with period ω, such that $D^k\alpha(t)\in C([0,\omega])$ for $k\in\{0,1\}$, $D^2\alpha(t)\in L^1(0,\omega)$, and*

$$\sum_{k=0}^{1}|D^k\alpha(t)|_\infty+\sup_{t\geq 0}\int_t^{t+1}|D^2\alpha(\tau)|d\tau<\mu_\omega/2.$$

Then, there exists a function $u_\circ(x)\in W^{2,2}(\Omega)\cap J^{1,2}_*(\Omega)\cap L^p(\Omega)$, with $|||u_\circ|||:=|u_\circ|_p+|u_\circ|_{2,2}<c(\mu_\omega)$ and $\Phi(u_\circ)=\alpha(0)$, such that a solution (u,π) of system (3) corresponding to $u(x,0)=u_\circ$ is time-periodic with period ω and

$$|||u|||\leq c(\mu).$$

Moreover, the solution $u(x,t)$ is the unique time-periodic solution corresponding to the flux $\alpha(t)$.

It is interesting to stress that the existence of a time periodic solution is not obtained by means a theorem of fixed point. We employ a technique introduced by J.Serrin in [21] for PDE, which is not different of some employed in ODE. The same technique is employed in [20]. Since the technique seems congenial for unbounded domains, it is employed in the papers [10,16,17]. In all these papers a time periodic solution is obtained as limit of unsteady solutions. We sketch the basic idea. Let $\alpha(t)$ be a periodic flux of period ω satisfying the hypotheses of the theorem. The existence problem is solved by determining an initial data $u_\circ$ in such a way that the solution corresponding to $u_\circ$ and to the periodic flux $\alpha(t)$ is a time periodic solution of period ω. Such an initial data $u_\circ$ is the limit point of any solution (v,q) of system (3), whose existence is ensured by the first existence theorem, provided that $|||v_\circ||| \leq \mu_\omega/2$ [†], for a suitable constant μ_ω. Setting

$$S_{\mu_\omega} = \{v(x)\colon |v|_p + |v|_{2,2} \leq \mu_\omega/2, \text{ for some } p \in (\frac{3}{2}, 2)\},$$

we can write *in a suitable sense*

$$u_\circ = S_{\mu_\omega}\text{-} \lim_{n\to\infty} v(x, n\omega, v_\circ), \quad \text{for any } v_\circ \in S_{\mu_\omega},$$

where $v(x,n\omega,v_\circ)$ is the solution corresponding to the initial data $v_\circ$. The solution (u,π) of system (3), corresponding to the limit point $u_\circ$ and an ω-periodic flux $\alpha(t)$, is an attractor of the solutions corresponding to initial data in S_{μ_ω} properly enclosed in a basin of attraction of $u(x,t,u_\circ)$. Actually this property makes $u(x,t,u_\circ)$ ω-periodic. Indeed, we have

$$S_{\mu_\omega}\text{-} \lim_n v(x, t+n\omega, v_\circ) - u(x,t) = 0,$$

hence

$$\begin{aligned} |u(t+\omega) - u(t)|_{S_{\mu_\omega}} = {} & S_{\mu_\omega}\text{-}\lim_n v(x, t+(n+1)\omega, v_\circ) - u(x, t+\omega) \\ & + S_{\mu_\omega}\text{-}\lim_n v(x, t+n\omega, v_\circ) - u(x,t) \\ & + S_{\mu_\omega}\text{-}\lim_n v(x, t+(n+1)\omega, v_\circ) - v(x, t+n\omega, v_\circ) = 0, \end{aligned}$$

which proves the periodicity of $u(x,t)$.

[†] we recall that $|||\cdot||| = |\cdot|_p + |\cdot|_{2,2}$

We conclude this section with the spatial behavior of $u(x,t)$ with respect to x uniformly in t, where $u(x,t)$ is the solution of the IBVP (3). The existence of the solution $u(x,t)$ starts from representation

$$u(x,t) = U(x,t) + \alpha(t)b(x).$$

The auxiliary function $\alpha(t)b(x)$ *translates* the flux data in a "body force". Function $b(x)$ is the solution to the Stokes problem

$$\begin{aligned} &\Delta b = \nabla \tilde{q}\,,\ \nabla \cdot b = 0, \quad \text{in } \Omega, \\ &b(x) = 0, \quad \text{on } \partial\Omega,\ \lim_{|x|\to\infty} b(x) = 0, \\ &\Phi(b) = 1, \end{aligned} \tag{8}$$

Function $b(x)$ has the following properties:

$$|(1+|x|)\nabla b(x)| + |b(x)| \leq c(1+|x|)^{-2}, \forall x \in \Omega.$$

As in the case of steady solutions to the Navier-Stokes equations in aperture domains, in quite natural way one poses the question: *what has the spatial behavior of the solution $u(x,t)$ with respect to x*
The difficulties are connected with the fact that we have not a representation formula of $u(x,t)$ by means a Green function with related asymptotic estimates in (x,t). However the fact that Ω_+ (Ω_-) can be seen as a half-space helps in our aims. Before giving the result, we recall that the pressure π has different limit at infinity. We set in Ω_+ $\lim_{|x|\to\infty} \pi(x,t) = \pi_+$ and in Ω_- $\lim_{|x|\to\infty} \pi(x,t) = \pi_-$ In [18] it is proved the following

Theorem 3.2. *Let $u_\circ(x) \in W^{2,2}(\Omega)$ with $|u_0(x)| \leq U_0(1+|x|)^{-s}$, for some $s \geq 1$. Let $D^k\phi(t) \in L^\infty(0,T) \cap C([0,T))$, $k \in \{0,1\}$, $D^2\phi(t) \in L^2(0,T)$. Assume that $u_\circ(x)$ and $\phi(t)$ satisfy the compatibility and smallness conditions employed in the existence theorem. Then, system (3) has a unique regular solution (u,π) such that, uniformly in (x,t),*

$$\begin{aligned} &|u(x,t)| \leq \frac{c(U_0,\mu)}{(1+|x|)^\nu}, \\ &|\nabla u(x,t)| \leq \left[\frac{1}{\sqrt{t^a}} + 1\right] \frac{c(U_0,\mu)}{(1+|x|)^\nu}, \\ &|\pi(x,t) - \pi_+(t)| \leq \left[\frac{1}{\sqrt{t^a}} + 1\right] \frac{c(U_0,\mu)}{(1+|x|)^\nu} \left[\log^a\left(\frac{|x|}{\sqrt{t}} + 1\right) + \log(|x|+e)\right], \\ &|\pi(x,t) - \pi_-(t)| \leq \left[\frac{1}{\sqrt{t^a}} + 1\right] \frac{c(U_0,\mu)}{(1+|x|)^\nu} \left[\log^a\left(\frac{|x|}{\sqrt{t}} + 1\right) + \log(|x|+e)\right], \end{aligned}$$

with $a = 0$, if $U_0 = 0$; otherwise $a = 1$, where $\nu = \min\{s, 2\}$.

References

1. Borchers W, Pileckas K. *Arch. Rational Mech. Anal* **120** (1992), 1-49.
2. F. Crispo and P. Maremonti, *Math. Meth. Appl. Sc.* Publ. online in Wiley InterScience (19/06/2007).
3. F. Crispo and A. Tartaglione, *Math. Methods Appl. Sci.* **30** (2007), 1375-1401
4. R. Farwig and H. Sohr, *Math. Nachr.* **170**, (1994) 53-77.
5. R. Farwig and H. Sohr, *Analysis,* **16**, (1996) 1-26.
6. R. Farwig, *Manuscripta Mathematica* **89**, (1996) 139-158.
7. M. Franzke, Preprint 2139, Department of Mathematics, Darmstadt University of Technology 2001.
8. M. Franzke, *Ann. Univ. Ferrara* - Sez.VII - Sc.Mat. **46**, (2000) 161-173.
9. G.P. Galdi, *Springer-Verlag* 1994; **38-39**.
10. G.P. Galdi and H. Sohr, *Arch. Ration. Mech. Anal.*, **172** (2004), 363-406.
11. J. Heywood, *Acta Math.* **136** (1976), 61–102.
12. J. Heywood, *Indiana University Math. J.* **29** (1980), 639–681.
13. T. Hishida, *Galdi G. P. (ed.) et al., Birkhuser. Advances in Mathematical Fluid Mechanics.* (2004), 79-123.
14. T. Kubo and Y. Shibata, *Quaderni di Matematica* **15** (2004), 149–220.
15. T. Kubo *Math. Methods Appl. Sci.* **28** (2005), 1341–1357.
16. P. Maremonti, *Nonlinearity,* **4** (1991), 503-529.
17. P. Maremonti and M. Padula, *Zap. Nauchn. Sem S.-Peterburg Otdel Mat. Inst. Steklov* (POMI), **233** (1996), 142-182.
18. P. Maremonti, M. Padula and V.A. Solonnikov Spatial behavior of solutions of the Navier-Stokes equations in aperture domains. To appear.
19. O.A. Ladyzhenskaya and V.A. Solonnikov, *Vestnik Leningrad Univ. Math.,* **10** (1977), 271-279.
20. S. Rionero, *Meccanica,* **3** (1968), 207-213.
21. J. Serrin. *Arch. Rational Mech. Anal.*, **3** (1959), 120-122.
22. V.A. Solonnikov, *Collé de France Seminar, Vol. 4, Pitman Research Notes in Mathematics* **84** (1983), 240-349.
23. V.A. Solonnikov, *in Mathematical topics in fluid mechanics. Proc. Summer Course, Lisbon, Portugal, September 9-13, 1991; Pitman Res. Notes Math. Ser.* **274** (1993), 117-162.

UNSTEADY SOLUTIONS OF PDEs GENERATED BY STEADY SOLUTIONS*

L. MARGHERITI, M. P. SPECIALE

Department of Mathematics, University of Messina
Salita Sperone 31, 98166 Messina, Italy
E-mail: margheriti@mat520.unime.it, speciale@mat520.unime.it

Invariant solutions of PDEs are found by solving a reduced system involving one independent variable less. When the solutions are invariant with respect to the so–called projective group, the reduced system is simply the steady version of the original system. This feature enables us to generate unsteady solutions when steady solutions are known. Various examples of physical relevance are provided.

Keywords: Lie group analysis; Invariant solutions.

1. Introduction

Lie group analysis[1–5] is a powerful tool for studying systematically differential equations (DEs). The Lie symmetries admitted by DEs may be used for finding invariant solutions, but also for introducing invertible variable transformations mapping DEs to a more convenient form.[4,6,7] In this paper, special Lie symmetries are used to determine unsteady solutions of PDEs by using steady solutions of the same PDEs and some identifications.

Let us consider a (system of) differential equation(s) $\Delta\left(\mathbf{x},\mathbf{u},\mathbf{u}^{(r)}\right)=0$, where $\mathbf{x}\in\mathbb{R}^m$ denotes the set of independent variables, $\mathbf{u}(\mathbf{x})\in\mathbb{R}^N$ the set of dependent variables and $\mathbf{u}^{(r)}$ the set of partial derivatives. A one–parameter Lie group of point transformations generated by the vector field

$$\Xi=\sum_{i=1}^{m}\xi_i(x_j,u_B)\frac{\partial}{\partial x_i}+\sum_{A=1}^{N}\eta_A(x_j,u_B)\frac{\partial}{\partial u_A}$$

leaves invariant the given system if its r–order prolongation $\Xi^{(r)}$ acting on the system is zero along the solutions.[1–5] This condition provides an

*Dedicated to T. Ruggeri on the occasion of his 60th birthday.

overdeterminated set of linear differential equations for the infinitesimal generators ξ_i and η_A, whose integration gives the admitted Lie symmetries.

An invariant solution $\mathbf{u} = \mathbf{\Theta}(\mathbf{x})$ satisfies the invariant surface condition

$$\xi_1(\mathbf{x}, \mathbf{u})\frac{\partial \Theta_A}{\partial x_1} + \xi_2(\mathbf{x}, \mathbf{u})\frac{\partial \Theta_A}{\partial x_2} + \dots \xi_m(\mathbf{x}, \mathbf{u})\frac{\partial \Theta_A}{\partial x_m} = \eta_A(\mathbf{x}, \mathbf{u}) \qquad (1)$$

for $A = 1, 2, \dots, N$, and the system itself. Solving (1) we may write

$$u_A = \psi_A(X_1(\mathbf{x}, \mathbf{u}), \dots, X_{m-1}(\mathbf{x}, \mathbf{u})) \qquad A = 1, 2, \dots, N$$

that, substituted into the system, gives a reduced system of PDE's with $m-1$ independent variables (*similarity variables*). In some cases the *reduced system*, if some identifications are made, coincides with the steady version of the original system. In this case, the solutions of the reduced system can be considered as the steady solutions of the original system, and the functional representation of the invariant solutions allows to build unsteady solutions from steady ones. Let us give some examples where the reduced equation can be viewed as the steady version of the equation at hand.

The linear heat equation

$$\frac{\partial u}{\partial t} = \frac{\partial^2 u}{\partial x^2} \qquad (2)$$

admits, among the others, the Lie point symmetry generated by

$$\Xi = t^2\frac{\partial}{\partial t} + tx\frac{\partial}{\partial x} - \left(\frac{x^2}{4} + \frac{t}{2}\right)u\frac{\partial}{\partial u}. \qquad (3)$$

The corresponding invariant solution is given by

$$u = t^{-\frac{1}{2}} \exp\left(-\frac{x^2}{4t}\right)U(\omega), \qquad \omega = \frac{x}{t}, \qquad (4)$$

where $U(\omega)$ satisfies the reduced equation

$$0 = \frac{d^2 U}{d\omega^2}. \qquad (5)$$

By identifying ω with x and U with u, (5) can be viewed as the steady version of (2). A solution of (5) has the same structure as the steady solutions of (2) and provides, through (4), an unsteady solution of the heat equation.

Analogous considerations can be made for the Hopf's equation

$$\frac{\partial u}{\partial t} + u\frac{\partial u}{\partial x} = \frac{\partial^2 u}{\partial x^2}, \qquad (6)$$

that admits the Lie point symmetry generated by

$$\Xi = t^2\frac{\partial}{\partial t} + tx\frac{\partial}{\partial x} + (x - tu)\frac{\partial}{\partial u}, \qquad (7)$$

from which we have that the invariant solutions are given by

$$u = \frac{x}{t} + \frac{1}{t}U(\omega), \qquad \omega = \frac{x}{t}, \tag{8}$$

where $U(\omega)$ satisfies the reduced equation

$$U\frac{dU}{d\omega} = \frac{d^2U}{d\omega^2}. \tag{9}$$

Also in this case, equation (9) can be viewed as the steady version of (6), and relation (8) states how to generate unsteady solutions from steady ones.

The Lie symmetries corresponding to the operators (3) or (7) characterize the projective transformation, because the finite form of the transformation for the independent variables is the projective transformation.

2. Galilean systems and the projective group

Usually, when one consider systems of PDEs, the infinitesimal generators of the projective group are t^2 for the time variable, x_it for the generic spatial cartesian coordinate x_i, $\alpha_A u_A t$ (α_A suitable constant) for a field variable not representing a velocity, and $x_i - v_i t$ for the generic component of the velocity (if any). Hence, in order to investigate whether a system of PDEs admits a group of symmetries whose invariant solutions link the unsteady equations to the corresponding steady ones, some considerations are in order. Many physical systems are mathematically modelled by first order quasilinear PDEs that in non–relativistic situations must be invariant with respect to Galilei's transformations. It is known[8–10] that a Galilean system involving N field variables u_i and m space variables x_j has the form

$$\frac{\partial u_i}{\partial t} + \sum_{j=1}^{m}\left(v_j\frac{\partial u_i}{\partial x_j} + \sum_{k=1}^{N} b^j_{ik}(\mathbf{c})\frac{\partial u_k}{\partial x_j}\right) - G_i(\mathbf{c}) = 0, \tag{10}$$

where $u_i = v_i$ if $1 \le i \le m$, and $u_i = c_{i-m}$ if $m+1 \le i \le N$ (v_i are the components of velocity, and c_i are the components of the field variables not representing a velocity), whereas b^j_{ik} and G_i are functions depending on the indicated arguments. The equations of mathematical physics should also remain invariant when we change the units of the independent physical quantities. Hence we require that the Galilean system (10) be invariant with respect to the stretching group[11,12]

$$\begin{aligned} &t^\star = \lambda^\gamma t, \qquad x_i^\star = \lambda x_i, \qquad v_i^\star = \lambda^{1-\gamma} v_i, \qquad c_i^\star = \lambda^{\beta_i} c_i, \\ &{b^j_{ik}}^\star(\mathbf{c}^\star) = \lambda^{\gamma_{ik}} b^j_{ik}(\mathbf{c}), \qquad g_i^\star(\mathbf{c}^\star) = \lambda^{\delta_i} g_i(\mathbf{c}), \end{aligned} \tag{11}$$

where $\lambda \neq 0$ is a parameter, and γ, β_i, γ_{ik} and δ_i constants such that

$$\delta_i = \begin{cases} \beta_0 - \gamma, & i = 1, \ldots, m, \\ \beta_{i-m} - \gamma, & i = m+1, \ldots, N, \end{cases} \qquad \beta_0 = 1 - \gamma,$$

$$\gamma_{ik} = \begin{cases} \beta_0, & i, k = 1, \ldots, m, \\ 2\beta_0 - \beta_{k-m}, & i = 1, \ldots, m, \; k = m+1, \ldots, N, \\ \beta_{i-m}, & i = m+1, \ldots, n, \; k = 1, \ldots, m, \\ \beta_0 + \beta_{i-m} - \beta_{k-m}, & i, k = m+1, \ldots, N. \end{cases}$$

In order the Galilean system to be invariant with respect to the stretching group, the involved constitutive functions should have the form:[11,12]

$$b^j_{ik}(\mathbf{c}) = c_{N-m}^{\gamma_{ik}/\beta_{N-m}} B^j_{ik}(\mathbf{w}), \qquad g_i(\mathbf{c}) = c_{N-m}^{\delta_i/\beta_{N-m}} G_i(\mathbf{w}),$$
$$w_i = \frac{c_i}{c_{N-m}^{\beta_i/\beta_{N-m}}}, \qquad i = 1, \ldots, N-m-1.$$

If the Galilean system (10), admitting also (11), possesses invariant solutions with respect to the projective group, these have the form

$$\begin{aligned} v_i &= \frac{x_i}{t} + \frac{V_i(\boldsymbol{\omega})}{t}, & i &= 1, \ldots, m, \\ c_{i-m} &= t^{\alpha_{i-m}} C_{i-m}(\boldsymbol{\omega}), & i &= m+1, \ldots, N, \end{aligned} \tag{12}$$

where $\boldsymbol{\omega} = \frac{\mathbf{x}}{t}$ denotes the set of the new independent variables. It follows

$$w_i = \frac{c_i}{c_{N-m}^{\beta_i/\beta_{N-m}}} = \frac{t^{\alpha_i} C_i}{t^{\alpha_{N-m}\beta_i/\beta_{N-m}} C_{N-m}^{\beta_i/\beta_{N-m}}}, \qquad i = 1, \ldots, N-m-1,$$

and, since we want the time t to disappear from the constitutive functions evaluated in the considered solution, it has to be:

$$\alpha_s = \frac{\alpha_{N-m}}{\beta_{N-m}} \beta_s, \qquad s = 1, \ldots, N-m.$$

Setting $\mathbf{U} = (\mathbf{V}, \mathbf{C})$, the reduced system is

$$\sum_{j=1}^{m} \left(V_j \frac{\partial U_i}{\partial z_j} + \sum_{k=1}^{N} C_{N-m}^{\gamma_{ik}/\beta_{N-m}} B^j_{ik}(\mathbf{w}) \frac{\partial U_k}{\partial z_j} \right) - C_{N-m}^{\delta_i/\beta_{N-m}} G_i(\mathbf{w}) = 0, \tag{13}$$

$(i = 1, \ldots, N)$, provided that

$$\begin{aligned} &\sum_{j=1}^{m} B^j_{ij} = \begin{cases} 0, & i = 1, \ldots, m, \\ -\alpha_{i-m} C_{i-m} C_{N-m}^{-\beta_{i-m}/\beta_{N-m}}, & i = m+1, \ldots, N, \end{cases} \\ &\alpha_s = \frac{\beta_s}{\gamma - 1}, \qquad s = 1, \ldots N-m. \end{aligned} \tag{14}$$

Furthermore, if the system is non–homogenous there are the constraints:

$$\alpha_s = \frac{3}{2\gamma - 1}\beta_s \quad s = 1, \dots N - m \quad \text{if } G_i \neq 0 \text{ for some } i \in \{1, \dots, m\},$$

$$\alpha_s = \frac{2}{\gamma}\beta_s, \quad s = 1, \dots N - m, \quad \text{if } G_i \neq 0 \text{ for some } i \in \{m+1, \dots, N\}.$$

If both previous cases occur, then $\gamma = 2$ and $\alpha_s = \beta_s$ $(s = 1, \dots, N - m)$.

3. Applications

First, let us consider the Euler equations:

$$\begin{aligned} &\frac{\partial \mathbf{v}}{\partial t} + (\mathbf{v} \cdot \nabla)\mathbf{v} + p^{-1/\Gamma} s^{-1} \nabla p = 0, \\ &\frac{\partial s}{\partial t} + \mathbf{v} \cdot \nabla s = 0, \qquad \frac{\partial p}{\partial t} + \mathbf{v} \cdot \nabla p + \Gamma p \nabla \cdot \mathbf{v} = 0, \end{aligned} \tag{15}$$

where $\mathbf{v}(t, \mathbf{x})$ is the velocity, $s(t, \mathbf{x})$ the entropy and $p(t, \mathbf{x})$ the pressure. Moreover, the density is $\rho = p^{1/\Gamma} s$, Γ being the adiabatic index. The Euler equations fall into the general form considered above; if $\Gamma = (m+2)/m$ (m is the number of space variables), they admit the symmetry generated by

$$\Xi = t^2 \frac{\partial}{\partial t} + \sum_{i=1}^{m} \left(x_i t \frac{\partial}{\partial x_i} + (x_i - v_i t) \frac{\partial}{\partial v_i} \right) - (m+2) p t \frac{\partial}{\partial p}.$$

The corresponding invariant solutions are given by

$$\mathbf{v} = \boldsymbol{\omega} + \frac{1}{t}\mathbf{V}(\boldsymbol{\omega}), \quad s = S(\boldsymbol{\omega}), \quad p = t^{-(m+2)} P(\boldsymbol{\omega}), \quad \boldsymbol{\omega} = \frac{\mathbf{x}}{t},$$

and the reduced system is the steady version of the original system:

$$\begin{aligned} &(\mathbf{V} \cdot \widetilde{\nabla})\mathbf{V} + P^{-1\Gamma} S^{-1} \widetilde{\nabla} P = 0, \\ &\mathbf{V} \cdot \widetilde{\nabla} S = 0, \qquad \mathbf{V} \cdot \widetilde{\nabla} P + \Gamma P \widetilde{\nabla} \cdot \mathbf{V} = 0, \end{aligned}$$

where $\widetilde{\nabla} = (\partial_{\omega_1}, \dots, \partial_{\omega_m})$.

As a second example let us consider the 2D equations of ideal magneto-gasdynamics with a magnetic field transverse to the plane of motion:

$$\begin{aligned} &\frac{\partial p}{\partial t} + v_1 \frac{\partial p}{\partial x_1} + v_2 \frac{\partial p}{\partial x_2} + \Gamma p \left(\frac{\partial v_1}{\partial x_1} + \frac{\partial v_2}{\partial x_2} \right) = 0, \\ &p^{1/\Gamma} s \left(\frac{\partial v_i}{\partial t} + v_1 \frac{\partial v_i}{\partial x_1} + v_2 \frac{\partial v_i}{\partial x_2} \right) + \frac{\partial p}{\partial x_i} + \mu h \frac{\partial h}{\partial x_i} = 0, \quad i = 1, 2, \\ &\frac{\partial s}{\partial t} + v_1 \frac{\partial s}{\partial x_1} + v_2 \frac{\partial s}{\partial x_2} = 0, \\ &\frac{\partial h}{\partial t} + v_1 \frac{\partial h}{\partial x_1} + v_2 \frac{\partial h}{\partial x_2} + h \left(\frac{\partial v_1}{\partial x_1} + \frac{\partial v_2}{\partial x_2} \right) = 0. \end{aligned} \tag{16}$$

If $\Gamma = 2$ (see Ref. 13) the system (16) admits the projective group

$$\Xi = t^2\frac{\partial}{\partial t} + \sum_{i=1}^{2}\left(x_i t\frac{\partial}{\partial x_i} + (x_i - v_i t)\frac{\partial}{\partial v_i}\right) - 4pt\frac{\partial}{\partial p} - 2ht\frac{\partial}{\partial h},$$

and we have the invariant solutions

$$\begin{aligned} &p = t^{-4}P(\omega_1,\omega_2), \qquad s = S(\omega_1,\omega_2), \qquad h = t^{-2}H(\omega_1,\omega_2), \\ &v_i = \omega_i + t^{-1}V_i(\omega_1,\omega_2), \qquad \omega_i = \frac{x_i}{t}, \quad i = 1,2, \end{aligned} \tag{17}$$

where the reduced system is the steady version of (16), *i.e.*,

$$\begin{aligned} &V_1\frac{\partial P}{\partial \omega_1} + V_2\frac{\partial P}{\partial \omega_2} + 2P\left(\frac{\partial V_1}{\partial \omega_1} + \frac{\partial V_2}{\partial \omega_2}\right) = 0, \\ &P^{1/2}S\left(V_1\frac{\partial V_i}{\partial \omega_1} + V_2\frac{\partial V_i}{\partial \omega_2}\right) + \frac{\partial P}{\partial \omega_i} + \mu H\frac{\partial H}{\partial \omega_i} = 0, \quad i = 1,2, \\ &V_1\frac{\partial S}{\partial \omega_1} + V_2\frac{\partial S}{\partial \omega_2} = 0, \qquad V_1\frac{\partial H}{\partial \omega_1} + V_2\frac{\partial H}{\partial \omega_2} + H\left(\frac{\partial V_1}{\partial \omega_1} + \frac{\partial V_2}{\partial \omega_2}\right) = 0. \end{aligned} \tag{18}$$

Acknowledgments

The authors would thank F. Oliveri for the useful suggestions.

Bibliography

1. L. V. Ovsiannikov, *Group Analysis of differential equations* (Academic Press, New York, 1982).
2. N. H. Ibragimov, *Transformation groups applied to mathematical physics* (D. Reidel Publishing Company, Dordrecht, 1985).
3. P. J. Olver, *Applications of Lie groups to differential equations* (Springer, New York, 1986).
4. G. W. Bluman, S. Kumei, *Symmetries and differential equations* (Springer, New York, 1989).
5. P. J. Olver, *Equivalence, Invariants and Symmetry* (Cambridge Univ. Press, New York, 1995).
6. A. Donato, F. Oliveri, *Applicable Analysis* **58**, 313–323 (1995).
7. A. Donato, F. Oliveri, *J. Math. Anal. Appl.* **188**, 552–568 (1994).
8. S .M. Shugrin, *Differential Equations* **16**, 1402–1413 (1981).
9. T. Ruggeri, *Continuum Mech. Termodyn.* **1**, 3–20 (1989).
10. F. Oliveri, In *Nonlinear Hyperbolic equations: theory and computational aspects*, A. Donato & F. Oliveri eds., *Notes on Num. Fluid Mech.* **43**, Vieweg, Wien, 1993.
11. A. Donato, *Int. J. Non-linear Mech.* **22**, 307–314 (1987).
12. W. F. Ames, A. Donato, *Int. J. Non-linear Mech.* **23**, 167–174 (1988).
13. J.C. Fuchs, E.W. Richter, *J. Phys.A: Math. Gen.* **20**, 3135–3157 (1987).

ON TWO-PULSE INTERACTION IN A CLASS OF MODEL ELASTIC MATERIALS

A. MENTRELLI

Research Centre of Applied Mathematics (CIRAM), University of Bologna, Italy

C. ROGERS

The Polytechnic University of Hong Kong, Hong Kong & School of Mathematics and Statistics, The University of New South Wales, Sydney, NSW 2052, Australia & Australian Research Council Centre of Excellence for Mathematics and Statistics of Complex Systems

W. K. SCHIEF

Institut für Mathematik, Technische Universität Berlin, Straße des 17. Juni 136, D-10623 Berlin, Germany & Australian Research Council Centre of Excellence for Mathematics and Statistics of Complex Systems, School of Mathematics and Statistics, The University of New South Wales, Sydney, NSW 2052, Australia

A class of nonlinear stress-strain laws is introduced which may be used to model a type of material response encountered in superelastic materials such as nitinol. Here, pulse interaction and shock transmission in media with particular such model constitutive laws are investigated for a variety of initial data.

Keywords: pulse; nonlinear; elasticity

1. Introduction

Loewner,[1] in 1952, introduced a class of infinitesimal Bäcklund transformations (BTs) applied to the hodograph equations of gasdynamics with a view to their asymptotic reduction in subsonic régimes to the Cauchy-Riemann equations. Such reduction is available for appropriate model multi-parameter gas laws. Subsequently[2,3] a reinterpretation and generalisation of the Loewner class of infinitesimal BTs was shown to lead to a linear representation for a master class of 2+1-dimensional integrable nonlinear equations which may be parametrised in terms of a triad of eigenfunctions. The $\bar{\partial}$-dressing method for the generalised Loewner system was described in.[3]

Here, our concern is with the application of Loewner-type infinitesimal

BTs in the context of $1+1$-dimensional nonlinear elastodynamics. Again a solitonic connection is made. Thus, it emerges that the hodograph system in this case may be asymptotically reduced to a classical wave equation provided that the model nonlinear stress-strain laws have signal speed determined by particular travelling wave solutions of the integrable hyperbolic sinh-Gordon equations.

The soliton connection is exploited to construct model nonlinear constitutive laws which allow an interior change of concavity corresponding to a change of soft to hard elastic behaviour (or vice versa). Such response is encountered, importantly, in superelastic materials like nickel-titanium as used extensively in shape memory alloys with a wide range of medical device applications.[4,5] The class of novel model stress-strain laws to be discussed here involve elliptic functions and integrals thereof.[6] Pulse interaction in materials with these constitutive laws is investigated for appropriate initial data.

2. The infinitesimal BTs. A sinh-Gordon connection

Loewner,[1] in a gasdynamics setting, introduced a class of infinitesimal BTs

$$\mathbb{B}_\epsilon : \quad \mathbf{\Omega}_{\sigma\epsilon} = \mathbf{\Lambda}\mathbf{\Omega}_\sigma + \mathbf{A}\mathbf{\Omega} + \mathbf{C}\mathbf{\Omega}_\epsilon, \quad \mathbf{\Omega}_{\tau\epsilon} = \mathbf{\Lambda}\mathbf{\Omega}_\tau + \mathbf{B}\mathbf{\Omega} + \mathbf{D}\mathbf{\Omega}_\epsilon, \tag{1}$$

acting on a matrix equation

$$\mathbf{\Omega}_\sigma = \mathbf{S}\mathbf{\Omega}_\tau. \tag{2}$$

Thus, hodograph systems represented by (2) were sought which are asymptotic to an appropriate canonical form via the above infinitesimal BTs as the parameter $\epsilon \to \infty$. The compatibility of the system (1) augmented by (2) imposes six algebraic and differential constraints on the six matrices $\mathbf{A}, \mathbf{B}, \mathbf{C}, \mathbf{D}, \mathbf{\Lambda}$ and $\mathbf{S}$.[1]

Here, we shall be concerned with the application of a subclass of the infinitesimal BTs (1) to systems of the type (2) with

$$\mathbf{S} = \begin{pmatrix} 0 & h \\ \delta h^{-1} & 0 \end{pmatrix}, \qquad \delta = \text{const.}$$

A canonical reduction of the above-mentioned six constraints has been shown to be the $2+1$-dimensional integrable sinh-Gordon system[2,3,6]

$$\begin{aligned} (e^{-2(\alpha-\beta)}\alpha_{\sigma\epsilon})_\sigma &= (\delta e^{-2(\alpha-\beta)}\alpha_{\tau\epsilon})_\tau, \\ (\delta^{-1}e^{2(\alpha-\beta)}\beta_{\sigma\epsilon})_\sigma &= (e^{2(\alpha-\beta)}\beta_{\tau\epsilon})_\tau \end{aligned} \tag{3}$$

with $h = e^{2(\alpha-\beta)}$.

3. Application in nonlinear elasticity. A novel class of model stress-strain laws

In what follows, we consider the uniaxial deformation $x = x(X,t)$ of a nonlinear elastic medium with stress-strain law

$$T = \Sigma(e), \tag{4}$$

where T is the stress and $e = x_X - 1$ is the strain. In the above, x and X denote, in turn, Eulerian and Lagrangian coordinates. The governing Lagrangian equations of motion consist of the compatibility condition

$$e_t = v_X, \tag{5}$$

where $v = x_t$ is the velocity, together with the momentum equation

$$T_X = \rho_0 v_t. \tag{6}$$

Here, ρ_0 is the density of the elastic medium in its undeformed state. Substitution of (4) into (6) yields

$$A^2 e_X = v_t, \tag{7}$$

where $A = \sqrt{\rho_0^{-1} d\Sigma/de}$ is the signal speed.

On introduction of the hodograph transformation, wherein X, t are taken as new dependent variables and v, e as the new independent variables, it is seen that

$$v_X = Jt_e, \quad v_t = -JX_e, \quad e_X = -Jt_v, \quad e_t = JX_v, \tag{8}$$

where $J = J(v, e; X, t)$ is the Jacobian and it is required that $0 < |J| < \infty$. Use of the relations (8) in (5) and (7) produces the hodograph system

$$\mathbf{\Omega}_\sigma = \mathbf{S}\mathbf{\Omega}_\tau, \tag{9}$$

where

$$\mathbf{S} = \begin{pmatrix} 0 & A \\ A^{-1} & 0 \end{pmatrix}, \quad \mathbf{\Omega} = \begin{pmatrix} X \\ t \end{pmatrix}, \quad \frac{d\sigma}{de} = A, \quad \tau = v.$$

Here, the class of infinitesimal BTs of the preceding section is applied to the above hodograph system. Thus $h = A$ and hence $\alpha - \beta$ depends only on the single variable σ. If we impose the stronger requirement that α and β individually depend on σ alone, then the system (3) reduces to

$$g^*_{\sigma\epsilon} = \frac{1}{\rho^2} \sinh(g^* + \mu^*)$$

with $g^* = \ln h$ and a suitable choice of certain constants of integration. This integrable hyperbolic sinh-Gordon equation has been discussed elsewhere in the literature in other connections.[7,8]

Accordingly, nonlinear constitutive laws (4) are investigated for which the signal speed A is such that

$$[\ln (A/A_0)]_{\sigma\epsilon} = \frac{1}{\rho^2} \sinh[\ln(A/A_0) + \mu], \tag{10}$$

where $A_0 = A|_{e=0}$. If we set $g = \ln(A/A_0) + \mu$ in (10) then we obtain

$$g_{\sigma\epsilon} = \frac{1}{\rho^2} \sinh g. \tag{11}$$

The latter admits an auto-Bäcklund transformation

$$\mathbb{B}_\beta : \quad \left(\frac{\tilde{g} - g}{2}\right)_\sigma = \frac{\beta}{\rho} \sinh\left(\frac{\tilde{g} + g}{2}\right), \quad \left(\frac{\tilde{g} + g}{2}\right)_\epsilon = \frac{1}{\beta\rho} \sinh\left(\frac{\tilde{g} - g}{2}\right),$$

where β is a Bäcklund parameter. The vacuum solution $\tilde{g} = 0$ of (11) corresponds to a canonical Hooke's law with constant signal speed. Insertion of this seed Hookean solution into $\mathbb{B}_\beta$ and integration produces the solution

$$g = \ln(A/A_0) + \mu = 4 \tanh^{-1}\left[\exp\left(\frac{\beta\sigma}{\rho} + \frac{1}{\beta\rho}\epsilon + \gamma\right)\right],$$

where γ is an arbitrary phase constant. Thus, the signal speed A for the model stress-strain laws generated via this action of a single BT is given by

$$A/A_0 = M \coth^2\left(\frac{\beta\sigma}{2\rho} + \frac{\epsilon}{2\beta\rho} + \frac{\gamma}{2}\right), \tag{12}$$

where $M = A_\infty/A_0 = e^{-\mu}$. Moreover, the invariance of (11) under $g \to -g$ shows that (10) also admits the companion solution

$$A/A_0 = M \tanh^2\left(\frac{\beta\sigma}{2\rho} + \frac{\epsilon}{2\beta\rho} + \frac{\gamma}{2}\right). \tag{13}$$

It is noted that as $\epsilon \to \infty$, the relations (12) and (13) yield $A \to A_\infty$. The corresponding stress-strain laws are given parametrically in terms of the strain measure σ by

$$T = \rho_0 \int A(\sigma) d\sigma, \quad e = \int A^{-1}(\sigma) d\sigma. \tag{14}$$

These model laws were originally obtained in another manner[9] and were extensively applied in the study of uniaxial pulse propagation and reflexion in bounded nonlinear elastic media.[9–11] However, as noted in,[10] a limitation of these model laws is that they are restricted in their application in that they can only be used to approximate the response of nonlinear elastic materials in which the signal speed A has either a monotonically increasing or decreasing dependence on the strain e. Thus, the models cannot accommodate an interior inflexion point associated with a change from soft elastic behaviour ($d^2T/de^2 < 0$) to hard elastic behaviour ($d^2T/de^2 > 0$) or vice versa. Such behaviour is observed, importantly, in superelasticity in nickel-titanium.[4,5] In,[6] the connection with the integrable sinh-Gordon equation has been used to construct a novel class of model stress-strain laws which indeed exhibit an internal change of concavity. Here, our main purpose is to examine the nonlinear interaction of two-pulse solutions propagating in such materials.

4. The model laws

Travelling wave solutions of the hyperbolic sinh-Gordon equation (10) with $A = A(c)$, where $c = \sigma + \lambda\epsilon$, are given by, on integration,

$$\left(\frac{A_c}{A}\right)^2 = \frac{1}{\rho^2\lambda}\left[\frac{A}{A_\infty} + \frac{A_\infty}{A}\right] + \mathbb{C}, \tag{15}$$

where $\mathbb{C}$ is an arbitrary constant. In particular, if $\mathbb{C} = \pm 2/(\rho^2\lambda)$ then we retrieve a signal speed relation of the type $A_c = \tilde{\mu}A^{1/2} + \tilde{\nu}A^{3/2}$ as discussed in.[9–12] In general, (15) admits solutions in terms of the Jacobi elliptic functions dn and sn as considered below in terms of the variable $\mathcal{A} = A^{1/2}$:

$$^{(i)}\mathcal{A} = \mathbb{A}\mathrm{dn}(\alpha\sigma + \delta, k)$$

Here, $\delta := \alpha\lambda\epsilon + \delta_0$. Substitution in (15) reveals that

$$\mathbb{C} = 4\alpha^2(2 - k^2), \quad \lambda = -\mathbb{A}^2/4\rho^2\alpha^2 A_\infty, \quad \mathbb{A}^2 = A_\infty/k',$$

where $k'^2 = 1 - k^2$ and $0 \leq k < 1$. Integration of the pair (14) shows that the associated model stress-strain laws are given parametrically in terms of σ by the relations

$$\begin{aligned} T &= \rho_0\frac{\mathbb{A}^2}{\alpha}\mathrm{E}(\alpha\sigma + \delta, k) + T_0, \\ e &= \frac{1}{\alpha\mathbb{A}^2 k'^2}[\mathrm{E}(\alpha\sigma + \delta, k) - k^2\mathrm{sn}(\alpha\sigma + \delta, k)\mathrm{cd}(\alpha\sigma + \delta, k)] + e_0. \end{aligned} \tag{16}$$

In the above, $\mathrm{E}(\sigma, \mathrm{k})$ denotes Jacobi's epsilon function.

$$^{(ii)}\mathcal{A} = \mathbb{D}\mathrm{sn}(\alpha\sigma + \delta, k)$$

Here, substitution in (15) shows that

$$\mathbb{C} = -4\alpha^2(1+k^2), \quad \lambda = \mathbb{D}^2/4k^2\alpha^2\rho^2 A_\infty, \quad \mathbb{D}^2 = kA_\infty.$$

Hence, in this case, the model constitutive laws are given parametrically by the relations

$$T = \rho_0 \frac{\mathbb{D}^2}{\alpha k^2}[\alpha\sigma - \mathrm{E}(\alpha\sigma+\delta,k)] + T_0,$$
$$e = \frac{1}{\alpha\mathbb{D}^2}[\alpha\sigma - \mathrm{dn}(\alpha\sigma+\delta,k)\mathrm{cs}(\alpha\sigma+\delta,k) - \mathrm{E}(\alpha\sigma+\delta,k)] + e_0.$$

The model laws with signal speed

$$A = A_\infty \tanh^2(\alpha\sigma+\delta) \tag{17}$$

are retrieved in the limit $k \to 1$. The corresponding stress-strain laws have been applied to model real material dynamic response in.[9–11] It is noted that reciprocal model laws with $(i)^*$ $\mathcal{A} = \mathbb{B}^*\mathrm{nd}(\alpha\sigma+\delta,k)$ and $(ii)^*$ $\mathcal{A} = \mathbb{D}^*\mathrm{ns}(\alpha\sigma+\delta,k)$ are readily constructed.[6]

5. Integration of the hodograph system

In general, the stress-strain laws for real materials do not allow the hodograph system (9) to be integrated. The advantage of the model laws investigated in[2,4] is that not only do they permit integration of the hodograph equations but also that they may be used to approximate closely actual stress-strain relations for a wide variety of real materials. However, a limitation of these model laws is that they cannot accommodate changes in the concavity of the stress-strain law. Such complex elastic response is of current interest in superelasticity.[5,6] Here, the model laws involving elliptic functions also allow explicit integration of the hodograph system, but also may admit changes in concavity. The cases to be investigated are (i) $\mathcal{A} = \mathbb{A}\mathrm{dn}(\alpha\sigma+\delta,k)$ and (ii) $\mathcal{A} = \mathbb{D}\mathrm{sn}(\alpha\sigma+\delta,k)$.

Let us consider solutions of the hodograph system (9) in the separable form

$$X = \hat{X}(\sigma,\hat{\lambda})e^{\hat{\lambda}v}, \quad t = \hat{t}(\sigma,\hat{\lambda})e^{\hat{\lambda}v}, \tag{18}$$

whence

$$\begin{pmatrix}\hat{X}\\ \hat{t}\end{pmatrix}_\sigma = \hat{\lambda}\begin{pmatrix}0 & A(\sigma)\\ A^{-1}(\sigma) & 0\end{pmatrix}\begin{pmatrix}\hat{X}\\ \hat{t}\end{pmatrix}. \tag{19}$$

It is observed that (18) implies that $\sigma = \sigma(X/T;\hat{\lambda})$ corresponding to similarity solutions of the original strain equation. Now, application of the gauge transformation $\phi = \phi_\circ \psi$ to the Schrödinger equation

$$\phi_{\sigma\sigma} + u\phi = \hat{\mu}\phi \tag{20}$$

yields

$$\psi_{\sigma\sigma} + (\ln \phi_\circ^2)_\sigma \psi_\sigma + (u + \frac{\phi_{\circ\sigma\sigma}}{\phi_\circ} - \hat{\mu})\psi = 0. \tag{21}$$

But, (19) shows that

$$\hat{t}_{\sigma\sigma} + (\ln A)_\sigma \hat{t}_\sigma - \hat{\lambda}^2 \hat{t} = 0, \tag{22}$$

which may be aligned with (21) by setting

$$\hat{t} = \psi, \quad A = \phi_\circ^2, \quad \hat{\mu} - \hat{\mu}_\circ = \hat{\lambda}^2$$

if we choose $\phi_\circ$ to be a solution of the Schrödinger equation (20) with parameter $\hat{\mu}_\circ$. The reduction (19) of the hodograph system is therefore equivalent to the 'modified Schrödinger equation' (22) with

$$\hat{X} = A\hat{t}_\sigma/\hat{\lambda} = \phi_\circ^2 \psi_\sigma/\hat{\lambda}.$$

We now put $u = -2\wp$, where $\wp(\sigma)$ is the Weierstrass elliptic function determined as the unique even solution of

$$\wp'^2 = 4\wp^3 - g_2\wp - g_3 = 4(\wp - e_1)(\wp - e_2)(\wp - e_3)$$

with arbitrary invariants g_2, g_3 which define the roots e_1, e_2, e_3. Insertion in (20) produces the particular Lamé equation

$$\phi_{\sigma\sigma} - 2\wp\phi = \hat{\mu}\phi$$

with, if $\hat{\mu} \neq e_i$, two linearly independent solutions $\phi_\pm$ given by

$$\phi_\pm(\sigma) = \frac{\sigma^*(\sigma \pm a)}{\sigma^*(\sigma)} e^{\mp\sigma^*\zeta(a)}, \quad \wp(a) := \hat{\mu},$$

where ζ and σ^* here denote the Weierstrass functions related to $\wp$ by[a] $\zeta' = -\wp$ and $\sigma^{*'} = \zeta\sigma^*$. If $\hat{\mu} = e_i$ for some $i \in \{1, 2, 3\}$ then the above solutions are linearly dependent and may be replaced by

$$\phi_1^i(\sigma) = \sqrt{\wp(\sigma) - e_i}, \quad \phi_2^i(\sigma) = \sqrt{\wp(\sigma) - e_i}\big[\zeta(\sigma + \omega_i) + e_i\sigma\big].$$

In this connection, it is noted that

$$\sigma^*(\sigma + \omega_i) \sim \sigma^*(\sigma)\sqrt{\wp(\sigma) - e_i}\, e^{\sigma\zeta(\omega_i)}$$

[a]Here, to avoid a clash of notation, we have denoted the Weierstrass sigma function by σ^* rather than the usual σ.

and

$$\wp(\sigma + \omega') = e_3 + \frac{(e_3 - e_1)(e_3 - e_2)}{\wp(\sigma) - e_3},$$

where, if 2ω and $2\omega'$ denote the complex periods of $\wp$, then $\omega_1 = \omega$, $\omega_2 = \omega + \omega'$ and $\omega_3 = \omega'$. The following identities hold:

$$\wp(\sigma + 2\omega_k) = \wp(\sigma), \quad \wp(\omega_k) = e_k, \quad \wp'(\sigma + 2\omega_k) = \wp'(\sigma), \quad \wp'(\omega_k) = 0.$$

The connection with the Jacobi elliptic functions is given by

$$\begin{aligned}\wp(\sigma) = e_3 + \frac{e_1 - e_3}{\mathrm{sn}^2(\gamma\sigma, k)} &= e_3 + (e_2 - e_3)\mathrm{sn}^2[\gamma(\sigma - w'), k] \\ &= e_2 + (e_3 - e_2)\mathrm{cn}^2[\gamma(\sigma - w'), k] \\ &= e_1 + (e_3 - e_1)\mathrm{dn}^2[\gamma(\sigma - w'), k],\end{aligned}$$

where

$$\gamma^2 = e_1 - e_3, \quad k^2 = \frac{e_2 - e_3}{e_1 - e_3}.$$

The modified Lamé equation (22) may be written as

$$\hat{t}_{\sigma\sigma} + (\ln \phi_\circ^2)_\sigma \hat{t}_\sigma - (\hat{\mu} - \hat{\mu}_\circ)\hat{t} = 0$$

which, with the choice

$$\phi_\circ^2 \sim e_3 - e_i + \frac{(e_3 - e_1)(e_3 - e_2)}{\wp(\sigma) - e_3}, \quad \hat{\mu}_\circ = e_i, \quad \wp(a) = \hat{\mu} \notin \{e_1, e_2, e_3\},$$

becomes

$$\hat{t}_{\sigma\sigma} + [\ln(\wp(\sigma + \omega') - e_i)]_\sigma \hat{t}_\sigma - [\wp(a) - e_i]\hat{t} = 0.$$

If $i = 1, 2, 3$, this delivers, in turn

$$\hat{t}_{\sigma\sigma} + (\ln \mathrm{en}_i^2(\gamma\sigma, k))_\sigma \hat{t}_\sigma - (\wp(a) - e_i)\hat{t} = 0,$$

where $\mathrm{en}_1 = \mathrm{dn}$, $\mathrm{en}_2 = \mathrm{cn}$, $\mathrm{en}_3 = \mathrm{sn}$, with linearly independent solutions

$$\hat{t}_\pm = \frac{\sigma^*(\sigma \pm a)}{\sigma^*(\sigma)} e^{\mp\sigma\zeta(a)} \frac{\sqrt{\wp(\sigma \pm a) - e_3}}{\sqrt{\wp(\sigma) - e_3}} / \phi_i, \qquad i = 1, 2, 3.$$

If $\hat{\mu} = e_k$ then the general solution reads

$$\hat{t} = \frac{\phi_k}{\phi_i} \int \frac{c}{\phi_k^2} d\sigma.$$

In particular, if $i = 1, 3$, the signal speeds $A = \phi_i^2$ corresponding to the above solutions are given, in turn, by

$$\phi_1^2 \sim \mathrm{dn}^2(\gamma\sigma, k), \quad \phi_3^2 \sim \mathrm{sn}^2(\gamma\sigma, k).$$

The corresponding results for $A = \mathrm{nd}^2(\alpha\sigma + \delta, k)$ and $A = \mathrm{ns}^2(\alpha\sigma + \delta, k)$ are readily generated.

It is remarked that, importantly, (18) represents a single mode of more general solutions obtained via the Fourier synthesis

$$X = \int \hat{X}(\sigma; \hat{\lambda}) e^{\hat{\lambda} v} d\hat{\lambda}, \quad t = \int \hat{t}(\sigma; \hat{\lambda}) e^{\hat{\lambda} v} d\hat{\lambda}.$$

6. Pulse interaction. Numerical results

Here, we consider pulse interaction in the class of materials with model nonlinear stress-strain law given parametrically by the relation (16) corresponding to the signal speed $A = \mathbb{A}^2\mathrm{dn}^2(\alpha\sigma + \delta, k)$. Numerical investigations have been performed in order to study the interaction of two pulses under certain initial data. The interaction of the two pulses is investigated by comparison of the profiles of the strain and velocity fields at different times with the superposition of the corresponding profiles that would be generated by each single pulse separately.

Numerical solutions of the system given by (5) and (7) with the above signal speed have been calculated with a general-purpose code solving hyperbolic systems of nonlinear balance laws. This code is appropriate for solving numerically a system of conservation and balance laws that can be written in conservative form. The algorithm employed is a modified version of the *Uniformly Accurate Central Scheme of Order 2* (UCS2), developed by Liotta, Romano and Russo[13] as used in.[14]

The behaviour of the stress-strain law is shown in Fig. 1(a) for two different values of k ($k = 0.72$ and $k = 0.92$). The initial data of the

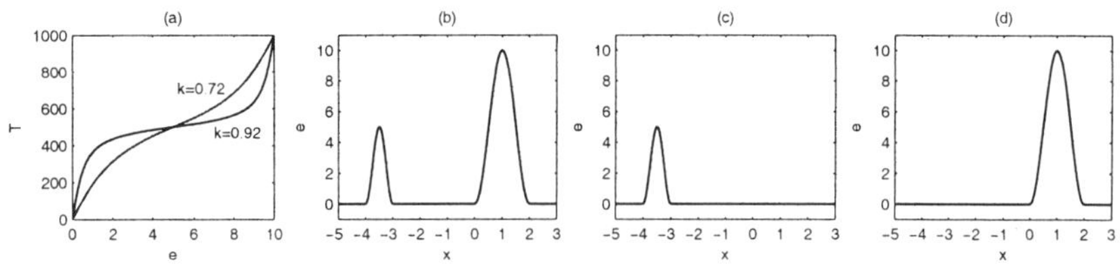

Fig. 1. Stress-strain law for different values of k (a); initial data for the strain, e_0, with two pulses (b), with only the left pulse (c) and with only the right pulse (d).

strain field, $e_0(X) \equiv e(X, t = 0)$, adopted in the calculations are shown in Fig. 1(b,c,d): the initial profile with two distinct pulses, shown in Fig. 1(b), is studied together with initial data characterised by each of the two pulses taken separately, as shown in Fig. 1(c) and Fig. 1(d), respectively, in order to investigate the effects of the nonlinearity of the model laws on the pulse interaction. In each case, the initial profile of the velocity field is assumed to be uniformly zero: $v_0(X) \equiv v(X, t = 0) = 0$.

The results show that when $k = 0.92$ (Fig. 2), the two pulses interact in a strongly nonlinear way, while in the case with $k = 0.72$ (Fig. 3) the behaviour is close to the linear one and the strain and velocity profiles obtained with the two-pulses initial data slightly differ from those generated by the superposition of the profiles obtained with each of the pulses taken alone. As one can see in Fig. 1(a), in the latter case ($k = 0.72$), the stress-strain law exhibits a behaviour quite close to the linear one. In this case, the signal speed is given by (17). It should be recalled here that, according

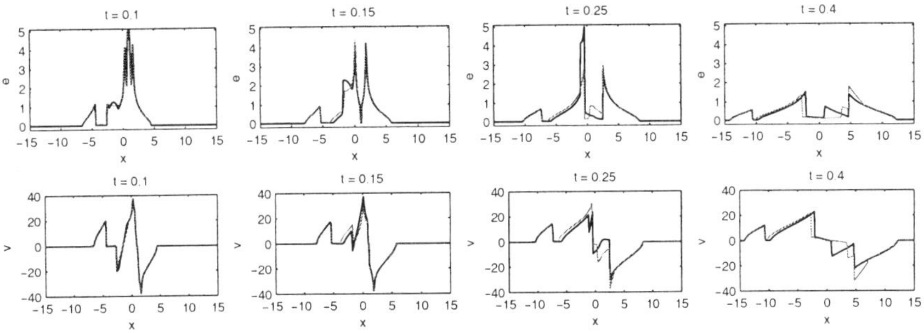

Fig. 2. Evolution of strain (top) and velocity (bottom) fields for initial data with two pulses (thick line), compared to the evolutions corresponding to initial data with the two pulses taken separately (thin lines) for the stress-strain law with $k = 0.92$.

to the theoretical results obtained by Liu,[16,17] the solution of the system under investigation with the given initial data, converges for large time to the solution of the corresponding Riemann problem, i.e. a problem for which the initial data are obtained replacing the pulses with a discontinuity connecting the values of the fields to the right and to the left of the pulses. Since in the present situation these initial data are constant null states, the solution is asymptotically converging to the null solution in the case of the two-pulse initial data as well as in the case of one-pulse initial data.

Acknowledgements

This work is dedicated to our great friend Professor Tommaso Ruggeri; Andrea would also like to thank him for the precious advice, not re-

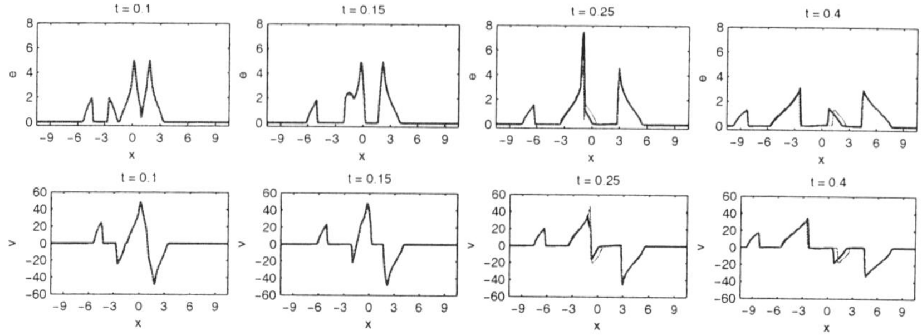

Fig. 3. Evolution of strain (top) and velocity (bottom) fields for initial data with two pulses (thick line), compared to the evolutions corresponding to initial data with the two pulses taken separately (thin lines) for the stress-strain law with $k = 0.72$.

stricted just to mathematics. We received partial support by MIUR/PRIN Project *Nonlinear Propagation and Stability in Thermodynamical Processes of Continuous Media*, Coordinator: T. Ruggeri (A.M. and T.R.) and GNFM/INdAM *Young Researcher Project*, Coordinator: A. Mentrelli (A.M.).

References

1. C. Loewner, *J. Anal. Math.* **2**, 219 (1952).
2. B. G. Konopelchenko, C. Rogers, *Phys. Lett. A* **158** 391 (1991).
3. B. G. Konopelchenko, C. Rogers, *J. Math. Phys.* **34**, 214 (1993).
4. T. W. Duerig, A. R. Pelton, D. Stöckel, *Metall* **9**, 569 (1996).
5. T. W. Duerig, *Mat. Res. Soc. Symp.* **360**, 497 (1995).
6. C. Rogers, W. K. Schief, K. W. Chow, *Theor. Math. Phys.* **151**, 711 (2007).
7. O. Babelon, D. Bernard, *Int. J. Mod. Phys. A* **8**, 507 (1993).
8. J. Dorfmeister, J. Inoguchi, M. Toda, *Contemporary Mathematics, AMS* **308**, 1 (2002).
9. H. M. Cekirge, E. Varley, *Philos. Trans. Roy. Soc. London Ser. A* **273**, 261 (1973).
10. J. Y. Kazakia, E. Varley, *Philos. Trans. Roy. Soc. London Ser. A* **277**, 191 (1974).
11. J. Y. Kazakia, E. Varley, *Philos. Trans. Roy. Soc. London Ser. A* **277**, 239 (1974).
12. B. R. Seymour, E. Varley, *SIAM J. Appl. Math.* **42**, 804 (1982).
13. F. Liotta, V. Romano, G. Russo, *SIAM J. Numer. Anal.* **38**, 1337 (2000).
14. A. Mentrelli, T. Ruggeri, *Suppl. Rend. Circ. Mat. Palermo* **II/78**, 201 (2006).
15. H. Nessyahu, E. Tadmor, *J. Comput. Phys.* **87**, 408 (1990).
16. T. -P. Liu, *Comm. Pure Appl. Math.* **30**, 7676 (1977).
17. T. -P. Liu, *Commun. Math. Phys.* **55**, 163 (1977).

INTERACTION BETWEEN A SHOCK AND AN ACCELERATION WAVE IN A PERFECT GAS FOR INCREASING SHOCK STRENGTH

A. MENTRELLI

Research Centre of Applied Mathematics (CIRAM), University of Bologna, Italy
E-mails: mentrelli@ciram.unibo.it

M. SUGIYAMA and N. ZHAO

Graduate School of Engineering, Nagoya Institute of Technology, Japan
E-mails: sugiyama@nitech.ac.jp, zhaonanr@yahoo.com

The interaction between shocks and acceleration waves in a perfect gas is studied following the general theory by Boillat and Ruggeri (Proc. Roy. Soc. Edinburgh **83A**, 17, 1979). Special attention is devoted to the analysis of the effects of varying the shock strength on the jump of the shock acceleration and on the amplitudes of the reflected/transmitted waves. Numerical calculations are in perfect agreement with theoretical results. This paper is based on a more comprehensive one by Mentrelli, Ruggeri, Sugiyama and Zhao (Wave Motion, in press, doi:10.1016/j.wavemoti.2007.09.005, 2007).

Keywords: shock waves; discontinuity waves; wave interaction; Euler fluid

1. Introduction

Shocks and rarefaction waves are commonly encountered when studying the solutions of nonlinear hyperbolic systems; discontinuity waves (or *acceleration waves*) are other important solutions of this kind of systems. The study of the interaction between shocks and discontinuity waves was first approached by Jeffrey[1] and Brun,[2] but the first comprehensive study including the case of characteristic shocks and a *weak shock* analysis was performed by Boillat and Ruggeri,[3] who proved that the interaction between a shock $\mathcal{S}^{(k)}$ and a discontinuity wave $\mathcal{D}^{(i)}$ has an important feature[a]: when

[a]We call $\mathcal{D}^{(i)}$ a discontinuity wave traveling with velocity equal to $\lambda^{(i)}$, and we call $\mathcal{S}^{(k)}$ a shock traveling with a velocity s such that $\lambda_0^{(k)} < s < \lambda_1^{(k)}$, being $\lambda_0^{(k)}$ and $\lambda_1^{(k)}$ the $k-th$ eigenvalue evaluated, respectively, in the unperturbed and perturbed state.[4]

the waves belong to different families (i.e. $k \neq i$) the interaction shows a typical nonlinear behavior, being the jump of the shock acceleration due to the interaction non-vanishing even for a weak shock.

The system of Euler equations, widely used in gas dynamics, is hyperbolic and much attention has been dedicated to the study of its solutions. Nevertheless, there is a lack in the understanding of the effects of the interaction between shock and acceleration waves even for this case; to the authors' knowledge the only papers devoted to this problem are those by Ruggeri[5] (for ideal gases) and Pandey *et al.*[6] (for van der Waals gases) restricted to the study of the interaction involving only characteristic shocks.

The aim of this paper is thus to achieve a comprehension of the consequences of the interaction process between shock and acceleration waves in a one-dimensional ideal gas. The results put into evidence the role played by the nonlinearity: it turns out that the reflection/transmission effects, typical of the nonlinear problems, are of primary importance. Moreover, a striking effect of the nonlinearity is appreciable for the interaction between a shock $\mathcal{S}^{(k)}$ and a discontinuity wave $\mathcal{D}^{(i)}$ with $k \neq i$: in this circumstance the jump of the shock acceleration is not vanishing even for a weak shock.

Numerical simulations have also been performed by means of a code developed by some of the authors.[7] The numerical results are perfectly consistent with the theory and a selection of them will be presented.

2. The system of Euler equations and and the interaction between shocks and acceleration waves

The system of Euler equations in the 1-D case takes the form $\mathbf{u}_t + \mathbf{F}_x(\mathbf{u}) = 0$, where $\mathbf{u} \equiv (\rho, \rho v, E)$, $\mathbf{F} \equiv \left(\rho v, \rho v^2 + p, (E+p)\, v\right)$ where ρ is the mass density, v the velocity, E the total energy density ($E = \rho e + \frac{1}{2}\rho v^2$ where e is the specific internal energy), p the pressure and the subscripts denote partial differentiation. The equation of state usually given to close the system is, for an ideal gas: $p = k\rho T/m$ and $e = p/\left(\rho\,(\gamma - 1)\right)$, where γ is the ratio of specific heats, T is the absolute temperature, k and m are, respectively, the Boltzmann's constant and the molecular mass of the gas.

According to the theory developed in Ref.,[3] when a shock and an acceleration wave interact, the latter is partially transmitted and partially reflected across the shock front and the latter experiences a jump in its acceleration. Letting τ_* be the impact time, the theory provides the number of transmitted and reflected waves, as well as their amplitude at $t = \tau_*$ and the jump of the acceleration of the shock, $[\dot{s}] := \dot{s}_{\tau_*^+} - \dot{s}_{\tau_*^-}$. Following Ref.,[3] two cases are possible (see Fig. 1): the interaction $\mathcal{S}^{(k)} \to \mathcal{D}^{(i)}$ with $k \geq i$

(Case I) and the interaction $\mathcal{D}^{(i)} \to \mathcal{S}^{(k)}$ with $i \geq k$ (Case II). In the one-

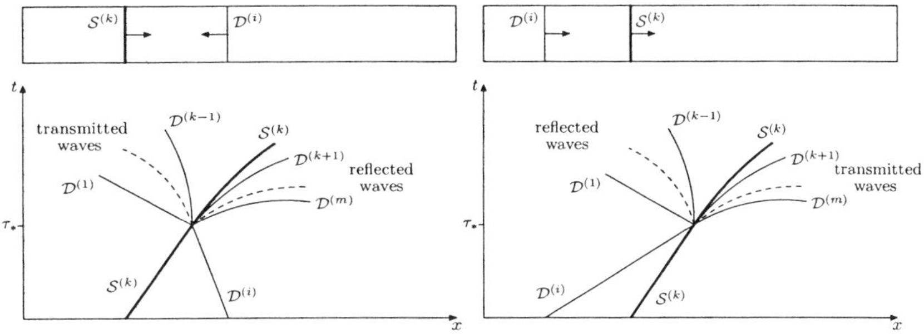

Fig. 1. (left) Propagation of $\mathcal{S}^{(k)}$ and $\mathcal{D}^{(i)}$ and its representation in the $x - t$ plane. Without loss of generality, the velocities of $\mathcal{S}^{(k)}$ and $\mathcal{D}^{(i)}$ are supposed to be positive and negative, respectively; (right) Propagation of $\mathcal{D}^{(i)}$ and $\mathcal{S}^{(k)}$ and its representation in the $x-t$ plane. In this case the velocities of $\mathcal{D}^{(i)}$ and $\mathcal{S}^{(k)}$ are supposed to be positive.

dimensional Euler fluid, five different interaction patterns exist, belonging to either Case I or Case II. These are summed up in Table 1.

Table 1. Interactions between shock and acceleration waves in the Euler fluid.

Case	*Interaction patterns*	*waves after interaction*
A	$\mathcal{S}^{(3)} \to \mathcal{D}^{(1)}$, or $\mathcal{D}^{(3)} \to \mathcal{S}^{(1)}$	two transmitted waves
B	$\mathcal{D}^{(3)} \to \mathcal{S}^{(3)}$, or $\mathcal{D}^{(1)} \to \mathcal{S}^{(1)}$	two reflected waves
C	$\mathcal{S}^{(3)} \to \mathcal{D}^{(2)}$, or $\mathcal{D}^{(2)} \to \mathcal{S}^{(1)}$	two transmitted waves
D	$\mathcal{D}^{(3)} \to \mathcal{S}^{(2)}$, or $\mathcal{S}^{(2)} \to \mathcal{D}^{(1)}$	one transmitted w., one reflected w.
E	$\mathcal{S}^{(3)} \to \mathcal{D}^{(3)}$, or $\mathcal{S}^{(1)} \to \mathcal{D}^{(1)}$	two transmitted waves

3. Analysis of the interaction $\mathcal{S}^{(3)} \to \mathcal{D}^{(1)}$

A thorough analysis of the interaction between shock and acceleration waves in the Euler fluid may be found in Ref.[8] In this section, an excerpt of the results obtained for Case A is presented. In this case, two transmitted waves ($\mathcal{D}^{(1)}$ and $\mathcal{D}^{(2)}$) are generated at the impact time. The theory, once the jump of density associated to the incident wave, $[\rho_x]_0^{(1)}$, is given, allows to determine the quantities $[\dot{s}]$, $[\rho_x]^{(1)}$, $[\rho_x]^{(2)}$ (representing, respectively, the jump of shock acceleration and the jump of the derivative of the density associated to $\mathcal{D}^{(1)}$ and $\mathcal{D}^{(2)}$) as explicit functions of the Mach number M_0. In order to analyze the effects of the Mach number on the propagation of the *fastest* transmitted wave, the jump of the acceleration of this wave, $G^{(1)}$, is introduced. All the quantities are replaced by suitable dimensionless ones ($[\dot{s}] \to [\dot{s}]/G_0^{(1)}$, $G^{(1)} \to G^{(1)}/G_0^{(1)}$, $[\rho_x]^{(1)} \to [\rho_x]^{(1)} / [\rho_x]_0^{(1)}$, $[\rho_x]^{(2)} \to [\rho_x]^{(2)} / [\rho_x]_0^{(1)}$, being $G_0^{(1)}$ the jump of the acceleration of the incident wave) and their behavior is shown in Fig. 2. In order to discuss the effects of a

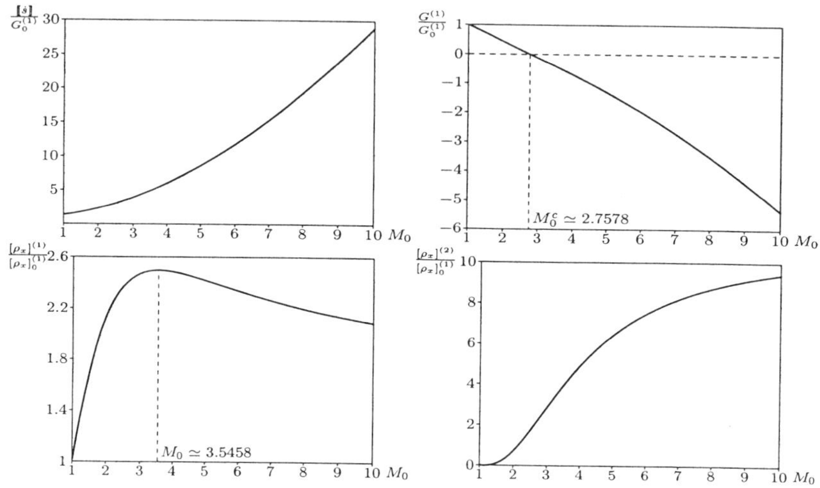

Fig. 2. Case A (Interaction $\mathcal{S}^{(3)} \to \mathcal{D}^{(1)}$). Dependence of (a) $[\dot{s}]/G_0^{(1)}$; (b) $G^{(1)}/G_0^{(1)}$; (c) $[\rho_x]^{(1)} / [\rho_x]_0^{(1)}$; (d) $[\rho_x]^{(2)} / [\rho_x]_0^{(1)}$ on the Mach number M_0 for a monatomic gas.

varying the Mach number M_0 on the interaction process, it is interesting to analyze the situations of incident *weak shock* ($M_0 \to 1$) and *strong shock* ($M_0 \to \infty$). In the *weak shock* situation, the interaction between an incident wave $\mathcal{D}^{(1)}$ and even a very weak shock $\mathcal{S}^{(3)}$ generates a non-trivial discontinuity in the shock acceleration, as seen in Fig. 2(a). This is in agreement with the results in Ref.,[3] according to which the jump of the shock acceleration is null, as $M_0 \to 1$, only when $\mathcal{S}^{(k)}$ and $\mathcal{D}^{(i)}$ belong to the same family. It is also remarkable that the wave $\mathcal{D}^{(2)}$ should be quite difficult to observe while the wave $\mathcal{D}^{(1)}$ is transmitted being just weakly altered. In the *strong shock* situation, it turns out that the jump of the shock acceleration, $[\dot{s}]$, becomes arbitrarily large as the Mach number M_0 increases. Moreover, $\left|[\rho_x]^{(1)}\right| > \left|[\rho_x]_0^{(1)}\right|$ for every Mach number. This means than an incident compression (expansion) wave generates a compression (expansion) transmitted wave $\mathcal{D}^{(1)}$ and the *strength* of the latter is always greater than that of the incident wave. A striking feature of the behavior of $G^{(1)}/G_0^{(1)}$, shown in Fig. 2(b), is that this quantity changes sign when the Mach number M_0 come across a *threshold* value $M_0^c \simeq 2.7578$ (for $\gamma = 5/3$). It may be proved that $G^{(1)}/G_0^{(1)}$ changes sign when $v = c$, that is when the velocity in the perturbed state equals the sound velocity and becomes (for increasing Mach number) supersonic. In order to test these

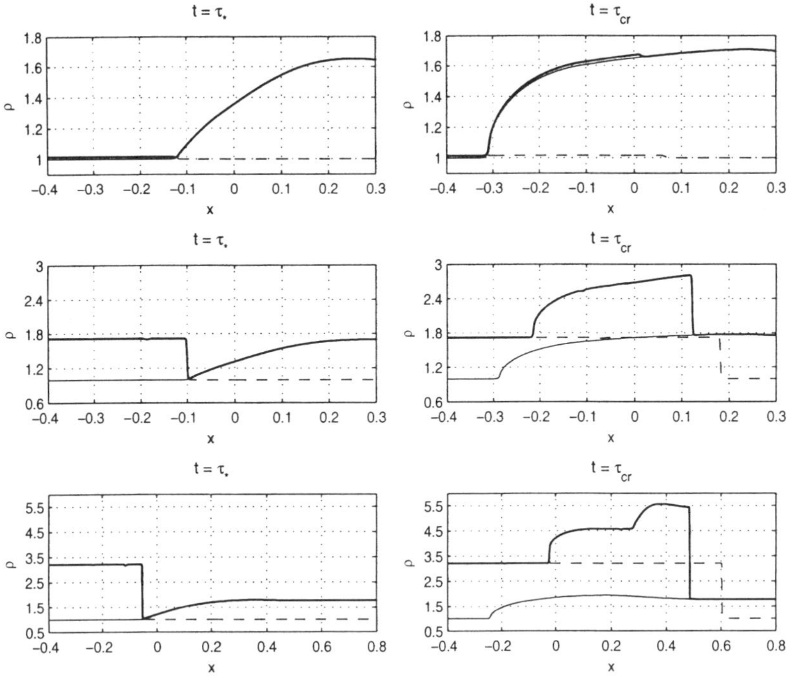

Fig. 3. Case A: Density fields at different instants t with $[\rho_x]_0^{(1)} > 0$ (compression wave) and $M_0 = 1.01$ (first row), $M_0 = 1.5$ (second row), $M_0 = 3.5$ (third row).

results, numerical experiments have been performed. The results in Fig. 3 show that, according to Fig. 2(a), even for a *weak* shock ($M_0 = 1.01$) the propagation of the shock is appreciably affected by the interaction: it may be easily seen comparing the density profiles (thick continuous line) to the corresponding ones obtained for a shock propagating without interaction (thin dashed line) that the shock is decelerated, that is $[\dot{s}] < 0$. It is worth recalling here that the sign of the shock acceleration cannot change, thus allowing us to say that if the shock experiences a deceleration at $t = \tau_*$ (i.e. $[\dot{s}] < 0$), it cannot accelerate for any $t > \tau_*$. This observation is crucial since it allows us to infer the sign of $[\dot{s}]$ comparing the position of the shock after the interaction to the one obtained without interaction.[8,9] An analogous conclusion may be drawn by comparing the thick and dashed lines for the cases with Mach numbers $M_0 = 1.5$ and $M_0 = 3.5$, all shown in Fig. 3.

4. Conclusions

The results presented for the the case of the $\mathcal{S}^{(3)} \rightarrow \mathcal{D}^{(1)}$ interaction show that under the *weak shock* condition ($M_0 \rightarrow 1$) the incident wave $\mathcal{D}^{(1)}$ is completely transmitted across the shock, while the jump of the shock acceleration is not vanishing, thus emerging the non-linear character of the interaction process. On the other hand, on the *strong shock* condition ($M_0 \rightarrow \infty$) two transmitted waves emerge after the interaction, the amplitude of which is bounded. The jump of the shock acceleration, for a strong shock, is instead unbounded and it grows as the Mach number increases.

Acknowledgement

This work was developed during the stay of T.R. in Nagoya as visiting professor with a fellowship of JSPS and was partially supported by: MIUR/PRIN Project *Non-linear Propagation and Stability in Thermodynamical Processes of Continuous Media*, Coordinator: T. Ruggeri (A.M. and T.R.); GNFM/INdAM *Young Researcher Project*, Coordinator: A. Mentrelli (A.M.); Daiko Foundation (No. 10100) (M.S. and N.Z.).
Andrea, Masaru and Nanrong would like to dedicate this paper to Tommaso with high esteem and gratitude.

References

1. A. Jeffrey, *Applicable Anal.* **3**, 79 (1973); *Applicable Anal.* **3**, 359 (1973/74).
2. L. Brun, in *Mechanical Waves in Solids* (Spinger, Wien, 1975).
3. G. Boillat, T. Ruggeri, *Proc. Roy. Soc. Edinburgh* **83A**, 17 (1979).
4. T. Ruggeri, in *Nonlinear Wave Motion* (Longman, New York, 1989).
5. T. Ruggeri, *Applicable Analysis* **11**, 103 (1980).
6. M. Pandej, V. D. Sharma, *Wave Motion* **44**, 346 (2007).
7. A. Mentrelli, T. Ruggeri, *Suppl. Rend. Circ. Mat. Palermo* **II/78**, 201 (2006).
8. A. Mentrelli, T. Ruggeri, M. Sugiyama, N. Zhao, *Wave Motion*, doi:10.1016/j.wavemoti.2007.09.005 (2007).
9. A. Muracchini, T. Ruggeri, L. Seccia, *Nuovo cimento D* **16**, 15 (1994).

ON A PARTICLE-SIZE SEGREGATION EQUATION

C. MINEO and M. TORRISI

Department of Mathematics and Computer Sciences, University of Catania, Viale A. Doria, 6, 95125 Catania, Italy
E-mail:mineo@dmi.unict.it
torrisi@dmi.unict.it

In an attempt to better understand the particle-size segregation phenomena we consider an avalanche of a mixture of a granular material of different size along an inclined chute. By taking into account the essential mechanisms we analyze a segregation remixing equation in some cases of physical interest.

Keywords: Segregation; avalanches; granular media, nonlinear diffusion, generalized Burgers equation.

1. Introduction

The segregation phenomenon, appearing in granular mixture flows, consists in the separation of the particles of different size. By changing the spatial composition of the flow, segregation has significant consequences on the behavior of mixture.

A granular mixture, initially homogeneous, in presence of the gravitational field is brought in a mixture where the distribution of the grains is inversely graded. That is, the small particles are fallen to the bottom while the larger ones are drifted to the top of the sheared layer.

In shear phenomena of a granular mixture the small particles take easily place in the voids which continuously are created and annihilated during the process.

Segregation occurs in the course of many natural phenomena such as rock falls, debris flows, avalanches, sub aqueous grain flow as well as during some technological processes related to the pharmaceutical, bulk chemical, food and agricultural industries. Many processes involving granular material mixture are concerned with particles of varying sizes, so an accurate model of the mechanics of the segregation may be advantageous. The change in the even dispersion of different size particles may have some important effects. In some applications, the result of this flow property may be undesirable, as it modifies the homogeneity of the mixture. In other processes instead, the separation of different sized particles could be an integral part of system operations. These reasons, among others encountered in many disciplines, justify the study of the mechanism of non-uniform granular flows.

We consider a granular avalanche that downslopes along a chute, inasmuch as particle segregation on an incline is characteristic of a non-uniform, gravity driven granular flow. In mixture with varying particle sizes, it has been observed that larger particles will generally go to the top of the flow regime, while smaller particles will go at the bottom.

Within this flow, the separation of grains of different size is due to a process called *kinetic sieving* generated by the smaller particles which fall under gravity in the voids between the grains. Competing against this process is *diffusive remixing* which is caused by random motions of the particles as they collide and shear over one another.

Several attempts have been done to study some of segregation mechanisms. In the paper of S. B. Savage and C. K. K. Lun[1] , an attempt is done to isolate and analyze some of the *essential* segregation mechanisms. In that paper is considered a binary mixture of small and large spherical particles of equal mass density and the problem of a steady to two-dimensional flow along a inclined chute is studied. V. N. Dolgunin and A. A. Ukolov[2] , show a model of segregation for a particle rapidly gravity flow based on a general equation of species transfer, taking in account convection transfer, quasi-diffusional mixing and particle segregation. Recently J.M.N.T. Gray and A.R. Thornton[3] have proposed a theory for particle size segregation in shallow granular in free surface flow. Eventually J.M.N.T. Gray and V. A. Chugunov[4] starting from Ref. 3 used a simple binary mixture theory to model the particle size segregation and diffusive remixing of large and small particle in a gravity driven flows.

The plan of this paper is the following. In section 2 the basic equations for the mixture are written by using the mass and momentum balance laws. In section 3 we introduce the constitutive relations that take into account the essential features of the segregation phenomena. In section 4 the segregation-remixing equation and its associated boundary conditions are written. In section 5 some case of physical interest are taken in consideration. The conclusions are given in section 6.

2. Basic equations

In order to construct a model for gravity driven segregation and diffusive remixing in a granular avalanche we consider this one as a binary mixture composed of *large* and *small* particles.

In this approach it is implicitly assumed that, in average, the interstitial pore space remain and that the unit of mixture volume can be subsumed in the volume fractions ϕ^l and ϕ^s of large and small particles respectively and

$$\phi^l + \phi^s = 1.$$

We write, according with the literature concerned with this field (see, e.g. S. B. Savage and K. Hutter[6] , D.A. Drew and S.L. Passman[7]), the mass balance equation and the momentum balance equation for each of the constituents

$$\partial_t \rho^\mu + \nabla \cdot (\rho^\mu \mathbf{u}^\mu) = 0, \tag{1}$$

$$\partial_t(\rho^\mu \mathbf{u}^\mu) + \nabla \cdot (\rho^\mu \mathbf{u}^\mu \otimes \mathbf{u}^\mu) = -\nabla p^\mu + \rho^\mu \boldsymbol{g} + \boldsymbol{\beta}^\mu, \quad \mu = l, s \tag{2}$$

where we denote with $\rho^\mu, \mathbf{u}^\mu$, respectively, partial densities and partial velocities per unit mixture volume for each of $\mu = l, s$, while, p^μ are partial pressures, $\rho^\mu \boldsymbol{g}$ is the gravitational force and $\boldsymbol{\beta}^\mu$ is the force exerted on the phase μ by other constituent. $\otimes$ denotes the dyadic product.

The quantities $\boldsymbol{\beta}^l$ and $\boldsymbol{\beta}^s$ as $\boldsymbol{\beta}^l = -\boldsymbol{\beta}^s$, of course, disappear when we consider the bulk mass and momentum balance equations obtained by summing the balance laws of the single constituents and defining the bulk density ρ, bulk velocity $\boldsymbol{u}$ and bulk pressure p, as

$$\rho = \rho^l + \rho^s, \qquad \rho \boldsymbol{u} = \rho^l \boldsymbol{u}^l + \rho^s \boldsymbol{u}^s, \qquad p = p^l + p^s$$

A crucial point of a mixture theory is how partial quantities defined per unit mixture volume, are linked to measurable intrinsic quantities, defined per unit constituent volume. As known[5] the partial and intrinsic densities and velocity fields are related by

$$\rho^\mu = \phi^\mu \rho^{\mu\star} \qquad \boldsymbol{u}^\mu = \boldsymbol{u}^{\mu\star}$$

where the superscript * denotes an intrinsic variable.

The partial and intrinsic pressures, instead, can be related by any functional form which satisfies the condition

$$p = p^l + p^s.$$

Now we consider the chute, where take place the avalanche phenomenon, inclined at an angle ζ to the horizontal.

We take a reference system $Oxyz$ with the plane xy coinciding with the chute, $x-$axis pointing down and the $z-$axis as the pointing upward normal.

With respect to the system $Oxyz$ the constituent velocities $\boldsymbol{u}^\mu$ and the bulk velocity $\boldsymbol{u}$ have components (u^μ, v^μ, w^μ) and (u, v, w) respectively.

If we assume that the $\rho^{\mu\star}$ are equal to the same constant

$$\rho^{l\star} = \rho^{s\star} = const. \tag{3}$$

then the bulk density is constant and we get:

$$\boldsymbol{\nabla} \cdot \mathbf{u} = 0. \tag{4}$$

So the mixture is incompressible. Moreover assuming that the normal accelerations therms are negligible (as will be usual in the following) from the $z-$component of the bulk momentum balance we get:

$$p_z = -\rho g \cos \zeta. \tag{5}$$

Taking into account that ρ is constant and that the free surface is traction free we are able to integrate the previous equation through the avalanche depth h to get that the bulk pressure is lithostatic:

$$p = \rho g(h - z) \cos \zeta. \tag{6}$$

Relations (4) and (6) show that the present theory is in a good agreement with most of the current avalanche theory models in fact the incompressibility of the bulk velocity and the lithostaticity of the bulk pressure are two key assumptions for them.

3. The constitutive relations

According with Gray and Thornton[3] we introduce a new pressure scaling in which the partial pressure is related to the bulk pressure by

$$p^\mu = f^\mu p \qquad (\mu = l, s). \tag{7}$$

The factor f^μ determine the fraction of the overburden pressure supplied by each of the single components. Often in the standard mixture theories is assumed $f^\mu = \phi^\mu$ but, here, there are the perturbations away from ϕ^μ that play a key role in the segregation.

f^μ must satisfy the following constraints:

$$f^l + f^s = 1 \quad f^\mu = 1 \text{ when } \phi^\mu = 1 \quad \forall \mu.$$

In the wide class of functions $f^\mu = f^\mu(\phi^l, \phi^s)$ satisfying the previous conditions, we focused our attention with

$$f^l = \phi^l + b\phi^s\phi^l \qquad f^s = \phi^s - b\phi^s\phi^l \tag{8}$$

where b represents the magnitude of the adimensional perturbation away from ϕ^μ.

The dynamic flows of water and other viscous fluids through porous solids is commonly modeled by using Darcy's law. Since experimental investigations of the segregation processes show a certain similarity with the percolation of aforesaid fluids, in the present model the form of interaction drag $\boldsymbol{\beta}^\mu$ is assumed as a combination of a linear velocity-dependent drag, a grain-grain interaction force $p\nabla f^\mu$ and an additional remixing force $-\rho d\nabla\phi^\mu$. This force seeks to drive grains of the phase μ toward areas of lower concentration.

So the interaction drag is written as

$$\boldsymbol{\beta}^\mu = p\nabla f^\mu - \rho^\mu c(\boldsymbol{u}^\mu - \boldsymbol{u}) - \rho d\nabla\phi^\mu \quad (\mu = l, s) \tag{9}$$

where c is the linear drag coefficient, $\mathbf{u}$ and d are respectively the bulk velocity and the strength of the diffusive force. In view of the further developments, taking into account the normal component of the constituent momentum balance equation and neglecting the acceleration therms, we derive

$$\phi^\mu w^\mu = \phi^\mu w + (f^\mu - \phi^\mu)\cos\zeta - \left(\frac{d}{c}\right)\partial_z\phi^\mu \qquad (\mu = l, s) \tag{10}$$

that is

$$w^l = w + q\phi^s - D\partial_z \ln\phi^l, \qquad w^s = w + q\phi^l - D\partial_z \ln\phi^s, \tag{11}$$

where the coefficients:

$$q = \left(\frac{b}{c}\right) g\cos\zeta, \qquad D = \left(\frac{d}{c}\right) \tag{12}$$

are the mean segregation velocity and diffusivity respectively. So in the present model there are only two constitutive parameters to identify. That is the model is identified once, from experimental observations, we are able to get the values of q and D.

From physical point of view:

- q determines the maximum percolation velocity of grains,
- D determines the strength of the remixing.

4. A segregation-remixing equation

In order to write the segregation-remixing equation, by following Gray and Chugunov[4] , we observe that the velocities induced by the particle-size segregation and diffusive remixing can be assumed of the same order of the normal component of the bulk velocity. Moreover due to the shallowness of the avalanche the lateral velocity components are much larger than the normal velocity.

So we can assume that the lateral components of velocities of each constituent are equal to bulk velocity:

$$u^{\mu} = u,$$
$$v^{\mu} = v \qquad (\mu = l, s).$$

Taking into account the previous assumptions, by substituting partial densities

$$\rho^{\mu} = \phi^{\mu} \rho^{\mu\star}$$

and the normal constituent velocities

$$w^{l} = w + q\phi^{s} - D\partial_z \ln \phi^{l}, \qquad w^{s} = w + q\phi^{l} - D\partial_z \ln \phi^{s}, \tag{13}$$

in the small particle mass balance equation

$$\partial_t \rho^{s} + \boldsymbol{\nabla} \cdot (\rho^{s} \mathbf{u}^{s}) = 0, \tag{14}$$

we get the *segregation-remixing equation* which, taking into account the incompressibility of the bulk velocity, reads:

$$\partial_t \phi^{s} + \mathbf{u} \cdot \boldsymbol{\nabla} \phi^{s} - \partial_z (q\phi^{s}\phi^{l}) = \partial_z (D\partial_z \phi^{s}) \tag{15}$$

- first two terms sweep the local concentration along with the bulk flow,
- third term is responsible for the particle size segregation,
- fourth term is responsible for the diffusive remixing.

After having eliminated ϕ^{l}, the equation (15) is put in a non dimensional form by using the following scaling transformation:

$$\begin{cases} (x, y, x) = L\left(\tilde{x}, \tilde{y}, \frac{H}{L}\tilde{x}\right) \\ (u, v, w) = U\left(\tilde{u}, \tilde{v}, \frac{H}{L}\tilde{w}\right) \\ t = \frac{L}{U}\tilde{t} \end{cases} \tag{16}$$

where H and L are the thickness and the length of the avalanche, while U the typical downslope velocity.

So, by dropping the superscript s and the tildes, we write

$$\partial_t\phi + u\partial_x\phi + v\partial_y\phi + w\partial_z\phi - \partial_z(S_r\phi(1-\phi)) = \partial_z(D_r\partial_z\phi) \tag{17}$$

where

$$S_r = \frac{qL}{HU}, \qquad D_r = \frac{DL}{H^2U}$$

are the segregation and diffusive remixing numbers.

4.1. *Boundary conditions*

Assumed that there is not erosion or deposition (see V. N. Dolgunin and A. A. Ukolov[2] , J.M.N.T. Gray and V. A. Chugunov[4]) the appropriate condition at the surface and base of the avalanche flow are that there is no flux of small particles across the boundary. These assumptions lead (see e.g P. Chadwick[8]) to

$$S_r\phi(1-\phi) + D_r\partial_z\phi = 0 \tag{18}$$

at the surface $z = z_s(x, y, t)$ and at the base of the flux $z = z_b(x, y, t)$.

5. Some problems for a segregation-remixing equation

In order to solve some problems of physical interest we consider, here, some additional restrictions to the segregation equation.

By assuming that $z = z_s(x, y, t) = 0$ and $z = z_b(x, y, t) = 1$, in agreement with Ref. 4 and Ref. 1, we can consider:

$$u = u(z) - \frac{w(z)}{2}x, \quad v = v(z) - \frac{w(z)}{2}y, \quad w = w(z) \qquad 0 < z < 1 \tag{19}$$

where $w(z)$ could be a non negative function such that

$$w(0) = w(1) = 0. \tag{20}$$

A physically suitable example of this class of functions could be

$$w(z) = w_0 z(1-z) \tag{21}$$

where $w_0 \geq 0$.

Moreover it is assumed that there are no lateral gradients in the small particle concentration

$$\partial_x\phi = 0 \qquad \partial_y\phi = 0. \tag{22}$$

Then, under the previous assumptions, the segregation remixing equation takes the form

$$\partial_t \phi - (S_r - w(z))\partial_z \phi + 2S_r\, \phi \partial_z \phi = D_r \partial_{zz}\phi \qquad (23)$$

while we are able to associate the following initial and boundary conditions

$$\begin{cases} t = 0 \quad \phi = \phi_0(z), \\ z = 0 \qquad S_r\phi(1-\phi) + D_r\partial_z\phi = 0, \\ z = 1 \qquad S_r\phi(1-\phi) + D_r\partial_z\phi = 0. \end{cases} \qquad (24)$$

In order to get solutions of the equation (23) we make the following transformations of variables

$$\begin{cases} t = \frac{D_r}{(S_r)^2}\tau \\ z = \frac{D_r}{S_r}\zeta \\ \phi = \frac{1}{2}\left(1 - \frac{w(\zeta)}{S_r} + \psi\right). \end{cases} \qquad (25)$$

So, our equation becomes the following *generalized Burgers equation*

$$\partial_\tau \psi + \frac{w''}{S_r} - \frac{w'}{S_r}\psi + \psi\partial_\zeta\psi = \partial_{\zeta\zeta}\psi \qquad (0 < \zeta < \zeta_1 \equiv \frac{S_r}{D_r}) \qquad (26)$$

while the initial and boundary conditions are transformed as follows:

$$\text{at } \ \tau = 0 \qquad \psi = \psi_0 \equiv 2\phi_0 + \frac{w(\zeta)}{S_r} - 1 \qquad (27)$$

$$\left[1 - \left(\frac{w(\zeta)}{S_r} - \psi\right)^2 + 2\partial_\zeta\psi + 2\frac{w'(\zeta)}{S_r}\right]_{\zeta=0,\zeta_1} = 0 \qquad (28)$$

It is a simple matter to see that for $w = w_0$ the equation (26) reduces to the Burgers one.

For non constant functions $w(\zeta)$, instead, we do not know solutions or results apart the steady solution $\psi = \frac{w(\zeta)}{S_r}$ that bring to the trivial solution $\phi = const$ for the equation (23).

5.1. *A special case*

If we consider $w(z) = 0$ we fall in the special case considered in Ref. 4 and the equation (26) becomes:

$$\partial_\tau \psi + \psi \partial_\zeta \psi = \partial_{\zeta\zeta} \psi \qquad (0 < \zeta < \zeta_1 \equiv \frac{S_r}{D_r}) \tag{29}$$

By applying the Cole-Hopf transformation

$$\psi = -\frac{2}{\chi} \partial_\zeta \chi \tag{30}$$

the equation (26) and its initial and boundary conditions (27) and (28) assume the following linear form

$$\partial_\tau \chi = \partial_{\zeta\zeta} \chi + \alpha(t) \chi \tag{31}$$

$$\chi|_{\tau=0} = \chi_0(\zeta) = exp\left(\int_0^\zeta -\frac{1}{2} \psi_0(\zeta') d\zeta' \right) \tag{32}$$

$$\partial_{\zeta\zeta} \chi - \frac{1}{4} \chi = 0|_{\zeta=0,\ \zeta_1}. \tag{33}$$

The function $\alpha(t)$ is a transformation parameter which will be determined by requiring the compatibility with the boundary condition.

By looking for stationary solution from (31) we get

$$\partial_{\zeta\zeta} \chi + \alpha(t) \chi = 0 \tag{34}$$

$$\partial_{\zeta\zeta} \chi - \frac{1}{4} \chi = 0|_{\zeta=0,\ \zeta_1} \tag{35}$$

then :

$$\alpha = -\frac{1}{4}.$$

So the equation (34) assumes the form

$$\partial_{\zeta\zeta} \chi - \frac{1}{4} \chi = 0 \tag{36}$$

Moreover the boundary conditions imply

$$\text{at} \quad \zeta = 0,\ \zeta_1 \quad \partial_\tau \chi = 0$$

then the boundary values of χ remain constant and equal to their initial values, that is

$$\zeta = 0: \quad \chi = 1 \tag{37}$$

$$\zeta = \zeta_1: \quad \chi = \chi_0(\zeta_1) = exp\left(\int_0^{\zeta_1} -\frac{1}{2} \psi_0(\zeta')\, d\zeta' \right). \tag{38}$$

The equation (31), with the initial condition (32) and boundary conditions (37)and (38), after having put $\alpha = -\frac{1}{4}$, can be solved, easily, by separable variable method.

6. Conclusions

In this short paper starting from a recent work of J.M.N.T. Gray and V. A. Chugunov[4] we have written the segregation remixing equation for a problem where the normal component to the chute of the bulk velocity is taken into account. By introducing a transformation of dependent and independent variables the segregation remixing equation is mapped in a *generalized Burgers equation.* Unfortunately the solution of this problem is far from trivial. However this equation shows that this problem contains as special case the uniform flows problem studied in Ref. 4.

The search of solutions for the class of *generalized Burgers equations*, as (26), will be the subject of a longer paper in order to get, at least, some means to check numerical schemes devoted to improve the study of the problem considered.

Acknowledgments

This paper is dedicated to Prof. Tommaso Ruggeri on the occasion of his 60th birthday. The authors acknowledge the financial support from P.R.A. (ex 60%) of University of Catania and from G.N.F.M. of INdAM. M.T. has been supported from M.I.U.R. through the PRIN 2005/2007: *Nonlinear Propagation and Stability in Thermodynamical Processes of Continuous Media.*

References

1. S.B. Savage and C.K. Lun, *J. Fluid Mech.*, **189**, 311, (1988).
2. V. N. Dolgunin and A. A. Ukolov , *Powder Tecnology*, **83**, 95, (1995).
3. J.M.N.T. Gray and A.R. Thornton , *Proc. R. Soc. A*, **441**, 1447, (2005).
4. J.M.N.T. Gray and V. A. Chugunov, *J. Fluid Mech.*, **569**, 365, (2006).
5. L. W. Morland, *Surveys Geophys.*, **13**, 209, (1992).
6. S. B. Savage and K. Hutter. *J. Fluid Mech.*, **199**, 117, (1989).
7. D. A. Drew and S. L. Passman, *Theory of multicomponent fluids*, Springer Verlag, New York, (1998).
8. P. Chadwick, *Continuum mechanics. Concise theory and problems*, Allen and Unwin, Dover, (1976).

MODELING OF LINEAR STABILITY OF PLANAR DETONATION WAVES IN EXTENDED KINETIC THEORY

R. MONACO AND M. PANDOLFI BIANCHI

Dipartimento di Matematica, Politecnico di Torino,
10129 Turin, Italy
E-mails: roberto.monaco@polito.it, miriam.pandolfi@polito.it

A. J. SOARES

Departamento de Matemática, Universidade do Minho,
4710-057 Braga, Portugal
E-mail: ajsoares@math.uminho.pt

The classical linear stability of the steady detonation wave with finite reaction zone is here studied within kinetic theory extended to chemically reacting gases with a two-way bimolecular reaction. The Euler equations and the Rankine-Hugoniot conditions of the kinetic model are linearized through a normal mode expansion around the steady detonation solution which is known from a previous paper. The linear stability is formulated by means of a non closed system of unknown perturbations with initial data at von Neumann state. A dispersion relation at the end of the reaction zone assures its determinacy.

Keywords: Reactive Boltzmann equation; detonation solution; linear stability.

1. Introduction

Kinetic approaches to the steady detonation problem, based on ZND description, have been developed mainly in [1] for discrete velocity models extended to reversible reactions, and in [2] for the full reactive Boltzmann equation. In the present paper, the detonation stability problem is formulated in the kinetic frame for a steady one dimensional wave propagating in a 4-component gas with bimolecular reactions. The linear stability is a widely investigated topic after the pioneering theoretical studies of Erpenbeck [3], who used a Laplace transform method to determine whether the detonation is stable or unstable to small disturbances induced in the reaction zone. Further contributions to linear stability spectra are based on a normal mode approach combined with a numerical shooting algorithm, first developed by Lee and Stewart [4] and then modified by other authors (see the very recent paper [5] and its related bibliography).

In the kinetic frame, the linear stability has been investigated in the context of discrete models of Boltzmann equation in paper [6] for the solution to the steady detonation problem of paper [1].

Aim of this present work is to formulate the linear stability in the context of full Boltzmann equation extended to a quaternary mixture with reversible bimolecular reaction for the steady detonation wave of paper [2]. According to papers [4] and [5], the stability is here formulated in terms of an initial value problem defined by the linearized reactive Euler equations with initial data stated by the Rankine-Hugoniot (RH) conditions at the von Neumann state. Such system provides the evolution of complex disturbances in dependence on complex growth rate parameters. The determinacy is assured by an acoustic radiation condition at the end of the reaction zone which imposes that no acoustic wave emanates from the rear boundary to interfere with the shock wave propagation.

2. Gas system

Consider a mixture of four constituents A_i, with molecular masses m_i and formation heats E_i, chemical reaction $A_1 + A_2 \rightleftharpoons A_3 + A_4$ such that $m_1 + m_2 = m_3 + m_4$ and $E = E_3 + E_4 - E_1 - E_2$, assumed as positive (endothermic forward reaction). The one–particle distribution $f_i(t, \boldsymbol{x}, \boldsymbol{v})$, $t \in \mathbb{R}_+$, $\boldsymbol{x} \in \mathbb{R}^3$, $\boldsymbol{v} \in \mathbb{R}^3$, for each A_i satisfies the reactive Boltzmann equation

$$\partial_t f_i + \boldsymbol{v} \cdot \nabla f_i = \mathcal{Q}_i[\boldsymbol{f}] + \mathcal{R}_i[\boldsymbol{f}], \quad i = 1, \ldots, 4, \tag{1}$$

where $\boldsymbol{f} = (f_1, f_2, f_3, f_4)$ and $\mathcal{Q}_i[\boldsymbol{f}]$, $\mathcal{R}_i[\boldsymbol{f}]$ are the elastic and reactive collision terms whose explicit form is detailed in [2]. The chemical kinetics of the gas system is based on the properties

$$\int_{\mathbb{R}^3} \mathcal{R}_1[\boldsymbol{f}](\boldsymbol{v})d\boldsymbol{v} = \int_{\mathbb{R}^3} \mathcal{R}_2[\boldsymbol{f}](\boldsymbol{v})d\boldsymbol{v} = -\int_{\mathbb{R}^3} \mathcal{R}_3[\boldsymbol{f}](\boldsymbol{v})d\boldsymbol{v} = -\int_{\mathbb{R}^3} \mathcal{R}_4[\boldsymbol{f}](\boldsymbol{v})d\boldsymbol{v}, \tag{2}$$

which assures the correct exchanges τ_i for the bimolecular reaction, that is

$$\tau_i = \sigma_i \int_{\mathbb{R}^3} \mathcal{R}_1[\boldsymbol{f}](\boldsymbol{v})d\boldsymbol{v}, \; i = 1, \ldots, 4, \text{ with } \quad \sigma_1 = \sigma_2 = -\sigma_3 = -\sigma_4 = 1, \tag{3}$$

and $\tau_1 = \tau_{\text{forward}} - \tau_{\text{backward}}$ thanks to the explicit form of $\mathcal{R}_i[\boldsymbol{f}]$.
The *mechanical equilibrium* is characterized by the Maxwellian distributions

$$\widetilde{f}_i(\boldsymbol{v}) = n_i \left(\frac{m_i}{2\pi k_B T}\right)^{\frac{3}{2}} \exp\left(-\frac{m_i(\boldsymbol{v} - \boldsymbol{u})^2}{2k_B T}\right), \tag{4}$$

where the number densities n_i are completely uncorrelated, whereas the *chemical equilibrium* is characterized by the mass-action-law

$$\frac{n_1^{\mathrm{eq}} n_2^{\mathrm{eq}}}{n_3^{\mathrm{eq}} n_4^{\mathrm{eq}}} = \left(\frac{m_1 m_2}{m_3 m_4}\right)^{3/2} \exp\left(\frac{E}{kT}\right), \tag{5}$$

E being the binding energy given in units of the gas thermal energy kT.

Reactive Euler equations

The evolution equations for constituent mass densities and conservation equations for momentum and total energy of the mixture are derived as

$$\partial_t \varrho_i + \nabla \cdot (\varrho_i \boldsymbol{u}_i) = m_i \tau_i, \quad i = 1, \ldots, 4, \tag{6}$$

$$\partial_t (\varrho \boldsymbol{u}) + \nabla \cdot (\varrho \boldsymbol{u} \otimes \boldsymbol{u} + P) = 0, \tag{7}$$

$$\partial_t \Big(\frac{3p}{2} + \sum_{i=1}^{4} n_i E_i + \frac{\varrho \boldsymbol{u}^2}{2} \Big) + \nabla \cdot \Big[\boldsymbol{q} + P\boldsymbol{u} + \Big(\frac{3p}{2} + \sum_{i=1}^{4} n_i E_i + \frac{\varrho \boldsymbol{u}^2}{2} \Big) \boldsymbol{u} \Big] = 0, \tag{8}$$

where the observables are defined by

$$\varrho_i = \int m_i f_i d\boldsymbol{v} = m_i n_i, \quad \text{with} \quad \varrho = \sum_{i=1}^{4} \varrho_i \quad \text{and} \quad n = \sum_{i=1}^{4} n_i, \tag{9a}$$

$$\varrho_i u_k^i = \int m_i v_k f_i d\boldsymbol{v}, \quad \text{with} \quad \varrho u_k = \sum_{i=1}^{4} \varrho_i u_k^i, \tag{9b}$$

$$P_{kl}^i = \int m_i (v_k - u_k)(v_l - u_l) f_i \, d\boldsymbol{v}, \quad \text{with} \quad P_{kl} = \sum_{i=1}^{4} P_{kl}^i, \tag{9c}$$

$$p_i = \frac{1}{3} \int m_i (\boldsymbol{v} - \boldsymbol{u})^2 f_i \, d\boldsymbol{v}, \quad \text{with} \quad p = \sum_{i=1}^{4} p_i, \tag{9d}$$

$$q_k^i = \int \frac{1}{2} m_i (\boldsymbol{v} - \boldsymbol{u})^2 (v_k - u_k) f_i \, d\boldsymbol{v}, \text{ with } q_k = \sum_{i=1}^{4} \big[q_k^i + E_i n_i (u_k^i - u_k) \big]. \tag{9e}$$

The closure of Eqs. (6-8) is achieved referring them to the Maxwellians proper of a flow regime for which the mechanical relaxation time is smaller than the chemical one, consistently with the detonation process. Accordingly, the reactive Euler equations are derived in one space dimension as

$$\partial_t \varrho_i + u \, \partial_x \varrho_i + \varrho_i \, \partial_x u = m_i \widetilde{\tau}_i, \tag{10}$$

$$\partial_t u + u \, \partial_x u + \frac{1}{\varrho} \, \partial_x p = 0, \tag{11}$$

$$\partial_t p + u \, \partial_x p + \frac{5}{3} \, p \, \partial_x u = \eta, \tag{12}$$

where $\eta = -\frac{2}{3}\sum_{i=1}^{4}\widetilde{\tau}_i E_i$ and $\widetilde{\tau}_i$ represents the reaction rate τ_i of Eq.(3), referred to mechanical equilibrium. In paper [2] a proper choice of forward and backward reactive cross sections leads to a final form of $\widetilde{\tau}_1$, i.e. $\widetilde{\tau}_1 = \left[\varrho_3\varrho_4 - \varrho_1\varrho_2\left(\frac{m_3m_4}{m_1m_2}\right)^{5/2}\exp\left(-\frac{E}{kT}\right)\right]\widehat{S}$, where

$$\widehat{S}=\frac{4\pi\beta}{kT(m_3+m_4)}\left[\frac{\chi}{\pi}\sqrt{\frac{2\pi kT}{\mu_{13}}}\exp\Big(-\frac{\mu_{34}\chi^2}{2kT}\Big)+\Big(\frac{kT}{\mu_{34}}-\chi^2\Big)\Big[1-\mathrm{erf}\Big(\sqrt{\frac{\mu_{34}}{2kT}}\,\chi\Big)\Big]\right],$$

χ is a threshold velocity, and β a collision frequency scale factor.

Steady detonation waves

The lead element of the ZND detonation wave is a non-reactive shock propagating with constant velocity s towards the unreacted mixture in the given initial state I. The shock front is followed by a finite reaction zone which connects the von Neumann state N, where pressure, density and temperature jump to very high values triggering an exothermic reaction, to the final equilibrium state S. Inside this zone, the explosive mixture is in strong chemical disequilibrium but in mechanical equilibrium. The kinetic model is stated by the reactive Euler equations in the shock attached steady frame

$$(u-s)\frac{d\varrho_i}{dx}+\varrho_i\frac{du}{dx}=m_i\widetilde{\tau}_i,\quad i=1,\dots,4, \tag{13}$$

$$(u-s)\frac{du}{dx}+\frac{1}{\varrho}\frac{dp}{dx}=0, \tag{14}$$

$$(u-s)\frac{dp}{dx}+\frac{5p}{3}\frac{du}{dx}=\eta. \tag{15}$$

A maximal set of conservation laws can be deduced from Eqs.(13-15) for partial number densities, momentum and total energy. Its integration through the shock leads to RH conditions which allow to determine the wave structure, i.e. all intermediate, N and S states in the reaction zone.

3. Linear stability problem

The detonation stability is classically studied assuming that a small rear boundary perturbation, instantaneously assigned, induces a distortion on the shock wave whereas subsequent rear oscillations do not affect the wave propagation [4]. In the reaction zone the evolution of the perturbations is studied linearizing Eqs.(10-12) with RH conditions at von Neumann state, through a normal mode expansion around the steady detonation solution.

Let x denote the *wave coordinate* which measures the distance from the perturbed shock, $x = x^\ell - [st + \psi^\ell(t)]$, with $\psi^\ell(t)$ the induced distortion on the shock position. In this frame Eqs.(10-12) assume the matrix form

$$\partial_t \boldsymbol{z} + \left[A(\boldsymbol{z}) - \left(s + \frac{d\psi^\ell}{dt} \right) I \right] \partial_x \boldsymbol{z} = \boldsymbol{c}(\boldsymbol{z}), \tag{16}$$

where $\boldsymbol{z} = \left[\varrho_1\, \varrho_2\, \varrho_3\, \varrho_4\, u\, p \right]^{\mathrm{T}}$ is the unknown state vector, and

$$A = \begin{bmatrix} u & 0 & 0 & 0 & \varrho_1 & 0 \\ 0 & u & 0 & 0 & \varrho_2 & 0 \\ 0 & 0 & u & 0 & \varrho_3 & 0 \\ 0 & 0 & 0 & u & \varrho_4 & 0 \\ 0 & 0 & 0 & 0 & u & 1/\varrho \\ 0 & 0 & 0 & 0 & 5p/3 & u \end{bmatrix}, \quad \boldsymbol{c} = \begin{bmatrix} m_1 \widetilde{\tau}_1 \\ m_2 \widetilde{\tau}_2 \\ m_3 \widetilde{\tau}_3 \\ m_4 \widetilde{\tau}_4 \\ 0 \\ \eta \end{bmatrix}. \tag{17}$$

The RH conditions written in terms of the state variables become

$$\varrho_i \left(u - s - \frac{d\psi^\ell}{dt} \right) = -\varrho_{i0} \left(s + \frac{d\psi^\ell}{dt} \right), \quad i = 1, \ldots, 4 \tag{18}$$

$$\left(u - s - \frac{d\psi^\ell}{dt} \right) u \sum_i \varrho_i + p = p_0, \tag{19}$$

$$\left(u - s - \frac{d\psi^\ell}{dt} \right) \left[u^2 \sum_i \varrho_i + 2 \sum_i \frac{E_i}{m_i} \varrho_i + 3p \right] + 2pu$$
$$= -\left(3p_0 + 2 \sum_i \frac{E_i}{m_i} \varrho_{i0} \right) \left(s + \frac{d\psi^\ell}{dt} \right). \tag{20}$$

Assuming that the shock distortion and state variables perturbations have an exponential time dependence, the governing equations (16) and RH conditions (18-20) are linearized through a normal mode expansion around the known steady solution $\boldsymbol{z}^*(x)$ in the form

$$\boldsymbol{z}(x,t) = \boldsymbol{z}^*(x) + \exp(at)\, \overline{\boldsymbol{z}}(x), \quad \psi^\ell(t) = \overline{\psi} \exp(at), \quad a, \overline{\psi} \in \mathbb{C}, \tag{21}$$

Above, the bar refers to disturbances and growth rate $Re\, a$ determines the stability behaviour of the steady solution. The linearization leads to

$$(A^* - sI) \frac{d\overline{\boldsymbol{z}}}{dx} + (aI + C^*)\, \overline{\boldsymbol{z}} - a \boldsymbol{b}^* \overline{\psi} = 0, \tag{22}$$

where $\boldsymbol{b}^* = d\boldsymbol{z}^*/dx$, $A^* = A(\boldsymbol{z}^*)$, and C^* is a matrix defined in terms of the steady state. The linearized RH conditions are

$$(A^*_{\mathrm{VN}} - sI)\, \overline{\boldsymbol{z}}_{\mathrm{VN}} = a\overline{\psi} \left(K^*_{\mathrm{VN}} \boldsymbol{z}^*_{\mathrm{VN}} - \boldsymbol{z}_0 \right), \tag{23}$$

where the matrix $K^*_{\rm VN}$ is such that the non-zero elements are

$$K^*_{ii} = 1,\ i = 1, \ldots, 4 \text{ and } i = 6,\ K^*_{55} = \frac{\sum_i \varrho_{i0}}{\sum_i \varrho^*_{i\rm VN}},\ K^*_{64} = -\frac{\sum_i \varrho_{i0} u^*_{\rm VN}}{3}. \quad (24)$$

The state vector $\boldsymbol{z}_0$ is assigned in the initial state I, ahead of the wave, where the mixture is quiescent. The RH conditions (23) relate the value of the disturbances $\overline{\boldsymbol{z}}_{\rm VN}$ behind the shock to their zero value at the initial state ahead of the shock. Normalizing the disturbances through the position $\overline{\boldsymbol{\zeta}} = \overline{\boldsymbol{z}}/\overline{\psi}$, the stability problem is formulated in the reaction zone as

$$(A^* - sI)\frac{d\overline{\boldsymbol{\zeta}}}{dx} + (aI + C^*)\,\overline{\boldsymbol{\zeta}} - a\boldsymbol{b}^* = 0 \quad (25)$$

$$\overline{\boldsymbol{\zeta}}_{\rm VN} = a(A^*_{\rm VN} - sI)^{-1}\left(K^*_{\rm VN}\boldsymbol{z}^*_{\rm VN} - \boldsymbol{z}_0\right), \quad \text{for} \quad x = x_{\rm VN}. \quad (26)$$

Since the growth rate $Re\,a$ cannot be known *a priori*, this problem is not closed in general. Its determinacy is assured by an algebraic condition which corresponds to the dispersion relation of the normal mode expansions (21), obtained resorting to acoustic arguments. A rear boundary condition applied at the equilibrium final state where the chemical process is exhausted, leads to the acoustic radiation condition [7] of inert flows, i.e.

$$\overline{\zeta}_6 = \gamma \varrho^*_{\rm eq} c^*_{\rm eq} \overline{\zeta}_5, \quad \text{for} \quad x = x_{\rm eq}, \quad (27)$$

where $c^*_{\rm eq}$ is the isentropic sound speed and γ the ratio of specific heats.

In conclusion, the linear stability problem is stated in a closed form in terms of complex disturbances $\overline{\boldsymbol{\zeta}}$, growth rate $Re\,a$ and frequency $Im\,a$, by means of stability equations (25) with initial data (26) and dispersion relation (27). Since $\overline{\boldsymbol{\zeta}}, a \in \mathbb{C}$, the problem splits in twelve ODEs and twelve shock conditions at the von Neumann state, plus two algebraic conditions at the equilibrium final state.

A numerical shooting technique to compute eigenvalues a and eigenfuctions $\overline{\boldsymbol{\zeta}}$ will be implemented in a forthcoming paper devoted to numerical simulations and comparisons with results of classical theory.

References

1. M. Pandolfi Bianchi, A.J. Soares, *Phys. Fluids*, **8**, 3423–3432 (1996).
2. F. Conforto, R. Monaco, F. Schürrer, I. Ziegler, *J. Phys. A: Math. Gen.*, **36**, 5381–5398 (2003).
3. J.J. Erpenbeck, *Phys. Fluids*, **5**, 604–614 (1962).
4. H.I. Lee, D.S. Stewart, *J. Fluid Mech.*, **216**, 103–132 (1990).
5. V. Gorchkov, C.B. Kiyanda, M. Short, J.J. Quirck, *Proc. Combustion Inst.*, **31**, 2397-2405 (2006).
6. M. Pandolfi Bianchi, A.J. Soares, *Riv. Mat. Univ. Parma*, **3**, 201-215 (2000).
7. M. Short, D.S. Stewart, *J. Fluid Mech.*, **368**, 229–262 (1998).

ON SLOW PROCESSES IN PIEZOTHERMOELASTIC PLATES

A. MONTANARO
DMMSA, University of Padua,
Padua, Italy
E-mail: montanaro@dmsa.unipd.it
www.dmsa.unipd.it/montanaro

Here we present the results of the work Ref. 1. We consider a plate infinite in extent and bounded by two parallel planes. The plate is filled by a heat-conducting piezoelectric material with the symmetry of a hexagonal crystal class $C_{6\nu} = 6mm$, so that e.g. ferroelectric ceramics are included. We assume that the panel is subject to a thermic exposure on the upper face which varies very slowly with time. On the lower face the displacement is prescribed, as when e.g. the panel is welded above a fixed flat body. We study processes which are homogeneous on each plane parallel to the boundary planes, i.e. they depend only on the thickness coordinate, and moreover they vary very slowly with time. Hence they form a family of equilibrium configurations, for the thermopiezoelastic panel, indexed by a parameter τ, which is a function of time with everywhere small time-derivative; they can be identified with quasi-static processes of the body, induced in particular by a given thermic exposure on the upper bounding face. Thus we formulate some appropriate boundary-value problems involving the equilibrium field equations. Such problems are completely solved by means of their explicit solutions. In particular, we show that whatever temperature is given at the upper face, by the thermic exposure, the temperature at the lower face can be controlled by the difference of electric potential between the two bounding planes.

1. The linear theory of thermo piezoelectricity

1.1. *Constitutive equations*

By using the matrix notation

ij or kl	11	22	33	23 or 32	31 or 13	12 or 21
p, q	1	2	3	4	5	6

we assume the following constitutive equations respectively for the Cauchy stress, electric displacement vector, heat flux vector and specific entropy:

$$t_p = c_{pq} S_q + e_{ip} \phi_{,i} - \beta_p T , \qquad D_i = e_{iq} S_q - \epsilon_{ik} \phi_{,i} + \tilde{\omega}_i T , \quad (1)$$

$$q_i = \kappa_{il} T_{,l} - \kappa_{il}^{E} \phi_{,l}, \qquad \eta = \eta_0 + \frac{\gamma}{T_0} T + \frac{1}{\rho_o} (\beta_p S_p - \tilde{\omega}_i \phi_{,i}), \quad (2)$$

where $S_p = \frac{1}{2}(u_{i,j} + u_{j,i})$, $t_{ij} = t_p$ when $i, j = 1, 2, 3$, $p = 1, \ldots, 6$, ϕ is the electric potential, and $c_{pq} = c_{qp}$, $\kappa_{kl} = \kappa_{lk}$, $\kappa_{kl}^{E} = \kappa_{lk}^{E}$.

1.2. *Balance laws of equilibrium*

In the absence of external sources, the local equations of equilibrium corresponding to the (i) balance law of linear momentum, (ii) Maxwell's equation, and (iii) balance law of conservation of energy, respectively write as

$$t_{kl,k} = 0, \qquad D_{k,k} = 0, \qquad q_{k,k} = 0. \quad (3)$$

1.3. *Field equations of equilibrium*

By replacing the above constitutive equations in (3), in the homogeneous case, we obtain the linearized field equations of equilibrium

$$c_{klij} u_{i,jk} + e_{ijl} \phi_{,ij} - \beta_{kl} T_{,k} = 0 \qquad (l = 1, 2, 3), \quad (4)$$

$$e_{kji} u_{j,ik} - \epsilon_{kj} \phi_{,jk} + \tilde{\omega}_k T_{,k} = 0, \quad -\kappa_{kj} T_{,jk} + \kappa_{jk}^{E} \phi_{,jk} = 0. \quad (5)$$

1.4. *Field equations for ferroelectric ceramics*

When the arrays for the material with hexagonal symmetry in class $C_{6\nu} = 6mm$ are substituted in the above constitutive equations we obtain (see Refs. 2, p.58, and 3, p.705)

$$\begin{aligned}
t_1 &= c_{11} u_{1,1} + c_{12} u_{2,2} + c_{13} u_{3,3} + e_{31} \phi_{,3} - \beta_1 T, \\
t_2 &= c_{12} u_{1,1} + c_{11} u_{2,2} + c_{13} u_{3,3} + e_{31} \phi_{,3} - \beta_2 T, \\
t_3 &= c_{13} u_{1,1} + c_{13} u_{2,2} + c_{33} u_{3,3} + e_{33} \phi_{,3} - \beta_3 T, \\
t_4 &= c_{44} (u_{3,2} + u_{2,3}) + e_{15} \phi_{,2}, \\
t_5 &= c_{44} (u_{3,1} + u_{1,3}) + e_{15} \phi_{,1}, \quad t_6 = c_{66} (u_{1,2} + u_{2,1}),
\end{aligned} \quad (6)$$

$$\begin{aligned}
D_1 &= e_{15} u_{3,1} + e_{15} u_{1,3} - \epsilon_{11} \phi_{,1} + \tilde{\omega}_1 T, \\
D_2 &= e_{15} (u_{3,2} + u_{2,3}) - \epsilon_{11} \phi_{,2} + \tilde{\omega}_2 T, \\
D_3 &= e_{31} u_{1,1} + e_{31} u_{2,2} + e_{33} u_{3,3} - \epsilon_{33} \phi_{,3} + \tilde{\omega}_3 T.
\end{aligned} \quad (7)$$

When these constitutive relations are substituted in the equilibrium field equations (4)-(5) we obtain

$$-\beta_1 T_{,1} + c_{11}u_{1,11} + (c_{12}+c_{66})u_{2,12} + (c_{13}+c_{44})u_{3,13} + c_{66}u_{1,22} + c_{44}u_{1,33} + (e_{31}+e_{15})\phi_{,13} = 0 \qquad (8)$$

$$-\beta_2 T_{,2} + c_{66}u_{2,11} + (c_{66}+c_{12})u_{1,12} + c_{11}u_{2,22} + + (c_{13}+c_{44})u_{3,23} + c_{44}u_{2,33} + (e_{31}+e_{15})\phi_{,23} = 0 \qquad (9)$$

$$-\beta_3 T_{,3} + c_{44}u_{3,11} + (c_{44}+c_{13})u_{1,31} + c_{44}u_{3,22} + (c_{44}+c_{13})u_{2,23} + c_{33}u_{3,33} + + e_{15}\phi_{,11} + e_{15}\phi_{,22} + e_{33}\phi_{,33} = 0 \qquad (10)$$

$$e_{15}u_{3,11} + (e_{15}+e_{31})u_{1,13} + e_{15}u_{3,22} + (e_{15}+e_{31})u_{2,32} + e_{33}u_{3,33} + \tilde{\omega}_l T_{,l} - \epsilon_{11}\phi_{,11} - \epsilon_{11}\phi_{,22} - \epsilon_{33}\phi_{,33} = 0 \qquad (11)$$

$$-\kappa_{ll} T_{,ll} + \kappa_{ll}^{E}\,\phi_{,ll} = 0\,. \qquad (12)$$

2. Quasi-statics

Having in mind the study of the sun exposure of a panel settled to a rigid underlying body, the boundary conditions (*b.c.*) that we must add to the above five second-order differential equations must include the prescriptions of temperature and of vanishing normal stress on the upper bounding plane, and the condition of assigned displacement on the lower bounding plane. In addition, other conditions must be added in order to have ten independent *b.c.*, which will determine a unique solution to the resulting boundary-value problem.

We assume that the prescriptions on the boundary are functions of a parameter τ which depends slowly on time: $\tau = \tau(t)$, $|\tau'(t)|$ small.

We refer to equations (8)-(12) as the *quasi-static equations*, and to such equations joined with a set of prescriptions on the boundary as a *boundary-value problem of quasi-statics.*

2.1. *Problem B.1.3 [C.1.3]*

2.1.1. *Statement of the problems*

We consider a panel $\mathcal{P}$, filled by a material of class $C_{6\nu} = 6mm$, which is bounded by the parallel planes $x_1 = \pm h$. The panel is coated by an electrode on the plane $x_1 = h$, which is infinitesimally thin, so that all its mechanical effects may be ignored [is coated by electrodes on the planes $x_1 = \pm h$, which are infinitesimally thin, so that all their mechanical effects may be ignored]. Remind that x_3 is the poling direction. Referring to thickness equilibrium processes $(T,\, \phi,\, \mathbf{u})$, superimposed to the natural equilibrium configuration, of the form

$$T = T(x_1,\, \tau)\,, \qquad \phi = \phi(x_1,\, \tau)\,, \qquad u_i = u_i(x_1,\, \tau)\,, \tag{13}$$

the field equations (8)-(12) reduce to

$$-\beta_1 T_{,1} + c_{11} u_{1,11} = 0 \tag{14}$$

$$c_{66} u_{2,11} = 0 \tag{15}$$

$$c_{44} u_{3,11} + e_{15} \phi_{,11} = 0 \tag{16}$$

$$e_{15} u_{3,11} + \tilde{\omega}_l T_{,l} - \epsilon_{11} \phi_{,11} = 0 \tag{17}$$

$$-\kappa_{11} T_{,11} + \kappa_{11}^{E} \phi_{,11} = 0\,. \tag{18}$$

Problem $B.1.3$ [$C.1.3$] is formulated as follows:

To find the particular solution of the form (13) to the field equations (14)-(18), which satisfies the ten boundary conditions below, where Θ, Φ, Π_i, Υ_i, Δ [Φ_2], Ω *are given smooth functions of* τ.

1 Temperature prescribed $\quad T = \Theta \quad$ at $\quad x_1 = h\,.$
2 Electric potential prescribed $\quad \phi = \Phi\,, \quad$ at $\quad x_1 = h\,.$
3 Normal stress prescribed $\quad t_1 = \Pi_1\,,\ t_6 = \Pi_2,\ t_5 = \Pi_3, \quad$ at $\quad x_1 = h.$
4 Displacement $u_i\ (i = 1,\, 2,\, 3)$ prescribed $u_i = \Upsilon_i$ at $x_1 = -h\,.$
5 [**6**] Normal electric displacement [Electric potential] prescribed $-D_1 = \Delta \quad [\phi = \Phi_2]$ at $x_1 = -h.$
7 Normal heat flux prescribed $\quad -q_1 = \Omega \quad$ at $\quad x_1 = -h\,.$

2.1.2. *General solution*

Put $c = c_{44}$ $e = e_{15}$ $\omega = \tilde{\omega}_1$ $\epsilon = \epsilon_{11}$ $\kappa = \kappa_{11}$ $\kappa' = \kappa_{11}'^{E}$ $\beta = \beta_1\,,$ $K = k/k'$, $A = \omega/e$, $B = K\big(e/c + \epsilon/e\big)$, $a = A/B\,,$ $V = -Kea\,.$ Then the general solution to equations (14)-(18) is

$$T(x_1,\, \tau) = T_1 e^{a x_1} + T_2 \tag{19}$$

$$\phi(x_1,\, \tau) = K T_1 e^{a x_1} + F_1 x_1 + F_2 \tag{20}$$

$$u_1(x_1,\, \tau) = a^{-1} c_{11}^{-1} \beta T_1 e^{a x_1} + U_{11} x_1 + U_{12} \tag{21}$$

$$u_2(x_1, \tau) = U_{21}x_1 + U_{22}\,, \; u_3(x_1, \tau) = a^{-1}c^{-1}VT_1e^{ax_1} + U_{31}x_1 + U_{32} \quad (22)$$

where T_1, T_2, F_1, F_2, $U_{\alpha 1}$, $U_{\alpha 2}$, $(\alpha = 1, 2, 3)$ are arbitrarily chosen smooth functions of τ.

2.1.3. *On the solution of Problem B.1.3*

By choosing $\Omega = 0$ one finds

$$T_1 = e^{-ah}\Big(T_2 + \Theta\Big)\,, \; T_2 = \frac{c\Delta + e\Pi_3}{c\omega}\,, \; F_1 = 0\,, \; F_2 = \Phi - Ke^{ah}T_1\,, \quad (23)$$

$$U_{31} = \frac{\Pi_3}{c_{44}k}\,, \quad U_{32} = \Upsilon_3 + \frac{h}{c_{44}}\Pi_3 + e^{ah}KET_1\,. \quad (24)$$

Remark 2.1. By replacing (23) in (20), and evaluating the resulting equality at $x_1 = -h$, we find the following linear expression for $\phi(-h)$:

$$\phi(-h) = K\Big(\frac{c\Delta + e\Pi_3}{c\omega} + \Theta\Big)\Big(e^{-2ah} - 1\Big) + \Phi\,. \quad (25)$$

Hence, given any three quantities in $\{\Delta, \Pi_3, \Theta, \Phi\}$, the fourth can be choosen in such a way to control $\phi(-h)$.

Remark 2.2. By replacing (23) in (19), and evaluating the resulting equality at $x_1 = -h$, we find the following linear expression for $T(-h)$:

$$T(-h) = \Theta e^{-2ah} + \Big(e^{-2ah} + 1\Big)\frac{c\Delta + e\Pi_3}{c\omega}\,. \quad (26)$$

Hence, given any two quantities in $\{\Delta, \Pi_3, \Theta\}$, the third can be choosen in such a way to control $T(-h)$.

In particular we note that, by putting $\Delta = 0 = \Pi_3$ in (25), we obtain

$$\phi(-h) = \Phi + K\Big(e^{-2ah} - 1\Big)\Theta\,, \quad (27)$$

which yields the electric potential in the plane $x_1 = -h$ in terms of the boundary temperature Θ at $x_1 = h$.

2.1.4. *On the solution to Problem C.1.3*

One finds the following expressions for the coefficients of the general solution in terms of the boundary data. In particular, one finds

$$T_1 = \mathcal{A}\Omega + \mathcal{B}\Big(\Phi - \Phi_2\Big)\,, \quad T_2 = \Theta - e^{ah}T_1\,, \quad (28)$$

where

$$\mathcal{A} = \frac{2h}{K^2k(e^{ah} - e^{-ah})}, \qquad \mathcal{B} = \frac{Kk}{2h}\mathcal{A},$$

$$F_1 = -\frac{\Omega}{kK}, \qquad F_2 = \Phi + \frac{h}{Kk}\Omega - Ke^{ah}T_1. \tag{29}$$

Hence, the explicit expressions of T and ϕ in terms of the boundary data write as

$$T(x_1) = \Theta + \Big(\mathcal{A}\Omega + \mathcal{B}(\Phi - \Phi_2)\Big)\Big(e^{ax_1} - e^{ah}\Big) \tag{30}$$

$$\phi(x_1) = K\Big(\mathcal{A}\Omega + \mathcal{B}(\Phi - \Phi_2)\Big)(e^{ax_1} - e^{ah}) + \Phi + \frac{\Omega}{kK}(h - x_1). \tag{31}$$

Remark 2.3. Note that (30) yields

$$T(-h) = \Theta + \Big(\mathcal{A}\Omega + \mathcal{B}(\Phi - \Phi_2)\Big)\Big(e^{-ah} - e^{ah}\Big). \tag{32}$$

This equality shows that the temperature $T(-h)$ at $x_1 = -h$ can be controlled by, e.g., the electric potential difference $\Phi - \Phi_2$. More in general, given three quantities choosen in $\{\Theta, \Omega, \Phi, \Phi_2\}$, the fourth can be choosen in such a way to control $T(-h)$.

References

1. A. Montanaro, *Control Problems for Thermopiezoelastic Plates under Quasi-static Thermic Exposure* (Being presented, 2007).
2. H.F. Tiersten, *Linear Piezoelectric Plate Vibrations*, (Plenum Press – New York, 1969).
3. W. Q. Chen, *J. Applied Mechanics* **67**, 705 (2000).

PROBLEMS OF STABILITY AND WAVES IN BIOLOGICAL SYSTEMS

G. MULONE

Dipartimento di Matematica e Informatica,
Città Universitaria, Viale A. Doria, 6, 95125, Catania, Italy
E-mail: mulone@dmi.unict.it

The reduction method is used to obtain some optimal stability results in biomathematics: an epidemic model with diffusion and the May-Leonard system for competition between three species with diffusion. The existence of travelling waves solutions is proved for the epidemic model.

Keywords: Lyapunov stability; biological systems; travelling waves.

1. Introduction

Let us consider the ODE system

$$\dot{U} = AU + N(U), \tag{1}$$

where U is a vector of $\mathbb{R}^n$, $A = (a_{ij})$, $i, j = 1, 2, \ldots, n$, is a constant matrix, N is a nonlinear operator, sufficiently smooth, vanishing at 0, so that $\tilde{U} = 0$ is a solution of (1). The linear instability of this solution can be studied with the eigenvalues-eigenvectors method and a critical instability parameter, say R_L, is obtained: above R_L the solution is unstable, below R_L it is stable against infinitesimal perturbations. The nonlinear stability of zero solution can be studied with the classical energy method, by introducing the energy (Lyapunov function) $E = \frac{1}{2}|U|^2$. This method furnishes a critical nonlinear stability parameter, say R_E, below which the zero solution is nonlinearly stable. Generally we have $R_E \leq R_L$. If $R_E = R_L$ we say that the Lyapunov function is *optimal* and we obtain necessary and sufficient stability conditions. The nonlinear stability of the zero solution can also be studied by the *canonical reduction method* based on the eigenvalues-eigenvectors method (see Ref. 1,7,8). Indeed, by means of a change of dependent variables, with a *transformation* matrix Q, we obtain new canonical field $V = Q^{-1}U$ and a *new system*

$$\dot{V} = BV + \bar{N}(V), \tag{2}$$

where $B = Q^{-1}AQ$ is a *similar matrix* to A (it is in a diagonal or a general Jordan form), $\bar{N}(V) = Q^{-1}N(QV)$. As it is well known, similar operators define ordinary differential equations that have the same dynamical properties (see Ref. 1,8,9), and system (2) is topologically equivalent to (1). We thus define the *optimal* Lyapunov function

$$E := \frac{1}{2}|V|^2 = \frac{1}{2}[V_1^2 + V_2^2 + \cdots + V_n^2], \tag{3}$$

which gives the critical linear and (at least local) nonlinear stability thresholds; we also obtain a computable radius of attraction of initial data. In particular cases we may obtain global stability.

This method can be generalized (see Ref. 12,13) to reaction-diffusion systems

$$U_t = D\Delta U + LU + N(U) \tag{4}$$

(where D and L are constant matrices and Δ is the Laplacian), fluid dynamics (see Ref. 11,15) and flows in porous media, Ref. 16, by introducing a change of dependent variables, connected with *first* eigenvalue of the operator $D\Delta + L$ (with suitable boundary conditions) and defining as optimal Lyapunov function (at least for the linear problem) the classical energy of the new canonical fields:[12,13,16]

$$E(t) = \frac{1}{2}\|V\|^2, \tag{5}$$

where the symbols $(\, , \,)$ and $\| \cdot \|$ denote the usual scalar product and norm in a Hilbert space $\mathcal{H}$ (usually $\mathcal{H} = L^2(\Omega)$) where $\Omega \subseteq \mathbb{R}^m$ is the space domain. Here we apply this method to stability of the endemic equilibrium for an epidemic model with diffusion introduced in Ref. 18, and for a three-competing species model (the May Leonard system with diffusion[2,14,17]).

There is a vast number of phenomena in biology in which a key element to a developmental process seems to be the appearance of a travelling wave of chemical concentration, mechanical deformation, electrical signal and so on, (see Ref. 3,6,19,20). In what follows, we look for constant shape travelling wavefront solutions in the epidemic model (6) with Neumann boundary conditions and follow Dunbar[4,5] which studied a similar problem for a prey-predator model. By assuming that the reproduction number R_0 (see sect.

2) satisfies the inequality $R_0 > 1$, we show that there exist travelling wavefront solutions that approach the steady state $(\frac{1}{R_0}, \lambda(1-\frac{1}{R_0}))$, both in an oscillatory manner or they are monotonic.

2. Optimal Lyapunov function for an epidemic model with diffusion

Mulone et al.[18] have considered the epidemic model with diffusion and cross-diffusion

$$\begin{aligned} S_t &= aS_{xx} + cI_{xx} + \mu - \mu S - \beta SI \\ I_t &= cS_{xx} + aI_{xx} - (\mu+\epsilon)I + \beta SI \end{aligned} \tag{6}$$

in $(0, L) \times (0, \infty)$, where $S(t, x)$ and $I(t, x)$ are the densities of susceptible individuals and infectious individuals of a biological population at the spatial position x and the time t, respectively, μ is the recruitment rate of the population and the per capita death rate of the population ($1/\mu$ is the mean lifetime), β is the disease transmission coefficient (contact rate) and ϵ is the disease-induced death rate ($1/\epsilon$ is the average infectious period), a is a diffusion term and c is a cross-diffusion term, $0 \le c < a$. Rescaling the variables by introducing $t^* = t(\mu+\epsilon)$, $x^* = \sqrt{\frac{\mu+\epsilon}{a}}x$, $R_0 = \frac{\beta}{\mu+\epsilon}$, $p = \frac{c}{a}$, $\lambda = \frac{\mu}{\mu+\epsilon}$, we obtain (omitting the asterisks)

$$\begin{aligned} S_t &= S_{xx} + pI_{xx} + \lambda(1-S) - R_0 SI \\ I_t &= pS_{xx} + I_{xx} - I + R_0 SI. \end{aligned} \tag{7}$$

The equilibria are the disease-free $H_0 = (1, 0)$, for any R_0, and, for $R_0 > 1$, the endemic $H_1 = (\frac{1}{R_0}, \lambda(1-\frac{1}{R_0}))$. We note that $0 \le p < 1$ and $0 < \lambda < 1$. In Ref. 18 the stability of the disease free equilibrium has been studied in the case of Dirichlet boundary conditions. To consider the stability of $(1, 0)$, the basic reproduction number R_0 is used as the threshold quantity that determines *whether a disease can invade* a population. For this model the threshold quantity is the contact rate β times the average death-adjusted infectious period $\frac{1}{\mu+\epsilon}$. The critical instability reproduction number is found to be

$$R_{0c} := 1 + \xi - \frac{p\xi(1+\xi+p\xi)}{\lambda+\xi+p\xi}$$

where $-\xi$ is the *first* eigenvalue of the Laplacian (here f_{xx}) with Dirichlet b. c., $\xi = \frac{\pi^2}{L^2}$ (in the case of Neumann boundary conditions, $\xi = 0$ and

$R_{0c} = 1$, as in the kinetic case, see Ref. 10). By introducing the *classical energy* (see e.g. Ref. 22) of perturbation (s, i) to H_0:

$$E_0 = \frac{1}{2}[\|s\|^2 + \|i\|^2], \tag{8}$$

the critical energy reproduction number R_{0E_0} is below the linearized instability reproduction number R_{0c} (see Ref. 18). By using the reduction method, in Ref. 18, the following theorem has been proved.

Theorem 2.1. *If $\epsilon < \beta$ and $R_0 < R_{0c}$, the disease free equilibrium* $(1,0)$ *of (7) is nonlinearly stable.*

Moreover, a known radius of attraction for the initial data has been given, see Ref. 18.

Here we shortly study the stability of the endemic equilibrium H_1 by assuming Dirichlet boundary conditions (in the case of Neumann boundary conditions, the same results of the kinetic model can be obtained). The perturbations equations to H_1 are found to be

$$\begin{aligned} s_t &= s_{xx} + p i_{xx} - R_0 \lambda s - i - R_0 s i \\ i_t &= p s_{xx} + i_{xx} + \lambda (R_0 - 1) s + R_0 s i \,. \end{aligned} \tag{9}$$

It can be seen that if p is sufficiently small, $(p^2-1)\xi + p \leq 0$ or if $\lambda > (p^2-1)\xi + p$, the endemic equilibrium is linearly stable for any $R_0 > 1$. By using the classical energy $E_0 = \dfrac{\lambda(R_0-1)}{2}\|s\|^2 + \dfrac{1}{2}\|i\|^2$, we obtain the energy equation:

$$\dot{E}_0 = -R_0\lambda^2(R_0-1)\|s\|^2 - [\lambda(R_0-1)\|s_x\|^2 + p(\lambda(R_0-1)+1)(i_x, s_x) + \|i_x\|^2].$$

From this we have conditional nonlinear stability if $p^2(\lambda(R_0-1)+1)^2 - 4\lambda(R_0-1) < 0$. With this energy we do not have stability conditions for any $R_0 > 1$. Indeed, if we consider $p = 0.5$ and $\lambda = 0.8$, we have nonlinear stability, according to E_0, if R_0 is in the interval $[1.089, 18.41]$. For $R_0 \in (1, \infty) - [1.089, 18.41]$, we can search for another energy with an optimal coupling Lyapunov parameter; instead, we use the reduction method to obtain a new energy E (equivalent to the classical one) which gives nonlinear stability. For example, if $L = \pi$, and $R_0 = 1.01$, proceeding as in Ref. 18, we consider the matrix (where now $\xi = 1$)

$$A = \begin{pmatrix} -\xi - R_0\,\lambda & -p\xi - 1 \\ -p\xi + \lambda\,(R_0 - 1) & -\xi \end{pmatrix}.$$

Its eigenvalues are $\alpha_1 = -2.353$ and $\alpha_2 = -0.454$. A transformation matrix and its inverse are

$$Q = \begin{pmatrix} -0.939 & 0.742 \\ -0.341 & -0.669 \end{pmatrix}, \quad Q^{-1} = \begin{pmatrix} -0.758 & -0.840 \\ 0.386 & -1.0640 \end{pmatrix}$$

By introducing the change of variables $(\phi, \psi)^T = Q(s, i)^T$, we easily obtain the new system (equivalent to (9))

$$\begin{aligned} \phi_t =& 1.524\phi_{xx} - .058\psi_{xx} - .828\phi - .058\psi - .026\phi^2 - .031\phi\psi + .413\psi^2 \\ \psi_t =& .434\phi_{xx} + .475\psi_{xx} + .434\phi + .020\psi - .470\phi^2 - .550\phi\psi + .729\psi^2 . \end{aligned} \tag{10}$$

By writing the energy equation and following Ref. 18, section 4, we easily obtain that the condition $E(0) < 8.34 \times 10^{-4}$ implies nonlinear exponential stability.

The problem of global stability (showed in the kinetic case, see Ref. 10) remains open.

3. Optimal stability in the May-Leonard system with diffusion

An application of the reduction method to a three component reaction-diffusion system is given (for more details, see Ref. 17). The (symmetric) May-Leonard system with diffusion is given by

$$\begin{aligned} u_{1,t} &= \mu\Delta u_1 + u_1(1 - u_1 - \alpha u_2 - \beta u_3) \\ u_{2,t} &= \mu\Delta u_2 + u_2(1 - \beta u_1 - u_2 - \alpha u_3) \\ u_{3,t} &= \mu\Delta u_3 + u_3(1 - \alpha u_1 - \beta u_2 - u_3) \\ &\qquad\qquad\qquad \text{in} \quad \Omega \times (0, \infty) \end{aligned} \tag{11}$$

with assigned boundary conditions (of Dirichlet or Neumann type) on $\partial\Omega \times (0, \infty)$, where Ω is a bounded domain of $\mathbb{R}^3$, u_i denotes the density of the ith competitor, the positive constants μ, α and β are the diffusion rate and the competition coefficients, with $\alpha < 1 < \beta$, (the stability of other three-competing species has also been studied in Ref. 21).

It is easy to check that the unique positive constant equilibrium of system (11) is given by $M_1 = (\bar{u}, \bar{u}, \bar{u}) = (\frac{1}{h}, \frac{1}{h}, \frac{1}{h})$, with $h = 1 + \alpha + \beta$ (it is obtained with zero Neumann conditions or assigned Dirichlet conditions). In order to study its stability, we write the perturbation equations to M_1:

$$\begin{aligned} U_{1,t} &= \mu\Delta U_1 - (\bar{u} + U_1)(U_1 + \alpha U_2 + \beta U_3) \\ U_{2,t} &= \mu\Delta U_2 - (\bar{u} + U_2)(\beta U_1 + U_2 + \alpha U_3) \\ U_{3,t} &= \mu\Delta U_3 - (\bar{u} + U_3)(\alpha U_1 + \beta U_2 + U_3) . \end{aligned} \tag{12}$$

It is easy to check that the linear instability condition is given by

$$\frac{\alpha+\beta-2}{2h}-\mu\xi>0,$$

where $-\xi$ is the *first eigenvalue* of the $\mu I\Delta U + LU$ (I is the 3×3 identity matrix) with the appropriate boundary conditions. As concerns the nonlinear stability, we note that, by using the classical energy $E_0 = \frac{1}{2}[\|U_1\|^2 + \|U_2\|^2 + \|U_3\|^2]$, we have conditional nonlinear stability whenever

$$\frac{\alpha+\beta-1}{h}-\mu\xi<0.$$

Thus we do not have the same stability region of linearized case. In order to have the same stability region, we use the reduction method. For this, we define the matrix $A_\xi = -\mu\xi I + J(M_1)$, where I is the 3×3 identity matrix and $J(M_1)$ is given by the Jacobian matrix

$$J(M_1)=\begin{pmatrix} -\bar{u} & -\alpha\bar{u} & -\beta\bar{u} \\ -\beta\bar{u} & -\bar{u} & -\alpha\bar{u} \\ -\alpha\bar{u} & -\beta\bar{u} & -\bar{u} \end{pmatrix}.$$

The eigenvalues λ_i $(i=1,2,3)$ of the matrix A_ξ are

$$\lambda_1=-\mu\xi-1,\quad \lambda_{2,3}=\frac{\alpha+\beta-2}{2h}-\mu\xi\pm i\frac{\sqrt{3}}{2}\frac{\beta-\alpha}{h}.$$

The new equivalent nonlinear system associated to (12) is found to be

$$\begin{aligned} V_{1,t} &= \mu\Delta V_1 - V_1 + \bar{N}_1 \\ V_{2,t} &= \mu\Delta V_2 + \frac{\alpha+\beta-2}{2h}V_2 + \frac{\sqrt{3}}{2}\frac{\beta-\alpha}{h}V_3 + \bar{N}_2 \\ V_{3,t} &= \mu\Delta V_3 - \frac{\sqrt{3}}{2}\frac{\beta-\alpha}{h}V_2 + \frac{\alpha+\beta-2}{2h}V_3 + \bar{N}_3, \end{aligned} \tag{13}$$

where N_i are the nonlinear terms (see Ref. 17). By using the new energy

$$E_1=\frac{1}{2}[\|V_1\|^2+\|V_2\|^2+\|V_3\|^2],$$

we obtain

$$\frac{\alpha+\beta-2}{2h}-\mu\xi<0,$$

and we reach the coincidence of the linear and conditional nonlinear stability regions.

4. Travelling waves in the epidemic model with diffusion

Let us consider the epidemic model described by system (7). It is known (see Ref. 10) that in the kinetic case, i.e. in the ODEs case, H_0 is stable if $R_0 < 1$, H_0 is unstable and H_1 is stable (globally in the quadrant $\mathrm{Int}(\mathbb{R}^2_+)$) if $R_0 > 1$. We shall consider system (7) with Neumann boundary conditions. We look for wavefront solutions:

$$S(x,t) = s(z), \quad I(x,t) = i(z), \quad z = x + ct,\ c > 0\,,$$

where c is the wave speed to be determined. We obtain the ODE system

$$\begin{aligned} cs' &= s'' + pi'' + \lambda(1-s) - R_0 si, \\ ci' &= ps'' + i'' + R_0 si - i \end{aligned} \tag{14}$$

and the dynamical system in $\mathbb{R}^4$

$$\begin{aligned} s' &= v\,, \quad i' = w \\ v' &= \frac{cv - \lambda(1-s) + R_0 si - p(cw - R_0 si + i)}{1-p^2} \\ w' &= \frac{cw - R_0 si + i - p(cv - \lambda(1-s) + R_0 si)}{1-p^2}\,. \end{aligned} \tag{15}$$

In the (s,i,v,w) phase-space, the steady solutions are $A_0 = (1,0,0,0)$ and $A_1 = (\frac{1}{R_0}, \lambda(1-\frac{1}{R_0}), 0, 0)$, with A_0 stable if $R_0 < 1$, and A_0 unstable, A_1 stable, if $R_0 > 1$. Thus, there is the possibility of travelling wave from unstable steady state A_0 to the stable one in the hypothesis $R_0 > 1$. We should look for solutions $(s(z), i(z))$ of (15) with boundary conditions

$$\begin{aligned} &s(-\infty) = 1,\ i(-\infty) = 0, \\ &s(+\infty) = \frac{1}{R_0},\ i(+\infty) = \lambda(1-\frac{1}{R_0}). \end{aligned} \tag{16}$$

First, we linearize the system about the singular point $A_0 = (1,0,0,0)$. If $\mathbf{f}(s,i,v,w)$ is the vector representing the right hand side of (15), the Jacobian of $\mathbf{f}(s,i,v,w)$ with respect to s, i, v, w is given by

$$J = \begin{pmatrix} 0 & 0 & 1 & 0 \\ 0 & 0 & 0 & 1 \\ \dfrac{\lambda + R_0 i + pR_0 i}{1-p^2} & \dfrac{R_0 s + pR_0 s - p}{1-p^2} & \dfrac{c}{1-p^2} & \dfrac{-pc}{1-p^2} \\ \dfrac{-R_0 i - p\lambda - pR_0 i}{1-p^2} & \dfrac{-R_0 s + 1 - pR_0 s}{1-p^2} & \dfrac{-pc}{1-p^2} & \dfrac{c}{1-p^2} \end{pmatrix}.$$

We determine the eigenvalues in A_0. In the case $p = 0$ (no cross-diffusion), the eigenvalues are:

$$\sigma_{1,2} = (c \pm \sqrt{c^2 - 4(R_0 - 1)})/2,\ \sigma_{3,4} = (c \pm \sqrt{c^2 + 4\lambda})/2.$$

If $R_0 > 1,\ 0 < c < 2\sqrt{R_0 - 1}$ there is a three-dimensional unstable manifold and one-dimensional stable manifold based at A_0. There is also a two-dimensional unstable manifold associated to the eigenvalues $\sigma_{1,2}$. In particular, the critical point is a spiral point on the unstable manifold. Therefore a trajectory approaching A_0 must have $i(z) < 0$ for some z. This violates the requirement that waves be non-negative (see Ref. 5). Thus the inequalities $R_0 > 1,\ 0 < c < 2\sqrt{R_0 - 1}$ imply that travelling wave solutions do not exist. Thus, we require:

$$R_0 > 1,\ c > 2\sqrt{R_0 - 1}\,,$$

in this case all the eigenvalues are real and there is a three-dimensional unstable manifold and a one-dimensional stable manifold at A_0. Now the system gives rise to travelling wavefront solutions which can also display oscillatory behaviour. The proof of existence of these waves involves a careful analysis of the phase space system to show that there is a trajectory, lying in the positive quadrant, which joins the relevant singular points and can be done in the same way as in Ref. 4,5.

A key point in the proof of the existence of travelling waves is the introduction of a suitable Lyapunov function for (15) in a neighborhood of the fixed point $A_1 = (\frac{1}{R_0}, \lambda(1 - \frac{1}{R_0}), 0, 0)$ and the use of the Lasalle invariance principle. In the case $p = 0$, we define

$$\begin{aligned} V = {} & \frac{c(R_0 s - 1)^2}{2} + cR_0 I_e \left(\frac{i}{I_e} - 1 - \ln\frac{i}{I_e}\right) \\ & + R_0^2 v(S_e - s) + R_0 w(\frac{I_e}{i} - 1)\,, \end{aligned} \tag{17}$$

where $S_e = \frac{1}{R_0}$, $I_e = \lambda(1 - \frac{1}{R_0})$. Now, we linearize the system about the singular point A_1 and compute its eigenvalues. They are given by the solutions of the equation $\sigma^2(\sigma - c)^2 - \lambda R_0 \sigma(\sigma - c) + \lambda(R_0 - 1) = 0$. It can be proved that there exists $\lambda^* = \frac{4(R_0 - 1)}{R_0^2}$ such that, for $\lambda < \lambda^*$ the wavefront solutions (s, i) of (15), with boundary conditions (16), approach the steady state $(\frac{1}{R_0}, \lambda(1 - \frac{1}{R_0}))$ in an oscillatory manner while for $\lambda \geq \lambda^*$ they are monotonic. For example, Fig. 1 illustrates the monotonic solutions behaviour.

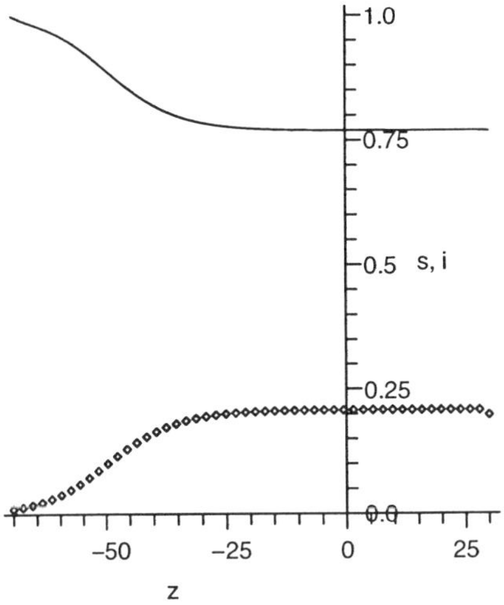

Fig. 1. A monotonic travelling wave. $\lambda = 0.9, R_0 = 1.3; c = 2.19, H_0 = (1, 0); H_1 = (0.77, 0.21)$. The continuous curve refers to $s(z)$ and the other curve to $i(z)$.

If $p \neq 0$ (cross-diffusion), the analysis is much more difficult, however, numerical computations show that there exist travelling wavefront solutions that approach the steady state $(\frac{1}{R_0}, \lambda(1 - \frac{1}{R_0}))$ both in an oscillatory manner or they are monotonic. For example, if $p = 0.1$, $R_0 = 3$, $c \geq c^* = \sqrt{27.98}$ wave front solutions that approach the steady state H_1 in an oscillatory manner exist. When p increases from 0.1 to 1 the minimum velocity c^* for the existence of a travelling wave decreases and this is in agreement with the destabilizing effect of the cross diffusion term p.

It will interesting to see what kind of initial data for the PDE system will develop to a travelling wave and the study of the stability of the wave front solutions. This will be done in a forthcoming paper.

Acknowledgments

The paper is dedicated to Tommaso Ruggeri on the occasion of his birthday. The research has been partially supported by the University of Catania under a local PRA contract, by the Italian Ministry for University and Scientific Research, PRIN: "*Problemi matematici non lineari di propagazione e stabilità nei modelli del continuo*", and GNFM of INDAM.

References

1. E. Beltrami, *Mathematics for Dynamic Modeling*, 2nd edn. (Academic Press, San Diego, 1997).
2. R.S. Cantrell and C. Cosner, *Spatial ecology via reaction-diffusion equations* (Wiley, Chichester, UK, 2003).
3. D.A.T. Cummings, R.A. Irizarry, N.E. Huang, T.P. Endy, A. Nisalak, K. Ungchusak and D.S. Burke, *Nature* **427**, 344–347 (2004).
4. S.R. Dunbar, *J. Math. Biol.*, **17**, 11- 32, (1983).
5. S.R. Dunbar, *Trans. Amer. Math. Soc.*, **268**, 557-594, (1984).
6. B.T. Grenfell, O.N. Bjørnstad and J. Kappey, *Nature* **414**, 716–723 (2001).
7. J.K. Hale, *Ordinary differential equations* (R. E. Krieger Publ. Co., Huntington, New York, 1980).
8. P. Hartman, *Ordinary differential equations.* Reprint of the 2nd edn. (Birkhäuser, Boston, 1982).
9. M.W. Hirsch and S. Smale, *Differential equations, dynamical systems, and linear algebra* (Academic Press, New York, 1974).
10. H.W. Hethcote, *SIAM Review* **42**, 599-653, (2000).
11. S. Lombardo and G. Mulone, *Nonlinear, Anal.* **63** /5-7, e2091-e2101, 2005 (doi:10.1016/j.na.2004.09.003).
12. S. Lombardo, G. Mulone and M. Trovato, *Rend. Circolo Mat. Palermo*, ser. II, Sppl. **78**, 173–185 (2006).
13. S. Lombardo, G: Mulone and M. Trovato, Nonlinear stability in reaction-diffusion systems via optimal Lyapunov functions (submitted).
14. R.M. May and W.J. Leonard, *SIAM J. Appl. Math.* **29**, 243–253 (1975).
15. G. Mulone, *Far East J. Appl. Math.* **15**, n.2, 117–134 (2004).
16. G. Mulone and B. Straughan, *ZAMM* **86**, n. 7, 507-520 (2006).
17. G. Mulone and B. Straughan, Nonlinear stability of multi-species ecological problems (submitted).
18. G. Mulone, B. Straughan and W. Wang, *Stud. Appl. Math.* **118**, 117–132 (2007).
19. J.D. Murray, *Mathematical biology. I. An introduction.* 3rd edn. Interdisciplinary Applied Mathematics, 17 (Springer-Verlag, New York, 2002).
20. J.D. Murray, *Mathematical biology. II. Spatial models and biomedical applications.* 3rd edn. Interdisciplinary Applied Mathematics, 18 (Springer-Verlag, New York, 2003).
21. S. Rionero, Long time behaviour of three competing species and mutualistic communities, in *Proc. Asymptotic Methods in Nonlinear Wave Phenomena*, Palermo 5-7 June 2006, T. Ruggeri, M. Sammartino Eds., World Scientific, Singapore 171–185 (2007).
22. B. Straughan, *The Energy Method, Stability, and Nonlinear Convection*, 2nd edn. (Springer-Verlag: New-York, 2004).

MULTIPLE COLD AND HOT SECOND SOUND SHOCKS IN HE II *

AUGUSTO MURACCHINI AND LEONARDO SECCIA

Department of Mathematics and Research Center of Applied Mathematics, University of Bologna, Via Saragozza, 8, 40123-Bologna, Italy.
E-mail: muracchi@ciram.unibo.it

The propagation of a temperature pulse in superfluid helium is studied by the simple waves theory. We determine the shape change of this pulse, initially represented by a gaussian profile, using a generalized non-linear Cattaneo model proposed, in the framework of Extended Thermodynamics, by Ruggeri and co-workers in the case of a rigid conductor. Hence, we prove the existence of several families of multiple hot and cold shocks (i.e. usual double shocks, double shocks only ahead or behind the wave profile, and very strange quadri-shocks), depending on the relation among the unperturbed temperature of Helium II and three special temperatures (characteristic temperatures) and, in some cases, on the wave's amplitude.

1. Introduction

In the present paper, a realistic initial regular temperature pulse, of gaussian type, is considered in the one dimensional case, studying the evolution of a simple plane wave travelling in superfluid helium. Since our model is non linear, the original profile is deformed during its evolution and a critical time occurs for which its slope becomes infinite and a shock wave arises. We are able to prove that three characteristic temperatures exist, playing an essential role in the shape changes of travelling waves. So, unlike the case of crystals, several new possible changes in the simple wave structure become visible, as it was pointed out in our previous papers,[1,2] although for different types of shock waves. It is also possible to better justify the experimental evidence of double shocks in a neighborhood of 1.8 K.[3] Finally, the existence of very special quadri-shocks, as limiting case of two different

---

* *Al nostro amico e maestro Tommaso Ruggeri per il suo 60º compleanno.*

families of usual double shocks, is discussed.†

2. The model

We use a mathematical model based on a differential non linear system constituted by an energy conservation equation and an evolution equation for the heat flux. It was before used with satisfactory results in the case of heat propagation in a rigid conductor.[4] This system is

$$\frac{\partial(\rho\varepsilon)}{\partial t} + \operatorname{div}\mathbf{q} = 0\,, \qquad \frac{\partial\,(\alpha\mathbf{q})}{\partial t} + \operatorname{grad}\nu = -\frac{\nu'}{\kappa}\,\mathbf{q}\,. \tag{1}$$

Here ρ is the constant mass density, ε the internal energy density, $\mathbf{q}$ the heat flux vector, α (the so called thermal inertia) and ν constitutive scalars depending on the temperature θ and κ the heat conductivity.

Choosing as field variables θ and $\mathbf{q}$, all the constitutive functions appearing in (1), i.e. $\alpha(\theta)$, $\nu(\theta)$, $\varepsilon(\theta)$ and $\kappa(\theta)$, may be determined from measurements of the equilibrium quantities $\varepsilon \equiv \varepsilon(\theta)$, $\kappa \equiv \kappa(\theta)$, $U_E \equiv U_E(\theta)$ ($U_E(\theta)$ represents the velocity of small perturbations propagating into an equilibrium state and $\varepsilon'(\theta) = c_v(\theta)$ is the specific heat).

This model can properly describe the phenomenology of second sound propagation since the thermal inertia α is a function of the temperature and, in particular, it can be used also in the superfluid case taking into account the results of the paper[5] where the differential system of a binary mixture of Euler's fluids is written as a system for a single heat conducting fluid. Following this way, it is possible to develop an unified macroscopic theory valid both for He II and crystals.[1,2,6] For more details, the interested reader can refer to the papers.[5,7]

3. Evolution of the simple thermal wave

Let us consider one dimensional waves [$\theta \equiv \theta(x,t)$, $q \equiv q(x,t)$] propagating along the positive x direction in an initially unperturbed state [i.e. $(\theta, q) = (\theta_0, 0)$]. On the ground of the experimental data, it is reasonable to consider infinite the heat conductivity κ for liquid helium below 2.2 K.[8] In this way, the right hand side of $(1)_2$ becomes zero and so the system (1) is conservative and the simple wave theory is applicable.

†In the case of usual double shocks, the two points of the wave profile for which its slope becomes infinite at the same critical time are on opposite sides of the wave peak; instead in the new double shocks here identified, both the points lie on the same side of the peak, either in front or behind.

A solution is a simple wave if the fields θ, q depend on x and t through only a variable φ

$$\theta(x,t) = \theta(\varphi), \quad q(x,t) = q(\varphi) \ , \tag{2}$$

with φ given as a function of x and t in the implicit form $\varphi = x - \lambda(\varphi)t$, with $\lambda(\varphi)$ to be determined. Then, by (2), one obtains

$$\theta(\varphi) = \Theta(\varphi), \quad q(\varphi) = Q(\varphi) \ , \tag{3}$$

where Θ and Q are the initial data of θ and q respectively: $\theta(x,0) = \Theta(x)$, $q(x,0) = Q(x)$. As it is well known, λ is the solution of the characteristic polynomial associated to the system (1)

$$\rho c_v \alpha \lambda^2 + \lambda \alpha' q - \nu' = 0 \ , \quad (\alpha' = d\alpha/d\theta) \tag{4}$$

Since in the present case the simple wave propagates in the x direction, then the positive eigenvalue solution of (4) is taken and it is easy to show that $\lambda(\theta_0, 0) = U_E(\theta_0)$.

Inserting (2) in (1) we can deduce that q must be function of θ through the differential equation

$$\frac{dq}{d\theta} = \rho c_v(\theta)\lambda(q,\theta), \quad q(\theta_0) = 0. \tag{5}$$

Therefore, from (4) and (5) one finds that $\lambda(\theta, q)$ depends only on the temperature i.e.

$$\lambda(\theta, q) = \lambda(\theta, q(\theta)) = \hat{\lambda}(\theta). \tag{6}$$

The free component θ of the field is obtained by $\varphi = x - \lambda(\varphi)t$ and (3) as an implicit function of x and t

$$\theta = \Theta\left(x - \hat{\lambda}(\theta)t\right). \tag{7}$$

To obtain $\theta = \theta(x,t)$, we need to invert (7). We define local critical time, for a generic point x of the initial profile of the simple wave, the time for which (7) cannot be inverted and the critical time the minimum possible value among all local critical times in the future of the profile i.e., respectively,

$$\hat{t}_{cr}(x) = -\frac{1}{d\hat{\lambda}(\Theta(x))/dx} \ , \quad t_{cr} = \inf_x \{\hat{t}_{cr}(x) > 0\}. \tag{8}$$

Let x_{cr}^o be the value of x corresponding to the lower limit of the function $\hat{t}_{cr}(x)$ appearing in (8). So x_{cr}^o represents the point of the initial profile which is mapped, at the critical time, into the point x_{cr} (critical distance), where the wave degeneration occurs. In correspondence to x_{cr}^o we define the critical temperature $\theta_{cr} = \Theta(x^o_{cr})$. Since the point (x^o_{cr}, θ_{cr}) has a velocity $\hat{\lambda}(\theta_{cr})$, the critical distance x_{cr} is given by $x_{cr} = x^o_{cr} + \hat{\lambda}(\theta_{cr})t_{cr}$.

A gaussian curve is chosen as initial temperature profile $\Theta(x)$, representing for a hot wave the increase (or decrease for a cold one) of the equilibrium temperature θ_0 of He II to the perturbed temperature θ_1

$$\Theta(x) = (\theta_1 - \theta_0)e^{-8\left(\frac{x-a}{\delta}\right)^2} + \theta_0. \tag{9}$$

Here, a is the point where the curve attains the maximum value θ_1 (minimum for the cold waves) and the thickness δ of the gaussian is defined as the distance between the intersection points, with the line $\theta = \theta_0$, of the tangent lines in the inflections points. As δ depends on the time duration Δt of the initial pulse and the points lying at the bottom of the gaussian curve, corresponding to a temperature close to θ_0, have a velocity $\hat{\lambda}(\theta_0) = U_E(\theta_0)$, it seems natural to choose the constant δ and a (depending on θ_0) as $\delta = U_E(\theta_0)\Delta t$, $a = \delta/2$. In such way the initial profile (9) is completely determined by θ_0, Δt and the amplitude $\Delta\theta = \theta_1 - \theta_0$. Then, the equation (7), yields

$$\theta = (\theta_1 - \theta_0)e^{-\frac{8}{\delta^2}\left(x-\hat{\lambda}(\theta)t-a\right)^2} + \theta_0. \tag{10}$$

4. Shape change of the second sound wave

In this section we study the wave's shape and we evaluate the critical distance and the critical time. At first, let us recall that the wave degeneration occurs in the spatial interval for which $d\hat{\lambda}/dx < 0$ (see (8)). Since $\hat{\lambda}$ is a function of x through θ, it is evident that the breakdown of the solution occurs in the front part or in the back part of the gaussian profile if $\hat{\lambda}$ is, respectively, an increasing or a decreasing function of θ.

4.1. *Small amplitude waves*

To investigate the sign of $d\hat{\lambda}/d\theta$, in the case of small amplitude waves, we need to evaluate it in the equilibrium state (i.e. when $\theta = \theta_0$, $q = 0$), since if the amplitude $\mid \theta_1 - \theta_0 \mid$ is very small one has $\hat{\lambda}' \simeq \hat{\lambda}'_0$, $(\lambda' = d\lambda/d\theta)$.

This was done in previous papers,[2,4] where it is proved that

$$\left(\frac{d\hat{\lambda}}{d\theta}\right)_0 = \frac{3}{2}\left(\frac{U_E}{\Phi}\frac{d\Phi}{d\theta}\right)_0 , \quad \text{with} \quad \Phi(\theta) = U_E c_v^{1/6}\theta^{1/3} \tag{11}$$

and then the derivative changes sign where the shape function Φ has extrema (the subscript 0 indicates that the expression (11) is evaluated for

$\theta = \theta_0$).[‡] Performing a numerical analysis one finds that the principal differences of He II with respect to the crystals as NaF and Bi[4] is that there exist three characteristic temperatures $\tilde{\theta}$ for which the shape function Φ reaches the extrema and $d\hat{\lambda}/d\theta$, evaluated at equilibrium, changes sign. At equilibrium, these temperatures are [§]

$$\widetilde{\theta}_1 \simeq 0.47\,\mathrm{K}, \quad \widetilde{\theta}_2 \simeq 0.95\,\mathrm{K}, \quad \widetilde{\theta}_3 \simeq 1.87\,\mathrm{K}. \tag{12}$$

In particular, the function Φ shows as well as two maxima a minimum too, on the contrary of the crystals case where only a maximum appears; so, the existence of a critical temperature $\widetilde{\theta}_2$, corresponding to the minimum point of $\Phi(\theta)$, leads to new shape changes of second sound. From the behavior of Φ, taking into account $(8)_1$ and $(8)_2$, a hot (cold) wave at the critical time degenerates in the front (back) part if $\theta_0 < \tilde{\theta}_1$ and $\tilde{\theta}_2 < \theta_0 < \tilde{\theta}_3$, and in the back (front) part if $\tilde{\theta}_2 > \theta_0 > \tilde{\theta}_1$ and $\theta_0 > \tilde{\theta}_3$.

Starting from these observations, we have also proved in[7] that our model, in the limit of small (non vanishing) wave amplitudes, gives shape changes qualitatively analogous to those obtained in.[10]

4.2. *Arbitrary amplitude waves*

For studying arbitrary amplitude waves, or if we are in a neighborhood of the characteristic temperatures, $d\hat{\lambda}/d\theta$ must be evaluated in a non equilibrium state. To this goal, we estimate it numerically, for example in the case of a gaussian curve with $\Delta t = 50\,\mu s$,[¶] finding the critical time of the simple wave as a function of the unperturbed temperature and of the initial wave amplitude. In particular, we determine in the plane $(\theta_0, \Delta\theta)$ the curves of the threshold amplitude for which there exist two points, one in the front and one in the back part of the wave profile, such that the local critical time has the same minimum: so the double shock arises (see Fig. 1). In particular, we underline that:

- there are three curves passing through the characteristic temperatures and dividing the $(\theta_0, \Delta\theta)$ plane in seven different regions where the wave degenerates alternatively in the front or in the back part;
- the two curves through $\widetilde{\theta}_2$ and $\widetilde{\theta}_3$ present a very strange behavior, since they coalesce in a point approximatively given by $\theta_0 \simeq 0.8$ K and $\Delta\theta \simeq 1$ K (the point A in the Fig. 1);

[‡]The general behavior of (11) is studied in.[2]
[§]It must be underlined that they have been experimentally identified[8,9] in He II and important differences in wave propagation appear in correspondence.
[¶]This value is in accord to many experimental data: see, for example, in.[9]

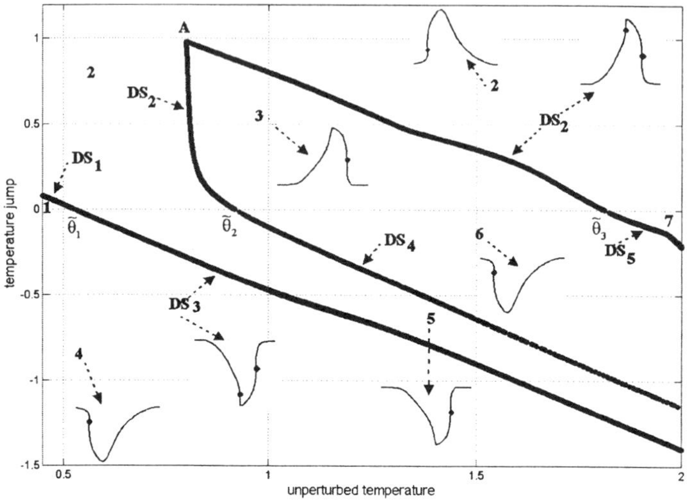

Fig. 1. The curve of the threshold amplitudes in the case of superfluid helium. The points lying on the curve correspond to the amplitudes for which the double shock occurs.The singular point A is shown and different regions and curves are pointed out. The wave profile evolution is sketched: note that in region 1 and on the curve DS_1 the behavior is like in the region 3 and on the curve DS_2, respectively. Moreover, in region 7 and on the curves DS_4, DS_5 the behavior is like in the region 5 and on the curve DS_3, respectively.

• there exists an unperturbed temperature range between about 0.8 K and 2 K such that, if one changes the temperature jump, more than one double shock can arise, almost in principle: this feature is different from the case of NaF crystal where, for a fixed value θ_0, always a double shock only occurs.

A first important consequence of these results is that, at least theoretically, there exist, for suitable amplitudes, double shock waves in all the temperature range represented in Fig. 1. Another interesting observation

temperature ranges	hot double shocks	cold double shocks
$\theta_0 < \tilde{\theta}_1$	1	0
$\tilde{\theta}_1 < \theta_0 < 0.8$	0	1
$0.8 < \theta_0 < \tilde{\theta}_2$	2	1
$\tilde{\theta}_2 < \theta_0 < \tilde{\theta}_3$	1	2
$\tilde{\theta}_3 < \theta_0$	0	3

Table I. In the horizontal lines one finds, in correspondence of the meaningful temperature ranges, the number of possible hot and cold double shocks, appearing if the temperature jump passes from negative to positive values.

regards the not complete symmetry between hot and cold double shocks, as resumed in the table I.

In what follows, let's summarize the complex behavior of the point of the profile where the shock arises.

When the amplitude increases, different shape changes could appear. If one studies $(\theta_{cr} - \theta_0)/\Delta\theta$ vs. the unperturbed temperature θ_0 for different $\Delta\theta$ and a fixed Δt, a very particular behavior appears, permitting us to better characterize the special point A shown in the figure 1. The plot of

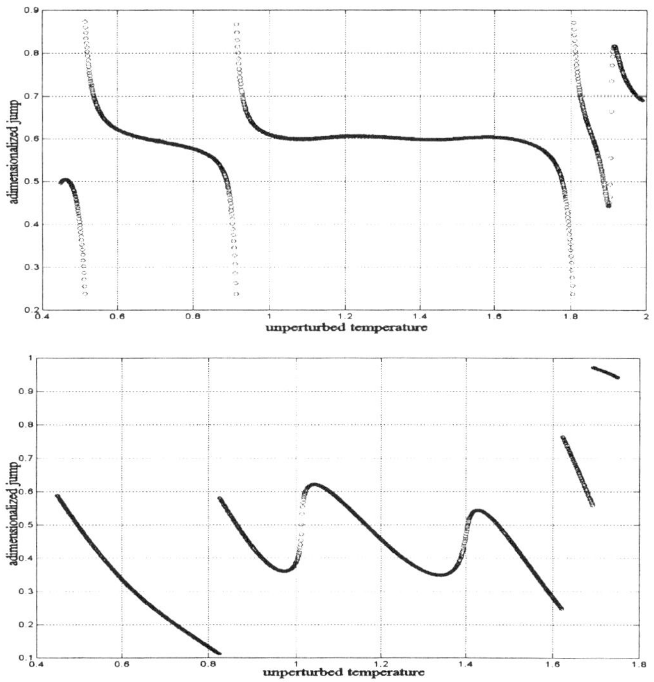

Fig. 2. Upper plot: jump for $\Delta\theta = 0.01$ K, with three discontinuities clearly present. Lower plot: jump for $\Delta\theta = 0.25$ K, with a new discontinuity at about 1.7 K.

$(\theta_{cr} - \theta_0)/\Delta\theta$ vs. the unperturbed temperature θ_0 is given in the figure 2, in correspondence of two fixed values of $\Delta\theta$ (0.01 K, 0.25 K) and with a constant value of Δt (50 μs). In the upper plot of Fig. 2 we can see that only three discontinuities are present. They characterize the three usual double shocks for a wave of initial amplitude $\Delta\theta = 0.01$ K in the whole range of unperturbed temperatures, as it is possible to verify considering in Fig. 1 a temperature jump equal to 0.01 K. In the lower plot of Fig. 2 a new

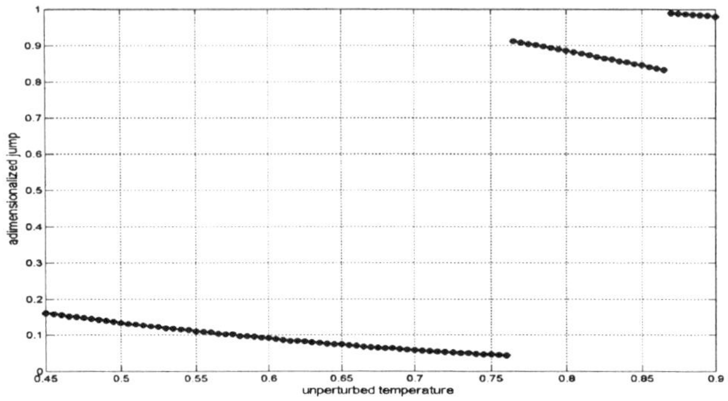

Fig. 3. Jump for $\Delta\theta = 1.1$ K, with two discontinuities clearly present.

situation appears: in fact, since here $\Delta\theta = 0.25$ K, we should expect two discontinuities, due to the presence of only two usual double shocks (see again Fig. 1), but a new discontinuity is shown at $\theta_0 \simeq 1.7$ K. The reason is the appearance of a new type of double shock, for which the breakdown of the wave profile happens in the same moment in two points of the back part. Now if the initial amplitude of the wave is increased (for example $\Delta\theta = 0.29$ K), a new discontinuity appears just before 1 K, revealing again the onset of a double shock of new type similar to the previous one, but with the difference that the breakdown of the wave profile occurs in the same moment in two points of the front part. Then, increasing again the initial amplitude other new discontinuities of these types appear, until about 1 K (the point A in the Fig. 1) where a striking quadri-shock becomes visible: it is characterized by four points of breakdown in the profile taking place at the same critical time, two in the front part and two in the back one of the wave (for more details, see[7]). It is interesting to observe that, if one increases the initial amplitude over 1 K, two new discontinuities are still present, as shown in figure 3; in particular, two double shocks of new type appear in the back part of the wave profile at $\theta_0 \simeq 0.76$, $\theta_0 \simeq 0.86$.

The plot of $(\theta_{cr} - \theta_0)/\Delta\theta$ vs. the unperturbed temperature θ_0 is given again in the figure 4, in correspondence of two negative fixed values of $\Delta\theta$ (-0.005 K, -0.01 K) and with the constant value $\Delta t = 50$ μs. In the upper plot of Fig. 4 only three discontinuities are present. They characterize the three usual cold double shocks for a wave of initial amplitude $\Delta\theta = -0.005$ K in the whole range of unperturbed temperatures (see in Fig. 1 a temperature jump equal to -0.005 K). In the lower plot of Fig. 4 a new situation appears: in fact, since here $\Delta\theta = -0.01$ K, we should expect again three discontinuities, but instead a new discontinuity takes place at $\theta_0 \simeq 1.92$ K. This is because of the appearance of a new type of cold double shock, for which the breakdown of the wave profile happens in the same moment in two points of the front part. If the wave initial amplitude

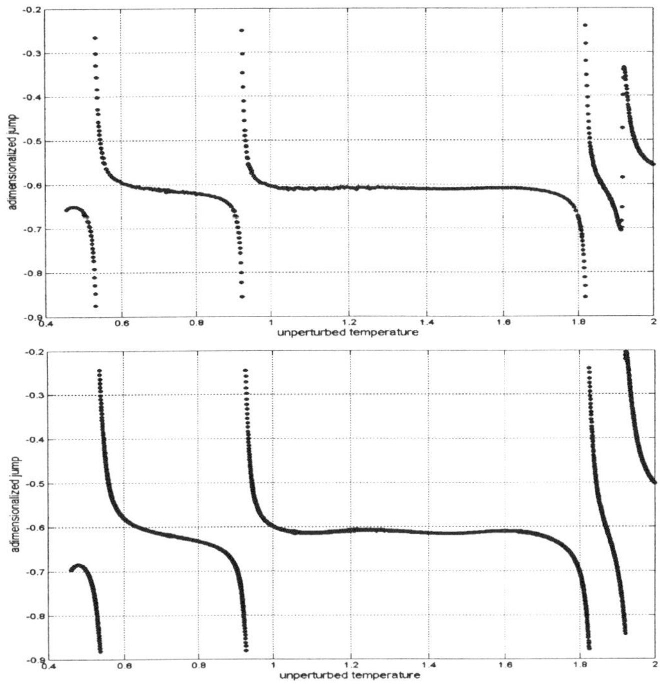

Fig. 4. Upper plot: jump for $\Delta\theta = -0.005$ K, with three discontinuities clearly present. Lower plot: jump for $\Delta\theta = -0.01$ K, with a new discontinuity at about 1.92 K.

decreases (for example $\Delta\theta = -0.4$ K), two discontinuities arise showing two usual cold double shocks (see Fig. 1). Then, further decreasing the initial amplitude until to $\Delta\theta = -0.6$ K, we should expect again two discontinuities, but a new one is present at about 1.95 K, due to the onset of a new type of cold double shock in the back part of the wave profile.

Finally, we underline that in[7] a full discussion about the optimal choice of the free parameters is performed, in order to reveal, in experimental way, the usual double shocks and the new hot multiple shocks. Here we restrict ourselves to give, in the table II, some possible numerical values in order to find both hot and cold multiple shocks.

He II

	θ_0	$\Delta\theta$	t_{cr}	x_{cr}
b	1.694	0.25	183.1	0.0041
a	0.993	0.29	60.43	0.0021
b	1.6564	0.29	128.6	0.0031
c	0.8023	0.9805	40.89	0.0024
b	0.765	1.1	27.02	0.0015
b	0.8665	1.1	12.08	0.0007
a	1.9221	-0.01	$4.32 \cdot 10^5$	7.89
b	1.948	-0.6	19.46	0.0004

Table II. The critical time and distance, for He II, in correspondence with some choices of initial profile parameters. The temperatures are in Kelvin degrees, the critical distance in centimeters, t_{cr} in μs and $\Delta t = 50\,\mu$s. The values corresponding to the letters a, b, c are referred, respectively, to a front double shock, to a back double shock and to the quadri-shock.

References

1. A. Muracchini, T. Ruggeri and L. Seccia, Second sound propagation in superfluid helium, in: *Proceedings "WASCOM 2001" 11th Conference on Waves and Stability in Continuous Media, Porto Ercole, 2001.* World Scientific, 347 (2002).
2. A. Muracchini, T. Ruggeri and L. Seccia, *Z. Angew. Math. Phys.* **57**, 4, 567 (2006).
3. T. N. Turner, *Physica* **107**B, 701 (1981); *Phys. Fluids* **26** (11), 3227 (1983).
4. T. Ruggeri, A. Muracchini and L. Seccia, *Phys. Rev. Lett.* **64** (22), 2640 (1990); *Il Nuovo Cimento* **16**D, N.1 15 (1994); *Phys. Rev. B* **54** (1), 332 (1996).
5. T. Ruggeri, The binary mixtures of Euler fluids: a unified theory of second sound phenomena, in: *Continuum Mechanics and Applications in Geophysics and Environment*. Springer-Verlag, Berlin 79 (2001).
6. A. Muracchini, L. Seccia, *Rend. Circ. Mat. Palermo* (2) Suppl. **78**, 227 (2006).
7. L. Seccia, T. Ruggeri and A. Muracchini, Second sound and multiple shocks in superfluid helium. Submitted to *Z. Angew. Math. Phys.*
8. S. J. Putterman, *Superfluid Hydrodynamics*, North Holland, Amsterdam (1974).
9. J. Cummings, D.W. Schmidt and W.J. Wagner, *Phys. Fluids* **21** (5), 713 (1978).
10. I.M. Khalatnikov, *An Introduction to the Theory of Superfluidity*, Benjamin, New York (1965).

Acknowledgments - This paper was supported by MIUR PRIN (Progetto di Ricerca di Interesse Nazionale: *Nonlinear Propagation and Stability in Thermodynamical Processes of Continuous Media*. Coordinator: Prof. Tommaso Ruggeri) and by GNFM-INdAM.

EXTENDED HYDRODYNAMIC MODEL FOR THE COUPLED ELECTRON-PHONON SYSTEM IN SILICON SEMICONDUCTORS

O. MUSCATO*, V. DI STEFANO** and C. MILAZZO*

**Dipartimento di Matematica e Informatica, Università degli Studi di Catania, V.le A. Doria, 6 - 95125 - Catania, Italia*

***Dipartimento di Matematica, Università degli Studi di Messina, Salita Sperone, 31 - 98166 - Messina, Italia*

E-mail:{muscato,vdistefano,cmilazzo}@dmi.unict.it

In this paper we introduce an extended hydrodynamic model for the description of heat generation in thin body silicon semiconductor devices. The hydrodynamic model is obtained by taking the moments of the Bloch-Boltzmann-Peierls equations, and by using the Maximum entropy principle for the closure of the system.

Keywords: Semiconductor devices, Boltzmann-Poisson system for semiconductors, Hydrodynamic model, Maximum Entropy Principle.

1. Introduction

In recent years, the dimensions of semiconductor devices have been scaled down reaching submicrometric order. As a result, large electric fields near the drain region generate hot or energetic electrons, and a very large quantity of heat. From a physical point of view, self-heating starts when the nearly-free electrons in a conduction band of a semiconductor are accelerated by the electric field. The electrons gain energy from the field, then lose it by inelastically scattering with boundaries and phonons, heating up the lattice through the mechanism known as Joule heating. The heat conduction can be viewed as the transport of phonons inside the device, which can be modeled by means of kinetic equations. To solve these equations is a daunting task, also from the numerical viewpoint. An alternative is to introduce hydrodynamic models which can be derived by taking moments from the kinetic equations for electrons and phonons. The main problems of this approach is the closure for the high-order moments, as well as the closure of the production terms.

The aim of this paper is to tackle this problem by coupling self-consistently the electron current and the heat transport, and to propose a closure based on the application of the Maximum Entropy Principle of Extended thermodynamics.

2. Kinetic equations for electrons and phonons

The most complete description of a system formed by electrons (e) and phonons (p), is based on the Bloch-Boltzmann-Peierls (BBP) equations[1] . In principle the interactions in such a system are p-p, p-e and e-e. For the sake of simplicity, we shall consider the electrons as a rarefied gas in a sea of phonons, which means that the e-e interactions can be neglected.
Let be $f(t, \mathsf{x}, \mathsf{k})$ the probability density to find an electron at time t, position x, with wave vector k, $N_g(t, \mathsf{x}, \mathsf{q})$ the probability density to find a phonon at time t, position x with wave vector q of type g (i.e. the branch g of the phonon spectrum). The BBP equations read[1-3]

$$\frac{\partial f}{\partial t} + \mathsf{v}(\mathsf{k}) \cdot \nabla_{\mathsf{x}} f - \frac{e}{\hbar} \mathsf{E} \cdot \nabla_{\mathsf{k}} f = \sum_g \mathcal{Q}_g^{ep}[f, N_g] \tag{1}$$

$$\frac{\partial N_g}{\partial t} + \mathsf{u}_g(\mathsf{q}) \cdot \nabla_{\mathsf{x}} N_g = \mathcal{Q}_g^{pe}[f, N_g] + \mathcal{Q}_g^{pp}[N_g] \,, \quad g = \{g_1, g_2, g_3, ...\} \tag{2}$$

where e is the absolute value of the electron charge, $\hbar$ the Planck constant divided by 2π, v and u_g are respectively the electron and phonon group velocity. The formula of the collisional operators $\mathcal{Q}_g^{pp}$, $\mathcal{Q}_g^{pe}$, $\mathcal{Q}_g^{ep}$ can be found in[2] . Finally the electric field $\mathsf{E}(t, \mathsf{x})$ satisfies the Poisson equation

$$\Delta(\epsilon\phi) = e\left[n(t, \mathsf{x}) - N_D(\mathsf{x}) + N_A(\mathsf{x})\right]$$

$$\mathsf{E} = -\nabla_{\mathsf{x}}\phi \tag{3}$$

where $\phi(t, \mathsf{x})$ is the electric potential, n is the electron density, N_D and N_A, respectively, are the donor and acceptor densities, ϵ the dielectric constant. The semiconductor band structure enters into the dispersion relation, which relates the electron energy and its wave vector. In the following we have considered the analytic band approximation, i.e.

$$\varepsilon(\mathsf{k})\left[1 + \alpha\varepsilon(\mathsf{k})\right] = \gamma = \frac{\hbar^2 \mathsf{k}^2}{2m^\star},$$

where α is the non parabolicity factor. Moreover linear dispersion relation for the acoustic phonons and constant dispersion relation for optical phonons have been considered.

3. Moment equations

Starting from the transport equations (1), (2) one can obtain balance equations for macroscopic quantities associated to the flow. To this end it is sufficient to multiply (1) by a weight function $\psi_A(\mathsf{k})$, (2) by $\phi_B(\mathsf{q})$ and integrate on $\mathbb{R}^3$. For the electrons we have chosen an 8-moments model, for the phonon flow we have chosen a 4-moments model, then we have obtained the following balance equations

$$\frac{\partial n}{\partial t} + \frac{\partial (nV^i)}{\partial x^i} = 0 \tag{4}$$

$$\frac{\partial (nP^i)}{\partial t} + \frac{\partial (nU^{ij})}{\partial x^j} + neE_j = nC_P^i \tag{5}$$

$$\frac{\partial (nW)}{\partial t} + \frac{\partial (nS^i)}{\partial x^i} + neV_iE^i = nC_W \tag{6}$$

$$\frac{\partial (nS^i)}{\partial t} + \frac{\partial (nF^{ij})}{\partial x^j} + neE_jG^{ij} = nC_W^i \tag{7}$$

$$\frac{\partial W_{ac}}{\partial t} + \frac{\partial Q^i}{\partial x^i} = \mathcal{C} \tag{8}$$

$$\frac{\partial Q^i}{\partial t} + \frac{\partial N^{ij}}{\partial x^j} = \mathcal{C}^i \tag{9}$$

$$\frac{\partial W_{op}}{\partial t} = \mathcal{C}_{op} \tag{10}$$

where V^i is the average electron velocity, W is the average electron energy, S^i is the average electron energy flux, P^i is the average electron crystal momentum, U^{ij} is the flow of electron crystal momentum, F^{ij} is the flux of electron energy flux, C_P^i is the production of electron crystal momentum, C_W is the production of electron energy, C_W^i is the production of electron energy flux, W_{ac} is the acoustic phonon energy density, Q^i is the acoustic phonon energy-flux density, N^{ij} is the acoustic phonon flux of energy-flux density, $\mathcal{C}$ is the production of acoustic phonon energy density, $\mathcal{C}_i$ is the production of acoustic phonon energy-flux density, W_{op} and $\mathcal{C}_{op}$ are the optical phonon energy density and the production of optical phonon energy density respectively.

4. Maximum entropy principle and closure relations for the fluxes

The moment equations (4)-(10) do not constitute a set of closed relations. If we assume as fundamental variables n, V^i, W, S^i, W_{ac}, Q^i, W_{op}, which have a direct physical interpretation, the closure problem consists in expressing the high-order fluxes U^{ij}, G^{ij}, F^{ij} and the moments of the collisional terms as function of the fundamental variables. A good, physically motivated way to obtain constitutive equations is by means of the Maximum Entropy Principle (MEP)[4-7] . It offers a procedure to provide an approximation of the electron and phonon distribution functions, in terms of a finite number of moments. For the electron distribution function Anile et al.[8] , by using a suitable approximation, obtained at lower order the following closures for the fluxes

$$U_{ij} = U^{(0)}(W)\delta_{ij}\,, \quad F_{ij} = F^{(0)}(W)\delta_{ij}\,, \quad G_{ij} = G^{(0)}(W)\delta_{ij}. \tag{11}$$

The phonon distribution function has been obtained in[9] without any approximation procedure. In this case the closure for the flux N^{ij} is

$$N^{ij} = \frac{1}{3}v_s^2 W_{ac}\delta_{ij} + \frac{1}{2}v_s^2 W_{ac}(3\chi - 1)\left(\frac{Q_i Q_j}{Q^2} - \frac{1}{3}\delta_{ij}\right) \tag{12}$$

$$\chi = \frac{5}{3} - \frac{4}{3}\sqrt{1 - \frac{3}{4}\left(\frac{Q}{v_s W_{ac}}\right)^2} \tag{13}$$

which is valid supposing $Q = |Q_i| < v_s W_{ac}$, where v_s is the sound speed.

5. Characterization of the hydrodynamic model

In this section we shall investigate the mathematical properties of the closed system (4)-(10). Our system can be written as

$$\frac{\partial}{\partial t}\Psi^{(0)}(\mathbf{U}) + \sum_{i=1}^{3} \frac{\partial}{\partial x_i}\Psi^{(i)}(\mathbf{U}) = B(\mathbf{U}), \tag{14}$$

where $\mathbf{U}$ is the vector of the fundamental variables. It can be represented in the form of a quasi-linear system of PDEs

$$A^{(0)}\frac{\partial \mathbf{U}}{\partial t} + \sum_{i=1}^{3} A^{(i)}\frac{\partial \mathbf{U}}{\partial x^i} = B(\mathbf{U}) \tag{15}$$

by introducing the Jacobian matrix of the fluxes

$$A^{(i)} = \nabla_{\mathbf{U}}\Psi^{(i)}, \quad i = 0, 1, 2, 3 \quad .$$

It can be proved, by using the entropy principle and the concavity of the entropy density, that the system (14) can be transformed into a symmetric hyperbolic form, inside a given neighborhood of the local equilibrium configuration[4] . This property assures regular solutions with finite speeds of propagation and also the applicability of numerical schemes for hyperbolic systems. Since some of the constitutive equations of our system have been obtained through a series expansion, the question of the hyperbolicity requires more care. We recall that a quasi-linear system of PDEs is said to be *hyperbolic in the t-direction* if $\det(A^{(0)}) \neq 0$ and the eigenvalues problem

$$\det\left(\sum_{i=1}^{3} n_i A^{(i)}(\mathbf{U}) - \lambda A^{(0)}(\mathbf{U})\right) = 0 \tag{16}$$

has real eigenvalues and the eigenvectors span $\mathbb{R}^{16}$ *for all* the unit vectors $\mathbf{n}$. The first condition on the hyperbolicity is easily satisfied because

$$\det(A^{(0)}) = n^7 (m^\star)^3 \neq 0 \quad .$$

If we consider eqs.(11), by using suitable dimensionless variables, the characteristic polynomial gives:

$$\lambda^2 = 0 \tag{17}$$
$$m^\star \lambda^4 - \lambda^2 \left[U + m^\star F' - WU' + 2\alpha(WF' - F)\right] + F'U - FU' = 0 \tag{18}$$

where $U = U^{(0)}, F = F^{(0)}$ and the prime denotes derivation respect to W. The roots of eq.(18) are real, distinct and different from zero if the following inequalities hold

$$g_1(W) = \left[U + F' - WU' + 2\alpha(WF' - F)\right]^2 + 4\,(FU' - F'U) > 0\,, \tag{19}$$
$$g_{2\pm}(W) = U + F' + 2\alpha(WF' - F) - WU' \pm \sqrt{g_1(W)} > 0 \tag{20}$$

which are fulfilled in standard regimes. The eingenspace associated to λ=0 has two independent eigenvectors providing $F' \neq 0$.

In the case of parabolic band approximation, where $\alpha = 0$, the eqs. (18) and the condition $F' \neq 0$ reduce to

$$\lambda_{\pm\pm} = \pm\frac{1}{3}\sqrt{(10 \pm 2\sqrt{10})W}\,, \quad F' = \frac{20}{9}W > 0\,.$$

The characteristic polynomial for phonons gives:

$$\lambda^6 = 0$$

$$\lambda^2 - \lambda\left(\frac{Q}{v_s^2 W_{ac}} \frac{1}{\sqrt{1 - \frac{3}{4}\left(\frac{Q}{v_s W_{ac}}\right)^2}}\right) - \left(\frac{5}{3} - \frac{4}{3}\frac{1}{\sqrt{1 - \frac{3}{4}\left(\frac{Q}{v_s W_{ac}}\right)^2}}\right) = 0$$

The roots are real, distinct and different from zero supposing $Q < v_s W_{ac}$, and read

$$\lambda_{\pm} = \frac{1}{2}\left(\frac{Q}{W_{ac}}\frac{1}{\sqrt{1-\frac{3}{4}\left(\frac{Q}{v_s W_{ac}}\right)^2}} \pm \sqrt{\Delta}\right)$$

where

$$\Delta = \frac{3\left(\frac{Q}{W_{ac}}\right)^2 + 20v_s^2\left[1-\frac{3}{4}\left(\frac{Q}{v_s W_{ac}}\right)^2\right] - 16v_s^2\sqrt{1-\frac{3}{4}\left(\frac{Q}{v_s W_{ac}}\right)^2}}{3\left[1-\frac{3}{4}\left(\frac{Q}{v_s W_{ac}}\right)^2\right]}.$$

The eingenspace associated to $\lambda = 0$ has six independent eigenvectors. Finally we have stated that the above system is hyperbolic for the parabolic and quasi-parabolic band approximations.
The closure of the production terms will be the subject of future work.

References

1. J.M. Ziman, *Electrons and Phonons*, Claredon Press, Oxford, (1967)
2. A. Rossani, "Generalized kinetic theory of electrons and phonons", Physica A, **305**, 323-329, (2002)
3. Ch. Auer, F. Schürrer and W. Koller, "A semi-continuos formulation of the Bloch-Boltzmann-Peierls equations", SIAM J. Appl. Math, **64**(4), 1457-1475, (2004)
4. I. Müller and T. Ruggeri, *Rational Extended thermodynamics*, Springer-Verlag, Berlin, (1998)
5. D. Jou, J. Casas-Vázquez and G. Lebon, *Extended irreversible thermodynamics*, Springer-Verlag, Berlin, (2001)
6. C.D. Levermore, "Moment Closure Hierarchies for Kinetic Theories" J. Stat. Phys, **83** , 331–407, (1996)
7. W. Dreyer, "Maximation of the entropy in non-equilibrium", J. Phys. A, **20**, 6505-6517,(1987)
8. A.M. Anile V. and Romano, "Nonparabolic band transport in semiconductors: closure of the moment equations", Cont. Mech. Therm., **11**, 307-325, (1999)
9. W. Larecki, "Symmetric conservative form of low-temperature phonon gas hydrodynamics", Nuovo Cimento, **14**D, N.2, 141-176, (1992)

DIFFERENTIAL EQUATIONS AND LIE SYMMETRIES*

F. OLIVERI[1], G. MANNO[2], R. VITOLO[2]

[1] *Department of Mathematics, University of Messina*
Salita Sperone 31, 98166 Messina, Italy,
[2] *Department of Mathematics "E. De Giorgi", University of Salento*
Via per Arnesano, 73100 Lecce, Italy
E-mail: oliveri@mat520.unime.it; gianni.manno@unile.it; raffaele.vitolo@unile.it

In a recent paper, within the framework of the inverse Lie problem, the definitions of strongly and weakly Lie remarkable equations have been introduced. The former are uniquely determined by their Lie point symmetries, whereas the latter are equations which do not intersect other equations admitting the same symmetries. After reviewing the basic theorems, some examples of strongly and weakly Lie remarkable equations are given. Moreover, the strongly Lie remarkable equations admitted by relevant algebras of vector fields on $\mathbb{R}^k$ (such as the isometric, affine or conformal algebra) are determined.

Keywords: Lie group analysis; Inverse Lie problem; Lie remarkable equations.

1. Introduction

Differential equations (DEs) provide a useful language to express the laws of nature, as well as to characterize complex geometric objects. After Sophus Lie, a powerful method for investigating the properties of DEs, either ordinary or partial, as well as determining their solutions, is given by the theory of symmetries.[1–7] Symmetries of DEs are (finite or infinitesimal) transformations of the independent and dependent variables and derivatives of the latter with respect to the former, with the further property of sending solutions into solutions. Among symmetries, there is a distinguished class, that of symmetries coming from a transformation of the independent and dependent variables: point symmetries.

The problem of finding the symmetries of a DE has associated a natural "inverse" problem, namely, the problem of finding the most general form of a DE admitting a given Lie algebra as subalgebra of infinitesimal point

*Dedicated to T. Ruggeri on the occasion of his 60th birthday.

symmetries. This problem may be addressed by classifying all possible realizations of the given Lie algebra as algebra of vector fields on the base manifold and then finding the differential invariants I_i of the realization under consideration.[8,9] In fact, it is well known[5] that, under suitable hypotheses of regularity, the most general DE admitting a given Lie algebra as subalgebra of point symmetries is locally given by $F^{\mu}(I_1, I_2, \ldots, I_k) = 0$ where F^{μ} are arbitrary smooth functions.

Within the framework of inverse Lie problem, it has been introduced the notion of *Lie remarkable equation*,[10] *i.e.*, of a DE uniquely determined by its Lie point symmetries. This definition has been reformulated from a geometric viewpoint,[11] strongly and weakly Lie remarkable equations have been distinguished, and necessary and sufficient conditions for their detection have been established.

2. Theoretical framework

Here we recall some basic facts regarding jet spaces[4,12] and the properties of DEs determined by their Lie point symmetries.[11]

All manifolds and maps are supposed to be C^{∞}. If E is a manifold then we denote by $\chi(E)$ the Lie algebra of vector fields on E. Also, for the sake of simplicity, all submanifolds of E are *embedded* submanifolds.

Let E be an $(n+m)$-dimensional smooth manifold and L an n-dimensional embedded submanifold of E. Let (V, y^A) be a local chart on E. The coordinates (y^A) can be divided in two sets, $(y^A) = (x^{\lambda}, u^i)$, $\lambda = 1, \ldots, n$ and $i = 1, \ldots, m$, such that the submanifold L is locally described as the graph of a vector function $u^i = f^i(x^1, \ldots, x^n)$. In what follows, Greek indices run from 1 to n and Latin indices run from 1 to m unless otherwise specified.

The set of equivalence classes $[L]^r_p$ of submanifolds L having at $p \in E$ a contact of order r is said to be the *r-jet of n-dimensional submanifolds of E* (also known as extended bundles[4]), and is denoted by $J^r(E, n)$. If E is endowed with a fibring $\pi : E \to M$ where $\dim M = n$, then the r-th order jet $J^r\pi$ of local sections of π is an open dense subset of $J^r(E, n)$. We have the natural maps $j_r L \colon L \to J^r(E, n)$, $p \mapsto [L]^r_p$, and $\pi_{k,h} \colon J^k(E, n) \to J^h(E, n)$, $[L]^k_p \mapsto [L]^h_p$, $k \geq h$.

The set $J^r(E, n)$ is a smooth manifold whose dimension is

$$\dim J^r(E, n) = n + m \sum_{h=0}^{r} \binom{n+h-1}{n-1} = n + m \binom{n+r}{r}, \qquad (1)$$

whose charts are $(x^{\lambda}, u^i_{\boldsymbol{\sigma}})$, where $u^i_{\boldsymbol{\sigma}} \circ j_r L = \partial^{|\boldsymbol{\sigma}|} f^i / \partial x^{\boldsymbol{\sigma}}$, where $0 \leq |\boldsymbol{\sigma}| \leq$

r. On $J^r(E,n)$ there is a distribution, the contact distribution, which is generated by the vectors

$$D_\lambda \overset{\text{def}}{=} \frac{\partial}{\partial x^\lambda} + u^j_{\boldsymbol{\sigma}\lambda}\frac{\partial}{\partial u^j_{\boldsymbol{\sigma}}} \quad \text{and} \quad \frac{\partial}{\partial u^j_{\boldsymbol{\tau}}},$$

where $0 \leq |\boldsymbol{\sigma}| \leq r-1$, $|\boldsymbol{\tau}| = r$ and $\boldsymbol{\sigma}\lambda$ denotes the multi-index $(\sigma_1,\ldots,\sigma_{r-1},\lambda)$. The vector fields D_λ are the (truncated) total derivatives. Any vector field $X \in \chi(E)$ can be lifted to a vector field $X^{(r)} \in \chi(J^r(E,n))$ which preserves the contact distribution. In coordinates, if $\Xi = \Xi^\lambda\partial/\partial x^\lambda + \Xi^i\partial/\partial u^i$ is a vector field on E, then its k-lift $\Xi^{(k)}$ has the coordinate expression

$$\Xi^{(k)} = \Xi^\lambda \frac{\partial}{\partial x^\lambda} + \Xi^i_{\boldsymbol{\sigma}} \frac{\partial}{\partial u^i_{\boldsymbol{\sigma}}}, \tag{2}$$

where $\Xi^j_{\boldsymbol{\tau},\lambda} = D_\lambda(\Xi^j_{\boldsymbol{\tau}}) - u^j_{\boldsymbol{\tau},\beta}D_\lambda(\Xi^\beta)$ with $|\boldsymbol{\tau}| < k$.

A *differential equation* $\mathcal{E}$ *of order* r *on* n*-dimensional submanifolds of a manifold* E is a submanifold of $J^r(E,n)$. The manifold $J^r(E,n)$ is called the *trivial equation*. An *infinitesimal point symmetry* of $\mathcal{E}$ is a vector field of the type $X^{(r)}$ which is tangent to $\mathcal{E}$.

Let $\mathcal{E}$ be locally described by $\{F^i = 0\}$, $i = 1,\ldots,k$ with $k < \dim J^r(E,n)$. Then finding point symmetries amounts to solve the system

$$\Xi^{(r)}\left(F^i\right) = 0\Big|_{F^i=0}$$

for some $\Xi \in \chi(E)$. We denote by $\operatorname{sym}(\mathcal{E})$ the Lie algebra of infinitesimal point symmetries of the equation $\mathcal{E}$.

By an r*-th order differential invariant* of a Lie subalgebra $\mathfrak{s}$ of $\chi(E)$ we mean a smooth function $I\colon J^r(E,n) \to \mathbb{R}$ such that for all $\Xi \in \mathfrak{s}$ we have $\Xi^{(r)}(I) = 0$.

The problem of determining the Lie algebra $\operatorname{sym}(\mathcal{E})$ is said to be the *direct Lie problem*. Conversely, given a Lie subalgebra $\mathfrak{s} \subset \chi(E)$, we consider the *inverse Lie problem*, *i.e.*, the problem of characterizing the equations $\mathcal{E} \subset J^r(E,n)$ such that $\mathfrak{s} \subset \operatorname{sym}(\mathcal{E})$.[13,14]

In what follows, we will devote ourselves to the analysis of the inverse Lie problem. We start by the definition and main properties, contained in Ref. 11, of DEs which are uniquely determined by their point symmetries, that we call *Lie remarkable* DEs.

Definition 2.1. Let E be a manifold, $\dim E = n+m$, and let $r \in \mathbb{N}$, $r > 0$. An l-dimensional equation $\mathcal{E} \subset J^r(E,n)$ is said to be

(1) *weakly Lie remarkable* if $\mathcal{E}$ is the only maximal (with respect to the inclusion) l-dimensional equation in $J^r(E,n)$ passing at any $\theta \in \mathcal{E}$ admitting $\mathrm{sym}(\mathcal{E})$ as subalgebra of the algebra of its infinitesimal point symmetries;
(2) *strongly Lie remarkable* if $\mathcal{E}$ is the only maximal (with respect to the inclusion) l-dimensional equation in $J^r(E,n)$ admitting $\mathrm{sym}(\mathcal{E})$ as subalgebra of the algebra of its infinitesimal point symmetries.

Of course, a strongly Lie remarkable equation is also weakly Lie remarkable. Some direct consequences of our definitions are due. For each $\theta \in J^r(E,n)$ denote by $S_\theta(\mathcal{E}) \subset T_\theta J^r(E,n)$ the subspace generated by the values of infinitesimal point symmetries of $\mathcal{E}$ at θ. Let us set $S(\mathcal{E}) \stackrel{\text{def}}{=} \bigcup_{\theta \in J^r(E,n)} S_\theta(\mathcal{E})$. In general, $\dim S_\theta(\mathcal{E})$ may change with $\theta \in J^r(E,n)$. The following inequality holds:

$$\dim \mathrm{sym}(\mathcal{E}) \geq S_\theta(\mathcal{E}),\ \forall \theta \in J^r(E,n), \tag{3}$$

where $\dim \mathrm{sym}(\mathcal{E})$ is the dimension, as real vector space, of the Lie algebra of infinitesimal point symmetries $\mathrm{sym}(\mathcal{E})$ of $\mathcal{E}$. If the rank of $S(\mathcal{E})$ at each $\theta \in J^r(E,n)$ is the same, then $S(\mathcal{E})$ is an involutive (smooth) distribution.

A submanifold N of $J^r(E,n)$ is an *integral submanifold* of $S(\mathcal{E})$ if $T_\theta N = S_\theta(\mathcal{E})$ for each $\theta \in N$. Of course, an integral submanifold of $S(\mathcal{E})$ is an equation in $J^r(E,n)$ which admits all elements in $\mathrm{sym}(\mathcal{E})$ as infinitesimal point symmetries. The points of $J^r(E,n)$ of maximal rank of $S(\mathcal{E})$ form an open set of $J^r(E,n)$.[11] It follows that $\mathcal{E}$ can not coincide with the set of points of maximal rank of $S(\mathcal{E})$. In Ref. 11 the following theorems have been proved.

Theorem 2.1. *A necessary condition for the differential equation $\mathcal{E}$ to be strongly Lie remarkable is that*

$$\dim \mathrm{sym}(\mathcal{E}) > \dim \mathcal{E},$$

whereas a necessary condition for the differential equation $\mathcal{E}$ to be weakly Lie remarkable is that

$$\dim \mathrm{sym}(\mathcal{E}) \geq \dim \mathcal{E}.$$

Sufficient conditions have been also established, that reveal useful when computing examples and applications.

Theorem 2.2. *If $S(\mathcal{E})|_{\mathcal{E}}$ is an l-dimensional distribution on $\mathcal{E} \subset J^r(E,n)$, then $\mathcal{E}$ is a weakly Lie remarkable equation. Let $S(\mathcal{E})$ be such that for any $\theta \notin \mathcal{E}$ we have $\dim S_\theta(\mathcal{E}) > l$. Then $\mathcal{E}$ is a strongly Lie remarkable equation.*

Finally, the next theorem gives the relationship between Lie remarkability and differential invariants.

Theorem 2.3. *Let $\mathfrak{s}$ be a Lie subalgebra of $\chi(J^r(E,n))$. Let us suppose that the r-prolongation subalgebra of $\mathfrak{s}$ acts regularly on $J^r(E,n)$ and that the set of r-th order functionally independent differential invariants of $\mathfrak{s}$ reduces to a unique element $I \in C^\infty(J^r(E,n))$. Then the submanifold of $J^r(E,n)$ described by $\Delta(I) = 0$ (in particular $I = k$ for any $k \in \mathbb{R}$), with Δ an arbitrary smooth function, is a weakly Lie remarkable equation.*

3. Examples

By using theorem 2.2 several strongly and weakly Lie remarkable equations have been characterized.[11] We recall some of them.

(1) The equation of minimal surfaces in $\mathbb{R}^{m+2}$

$$(1 + |\mathbf{u}_y|^2)\mathbf{u}_{xx} - 2|\mathbf{u}_x| \cdot |\mathbf{u}_y|\mathbf{u}_{xy} + (1 + |\mathbf{u}_x|^2)\mathbf{u}_{yy} = \mathbf{0}, \qquad \mathbf{u} \in \mathbb{R}^m$$

is nor strongly neither weakly Lie remarkable for $m = 2$ and $m = 3$, whereas it is weakly Lie remarkable for $m = 1$ and $m = 4$, provided we remove singular equations.

(2) The equation of unparametrized geodesic on a complete simply connected Riemannian 2–dimensional manifold E is strongly Lie remarkable if and only if E has constant Gaussian curvature.

(3) The second order Monge–Ampère equation

$$u_{xx}u_{yy} - u_{xy}^2 = \kappa, \qquad \kappa \text{ constant},$$

is weakly Lie remarkable if $\kappa \neq 0$, whereas it is strongly Lie remarkable if $\kappa = 0$ (equation of surfaces with vanishing Gaussian curvature).

(4) The third order Monge–Ampère equation[15]

$$(u_{ttx}u_{xxx} - u_{txx}^2) + \lambda(u_{ttt}u_{xxx} - u_{ttx}u_{txx}) + \lambda^2(u_{ttt}u_{txx} - u_{ttx}^2) = \mu,$$

where λ and μ are constants, as well as the homogeneous fourth order Monge-Ampère equation[15]

$$u_{tttt}(u_{ttxx}u_{xxxx} - u_{txxx}^2) + 2u_{tttx}u_{ttxx}u_{txxx} - u_{ttxx}^3 - u_{tttx}^2u_{xxxx} = 0$$

are weakly Lie remarkable, provided we remove singular subsets.

(5) The equation introduced by Bateman[16]

$$u_x^2u_{tt} - 2u_xu_tu_{xt} + u_t^2u_{xx} = 0,$$

and playing a role in the Painlevé analysis of 2–dimensional nonintegrable PDEs, as well as its generalization to the n–dimensional case[17]

$$\det \begin{bmatrix} 0 & u_{x^0} & u_{x^1} & \dots u_{x^{n-1}} \\ u_{x^0} & u_{x^0x^0} & u_{x^0x^1} & \dots u_{x^0x^{n-1}} \\ u_{x^1} & u_{x^0x^1} & u_{x^1x^1} & \dots u_{x^1x^{n-1}} \\ \dots & \dots & \dots & \dots\dots \\ u_{x^{n-1}} & u_{x^0x^{n-1}} & u_{x^1x^{n-1}} & \dots u_{x^{n-1}x^{n-1}} \end{bmatrix} = 0$$

(where x^0 is the time) are strongly Lie remarkable; in fact, for every n the Bateman equation admits a Lie algebra of dimension n^2, and the second order prolongation generates a distribution of maximal rank provided singular subsets are excluded.

4. Equations determined by Lie algebras of vector fields on $\mathbb{R}^{n+1}$

Strongly Lie remarkable equations can be determined in a general way by considering relevant Lie algebras. In this section we consider the strongly Lie remarkable equations associated with isometric, affine and conformal algebra of $\mathbb{R}^k$, where $\mathbb{R}^k$ is provided with metrics of various signatures. Since we start from concrete algebras and not from abstract ones, we do not have the problem of realizing them as vector fields. Let us consider scalar equations, *i.e.*, $m = 1$. Denote by

$$\mathcal{I}(\mathbb{R}^{n+1}), \qquad \mathcal{A}(\mathbb{R}^{n+1}), \qquad \mathcal{C}(\mathbb{R}^{n+1}),$$

the isometric, affine and conformal algebra of $\mathbb{R}^{n+1}$, respectively, with reference to the metric

$$g = \sum_{i=1}^{n} k_i dx^i \otimes dx^i + du \otimes du, \tag{4}$$

where k_i $(i = 1, \dots, n)$ are non–vanishing real constants. The algebras $\mathcal{I}(\mathbb{R}^{n+1})$ and $\mathcal{C}(\mathbb{R}^{n+1})$ depend on k_i. For instance, $\mathcal{I}(\mathbb{R}^4)$ can represent both the Euclidean and Poincaré algebra for suitable values of k_i.

The algebra $\mathcal{I}(\mathbb{R}^{n+1})$ has dimension equal to $(n+1)(n+2)/2$.

On the contrary, it is

$$\dim J^1(\mathbb{R}^{n+1}, n) = 2n + 1,$$
$$\dim J^2(\mathbb{R}^{n+1}, n) = 2n + 1 + \frac{n(n+1)}{2}, \quad \dots,$$

whereupon strongly Lie remarkable equations can be of order 1.

If $n = 2$, by setting $x = x^1$, $y = x^2$, we have the vector fields:

$$\Xi_1 = \frac{\partial}{\partial x}, \qquad \Xi_2 = \frac{\partial}{\partial y}, \qquad \Xi_3 = \frac{\partial}{\partial u},$$
$$\Xi_4 = k_2 y\frac{\partial}{\partial x} - k_1 x\frac{\partial}{\partial y}, \ \Xi_5 = u\frac{\partial}{\partial x} - k_1 x\frac{\partial}{\partial u}, \ \Xi_6 = u\frac{\partial}{\partial y} - k_2 y\frac{\partial}{\partial u}.$$

The rank of distribution of first order prolongations is 5 on $J^1(\mathbb{R}^3, 2)$ except for a 4-dimensional submanifold where the rank decreases.

Such a submanifold will be the strongly Lie remarkable equation which we are looking for:

$$1 + \frac{u_x^2}{k_1} + \frac{u_y^2}{k_2} = 0, \tag{5}$$

which describes the vanishing of the infinitesimal area element.

The generalization to arbitrary n is straightforward. The strongly Lie remarkable equation in this case is

$$1 + \sum_{i=1}^{n} \frac{u_{x^i}^2}{k_i} = 0. \tag{6}$$

Of course, in order these equations to be nonempty, we have to require that not all k_i are positive.

The algebra $\mathcal{A}(\mathbb{R}^{n+1})$ has dimension $n^2 + 3n + 2$, and strongly Lie remarkable equation can be of order 2 or 3 in the case $n = 2$, and of order 2 if $n \geq 3$. The infinitesimal generators of this algebra are

$$\frac{\partial}{\partial a}, \qquad a\frac{\partial}{\partial a}, \qquad a\frac{\partial}{\partial b}, \qquad \forall\, a, b \in \{x^1, x^2, \ldots, x^n, u\}.$$

We see that the strongly Lie remarkable second order equation in $J^2(\mathbb{R}^3, 2)$ is the homogeneous Monge-Ampère equation

$$u_{xx}u_{yy} - u_{xy}^2 = 0.$$

Moreover, there exists[18] a strongly Lie remarkable equation in $J^3(\mathbb{R}^3, 2)$ which has the following local expression:

$$\begin{aligned}
&u_{xx}^3 u_{yyy}^2 + u_{xxx}^2 u_{yy}^3 + 6u_{xx}u_{xxx}u_{xy}u_{yy}u_{yyy} - 6u_{xxx}u_{xxy}u_{xy}u_{yy}^2 \\
&-6u_{xx}u_{xxx}u_{xyy}u_{yy}^2 - 6u_{xx}^2 u_{xy}u_{xyy}u_{yyy} - 6u_{xx}^2 u_{xxy}u_{yy}u_{yyy} \\
&-8u_{xxx}u_{xy}^3 u_{yyy} + 9u_{xx}u_{xxy}^2 u_{yy}^2 + 9u_{xx}^2 u_{xyy}^2 u_{yy} \\
&+12u_{xxx}u_{xy}^2 u_{xyy}u_{yy} + 12u_{xx}u_{xxy}u_{xy}^2 u_{yyy} - 18u_{xx}u_{xxy}u_{xy}u_{xyy}u_{yy} = 0.
\end{aligned}$$

By considering the equations which live in $J^2(\mathbb{R}^{n+1}, n)$ with $n \geq 3$, for each n, we still have a strongly Lie remarkable equation, namely

$$\det\left(\frac{\partial^2 u}{\partial x^i \partial x^j}\right) = 0,$$

that is, the second order homogeneous Monge–Ampère equation in n variables.[19]

The algebra $\mathcal{C}(\mathbb{R}^{n+1})$ has dimension equal to $(n+2)(n+3)/2$; therefore, we may look for second order strongly Lie remarkable equations.

For $n = 2$, by setting $x = x^1$ and $y = x^2$, the algebra is spanned by isometries and by the vector fields

$$\begin{cases} \frac{1}{2}\frac{k_1x^2 - k_2y^2 - u^2}{k_1}\frac{\partial}{\partial x} + xy\frac{\partial}{\partial y} + xu\frac{\partial}{\partial u} \\ xy\frac{\partial}{\partial x} + \frac{1}{2}\frac{k_2y^2 - k_1x^2 - u^2}{k_1}\frac{\partial}{\partial y} + yu\frac{\partial}{\partial u} \\ xu\frac{\partial}{\partial x} + yu\frac{\partial}{\partial y} + \frac{1}{2}(u^2 - k_1x^2 - k_2y^2)\frac{\partial}{\partial u} \\ x\frac{\partial}{\partial x} + y\frac{\partial}{\partial y} + u\frac{\partial}{\partial u}. \end{cases}$$

By considering the metric (4) on $\mathbb{R}^3$, the (scalar) mean curvature H of a generic surface $u = u(x, y)$ is

$$H = \frac{1}{2}\frac{(k_2 + u_y^2)u_{xx} - 2u_xu_yu_{xy} + (k_1 + u_x^2)u_{yy}}{(k_1k_2 + k_2u_x^2 + k_1u_y^2)^{\frac{3}{2}}}, \tag{7}$$

and the Gaussian curvature G is

$$G = \frac{u_{xx}u_{yy} - u_{xy}^2}{(k_1k_2 + k_2u_x^2 + k_1u_y^2)^2}. \tag{8}$$

Then, by analyzing the rank of the matrix of 2-prolongations of the vector fields of $\mathcal{C}(\mathbb{R}^3)$, the unique second order equation which is strongly Lie remarkable is given by

$$G = H^2,$$

characterizing a surface whose Gaussian curvature is equal to the squared mean curvature. On the contrary. for $n > 2$ there are not second order strongly Lie remarkable equations.

5. Open problems

The property of Lie remarkability gives severe restrictions to the structure of differential equations admitting it. Various open questions are worth of being investigated. For instance, do the invariant solutions of a strongly

Lie remarkable equation provide all solutions? This is certainly true for homogeneous Monge–Ampère equation $u_{xx}u_{yy} - u_{xy}^2 = 0$,[20] and for the n–dimensional Bateman equation, that admits the general implicit solution[17]

$$\sum_{i=0}^{n-1} x^i f_i(u) = c,$$

where $f_i(u)$ $(i = 0, \ldots, n-1)$ are arbitrary functions, and c is constant.

Another open question, concerned with ODEs, has a direct link with the notion of complete symmetry group:[21,22] can a Lie remarkable ODE always be solved by quadratures?

Finally, since in mathematical physics we often have classes of differential equations, *i.e.*, equations having an assigned differential structure but involving arbitrary functions with assigned functional dependencies, it can be interesting to look for classes of DEs uniquely determined by their symmetries (in this case equivalence symmetries[23]). For instance, the nonlinear wave equation

$$u_{tt} - u_{xx} = f(t, x, u, u_t, u_x), \tag{9}$$

where f is an arbitrary function of its arguments, admits[24] the following operator of the equivalence group

$$\Xi = [\alpha(t+x) + \beta(t-x) + c_0]\frac{\partial}{\partial t} + [\alpha(t+x) - \beta(t-x)]\frac{\partial}{\partial x} + \gamma(t,x,u)\frac{\partial}{\partial u} +$$
$$[\gamma_{tt} - \gamma_{xx} + (u_t^2 - u_x^2)\gamma_{uu} + 2u_t\gamma_{tu} - 2u_x\gamma_{xu} + f\gamma_u - 2f\alpha' - 2f\beta']\frac{\partial}{\partial f},$$

where $\alpha(t+x)$, $\beta(t-x)$, $\gamma(t,x,u)$ are arbitrary functions of the indicated arguments, and c_0 is a constant. In the second order jet space augmented with f, the second order prolongations of the admitted equivalence symmetries generate a distribution of rank 9, and the rank decreases only on the manifold described by the equation itself, and on another submanifold not admitting the whole symmetries. Roughly speaking we could say that the class of equations (9) is strongly Lie remarkable, but it has to be stressed that at this stage this is only a naive statement, since much work is needed before extending the definitions and the theorems on Lie remarkability to equivalence transformations.

References

1. L. V. Ovsiannikov, *Group analysis of differential equations* (Academic Press, New York, 1982).

2. N. H. Ibragimov, *Transformation groups applied to mathematical physics* (D. Reidel Publishing Company, Dordrecht, 1985).
3. G. W. Bluman, S. Kumei, *Symmetries and differential equations* (Springer, New York, 1989).
4. P. J. Olver, *Applications of Lie Groups to Differential Equations*, 2nd ed. (Springer, New York, 1991).
5. P. J. Olver, *Equivalence, Invariants, and Symmetry* (Cambridge Univ. Press, New York, 1995).
6. N. H. Ibragimov, *Handbook of Lie group analysis of differential equations*, 3 volumes (CRC Press, Boca Raton, 1994, 1995, 1996).
7. G. Baumann, *Symmetry analysis of differential equations with Mathematica* (Springer, Berlin, 2000).
8. W. I. Fushchych, I. Yehorchenko, *Acta Appl. Math.* **28**, 69–92 (1992).
9. G. Rideau, P. Winternitz, *J. Math. Phys.* **31**, 1095–1105 (1990).
10. F. Oliveri, *Note di Matematica* **23**, 195–216 (2005).
11. G. Manno, F. Oliveri, R. Vitolo, *J. Math. Anal. Appl.* **332**, 767–786 (2007).
12. D. J. Saunders, *The Geometry of Jet Bundles* (Cambridge Univ. Press, 1989).
13. G. W. Bluman, J. D. Cole, *Similarity methods of differential equations* (Springer, New York, 1974).
14. W. I. Fushchych, *Collected Works* (Kyiv, 2000).
15. G. Boillat, *C. R. Acad. Sci. Paris Sér. I. Math.* **315**, 1211–1214 (1992).
16. H. Bateman, *Partial differential equations of mathematical physics* (Cambridge Univ. Press, 1992).
17. D. B. Fairlie, *Lett. Math. Phys.* **49**, 213–216 (1985).
18. G. Manno, F. Oliveri, R. Vitolo, *Theor. Math. Phys.* **151**, 843–850 (2007).
19. G. Boillat, *C. R. Acad. Sci. Paris* **313**, 805–808 (1991).
20. V. Rosenhaus, *Algebras, Groups and Geometries* **5**, 137–150 (1988).
21. J. Krause, *J. Math. Phys.* **35**, 5734–5748 (1994).
22. K. Andriopulos, P. G. L. Leach, G. P. Flessas, *J. Math. Anal. Appl.* **262**, 256–273 (2001).
23. I. G. Lisle, *Equivalence transformations for classes of differential equations* (Ph.D. thesis, Univ. of British Columbia, 1992).
24. R. Tracinà, *Private communication* (2007).

A PARTICLE–MESH NUMERICAL METHOD FOR ADVECTION–REACTION–DIFFUSION EQUATIONS WITH APPLICATIONS TO PLANKTON MODELLING*

F. PAPARELLA[1], F. OLIVERI[2]

[1] *Department of Mathematics "E. De Giorgi", University of Salento*
Via per Arnesano, 73100 Lecce, Italy
[2] *Department of Mathematics, University of Messina*
Salita Sperone 31, 98166 Messina, Italy
E-mail: francesco.paparella@unile.it; oliveri@mat520.unime.it

We present a new method for the numerical solution of advection-reaction-diffusion equations. The method is lagrangian for the advection-reaction part, and uses an auxiliary eulerian grid for the diffusive operator. We discuss the application of the method to the modelling of planktonic populations.

Keywords: Advection–reaction–diffusion equations; plankton modelling.

1. Introduction

Any believable model of plankton dynamics should involve at least the three following key factors. *Population dynamics*: planktonic communities are complex ecosystems where tens or hundreds of different species interact through non–trivial food networks; a minimalistic model should at least include the predator–prey dynamics between zooplankton and phytoplankton; more realistic models also explicitly track the concentration of nutrients.[1,2] *Large–scale transport*: on large scales (from a few kilometers to planetary scale) plankton acts as a tracer carried by vortices and ocean currents; communities colonizing nearby water patches may be separated by large distances at later times, and, conversely, initially distant populations may be brought into contact by the oceanic transport. *Small–scale mixing*: plankton is subject to the mixing effects of small–scale turbulence and wave–induced drift; swimming patterns of individual animals, such as nictemeral migrations across the water column, also induce mixing.

*Dedicated to T. Ruggeri on the occasion of his 60th birthday.

Then, the mathematical form of a N–species plankton model is that of a set of advection–reaction–diffusion equations:

$$\frac{\partial c_i}{\partial t} = -\underbrace{J(\psi, c_i)}_{advection} + \underbrace{f_i(c_1, \ldots, c_i, \ldots, c_N)}_{reaction} + \underbrace{D_i \Delta c_i}_{diffusion} , \qquad (1)$$

where c_i is the concentration of the i–th plankton species, ψ is a known streamfunction, the jacobian operator J is defined by $J(A, B) = \partial_x A \partial_y B - \partial_x B \partial_y A$, and the numbers D_i are given diffusion coefficients.

Although equations (1) are a rather general formulation of the problem, several assumptions and approximations have already been made. The large–scale flow is assumed to be two–dimensional and incompressible. This is a reasonable approximation for mesoscale and basin–scale dynamics in most areas of the ocean.[3] The small–scale mixing processes are modeled by a Laplacian operator with a constant diffusion coefficient. In some cases a nonlinear diffusion operator might be more realistic, for example in the presence of swimming of the individuals,[4] or slumping of density fronts;[5] however, at the exploratory level of this paper, we prefer to limit ourselves to laplacian diffusion.

2. The Numerical Method

In the absence of diffusion, equations (1) are easily solved by the method of characteristics. A simple and effective numerical scheme, already used in plankton studies,[6] is the following. The initial positions $\{(x_p, y_p)\}_{p=1,\cdots,N_p}$ are assigned to a set of N_p particles with uniformly random distribution within the computational domain. Each particle represents a fluid parcel, which moves according to the law

$$\dot{x}_p = -\frac{\partial \psi}{\partial y}(x_p, y_p, t); \qquad \dot{y}_p = \frac{\partial \psi}{\partial x}(x_p, y_p, t). \qquad (2)$$

The concentration fields are sampled at the particle positions: $c_{p,i}(t)$ is the concentration of the i–th field at the position (x_p, y_p) at time t. They obey to the set of equations

$$\dot{c}_{p,i} = f_i(c_{p,1}, \ldots, c_{p,i}, \ldots, c_{p,N}). \qquad (3)$$

Finally, equations (2,3) may be discretized in time with one of the several well–known schemes for ODEs (for example, we use the second–order Runge–Kutta scheme). We recall that incompressible flows conserve the Lebesgue measure. This implies that an initially uniform distribution of particles moving according to (2) will remain uniform at any later time. As a consequence there will be no undersampled or oversampled regions of the domain.

Adding a diffusion operator while maintaining the purely lagrangian nature of this scheme is not straightforward. Although particle–based methods for reaction–diffusion equations do exist,[7,8] their application to plankton modelling and a comparison with the method presented here is the topic for a future work.

In this paper we prefer to abandon a purely lagrangian approach. We introduce an auxiliary eulerian grid with meshes of fixed size $\Delta \times \Delta$ to approximate the diffusion operator. The p–th particle contributes to the value of the concentration fields on the k–th grid node, located at (x_k, y_k), with a weight

$$w_p(x_k, y_k) = \int_{x_k-\Delta/2}^{x_k+\Delta/2} \int_{y_k-\Delta/2}^{y_k+\Delta/2} \mathcal{P}(x - x_p, y - y_p)\, dx\, dy,$$

where the *cloud* $\mathcal{P}$ is a function having the following properties: $\mathcal{P}(x, y) \geq 0$; $\mathcal{P}(-x, y) = \mathcal{P}(x, -y) = \mathcal{P}(x, y)$; $\max_{x,y} \mathcal{P} = \mathcal{P}(0, 0)$; $\int_{-\infty}^{\infty} \int_{-\infty}^{\infty} \mathcal{P}(x, y)\, dx\, dy = 1$.

We compute interpolated concentration fields $\tilde{c}_i$ on the eulerian grid nodes with the following weighted averages

$$\tilde{c}_i(x_k, y_k) = \frac{\sum_p w_p(x_k, y_k)\, c_{p,i}}{\sum_p w_p(x_k, y_k)}.$$

This is a *cloud–in–cell* interpolation scheme.[9] If P has a compact support (not larger than a few meshes), most of the terms in the above sums are zero, and this makes the algorithm $O(N_p)$ in execution time. If we define $\mathcal{P}(x, y) = \{(x, y) \in [-\Delta/2, \Delta/2]^2 : 1/\Delta^2;\ \text{otherwise}:\ 0\}$ then each particle generates only four non–zero weights. Taking $x_k \leq x_p \leq x_k + \Delta$ and $y_k \leq y_p \leq y_k + \Delta$, their explicit expression are

$$\begin{aligned}
w_p(x_k, y_k) &= \Delta^{-2}(x_k + \Delta - x_p)(y_k + \Delta - y_p), \\
w_p(x_k + \Delta, y_k) &= \Delta^{-2}(x_p - x_k)(y_k + \Delta - y_p), \\
w_p(x_k, y_k + \Delta) &= \Delta^{-2}(x_k + \Delta - x_p)(y_p - y_k), \\
w_p(x_k + \Delta, y_k + \Delta) &= \Delta^{-2}(x_p - x_k)(y_p - y_k).
\end{aligned}$$

The concentration fields interpolated on the grid are subject to a Jacobi sweep:

$$\begin{aligned}
&\tilde{c}_i(x_k, y_k) = \\
&\quad = \frac{\tilde{c}_i(x_k - \Delta, y_k) + \tilde{c}_i(x_k, y_k - \Delta) + \tilde{c}_i(x_k + \Delta, y_k) + \tilde{c}_i(x_k, y_k + \Delta)}{4},
\end{aligned}$$

which is a second–order, finite–difference discretization of the heat equation.

Finally, after each time step in the numerical solution of (2)–(3), the concentration fields carried by the particles are relaxed toward their diffused counterparts known on the eulerian grid:

$$\dot{c}_{p,i} = \frac{4D_i}{\Delta^2}\left(\tilde{c}_i(x_p, y_p) - c_p\right), \tag{4}$$

where $\tilde{c}_i(x_p, y_p)$ is the value of the i–th gridded concentration field interpolated at the position of the p–th particle. We use bilinear interpolation:

$$\tilde{c}_i(x_p, y_p) = w_p(x_k, y_k)\tilde{c}_i(x_k, y_k) + w_p(x_k + \Delta, y_k)\tilde{c}_i(x_k + \Delta, y_k) +$$
$$w_p(x_k, y_k + \Delta)\tilde{c}_i(x_k, y_k + \Delta) + w_p(x_k + \Delta, y_k + \Delta)\tilde{c}_i(x_k + \Delta, y_k + \Delta)$$

and an Euler integration step for the temporal part of (4).

A remarkable feature of the numerical method embodied by (2)–(4) is the possibility of taking arbitrarily small values for the diffusivities D_i without destabilizing the scheme and to recover the non–diffusive case in the limit $D_i \to 0$.

3. Biological applications

A simple example of predator–prey model was given by May:[10]

$$\begin{aligned} \dot{P} &= P(1-P) - aZ(1-\exp(-\lambda_1 P)), \\ \dot{Z} &= -\gamma Z + bZ(1-\exp(-\lambda_2 P)). \end{aligned} \tag{5}$$

Here P is the concentration of preys and Z is the concentration of predators. Unlike the classical Lotka–Volterra model, these equations admit a structurally stable limit cycle in a wide range of parameters. In the following we use the values $a = 1$, $b = 2.5$, $\gamma = 1.5$, $\lambda_1 = \lambda_2 = 4$, which make the limit cycle about 10 non–dimensional time–units long. To illustrate the importance of a small amount of diffusion in this class of problems, we solve the equations (1) with the reaction terms given by (5), and the streamfunction

$$\psi = \psi_0 \sin(kx)\sin(ky), \tag{6}$$

with $\psi_0 = 0.1$ and $k = 1.5$. This is a cellular flow of 3×3 steady vortices on the domain $[0, 2\pi]^2$. The initial conditions are $P(x, y) = 1$ and $Z(x, y) = \{(x, y) \in [7/10\pi, 13/10\pi] \times [5/6\pi, 7/6\pi] : 0.1;\ \text{otherwise} : 0\}$; that is, the preys are initially uniformly distributed with a concentration equal to the carrying capacity of the system, and the predators are confined in a rectangular patch all contained inside the central vortex. Figure 1 shows

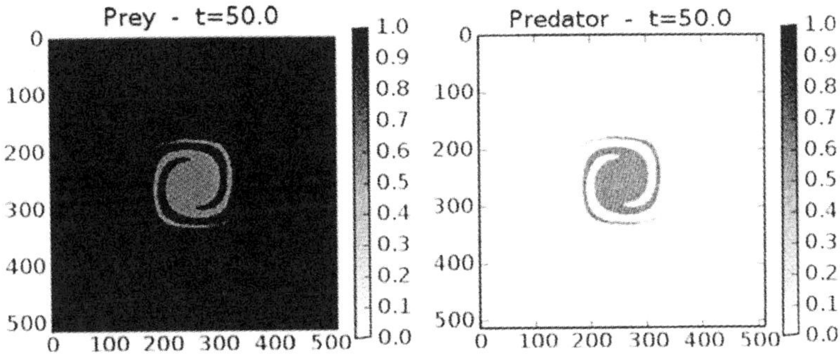

Fig. 1. Predator–prey model with a cellular flow and no diffusion: the predators are unable to escape the central vortex. The prey in the other 8 vortices is unaffected.

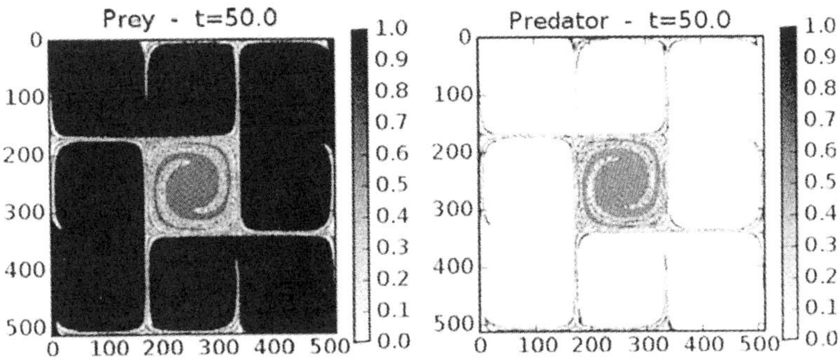

Fig. 2. Predator–prey model with a cellular flow and small diffusivity ($D = 10^{-7}$ for both species): the predators escape the central vortex from each of its four stagnation points and invade the other vortices.

the situation after 50 non–dimensional time units in a simulation with no diffusion: because fluid parcels in this case follow the streamlines, which are closed, the predators are unable to leave the central vortex. With the addition of a very small diffusivity ($D = 10^{-7}$ for both species in the simulations of figure 2) the predators, within a few turnover times, are able to escape the central vortex from each of its four stagnation points and invade the other vortices.

An outstanding open problem in the study of plankton is the absence of a limit–cycle behavior in the available observations: while the relevant population models exhibit a limit cycle (or even a chaotic attractor) in a wide range of parameters, satellite images of clorophyll concentration do not show such oscillations.[12] A possible explanation is that, because of the mixing effect of ocean currents, nearby water patches oscillate with very different phases; coarse–grained satellite observations average them out, disguising the oscillations.

We performed several simulations with the renovating random wave streamfunction, a well–known homogeneously stirring flow which induces chaotic lagrangian trajectories.[11] In the absence of diffusion the coarse–graining explanation is upheld. However, when diffusion is not zero, we observe the sinchronization of phases among nearby particles: eventually, we reach a spatially uniform state which follows the limit cycle in time. The synchronization time increases steeply as diffusivity decreases. For low diffusivities, there is a large interval of times where synchronization is not evident, but stirring has already created enough fine structure that coarse graining observations would disguise the limit cycle (for further details on these simulations see `http://smaug.unile.it/plankton`).

At this stage it is not clear whether there is a small, but non–zero, diffusivity threshold below which complete synchronization is never attained. Furthermore, we have to stress that our simulations used the same diffusivities for both species. With different diffusivities, Turing–like instabilities might destabilize a spatially homogeneous situation. These lines of research will be pursued in future works.

References

1. A. M. Edwards & J. Brindley, *Dynamics and stability of Systems* **11**, 347–370 (1996).
2. A. M. Edwards & J. Brindley, *Bull. Math. Biol.* **61**, 303–339 (1999).
3. J. Pedlosky, *Ocean Circulation Theory*, (Springer, Berlin, 1996).
4. A. W. Epstein, R. C. Beardsley, *Deep–Sea Res. II* **48**, 395–418 (2001).
5. R. Ferrari & F. Paparella, *J. Phys. Ocean.* **33**, 2214–2223 (2003).
6. C. Pasquero, A. Bracco, A. Provenzale, in *Shallow Flows* (Balkema Publishers, Leiden, NL., 2004).
7. P. Degond & F. J. Mustieles, *SIAM J. Sci. Stat. Comput.* **11**, 293–310 (1990).
8. G. Gambino, M. C. Lombardo, M. Sammartino, *Proccedings WASCOM 2007, this volume.*
9. R.W Hockney & J.W Eastwood, *Computer Simulation Using Particles*, (Taylor & Francis, London, 1988).
10. R. M. May, *Science* **177**, 900–902 (1972).
11. W. R. Young, *Stirring and Mixing: Proceedings of the 1999 Summer Program in Geophysical Fluid Dynamics* (Woods Hole Oceanographic Institution, Woods Hole, MA, USA, 1999).
12. I. Koszalka, A. Bracco, C. Pasquero, A. Provenzale, *Theor. Pop. Biol.* **72**, 1–6 (2007).

BIFURCATION ANALYSIS OF EQUILIBRIA IN COMPETITIVE LOGISTIC NETWORKS WITH ADAPTATION

A. RAIMONDI and C. TEBALDI

Department of Mathematics, Politecnico di Torino,
Torino, Italy
E-mail: alessia.raimondi@polito.it
E-mail: claudio.tebaldi@polito.it

A general n-node network is considered for which, in absence of interactions, each node is governed by a logistic equation. Interactions among the nodes take place in the form of competition, which also includes adaptive abilities through a (short term) memory effect. As a consequence the dynamics of the network is governed by a system of n^2 nonlinear ordinary differential equations. As a first step, equilibria and their stability are investigated analytically for the general network in dependence of the relevant parameters, namely the strength of competition, the adaptation rate and the network size. The existence of classes of invariant subspaces, related to symmetries, allows the introduction of a reduced model, four dimensional, where n appears as a parameter, which give full account of existence and stability for the equilibria in the network.

Keywords: Complex Systems; Networks; Logistic Equation; Reduced Models; Stability.

1. Introduction

A network is a collection of parts for which interactions among them give rise to a collective behavior, with characteristics that are often not easily foreseen from the properties of the components. A network is therefore a kind of "prototype" for complex systems, see.[1]

No matter of which definition one has in mind, characteristic features of complex systems are "emergence" and "adaptation". The former is the process of obtaining some new structures and properties, usually in the evolution of the system. Emergent phenomena occur due to the pattern of interactions (non-linear and distributed) between the elements of the system over time. Adaptation is the capacity of learning from experience and adjust. The adaptive changes that occur are often relevant to achieving a goal or objective and certainly the property characterizes living entities, that are a fundamental area of application in the subject of complex systems.

In this work we consider a complex system described by a network exhibiting both such features.

2. The Model

We consider a general network of n interconnected nodes, in which the dynamics of every node is described by the logistic equation

$$\frac{dM_i(t)}{dt} = r_i[1 - \frac{M_i(t)}{k_i}]M_i(t), \tag{1}$$

where $M_i(t)$ is the non negative variable that denotes the activity level of node i at time t, while r_i and k_i are the positive parameters that denote, respectively, the intrinsic growth rate and the time asymptotic value of the activity level of node i. We can suppose, by a suitable change of variables, $k_i = 1$ without loss of generality. Having as main aim the discussion of the role of interactions among nodes, we suppose for all of them the same intrinsic growth rate $r_i = r$. Rescaling time, we can have $r = 1$ without loss of generality.

The interactions among nodes take place in the form of competition, i.e. for each node i we have, in the activity equation, the additional term

$$-\sum_{\substack{j=1\\ j\neq i}}^{n} \beta_{ij} M_i M_j \qquad 1 \le i \le n.$$

The non negative function β_{ij}, $j \neq i$, represents the strength of competition of node j to node i.

We want to consider also a mechanism of adaptation in competition, therefore we consider β_{ij} proportional to the level of activity of nodes j and i over the past:

$$\beta_{ij}(t) = C \int_{-\infty}^{t} M_i(u) M_j(u) K_{T_j}(t-u)du \qquad 1 \le i,j \le n, j \neq i. \tag{2}$$

The term $K_T(t) = e^{(-t/T)}/T$ is a delay kernel representing a short term memory effect.[2] Such a mechanism can be defined as "learning by doing".

Differentiating β_{ij} with respect to t, analogously with[3] and assuming the same adaptation rate for all nodes, i.e. $T_j = T$, we obtain:

$$\begin{cases} \frac{dM_i}{dt} = [1 - M_i]M_i - \sum_{\substack{j=1\\ j\neq i}}^{n} \beta_{ij} M_i M_j \\ \frac{d\beta_{ij}}{dt} = \frac{CM_iM_j - \beta_{ij}}{T} \qquad 1 \le i,j \le n, i \neq j. \end{cases} \tag{3}$$

Therefore the dynamics of the network is described by a set of n^2 ordinary differential equations:

(3.1) provide the activity level equation for each node;
(3.2) prescribe the rule of adaptation in competition.

The relevant parameters are: the interaction strength C, the adaptation rate T and the dimensions of the network n.

3. Reduced models and equilibria

In order to carry on analytically and in general the study of system (3), we start by considering the invariant subspaces related to symmetries.

(1) We have the following invariant subspace for system (3)

$$\beta_{ji} = \beta_{ij} \qquad 1 \leq i, j \leq n, j \neq i \tag{4}$$

which is also attractive. In (4) we can rewrite (3) as a system of $(n^2 + n)/2$ differential equations

$$\begin{cases} \frac{dM_i}{dt} = [1 - M_i]M_i - \sum_{\substack{j=1 \\ j \neq i}}^{n} \beta_{ij} M_i M_j \\ \frac{d\beta_{ij}}{dt} = \frac{CM_iM_j - \beta_{ij}}{T} \qquad 1 \leq i, j \leq n, i > j \\ \beta_{ji} = \beta_{ij} \qquad j \neq i. \end{cases} \tag{5}$$

Because of the attractivity of (4), such system gives complete account of all the attractors in system (3).

(2) We consider now, in general, the equilibria of system (3). They are solutions of the equations $M_i - M_i^2 - \sum_{\substack{j=1 \\ j \neq i}}^{n} CM_i^2 M_j^2 = 0$ with $i = 1, \ldots, n$, since $\beta_{ij} = CM_iM_j$ at equilibria.
Therefore, for sake of conciseness, in the following, only the variables M_i's are reported in order to characterized the equilibria. We notice that their existence does not depend on the adaptation parameter T.
We call interior equilibria the ones with positive M_i's.
In the logistic case ($C = 0$) we obtain the interior equilibrium R_0 such that $M_i = 1$, for $i = 1, \ldots, n$ and the class of equilibria S_0 such that one or more $M_i = 1$ are replaced by $M_i = 0$. One can easily see that R_0 is stable and the S_0's are unstable for any T and n.

(3) Furthermore, choosing arbitrarily $1 \leq h, k \leq n$ with $k \neq h$, for (3) we have the following invariant subspace

$$M_i = M, \quad \beta_{ij} = \beta \qquad 1 \leq i, j \leq n, j \neq i. \tag{6}$$

In this subspace system (3) becomes

$$\begin{cases} \frac{dM}{dt} = [1-M]M - (n-1)\beta M \\ \frac{d\beta}{dt} = \frac{CM^2-\beta}{T} \end{cases} \tag{7}$$

where n appears as parameter.
The equilibria are solutions of $(n-1)CM^3+M-1=0$, since $\beta = CM^2$. The cubic equation has been found to have one and only one positive root r for any $C>0$ and for any $n \geq 3$, then an interior equilibrium $R=(r)$ exists. It is found that $r < 1/\sqrt{C}$, if $C < n^2$, $r = 1/n$, if $C = n^2$ and $r > 1/\sqrt{C}$, if $C > n^2$, which are conditions relevant for the study of stability, as we will see in the following.
In the complete model, an interior equilibrium $R = (r, r, \dots, r)$ exists for any n.

(4) We can choose arbitrarily $1 \leq h, k, l \leq n$ with $k, l \neq h$ and $k \neq l$, and we have also the invariant subspace

$$M_i = M_k \quad i \neq h, k; \beta_{ij} = \beta_{hk} \quad i = h \vee j = h; \beta_{ij} = \beta_{kl} \quad i, j \neq h \tag{8}$$

of which (6) is a subspace. In (8), system (3) becomes

$$\begin{cases} \frac{dM_h}{dt} = M_h[1 - M_h - (n-1)\beta_{hk}M_k] \\ \frac{dM_k}{dt} = M_k[1 - M_k - \beta_{kh}M_h - (n-2)\beta_{kl}M_l] \\ \frac{d\beta_{hk}}{dt} = \frac{CM_hM_k - \beta_{hk}}{T} \\ \frac{d\beta_{kl}}{dt} = \frac{CM_k^2 - \beta_{kl}}{T} \end{cases} \tag{9}$$

where n appears as parameter.
In this subspace the equilibria are solutions of

$$\begin{aligned} &M_h[1 - M_h - (n-1)CM_hM_k^2] = 0 \\ &M_k[1 - M_k - CM_kM_h^2 - (n-2)CM_k^3] = 0 \\ &\beta_{hk} = CM_hM_k \\ &\beta_{kl} = CM_k^2. \end{aligned}$$

By the result in subspace (6),we can write

$$[(n-1)CM_k^3 + M_k - 1][(n-1)(n-2)C^2M_k^4 + (2n-3)CM_k^2 - CM_k + 1] = 0.$$

Then,

$$C^* = C^*(n) = \frac{8}{\sqrt{(2n-3)^2[\gamma(n)]^2 + 108(n-1)(n-2)} - (2n-3)\gamma(n)},$$

with $\gamma(n) = [32n(n-3)+63]$, exists, determining the number of equilibria.

(a) For $C < C^*(n)$, one equilibrium $R = (r, r)$ exists, where r is the only real root of $(n-1)CM_k^3 + M_k - 1 = 0$.
In the complete model it corresponds to the equilibrium $R = (r, r, \ldots, r)$.

(b) For $C > C^*(n)$, three equilibria $R = (r, r)$, $S = (b_1, s_1)$, $S^* = (b_2, s_2)$ exist, where s_1 and s_2, $(s_1 < s_2)$, are the only two real root of $(n-1)(n-2)C^2M_k^4 + (2n-3)CM_k^2 - CM_k + 1 = 0$ for any n and $b_\alpha = 1/[(n-1)Cs_\alpha^2 + 1]$ for $\alpha = 1, 2$, $(b_1 > b_2)$.
In the complete model, S and S^* correspond to $2n$ equilibria of the form

$$\begin{array}{l} S_h = (s_1, \ldots, s_1, b_1, s_1, \ldots, s_1) \\ S_h^* = (s_2, \ldots, s_2, b_2, s_2, \ldots, s_2) \end{array} \quad h = 1, \ldots, n$$

i.e. having the $M_i = s_\alpha$, for $\alpha = 1, 2$, if $i \neq h$ and $M_h = b_\alpha$, present together with $R = (r, \ldots, r)$.

(c) At $C = C^*(n)$, S and S^* disappear by a saddle-node bifurcation. Consequently, in the complete model, the n S_h's and the n S_h^*'s disappear by n saddle-node bifurcations.

(5) In the same line, if we choose $1 \leq h, k, l, t \leq n$ with $k, l, t \neq h$, $l, t \neq k$ and $t \neq l$ and n_h, n_k, n_l such that $n_h + n_k + n_l = n$, we have the following invariant subspace

$$\begin{array}{ll} M_i = M_h & i = 1, 2, \ldots, n_h \\ M_i = M_k & i = n_h + 1, \ldots, n_h + n_k \\ M_i = M_l & i = n_h + n_k + 1, \ldots, n_h + n_k + n_l \\ \beta_{ij} = \beta_{hk} & i = h \vee j = h \\ \beta_{ij} = \beta_{kl} & i = k \vee j = k, i, j \neq h \\ \beta_{ij} = \beta_{hl} & i = l \vee j = l, i, j \neq k \\ \beta_{ij} = \beta_{lt} & i \neq h, k \end{array} \tag{10}$$

of which (6) and (8) are subspaces.
We will prove that no other equilibria can be found in this subspace. In fact, equilibria are solutions of

$$M_h\left[1 - M_h - (n_h - 1)CM_h^3 - n_kCM_k^2M_h - n_lCM_l^2M_h\right] = 0 \quad (11)$$
$$M_k\left[1 - M_k - n_hCM_h^2M_k - (n_k - 1)CM_k^3 - n_lCM_l^2M_k\right] = 0 \quad (12)$$
$$M_l\left[1 - M_l - n_hCM_h^2M_l - n_kCM_k^2M_l - (n_l - 1)CM_l^3\right] = 0 \quad (13)$$

with $\beta_{hk} = M_hM_k$, $\beta_{kl} = M_kM_l$ and $\beta_{hl} = M_hM_l$.

Subtracting Eq. (13) from Eq. (12) we have

$$(M_l - M_k)[1 + n_h C M_h^2 + (n_k - 2) C M_k^2 + (n_l - 2) C M_l^2 + C(M_l - M_k)^2 + C M_l M_k] = 0$$

with the only solution $M_l = M_k$.
Subtracting Eq. (12) from Eq. (11) we have

$$(M_k - M_h)[1 + (n_h - 2) C M_h^2 + (n - n_h - 1) C M_k^3 + C(M_k - M_h)^2 + C M_h M_k] = 0.$$

Therefore, if $n_h = 1$ we have $M_k \neq M_h$ and we obtain the points S_h and S_h^* of the complete model, while if $n_h \geq 2$ we obtain $M_k = M_h$ and we have the point R of the complete model.
Further applying this procedure, it is possible to show that (9) gives all the equilibria of the complete model.

Figure 1 shows that C^* is an increasing function of n, $C^* \to (2^6 n/3^3)$ for $n \to \infty$; r decreases with C and n; β increases with C and decreases with n. We can notice that the activity level of all nodes becomes low in the equilibrium R at high C.

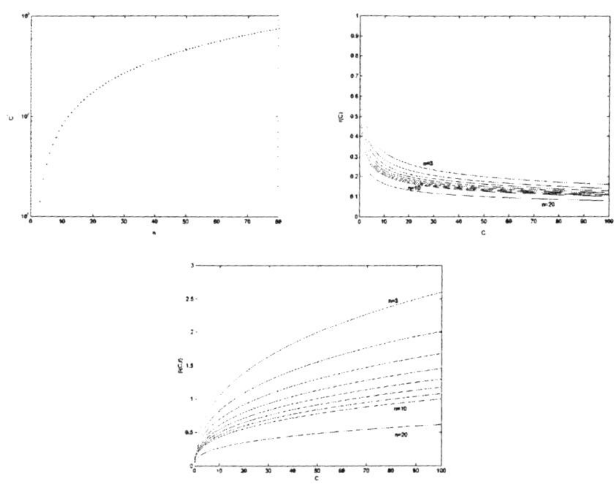

Fig. 1. $C^*(n)$, $r(C, n)$ and $\beta(C, n)$.

Figure 2 shows that s_1 decreases with C and increases with n, while s_2 is not a monotonic function of C but decreases with n; b_1 increases with C and decreases with n, b_2 decreases with C and increases with n; β_1 increases with C and decreases with n, while β_2 decreases with C and increases with n. We observe that for each S_h the activity level of node h becomes much higher than for the others, approaching 1 to high C.

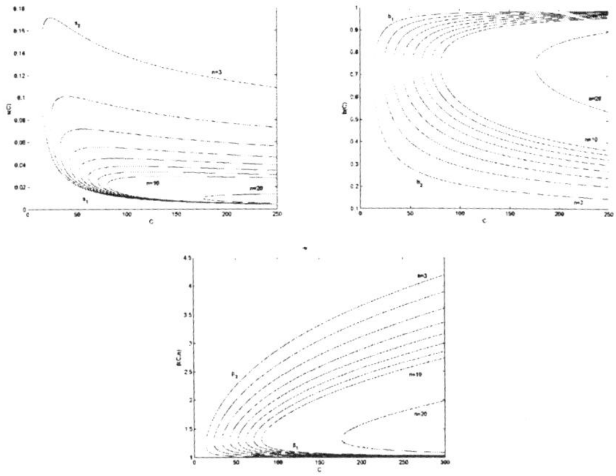

Fig. 2. $s_\alpha(C,n)$, $h_\alpha(C,n)$ for $\alpha = 1, 2$, $\beta_1(C,n) = Cb_1s_1$ and $\beta_2(C,n) = Cs_1^2$.

4. Stability of equilibria

4.1. *Stability in the reduced model*

The eigenvalues of Jacobian matrix DF_{red} are all the roots λ of the polynomial of degree 4,

$$Q(\lambda) = (2+\lambda T)^2 M_h^3 M_k^3 C^2 [p_{M_h} p_{M_k} + (n-2)p_{M_h} - (n-1)]$$

where $p_{M_h} = [\lambda^2 T + \lambda(1+M_hT)+1]/[(2+\lambda T)CM_h^3]$ and $p_{M_k} = [\lambda^2 T + \lambda(1+M_kT)+1]/[(2+\lambda T)CM_k^3]$.

4.1.1. *Stability of point R in the reduced model*

For the equilibrium R we have

$$Q(\lambda) = [\lambda^2 T + \lambda(1+rT-Cr^3T)+1-2Cr^3] \times \\ \times [\lambda^2 T + \lambda(1+rT+(n-1)Cr^3T)+1+2(n-1)Cr^3].$$

Therefore $R = (r,r)$ is asymptotically stable if and only if

$$r^2 \leq \frac{1}{C} \quad \text{or} \quad r^2 > \frac{1}{C} \quad \text{and} \quad T < \frac{1}{r(Cr^2-1)}.$$

Since it has been proven that $r^2 \leq 1/C$ if $C \leq n^2$, R is asymptotically stable for any T if $C \leq n^2$, only for $T < 1/[r(Cr^2-1)]$ if $C > n^2$.

4.1.2. *Stability of point S in the reduced model*

For the equilibrium $S = (b_1, s_1)$ we have $Q_2(\lambda) = [T\lambda^4 + \alpha\lambda^3 + \beta\lambda^2 + \gamma\lambda + \delta]$ where

$$\begin{aligned}
\alpha &= T[2 + (s_1 + b_1)T + (n-2)Cs_1^3 T] \\
\beta &= 1 + T[2 + s_1 + b_1 + 3(n-2)Cs_1^3] \\
&\quad + b_1 s_1 T^2[1 + (n-2)Cs_1^2 - (n-1)C^2 b_1^2 s_1^2] \\
\gamma &= 2[1 + (n-2)Cs_1^3] + T[b_1 + s_1 + 2(n-2)Cb_1 s_1^3 \\
&\quad - 4(n-1)C^2 b_1^3 s_1^3 + (n-2)Cs_1^3] \\
\delta &= 1 + 2Cs_1^3[(n-2) - 2(n-1)Cb_1^3].
\end{aligned}$$

Therefore S is asymptotically stable for any $C \geq C^*$ and for any $n \geq 3$ if and only if the quartic $Q_2(\lambda)$ has all the roots with negative real part.

The equilibrium $S^* = (b_2, s_2)$ is unstable for any $C \geq C^*$, $T > 0$ and n.

4.2. *Stability in the complete model*

The eigenvalues of the Jacobian matrix DF are $\lambda = -1/T$ with multiplicity $n^2 - n$ and all the roots λ of polynomial of degree $2n$,

$$Q(\lambda) = (2 + \lambda T)^n (M_1^3 \dots M_n^3)(-1)^n C^n \det \mathbf{P}$$

where $\mathbf{P}$ is a matrix $n \times n$ where its entries are $a_{ij} = p_k(\lambda) = [\lambda^2 T + \lambda(1 + M_k T) + 1]/[(2 + \lambda T)CM_k^3]$, $k = 1, 2, \dots, n$ if $i = j$ and $a_{ij} = 1$ if $i \neq j$.

4.2.1. *Stability of point R in the complete model*

For the equilibrium R we have

$$\begin{aligned}
Q(\lambda) = (-1)^n[\lambda^2 T + \lambda(1 + rT - Cr^3 T) + 1 - 2Cr^3]^{n-1} \times \\
\times[\lambda^2 T + \lambda(1 + rT + (n-1)Cr^3 T) + 1 + 2(n-1)Cr^3].
\end{aligned}$$

Therefore $R = (r, \dots, r)$ is asymptotically stable if and only if

$$r^2 \leq \frac{1}{C} \quad \text{or} \quad r^2 > \frac{1}{C} \quad \text{and} \quad T < \frac{1}{r(Cr^2 - 1)}.$$

Since we have exactly the same conditions obtained in the reduced model, also the stability properties of the equilibrium R are given by the reduced model.

4.2.2. *Stability of point S_h in the complete model*

For the equilibria S_h we have

$$\begin{aligned} Q(\lambda) &= [Q_1(\lambda)]^{n-2} \times Q_2(\lambda) \\ &= [\lambda^2 T + \lambda(1 + sT - Cs^3T) + 1 - 2Cs^3]^{n-2} \times \\ &\quad \times [T\lambda^4 + \alpha\lambda^3 + \beta\lambda^2 + \gamma\lambda + \delta] \end{aligned}$$

where $Q_2(\lambda)$ is the same that we have obtained in the reduced model.

Therefore any equilibria $S_h = (s_1, \ldots, s_1, b_1, s_1, \ldots, s_1)$ is asymptotically stable for any $C \geq C^*$ and for any $n \geq 3$ if and only if

$$s_1^2 \leq \frac{1}{C} \quad \text{or} \quad T < \frac{1}{s_1(Cs_1^2 - 1)}$$

and the quartic $Q_2(\lambda)$ has all the roots with negative real part.

If we compare these stability conditions with the ones in the reduced model, we notice that we have now additional conditions. However we have proven that the condition $s_1^2 \leq \frac{1}{C}$ is true for any $C > C^*$ and therefore the stability of equilibria S_h is determined only by the quartic $Q_2(\lambda)$, as in the reduced model, for any n.

By the Routh-Hurwitz theorem, the condition on the stability of $Q_2(\lambda)$ holds if $P(T) = a_1T^4 + a_2T^3 + a_3T^2 + a_4T + a_5 > 0$ where the coefficients a_k's, $k = 1, \ldots, 5$, are complicated functions of n, C, s_1, b_1.

The coefficients of $P(T)$ are in fact:

$$\begin{aligned} a_1 &= b_1s_1[-1 - (n-2)Cs_1^2 + (n-1)C^2b_1^2s_1^2][b_1 + s_1 + (n-2)Cs_1^3] \\ &\quad [4(n-1)C^2b_1^3s_1^3 - s_1(1 + (n-2)Cs_1^2) + b_1(-1 - 2(n-2)Cs_1^3)] \\ a_2 &= -8(n-1)^2C^4b_1^6s_1^6 + [s_1 + (n-2)Cs_1^3]^3 - 2(n-1)C^2b_1^4s_1^3 \\ &\quad [2 + 4s_1 + 3(n-2)Cs_1^3] + b_1^3[1 + 2(n-2)Cs_1^3 - 4(n-1)C^2s_1^4 \\ &\quad -10(n-2)^2(n-1)C^4s_1^9 - 2(n-1)(n-2)(2 + 5s_1)C^3s_1^6] \\ &\quad +b_1^2s_1[7 + 10(n-2)^2C^2s_1^5 + (n-2)(7 + 10s_1)Cs_1^2] \\ &\quad +b_1s_1^2[7 + 8(n-2)^3C^3s_1^7 + 2(n-2)(7 + 2s_1)Cs_1^2 \\ &\quad +(n-2)^2(7 + 12s_1)C^2s_1^4] \\ a_3 &= 4(n-1)C^2b_1^4s_1^3 + 4(n-1)(-1 + s_1)C^2b_1^3s_1^3 + b_1^2[5 + 8(n-2)Cs_1^3] \\ &\quad +2b_1s_1[7 + 9(n-2)^2C^2s_1^5 + 7(n-2)(1 + s_1)Cs_1^2] \\ &\quad +s_1^2[5 + 6(n-2)^3C^3s_1^7 + 2(n-2)(5 + s_1)Cs_1^2 + (n-2)^2(5 + 8s_1)C^2s_1^4] \\ a_4 &= 2[4(n-1)C^2b_1^3s_1^3 + b_1(4 + 5(n-2)Cs_1^3) + s_1(4 + 5(n-2)^2C^2s_1^5 \\ &\quad +(n-2)(4 + 3s_1)Cs_1^2)] \\ a_5 &= 4[1 + (n-2)Cs_1^3]. \end{aligned}$$

Because of the complexity of the polynomial $P(T)$, no analytical condition on T has been determined for the stability of S_h. Numerical results are reported in Fig. 3.

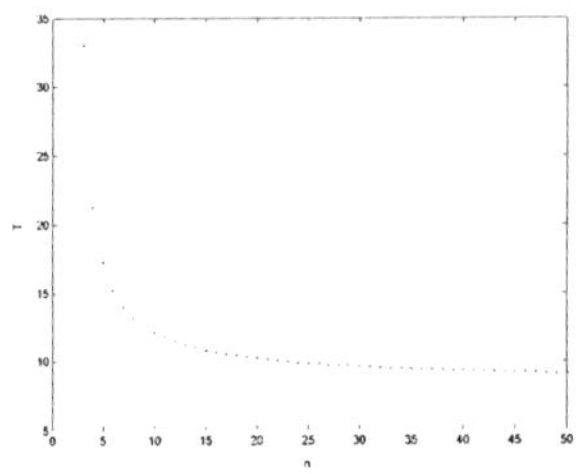

Fig. 3. The numerical calculation of the critical value of T, T_c, for different value of n.

We can see that T decreases with n. Increasing the network size, the S_h's appear for larger values of $C^*(n)$ but they become unstable for smaller values of $T_c(n)$.

The S_h^*'s are unstable for any $C \geq C^*$, $T > 0$ and for any n.

We finish briefly discussing what happens after destabilization of the equilibria, in the case $n = 3$ (9 equations).

For $9 < C < C^*$, R looses stability at a critical value $T_R(C, n)$ when a supercritical Hopf bifurcation takes place, giving rise to periodic behavior.

For $C > C^*$, instead, when R becomes unstable at T_R, a subcritical Hopf bifurcation takes place. Depending on the initial condition the solution goes to one of the S_h's, stable up to a threshold $T_S(C, n) > T_R$, at which a subcritical Hopf bifurcation takes place. For $T > T_S$ chaotic behavior appears. Interesting point is that the reduced model (9) is found to give complete account also of the appearance of time dependent behavior in the system. Detailed analysis of the time dependent behavior, with remarkable properties of synchronization, will be reported elsewhere.

References

1. A. L. Barabasi, *The new science of networks* (Perseus Publishing, 2002).
2. V. W. Noonburg, *SIAM J. Applied Mathematics* **49**, 1779 (1989).
3. D. Lacitignola and C. Tebaldi, *Mathematical Biosciences* **194**, 95 (2005).
4. C. Bortone and C. Tebaldi, *Dynamics of Continuos, Discrete and Impulsive Systems* , 379 (1998).
5. E. Barone and C. Tebaldi, *Mathematical Methods in the Applied Sciences* , 1179 (2000).

DIFFUSION INFLUENCE ON STABILITY - INSTABILITY

S.RIONERO

Department of Mathematics and Applications "R. Caccioppoli", Federico II
Naples 80126, Italy
rionero@unina.it

The diffusion influence on the linear stability-instability for a binary reaction-diffusion systen of P.D.Es., under Robin boundary conditions, is considered. Turing instability, with different and equal diffusivities, is analized.

Keywords: Diffusion, Stability-Instability, Turing effect, Equal diffusivities

1. Introduction

The role of diffusion on the solution's behaviour of P.D.Es., is very significant. In fact, although in a simple reaction-diffusion P.D.E. the diffusion has a smoothing and stabilizing effect, the situation is drastically different for a reaction-diffusion system of P.D.Es. This is because different diffusion rates can drive (linear) instability. The diffusion-driven instability is called Turing's instability since A. Turing, in 1952, argued this phenomenon.[1] From then this surprising phenomenon has been studied by many authors and is usually included in the recent literature[2–13] [a]. Generally the Turing's instability has been considered under Neumann boundary conditions. Further the possibility of having Turing's instability with equal diffusivities does not appear to be completely investigated. We here consider the case of the general Robin boundary conditions and present some results obtained recently.[14] In particular a case of Turing's instability with equal diffusivities is investigated. The plan of the paper is as follows. Section 2 is devoted to some preliminaries. In Section 3 the (linear) diffusion-driven stability-instability is studied. The case of equal diffusivities is considered in Section 4. The paper ends with a final remark (Section 5) in which is put in evidence that phenomena analogous to diffusion-driven instability occur

[a] Cfr. also the references therein.

also in Classical Mechanics.

2. Preliminaries

Let $\Omega \subset I\!R^3$ be a bounded (smooth) domain and Σ its boundary. The stability-instability analysis of an equilibrium spatially uniform state in Ω of two "substances" $S_i\,(i = 1,2)$ diffusing there often can be traced back to the stability-instability of the zero solution of a dimensionless system of P.D.Es. like

$$\begin{cases} \dfrac{\partial C_1}{\partial t} = a_{11}C_1 + a_{12}C_2 + \gamma_1 \Delta C_1 + f(C_1, C_2), \\ \\ \dfrac{\partial C_2}{\partial t} = a_{21}C_1 + a_{22}C_2 + \gamma_2 \Delta C_2 + g(C_1, C_2), \end{cases} \tag{1}$$

under the Robin boundary conditions

$$\beta_i C_i + (1 - \beta_i)\nabla C_i \cdot \mathbf{n} = 0 \qquad \text{on } \Sigma \tag{2}$$

with

$$\begin{cases} i, j = 1, 2; \quad a_{ij},\, \gamma_i \text{ constants} \\ \\ C_i = \text{concentrations of } S_i \; (i = 1, 2) \\ \\ \gamma_1 > 0,\, \gamma_2 > 0,\; \beta_i \in]0, 1[\\ \\ C_i \,:\, \mathbf{x} \in \Omega \Rightarrow C_i(\mathbf{x},\, t) \in I\!R, \end{cases} \tag{3}$$

$\mathbf{n}$ unit outward normal to Σ and f, g nonlinear functions such that

$$f(0,0) = g(0,0) = 0\,. \tag{4}$$

System (1)-(3) can be interpreted, both as system for diffusing biological communities and as an economy where the two species correspond to the investment capital and labor force.

We denote by

- $\langle \cdot, \cdot \rangle$ the scalar product in $L^2(\Omega)$
- $\langle \cdot, \cdot \rangle_\Sigma$ the scalar product in $L^2(\Sigma)$
- $\| \cdot \|$ the $L^2(\Omega)$-norm;
- $\| \cdot \|_\Sigma$ the $L^2(\Sigma)$-norm
- $W_i^{1,2}(\Omega,\, \beta_i)$ the functional space such that

$$\varphi \in W_i^{1,2}(\Omega,\, \beta_i) \to \{\varphi \in W_i^{1,2}(\Omega), (2) \text{ hold}\} \tag{5}$$

- $\bar{\alpha}_i(\Omega)$ the positive constant appearing in the inequality ($i = 1, 2$)

$$\|\nabla\varphi\|^2 + \frac{\beta_i}{1-\beta_i}\|\varphi\|_\Sigma^2 \geq \bar{\alpha}_i\|\varphi\|^2\,, \tag{6}$$

with $\Sigma = \partial\Omega$, holding in $W_i^{1,2}(\Omega,\,\beta_i)$, for Ω of class $C^p(p > 2)$ and having the cone property.

As it is well known $\bar{\alpha}_i(\Omega,\,\beta_i)$ is the lowest eigenvalue of

$$\Delta\varphi + \lambda\varphi = 0 \tag{7}$$

in $W_i^{1,2}(\Omega,\,\beta_i)$ (i.e. the principal eigenvalue of $-\Delta$).[5,11]

By setting

$$b_{11} = a_{11} - \bar{\alpha}_1\gamma_1, \quad b_{22} = a_{22} - \bar{\alpha}_2\gamma_2 \tag{8}$$

and introducing the scalings α, β such that

$$C_1 = \alpha u\,, \quad C_2 = \beta v\,, \tag{9}$$

(1) reduces to $\left(\mu = \dfrac{\alpha}{\beta}\right)$

$$\begin{cases} \dfrac{\partial u}{\partial t} = b_{11}u + \dfrac{a_{12}}{\mu}v + f^* + f_1^* \\ \\ \dfrac{\partial v}{\partial t} = \mu a_{21}u + b_{22}v + g^* + g_1^* \end{cases} \tag{10}$$

under the boundary conditions

$$\begin{cases} \beta_1 u + (1-\beta_1)\nabla u \cdot \mathbf{n} = 0 \\ \\ \beta_2 v + (1-\beta_2)\nabla v \cdot \mathbf{n} = 0 \end{cases} \quad \text{on } \Sigma \tag{11}$$

with

$$\begin{cases} f^* = \dfrac{1}{\alpha}f(\alpha\, u,\, \beta\, v)\,, & g^* = \dfrac{1}{\beta}g(\alpha\, u,\, \beta\, v) \\ \\ f_1^* = \gamma_1(\Delta u + \bar{\alpha}_1 u)\,, & g_1^* = \gamma_2(\Delta v + \bar{\alpha}_2 v) \end{cases} \tag{12}$$

Remark 2.1. We observe that: i) the cases $\beta_1 = \beta_2 = 1$ and $\beta_1 = \beta_2 = 0$, have been already considered;[10–12] ii) $\beta_1 = \beta_2 \Rightarrow \bar{\alpha}_1 = \bar{\alpha}_2$. Then β and $\bar{\alpha}$ will denote, respectively, the values $\beta_1 = \beta_2$ and $\bar{\alpha}_1 = \bar{\alpha}_2$; iii) a special case with $\beta_i = \beta_i(\mathbf{x})$ ($i = 1, 2$), $\mathbf{x} \in \Sigma$ has been considered recently.[17]

3. Linear stability-instability

Theorem 3.1. *Let*

$$I_1 = b_{11} + b_{22} < 0,\ A_1 = b_{11}b_{22} - a_{12}a_{21} > 0,\ b_{11}b_{22}a_{12}a_{21} > 0,\ a_{12}a_{21} > 0. \quad (13)$$

Then the zero solution of (10) is (linearly) asymptotically stable in the $L^2(\Omega)$-norm.

Proof. On choosing $\mu = \sqrt{\dfrac{a_{12}}{a_{21}}}$ it easily follows that

$$\frac{dE}{dt} = b_{11}\|u\|^2 + 2\sqrt{a_{12}a_{21}}\langle u, v\rangle + b_{22}\|v\|^2 + \langle u, f_1^*\rangle + \langle v, g_1^*\rangle \quad (14)$$

with $E = \frac{1}{2}(\|u\|^2 + \|v\|^2)$. By virtue of (6) and (13) one obtains

$$\frac{dE}{dt} \leq -|b_{11}|\|u\|^2 + 2\sqrt{a_{12}a_{21}}|\langle u, v\rangle| - |b_{22}|\|v\|^2 . \quad (15)$$

Since $\{A_1 = |b_{11}b_{22}| - a_{12}a_{21} > 0 \Rightarrow a_{12}a_{21} = \varepsilon^2|b_{11}b_{22}|\}$, with $0 < \varepsilon < 1$, one easily obtains $\dfrac{dE}{dt} \leq -dE$, with $d = 2(1-\varepsilon)\inf(|b_{11}|, |b_{22}|)$. □

From now on, we consider the case $\beta_1 \equiv \beta_2$ and denote with $\{\bar{\alpha}_n\}$, $\{\varphi_n\}$ respectively the sequence of the eigenvalues of (6) and the associate sequence of eigenfunctions. We assume that: 1) $\{\varphi_n\}$ is complete and orthogonal in $W^{1,2}(\Omega, \beta)$; 2) u, v and their first and second (spatial) derivatives can be expanded in a (Fourier) series absolutely and uniformly convergent in Ω according to

$$u = \sum_{n=1}^{\infty} u_n = \sum_{n=1}^{\infty} X_n(t)\varphi_n(\mathbf{x}), \quad v = \sum_{n=1}^{\infty} v_n = \sum_{n=1}^{\infty} Y_n(t)\varphi_n(\mathbf{x}) \quad (16)$$

derivable term by term. Then, linearizing (10) with $\mu = 1$, it turns out that

$$\begin{cases} \dfrac{dX_n}{dt} = b_{11}^{(n)} X_n + a_{12} Y_n \\ \\ \dfrac{dY_n}{dt} = a_{21} X_n + b_{22}^{(n)} Y_n \end{cases} \quad (17)$$

with

$$b_{11}^{(n)} = a_{11} - \gamma_1\bar{\alpha}_n, \quad b_{22}^{(n)} = a_{22} - \gamma_2\bar{\alpha}_n, \quad b_{11}^{(1)} = b_{11}, \quad b_{22}^{(1)} = b_{22}. \quad (18)$$

Setting

$$I_n = b_{11}^{(n)} + b_{22}^{(n)}, \quad A_n = b_{11}^{(n)} b_{22}^{(n)} - a_{12}a_{21}, \quad A^* = \inf A_n \quad (19)$$

it follows that:

Theorem 3.2. *Let* $\beta_1 \equiv \beta_2$ *and*

$$I_1 < 0\,, \qquad A^* > 0\,. \tag{20}$$

Then the zero solution of (10) is asymptotically stable.

Proof. Introducing the functional

$$V_n = \frac{1}{2}[A_n(\|u_n\|^2 + \|v_n\|^2) + \|b_{11}^{(n)} v_n - a_{21} u_n\|^2 + \|a_{12} v_n - b_{22}^{(n)} u_n\|^2 \tag{21}$$

one obtains[10]

$$\frac{dV_n}{dt} = -|I_n| A_n(\|u_n\|^2 + \|v_n\|^2)\,. \tag{22}$$

In view of

$$\begin{cases} \dfrac{k_1^{(n)}}{2}(\|u_n\|^2 + \|v_n\|^2) \le V_n \le \dfrac{k_2^{(n)}}{2}(\|u_n\|^2 + \|v_n\|^2) \\ \\ k_1^{(n)} = A_n\,, \quad k_2^{(n)} = A_n + 2[(b_{11}^{(n)})^2 + (b_{22}^{(n)})^2 + a_{12}^2 + a_{21}^2] \end{cases} \tag{23}$$

it follows that

$$V_n \le V_n(0) e^{-d_n t}\,, \quad d_n = \frac{2A_n|I_n|}{k_2^{(n)}} = 2\frac{k_1^{(n)}}{k_2^{(n)}}|I_n|\,. \tag{24}$$

In view of (24) each harmonic (u_n, v_n) of the perturbation (u, v) tends to zero for $t \to \infty$. There exists a positive δ such that:[14] $V = \sum_{n=1}^{\infty} V_n \le V(0)e^{-\delta t}$. □

Theorem 3.3. *Let* $\beta_1 \equiv \beta_2$ *and exists a* $\bar{n} \in \{1, 2, ..\}$ *such that either*

$$I_{\bar{n}} > 0 \tag{25}$$

or

$$A_{\bar{n}} < 0 \tag{26}$$

hold. Then the zero solution of (10) is linearly unstable.

Proof. In fact then at least one of the eigenvalues of (17) with $n = \bar{n}$, is positive or has positive real part. □

Immediate is the proof of the following theorems 4-5.

Theorem 3.4. *Let theorem 1 and either*

$$I_0 = a_{11} + a_{22} > 0 \tag{27}$$

or

$$A_0 = a_{11}a_{22} - a_{12}a_{21} < 0 \tag{28}$$

hold. Then the zero solution of (10) (linearly unstable in the absence of diffusion), is stabilized by diffusion.

Theorem 3.5. *Let $\beta_1 \equiv \beta_2$, theorem 2 and either (27) or (28) hold. Then the zero solution of (10) (linearly unstable in the absence of diffusion), is stabilized by diffusion.*

Theorem 3.6. *Let $\beta_1 \equiv \beta_2$ and $\{I_0 < 0,\ A_0 > 0\}$ hold. Then necessary and sufficient condition for the onset of Turing instability is that $A_n < 0$, for at least one $n \in \{1, 2, ..\}$.*

Proof. By virtue of $\{I_0 < 0,\ A_0 > 0\}$ the zero solution is (linearly) stable in the absence of diffusion and is destabilized by the perturbation (u_n, v_n) with n such that $A_n < 0$. □

Theorem 3.7. *Let $\{\beta_1 \equiv \beta_2,\ I_0 < 0,\ A_0 > 0\}$ hold. Then Turing instability can occur only if*

$$\gamma_1 \neq \gamma_2,\ a_{11}a_{22} < 0,\ a_{12}a_{21} < 0,\ \Delta = (a_{11}\gamma_2 + a_{22}\gamma_1)^2 - 4\gamma_1\gamma_2 A_0 > 0. \tag{29}$$

Proof. In view of

$$A_n = \bar{\alpha}_n^2\gamma_1\gamma_2 - (a_{11}\gamma_2 + a_{22}\gamma_1)\bar{\alpha}_n + A_0 < 0 \tag{30}$$

for at least one $n \in \{1, 2, ..\}$, with $\{A_0 > 0,\ I_0 = a_{11} + a_{22} < 0\}$, $(29)_1$–$(29)_2$ immediately follow. As concerns $(29)_3$, it comes immediately from $A_0 > 0$ and $(29)_2$. Finally $(29)_4$ is implied by (30). □

Theorem 3.8. *Let $\{\beta_1 \equiv \beta_2,\ I_0 < 0,\ A_0 > 0\}$ and (29) hold. Then the Turing instability occurs if exists a n such that*

$$0 < \frac{a_{11}\gamma_2 + a_{22}\gamma_1 - \sqrt{\Delta}}{2\gamma_1\gamma_2} < \bar{\alpha}_n < \frac{a_{11}\gamma_2 + a_{22}\gamma_1 + \sqrt{\Delta}}{2\gamma_1\gamma_2}. \tag{31}$$

We remark that, solving (30) with respect to γ_1, γ_2 one can find conditions on $\gamma_i (i = 1, 2)$ guaranteeing the Turing instability for a fixed $\bar{\alpha}_n$. In particular, in the case of the principal eigenvalue $(n = 1)$ one has

Theorem 3.9. *Let* $\{\beta_1 \equiv \beta_2,\ I_0 < 0,\ A_0 > 0\}$ *and*

$$a_{11} < 0,\ \gamma_2 < a_{22}/\bar{\alpha},\ \gamma_1 > \frac{|a_{11}|\gamma_2\bar{\alpha} + A_0}{(a_{22} - \gamma_2\bar{\alpha})\bar{\alpha}} \tag{32}$$

hold. Then the Turing instability occurs.

4. Turing instability with equal diffusivities

The results of the previous Sections are concerned with "two substances" diffusing in any direction. It remains to consider the cases in which it is not so. We confine ourselves only to the following two cases: 1) only one substance spreads in Ω; 2) a substance spreads (essentially) only in one direction, while the other spreads in more than one. In the case 1), without loss of generality, we assume $\gamma_1 = 0$. Then $b_{11} = a_{11}$, $A_n = -a_{11}\gamma_2\bar{\alpha}_n + A_0$ and theorems 3.1 3.6 continue to hold. As concerns theorems 3.7-3.9 (29) reduces to $a_{11} > 0$ and Turing instability occurs if and only if $\{I_0 < 0,\ A_0 > 0,\ \bar{\alpha}_n\gamma_2 > A_0/a_{11}\}$. More interesting appears the case 2). For the sake of concreteness we consider the case

$$\begin{cases} u_t = a_{11}u + a_{12}v + \gamma_1 u_{xx} + f \\ \\ v_t = a_{21}u + a_{22}v + \gamma_2(v_{xx} + v_{yy}) + g \end{cases} \qquad \Omega = (0,1) \times (0,1) \tag{33}$$

under the Neumann homogeneous boundary conditions

$$\nabla u \cdot \mathbf{n} = \nabla v \cdot \mathbf{n} = 0 \qquad \text{on } \Sigma\,. \tag{34}$$

Linearizing and looking for solutions such that

$$u = \sum_{n,m}^{1-\infty} X_{n,m}(t)\cos n\pi x \cos m\pi y,\ v = \sum_{n,m}^{1-\infty} Y_{n,m}(t)\cos n\pi x \cos m\pi y \tag{35}$$

with the double series uniformly and absolutely convergent and derivable term by term, one easily obtains

$$\begin{cases} \dfrac{dX_{n,m}}{dt} = (a_{11} - \gamma_1 n^2\pi^2)X_{n,m} + a_{12}Y_{n,m} \\ \\ \dfrac{dY_{n,m}}{dt} = a_{21}X_{n,m} + [a_{22} - \gamma_2\pi^2(n^2 + m^2)]Y_{n,m}\,. \end{cases} \tag{36}$$

Equations (36) are the starting point for generalizing theorems 3.1-3.9. We here confine ourselves to the onset of Turing instability with *equal diffusivities* ($\gamma_1 = \gamma_2 = \gamma$) for $n = 1$, i.e. $\{I_0 = a_{11}+a_{22} < 0,\ A_0 = a_{11}a_{22}-a_{12}a_{21} > 0\}$ and

$$A_{1m} == (1 + m^2)(\gamma\pi^2)^2 - [a_{11}(1 + m^2) + a_{22}](\gamma\pi^2) + A_0 < 0\,, \tag{37}$$

for at least one $m \in \{1, 2, ..\}$. Setting

$$\begin{cases} \Delta = [a_{11}(1+m^2) + a_{22}]^2 - 4(1+m^2)A_0 \\ \\ \Delta_1 = (a_{11}a_{22} - 2A_0)^2 - (a_{11}a_{22})^2 = 4A_0(A_0 - a_{11}a_{22}) \end{cases} \tag{38}$$

the following theorems hold

Theorem 4.1. *Let $\gamma_1 = \gamma_2 = \gamma$ and the zero solution of (33) – (34) be (linearly) stable in the absence of diffusion. The the Turing instability arises on the harmonic $\{u_{1m} = X_{1m} \cos \pi x \cos m\pi x,\ v_{1m} = Y_{1m} \cos \pi x \cos m\pi y\}$ if and only if*

$$\begin{cases} a_{22} < 0\,, \quad a_{11} > 0\,, \quad a_{12}a_{21} < 0 \\ \\ 1 + m^2 \in]1, \xi_1[\cup]\xi_2, \infty[\cap \left]\dfrac{|a_{22}|}{a_{11}}, \infty\right[\;; \quad \gamma \in]\eta_1, \eta_2[\\ \\ \xi_{1,2} = \dfrac{2A_0 - a_{11}a_{22} \pm \sqrt{\Delta_1}}{a_{11}}\,, \quad \eta_{1,2} = \dfrac{a_{11}(1+m^2) + a_{22} \pm \sqrt{\Delta}}{2\pi^2(1+m^2)} \end{cases} \tag{39}$$

Proof. Since$\{I_0 = a_{11} + a_{22} < 0,\ a_{11}(1+m^2) + a_{22} = m^2 a_{11} + I_0\}$, (37) can hold only if $a_{11} > 0$ and hence $a_{22} < 0$. Then for $\{a_{11} > 0,\ a_{22} < 0,\ a_{11} < -a_{22}\}$ for $m^2 > \dfrac{|a_{22}|}{a_{11}}$, one has $a_{11}(1+m^2) + a_{22} > 0$. Since $A_0 > 0$, one obtains $a_{12}a_{21} < 0$. Obviously (37) can be verified if and only if

$$\Delta = a_{11}^2(1+m^2)^2 + 2(a_{11}a_{22} - 2A_0)(1+m^2) + a_{22}^2 > 0\,. \tag{40}$$

But $\{A_0 > 0,\ a_{11}a_{22} < 0\} \Rightarrow \Delta_1 > 0$, therefore (40) is implied by $(39)_5$ and hence (37) is implied by $(39)_6$. □

Theorem 4.2. *Let the assumptions of Theorem 4.1 hold. Then for $\gamma_1 = \gamma_2 \in \left]0, \dfrac{a_{11}}{\pi^2}\right[$, the Turing instability occurs.*

Proof. In fact $\lim\limits_{m \to \infty} \eta_{1,2} = \dfrac{a_{11} \pm a_{11}}{2\pi^2}$. □

Remark 4.1. It is easily verified that theorem 4.1 continues to hold also when to (33) are appended the boundary conditions

$$\begin{cases} \nabla u \cdot \mathbf{n} \equiv \nabla v \cdot \mathbf{n} = 0 & \text{for } x = 0, 1 \\ \\ u = v = 0 & \text{for } y = 0, 1\,. \end{cases} \tag{41}$$

In fact, in this case

$$u = \sum_{n,m} X_{n,m}(t) \cos n\pi x \sin m\pi y\,, \qquad v = \sum_{n,m} Y_{n,m}(t) \cos n\pi x \sin m\pi y$$

and (36) continue to hold.

5. Final remark

Let us consider the stability of the zero solution of the classical binary system of O.D.Es.

$$\begin{cases} \dfrac{d^2x}{dt^2} - 2\omega\dfrac{dy}{dt} + \gamma\dfrac{dx}{dt} = \rho_1 x \\ \\ \dfrac{d^2y}{dt^2} + 2\omega\dfrac{dx}{dt} + \gamma\dfrac{dy}{dt} = \rho_2 y \end{cases} \tag{42}$$

with ω, γ, ρ_1, ρ_2 positive constants, with $2\omega > \sqrt{\rho_1} + \sqrt{\rho_2}$.

Equations (42) govern the motion of a material point, of unitary mass, in the xy plane, under the action of the conservative, gyrostatic and dissipative forces $(\rho_1 x,\, \rho_2 y)$, $2\omega\left(\dfrac{dy}{dt}, \dfrac{dx}{dt}\right)$, $-\gamma\left(\dfrac{dx}{dt}, \dfrac{dy}{dt}\right)$ respectively. The zero solution is destabilized by the dissipative force since is stable in its absence.[18–20] This result (known in Classical Mechanics well before 1952 and analogous to the Turing instability) was recalled to me by prof. F. Cardin in a private conversation, at the coffee break time, after my talk.

Acknowledgments

To Professor T. Ruggeri on the occasion of his 60th birthday.
This work has been performed under the auspices of the G.N.F.M. of I.N.D.A. M. and M.I.U.R. (P.R.I.N. 2005): "Nonlinear Propagation and Stability in Thermodynamical Processes of Continuous Media".

References

1. A. Turing, The chemical basis of morphogenesis, *Ph. Tran. Royal Soc. (B)* **237** (1952).
2. Wei-Ming Ni, Diffusion, Cross-Diffusion, and their Spike-Layer Steady States. In *Notices of the AMS*, January 1998.
3. Liancheng Wang, Michel Y. Li, Diffusion- Driven Instability in Reaction-Diffusion Systems *Journal of Mathematical Analysis and Application* **254**,138 (2001).
4. Sándor Kovács, Turing bifurcation in a system with cross diffusion, *Nonlinear Anal.* **59**, 567 (2004).

5. Wei Ming Ni, Moxun Tang, Turing patterns in the Lengel-Epstein system for the CIMA reaction. *Transaction of the AMS* **357**, 10, 3953 (2005).
6. V. Capasso, A. Di Liddo, Global attractivity for reaction-diffusion systems. The case of nondiagonal diffusion matrices, *Journal of Mathematical Analysis and Application* **188**, 510 (1993).
7. V. Capasso, A. Di Liddo, Asymptotic behaviour of reaction-diffusion systems in population and epidemic models, *J. Math. Biol.* **32**, 453 (1994).
8. A. Okubo, S.A. Levin, *Diffusion and ecological problems: modern perspectives.* Second Edition. Interdisciplinary Applied Mathematics, vol. 14. (Springer-Verlag, New York, 2001).
9. J.D. Murray, *Mathematical Biology. I, II* . Third Edition. (Interdisciplinary Appl.Math., vol.17. Springer-Verlag, New York, 2002).
10. S. Rionero, A rigorous reduction of the L^2-stability of the solutions to a nonlinear binary reaction-diffusion system of P.D.Es. to the stability of the solutions to a linear binary system of O.D.Es., *Journal of Mathematical Analysis and Application* **319**, 372 (2006).
11. S. Rionero, A nonlinear L^2-stability analysis for two-species population dynamics with dispersal, *Math. Biosc. Eng.* **3**, n. 1, 189 (2006).
12. S. Rionero, L^2-stability of the solutions to a nonlinear binary reaction-diffusion system of P.D.Es. *Rend.Mat.Acc.Lincei*, s.9, **16** (2005).
13. J.N. Flavin, S. Rionero, Cross-diffusion influence, *IMAMAT*, 1 (2007)
14. S. Rionero, Diffusion Driven Stability and Turing Effect Under Robin Boundary data (To appear)
15. R.S. Cantrell, C. Cosner, *Spatial Ecology via Reaction-Diffusion Equations* (Wiley, 2003).
16. S. Senn, On a nonlinear elliptic eigenvalue problem with Neumann boundary conditions with an application to population genetics, *Comm. in P.D.Es.* **8(11)**, 119 (1983)
17. F. Capone, M. Piedisacco, S. Rionero, Nonlinear stability for reaction-diffusion Lotka-Volterra model with Beddington-De Angelis functional response, *Rend. Acc. Sc. fis. mat. Napoli* **LXXIII** (2006).
18. T. Levi Civita, U. Amaldi *Lezioni di Meccanica Razionale. Parte Prima*, Vol. II, pp. 471-474 (Zanichelli Editore, Bologna, 1951).
19. R. Krechetnikov, J.E. Marsden *Dissipation-induced instabilities in finite dimensions* (California Institute of Tecnology, Pasadena, CA91125).
20. A. Block, P.S. Krishnaprasad, J.E. Marsden, T.S. Ratiu, Dissipation induced instability, *Analyse Nonlineaire Annales Institute H. Poincaré* **11**, 37 (1994).

STABILITY PROPERTIES OF THE SOLUTIONS OF A REACTION DIFFUSION EQUATION WITH ROBIN BOUNDARY CONDITION

S. RIONERO

Dipartimento di Matematica ed Applicazioni "R. Caccioppoli",
Università degli Studi di Napoli Federico II.
Naples, Italy
E-mail: rionero@unina.it.

M. VITIELLO

Dipartimento di Matematica, Politecnico di Bari, II Facoltà di Ingegneria Taranto
Bari, Italy
E-mail: vitiello@dm.uniba.it.

Equation (1) is considered under Robin boundary data in a bounded domains. Existence of equilibria and longtime behaviour of solutions, under appropriate assumptions on F anf f, are established.

1. Introduction

Let $\Omega \in R^3$ be a bounded smooth domain. We consider the nonlinear reaction diffusion equation

$$u_t = \Delta F(\mathbf{x}, u) + f(\mathbf{x}, u, \nabla u). \tag{1}$$

Equation (1) models a variety of phenomena such as chemical reactions, heat conduction, electrons flow and population dynamics. Recently for $f = f(\mathbf{x}, u)$, (1) has been studied intensively [1]- [8]. Precisely in [1]- [5], under Dirichlet boundary conditions, either in the case of Ω-bounded or unbounded, the longtime behaviour of solutions of (1) has been studied. In the present paper we consider Robin boundary conditions and look for conditions on F and f guaranteing the existence of equilibria and their stability.

The paper is structured as follows, Section 2 deals with some preliminaries, followed by a review of results concerning the basic existence theory for equilibria (Section 3). In Section 4 we obtain conditions guarenteing the stability of steady states. The instability is studied in Section 5. A pointwise estimate is obtained in Section 6.

2. Statement of the problem

Let $\Omega \in R^3$ be an open bounded set with $C^{2+p}(p > 0)$ boundary $\partial\Omega$, having the interior cone property. We consider the initial boundary value problem

$$\begin{cases} u_t = \Delta F(\mathbf{x}, u) + f(\mathbf{x}, u, \nabla u), & (\mathbf{x}, t) \in \Omega \times R^+ \\ u(\mathbf{x}, 0) = u_0(\mathbf{x}), & \mathbf{x} \in \Omega \\ \beta u(\mathbf{x}, t) + (1-\beta)\nabla u(\mathbf{x}, t) \cdot \mathbf{n} = 0, \ \mathbf{x} \in \partial\Omega, & t \in R^+, \end{cases} \tag{2}$$

where $\mathbf{n}$ denotes the outward unit normal vector to $\partial\Omega$, β is a constant $\in [0, 1]$, $F \in C^2(R \times \Omega)$, $f \in C(\Omega \times R \times R^n)$ and $u_0 \in C(\overline{\Omega})$.
The associated steady boundary value problem is given by

$$\begin{cases} \Delta F(\mathbf{x}, U) + f(\mathbf{x}, U, \nabla U) = 0 \ \mathbf{x} \in \Omega \\ \beta U + (1-\beta)\nabla U \cdot \mathbf{n} = 0 \qquad \mathbf{x} \in \partial\Omega. \end{cases} \tag{3}$$

On setting

$$\begin{cases} u = U + v \\ L(U, v) = F(\mathbf{x}, U + v) - F(\mathbf{x}, U) \\ g(U, v, \nabla U, \nabla v) = f(\mathbf{x}, U + v, \nabla U + \nabla v) - f(\mathbf{x}, U, \nabla U) \end{cases}$$

then, in view of (2)-(3) it follows that

$$\begin{cases} v_t = \Delta L(U, v) + g(U, v, \nabla U, \nabla v) & (\mathbf{x}, t) \in \Omega \times R^+ \\ v(\mathbf{x}, 0) = v_0(\mathbf{x}) & \mathbf{x} \in \Omega \\ \beta v(\mathbf{x}, t) + (1-\beta)\nabla v(\mathbf{x}, t) \cdot \mathbf{n} = 0, \ \mathbf{x} \in \partial\Omega, & t \in R^+. \end{cases} \tag{4}$$

Now we assume that exist two positive constants m and m_1 such that

$$2m \leq \frac{\partial F(\mathbf{x}, U)}{\partial u} \leq m_1 \qquad a.e. \qquad in \ \ \Omega \tag{5}$$

and hence there exists a positive constant ϵ such that [3]- [9]

$$|v| < \epsilon \Rightarrow m \leq \frac{\partial F(\mathbf{x}, U + v)}{\partial u} \leq m^* \quad with \qquad m^* = m + m_1 \quad a.e. \ \ in \ \ \Omega.$$

Moreover, denoting by $\|\cdot\|$ be the $L^2(\Omega)$-norm, we assume that exist two positive constant μ and k such that [3]- [4]

$$\left\|g^{n+1}\right\| \leq \mu \left\|v^{n+1}\right\|^{2k+1}. \tag{6}$$

In the sequel we will denote by $W^{1,2}(\Omega, \partial\Omega, \beta)$ the subspace of $W^{1,2}(\Omega)$ such that $\Psi \in W^{1,2}(\Omega, \partial\Omega, \beta) \Rightarrow$

$$\left\{ \Psi \in W^{1,2}(\Omega) \cap W^{1,2}(\partial\Omega); \beta\Psi + (1-\beta)\frac{d\Psi}{d\mathbf{n}} = 0 \ \ on \ \ \partial\Omega \times R^+ \right\}.$$

Recall that in $W^{1,2}(\Omega, \partial\Omega, \beta)$ the following inequality holds [10]- [11]

$$\|\nabla\Psi\|^2 + \frac{\beta}{1-\beta}\int_{\partial\Omega} \Psi^2 d\Sigma \geq \lambda \|\Psi\|^2 . \tag{7}$$

As it is well known, λ is the lowest eigenvalue in $W^{1,2}(\Omega, \partial\Omega, \beta)$ of

$$\Delta\Psi + \lambda\Psi = 0.$$

3. A remark on existence of steady state

This section is devoted to the existence of solutions of (3). We observe that f is a Caratheodory function (i.e.$f(\cdot, u, \nabla u)$ is measurable in Ω for any $(u, \nabla u) \in R \times R^n$ for almost every $x \in \Omega$) such that $u = 0 \Rightarrow f = 0$; then from (6), for $n = 0$ and $k = 0$; it follows that

$$||f(\mathbf{x}, u, \nabla u)|| \leq \mu||u||. \tag{8}$$

Lemma 3.1. *Suppose that $f(x, u, \nabla u)$ has an $L^2(\Omega)$ bound (i.e. there exists a positive constant H such that $\int_\Omega f^2 d\Omega < H$) and that (8) and (5) hold. Then exists a solution u to the problem (3).*

Proof. Let T be the map which associates to each v the unique solution u of the following linear problem (by letting $u = T(v)$)

$$\begin{cases} \nabla[F'(x,v)\nabla u] + f(x,v,\nabla v) = 0 & \mathbf{x} \in \Omega \\ \beta u + (1-\beta)\nabla u \cdot \mathbf{n} = 0 & \mathbf{x} \in \partial\Omega. \end{cases} \tag{9}$$

Notice that the solvability of problem (9) is guaranteed by the linear existence results and that T satisfies the hypothesis of Schauder's Fixed Point Theorem. In fact multiplying $(9)_1$ by u and using divergence theorem, we have

$$\int_{\partial\Omega} F'(x,v)u\nabla u\cdot\mathbf{n}d\Sigma - \int_\Omega F'(x,v)(\nabla u)^2 d\Omega + \int_\Omega f(x,v,\nabla v)u d\Omega = 0. \tag{10}$$

From (5), boundary condition $(9)_2$ and Cauchy inequality, it follows that

$$m||\nabla u||^2 + \frac{m\beta}{1-\beta}\int_{\partial\Omega} u^2 d\Sigma \leq \frac{1}{2}||f(x,v,\nabla v)||^2 + \frac{1}{2}||u||^2. \tag{11}$$

Since in view of (7) it turns out that

$$||u||^2 \leq \frac{1}{(2m\lambda - 1)}||f||^2. \tag{12}$$

Since f bounded, by (12), $\{u_n\}$ is bounded in $L^2(\Omega)$, then there exists a subsequence (we relabel $\{u_n\}$) such that $u_n \to w$ in $L^2(\Omega)$. From the uniqueness of the solution to the equation (9) we have $u = w$, therefore T is continuous. A similar argument shows that T is compact. [6] □

Let now introduce the *Approximated Problem* of (3)

$$\begin{cases} \nabla[F'(x,U_n)\nabla U_n] + f_n(x,U_n,\nabla U_n) = 0 & in \ \ \Omega \\ \beta U_n + (1-\beta)\nabla U_n \cdot \mathbf{n} = 0 & on \ \ \partial\Omega \end{cases} \tag{13}$$

where f_n (for $n > 0$) is the truncation of f by $\pm n$.
Now our aim is to prove such estimates for solution U_n of the problem (13).

Lemma 3.2. *Under the assumption of theorem 3.1, there exists a subsequence of the approximating solution $\{U_n\}$ to problem (13) which is $L^2(\Omega)$ bounded, $W^{2,p}(\Omega)$ bounded and $L^2(\partial\Omega)$ bounded.*

Proof. Following the procedure used in Lemma 3.1, from (13) we have

$$m||\nabla U_n||^2 + \frac{m\beta}{1-\beta}\int_{\partial\Omega} U_n^2 d\Sigma \leq ||f_n(x,U_n,\nabla U_n)|| \ \ ||U_n||. \tag{14}$$

From estimate (8) and (7) we obtain

$$\lambda m||U_n||^2 \leq \mu||U_n||^2 \leq \mu||U_n||^2 + K \qquad K = positive \ \ constant.$$

Hence by choosing $\mu \leq \lambda m$, we obtain

$$||U_n||^2 \leq M \quad with \quad M = \frac{K}{\lambda m - \mu} > 0. \tag{15}$$

Moreover by (14), (8) and (15) we have

$$||\nabla U_n||^2 \leq \frac{M\mu}{m} \quad and \quad \int_{\partial\Omega} U_n^2 d\Sigma \leq \mu M \frac{1-\beta}{m\beta}.$$

□

Theorem 3.1. *Suppose that (5) and (8) hold. Then there exists a solution $U \in L^2(\Omega)$ to problem (3).*

Proof. Let us consider the approximated problem (13), Lemma 3.1 implies the existence of weak solutions. Moreover Lemma 3.2 allows us to pass to the limit in each term of (13) and applying Vitali's theorem, we obtain the existence of at least one solution. □

4. Stability criteria

Theorem 4.1. *Let (5) and (6) hold. Then U is asymptotically exponentially stable.*

Proof. Let us introduce the functional

$$E(t) = \frac{1}{2}\int_\Omega v^2 d\Omega, \tag{16}$$

along the solutions of (4) we obtain

$$\dot{E}(t) = \int_\Omega v v_t d\Omega = -\int_\Omega \nabla v \nabla L d\Omega + \int_\Omega g v d\Omega + \int_{\partial\Omega} v \left(\frac{\partial L}{\partial v}\right) \nabla v \cdot \mathbf{n} d\Sigma. \tag{17}$$

In view of $(4)_3$ and (5) it follows that

$$\dot{E}(t) \leq -m \left[\int_\Omega (\nabla v)^2 d\Omega + \frac{\beta}{1-\beta}\int_{\partial\Omega} v^2 d\Sigma\right] + \int_\Omega g v d\Omega. \tag{18}$$

By virtue of (7), Holder inequality and (16); it turns out that

$$\dot{E}(t) \leq -2(m\lambda - 2^k \mu E^k(t))E(t). \tag{19}$$

It easly follows that $E^k(0) < \dfrac{m\lambda}{2^k \mu}$ implies

$$\dot{E}(t) \leq -2\gamma E(t), \tag{20}$$

with $\gamma = m\lambda\left(1 - \dfrac{2^k\mu}{m\lambda}E^k(0)\right)$. □

5. Instability result

Assuming

$$\frac{\partial F(\mathbf{x}, U)}{\partial u} \leq -2m, \qquad a.e. \ \ in \ \ \Omega \tag{21}$$

it turns out that, following the procedure used in [5]- [7], there exists a positive constant ϵ such that

$$|v| < \epsilon \Rightarrow \frac{\partial F(\mathbf{x}, U+v)}{\partial u} \leq -m. \qquad a.e. \quad in \ \ \Omega$$

Theorem 5.1. *Let (6) and (21) hold true. Then U is unstable with respect to the $L^2(\Omega)$-norm.*

Proof. From (17), by virtue of (21), $(4)_3$ and Holder inequality it turns out that

$$\dot{E}(t) \geq m\left[\|\nabla v\|^2 + \frac{\beta}{1-\beta}\int_{\partial\Omega} v^2 d\Sigma\right] - \|g\|\,\|v\|\,.$$

Then (6) and (7) imply

$$\dot{E}(t) \geq 2\lambda m E(t) - \mu 2^{k+1} E^{k+1}.$$

Integrating, one obtains

$$E^k(t) \geq \frac{2mk\lambda E^k(0)e^{2mk\lambda t}}{2mk\lambda + 2^{k+1}k\mu E^k(0)e^{2mk\lambda t}}$$

and the instability is implied by

$$\lim_{t\to\infty} E^k(t) \geq \frac{2m\lambda}{2^{k+1}\mu}.$$

□

6. An essential sup estimate

Denoting by D the largest subdomain of Ω on each point of which, at time t, $|v(\mathbf{x}, \mathbf{t})|$ is bigger than $\epsilon(t) > 0$, the steady state U will be said to be weakly pointwise attractive if

$$\lim_{t\to\infty} \mu(\epsilon, t) = 0 \qquad \forall\epsilon > 0$$

where $\mu(\epsilon, t) = \mu(D(t))$ is the Lebesgue measure of $D(t)$. [3]

Theorem 6.1. *Let the assumptions of Theorem 4.1 hold. Then U is asymptotically stable with respect to pointwise norm (a.e.).*

Proof. In fact the following inequality holds [3]

$$\mu[D(t)] \leq \|v(\mathbf{x}, t)\|_2^{\frac{2}{3}}\,. \tag{22}$$

Then from (20) and (22) it follows that

$$\lim_{t\to\infty} \mu[D(t)] \leq \lim_{t\to\infty} \left[2E(0)e^{-2\gamma t}\right]^{\frac{1}{3}}$$

We introduce now the Liapunov functional

$$V_n(t) = \frac{1}{2(n+1)}\int_\Omega v^{2(n+1)} d\Omega. \tag{23}$$

Evaluating the time derivative of (23) along the solutions of (4), by recursive arguments, one obtains easily that

$$\dot{V}_n(t) \leq -\frac{2m\lambda(2n+1)}{(n+1)} V_n(t) + \mu^{\frac{1}{n+1}}(2n+1)^{k+1} V_n^{k+1}(t)$$

and hence

$$V_n^k(0) < \frac{m\lambda(2n+1)}{\mu^{\frac{1}{n+1}} 2^k (n+1)^{k+2}}$$

implies

$$V_n(t) \le V_n(0).$$

On raising both sides to the power of $\{2(n+1)\}^{-1}$ and, by the convergence dominate theorem, taking the limit as $n \to \infty$, one finds that

$$ess \sup_{\mathbf{x}\in\Omega} |v(\mathbf{x},t)| \le ess \sup_{\mathbf{x}\in\Omega} |v_0(\mathbf{x})|. \qquad \square$$

References

1. J. N. Flavin and S. Rionero, *Qualitative Estimates for Partial Differential Equation, An Introduction*, CRC Press, Boca Raton, FL, (1995).
2. J. Smoller, *Shock waves and Reaction-Diffusion Equations*, Springer-Verlag, n.258 of "A Series of Comprehensive Studies in Mathematics", (1983).
3. S. Rionero, *Asymptotic properties of solutions to nonlinear possibly degenerated parabolic equations in unbounded domains*, Math. and Mech. of Solids, (2003).
4. S. Rionero, *Asymptotic Properties of Solutions to Nonlinear Possibly Degenerated Parabolic Equations in Unbounded Domains*, Proc. Wascom 2003, 12th Conference on Waves and Stability in Continuous Media. (2003).
5. F. Capone, S. Rionero and I. Torcicollo, *On the stability of solutions of the remarkable equation* $u_t = \Delta F(x,u) - g(x,u)$, Suppl. Rend. Cir. Mat. Palermo, serie II, n.45, (1996).
6. Lawrence C. Evans, *Partial differential Equations*, 1992.
7. R. S. Cantrell, C. Cosner, *Diffusive logistic equations with indefinite weights: population models in distrupted environments*, J. Math. Anal. vol 22, n4, 1043-1064, (1991).
8. S. Rionero, *A rogorous reduction of the* L^2*-stability of the solutions to a nonlinear binary reaction-diffusion system of P.D.Es to the stability of the solutions to a linear binary system of O.D.Es.*, Journal of Math. Analysis and Application 319, 372-392, (2006).
9. I. Torcicollo and M. Vitiello, *A note on the nonlinear pointwise stability in the exterior of a sphere*, Rend. Acc. Sci. Fis. Mat. Napoli, Vol. LXX, pp.111-117, (2003).
10. K. Rektoris, *Varational methods in Mathematics*, Science and Engineering D. Reidd Pubb. Co., (1980).
11. Senn, *On a nonlinear elliptic eigenvalue problem with Neumann boundary conditions, with an application to population dynamics*, Comm. Partial Diff. Eq. 8, n.11, 1199-1228, (1983).

NONLINEAR WAVE PROPAGATION IN AN ELASTIC SOIL

V. ROMANO and M. RUGGIERI
Dipartimento di Matematica e Informatica,
Università di Catania,
viale A. Doria, 6, 95125 Catania, Italy
romano@dmi.unict.it
ruggieri@dmi.unict.it

Assuming the stress-strain relationship proposed by Kondner [1,2], a 1-D analysis of wave propagation in soil subjected to axial load is carried out. Simple waves, shocks and rarefaction waves are studied. The stable shocks are selected with the Lax conditions.

1. The mathematical model

The classical balance equations for an elastic non-linear material in the one dimensional case along a direction parallel to the gravity acceleration are [3,4]

$$\rho_0 v_t = S(e)_x + \rho_0 g, \tag{1}$$

$$e_t - v_x = 0, \tag{2}$$

where x is the spatial lagrangian coordinate, t the time, ρ_0 the density of the soil in the reference frame, v the significant component of velocity, S the significant component of the Piola-Kirchoff stress tensor, e the significant component of the displacement gradient, g the modulus of gravity acceleration. We assume that the mass density is constant and normalized to one.

The system can be written in compact form as follows

$$\frac{\partial U}{\partial t} + A\frac{\partial U}{\partial x} = G \tag{3}$$

where

$$U = (e, v)^T, \qquad A = \begin{pmatrix} 0 & -1 \\ -S'(e) & 0 \end{pmatrix}, \qquad G = (g, 0)^T,$$

the superscript T indicating transposition.

If we exclude the trivial case $S(e)$ = constant, the eigenvalue problem

$$\det(A - \lambda I) = 0 \tag{4}$$

has two distinct eigenvalues $\lambda_\pm = \pm\sqrt{S'(e)}$. Therefore the system (3) is hyperbolic provided that $S'(e) > 0$.

The stress-strain behavior in soils depends on a number of different factors including density, strain condition etc. In the sequel, we will consider the non linear constitutive stress-strain relation proposed by Kondner [1,2] represented by

$$S(e) = \frac{e}{a + be}. \tag{5}$$

More precisely in (5) one should subtract the transversal component of the stress which will be neglected because constant and therefore unimportant in the analysis.

The physical range of e allowed by (5) is $e > -\frac{b}{a}$. This interval has been selected by requiring that it contains the case $e = 0$. The physical meaning of a and b is given by the relations

$$\lim_{e\to+\infty} S = \sigma_{lim} = \frac{1}{b}, \qquad \lim_{e\to 0} \frac{dS}{de} = \frac{1}{a} = E_i \tag{6}$$

with $\sigma_{lim} > 0$ ultimate resistance and $E_i > 0$ initial tangent deformation modulus. The stress strain relation is completed by assigning the value of the constants a, $b \in \mathbb{R}^+$, which can be determined experimentally by laboratory tests carried out on soil samples.

The hyperbolicity conditions $\frac{\partial S}{\partial e} > 0$ is satisfied because

$$\frac{\partial S}{\partial e} = \frac{a}{(a + be)^2} \tag{7}$$

and we have the eigenvalues

$$\lambda_- = -(a + be)^{-1}\sqrt{a}, \qquad \lambda_+ = (a + be)^{-1}\sqrt{a}.$$

which are real and distinct implying that the system (3) is strictly hyperbolic.

It is also simple matter to see that the system is genuinely nonlinear.

The physical situation we are going to investigate is represented by a one dimensional problem describing the dynamics of a soil with an axial loading along the vertical direction.

2. Travelling waves

In this section we look for travelling waves admitted by the system (3), that is solutions of the form

$$v = \psi_1(x - \lambda t) = \psi_1(\xi), \quad e = \psi_2(x - \lambda t) = \psi_2(\xi), \tag{8}$$

with $\lambda \in \mathbb{R}$ and $\xi = x - \lambda t$ similarity variable.

By inserting (8) into the system (3), one obtains that ψ_1, ψ_2 must satisfy the conditions

$$\psi_1' + \lambda\psi_2' = 0, \quad \lambda\psi_1' + \frac{a}{(a + b\psi_2)^2}\psi_2' = g \tag{9}$$

where the prime denotes derivatives which respect to ξ.

From $(9)_1$ we have

$$\psi_1 + \lambda\psi_2 = c = \text{constant} \tag{10}$$

while ψ_2 is solution of the ODE

$$\left[\lambda^2 - \frac{a}{(a + b\psi_2)^2}\right]\psi_2' = g \tag{11}$$

which after integration gives

$$b\lambda^2\psi_2^2 + [\lambda^2 a - b(g\xi + k_0)]\psi_2 + \frac{a}{b} - a(g\xi + k_0) = 0. \tag{12}$$

The discriminant of (12) reads

$$\triangle = [b(g\xi + k_0) - a\lambda^2]^2 - 4a\lambda^2[1 - b(g\xi + k_0)].$$

In order to have real solutions $\triangle$ must be non negative. The discriminant of $\triangle$ considered as a polynomial in ξ is

$$\Delta_1 = 16b^2g^2\lambda^2 a$$

which is positive since $a \in \mathbb{R}^+$.

As a conseguence the solutions of (12) are given by

$$\psi_2^\pm = \frac{b(g\xi + k_0) - \lambda^2 a \pm \sqrt{[b(g\xi + k_0) - a\lambda^2]^2 - 4a\lambda^2[1 - b(g\xi + k_0)]}}{2b\lambda^2}, \qquad \xi \in \mathbb{R}\backslash[\xi_1, \xi_2]$$

where

$$\xi_1 = -\frac{a\lambda^2 + 2\lambda\sqrt{a} + bk}{bg}, \qquad \xi_2 = \frac{-a\lambda^2 + 2\lambda\sqrt{a} - bk}{bg}$$

are the solutions of $\triangle = 0$.

We observe that ψ_1^+ remains limited when $\xi \to -\infty$ while ψ_1^- remains limited when $\xi \to +\infty$. Therefore according to the domain to which ξ belongs at the initial time, only one of the two solutions is physically acceptable. In particular, for traveling wave propagating towards $+\infty$ ($\lambda > 0$), one takes the ψ_1^+; instead for for traveling wave propagating towards $-\infty$ ($\lambda < 0$), one takes the ψ_1^-.

3. Shock Waves

The Rankine-Hugoniot conditions for the set of balance equations under consideration are

$$s\,[e] + [v] = 0, \quad s\,[v] + [S(e)] = 0 \tag{13}$$

where $[u] = u^+ - u^-$ is the jump of u across the discontinuity surface, u^+ and u^- being the values the right and on the left respectively with respect to the chosen reference frame, s is the speed of the shock.

The jump conditions give

$$s = \pm\sqrt{\frac{a}{(a+be^+)(a+be^-)}}, \quad v^+ - v^- = \pm\sqrt{\frac{a(e-e^-)^2}{(a+be)(a+be^-)}} \tag{14}$$

The reality condition for the square root leads to consider two regions

$$R_I\,:\,\left\{a+be^- > 0, a+be^+ > 0\right\}, \; R_{II}\,:\,\left\{a+be^- < 0, a+be^+ < 0\right\}$$

which are not connected. Since the equilibrium state $e = 0$ does not belong to the region R_{II}, we retain only the region R_I as physically significant.

In order to select the stable shocks, the Lax conditions must to be satisfied. In the present case, for 1-shocks, which correspond to λ_-, and 2-shocks, which correspond to λ_+, the Lax conditions read respectively

$$-\sqrt{S'(e^-)} < -\,(a+be)^{-1}\sqrt{a} < -\sqrt{S'(e^+)} \tag{15}$$

$$\sqrt{S'(e^-)} < (a+be)^{-1}\sqrt{a} < \sqrt{S'(e^+)}. \tag{16}$$

In the 1-shock case, from (15), by taking into account (7) and $S'' = \frac{-2ab}{(a+be)^3} < 0$ it follows that

$$e^+ > e^- \tag{17}$$

and $(14)_2$ leads to

$$v^+ > v^- \tag{18}$$

Therefore the set of states which can be connected to (v^-, e^-) through a 1-shock is given by the curve

$$S^{(-)}: \ v = v^- + (e - e^-)\sqrt{\frac{a}{(a+be)(a+be^-)}}, \quad e > e^-. \tag{19}$$

In the 2-shock case, from (16) it follows that

$$e^+ < e^- \tag{20}$$

and $(14)_2$ leads again to (18). Therefore the set of states which can be connected to (v^-, e^-) through a 2-shock is given by the curve

$$S^{(+)}: \ v = v^- - (e - e^-)\sqrt{\frac{a}{(a+be)(a+be^-)}}, \quad e < e^-. \tag{21}$$

$S^{(-)}$ and $S^{(+)}$ are called 1-shock wave curve and 2-shock wave curve, respectively.

Remark. Relations (17)-(20) express the physical fact that *the shock propagates from the region with higher deformation toward the region with lower deformation.* This condition plays the same role of the compressive rule for shocks in gas dynamics

4. Rarefaction Waves

Finally we look for rarefaction waves which are represented by continuous solution of the form $U = U(x/t)$.

It is convenient perform the change of dependent variables $v \mapsto \tilde{v} + gt, \quad e \mapsto \tilde{e}$, which transforms the system (3) into an autonomous one

$$\rho_0 v_t - S_x = 0, \tag{22}$$

$$e_t - v_x = 0. \tag{23}$$

where for the sake of simplicity we have dropped the tilde in the new variables.

Looking for solutions of (22),(23) of the form $U = U(\phi(x,t))$, with $\phi(x,t) = x/t$, we get

$$e'\phi + v' = 0, \quad v'\phi + \frac{ae'}{(a+be)^2} = 0 \tag{24}$$

where the prime denotes derivation with respect to ϕ. Therefore

$$\frac{dv}{de} = \pm\frac{\sqrt{a}}{a+be},$$

and [5] the following rarefaction waves passing through $(\overline{e}, \overline{v})^T$ are found

$$R^{(-)}: \quad v = \overline{v} + \frac{\sqrt{a}}{b} \log \frac{a + be}{a + b\overline{e}} \tag{25}$$

$$R^{(+)}: \quad v = \overline{v} - \frac{\sqrt{a}}{b} \log \frac{a + be}{a + b\overline{e}} \tag{26}$$

They are called 1-rarefaction wave curve and 2-rarefaction wave curve, respectively.

Imposing

$$\frac{x}{t} = \pm \frac{\sqrt{a}}{a + be}$$

one has

$$e(x,t) = -\frac{\sqrt{a}}{b}\frac{x}{t} - \frac{a}{b}, \quad x < 0, \quad t > 0 \tag{27}$$

for 1-rarefaction wave and

$$e(x,t) = \frac{\sqrt{a}}{b}\frac{x}{t} - \frac{a}{b}, \quad x > 0, \quad t > 0 \tag{28}$$

for 2-rarefaction wave.

One example of application of the previous results is the resolution of the Riemann problem [6].

Acknowledgments

The authors acknowledge the financial support by P.R.A. University of Catania (ex 60 %). M.R. acknowledges also the financial support by MIUR-PRIN 2005/07 through the project "Nonlinear Propagation and Stability in Thermodynamical Processes of Continuous Media".

References

1. R. L. Kondner, *Journal of the Soil Mechanics and Foundations Division. ASCE*, **89** SM1, 115 (1963).
2. R. L. Kondner and J.S. Zelasko, *Proceedings 2nd Pan-American Conference on Soil Mechanics and Foundations Engineering, Brazil*, **1** 289 (1963).
3. M. E. Gurtin, *An introduction to Continuum Mechanics*, Academic Press (1981).
4. C. M. Dafermos, *Hyperbolic Conservation Laws in Continuum Physics*, Springer (2000).
5. J. Smoller, *Shock Waves and reaction-diffusion equation*, Springer-Verlag (A series of Comprehensive Studies in Mathematics Vol. **258**), Berlin (1983).
6. V. Romano and M. Ruggieri, *Riemann problem for an elastic soil*, preprint.

MODELING THE HEATING OF SEMICONDUCTOR CRYSTAL LATTICE BASED ON THE MAXIMUM ENTROPY PRINCIPLE

V. ROMANO and M. ZWIERZ
Department of Mathematics and Computer Science, University of Catania, Catania 90125, Italy
E-mail: romano@dmi.unict.it, zwierz@dmi.unict.it

1. Introduction

In micro and nano-devices the presence of very high and rapidly varying electric fields is the cause of non linear phenomena like thermal heating of the carriers and the crystal lattice. The external electric field transfers energy to electrons and in turn to the crystal lattice through the scattering mechanisms. The self-heating can influence significantly the electrical behavior since the dissipated electrical energy causes a temperature rise over an extended area of the device resulting in increased power dissipation [1].

On account of these phenomena the distribution functions of electrons and phonons are no longer represented by equilibrium statistics and the isothermal approximation made in the standard drift-diffusion models is in several applications very questionable. The crystal temperature has to be added as an additional dynamical state variable and in a phenomenological approach this has been achieved including into the model also the energy equation for the lattice whose temperature is denoted with T_L

$$\rho c \frac{\partial T_L}{\partial t} = \operatorname{div}\left[k(T_L)\nabla T_L\right] + H \tag{1}$$

where ρ and c are the specific mass density and specific heat of material. $k(T_L)$ and H denote thermal conductivity and the local energy production [1]. Up to now only phenomenological functional forms of $k(T_L)$ and H have been proposed, for instance

$$\begin{array}{ll} k(T_L) = \frac{1}{a+bT_L+cT_L^2} & \text{Glassbrenner and Slack} \\ k(T_L) = 1.5486\frac{VA}{cmK}\left(\frac{T_L}{300K}\right)^{-\frac{4}{3}} & \text{Adler} \\ H = J_n E & \text{Gaur and Navon} \\ H = \operatorname{div}\left(\frac{E_c}{q}J_n\right) & \text{Adler} \end{array} \tag{2}$$

where $J_n = qn\mu_n + qD_n\text{grad}n + qnD_n^T\text{grad}T_L$. There are no explicit expressions for $k(T_L)$ and H based on a consistent theoretical derivation (for an attempt in the framework of linear irreversible thermodynamics see [2]).

Here the heating of the lattice is investigated, treating the energy transport inside the crystal with a phonon gas. A macroscopic model is obtained as moment system of the semiclassical Boltzmann equation with closure relations deduced via the maximum entropy principle [3,4], including all the relevant scattering mechanisms for silicon.

2. Model of electron and phonon transport in semiconductors based on the MEP

In the semiclassical kinetic approach the charge transport in semiconductors is described by the Boltzmann equations for electrons in each band and for phonons, coupled with the Poisson equation for the electric potential ϕ

$$\frac{\partial f}{\partial t} + v^i(\mathbf{k})\frac{\partial f}{\partial x^i} - \frac{eE^i}{\hbar}\frac{\partial f}{\partial k^i} = \mathcal{C}[f,g], \tag{3}$$

$$\frac{\partial g}{\partial t} + \frac{\partial \omega}{\partial q^i}\frac{\partial g}{\partial x^i} = S[g,f], \tag{4}$$

$$E_i = -\frac{\partial \phi}{\partial x_i}, \quad \epsilon\Delta\phi = -e(N_D - N_A - n). \tag{5}$$

where respectively $f(\mathbf{x},\mathbf{k},t)$ and $g(\mathbf{x},\mathbf{q},t)$ are the electron and phonon distribution functions, e the absolute value of the electron charge, $\mathbf{k}$ the electron wave vector, $\mathbf{q}$ the phonon wave vector, $\boldsymbol{E}$ the electric field, ϵ the permittivity of the semiconductor, N_D and N_A the donor and acceptor density and n the electron density. The electron velocity $\mathbf{v}(\mathbf{k})$ is related to the electron energy $\mathcal{E}$ (the so called band structure) by the relation $\mathbf{v}(\mathbf{k}) = \frac{1}{\hbar}\nabla_{\mathbf{k}}\mathcal{E}$.

In this article the energy band $\mathcal{E}$ is approximated by the *parabolic band*

$$\mathcal{E}(\mathbf{k}) = \frac{\hbar^2|\mathbf{k}|^2}{2m^*}, \qquad \mathbf{k} \in \mathbb{R}^3 \quad \Rightarrow \quad \mathbf{v}(\mathbf{k}) = \frac{\hbar}{m^*}\mathbf{k}. \tag{6}$$

where m^* is the effective mass and $\hbar$ is Planck's constant divided by 2π. For the case of the Kane dispersion relation the interested reader is referred to [5].

$\mathcal{C}[f,g]$ is the electron collision operator, which takes into account scattering of the electrons with acoustic and non–polar optical phonons. Neglecting the electron–electron scattering, the collision operator $\mathcal{C}[f,g]$ can be written as

$$\mathcal{C}\left[f,g\right] = \int_{\mathcal{B}} \left[P\left(\mathbf{k}',\mathbf{k}\right) f\left(\mathbf{k}'\right) - P\left(\mathbf{k},\mathbf{k}'\right) f\left(\mathbf{k}\right)\right] d\mathbf{k}'. \tag{7}$$

$P\left(\mathbf{k}',\mathbf{k}\right)$ is the transition probability per unit time from a state $\mathbf{k}$ to a state $\mathbf{k}'$. It for acoustic and non–polar optical phonon $P\left(\mathbf{k}',\mathbf{k}\right)$ is given respectively by

$$P\left(\mathbf{k},\mathbf{k}'\right) = \sum_{\mathbf{q}} K_{ac}\left(g\left(\mathbf{q}\right) + \frac{1}{2} \mp \frac{1}{2}\right) \delta_{\mathbf{k}\pm\mathbf{q},\mathbf{k}'} \delta\left[\mathcal{E}\left(\mathbf{k}'\right) - \mathcal{E}\left(\mathbf{k}\right)\right]$$

$$P\left(\mathbf{k},\mathbf{k}'\right) = \sum_{\mathbf{q}} K_{op}\left(g\left(\mathbf{q}\right) + \frac{1}{2} \mp \frac{1}{2}\right) \delta_{\mathbf{k}\pm\mathbf{q},\mathbf{k}'} \delta\left[\mathcal{E}\left(\mathbf{k}'\right) - \mathcal{E}\left(\mathbf{k}\right) \mp \hbar\omega\left(q\right)\right]$$

where K_{ac} and K_{op} are constants and δ is the Dirac delta function.

$\mathcal{S}[g,f]$ is the phonon collision operator, which takes into account electron-phonon and phonon–phonon scattering. It is given by

$$\begin{aligned}\mathcal{S}\left[g,f\right] = 2\sum_{\mathbf{k},\mathbf{k}'} s\left(q\right) f\left(\mathbf{k}\right) \{&\left[g\left(\mathbf{q}\right)+1\right] \delta_{\mathbf{k}',\mathbf{k}-\mathbf{q}} \delta\left[E\left(\mathbf{k}'\right) - E\left(\mathbf{k}\right) + \hbar\omega\left(q\right)\right] \\ &-g\left(\mathbf{q}\right) \delta_{\mathbf{k}',\mathbf{k}+\mathbf{q}} \delta\left[E\left(\mathbf{k}'\right) - E\left(\mathbf{k}\right) - \hbar\omega\left(q\right)\right]\} + \frac{g_{BE} - g}{\tau}\end{aligned} \tag{8}$$

where g_{BE} is the Bose-Einstein distribution, τ the phonon relaxation time, ω the so-called dispersion relation. For acoustic phonons we adopt the Debye approximation $\omega = cq$ with c Debye velocity. In case of the non-polar optical phonons we assume $\hbar\omega =$ constant (Einstein approximation) and keep the phonon distribution at equilibrium, that is $g = g_{BE}$. Note that in the above expressions equipartition of energy is not imposed.

2.1. *Macroscopic model and maximum entropy principle*

Macroscopic models can be obtained by taking the moments of the Boltzmann transport equations (3) and (4). Multiplying equations (3) and (4) by a sufficiently regular functions $\Psi^e_A(\mathbf{k})$ and $\Psi^p_B(\mathbf{q})$ respectively and integrating over the first Brillouin zone. Various models employ different expressions of $\Psi^e_A(\mathbf{k})$, $\Psi^p_B(\mathbf{q})$ and number of moments. We choose the following sets

$$\Psi^e_A(\mathbf{k}) = (1, \hbar\mathbf{k}, \mathcal{E}, \hbar\mathbf{k}\mathcal{E}), \quad \Psi^p_B(\mathbf{q}) = \left(\hbar\omega, \hbar\mathbf{q_i}, \frac{\hbar\omega}{\mathbf{q}^2}\mathbf{q_{<i}q_{j>}}\right) \tag{9}$$

obtaining the following balance equations for electrons and phonons

$$\begin{cases} \frac{\partial}{\partial t}n + \frac{\partial(nV^i)}{\partial x^i} = 0, \\ \frac{\partial}{\partial t}(nP^i) + \frac{\partial(nU^{ij})}{\partial x^j} + neE^i = nC^i_P, \\ \frac{\partial}{\partial t}(nW) + \frac{\partial(nS^j)}{\partial x^j} + neV_kE^k = nC_W, \\ \frac{\partial}{\partial t}(nS^i) + \frac{\partial(nF^{ij})}{\partial x^j} + neE_jG^{ij} = nC^i_S. \end{cases} \qquad \begin{cases} \frac{\partial u}{\partial t} + Q_k = P_u, \\ \frac{\partial p_i}{\partial t} + \frac{1}{3}\frac{\partial u}{\partial x_i} + \frac{\partial N_{\langle ik\rangle}}{\partial x_k} = P_i, \\ \frac{\partial N_{\langle ij\rangle}}{\partial t} + \frac{\partial M_{\langle ij\rangle k}}{\partial x_k} = P_{\langle ij\rangle}. \end{cases}$$

with

$$\begin{cases} n = \int_{\mathcal{B}} f d\mathbf{k} & \text{electron density,} \\ V^i = \frac{1}{n}\int_{\mathcal{B}} f v^i d\mathbf{k} & \text{average electron velocity,} \\ P^i = \frac{1}{n}\int_{\mathcal{B}} f\hbar k^i d\mathbf{k} = m^* V^i & \text{average electron momentum,} \\ W = \frac{1}{n}\int_{\mathcal{B}} \mathcal{E}(k) f d\mathbf{k} & \text{average electron energy,} \\ S^i = \frac{1}{n}\int_{\mathcal{B}} f v^i \mathcal{E}(k) d\mathbf{k} & \text{energy flux} \\ u = \int_{\mathcal{B}} \hbar\omega g \, d\mathbf{q} & \text{phonon energy density} \\ p_i = \int_{\mathcal{B}} \hbar q_i g \, d\mathbf{q} & \text{phonon momentum density} \\ N_{\langle ij\rangle} = \int_{\mathcal{B}} \frac{\hbar\omega}{q^2} q_{<i}q_{j>} g \, d\mathbf{q} & \text{deviatoric part of the phonon momentum flux} \end{cases}$$

In case of phonons without coupling with electrons general results can be found in [6] under the Debye approximation for the dispersion relation.

The previous sets of moment equations for electrons and phonons are not closed because more unknowns appear than the number of equations. The closure problem consists in expressing unknown fluxes U_{ij}, F_{ij}, G_{ij}, Q_k, $M_{\langle ij\rangle k}$ and production terms C^i_P, C_W, C^i_S, P_u, P_i, $P_{\langle ij\rangle}$ as functions of n, V^i , W, S^i, u, p_i and $N_{\langle ij\rangle}$.

According to MEP [3,4] if a given number of moments $M_A, A = 1, \ldots, N$ and $M_B, B = 1, \ldots, N^*$ are known, the distribution functions f_{ME} and g_{ME} for electron and phonon systems, which can be used to evaluate the unknown moments of f and g, correspond to the maximum of the entropy functional under the constrains

$$\int_{\mathcal{B}} \Psi^e_A(\mathbf{k}) f_{ME} d\mathbf{k} = M_A, \quad \int_{\mathcal{B}} \Psi^p_B(\mathbf{q}) g_{ME} d\mathbf{q} = M_B \tag{10}$$

where k_B is the Boltzmann constant. In the case under investigation one can approximate [7] f_{ME} with

$$\exp\left(-\frac{\lambda}{k_B} - \lambda^W \mathcal{E}\right)\left(1 - \boldsymbol{\lambda}^V \cdot \mathbf{v} - \boldsymbol{\lambda}^S \cdot \mathbf{v}\,\mathcal{E}\right) \tag{11}$$

with λ, λ^W, $\boldsymbol{\lambda}^V$, $\boldsymbol{\lambda}^S$ lagrangian multipliers associated with density, momentum, energy, energy flux [7,8] while the non-equilibrium acoustic phonon distribution g_{ME} is given (as in [6]) by

$$g_{ME} = g_{BE} + g_{BE}^{+} \left[\frac{3c^2 \hbar q}{4uk_B T_L} p_i l_i + \frac{15 \hbar c q}{8uk_B T_L} \left(N_{ij} l_i l_j - \frac{u}{3} \right) \right] \qquad (12)$$

with

$$g_{BE}^{+} = \frac{\exp\left(\frac{\hbar c q}{k_B T_L}\right)}{\left(\exp\left(\frac{\hbar c q}{k_B T_L}\right) - 1\right)^2}, \quad \mathbf{l} \in S^2 \text{ the unit sphere of } \mathbb{R}^3. \qquad (13)$$

At last we relate u and T_L by the equation of state $u = aT_L^4$ with a a coefficient related to the radiation constant.

After substituting f_{ME} and g_{ME} into the definition of fluxes and production terms, we obtain the same expressions for C_P^i, C_W and C_S^i as in [8]. For the phonons we find

$$P_u = nC_W, \quad P_i = F\tfrac{nV_i}{u} + G\tfrac{nS_i}{u} - \tfrac{p_i}{\tau_R}, \quad P_{\langle ij \rangle} = -\tfrac{N_{\langle ij \rangle}}{\tau}. \qquad (14)$$

where F and G depend on electron energy W and crystal lattice temperature T_L (the explicit expression can be found in [5,9]) and τ_R is the relaxation time for resistive processes. Under the scaling

$$t = \mathcal{O}\left(\frac{1}{\delta^2}\right), x_k = \mathcal{O}\left(\frac{1}{\delta}\right), \mathbf{V} = \mathcal{O}(\delta), \mathbf{S} = \mathcal{O}(\delta), \mathbf{p} = \mathcal{O}(\delta), \mathbf{N}_{\langle \mathbf{ij} \rangle} = \mathcal{O}(\delta).$$

the following equations for Energy transport and lattice heating model is deduced

$$\frac{\partial n}{\partial t} + \operatorname{div}(nV) = 0, \quad \frac{\partial (nW)}{\partial t} + \operatorname{div}(nS) + nqV\nabla\phi = nC_W, \qquad (15)$$

$$4aT_L^3 \frac{\partial T_L}{\partial t} = \operatorname{div}\left[k(T_L)\nabla T_L\right] + H. \qquad (16)$$

where

$$k(T_L) = \frac{4c^2 \tau_R}{3} a T_L^3,$$

$$H = nC_W(W, T_L) - c^2 \operatorname{div} \tau_R \left(\frac{F^{ac}(W, T_L)}{u} nV + \frac{G^{ac}(W, T_L)}{u} nS \right).$$

and

$$V = D_{11} \frac{\partial}{\partial x} \log n + D_{12} \frac{\partial}{\partial x} W + D_{13} \frac{\partial}{\partial x} \phi, \qquad (17)$$

$$S = D_{21} \frac{\partial}{\partial x} \log n + D_{22} \frac{\partial}{\partial x} W + D_{23} \frac{\partial}{\partial x} \phi \qquad (18)$$

with the coefficients D_{ij} depending on W and T_L [9]. Only the contribution of F and G due to the acoustic phonon is present in H.
The thermal conductivity can be expressed by

$$k(T) = \frac{c^2}{3} c_v \tau_R. \tag{19}$$

where $c_v = \frac{1}{\rho}\frac{\partial}{\partial t}e = 4aT_L^3$ denotes the specific heat per unit volume. The obtained expression for H cannot be identified with any relation known from the literature. The first part of H can be expressed in the relaxation time approximation and resembles the Wachutka model [2]. The second term comes from the interaction with the electrons. The form of divergence of current resembles the Adler model.

Acknowledgments

This work has been supported by the EU Marie Curie RTN project **CoMSON** (COupled Multiscale Simulation and Optimization in Nanoelectronics) grant n. MRTN-CT-2005-019417.

References

1. Selberherr, S. Analysis and Simulation of Semiconductor Devices, Springer–Verlag, Wien, 1984.
2. Wachutka, G. Rigorous thermodynamic treatment of heat generation and conduction in semiconducotr device modeling, IEEETrans. on Computer Aided Design, Vol. 9 n. 11 (1990), pages 1141–1149.
3. Janes, E. T. Information theory and statistical mechanics, Physical Review, Vol. 106, (1957) pages 620–.
4. N. Wu, The Maximum Entropy Method, Springer–Verlag, Berlin, 1997.
5. M. Zwierz and V. Romano, Modeling the heating of semiconductor crystal lattice based on the maximum entropy principle – Kane dispersion relation case, report (2007), `www.comson.org`
6. Dreyer, W. and Struchtrup, H., Heat pulse experiment – revisited, Continuum Mech. Thermodyn., Vol. 5, (1993), pages 3–50.
7. Anile, A. M. and Romano, V., Non parabolic band transport in semiconductors: closure of the production terms in the moment equations, Continuum Mech. Thermodyn. 11 (1999), pages 307– .
8. Romano, V., Non parabolic band transport in semiconductors: closure of the production terms in the moment equations, Continuum Mech. Thermodyn., Vol. 12, (2000), pages 31–51.
9. M. Zwierz and V. Romano, Modeling the heating of semiconductor crystal lattice based on the maximum entropy principle – parabolic band approximation case, report (2007), `www.comson.org`

SYMMETRY ANALYSIS OF A VISCOELASTIC MODEL

M. RUGGIERI and A. VALENTI

Dipartimento di Matematica e Informatica,
Università di Catania,
viale A. Doria 6, 95125 Catania, Italy

Symmetry analysis of two equivalent mathematical models describing one dimensional motion in a nonlinear dissipative medium is performed. The relationships between the symmetries of those models are explored.

Keywords: Lie Group, symmetry analysis, dissipative media.

1. Introduction

We consider the third order partial differential equation

$$w_{tt} = f(w_x)w_{xx} + [\lambda(w_x)\, w_{tx}]_x, \qquad f' \neq 0, \tag{1}$$

where, f and λ are smooth functions, $w(t,x)$ is the dependent variable and subscripts denote partial derivative with respect to the independent variables t and x. Primes, here and in what follows, denote derivative of a function with respect to the only variable upon which it depends.

Some mathematical questions related to equation (1), as the global existence, uniqueness and stability of solutions can be found in Refs. 1-2. Moreover, shear wave solutions are found in Ref. 3, where some explicit examples of blow up for boundary-values problems with smooth initial data are shown.

When $\lambda(w_x) = \lambda_0$, with λ_0 a positive constant, a symmetry analysis can be found in Refs. 4-5 and for $\lambda_0 = \varepsilon \ll 1$ approximate symmetries were studied in Ref. 6.

Equation (1) can describe the behaviour of one-dimensional viscoelastic medium in which nonlinearities appear not only in the elastic part of the stress but also in the viscoelastic one.

Moreover, equation (1) occurs in the more well-known setting of one-dimensional motion of a viscous isentropic gas, treated from the lagrangian point of view. In fact, by setting $w_x = u$ and $w_t = v$ equation (1) can be written as a 2 x 2 system of the form

$$u_t - v_x = 0, \tag{2}$$

$$v_t - f(u)\, u_x = [\lambda(u)\, v_x]_x\,, \tag{3}$$

where, u corresponds to the specific volume, $p(u) = \int^u f(s)\, ds$ is the pressure and v is the velocity.

The system (2)-(3), as it is well known, is equivalent to the equation (1), consequently, a symmetry of any one of them defines a symmetry of the other. More specifically, because of the nonlocal transformation connecting (1) and (2)-(3), it is possible for a point symmetry of (2)-(3) to yield a contact symmetry of (1) (for details see Ref. 7).

We perform the complete symmetry classification of the equation (1) and the system (2)-(3). After observing that the point symmetries of the system do not produce any contact symmetry of the equation, we are able to demonstrate that the symmetry classifications of the equation (1) and the system (2)-(3) are identical in the sense that, for any f and λ, a point symmetry admitted by (1) induces a point symmetry admitted by (2)-(3) and vice versa.

2. Symmetry classifications

2.1. *Symmetry classification of equation (1)*

In order to investigate on the symmetry classification of equation (1), we apply the classical Lie method and look for the one-parameter Lie group of infinitesimal transformations in (t, x, w)-space given by

$$\hat{t} = t + a\, \xi^1(t, x, w) + \mathcal{O}(a^2), \tag{4}$$

$$\hat{x} = x + a\, \xi^2(t, x, w) + \mathcal{O}(a^2), \tag{5}$$

$$\hat{w} = w + a\, \eta(t, x, w) + \mathcal{O}(a^2), \tag{6}$$

where a is the group parameter and the associated Lie algebra $\mathcal{L}$ is the set of vector fields of the form

$$X = \xi^1 \frac{\partial}{\partial t} + \xi^2 \frac{\partial}{\partial x} + \eta \frac{\partial}{\partial w}. \tag{7}$$

Then, we require that the transformation (4)-(6) leaves invariant the set of solutions of equation (1), in others words, we require that the transformed equation has the same form as the original one. So that, we introduce the third prolongation of the operator X, namely

$$X^{(3)} = X + \zeta_1 \frac{\partial}{\partial w_t} + \zeta_2 \frac{\partial}{\partial w_x} + \zeta_{11} \frac{\partial}{\partial w_{tt}} + \zeta_{22} \frac{\partial}{\partial w_{xx}} + \zeta_{221} \frac{\partial}{\partial w_{xxt}},$$

where we have set

$$\zeta_1 = D_t(\eta) - w_t\, D_t(\xi^1) - w_x\, D_t(\xi^2), \tag{8}$$

$$\zeta_2 = D_x(\eta) - w_t\, D_x(\xi^1) - w_x\, D_x(\xi^2), \tag{9}$$

$$\zeta_{11} = D_t(\zeta_1) - w_{tt}\, D_t(\xi^1) - w_{tx}\, D_t(\xi^2), \tag{10}$$

$$\zeta_{22} = D_x(\zeta_2) - w_{tx}\, D_x(\xi^1) - w_{xx}\, D_x(\xi^2), \tag{11}$$

$$\zeta_{221} = D_t(\zeta_{22}) - w_{xxt} D_t(\xi^1) - w_{xxx} D_t(\xi^2), \tag{12}$$

with operators D_t and D_x denoting total derivatives with respect to t and x.

The invariance condition reads:

$$X^{(3)}\left(w_{tt} - f(w_x)w_{xx} - [\lambda(w_x)w_{tx}]_x\right) = 0, \tag{13}$$

under the constraints that the variable w_{tt} has to satisfy the equation (1).

The *determining system* of (1), arising from the invariance condition (13), leads to the following result:

$$\xi^1 = a_8\, t + a_1, \tag{14}$$

$$\xi^2 = a_5\, x + a_2, \tag{15}$$

$$\eta = a_6\, w + a_7\, x + a_4\, t + a_3, \tag{16}$$

$$[(a_6 - a_5)w_x + a_7]f' + 2(a_8 - a_5)f = 0, \tag{17}$$

$$[(a_6 - a_5)w_x + a_7]\lambda' + (a_8 - 2\, a_5)\lambda = 0, \tag{18}$$

where a_i $(i = 1, 2, ..., 8)$ are constants.

Equations (14)-(18) allow us to find the infinitesimal generator of the symmetry transformations and, at the same time, give the functional dependence of the constitutive functions $f(w_x)$ and $\lambda(w_x)$ for which the equation does admit symmetries.

For arbitrary f and λ, we have that the Principal Lie Algebra $\mathcal{L}_{\mathcal{P}}$ of equation (1) is four-dimensional and it is spanned by the operators

$$X_1 = \frac{\partial}{\partial t}, \quad X_2 = \frac{\partial}{\partial x}, \quad X_3 = \frac{\partial}{\partial w}, \quad X_4 = t\frac{\partial}{\partial w}, \tag{19}$$

otherwise we obtain the symmetry classification summarized in Table 1.

Table 1. Symmetry classification of equation (1). f_0, λ_0, p, q, r and s are constitutive constants with f_0, $\lambda_0 > 0$, and $p, s \neq 0$.

Case	Forms of $f(w_x)$ and $\lambda(w_x)$	Extensions of $\mathcal{L}_\mathcal{P}$
I	$f = f_0\, e^{\frac{w_x}{p}}$ $\lambda = \lambda_0\, e^{\frac{w_x}{s}}$	$X_5 = 2\,(s-p)\,t\,\frac{\partial}{\partial t} + (s-2\,p)\,x\,\frac{\partial}{\partial x}$ $+ \left[(s-2\,p)\,w - 2\,p\,s\,x\right]\frac{\partial}{\partial w}$
II	$f = f_0\,(w_x+q)^{\frac{1}{p}}$ $\lambda = \lambda_0\,(w_x+q)^{\frac{1+r}{p}}$	$X_5 = 2\,r\,t\,\frac{\partial}{\partial t} + (1+2\,r)\,x\,\frac{\partial}{\partial x}$ $+ \{[1+2(p+r)]\,w + 2\,p\,q\,x\}\,\frac{\partial}{\partial w}$

2.2. *Symmetry classification of system (2)-(3)*

When we look for the one-parameter Lie group of infinitesimal transformations of the system (2)-(3) in the (t, x, u, v)-space, the associated Lie algebra $\bar{\mathcal{L}}$ is the set of vector fields of the form

$$\bar{X} = \bar{\xi}^1 \frac{\partial}{\partial t} + \bar{\xi}^2 \frac{\partial}{\partial x} + \eta^1 \frac{\partial}{\partial u} + \eta^2 \frac{\partial}{\partial v}, \tag{20}$$

where the coordinates $\bar{\xi}^1, \bar{\xi}^2, \eta^1, \eta^2$ are functions of t, x, u and v.

Making use of the classical Lie method, from the invariance conditions which follow by applying the second prolongation of the operator $\bar{X}$ to (2)-(3), we give rise to the following result:

$$\bar{\xi}^1 = a_8\, t + a_1, \tag{21}$$

$$\bar{\xi}^2 = a_5\, x + a_2, \tag{22}$$

$$\eta^1 = (a_6 - a_5)u + a_7, \tag{23}$$

$$\eta^2 = (a_6 - a_8)v + a_4, \tag{24}$$

$$[(a_6 - a_5)\,u + a_7]f' + 2\,(a_8 - a_5)f = 0, \tag{25}$$

$$[(a_6 - a_5)u + a_7]\lambda' + (a_8 - 2\,a_5)\lambda = 0. \tag{26}$$

For arbitrary f and λ, the Principal Lie Algebra $\bar{\mathcal{L}}_\mathcal{P}$ of the system (2)-(3) is three-dimensional and it is spanned by the operators

$$\bar{X}_1 = \frac{\partial}{\partial t}, \quad \bar{X}_2 = \frac{\partial}{\partial x}, \quad \bar{X}_3 = \frac{\partial}{\partial v}, \tag{27}$$

otherwise we obtain the symmetry classification summarized in Table 2.

3. Discussions of the symmetry classifications

By inspecting the two classifications obtained in the previous section, we can deduce easily that the point symmetries of the system (2)-(3) do not

Table 2. Symmetry classification of system (2)-(3). f_0, λ_0, p, q, r and s are constitutive constants with f_0, $\lambda_0 > 0$, and $p, s \neq 0$.

Case	Forms of $f(u)$ and $\lambda(u)$	Extensions of $\bar{\mathcal{L}}_{\mathcal{P}}$
I	$f = f_0\, e^{\frac{u}{p}}$ $\lambda = \lambda_0\, e^{\frac{u}{s}}$	$\bar{X}_4 = 2\,(s-p)\,t\,\frac{\partial}{\partial t} + (s-2p)\,x\,\frac{\partial}{\partial x}$ $-2\,p\,s\,\frac{\partial}{\partial u} - s\,v\,\frac{\partial}{\partial v}$
II	$f = f_0\,(u+q)^{\frac{1}{p}}$ $\lambda = \lambda_0\,(u+q)^{\frac{1+r}{p}}$	$\bar{X}_4 = 2\,r\,t\,\frac{\partial}{\partial t} + (1+2\,r)\,x\,\frac{\partial}{\partial x}$ $+2\,p\,(u+q)\,\frac{\partial}{\partial u} + (2\,p+1)\,v\,\frac{\partial}{\partial v}$

produce any contact symmetry of equation (1). Moreover, after observing that the following relations

$$\bar{\xi}^1 = \xi^1 = a_8\, t + a_1, \tag{28}$$

$$\bar{\xi}^2 = \xi^2 = a_5\, x + a_2, \tag{29}$$

hold, we will demonstrate the following statement:

Theorem. *For any f and λ, a point symmetry admitted by (1) induces a point symmetry admitted by (2)-(3) and vice versa.*

In fact:

- starting from the classification of the equation (1), in order to link the coordinates of the operators $\bar{X}$ to that ones of X, as in ref. 4 we require the invariance of the transformations $w_x = u$ and $w_t = v$ with respect to the operator

$$X^* = \eta\frac{\partial}{\partial w} + \zeta_1\frac{\partial}{\partial w_t} + \zeta_2\frac{\partial}{\partial w_x} + \eta^1\frac{\partial}{\partial u} + \eta^2\frac{\partial}{\partial v}. \tag{30}$$

  That is, we require

$$X^*(w_x - u)|_{w_x-u=0,\; w_t-v=0} = 0, \tag{31}$$

$$X^*(w_t - v)|_{w_x-u=0,\; w_t-v=0} = 0. \tag{32}$$

  From (31) and (32) we obtain immediately the other two coordinates of the operator $\bar{X}$, namely

$$\eta^1 = \zeta_2|_{w_x=u} = (a_6 - a_5)\, u + a_7, \tag{33}$$

$$\eta^2 = \zeta_1|_{w_t=v} = (a_6 - a_8)\, v + a_4. \tag{34}$$

- Viceversa, starting from the classification of the system (2)-(3), relations (31) and (32) can be written as

$$\zeta_2 = \eta^1\big|_{u=w_x} = (a_6 - a_5)\, w_x + a_7, \tag{35}$$

$$\zeta_1 = \eta^2\big|_{v=w_t} = (a_6 - a_8)\, w_t + a_4. \tag{36}$$

and taking (8) and (9) into account, we obtain

$$\eta_x + \eta_w\, w_x - a_5\, w_x = (a_6 - a_5)\, w_x + a_7, \tag{37}$$

$$\eta_t + \eta_w\, w_t - a_8\, w_t = (a_6 - a_8)\, w_t + a_4. \tag{38}$$

Relations (37) and (38) must be verified identically, so that the following conditions hold:

$$\eta_w = a_6, \quad \eta_t = a_4, \quad \eta_x = a_7. \tag{39}$$

Then the relation (16) of η follows in a simple way from (39).

4. Conclusions

We have demonstrated that the symmetry classifications of the equation (1) and the system (2)-(3) are identical in the sense that, for any f and λ, a point symmetry admitted by (1) induces a point symmetry admitted by (2)-(3) and vice versa. This implies that, in order to search for invariant solutions of (1) and (2)-(3) it is not necessary to reduce both of them, but it is more convenient to reduce only one of them. Obviously, it is convenient to reduce the equation (1) because of, once obtained its solutions, simply deriving we can obtain the corresponding solutions of (2)-(3).

Acknowledgements

Authors acknowledge the support by M.I.U.R. through P.R.I.N. 2005-2007: *NonLinear Propagation and Stability in Thermodynamical Processes of Continuous Media*, national coordinator Prof. T. Ruggeri, to whom this article is dedicated, on the occasion of his 60th birthday.

References

1. R. C. MacCamy, *Indiana Univ. Math. J.* **20**, 231 (1970).
2. C. M. Dafermos, *J. of Differential Equations* **6**, 71 (1969).
3. K. R. Rajagopal and G. Saccomandi, *Q. Jl Mech. Appl. Math.* **56**, 311 (2003).
4. M. Ruggieri and A. Valenti, *Proceedings of MOGRAN X*, N. H. Ibragimov et al. Eds., 175 (2005).
5. M. Ruggieri and A. Valenti, *Proceedings of WASCOM 2005*, R. Monaco, G. Mulone, S. Rionero and T. Ruggeri Eds., World Sc. Pub., Singapore, 481 (2006).
6. A. Valenti, *Proceedings of MOGRAN X*, N. H. Ibragimov et al. Eds., 236 (2005).
7. L. V. Ovsiannikov, *Group Analysis of Differential Equations*, Academic Press, New York (1982).

HYPERBOLIC MULTI-TEMPERATURE MODEL FOR MIXTURES OF EULER FLUIDS

S. SIMIĆ

Department of Applied Mechanics, Faculty of Technical Sciences
University of Novi Sad, Trg Dositeja Obradovića 6, 21000 Novi Sad, Serbia
E-mail: ssimic@uns.ns.ac.yu

In this review a macroscopic multi-temperature (MT) model for homogeneous mixtures of Euler fluids is proposed. Starting from usual set of governing equations a problem of modelling of homogeneous mixtures is analyzed and MT approach is motivated. Using the principles of extended thermodynamics and the theory of hyperbolic systems of balance laws the following questions are addressed: restriction on constitutive functions, hierarchical structure of hyperbolic subsystems and global existence of smooth solutions. As an application shock structure in binary non-reacting mixture is studied, as well as thermodynamic limit of MT model and different definitions of an average macroscopic temperature.

Keywords: Mixtures of fluids, extended thermodynamics.

1. Introduction

Starting from the model of homogeneous mixtures proposed by Truesdell [1], it is possible to develop several mathematical descriptions based upon different constitutive assumptions. Two main lines of modelling may be observed: one of them relies on the assumption that all the constituents have common — single temperature (ST), while another presumes that every constituent has its own temperature. Although the former assumption is widely adopted and agrees with experimental point of view, the latter is supported by subtle physical argumentation and has its roots both in continuum theories and kinetic theory of gases.

The aim of this study is to support the idea of introducing several temperatures in the theory of homogeneous mixtures. First, a process of modelling will be revisited and a new hyperbolic model will be proposed for mixtures of Euler fluids within the framework of extended thermodynamics. The second part of the review will be concerned with some applications like the shock structure problem in binary mixtures, relation to classical thermodynamics (thermodynamic limit) and a question of appropriate definition of an average temperature of the mixture.

The results presented in this article are obtained in collaboration with Prof. Tommaso Ruggeri, University of Bologna, Italy. Their extensive exposition may be found in our recent papers [2–5].

2. Mathematical modelling of homogeneous mixtures

Theory of homogeneous mixtures is well-established within continuum thermomechanics, as well as in the framework of kinetic theory of gases. In either case basic macroscopic equations are conservation laws of mass, momentum and energy.

2.1. *A route to multi-temperature model*

In order to motivate the introduction of MT model in the mixture theory we shall give a sketch of the modelling process. To that end let us start with the simplest possible approach — model for the "whole mixture"

$$\begin{aligned}
&\frac{\partial \rho}{\partial t} + \operatorname{div}(\rho \mathbf{v}) = 0, \\
&\frac{\partial (\rho \mathbf{v})}{\partial t} + \operatorname{div}(\rho \mathbf{v} \otimes \mathbf{v} - \mathbf{t}) = \mathbf{0}, \\
&\frac{\partial \left(\frac{1}{2}\rho v^2 + \rho\varepsilon\right)}{\partial t} + \operatorname{div}\left\{\left(\frac{1}{2}\rho v^2 + \rho\varepsilon\right)\mathbf{v} - \mathbf{t}\mathbf{v} + \mathbf{q}\right\} = 0.
\end{aligned} \tag{1}$$

In this model macroscopic thermodynamic fields of density ρ, velocity $\mathbf{v}$ and temperature T, $\mathbf{u} = (\rho, \mathbf{v}, T)$, describe the mixture in the "average" sense. When the mixture is treated as neither viscous ($\mathbf{t} = -p\mathbf{I}$, p being the pressure), nor heat-conducting ($\mathbf{q} = \mathbf{0}$), continuum is regarded as Euler fluid and Eq. (1) is reduced to Euler's equations of gas dynamics.

Although well-established, this model cannot cover important phenomena, like diffusion, which are typical for mixtures. To make description of these phenomena possible, the "structure" of the mixture have to be recognized, i.e. n constituents of the mixture ought to have an appropriate contribution to the mathematical model. This contribution is achieved through extension of the list of relevant state variales. This process could be motivated either by physical arguments, or by the need for certain mathematical structure of the model which appropriately correlates with them. As a result, different degrees of accuracy of physical description could be reached, as well as different mathematical levels of approximation.

Refinement of the mixture model could be performed in the following three steps:

(1) introduction of partial densities ρ_α, $\alpha = 1, \ldots, n$, and balance laws of masses for the constituents;
(2) introduction of velocities of constituents $\mathbf{v}_\alpha$ and corresponding balance laws of momenta;
(3) introduction of temperatures of constituents T_α and balance laws of energies.

The first step is typical for thermodynamics of irreversible processes (TIP) [6] and requires introduction of constitutive equations for diffusion fluxes even in the case of Euler fluids. The second step, made in the work of Müller [7], lies in the heart of extended thermodynamics of mixtures: by introducing balance laws for diffusion fluxes it removes the paradox of infinite speed of propagation of diffusion pulses predicted by classical Fick's law. The third step, although looking quite natural in this context, calls for sound physical or mathematical motivation. The idea of several temperatures, however odd it may look at first sight, is naturally imbedded into kinetic theory of gases [8,9]. It had been shown, both in kinetic and continuum approach, that MT assumption is quite reasonable when masses of constituents are disparate, like the masses of ions and electrons in plasmas [10,11].

The MT model of mixtures is finely tuned with the so-called metaphysical principles for mixtures established by Truesdell [1]. In accordance with these principles field variables for the "whole mixture", i.e. ρ, $\mathbf{v}$ and T, should be related to the fields describing the behaviour of constituents. Therefore, the following relations are established

$$\begin{aligned}
&\rho = \sum_{\alpha=1}^{n} \rho_\alpha; \quad \mathbf{v} = \frac{1}{\rho}\sum_{\alpha=1}^{n} \rho_\alpha \mathbf{v}_\alpha; \quad \mathbf{u}_\alpha = \mathbf{v}_\alpha - \mathbf{v}; \quad \left(\sum_{\alpha=1}^{n} \rho_\alpha \mathbf{u}_\alpha = \mathbf{0}\right); \\
&\varepsilon = \varepsilon^I + \frac{1}{\rho}\sum_{\alpha=1}^{n} \frac{1}{2}\rho_\alpha u_\alpha^2; \quad \varepsilon^I = \frac{1}{\rho}\sum_{\alpha=1}^{n} \rho_\alpha \varepsilon_\alpha; \qquad\qquad (2) \\
&\mathbf{t} = \sum_{\alpha=1}^{n} (\mathbf{t}_\alpha - \rho_\alpha \mathbf{u}_\alpha \otimes \mathbf{u}_\alpha); \ \mathbf{q} = \sum_{\alpha=1}^{n} \left\{ \mathbf{q}_\alpha + \rho_\alpha \left(\varepsilon_\alpha + \frac{1}{2}u_\alpha^2\right)\mathbf{u}_\alpha - \mathbf{t}_\alpha \mathbf{u}_\alpha \right\}.
\end{aligned}$$

Although they suffice for derivation the MT model for mixtures, it has to be noted that in (2) there is no explicit relation whatsoever between temperatures of the constituents and the average macroscopic temperature of the mixture. Taking into account Eqs. (2) the set of balance laws for one constituent, say n^{th} one, may be replaced by appropriate conservation laws (1) for the "whole mixture", so that MT model for the mixture now reads

$$
\begin{aligned}
&\frac{\partial\rho}{\partial t} + \operatorname{div}(\rho\mathbf{v}) = 0 \\
&\frac{\partial(\rho\mathbf{v})}{\partial t} + \operatorname{div}(\rho\mathbf{v}\otimes\mathbf{v} - \mathbf{t}) = \mathbf{0} \\
&\frac{\partial\left(\frac{1}{2}\rho v^2 + \rho\varepsilon\right)}{\partial t} + \operatorname{div}\left\{\left(\frac{1}{2}\rho v^2 + \rho\varepsilon\right)\mathbf{v} - \mathbf{t}\mathbf{v} + \mathbf{q}\right\} = 0 \\
&\frac{\partial\rho_b}{\partial t} + \operatorname{div}(\rho_b\mathbf{v}_b) = \tau_b \\
&\frac{\partial(\rho_b\mathbf{v}_b)}{\partial t} + \operatorname{div}(\rho_b\mathbf{v}_b\otimes\mathbf{v}_b - \mathbf{t}_b) = \mathbf{m}_b \\
&\frac{\partial\left(\frac{1}{2}\rho_b v_b^2 + \rho_b\varepsilon_b\right)}{\partial t} + \operatorname{div}\left\{\left(\frac{1}{2}\rho_b v_b^2 + \rho_b\varepsilon_b\right)\mathbf{v}_b - \mathbf{t}_b\mathbf{v}_b + \mathbf{q}_b\right\} = e_b
\end{aligned}
\tag{3}
$$

for $b = 1, \dots, n-1$. Source terms τ_α, $\mathbf{m}_\alpha$ and e_α reflect mutual interactions of the constituents and satisfy the conditions

$$
\sum_{\alpha=1}^{n}\tau_\alpha = 0; \qquad \sum_{\alpha=1}^{n}\mathbf{m}_\alpha = 0; \qquad \sum_{\alpha=1}^{n}e_\alpha = 0. \tag{4}
$$

Analysis of this model is the main issue of this study.

2.2. *Structure of the multi-temperature model*

Several questions could be raised with regard to the structure of mathematical model — closure problem being one of the most important. The answers will be searched for within the framework of rational extended thermodynamics [12] by exploiting basic principles of this theory.

Extended thermodynamics of MT mixtures. The set of usual state variables ρ, $\mathbf{v}$ and T is extended with ones describing the behaviour of the constituents, $\mathbf{u} = (\rho, \mathbf{v}, T, c_b, \mathbf{u}_b, \Theta_b)$, $\Theta_b = T_n - T_b$. Then, typically for extended thermodynamics, processes are described by balance laws $(3)_{4-6}$ implying the existence of source terms. Also, the local character of constitutive equations is imposed leading to the set of quasi-linear first order PDE's.

Two important restrictions have to be imposed to the model in this context. The first is invariance of balance laws with respect to Galilean transformations which reveals velocity dependence of source terms

$$\tau_b = \hat{\tau}_b; \quad \mathbf{m}_b = \hat{\tau}_b \mathbf{v} + \hat{\mathbf{m}}_b; \quad e_b = \hat{\tau}_b \frac{v^2}{2} + \hat{\mathbf{m}}_b \cdot \mathbf{v} + \hat{e}_b; \tag{5}$$

hat denotes terms which depend on objective quantities only. The second is compatibility with entropy inequality with convex entropy which determines the structure of source terms through residual inequality

$$\hat{\tau}_b = -\sum_{c=1}^{n-1} \varphi_{bc}(\mathbf{w}) \left(\frac{\mu_c - \frac{1}{2}u_c^2}{T_c} - \frac{\mu_n - \frac{1}{2}u_n^2}{T_n} \right); \tag{6}$$

$$\hat{\mathbf{m}}_b = -\sum_{c=1}^{n-1} \psi_{bc}(\mathbf{w}) \left(\frac{\mathbf{u}_c}{T_c} - \frac{\mathbf{u}_n}{T_n} \right); \quad \hat{e}_b = -\sum_{c=1}^{n-1} \theta_{bc}(\mathbf{w}) \left(-\frac{1}{T_c} + \frac{1}{T_n} \right),$$

where $\varphi_{bc}(\mathbf{w})$, $\psi_{bc}(\mathbf{w})$ and $\theta_{bc}(\mathbf{w})$ are positive semi-definite matrix functions of objective quantities $\mathbf{w}$ and μ_c are chemical potentials of the constituents.

In the sequel our attention will be focused on the mixture of Euler fluids, $\mathbf{t}_\alpha = -p_\alpha \mathbf{I}$, $\mathbf{q}_\alpha = \mathbf{0}$, where pressure of the mixture is determined by Dalton's law

$$p = \sum_{\alpha=1}^{n} p_\alpha. \tag{7}$$

Hierarchy of hyperbolic subsystems. Compatibility of the model with the entropy inequality is achieved by means of Lagrange multipliers [13] and construction of the main field $\mathbf{u}' = (\Lambda^\rho, \Lambda^\mathbf{v}, \Lambda^\varepsilon, \Lambda^{\rho_b}, \Lambda^{\mathbf{v}_b}, \Lambda^{\varepsilon_b})$. Main field may be used for transformation of governing equations into symmetric hyperbolic form which reveals hierarchical—nested structure of hyperbolic subsystems. To that end, the most important components are the ones corresponding to non-equilibrium variables

$$\begin{aligned} \Lambda^{\rho_b} &= \frac{1}{T_b}\left(\mu_b - \frac{1}{2}(\mathbf{u}_b + \mathbf{v})^2\right) - \frac{1}{T_n}\left(\mu_n - \frac{1}{2}(\mathbf{u}_n + \mathbf{v})^2\right); \\ \Lambda^{\mathbf{v}_b} &= \frac{\mathbf{u}_b}{T_b} - \frac{\mathbf{u}_n}{T_n} - \left(\frac{1}{T_b} - \frac{1}{T_n}\right)\mathbf{v}; \\ \Lambda^{\varepsilon_b} &= \frac{1}{T_n} - \frac{1}{T_b}. \end{aligned} \tag{8}$$

Assuming one or more main field components equal zero throughout the process, one may recognize the subset of governing equations which could be dropped leading to a principal subsystem [14]. The following results could be immediately obtained

(i) if $\Lambda^{\varepsilon_b} = 0$, i.e. when MT assumption is dropped $T_1 = \ldots = T_n = T$, energy balance laws could be discarded from the system and single-temperature model is obtained as a principal subsystem of a MT one;
(ii) if the diffusion between constituents is also neglected, $\mathbf{u}_b = \mathbf{0}$ leading to $\Lambda^{\mathbf{v}_b} = 0$, momentum balance laws may be discarded and principal subsystem characteristic for TIP is obtained.

If it is assumed that the mixture is consisted of non-reacting constituents, $\tau_\alpha = 0$, then preceding assumptions lead to an equilibrium subsystem. On the other hand, in the case of chemically reacting mixture conditions which lead to an equilibrium subsystem cannot be derived from Lagrangian multipliers in a straightforward manner, i.e. conditions $\Lambda^{\rho_b} = 0$ have to be adjoined with the law of mass action. In the sequel only non-reacting mixtures will be analyzed.

Qualitative analysis. Recently a group of results appeared discussing global existence of smooth solutions [15,16] and stability of constant states [17]. These qualitative results are deeply based upon Shizuta-Kawashima condition (K-condition) of genuine coupling. It is worth to note that under certain assumptions K-condition may be violated in ST mixture models. On the other hand, it satisfied in MT model without restrictions, providing another argument in favor of multi-temperature assumption.

3. Applications of hyperbolic multi-temperature model

Any theoretical model, however good it may be, need to be tested to prove its validity. Potential of the proposed hyperbolic model of MT mixtures (3) will be enlighted through the following three problems: shock structure in binary mixtures of Euler fluids, Maxwellian iteration and thermodynamic limit and discussion of average temperature of the mixture.

3.1. *Shock structure in binary mixtures of Euler fluids*

The simplest possible way to test the mixture model is to analyze binary mixture. Governing equations (3) will be rearranged using the vector of diffusion flux $\mathbf{J} = \rho_1 \mathbf{u}_1 = -\rho_2 \mathbf{u}_2$ and concentration variable $c = \rho_1/\rho$.

To close the system we shall assume that constituents are perfect gases which obey classical thermal and caloric equations of state $p_\alpha = (k/m_\alpha)\rho_\alpha T_\alpha$, $\varepsilon_\alpha = (k/m_\alpha)T_\alpha/(\gamma_\alpha - 1)$. Furthermore, we shall assume that mixture as a whole could be described by constitutive functions of the same form, i.e. $p = (k/m)\rho T$, $\varepsilon_I = (k/m)T/(\gamma - 1)$, where m, T and γ are average atomic mass, average temperature and average ratio of specific heats of the mixture, respectively. In such a way m, T and γ could be expressed form Eqs. (2) and (7)

$$T = c\frac{m}{m_1}T_1 + (1-c)\frac{m}{m_2}T_2; \tag{9}$$
$$\frac{1}{m} = \frac{c}{m_1} + \frac{1-c}{m_2}; \qquad \frac{1}{\gamma - 1} = \frac{c}{\gamma_1 - 1}\frac{m}{m_1}\frac{T_1}{T} + \frac{1-c}{\gamma_2 - 1}\frac{m}{m_2}\frac{T_2}{T}.$$

The set of of state variables is completed with the difference of temperatures of the constituents $\Theta = T_2 - T_1$, so the state vector reads $\mathbf{u} = (\rho, \mathbf{v}, T, c, \mathbf{J}, \Theta)$.

The structure of source terms is fixed by exploiting the results of kinetic theory [10]. Along with Eqs. (5)-(6) they provide the following relations

$$m_1 = \hat{m}_1 = -\frac{\rho_0 c_0 (1-c_0) T_0}{\tau_J}\left(\frac{1}{cT_1} - \frac{1}{(1-c)T_2}\right)\frac{J}{\rho}; \tag{10}$$
$$e_1 = \hat{m}_1 v + \hat{e}_1; \quad \hat{e}_1 = \frac{k}{m_1(\gamma_1 - 1)}\frac{\rho_0 c_0 T_0^2}{\tau_T}\frac{\Theta}{T_1 T_2},$$

where τ_J and τ_T are relaxation times for diffusion and temperature related by

$$\tau_T = \frac{2}{3(\gamma_1 - 1)}\frac{m_1 + m_2}{(1-c_0)m_1}\tau_J. \tag{11}$$

Interestingly enough, (11) implies that diffusion processes are attenuated faster than thermal non-equilibrium for a large class of mixtures.

Assuming the profile of the shock wave in the form of plane traveling wave $\mathbf{u} = \mathbf{u}(\xi) = \mathbf{u}(x - st)$, $\xi = x - st$ governing equations (3) are reduced to a set of ordinary differential equations. Shock structure then represents a heteroclinic orbit which asymptotically connects stationary points of the system. Problem is solved numerically for He-Ar and He-Xe mixtures for $c_0 = 0.3$ and Mach number $M_0 = 1.5$ in upstream equilibrium state.

Numerical solution, shown in Fig. 1, reveals that non-equilibrium profile of Θ is wider than the profile of c and J which is in accordance with the ratio of relaxation times, $(\tau_T/\tau_J)_{\text{He-Ar}} = 15.7$, $(\tau_T/\tau_J)_{\text{He-Xe}} = 48.3$. Average temperature is monotonically increasing function within the profile, whereas temperature of the heavier component has an overshoot—there is a region within a profile where its values are greater that downstream equilibrium temperature, the result typical for MT models. It can also be observed that greater mass difference implies stronger thermal imbalance so that it can be viewed as a main driving agent for temperature difference.

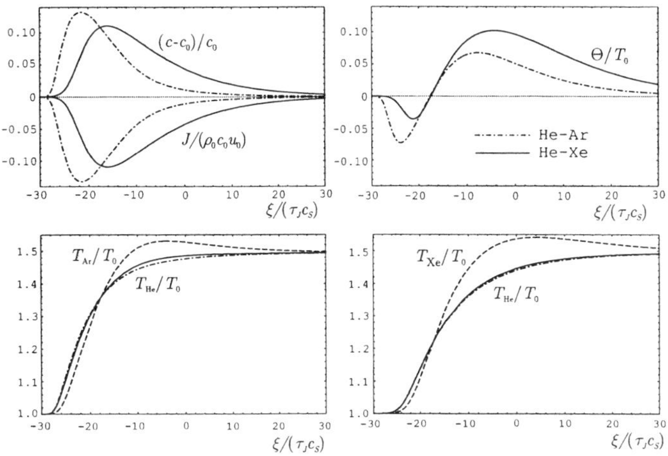

Fig. 1. Shock profiles in binary mixture: $u = v - s$, $c_S = \sqrt{\gamma_0 (k/m_0) T_0}$ sound speed in upstream equilibrium state.

3.2. *Thermodynamic limit of MT model*

A natural question could be raised about relation of extended model (3) to classical non-local constitutive equations. In this survey the answer will be obtained through Maxwellian iteration—an iterative procedure which could be regarded as expansion in powers of relaxation times. In the case of binary mixture these results have been recently presented by Ruggeri & Simic. It was shown that first iterates for diffusion flux and temperature difference have the following form

$$\mathbf{J}^{(1)} = L_{11} \operatorname{grad}\left(\frac{\mu_2 - \mu_1}{T}\right) + L \operatorname{grad}\left(\frac{1}{T}\right);$$
$$\Theta^{(1)} = L_\theta \left(\gamma_1 - \gamma_2\right) \operatorname{div}\mathbf{v}, \tag{12}$$

where L_{11}, L and L_θ are phenomenological coefficients. Diffusion flux $\mathbf{J}^{(1)}$ reproduces generalized Fick's law of diffusion, whereas $\Theta^{(1)}$ represents entirely new result showing that temperature difference is proportional to $\operatorname{div}\mathbf{v}$ in the first iteration. It is a matter of direct calculation to generalize this conclusion to multi-component mixtures of Euler fluids.

3.3. *Average temperature*

The necessity for defining an average macroscopic temperature in MT model of mixtures comes from the lack of possibility for measuring the temperatures of the constituents. Average temperature (9) has been defined with intention to express the constitutive equations for the mixture in the same form as for the constituents. However, this choice is not unique and other definitions of average temperature could also be put forward. One of the most striking motifs can be drawn from the analysis of homogeneous processes in binary mixture of Euler fluids. In this case spatial variations of state variables are neglected and governing equations are reduced to

$$\begin{aligned} \frac{d\rho_1}{dt} &= 0; \qquad \rho_1 \frac{d\mathbf{v}_1}{dt} = \mathbf{m}_1; \qquad \frac{d}{dt}\left(\frac{1}{2}\rho_1 v_1^2 + \rho_1\varepsilon_1\right) = e_1; \\ \frac{d\rho_2}{dt} &= 0; \qquad \rho_2 \frac{d\mathbf{v}_2}{dt} = -\mathbf{m}_1; \qquad \frac{d}{dt}\left(\frac{1}{2}\rho_2 v_2^2 + \rho_2\varepsilon_2\right) = -e_1. \end{aligned} \tag{13}$$

From Eqs. (13) one obtains conservation laws $\rho_1 = \text{const.}$, $\rho_2 = \text{const.}$ and $\rho\mathbf{v} = \text{const.}$, where $\mathbf{v} = \mathbf{0}$ can be put without loss of generality due to Galilean invariance. Using equations of state for ideal gas, Eqs. $(13)_{3,6}$ lead to energy conservation equation

$$\frac{1}{2}\rho_1 v_1^2 + \frac{1}{2}\rho_2 v_2^2 + \frac{k}{m_1}\frac{\rho_1 T_1}{(\gamma_1 - 1)} + \frac{k}{m_2}\frac{\rho_2 T_2}{(\gamma_2 - 1)} = E = \text{const.}, \tag{14}$$

where E is evaluated for the following equilibrium conditions $\mathbf{v}_1 = \mathbf{v}_2 = \mathbf{0}$, $T_1 = T_2 = T_0$

$$E = k\left(\frac{1}{m_1}\frac{\rho_1}{\gamma_1 - 1} + \frac{1}{m_2}\frac{\rho_2}{\gamma_2 - 1}\right)T_0. \tag{15}$$

Neglecting quadratic diffusive terms for processes not far from equilibrium, the following definition of average temperature could be derived from Eqs. (14)-(15)

$$T = T_0 = \frac{\rho_1 m_2(\gamma_2 - 1)T_1 + \rho_2 m_1(\gamma_1 - 1)T_2}{\rho_1 m_2(\gamma_2 - 1) + \rho_2 m_1(\gamma_1 - 1)}. \tag{16}$$

Two remarks have to be given. First, for processes near equilibrium, where higher order terms could be neglected, an average temperature (16) remains constant throughout the spatially homogeneous process. Second, this definition came from the energy equation, rather than from Dalton's law (7). In fact, it come from the assumption that intrinsic values of internal energy in MT and ST models are the same, $\varepsilon_{MT}^I = \varepsilon_{ST}^I$. This idea could be generalized to a definition of average temperature for multi-component mixtures

$$T = \frac{\sum_{\alpha=1}^{n} \frac{\rho_\alpha}{m_\alpha(\gamma_\alpha - 1)} T_\alpha}{\sum_{\beta=1}^{n} \frac{\rho_\beta}{m_\beta(\gamma_\beta - 1)}}. \tag{17}$$

This idea will be further developed in our forthcoming studies.

Acknowledgment

This paper is dedicated to Professor Tommaso Ruggeri, teacher and friend. Its preparation was supported by the Ministry of Science of Serbia within the project *Contemporary Problems of Mechanics of Deformable Bodies.*

References

1. C. Truesdell, *Rational Thermodynamics* (McGraw-Hill, New York, 1969).
2. T. Ruggeri and S. Simić, *Math. Methods Appl. Sci.* **30**, 827 (2007).
3. S. Simić and T Ruggeri, Shock Structure in a Hyperbolic Mixture of Binary Mixture of Non-Reacting Gases, in *Proc. of 1st International Congress of Serbian Society of Mechanics*, (Kopanoik, Serbia, 2007).
4. T. Ruggeri and S. Simić, Mixture of Gases with Multi-temperature: Maxwellian Iteration, in *Asymptotic Methods in Non Linear Wave phenomena*, eds. T. Ruggeri and M. Sammartino (World Scientific, Singapore, 2007), pp.186-194.
5. T. Ruggeri and S. Simić, Mixture of Gases with Multi-temperature: Identification of a Macroscopic Average Temperature, *Memorie dell'Accademia delle Scienze, Lettere ed Arti di Napoli*, Liguori Editore, Napoli (*in press*).
6. S.R. de Groot and P. Mazur, *Non-Equilibrium Thermodynamics* (North-Holland, Amsterdam, 1962).
7. I. Müller, *Arch. Rational Mech Anal.* **28**, 1 (1968).
8. S.C. Chapman and T.G. Cowling, *The Mathematical Theory of Non-Uniform Gases* (Cambridge University Press, London, 1961).
9. J.M. Burgers, *Flow Equations for Composite Gases* (Academic Press, New York, 1969).
10. T.K. Bose, *High-temperature gas dynamics* (Springer-Verlag, Berlin, 2004).
11. Ya.B. Zel'dovich and Yu.P. Raizer, *Physics of Shock Waves and High-Temperature Hydrodynamic Phenomena* (Dover Publications, Mineola, 2002).
12. I. Müller and T. Ruggeri, *Rational Extended Thermodynamics* (Springer-Verlag, New York, 1998).
13. I-S. Liu, *Arch. Rational Mech. Anal.* **46**, (1972).
14. G. Boillat and T. Ruggeri, *Arch. Rational Mech. Anal.* **137**, 305 (1997).
15. B. Hanouzet and R. Natalini, *Arch. Ration. Mech. Anal.* **163**, 89 (2003).
16. W.-A. Yong, *Arch. Ration. Mech. Anal.* **172**, 247 (2004).
17. T. Ruggeri and D. Serre, *Quart. Appl. Math* **62**, 163 (2004).

THE INCOMPRESSIBILITY ASSUMPTION ASSESSMENT IN FAST FLUIDIZED BED WAVE PROPAGATION

A. SORIA*, J. R. G. SÁNCHEZ-LÓPEZ and E. M. SALINAS-RODRÍGUEZ

Departamento de I. P. H., Universidad Autónoma Metropolitana-Iztapalapa, Apdo. 09340 México, D. F., Mexico

* *e-mail: asor@xanum.uam.mx*

A two-fluids linearized model was expressed for incompressible and compressible assumptions. The amplitude comparison showed significant incompressible first order waves and compressible second and third order waves with propagation speeds closely related to both the mixture flow velocity and the sound speed in the disperse medium.

Keywords: Compressibility effect; Fast fluidization; Pressure wave propagation; Sound speed; Wave hierarchy.

1. Introduction

Circulating fluidized beds (CFB) operate under the fast fluidization regime where solid particles are carried up by the fluid. In this regime the terminal velocity of the individual particles is surpassed by the up-flowing carrier fluid velocity and a continuous two-phase mixture flow is produced at the top exit. The usual fast fluidization models for incompressible fluid are improved by including a solid effective compressibility modulus as a part of the solid pressure force[1] . This modulus turns second order propagation speeds real, curing in this way the model ill-posedness. The compressible model is well-posed, even without the inclusion of this modulus.

In this study an incompressible and a compressible model were developed, within the same framework. The general model was introduced in Section 2, the method used was established in Section 3. The application was performed in Section 4, on an FCC catalyst dragged by water vapor at isothermal conditions and in Section 5 the main conclusions were stated.

2. Modeling the gas-solid flow in the riser

A simplified one-dimensional hydrodynamic behavior of powder swarms in a gas inside a vertical riser can be described by the following set of equations: The vapor and the solid phases continuity equations are

$$\frac{\partial}{\partial t}\left[\varepsilon_g \rho_g(p_g)\right] + \frac{\partial}{\partial z}\left[\varepsilon_g \rho_g(p_g) v_g\right] = 0, \tag{1}$$

$$-\rho_{s0}\frac{\partial}{\partial t}\varepsilon_g + \rho_{s0}\frac{\partial}{\partial z}\left[(1-\varepsilon_g)v_s\right] = 0, \tag{2}$$

respectively and the vapor and the solid phases momentum equations are

$$\begin{aligned}&\frac{\partial}{\partial t}\left[\varepsilon_g \rho_g(p_g) v_g\right] + \frac{\partial}{\partial z}\left[\varepsilon_g \rho_g(p_g) v_g^2\right] + \varepsilon_g\frac{\partial}{\partial z}p_g \\ &+\beta\varepsilon_g(1-\varepsilon_g)(v_g - v_s) + \beta_{gw}\rho_g(p_g)v_g^2 + \rho_g(p_g)g = 0,\end{aligned} \tag{3}$$

$$\begin{aligned}&\rho_{s0}\frac{\partial}{\partial t}\left[(1-\varepsilon_g)v_s\right] + \rho_{s0}\frac{\partial}{\partial z}\left[(1-\varepsilon_g)v_s^2\right] + (1-\varepsilon_g)\frac{\partial}{\partial z}p_g \\ &-\Phi'(1-\varepsilon_g)\frac{\partial}{\partial z}\varepsilon_g - \beta\varepsilon_g(1-\varepsilon_g)(v_g - v_s) + \beta_{sw}\rho_{s0}v_s^2 \\ &+(1-\varepsilon_g)\left[\rho_{s0} - \rho_g(p_g)\right]g = 0,\end{aligned} \tag{4}$$

respectively. Here $\rho_g(p_g)$ is the gas density, assumed to be known as a function of the gas pressure p_g, through a thermodynamic state equation, while ρ_{s0} is the (constant) solid density. ε_g and ε_s are the gas and solid volume fractions, respectively, where $\varepsilon_g + \varepsilon_s = 1$. v_g and v_s are the gas and solid average velocities, respectively. $\Phi' = d\Phi/d\varepsilon_s$ is the solid effective compressibility modulus derivative, $\beta = (3/4d_p)C_D\rho_g U_t$ is an interfacial drag coefficient while $\beta_{aw} = (2/D_t)\varepsilon_a f_{aw}$ is a wall drag coefficient for $a = g$, s. Considering that β, β_{gw} and β_{sw} are computed in a base state by standard methods, the balances (1) to (4) have a set of 5 unknowns and become a closed set when the function Φ' is specified, as done here by requiring the second order propagation speeds of the incompressible model be real. Therefore, Eqs. (1) to (4) can be taken as a closed set for the variables $\{\varepsilon_g, p_g, v_g, v_s\}$. This set can be linearized around a uniform base state, denoted by a zero subindex, such that $\varepsilon_g = \varepsilon_0 + \varepsilon$, $p_g = p_0 + p$, $v_g = v_{g0} + v_g$ and $v_s = v_{s0} + v_s$. Once the linearization is done the symbols for the velocities v_g and v_s are used for the correspondent perturbations. This gives the set of linearized equations

$$\mathbf{B}\frac{\partial \mathbf{u}}{\partial t} + \mathbf{C}\frac{\partial \mathbf{u}}{\partial z} + \mathbf{D}\mathbf{u} = \mathbf{0}, \tag{5}$$

where $\mathbf{u} = (\varepsilon \ \ p \ \ v_g \ \ v_s)^T$ is the perturbation vector and $\mathbf{B}$, $\mathbf{C}$, $\mathbf{D}$ are 4×4 coefficient matrices with elements $b_{11}, b_{12}, b_{21}, b_{33}, b_{44}$, $c_{11}, c_{12}, c_{13}, c_{21}, c_{24}, c_{32}, c_{33}, c_{41}, c_{42}, c_{44}$ and d_{jk}; $j = 3, 4$; $k = 1, ..., 4$ different from zero. The model (5) can be specialized for incompressible gas flow, where coefficients b_{12}, c_{12}, d_{32} and d_{42} become zero and for compressible gas flow, where the derivative $d\rho_g/dp_g$ is equal to s^{-2}; being s the vapor adiabatic sound speed.

3. Wave hierarchies for incompressible and compressible models

Assuming waving solutions for $\mathbf{u}$ in such a way that $\mathbf{u} = \mathbf{a}e^{i(\kappa z - \omega t)}$, Eq. (5) is transformed to $(-i\omega\mathbf{B} + i\kappa\mathbf{C} + \mathbf{D})\mathbf{u}{=}\mathbf{0}$. Performing similar algebraic operations on incompressible and compressible models, in order to get comparable higher degree polynomials on ε, the incompressible wave equation in the transformed domain can be set as

$$f_I(\omega, \kappa)\varepsilon \doteq [c_{32}f_2(\omega, \kappa) - c_{42}f_1(\omega, \kappa)]\,\varepsilon = 0 \tag{6}$$

and the compressible wave equation as

$$f_C(\omega, \kappa)\varepsilon \doteq [f_4(\omega, \kappa)f_I(\omega, \kappa) + f_2(\omega, \kappa)f_P(\omega, \kappa)]\,\varepsilon = 0 \tag{7}$$

where $f_I(\omega, \kappa)$ is also involved in Eq. (7) and where

$$\begin{aligned} f_1(\omega, \kappa) &= b_{11}b_{33}c_{24}\omega^2 - (b_{33}c_{11}c_{24} + b_{11}c_{24}c_{33})\omega\kappa + (c_{11}c_{24}c_{33})\kappa^2 \\ &\quad + (b_{11}c_{24}d_{33} + b_{21}c_{13}d_{34})i\omega \\ &\quad + (c_{13}c_{24}d_{31} - c_{11}c_{24}d_{33} - c_{13}c_{21}d_{34})i\kappa, \end{aligned} \tag{8}$$

$$\begin{aligned} f_2(\omega, \kappa) &= b_{21}b_{44}c_{13}\omega^2 - (b_{44}c_{13}c_{21} + b_{21}c_{13}c_{44})\omega\kappa \\ &\quad + (c_{13}c_{21}c_{44} - c_{13}c_{24}c_{41})\kappa^2 + (b_{11}c_{24}d_{43} + b_{21}c_{13}d_{44})i\omega \\ &\quad + (c_{13}c_{24}d_{41} - c_{11}c_{24}d_{43} - c_{13}c_{21}d_{44})i\kappa, \end{aligned} \tag{9}$$

$$\begin{aligned} f_3(\omega, \kappa) &= -b_{12}b_{33}\omega^2 - (b_{33}c_{12} - b_{12}c_{33})\omega\kappa + (c_{13}c_{32} + c_{12}c_{33})\kappa^2 \\ &\quad - d_{33}b_{12}i\omega - (c_{12}d_{33} + c_{13}d_{32})i\kappa, \end{aligned} \tag{10}$$

$$f_4(\omega, \kappa) = c_{13}c_{42}\kappa^2 - b_{12}d_{43}i\omega - (c_{13}d_{42} + c_{12}d_{43})i\kappa, \tag{11}$$

$$f_P(\omega, \kappa) = c_{42}f_3(\omega, \kappa) - c_{32}f_4(\omega, \kappa). \tag{12}$$

The correspondent incompressible and compressible wave equations in space-time domain are

$$\begin{aligned} f_I(t,z)\varepsilon &= A_{I2}\Big(\frac{\partial}{\partial t} + C_{21}\frac{\partial}{\partial z}\Big)\Big(\frac{\partial}{\partial t} + C_{22}\frac{\partial}{\partial z}\Big)\varepsilon \\ &\quad + A_{I1}\Big(\frac{\partial}{\partial t} + C_{11}\frac{\partial}{\partial z}\Big)\varepsilon = 0 \end{aligned} \tag{13}$$

$$f_C(t,z)\varepsilon = \\ A_{C4}\Big(\frac{\partial}{\partial t}+\eta_{41}\frac{\partial}{\partial z}\Big)\Big(\frac{\partial}{\partial t}+\eta_{42}\frac{\partial}{\partial z}\Big)\Big(\frac{\partial}{\partial t}+\eta_{43}\frac{\partial}{\partial z}\Big)\Big(\frac{\partial}{\partial t}+\eta_{44}\frac{\partial}{\partial z}\Big)\varepsilon \\ + A_{C3}\Big(\frac{\partial}{\partial t}+\eta_{31}\frac{\partial}{\partial z}\Big)\Big(\frac{\partial}{\partial t}+\eta_{32}\frac{\partial}{\partial z}\Big)\Big(\frac{\partial}{\partial t}+\eta_{33}\frac{\partial}{\partial z}\Big)\varepsilon \\ + A_{C2}\Big(\frac{\partial}{\partial t}+\eta_{21}\frac{\partial}{\partial z}\Big)\Big(\frac{\partial}{\partial t}+\eta_{22}\frac{\partial}{\partial z}\Big)\varepsilon = 0, \tag{14}$$

whose correspondent amplitudes, from Eqs. (6) and (7) are given by

$$A_{I2} = b_{11}b_{33}c_{24}c_{42} - b_{21}b_{44}c_{13}c_{32}, \tag{15}$$

$$A_{I1} = c_{42}(b_{11}c_{24}d_{33} + b_{21}c_{13}d_{34}) - c_{32}(b_{11}c_{24}d_{43} + b_{21}c_{13}d_{44}), \tag{16}$$

$$A_{C4} = -\frac{b_{21}c_{13}c_{42}}{d_{43}}b_{33}b_{44}, \tag{17}$$

$$A_{C3} = A_{I2} + b_{21}b_{44}c_{13}\Big(c_{32} - c_{42}\frac{d_{33}}{d_{43}}\Big) - b_{33}c_{42}\Big(b_{11}c_{24} + b_{21}c_{13}\frac{d_{44}}{d_{43}}\Big), \tag{18}$$

$$A_{C2} = A_{I1} + \Big(c_{32} - c_{42}\frac{d_{33}}{d_{43}}\Big)\Big(b_{11}c_{24}d_{43} + b_{21}c_{13}d_{44}\Big), \tag{19}$$

C_{ij} and η_{ij} are propagation speeds. A_{C3} is related to A_{I2} and A_{C2} is related to A_{I1}, as can be seen in Eqs. (18) and (19). In order to perform quantitative comparisons, a length scale L and a velocity scale U_m are introduced in order to define dimensionless independent variables and the following dimensionless amplitudes:

$$\widehat{A}_{I1} = \Big(\frac{L}{U_m}\Big)^2\frac{A_{I1}}{A_{C4}};\quad \widehat{A}_{I2} = \frac{L}{U_m}\frac{A_{I2}}{A_{C4}}; \\ \widehat{A}_{C2} = \Big(\frac{L}{U_m}\Big)^2\frac{A_{C2}}{A_{C4}};\quad \widehat{A}_{C3} = \frac{L}{U_m}\frac{A_{C3}}{A_{C4}}. \tag{20}$$

4. Discussion and results

The previous procedure is applied to an FCC catalyst carried up by a turbulent vapor flow ($Re_g \sim 2 \times 10^5$). The catalyst grains are assumed spherical and uniform size $d_p = 60$ μm. Since Re_s, based on the solid particle terminal velocity, is $Re_s \sim 0.12$, a local Stokes regime is occurring. A Stokes number $St \sim 0.13$ means that the catalyst motion is governed by the gas drag.

4.1. *Amplitudes*

Dimensionless amplitudes, Eqs. (20), are shown in Fig. 1(a) as functions of the solid volume fraction. A normalized $\widehat{A}_{C4} = 1$ was selected, so it is apparent that the fourth order wave behavior is at least 10^4 times smaller than all other wave hierarchies and can be neglected in the compressible wave model. On the other side, $\widehat{A}_{I1}$ is three orders of magnitude greater than the other wave amplitudes and $\widehat{A}_{I2}$ accounts only for the 0.274 % of the dynamical response. Therefore, the incompressibility assumption drives the model towards an overwhelming first order wave behavior.

The compressible wave behaviors are richer, since both $\widehat{A}_{C2}$ and $\widehat{A}_{C3}$ are same order of magnitude. They are also proportional since their ratio $\widehat{A}_{C2}/\widehat{A}_{C3}$ shows a main region, for $0.04 \leq \varepsilon_{s0} < 0.5$, where there is a linear behavior, thus both wave hierarchies should be taken into account.

4.2. *Propagation wave speeds*

The incompressible model propagation speeds remain close to the mixture mean flow velocity U_m. C_{11} is close to v_{s0}, C_{21} was set equal to C_{22} by selecting a solid compressibility modulus behavior $\Phi'(\varepsilon_0)$ that gives a zero discriminant for the relevant second degree polynomial roots. Other approaches to define the $\Phi'(\varepsilon_0)$ functions should give different C_{21} and C_{22} real propagation speed values. Nevertheless, the incompressible second order wave contribution is less than 0.3 %, as stated above. The set of compressible wave propagation speeds is shown in Fig. 1(b), where η_{22} and η_{32} lie between the gas and solid flow velocities for most of the domain.

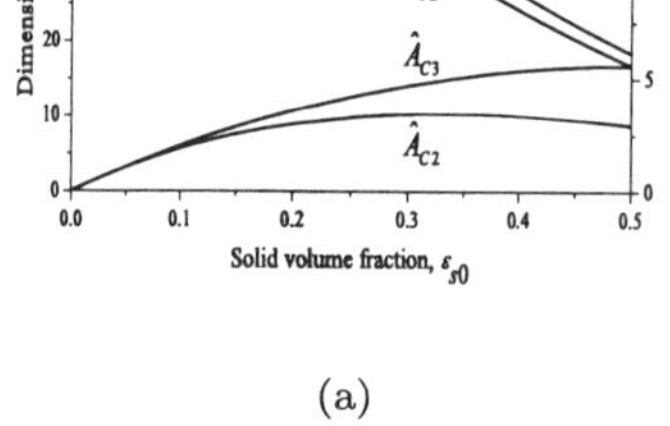

(a)

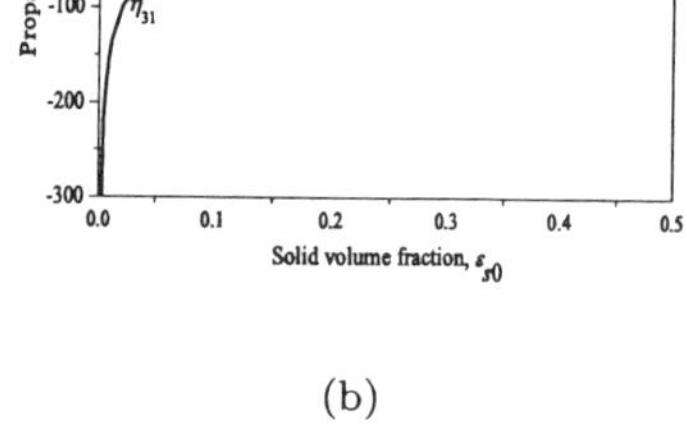

(b)

Fig. 1. (a) Dimensionless wave amplitudes for incompressible and compressible models. (b) Compressible model second and third order propagation speeds.

The compressible third order speeds η_{31} and η_{33} are the lower and upper bounds and can be understood as the sound speed propagating downwards and upwards in the bed, relative to U_m. The speed η_{21} is highly dependent upon the solid phase drag on the wall. A simplified expression for this speed, within 1 % error for $10^{-4} < \varepsilon_{s0} < 0.5$ is:

$$\eta_{21} \approx \frac{v_{g0} + v_{s0}}{2} - \frac{\varepsilon_{s0} g}{2 v_{s0} \beta_{sw}}. \tag{21}$$

It should be remarked that the incompressible model propagation speeds reappear into the compressible ones, thus $\eta_{32} = C_{11}$, $\eta_{42} = C_{21}$ and $\eta_{43} = C_{22}$. Also, the eigenvalues from the characteristic equation $\det(\mathbf{F} - \lambda \mathbf{1}) = \mathbf{0}$, where $\mathbf{F} = \mathbf{B}^{-1}\mathbf{C}$ [*c. fr.* Eq. (5)], are coincident with the compressible fourth order propagation speeds, $\lambda_{(j)} = \eta_{4j}$, $j = 1, ..., 4$. Therefore, the fourth order polynomial in the present approach is equivalent to the characteristic polynomial. Nevertheless, the compressible fourth order wave contribution is 10^4 times smaller than the second and third order ones.

5. Conclusions

The fluid incompressibility assumption in gas-solid fast fluidized beds drives the linear propagation analysis to significant first order waves with a propagation speed close to the solid flow velocity. The compressibility assumption renders second and third order wave hierarchies with competitive amplitudes and propagation speeds closely related to both the mixture flow velocity and the sound speed in the disperse medium. The solid phase drag on the wall introduces a second order propagation speed. The incompressible model propagation speeds reappear as some of the compressible and the fourth order propagation speeds are coincident with the eigenvalues of the characteristic polynomial.

Acknowledgements

The authors highly acknowledge the Consejo Nacional de Ciencia y Tecnología (CONACyT), Mexico, for financial support through 162476 Scholarship for one of them (JRGSL) and Grant 50379/2005.

References

1. D. Gidaspow, *Multiphase Flow and Fluidization: Continuum and Kinetic Theory Descriptions.* (Academic Press, 1994).

SOLUTIONS TO A SYSTEM OF PDES INVARIANT WITH RESPECT TO k LIE GROUPS*

M. P. SPECIALE, F. OLIVERI

Department of Mathematics, University of Messina
Salita Sperone 31, 98166 Messina, Italy
E-mail: speciale@mat520.unime.it; oliveri@mat520.unime.it

Lie point symmetries admitted by PDEs can be used to determine invariant solutions by solving a reduced system involving one independent variable less. For a system of PDEs with m independent variables we may consider a k–dimensional subalgebra of the Lie algebra of the admitted symmetries and get a reduced system with $m - k$ independent variables. Also, if we have a m–dimensional Abelian Lie subalgebra we may write the PDEs in an equivalent form for which some solutions can be found explicitly. An application to the Navier–Stokes–Fourier equations is considered.

Keywords: Lie group analysis, Invariant solutions.

1. Introduction

The explicit determination of exact solutions to systems of partial differential equations (PDEs) of physical interest is an important task. One of the most powerful methods in order to face this problem is based upon the study of their invariance with respect to one–parameter Lie groups of point transformations.[1–5] Given a system of differential equations $\Delta\left(\mathbf{x}, \mathbf{u}, \mathbf{u}^{(r)}\right) = 0$, where $\mathbf{x} \in \mathbb{R}^m$ denotes the set of independent variables, $\mathbf{u}(\mathbf{x}) \in \mathbb{R}^n$ the set of dependent variables and $\mathbf{u}^{(r)}$ the set of all partial derivatives of $\mathbf{u}$ with respect to $\mathbf{x}$ up to a fixed order r, it is invariant with respect to a one–parameter Lie group of transformations generated by the vector field

$$\Xi = \sum_{i=1}^{m} \xi_i(x_j, u_B)\frac{\partial}{\partial x_i} + \sum_{A=1}^{n} \eta_A(x_j, u_B)\frac{\partial}{\partial u_A},$$

if the r–prolongation of Ξ acting on Δ is zero along the solutions (see Refs. 1–5). This condition provides an overdeterminated set of linear differ-

*Dedicated to T. Ruggeri on the occasion of his 60th birthday.

ential equations for the infinitesimal generators ξ_i and η_A, whose integration gives the operators of the Lie symmetries admitted by the system; these operators span a Lie algebra that can be finite or infinite–dimensional.

A solution $\mathbf{u} = \boldsymbol{\Theta}(\mathbf{x})$ of the given system is an invariant solution with respect to a one–parameter Lie group of infinitesimal transformations, generated by the vector field Ξ, if it satisfies the invariant surface condition

$$\xi_1(\mathbf{x},\mathbf{u})\frac{\partial \Theta_A}{\partial x_1} + \cdots + \xi_m(\mathbf{x},\mathbf{u})\frac{\partial \Theta_A}{\partial x_m} - \eta_A(\mathbf{x},\mathbf{u}) = 0, \qquad A = 1,\ldots,n, \quad (1)$$

and the system itself. By solving (1) we may write

$$u_A = \psi_A(X_1(\mathbf{x},\mathbf{u}),\ldots,X_{m-1}(\mathbf{x},\mathbf{u})), \qquad A = 1,2,\ldots,n,$$

that, substituted into the given system, provides a reduced system of PDEs with one independent variable less. We can also consider a solution $\mathbf{u} = \boldsymbol{\Theta}(\mathbf{x})$ which is invariant with respect to $k \leq m$ admitted Lie groups; in this case the solution must be simultaneously invariant with respect to all the symmetries considered. By solving the k invariant surface conditions, and substituting the result in the system at hand, we finally get a reduced system involving only $m-k$ independent variables; if $k = m-1$ the reduced system is a set of ODEs, whereas if $k = m$ the reduced system is algebraic.

In order to determine the invariant solutions with respect to k Lie groups generating a k–dimensional Lie algebra, we need a set of functionally independent invariants $I_\alpha(\mathbf{x},\mathbf{u})$, *i.e.*, a set of functions defined by the conditions

$$\Xi_1 I_\alpha = \Xi_2 I_\alpha = \cdots = \Xi_k I_\alpha = 0.$$

For dimensional reasons, we have at most $m+n-k$ invariants. If

$$\operatorname{rank}\left(\frac{\partial I_\alpha(\mathbf{x},\mathbf{u})}{\partial \mathbf{u}}\right) = n,$$

by the implicit function theorem, we may express the field $\mathbf{u}$ in terms of the invariants I_α ($\alpha = 1,\ldots,m+n-k$) and obtain the functional form of the solution that, substituted into the original system, provides the reduced system. When $k = m-1$ and we have $n+1$ invariants, under the condition that we may solve n invariants with respect to the field variables, by choosing one invariant $\tau(\mathbf{x})$ as a new independent variable and the remaining ones (functions of τ) as the new dependent variables, the reduced system turns out to be ordinary.

2. Application: the Navier-Stokes-Fourier equations

Let us consider a viscous and heat-conducting monatomic gas in rotation about a fixed vertical axis with a constant angular velocity $\boldsymbol{\omega}$:

$$
\begin{aligned}
&\frac{\partial \rho}{\partial t}+\frac{\partial(\rho v_k)}{\partial x_k}=0,\\
&\frac{\partial(\rho v_i)}{\partial t}+\frac{\partial}{\partial x_k}\left\{\rho v_i v_k-\frac{2}{\alpha}\left(\frac{K}{m_0}\right)T\left[\frac{1}{2}\left(\frac{\partial v_i}{\partial x_k}+\frac{\partial v_k}{\partial x_i}\right)-\frac{1}{3}\frac{\partial v_r}{\partial x_r}\delta_{ik}\right]\right.\\
&\qquad\left.+\rho\left(\frac{K}{m_0}\right)T\delta_{ik}\right\}=\rho(f_i+i_i),\qquad i=1,2,3,\\
&\frac{\partial(\rho T)}{\partial t}+\frac{\partial}{\partial x_k}\left\{\rho T v_k-\frac{5}{2\alpha}\left(\frac{K}{m_0}\right)T\frac{\partial T}{\partial x_k}\right\}\\
&\qquad+\left\{\frac{2}{3}\rho T\delta_{lk}-\frac{4}{3\alpha}T\left[\frac{1}{2}\left(\frac{\partial v_l}{\partial x_k}+\frac{\partial v_k}{\partial x_l}\right)-\frac{1}{3}\frac{\partial v_r}{\partial x_r}\delta_{lk}\right]\right\}\frac{\partial v_l}{\partial x_k}=0.
\end{aligned}
\tag{2}
$$

i and **f** being the specific inertial (Coriolis and centrifugal forces) and external forces (gravity) acting on the gas, K the Boltzmann constant, m_0 the mass of a particle, and α a constant describing the interaction between Maxwellian molecules; moreover, the Einstein convention of sum over repeated indices has been used.

System (2) admits a 12–parameter group generated by the vector fields

$$
\begin{aligned}
\Xi_1 &= x_2\partial_{x_1}-x_1\partial_{x_2}+v_2\partial_{v_1}-v_1\partial_{v_2},\\
\Xi_2 &= 2t\partial_t+2(x_1+\omega x_2 t)\partial_{x_1}+2(x_2-\omega x_1 t)\partial_{x_2}+(2x_3-gt^2)\partial_{x_3}\\
&\quad+2\omega(v_2 t+x_2)\partial_{v_1}-2\omega(v_1 t+x_1)\partial_{v_2}-2gt\partial_{v_3}-2\rho\partial_\rho,\\
\Xi_3 &= -t\partial_t-\omega x_2 t\partial_{x_1}+\omega x_1 t\partial_{x_2}+gt^2\partial_{x_3}+\left[v_1-\omega(v_2 t+x_2)\right]\partial_{v_1}\\
&\quad+\left[v_2+\omega(v_1 t+x_1)\right]\partial_{v_2}+(v_3+2gt)\partial_{v_3}+\rho\partial_\rho+2\theta\partial_\theta,\\
\Xi_4 &= \partial_{x_3},\qquad \Xi_5=t\partial_{x_3}+\partial_{v_3},\qquad \Xi_6=\partial_t,\\
\Xi_7 &= \sin(\omega t)\partial_{x_1}+\cos(\omega t)\partial_{x_2}+\omega\cos(\omega t)\partial_{v_1}-\omega\sin(\omega t)\partial_{v_2},\\
\Xi_8 &= t\sin(\omega t)\partial_{x_1}+t\cos(\omega t)\partial_{x_2}\\
&\quad+\left[\omega t\cos(\omega t)+\sin(\omega t)\right]\partial_{v_1}-\left[\omega t\sin(\omega t)-\cos(\omega t)\right]\partial_{v_2},\\
\Xi_9 &= -\cos(\omega t)\partial_{x_1}+\sin(\omega t)\partial_{x_2}+\omega\sin(\omega t)\partial_{v_1}+\omega\cos(\omega t)\partial_{v_2},\\
\Xi_{10} &= -t\cos(\omega t)\partial_{x_1}+t\sin(\omega t)\partial_{x_2}\\
&\quad+\left[\omega t\sin(\omega t)-\cos(\omega t)\right]\partial_{v_1}+\left[\omega t\cos(\omega t)+\sin(\omega t)\right]\partial_{v_2},\\
\Xi_{11} &= \left[(2g-\omega^2(gt^2+2x_3))\sin(\omega t)-2g\omega t\cos(\omega t)\right]\partial_{x_1}\\
&\quad+\left[(2g-\omega^2(gt^2+2x_3))\cos(\omega t)+2g\omega t\sin(\omega t)\right]\partial_{x_2}\\
&\quad+2\omega^2(x_1\sin(\omega t)+x_2\cos(\omega t))\partial_{x_3}-\omega^2\left[\omega(gt^2+2x_3)\cos(\omega t)\right.
\end{aligned}
$$

$$
\begin{aligned}
&+ 2v_3\sin(\omega t)\big]\,\partial_{v_1} + \omega^2\left[\omega(gt^2+2x_3)\sin(\omega t) - 2v_3\cos(\omega t)\right]\partial_{v_2}\\
&+ 2\omega^2\left[(v_1\sin(\omega t)+v_2\cos(\omega t)) + \omega(x_1\cos(\omega t) - x_2\sin(\omega t))\right]\partial_{v_3},\\
\Xi_{12} = &\left[(2g\omega t\sin(\omega t) + \omega^2(gt^2-2x_3)+2g)\cos(\omega t)\right]\partial_{x_1}\\
&- \left[(2g-\omega^2(gt^2+2x_3))\sin(\omega t) - 2g\omega t\cos(\omega t)\right]\partial_{x_2}\\
&+ 2\omega^2\left[x_1\cos(\omega t) - x_2\sin(\omega t)\right]\partial_{x_3} + \omega^2\big[\omega(gt^2+2x_3)\sin(\omega t)\\
&- 2v_3\cos(\omega t)\big]\,\partial_{v_1} + \omega^2\left[\omega(gt^2+2x_3)\cos(\omega t) + 2v_3\sin(\omega t)\right]\partial_{v_2}\\
&+ 2\omega^2\left[\cos(\omega t)(v_1-\omega x_2) - \sin(\omega t)(v_2+\omega x_1)\right]\partial_{v_3}.
\end{aligned}
$$

To get a reduced set of ODEs we have to consider the 3–dimensional subalgebras of the admitted Lie algebra. Various possibilities arise that will be discussed in a forthcoming paper. Here we limit to report the results obtainable by considering the 3–dimensional subalgebra spanned by the vector fields $\{a\Xi_4+b\Xi_5, c\Xi_7+d\Xi_8, e\Xi_9+f\Xi_{10}\}$, where a, b, c, d, e and f are arbitrary constants. The invariant surface conditions lead to the solution:

$$
\begin{aligned}
v_1 &= U_1(t) + \omega x_2 - \Lambda_1\cos(\omega t)\frac{x_2\sin(\omega t) - x_1\cos(\omega t)}{(c+dt)(e+ft)} + \frac{dx_1}{c+dt},\\
v_2 &= U_2(t) - \omega x_1 - \Lambda_1\cos(\omega t)\frac{x_2\cos(\omega t) + x_1\sin(\omega t)}{(c+dt)(e+ft)} + \frac{fx_2}{e+ft},\\
v_3 &= U_3(t) + \frac{bx_3}{a+bt},\qquad \rho = R(t),\quad T = \widehat{T}(t),\qquad (\Lambda_1 = cf-de),
\end{aligned}
\tag{3}
$$

and the reduced ordinary system is

$$
\begin{aligned}
&\dot{U}_1 + \Lambda_1\cos(\omega t)\frac{U_1\cos(\omega t) - U_2\sin(\omega t)}{(c+dt)(e+ft)} - \omega U_2 + \frac{d}{c+dt}U_1 = 0,\\
&\dot{U}_2 - \Lambda_1\cos(\omega t)\frac{U_2\cos(\omega t) + U_1\sin(\omega t)}{(c+dt)(e+ft)} + \omega U_1 + \frac{f}{e+ft}U_2 = 0,\\
&\dot{U}_3 + g + \frac{b}{a+bt}U_3 = 0,\qquad \dot{R} + (\frac{b}{a+bt} + \frac{d}{c+dt} + \frac{f}{e+ft})R = 0,\\
&R\dot{\widehat{T}} + \frac{2}{3}\left(\frac{b}{a+bt} + \frac{d}{c+dt} + \frac{f}{e+ft}\right)R\widehat{T}\\
&\qquad - \frac{4}{9\alpha}\frac{\Lambda_1^2(a+bt)^2 + \Lambda_2^2(c+dt)^2 + \Lambda_3^2(e+ft)^2}{(a+bt)^2(c+dt)^2(e+ft)^2}\widehat{T} = 0,
\end{aligned}
\tag{4}
$$

where $\Lambda_2 = be - af$, $\Lambda_3 = bc - ad$, and the superposed dot denotes differentiation with respect to t. System (4) can be integrated and we finally obtain the following solution to (2):

$$v_1 = \frac{U_{10}\cos(\omega t) + dx_1}{c+dt} + \frac{U_{20}\sin(\omega t)}{e+ft}$$
$$- \Lambda_1 \cos(\omega t)\frac{x_2\sin(\omega t) - x_1\cos(\omega t)}{(c+dt)(e+ft)} + \omega x_2,$$
$$v_2 = \frac{U_{10}\sin(\omega t)}{c+dt} - \frac{U_{20}\cos(\omega t) + fx_2}{e+ft}$$
$$- \Lambda_1 \cos(\omega t)\frac{x_2\cos(\omega t) + x_1\sin(\omega t)}{(c+dt)(e+ft)} - \omega x_1,$$
$$v_3 = -\frac{gt(2a+bt)}{2(a+bt)} + \frac{U_{30}+bx_3}{a+bt}, \qquad R = \frac{\rho_0}{(a+bt)(c+dt)(e+ft)},$$
$$\widehat{T} = T_0 \frac{\left((a+bt)^{\Lambda_2\Lambda_3/b}(c+dt)^{\Lambda_1\Lambda_3/d}(e+ft)^{\Lambda_1\Lambda_2/f}\right)^{8/9\alpha\rho_0}}{((a+bt)(c+dt)(e+ft))^{2/3}},$$

where U_{10}, U_{20}, U_{30}, ρ_0 and T_0 are arbitrary constants.

By taking a 4–dimensional Lie subalgebra admitted by (2) and determining the invariants, we get an algebraic system. Various cases arise that will be discussed in a forthcoming paper. Here we consider the case where the 4–dimensional Lie subalgebra is Abelian. Instead of calculating the invariant solutions, we may introduce[6,7] a new set of independent and dependent variables and transform the system to an equivalent autonomous form. For instance, the vector fields $\{\Xi_2, \Xi_5, \Xi_8, \Xi_{10}\}$ span a 4–dimensional Abelian Lie algebra, whereupon, by introducing new independent and dependent variables such that

$$\begin{aligned} &\tau = \ln(t), \qquad \zeta_1 = \frac{x_1}{t}\cos(\omega t) - \frac{x_2}{t}\sin(\omega t),\\ &\zeta_2 = \frac{x_1}{t}\sin(\omega t) + \frac{x_2}{t}\cos(\omega t), \quad \zeta_3 = \frac{x_3}{t} + \frac{gt}{2},\\ &v_1 = \widehat{v}_1\cos(\omega t) + U_{20}\sin(\omega t) + \omega x_2 + a\frac{x_1}{t},\\ &v_2 = \widehat{v}_2\cos(\omega t) - U_{10}\sin(\omega t) - \omega x_1 + b\frac{x_2}{t},\\ &v_3 = \widehat{v}_3 - \frac{gt}{2} + \frac{x_3}{t}, \qquad \rho = \frac{\widehat{\rho}}{t}, \qquad T = \widehat{T}, \end{aligned} \tag{5}$$

we transform the source system (2) into autonomous form (that we omit to write here) that has the same form as the original system except for the inhomogeneity terms. By taking $\widehat{v}_1 = \widehat{v}_2 = \widehat{v}_3 = 0$, we get the solution:

$$v_1 = \frac{x_1}{t} + \omega x_2, \qquad v_2 = \frac{x_2}{t} - \omega x_1, \qquad v_3 = \frac{x_3}{t} - \frac{gt}{2},$$
$$\rho = t^{-3}\left[\psi(\zeta_1,\zeta_2,\zeta_3)\right]^{-1/2}, \quad T = T_0 t^{-2}\left[\psi(\zeta_1,\zeta_2,\zeta_3)\right]^{-1/2},$$

where T_0 (> 0) is constant and ψ is a positive solution of Laplace equation

$$\frac{\partial^2\psi}{\partial\zeta_1^2} + \frac{\partial^2\psi}{\partial\zeta_2^2} + \frac{\partial^2\psi}{\partial\zeta_3^2} = 0.$$

The same procedure can be used starting with the 4–dimensional Abelian Lie algebra spanned by the vector fields $\{\Xi_3, \Xi_4, \Xi_7, \Xi_9\}$, whereupon by introducing new independent and dependent variables such that

$$\begin{aligned}
&\tau = \ln(t), \quad \zeta_1 = x_1\cos(\omega t) - x_2 t\sin(\omega t),\\
&\zeta_2 = x_1\sin(\omega t) + x_2 t\cos(\omega t), \quad \zeta_3 = x_3 + \frac{gt^2}{2},\\
&v_1 = \frac{\widehat{v}_1\cos(\omega t) + \widehat{v}_2\sin(\omega t)}{t} + \omega x_2 + a\frac{x_1}{t},\\
&v_2 = \frac{\widehat{v}_2\cos(\omega t) - \widehat{v}_1\sin(\omega t)}{t} - \omega x_1 + b\frac{x_2}{t},\\
&v_3 = \widehat{v}_3 - gt, \qquad \rho = \widehat{\rho}, \qquad T = \widehat{T},
\end{aligned} \tag{6}$$

we get an autonomous system that, also in this case, has the same form as the original system except for the inhomogeneity terms. By assuming $\widehat{v}_1 = \widehat{v}_2 = \widehat{v}_3 = 0$, we get the following solution to (2):

$$\begin{aligned}
&v_1 = \omega x_2 \qquad v_2 = -\omega x_1, \qquad v_3 = -\frac{gt}{2},\\
&\rho = t\sqrt{\psi(\zeta_1, \zeta_2, \zeta_3)}, \qquad T = T_0 t\sqrt{\psi(\zeta_1, \zeta_2, \zeta_3)},
\end{aligned} \tag{7}$$

with T_0 (> 0) constant and ψ a positive solution of Poisson equation

$$\frac{\partial^2\psi}{\partial\zeta_1^2} + \frac{\partial^2\psi}{\partial\zeta_2^2} + \frac{\partial^2\psi}{\partial\zeta_3^2} + \frac{4\alpha m_0}{5KT_0} = 0.$$

Bibliography

1. L.V. Ovsiannikov. *Group analysis of differential equations* *Academic Press, New York, 1982(.
2. N.H. Ibragimov. *Transformation groups applied to mathematical physics* *D. Reidel Publishing Company, Dordrecht, 1985(.
3. P.J. Olver. *Applications of Lie groups to differential equations* (Springer, New York, 1986).
4. G.W. Bluman, S. Kumei. *Symmetries and differential equations* (Springer, New York, 1989).
5. P.J. Olver. *Equivalence, Invariants, and Symmetry* (Cambridge University Press, 1995).
6. A. Donato, F. Oliveri, *Applicable Analysis* **58**, 313–323 (1995).
7. A. Donato, F. Oliveri, *Transp. Th. Stat. Phys.* **25**, 303–322 (1996).

POISEUILLE FLOW OF A FLUID OVERLYING A POROUS MEDIUM

B. STRAUGHAN
Department of Mathematical Sciences, Durham University,
Science Laboratories, South Road,
Durham, DH1 3LE, UK
E-mail: brian.straughan@durham.ac.uk

We review work on the instability of Poiseuille flow when a fluid overlies a saturated porous medium. We investigate a two layer model where the porous medium is of Darcy type, and a three layer model where there is a transistion layer of Brinkman type.

1. Introduction

Allen and Khosravani [1], Habel *et al.* [2], Ewing and Weekes [3] study flow in an underground channel where the fluid also saturates at least part of the soil below. A study of such flows is of importance in obtaining drinking water supplies and for avoiding contamination. In this paper we review models for describing mathematically the flow of a fluid overlying a saturated porous medium.

Chang *et al.* [4] investigate the instability of Poiseuille flow when a Newtonian fluid overlies a porous medium saturated with the same fluid. The geometry for this problem is that of a Newtonian fluid occupying the domain $\mathbb{R}^2 \times \{z \in (0,d)\}$ with the saturated porous medium occupying the spatial domain $\mathbb{R}^2 \times \{z \in (-d_m, 0)\}$, as shown in figure 1. A pressure gradient in the $x-$ direction drives the flow and Chang *et al.* [4] studied linearized instability of this situation. A detailed description of the work of Chang *et al.* [4] and of other models pertaining to instability of Poiseuille flow in this situation may be found in chapter 6 of the book of Straughan [5].

The model of Chang *et al.* [4] consists of the Navier-Stokes equations in the domain $\mathbb{R}^2 \times (0,d) \times \{t > 0\}$ with the Darcy equations in the lower

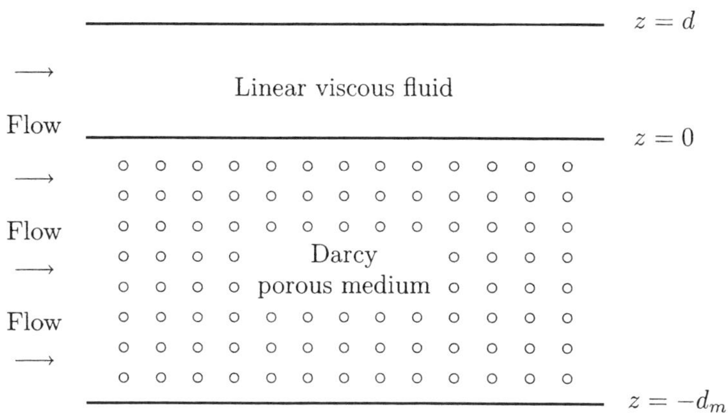

Figure 1. Poiseuille flow over a porous medium

domain $\mathbb{R}^2 \times (-d_m, 0) \times \{t > 0\}$ (see figure 1). Hence, they have

$$\frac{\partial u_i}{\partial t} + u_j \frac{\partial u_i}{\partial x_j} = -\frac{1}{\rho}\frac{\partial p}{\partial x_i} + \nu \Delta u_i, \qquad \frac{\partial u_i}{\partial x_i} = 0, \tag{1}$$

in $\mathbb{R}^2 \times \{z \in (0, d)\} \times \{t > 0\}$, and the Darcy equations, namely

$$\frac{1}{\Phi}\frac{\partial u_i^m}{\partial t} = -\frac{1}{\rho}\frac{\partial p^m}{\partial x_i} - \frac{\nu}{K} u_i^m, \qquad \frac{\partial u_i^m}{\partial x_i} = 0, \tag{2}$$

in $\mathbb{R}^2 \times \{z \in (-d_m, 0)\} \times \{t > 0\}$. In these equations u_i, p, ρ, ν denote fluid velocity, pressure, density and viscosity, u_i^m, p^m, K and Φ are the velocity and pressure in the porous medium, and the permeability and porosity, respectively.

The boundary conditions of Chang *et al.* [4] are no slip at the fixed upper surface and no flow out of the bottom of the porous layer, i.e.

$$u_i = 0, \quad z = d, \qquad w^m = 0, \quad z = -d_m. \tag{3}$$

At the interface they assume continuity of normal velocity and pressure, so

$$w = w^m, \quad p^m = p, \qquad z = 0, \tag{4}$$

and the Beavers and Joseph [6] condition

$$\frac{\partial u_\mu}{\partial z} = \frac{\alpha}{\sqrt{K}}(u_\mu - u_\mu^m), \quad \mu = 1, 2, \; z = 0. \tag{5}$$

For a constant pressure gradient dp/dx one obtains a basic profile for Poiseuille flow in the two - layer system in which the velocity is not continuous. The steady solution of Chang *et al.* [4] is

$$\bar{u}(z) = \frac{1}{2}A_1 z^2 + A_2 z + A_3, \ \bar{v} = \bar{w} = 0, \quad 0 \le z \le d, \tag{6}$$

$$\bar{u}^m(z) = -A_1 K, \ \bar{v}^m = \bar{w}^m = 0, \quad -d_m \le z \le 0, \tag{7}$$

where the (constant) coefficients A_1, A_2, A_3 are given by

$$A_1 = \frac{1}{\mu}\frac{dp}{dx}, \quad A_2 = \alpha A_1\sqrt{K} - \frac{\alpha A_1 d^2 + 2\alpha^2 A_1 d\sqrt{K}}{2(\alpha d + \sqrt{K})},$$

$$A_3 = -\frac{A_1 d^2\sqrt{K} + 2\alpha A_1 K d}{2(\alpha d + \sqrt{K})}.$$

2. Instability

To study instability Chang *et al.* [4] linearize the perurbation equations about the basic flow. they then introduce a normal mode form

$$u_i = u_i(z)\, e^{i(ax+by-act)}, \quad u_i^m = u_i^m(z_m)\, e^{i(a_m x_m + b_m y_m - a_m c_m t_m)},$$

$$p = \pi(z)\, e^{i(ax+by-act)}, \quad p^m = \pi^m(z_m)\, e^{i(a_m x_m + b_m y_m - a_m c_m t_m)},$$

and invoke Squire's theorem. In terms of stream functions and eigenfunctions $\psi, \psi_m, \phi, \phi_m$ related to the velocity fields by

$$u = \frac{\partial\psi}{\partial z}, \quad w = -\frac{\partial\psi}{\partial x}, \quad u^m = \frac{\partial\psi^m}{\partial z^m}, \quad w^m = -\frac{\partial\psi^m}{\partial x^m},$$

and

$$\psi = \phi(z)\, e^{ia(x-ct)}, \quad \psi^m = \phi^m(z_m)\, e^{ia_m(x_m - c_m t_m)},$$

Chang *et al.* [4] derive an Orr-Sommerfeld system. The resulting system consists of the eigenvalue equations

$$\begin{aligned} &(D^2 - a^2)^2\phi = ia\, Re(U - c)(D^2 - a^2)\phi - ia\, ReU''\phi, \quad z \in (0,1), \\ &\Big(\frac{1}{\delta^2} - Re^m \frac{ia_m c_m}{\Phi}\Big)(D_p^2 - a_m^2)^2\phi_m = 0, \quad z_m \in (-1,0). \end{aligned} \tag{8}$$

The notation is $D = d/dz$, $D_p = d/dz_m$ and $U(z)$ is $\bar{u}(z)$ rewritten with respect to a velocity scaling, see equation (2.8) of Chang *et al.* [4].

System (8) is solved numerically subject to the boundary conditions,

$$\phi = D\phi = 0, \quad \text{on } z = 1, \qquad \phi^m = 0, \quad \text{on } z_m = -1,$$

and the interface conditions

$$
\begin{aligned}
&Re\phi = Re^m\phi^m,\\
&D^2\phi - \frac{\alpha\hat{d}}{\delta}\,D\phi + \frac{\alpha\hat{d}^2 Re^m}{\delta Re}\,D_p\phi^m = 0,\\
&\Big(Re^m\frac{ia_m c_m}{\Phi} - \frac{1}{\delta^2}\Big)D_p\phi^m\\
&= \frac{Re}{\hat{d}^3 Re^m}\left[(D^2-a^2)D\phi - iaRe(U-c)D\phi + iaReU'\phi\right],
\end{aligned}
$$

on the surface $z = z_m = 0$.

Chang *et al.* [4] employ the D^2 Chebyshev tau numerical method of Dongarra *et al.* [7], see also chapters 6 and 9 of Straughan [5].

Chang *et al.* [4] report many numerical calculations. For suitable parameters they discover tri-modal neutral curves. For $\hat{d} = 0.11$ they found instability dominated by the porous layer whereas when $\hat{d} = 0.12$ the fluid layer is the controlling influence. As $\hat{d}$ was increased beyond $\hat{d} = 0.12$ a new instability mode was seen. For $\hat{d} = 0.121$ a third mode is seen but the fluid mode still dominates. By the time that $\hat{d} = 0.13$ the new (fluid) mode dominates, and continues to do so for larger $\hat{d}$. Chang *et al.* [4] interpret the new mode as a even-shear-mode of the Poiseuille flow.

3. Transistion layer

Hill and Straughan [8] argue that there must be a transistion layer between the fluid and the Darcy porous medium. This idea was proposed earlier by Nield [9], p. 45, who suggested using a Brinkman equation in the boundary layer region between the fluid and the Darcy porous medium. This scenario is also supported by Goharzadeh *et al.* [10] in their experimental work. Their experiments indicate that the thickness of the transistion zone is of the same order as the grain size of the material which comprises the porous medium.

Therefore, Hill and Straughan [8] develop an instability analysis for the Poiseuille flow problem for a fluid overlying a porous medium, but they adopt a three layer configuration with a Newtonian fluid saturating a porous layer below, and the porous layer is comprised of a Brinkman transistion layer overlying a Darcy porous layer.

The equations of Hill and Straughan [8] are (1), (2) in the layers contained between $z \in (0, d)$ and $z \in (-d_m, -\beta d_m)$, respectively. However, in the three dimensional layer contained between $z = -\beta d_m$ and $z = 0$ they suppose the Brinkman equations hold, namely

$$\frac{1}{\Phi_b} u^b_{i,t} = -\frac{1}{\rho} p^b_{,i} + \nu_e \Delta u^b_i - \frac{\nu}{K} u^b_i, \qquad u^b_{i,i} = 0. \tag{9}$$

The quantities $u^b_i, p^b, \nu_e, \Phi_b$ are the fluid velocity and pressure in the Brinkman layer and an effective viscosity and the porosity. Straughan [5] goes into detail about the form Φ_b may have since this depends on whether the porous material is composed of an easily dislodged substance like sand or somethng more rigid like a brick.

The boundary conditions of Hill and Straughan [8] assume no slip on $z = d$, no flow out of $z = -d_m$, continuity of normal and tangential stress on the interface $z = 0$, while on the Brinkman-Darcy interface $z = -\beta d_m$ they suppose continuity of normal stress and a Jones [11] condition

$$\frac{\partial u^b_\zeta}{\partial z} + \frac{\partial w^b}{\partial x_\zeta} = \frac{\alpha}{\sqrt{K}} \left(u^b_\zeta - u^m_\zeta \right), \quad z = -\beta d_m, \quad \zeta = 1, 2. \tag{10}$$

Hill and Straughan [8] derive the steady solution which is analogous to (6), (7) in the fluid and Darcy regions, but has an exponential behaviour in the transistion layer. They reduce the linearized perturbation equations to a two-dimensional form and in terms of velocity functions ϕ, ϕ^b, ϕ^m they derive the following Orr-Sommerfeld equations,

$$\begin{aligned}
&(D^2 - a^2)^2 \phi = Re(U - c) ia (D^2 - a^2)\phi - iaReU''\phi, \quad 0 < z < 1, \\
&\left(1 - \frac{ia_b c_b Re^b \delta^2}{\Phi_b} - \frac{\delta^2}{f^2}(D_b^2 - a_b^2)\right)(D_b^2 - a_b^2)\phi^b = 0, \quad -\beta < z < 0, \\
&\left(1 - \frac{ia_m c_m Re^m \delta^2}{\Phi}\right)(D_m^2 - a_m^2)\phi^m = 0, \quad -1 < z < -\beta.
\end{aligned} \tag{11}$$

Hill and Straughan [8] derive ten appropriate boundary conditions for (11) and solve the system numerically by the D^2-Chebyshev tau method. Their numerical findings are very interesting and substantially different from those of Chang *et al.* [4].

The whole question of modelling Poiseuille flow instability in a three layer situation is addressed in detail by Straughan [5]. He describes in detail other flow scenarios such as when the transistion layer is composed of a Forchheimer material, or even a Brinkman-Forchheimer one. Further details may be found in the book of Straughan [5].

We remark that nonlinear energy stabilty methods have not been too successful in analysing Poiseuille flows, cf. Straughan [12], [13]. However, this may be an area where new techniques such as that of Rionero [14], [15],

[16], [17], [18], [19], [20], could perhaps prove fruitful, see also Straughan [21].

Acknowledgments

This work was supported by a Research Project Grant of the Leverhulme Trust - Grant Number F/00128/AK.

References

1. M.B. Allen and A. Khosravani, *Adv. Water Resources* **15**, 125 (1992)
2. F. El-Habel, C. Mendoza and A.C. Bagtzoglou, *Adv. Water Resources* **25**, 455 (2002)
3. R.E. Ewing and S. Weekes, *Computational Mathematics* **202**, 75 (1998)
4. M.H. Chang, F. Chen and B. Straughan, *J. Fluid Mech.* **564**, 287 (1998)
5. B. Straughan, *Stability and wave motion in porous media.* Springer, New York (2008)
6. G.S. Beavers and D.D. Joseph, *J. Fluid Mech.* **30**, 197 (1967)
7. J.J. Dongarra, B. Straughan and D.W. Walker, *Appl. Numer. Math.* **22**, 399 (1996)
8. A.A. Hill and B. Straughan, *to be published.* (2008)
9. D.A. Nield, *J. Fluid Mech.* **128**, 37 (1983)
10. A. Goharzadeh, A. Khalili and B.B. Jorgensen, *Phys. Fluids* **17**, 057102 (2005)
11. I.P. Jones, *Proc. Camb. Phil. Soc.* **73**, 231 (1973)
12. B. Straughan, *Explosive instabilities in mechanics.* Springer, Heidelberg (1998)
13. B. Straughan, *The energy method, stability and nonlinear convection.* Springer, New York (2004)
14. S. Rionero, *Acc. Sc. fis. mat. Napoli* **71**, 53 (2004).
15. S. Rionero, *Nuovo Cimento B* **119**, 773 (2004).
16. S. Rionero, *Rend. Matem. Accad. Lincei* **16**, 227 (2005).
17. S. Rionero, *J. Math. Anal. Appl.* **319**, 377 (2006).
18. S. Rionero, *Mathematical Biosciences and Engineering* **3**, 189 (2006).
19. S. Rionero, *Rend. Circolo Matem. Palermo* **78**, 273 (2006).
20. S. Rionero, *J. Math. Anal. Appl.* bf 333, 1036 (2007).
21. B. Straughan, *Ricerche Matem.* In the press, (2008).

REGULARIZATION AND BOUNDARY CONDITIONS FOR THE 13 MOMENT EQUATIONS

HENNING STRUCHTRUP

ETH Zürich, Department of Materials, Polymer Physics, CH-8093 Zürich, Switzerland (on leave from University of Victoria, Canada, struchtr@uvic.ca)

MANUEL TORRILHON

ETH Zürich, Seminar for Applied Mathematics, CH-8092 Zürich, Switzerland

We summarize our recent contributions to the development of macroscopic transport equations for rarefied gas flows. A combination of the Chapman-Enskog expansion and Grad's moment method, termed as the order of magnitude method, yields the regularized 13 moment equations (R13 equations) which are of super-Burnett order. A complete set of boundary conditions is derived from the boundary conditions of the Boltzmann equations. The R13 equations are linearly stable and their results for Knudsen numbers below 0.5 stand in excellent agreement to DSMC simulations.

1. Introduction

Processes in rarefied gases are well described by the Boltzmann equation [1,2] which describes the evolution of the particle distribution function in phase space, i.e. on the microscopic level.

The relevant scaling parameter to characterize processes in rarefied gases is the Knudsen number Kn, defined as the ratio between the mean free path of a particle and a relevant length scale. If the Knudsen number is small, the Boltzmann equation can be reduced to simpler models, which allow faster solutions. Indeed, if $\mathrm{Kn} < 0.01$ (say), the hydrodynamic equations, the laws of Navier-Stokes and Fourier (NSF), can be derived from the Boltzmann equation, e.g. by the Chapman-Enskog method [1,2]. The NSF equations are macroscopic equations for mass density ρ, velocity v_i and temperature T, and thus pose a mathematically less complex problem than the Boltzmann equation.

Macroscopic equations for rarefied gas flows at Knudsen numbers above 0.01 promise to replace the Boltzmann equation with simpler equations that still capture the relevant physics. The Chapman-Enskog expansion is the classical method to achieve this goal, but the resulting Burnett and super-Burnett equations are unstable [3].

A classical alternative is Grad's moment method [4] which extends the set of variables by adding deviatoric stress tensor σ_{ij}, heat flux q_i, and possibly higher moments of the phase density. The resulting equations are stable but lead to spurious discontinuities in shocks, and for a given value of the Knudsen number it is not clear what set of moments one would have to consider [2].

Struchtrup and Torrilhon combined both approaches by performing a Chapman-Enskog expansion around a non-equilibrium phase density of Grad type [5,6] which resulted in the "regularized 13 moment equations" (R13 equations) which form a stable set of equations for the 13 variables $(\rho, v_i, T, \sigma_{ij}.q_i)$ of super-Burnett order.

An alternative approach to the problem was presented by Struchtrup in [7,8], partly based on earlier work by Müller et al. [9]. The Order of Magnitude Method, which is briefly outlined in Section 2, is based on a rigorous asymptotic analysis of the infinite hierarchy of the moment equations.

One of the biggest problems for all models beyond NSF is to prescribe suitable boundary conditions for the extended equations, which should follow from the boundary conditions for the Boltzmann equation. This task was recently tackled in [11], and our solution to the problem [12] will be briefly discussed in Section 3, which presents boundary conditions for the R13 equations.

Section 4 will briefly discuss the properties of the R13 equations, which are linearly stable, obey a H-theorem for the linear case, contain the Burnett and super-Burnett equations asymptotically, predict phase speeds and damping of ultrasound waves in excellent agreement to experiments, yield smooth and accurate shock structures for all Mach numbers, and exhibit Knudsen boundary layers and the Knudsen minimum in excellent agreement to DSMC simulations.

Lack of space forbids to present any details, the interested reader is referred to the cited literature, including the monograph [2].

2. The Order of Magnitude Method

The Order of Magnitude Method [7,8] considers not the Boltzmann equation itself, but its infinite system of moment equations. The method of finding the proper equations with order of accuracy λ_0 in the Knudsen number consists of the following three steps:

(1) Determination of the order of magnitude λ of the moments.
(2) Construction of moment set with minimum number of moments at order λ.
(3) Deletion of all terms in all equations that would lead only to contributions of orders $\lambda > \lambda_0$ in the conservation laws for energy and momentum.

Step 1 is based on a Chapman-Enskog expansion where a moment ϕ is expanded according to $\phi = \phi_0 + \mathrm{Kn}\phi_1 + \mathrm{Kn}^2\phi_2 + \mathrm{Kn}^3\phi_3 + \cdots$, and the leading order of ϕ is determined by inserting this ansatz into the complete set of moment equations. A moment is said to be of leading order λ if $\phi_\beta = 0$ for all $^\beta < \lambda$. This first step agrees with the ideas of [9]. Alternatively, the order of magnitude of the moments can be found from the principle that a single term in an equation cannot be larger in size by one or several orders of magnitude than all other terms [10].

In Step 2, new variables are introduced by linear combination of the moments originally chosen. The new variables are constructed such that the number of moments at a given order λ is minimal. This step gives an unambiguous set of moments at order λ.

Step 3 follows from the definition of the order of accuracy λ_0: A set of equations is said to be accurate of order λ_0, when stress and heat flux are known within the order $\mathcal{O}\left(\mathrm{Kn}^{\lambda_0}\right)$.

The order of magnitude method gives the Euler and NSF equations at zeroth and first order, and thus agrees with the Chapman-Enskog method in the lower orders [7]. The second order equations turn out to be Grad's 13 moment equations for Maxwell molecules [7], and a generalization of these for molecules that interact with power potentials [2,8]. At third order, the method was only performed for Maxwell molecules, where it yields the R13 equations [7], which read (θ is the temperature in energy units, μ is the viscosity of the gas)

$$\frac{D\rho}{Dt} + \rho\frac{\partial v_k}{\partial x_k} = 0$$

$$\rho\frac{Dv_i}{Dt} + \rho\frac{\partial \theta}{\partial x_i} + \theta\frac{\partial \rho}{\partial x_i} + \frac{\partial \sigma_{ik}}{\partial x_k} = 0$$

$$\frac{3}{2}\rho\frac{D\theta}{Dt} + \rho\theta\frac{\partial v_k}{\partial x_k} + \frac{\partial q_k}{\partial x_k} + \sigma_{kl}\frac{\partial v_k}{\partial x_l} = 0$$

$$\frac{D\sigma_{ij}}{Dt} + \frac{4}{5}\frac{\partial q_{\langle i}}{\partial x_{j\rangle}} + 2\sigma_{k\langle i}\frac{\partial v_{j\rangle}}{\partial x_k} + \sigma_{ij}\frac{\partial v_k}{\partial x_k} + \frac{\partial u^0_{ijk}}{\partial x_k} = -\rho\theta\left[\frac{\sigma_{ij}}{\mu} + 2\frac{\partial v_{\langle i}}{\partial x_{j\rangle}}\right]$$

$$\frac{Dq_i}{Dt}+\frac{5}{2}\sigma_{ik}\frac{\partial\theta}{\partial x_k}-\sigma_{ik}\theta\frac{\partial\ln\rho}{\partial x_k}+\theta\frac{\partial\sigma_{ik}}{\partial x_k}+\frac{7}{5}q_i\frac{\partial v_k}{\partial x_k}+\frac{7}{5}q_k\frac{\partial v_i}{\partial x_k}+\frac{2}{5}q_k\frac{\partial v_k}{\partial x_i}$$
$$-\frac{\sigma_{ij}}{\varrho}\frac{\partial\sigma_{jk}}{\partial x_k}+\frac{1}{2}\frac{\partial w^1_{ik}}{\partial x_k}+\frac{1}{6}\frac{\partial w^2}{\partial x_i}+u^0_{ijk}\frac{\partial v_j}{\partial x_k}=-\frac{5}{2}\rho\theta\left[\frac{q_i}{\kappa}+\frac{\partial\theta}{\partial x_i}\right]$$

$$w^2=-\frac{\sigma_{ij}\sigma_{ij}}{\rho}-12\frac{\mu}{p}\left[\theta\frac{\partial q_k}{\partial x_k}+\frac{5}{2}q_k\frac{\partial\theta}{\partial x_k}-\theta q_k\frac{\partial\ln\rho}{\partial x_k}+\theta\sigma_{ij}\frac{\partial v_i}{\partial x_k}\right]$$
$$u^0_{ijk}=-2\,\frac{\mu}{p}\left[\theta\frac{\partial\sigma_{\langle ij}}{\partial x_{k\rangle}}-\sigma_{\langle ij}\frac{\partial\ln\rho}{\partial x_{k\rangle}}+\frac{4}{5}q_{\langle i}\frac{\partial v_j}{\partial x_{k\rangle}}\right]$$

$$w^1_{ij}=-\frac{4}{7}\frac{\sigma_{k\langle i}\sigma_{j\rangle k}}{\rho}-\frac{24}{5}\frac{\mu}{p}\left[\theta\frac{\partial q_{\langle i}}{\partial x_{j\rangle}}+q_{\langle i}\frac{\partial\theta}{\partial x_{j\rangle}}-\theta q_{\langle i}\frac{\partial\ln\rho}{\partial x_{j\rangle}}\right.$$
$$\left.+\frac{5}{7}\theta\left(\sigma_{k\langle i}\frac{\partial v_{j\rangle}}{\partial x_k}+\sigma_{k\langle i}\frac{\partial v_k}{\partial x_{j\rangle}}-\frac{2}{3}\sigma_{ij}\frac{\partial v_k}{\partial x_k}\right)\right]$$

3. Boundary Conditions for R13

The computation of boundary conditions for the R13 equations is based on Maxwell's model for boundary conditions for the Boltzmann equation [1,2], which states that a fraction χ of the particles hitting the wall is thermalized, while the remaining $1-\chi$ particles are specularly reflected. Boundary conditions for moments follow by taking moments of the boundary conditions of the Boltzmann equation. To produce meaningful boundary conditions, one needs to obey the following rules:

(1) Continuity: In order to have meaningful boundary conditions for all accommodation coefficients χ in [0,1], only boundary conditions for tensors with an odd number of normal components should be considered [3,11,12].
(2) Consistency: Only boundary conditions for fluxes that actually appear in the equations should be considered [12].
(3) Coherence: The same number of boundary conditions should be prescribed for the linearized and the non-linear equations [12].

The application of Rules 1 and 2 is straightforward and yields the following set of boundary conditions (t and n denote tangential and normal tensor components, respectively, and $V_t=v_t-v_t^W$, $\beta=\frac{\chi}{2-\chi}\sqrt{\frac{2}{\pi\theta}}$, $P=\rho\theta+\frac{1}{2}\sigma_{nn}-\frac{1}{28}\frac{w_{nn}}{\theta}-\frac{1}{120}\frac{w_{kk}}{\theta}$)

$$\sigma_{nt}=-\beta\left[PV_t+\frac{1}{5}q_t+\frac{1}{2}u^0_{tnn}\right]$$

$$w_{tn} = \beta\left[P\theta V_t - \frac{1}{2}\theta u^0_{tnn} - \frac{11}{5}\theta q_t - PV_t^3 + 6P(\theta-\theta_W)V_t\right]$$

$$q_n = -\beta\left[2P(\theta-\theta_W) + \frac{5}{28}w_{nn} + \frac{1}{15}w_{kk} + \frac{1}{2}\theta\sigma_{nn} - \frac{1}{2}PV_t^2\right]$$

$$u^0_{nnn} = \beta\left[\frac{2}{5}P(\theta-\theta_W) - \frac{1}{14}w_{nn} + \frac{1}{75}w_{kk} - \frac{7}{5}\sigma_{nn} - \frac{3}{5}PV_t^2\right]$$

$$u^0_{ttn} + \frac{1}{2}u^0_{nnn} = -\beta\left[\theta\left(\sigma_{tt} + \frac{1}{2}\sigma_{nn}\right) + \frac{1}{14}\left(w_{tt} + \frac{1}{2}w_{nn}\right) - \frac{1}{2}PV_t^2\right]$$

θ_W and v^W denote temperature and velocity of the wall. The first condition above is the slip condition for the velocity, while the third equation is the jump condition for the temperature. In a manner of speaking, the other conditions can be described as jump conditions for higher moments.

When the R13 equations are considered for channel flows in their original form, it turns out that a different number of boundary conditions is required to solve the fully non-linear and the linearized equations. Since this would not allow a smooth transition between linear and non-linear situations, we formulated the third rule as given above.

Asymptotic analysis shows that some terms can be changed without changing the overall asymptotic accuracy of the R13 equations. This leads to the algebraization of several non-linear terms in the pde's which, after some algebra, leads to algebraic relations, termed as bulk equations, between the moments which serve as additional boundary conditions for the non-linear equations [12],

$$w_{nn} = \frac{136}{25}\frac{q_n^2}{p} - \frac{72}{35}\frac{\sigma_{tn}^2}{\rho}\quad,\quad u^0_{tnn} = \frac{32}{45}\frac{\sigma_{tn}q_n}{p}$$

4. Computations and Simulations

We summarize the most important features of the R13 equations which result from analytical considerations, and from analytical and numerical solutions:

The R13 equations:

- are derived in a rational manner by means of the order of magnitude method [7,8], or from a Chapman-Enskog expansion around non-equilibrium [5,6],
- are of third order in the Knudsen number [2,5–8],
- are linearly stable for initial and boundary value problems [5,6],
- contain Burnett and super-Burnett asymptotically [5,6],

- predict phase speeds and damping of ultrasound waves in excellent agreement to experiments [5],
- give smooth shock structures for all Mach numbers, with good agreement to DSMC simulations for Ma<3 [6],
- are accompanied by a complete set of boundary conditions [12],
- obey an H-theorem for the linear case, including the boundaries [13],
- exhibit the Knudsen paradox for channel flows [12,13],
- exhibit Knudsen boundary layers in good agreement to DSMC [14,15],
- are accessible to numerical simulations in higher dimensions [16],
- predict light scattering spectra in accordance to experiments [17]

With these properties and features, the R13 equations must be considered as the most successful macroscopic model for rarefied gas flows. The application of the R13 equations to a wider variety of one-, two, and three-dimensional rarefied gas flow problems is planned for the future.

Acknowledgments: H.S.: Support by the Natural Sciences and Engineering Research Council (NSERC) is gratefully acknowledged. I like to thank Prof. H.C. Öttinger (ETH Zürich) for his kind hospitality. M.T.: Support through the EURYI award of the European Science Foundation (ESF) is gratefully acknowledged.

References

1. C. Cercignani, *The Boltzmann Equation and its Applications*, Applied Mathematical Sciences 67, Springer, New York 1988
2. H. Struchtrup, *Macroscopic Transport Equations for Rarefied Gas Flows–Approximation Methods in Kinetic Theory*, Springer, Heidelberg 2005
3. A.V. Bobylev, Sov. Phys. Dokl. **27**, 29 (1982)
4. H. Grad, Comm. Pure Appl. Math. **2**, 331 (1949)
5. H. Struchtrup and M. Torrilhon, Phys. Fluids **15**, 2668 (2003)
6. M. Torrilhon and H. Struchtrup, J. Fluid Mech. **513**, 171 (2004)
7. H. Struchtrup, Phys. Fluids **16**, 3921 (2004)
8. H. Struchtrup, Multiscale Model. Simul. **3**, 221 (2005)
9. I. Müller, D. Reitebuch, W. Weiss, Cont. Mech. Thermodyn. **15**, 113 (2003)
10. H. Struchtrup, J. Stat. Phys. **125**, 565 (2006)
11. X. Gu and D. Emerson, J. Comput. Phys. **225**, 263 (2007)
12. M. Torrilhon and H. Struchtrup, submitted to J. Comp. Phys. (2007)
13. H. Struchtrup and M. Torrilhon, Phys. Rev. Lett. **99**, 014502 (2007)
14. H. Struchtrup, Cont. Mech. Thermodyn. **17**, 43-50 (2005)
15. H. Struchtrup and T. Thatcher, Cont. Mech. Thermodyn. **19**, 177 (2007)
16. M. Torrilhon, Multiscale Model. Sim. **5**, 695 (2006)
17. M. Torrilhon, Proceedings 25th Intl. Symposium on Rarefied Gas Dynamics, in press (2007)

PLANE WAVES AND VIBRATIONS IN THE THERMOELASTIC MIXTURES

M. SVANADZE

Faculty of Physics and Mathematics, Ilia Chavchavadze State University, Tbilisi, 0179, Georgia
E-mail: svanadze@gol.ge
www.iliauni.edu.ge

In this paper the diffusion model of the linear theory of thermoelasticity of binary mixtures is considered and the following results are obtained: the basic properties of wave numbers of the longitudinal and transverse plane waves are established. The interior boundary value problem (BVP) of steady vibration of binary mixtures of thermoelastic solids is formulated. The existence theorems of eigenoscillation frequencies (eigenfrequencies) of the BVP of steady vibrations are proved. The connection between plane waves and existence of eigenfrequencies is established.

Keywords: Plane waves; Vibration; Thermoelastic mixtures.

1. Introduction

The nonlinear theory of mixtures of thermoelastic solids was developed by Green and Steel [1]. A linear variant of the latter theory was proposed by Steel [2]. In [3], a wide class of BVPs of steady vibration of the linear theory of thermoelasticity for binary mixtures is investigated by means of the potential method. An extensive review of the results of the theory of mixtures can be found in [4, 5].

In this article the diffusion model of the linear theory of binary mixtures of thermoelastic solids (see [2]) is considered. In Section 2 the basic equations of this theory are summarized. In Section 3 some properties of wave numbers of the longitudinal and transverse plane waves are treated. In Section 4 the interior BVP of steady vibration of binary mixtures of thermoelastic solids is formulated, the existence theorems of eigenfrequencies of the aforementioned problem are proved. Finally, in Section 5 the connection between plane waves and eigenfrequencies is established.

2. Basic Equations

Let $\mathbf{x} = (x_1, x_2, x_3)$ be the point of the Euclidean three-dimensional space R^3, let t denote the time variable, $t \geq 0$. We denote by $\mathbf{u}'(\mathbf{x}, t)$ and $\mathbf{u}''(\mathbf{x}, t)$ partial displacements, and by θ temperature measured from the constant absolute temperature T_0. We assume that $\mathbf{u}' = (u_1', u_2', u_3')$ and $\mathbf{u}'' = (u_1'', u_2'', u_3'')$ are three-component vector functions.

In the absence of body forces and the heat supply the system of equations of the diffusion model of the linear dynamical theory of binary mixtures of thermoelastic solids can be written as follows [2]

$$a_1 \Delta \mathbf{u}' + b_1 \operatorname{grad} \operatorname{div} \mathbf{u}' + c \, \Delta \, \mathbf{u}'' + d \operatorname{grad} \operatorname{div} \mathbf{u}'' - \nu \, (\dot{\mathbf{u}}' - \dot{\mathbf{u}}'') - \alpha_1 \operatorname{grad} \theta = \rho_1 \, \ddot{\mathbf{u}}',$$

$$c \, \Delta \mathbf{u}' + d \operatorname{grad} \operatorname{div} \mathbf{u}' + a_2 \, \Delta \, \mathbf{u}'' + b_2 \operatorname{grad} \operatorname{div} \mathbf{u}'' + \nu \, (\dot{\mathbf{u}}' - \dot{\mathbf{u}}'') - \alpha_2 \operatorname{grad} \theta = \rho_2 \, \ddot{\mathbf{u}}'',$$

$$\alpha_0 \, \Delta \, \theta - \alpha_3 \, \dot{\theta} - \alpha_4 \operatorname{div} \dot{\mathbf{u}}' - \alpha_5 \operatorname{div} \dot{\mathbf{u}}'' = 0, \tag{1}$$

where Δ is the Laplacian, $\alpha_4 = \alpha_1 T_0 - \alpha_6, \alpha_5 = \alpha_2 T_0 + \alpha_6; a_1, a_2, b_1, b_2, c, d$ and $\alpha_0, \alpha_1, \alpha_2, \alpha_3, \alpha_6$ are elastic and thermoelastic constants, respectively, ρ_1 and ρ_2 are the partial density constants (positive constants), ν is the diffusion coefficient, $\nu > 0$; the dot denotes differentiation with respect to t, $\dot{\mathbf{u}} = \frac{\partial \mathbf{u}}{\partial t}$, $\ddot{\mathbf{u}} = \frac{\partial^2 \mathbf{u}}{\partial t^2}$.

We introduce the notation

$$a = a_1 + b_1, \quad b = a_2 + b_2, \quad c_0 = c + d, \quad d_1 = ab - c_0^2, \quad d_2 = a_1 a_2 - c^2,$$

$$p_1 = a \, \rho_2 + b \, \rho_1, \quad p_2 = a_1 \, \rho_2 + a_2 \, \rho_1, \quad q_1 = a + b + 2c_0, \quad q_2 = a_1 + a_2 + 2c,$$

$$D_1 = (a + c_0) \, \rho_2 - (b + c_0) \, \rho_1, \quad D_2 = (a_1 + c) \, \rho_2 - (a_2 + c) \, \rho_1.$$

In the sequel it is assumed that [2]: $a > 0, \quad b > 0, \quad a_l > 0, \quad d_l > 0, \quad l = 1, 2$.

3. Plane Waves

Suppose that plane waves corresponding to the wave number k and angular frequency ω are propagated in the x_1-direction through the binary mixture of thermoelastic solid. Then

$$\mathbf{u}'(\mathbf{x}, t) = \mathbf{A}' \exp \{i(kx_1 - \omega t)\}, \quad \mathbf{u}''(\mathbf{x}, t) = \mathbf{A}'' \exp \{i(kx_1 - \omega t)\},$$

$$\theta(\mathbf{x}, t) = A_0 \exp \{i(kx_1 - \omega t)\}, \tag{2}$$

where $\mathbf{A}' = (A_1', A_2', A_3')$ and $\mathbf{A}'' = (A_1'', A_2'', A_3'')$ are constant vectors, A_0 is constant value and $\omega > 0$.

Keeping in mind the condition (2) from (1) it follows that:

$$[-(a_1 + b_1\delta_{1l})k^2 + \beta_1]\, A_l' + [-(c + d\delta_{1l})k^2 - \beta_2]\, A_l'' - i\alpha_1\delta_{1l}kA_0 = 0,$$

$$[-(c + d\delta_{1l})k^2 - \beta_2]\, A_l' + [-(a_2 + b_2\delta_{1l})k^2 + \beta_3]\, A_l'' - i\alpha_2\delta_{1l}kA_0 = 0,$$

$$(-\alpha_0 k^2 + i\omega\alpha_3)A_0 - \omega k(\alpha_4 A_1' + \alpha_5 A_1'') = 0, \qquad l = 1, 2, 3, \tag{3}$$

where δ_{lj} is the Kronecker delta, $\beta_1 = i\omega\nu + \omega^2\rho_1$, $\beta_2 = i\omega\nu$, $\beta_3 = i\omega\nu + \omega^2\rho_2$. From (3) for A_1', A_1'' and A_0 we have the system:

$$(ak^2 - \beta_1)\, A_1' + (c_0k^2 + \beta_2)\, A_1'' + i\alpha_1 kA_0 = 0,$$

$$(c_0k^2 + \beta_2)\, A_1' + (bk^2 - \beta_3)\, A_1'' + i\alpha_2 kA_0 = 0,$$

$$\omega\alpha_4 kA_1' + \omega\alpha_5 kA_1'' + (\alpha_0 k^2 - i\omega\alpha_3)A_0 = 0. \tag{4}$$

Obviously, if k is a solution of the equation

$$\alpha_0 d_1 k^6 - (\alpha_0\beta_0 + i\omega\alpha_3 d_1 + i\omega\beta_4)k^4 + (\alpha_0\beta + i\omega\alpha_3\beta_0 + i\omega\beta_5)k^2 - i\omega\alpha_3\beta = 0, \tag{5}$$

then the system (4) has non-trivial solution (A_1', A_1'', A_0), where

$$\beta = \beta_1\,\beta_3 - \beta_2^2, \quad \beta_4 = \alpha_2\,(a\,\alpha_5 - c_0\,\alpha_4) + \alpha_1\,(b\,\alpha_4 - c_0\,\alpha_5),$$

$$\beta_0 = a\,\beta_3 + b\,\beta_1 + 2c_0\,\beta_2, \quad \beta_5 = \alpha_2\,(\beta_1\,\alpha_5 + \beta_2\,\alpha_4) + \alpha_1\,(\beta_3\,\alpha_4 + \beta_2\,\alpha_5).$$

Similarly, from (3) for A_l' and A_l'' $(l = 2, 3)$ we have the system

$$(a_1k^2 - \beta_1)\, A_l' + (ck^2 + \beta_2)\, A_l'' = 0, \quad ck^2 + \beta_2)\, A_l' + (a_2k^2 - \beta_3)\, A_l'' = 0. \tag{6}$$

Obviously, if k is a solution of the equation

$$d_2\, k^4 - (\omega^2\, p_2 + i\,\omega\,\nu\, q_2)\, k^2 + \omega^4\,\rho_1\,\rho_2 + i\,\omega^3\,\nu\,(\rho_1 + \rho_2) = 0, \tag{7}$$

then the system (6) has non-trivial solution (A_l', A_l'') $(l = 2, 3)$.

The relations (5) and (7) will be called the dispersion equations of longitudinal and transverse plane waves, respectively. Obviously, if k is real, then the corresponding plane wave has the constant amplitude, and if k is complex with $\operatorname{Im} k > 0$, then the plane wave is damped as $x_1 \to +\infty$.

Let k_1^2, k_2^2, k_3^2 and k_4^2, k_5^2 be roots of (5) and (7), respectively. We assume that $\operatorname{Im} k_j \geq 0$, $j = 1, 2, \cdots, 5$.

Remark. Through a binary mixture of thermoelastic solids propagates three longitudinal plane waves with wave numbers k_1, k_2, k_3 and four transverse plane waves (two by two waves with wave numbers k_4 and k_5).

Theorem 3.1. *If* $a > 0,\ d_1 > 0,\ \alpha_1 + \alpha_2 = 0,$ *then*

(a) $k_1^2 > 0,\ \mathrm{Im}\, k_j^2 \neq 0\ (j = 2,3)$ *for* $D_1 = 0,$
(b) $\mathrm{Im}\, k_l \neq 0\ (l = 1,2,3)$ *for* $D_1 \neq 0.$

Theorem 3.2. *f* $a_1 > 0,\ d_2 > 0,$ *then*

(a) $k_4^2 > 0,\ \mathrm{Im}\, k_5^2 \neq 0$ *for* $D_2 = 0,$
(b) $\mathrm{Im}\, k_j \neq 0\ (j = 4,5)$ *for* $D_2 \neq 0.$

Theorem 3.1 and 3.2 lead to the following result.

Corollary 3.1. *If* $\alpha_1 + \alpha_2 = 0$ *and* $D_1 D_2 \neq 0$, *then only damped plane waves propagate through the binary mixtures of thermoelastic solids.*

4. Existence of Eigenfrequencies

Let S be a closed surface surrounding a finite domain Ω in R^3, $\bar{\Omega} = \Omega \cup S$, $S \in C^{2,\lambda}$, $0 < \lambda \leq 1$. A vector function $\mathbf{u}$ is called *regular* in Ω if $u_j \in C^2(\Omega) \cap C^1(\bar{\Omega})$, $j = 1, 2 \cdots, l$.

The system of steady vibration equations of the linear theory of binary mixtures of thermoelastic solids can be written as follows [2, 3]

$$a_1 \Delta\, \mathbf{u}' + b_1 \text{ grad div } \mathbf{u}' + c\, \Delta\, \mathbf{u}'' + d \text{ grad div } \mathbf{u}'' + \beta_1 \mathbf{u}' - \beta_2 \mathbf{u}'' - \alpha_1 \text{ grad } \theta = \mathbf{0},$$

$$c \Delta\, \mathbf{u}' + d \text{ grad div } \mathbf{u}' + a_2 \Delta\, \mathbf{u}'' + b_2 \text{ grad div } \mathbf{u}'' - \beta_2 \mathbf{u}' + \beta_3 \mathbf{u}'' - \alpha_2 \text{ grad } \theta = \mathbf{0},$$

$$(\alpha_0\, \Delta + i\omega\alpha_3)\, \theta + i\omega\alpha_4 \,\mathrm{div}\, \mathbf{u}' + i\omega\alpha_5 \,\mathrm{div}\, \mathbf{u}'' = 0. \tag{8}$$

Let Ω be a domain of R^3 ocupied by a binary mixture of thermoelastic solids. We consider the following interior homogeneous BVP of steady vibration of binary mixtures of thermoelastic solids [5]: find a regular solution $\mathbf{U} = (\mathbf{u}', \mathbf{u}'', \theta)$ in Ω to the system (8) satisfying the boundary condition

$$\lim_{\Omega \ni \mathbf{x} \to \mathbf{z} \in S} U(x) \equiv U(z) = \mathbf{0}. \tag{9}$$

Lemma 4.1. *A regular solution* $\mathbf{U} = (\mathbf{u}', \mathbf{u}'', \theta)$ *of the BVP (8), (9) satisfies the conditions:* $\mathbf{u}'(\mathbf{x}) = \mathbf{u}''(\mathbf{x})$, $\theta(\mathbf{x}) = 0$, $D_1 \,\mathrm{div}\, \mathbf{u}'(\mathbf{x}) = 0$, $D_2 \,\mathrm{curl}\, \mathbf{u}'(\mathbf{x}) = \mathbf{0}$ *for* $\mathbf{x} \in \Omega$.

Lemma 4.2. *A regular solution* $\mathbf{u}'$ *to the system*

$$(a_1 + c)\Delta\,\mathbf{u}'(\mathbf{x}) + (b_1 + d)\,\mathrm{grad\,div}\,\mathbf{u}'(\mathbf{x}) + \omega^2\rho_1\mathbf{u}'(\mathbf{x}) = \mathbf{0},$$

$$(a_2 + c)\Delta\,\mathbf{u}'(\mathbf{x}) + (b_2 + d)\,\mathrm{grad\,div}\,\mathbf{u}'(\mathbf{x}) + \omega^2\rho_2\mathbf{u}'(\mathbf{x}) = \mathbf{0},$$

$$\alpha_3\,\theta + (\alpha_4 + \alpha_5)\,\mathrm{div}\,\mathbf{u}'(\mathbf{x}) = 0$$

satisfies the conditions

(a) $\mathbf{u}'(\mathbf{x}) = \mathbf{0}$ *for* $D_1 D_2 \neq 0$,
(b)

$$(\Delta + \omega^2\rho\, q_2^{-1})\,\mathbf{u}'(\mathbf{x}) = \mathbf{0}, \;\; \mathrm{div}\,\mathbf{u}'(\mathbf{x}) = 0 \;\; \textit{for} \;\; D_1 \neq 0, \quad D_2 = 0, \quad (10)$$

(c)

$$(\Delta + \omega^2\rho\, q_1^{-1})\,\mathbf{u}'(\mathbf{x}) = \mathbf{0}, \;\; \mathrm{curl}\,\mathbf{u}'(\mathbf{x}) = 0 \;\; \textit{for} \;\; D_1 = 0, \quad D_2 \neq 0, \quad (11)$$

(d)

$$q_1\Delta\mathbf{u}'(\mathbf{x}) + (q_2 - q_1)\mathrm{grad\,div}\,\mathbf{u}'(\mathbf{x}) + \omega^2(\rho_1 + \rho_2)\mathbf{u}'(\mathbf{x}) = \mathbf{0}, \qquad (12)$$

for $D_1 = D_2 = 0$, *where* $\mathbf{x} \in \Omega$.

Lemmas 4.1 and 4.2 yield the following theorems.

Theorem 4.1. *If* $D_1 D_2 \neq 0$, *then the homogeneous BVP (8), (9) has only the trivial solution in the class of regular vectors, that is there exists no eigenfrequency.*

Theorem 4.2. *If* $D_1 D_2 = 0$, *then the homogeneous BVP (8), (9) has a non-trivial solution* $\mathbf{U} = (\mathbf{u}', \mathbf{u}', 0)$ *in the class of regular vectors. In addition,*

(a) if $D_1 \neq 0$, $D_2 = 0$, *then the vector* $\mathbf{u}'$ *is a solution to the system (10) satisfying the boundary condition*

$$\mathbf{u}'(\mathbf{z}) = \mathbf{0}, \qquad \mathbf{z} \in S; \qquad (13)$$

the BVPs (8), (9) and (10), (13) have the same eigenfrequencies,
(b) if $D_1 = 0, D_2 \neq 0$, *then the vector* $\mathbf{u}'$ *is a solution to the BVP (11), (13); the BVPs (8), (9) and (11), (13) have the same eigenfrequencies,*
(c) if $D_1 = D_2 = 0$, *then the vector* $\mathbf{u}'$ *is a solution to the BVP (12), (13); the BVPs (8), (9) and (12), (13) have the same eigenfrequencies.*

5. Connection Between Plane Waves and Existence of Eigenfrequencies

By virtue of Theorems 3.1, 3.2, 4.1, 4.2 and Corollary 3.1 we have the following connection between plane waves and eigenfrequencies in the theory of binary mixtures of thermoelastic solids:

(i) If all plane waves propagating through a binary mixture of thermoelastic solids are damped, then the interior homogeneous steady vibration BVP have only the trivial solution, that is, there exist no eigenfrequencies.

(ii) If all longitudinal (transverse) plane waves propagating through a binary mixture of thermoelastic solids are damped, then in the interior homogeneous steady vibration BVP divergence (curl) of partial displacements vanishes.

(iii) If through a binary mixture of thermoelastic solids there propagates at least one plane wave with constant amplitude, then the existence of eigenfrequencies is possible in the interior homogeneous steady vibration BVP.

The above-mentioned connections (ii) and (iii) between plane waves and eigenfrequencies hold for isotropic elastic solids in the classical theory of elasticity, thermoelasticity, generalized thermoelasticity, micropolar theory of elasticity and thermoelasticity, theory of mixtures.

Acknowledgments

The designated project has been fulfilled by financial support of Georgian National Science Foundation (Grant # GNSF/ST06/3-033). Any idea in this publication is possessed by the author and may not represent the opinion of Georgian National Science Foundation itself.

This paper dedicated to Professor T. Ruggeri on the occasion of his 60th birthday.

References

1. A. E. Green and T. R. Steel, *Int. J. Engng. Sci.* **4**, 483 (1966).
2. T. R. Steel, *Quart. J. Mech. Appl. Math.* **20**, 57 (1967).
3. T. Burchuladze and M. Svanadze, *J. Thermal Stesses* **23**, 601 (2000).
4. Y. Y. Rushchitskii, *Elements of Mixture Theory* (Naukova Dumka, Kiev, 1991).
5. R. M. Bowen, *Theory of Mixtures*, in *Continuum Physics*, ed. A. C. Eringen (Academic Press, 1976), pp. 1–127.

ANALYSIS OF HEAT CONDUCTION PHENOMENA IN A ONE-DIMENSIONAL HARD-POINT GAS BY EXTENDED THERMODYNAMICS

S. TANIGUSHI[1], M. NAKAMURA[1], M. SUGIYAMA[1], M. ISOBE[1] and N. ZHAO[1,2]

[1] *Graduate School of Engineering, Nagoya Institute of Technology, Nagoya 466-8555, Japan.*

[2] *Department of Physics, Sichuan University, Chengdu 610064, China.*

E-mail: sugiyama@nitech.ac.jp

After explaining a one-dimensional hard-point gas, we show the following points: The first point is that extended thermodynamics (ET) can be adopted as a suitable phenomenological theory that can explain the simulation data of heat conduction in a one-dimensional hard-point gas. The agreement between the simulation data of the temperature profile and its theoretical prediction is fairly well. The second is that we specify the well-defined heat conductivity on the basis of ET, and then we revisit the system-size dependence of the heat conductivity. The last is the discussion about the higher moments.

1. Introduction

Studies of highly nonequilibrium and nonlinear phenomena such as shock wave phenomena with rapid change in time and steep gradient in space are strongly required nowadays in many scientific and engineering fields. Extended thermodynamics (ET) [1] is a theory by which we can analyze such phenomena deeply. Recently consistent-order extended thermodynamics (COET) [2] was proposed as a revised version of ET.

In this paper, firstly we show explicitly the usefulness of COET through studying heat conduction phenomena in a one-dimensional hard-point gas at rest both by molecular dynamics (MD) simulation and by COET. [3] The agreement between the simulation data of the temperature profile and its theoretical prediction is fairly well. Secondly we specify the well-defined heat conductivity on the basis of COET and analyze its system-size dependence in the thermodynamic limit. [3] Lastly we will discuss about the higher moments in COET by comparing their numerical results.

2. Hard-Point Gas Model

One-dimensional binary hard-point gas [4] is composed of N point-particles with alternating masses m_1 and m_2 moving in a one-dimensional segment on the x-axis ($0< x < L$) as shown in Fig.1. Interaction between two particles is made through elastic collision, and each particle moves with a constant velocity between two collisions.

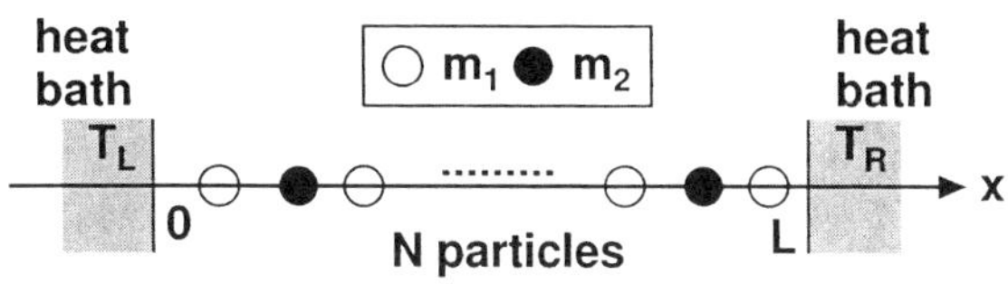

Fig. 1. Hard-point gas model.

The Maxwell boundary condition is adopted at two sides of the system. A colliding particle with the boundary wall at temperature T is reflected with the velocity c obeying the probability distribution function φ :

$$\varphi(c) = \frac{m|c|}{T} \exp\left(-\frac{mc^2}{2T}\right), \tag{1}$$

where m is the mass of the particle. The temperatures at the left and right sides in Fig.1, called wall temperatures, are fixed to be T_{L} and T_{R}, respectively.

When the average number density N/L is fixed, there are three independent parameters in the model, that is, N, $T_{\mathrm{L}}/T_{\mathrm{R}}$ and r ($\equiv m_2/m_1$). [4] Hereafter we will use dimensionless quantities in the units of mass m_1 and temperature T_{R}, and will show the numerical results with several total particle number N at $N/L = 1$, $r = 1.1$ and $T_{\mathrm{L}}/T_{\mathrm{R}} = 1.1$ as a typical example.

3. Molecular Dynamics Simulation of Heat Conduction in a Hard-Point Gas

By using efficient event-driven MD-simulation [5] of stationary heat conduction in a hard-point gas at rest we obtain stationary one-body distribution functions $f(x, c)$ for both of the two subsystems composed of the particles with the same mass (hereafter, referred to as m_1 and m_2-subsystem). Then we estimate the moments for each subsystem such as the number density

$n(x)$, the pressure $P(x)$, the heat flux $q(x)$:

$$n(x) = \int f(x,c)\mathrm{d}c, \tag{2}$$

$$P(x) = \int mc^2 f(x,c)\mathrm{d}c, \tag{3}$$

$$q(x) = \int \frac{m}{2}c^3 f(x,c)\mathrm{d}c, \tag{4}$$

where m is either m_1 or m_2. Moreover, the (kinetic) temperature is given by the relation:

$$T(x) = P(x)/n(x). \tag{5}$$

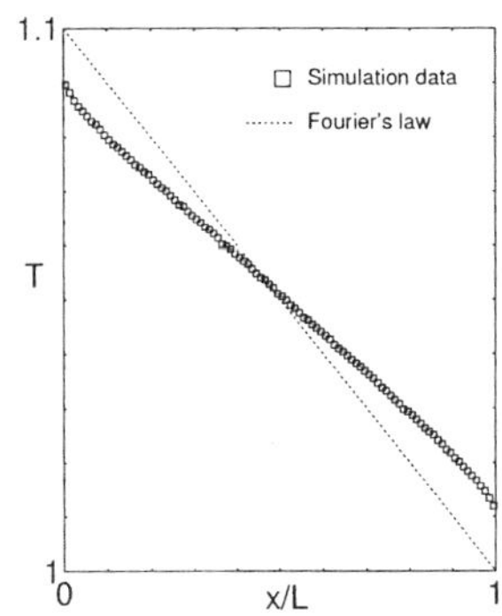

Fig. 2. Typical temperature profile.

Remarkable points obtained by the MD simulation are summarized as follows [3,6]: (Hereafter we show mainly the results of m_1-system.)
(A) Temperature profile is an S-shaped curve as shown in Fig.2.
(B) There appear temperature jumps at the boundaries of the system. See also Fig.2.
(C) It has been reported that the heat conductivity κ diverges as the system size L tends to infinity (in a thermodynamic limit: N/L =fixed) such that

$$\kappa \sim L^{\alpha}. \tag{6}$$

The estimated value of the exponent α is, however, scattered among researchers. [4,7,8]

Even though many previous simulation data were beyond the applicability range of linear irreversible thermodynamics, the heat conductivity has been estimated in a rough qualitative manner on the basis of linear irreversible thermodynamics. We need, therefore, suitable phenomenological theory that is valid beyond the local equilibrium assumption and that can explain (A)-(C) consistently.

4. Consistent-Order Extended Thermodynamics — Brief Summary

COET was proposed as a revised version of ET in the sense that it provides assignment of order of magnitude to the moments (G-moments). [2] In the context of COET, Bhatnagar-Gross-Krook (BGK) equation [9,10] with one relaxation time τ in the collision term has been adopted. The basic moment equations in the 3rd-order COET are derived as follows [2]:

$$\left.\begin{aligned} 0 &= \frac{\mathrm{d}P}{\mathrm{d}x}, \\ 0 &= \frac{\mathrm{d}G_3}{\mathrm{d}x}, \\ G_3 &= -\sqrt{\frac{3}{2}}P\tau\frac{\mathrm{d}\theta}{\mathrm{d}x} - 2\tau\frac{\mathrm{d}G_4}{\mathrm{d}x}, \\ G_4 &= -5G_3\tau\frac{\mathrm{d}\theta}{\mathrm{d}x}, \end{aligned}\right\} \text{(A)} \tag{7}$$

$$\left.\begin{aligned} G_6 &= -\sqrt{30}\theta G_3\tau\frac{\mathrm{d}\theta}{\mathrm{d}x}, \\ G_5 &= -\sqrt{5}\theta\tau\frac{\mathrm{d}G_4}{\mathrm{d}x} - \sqrt{6}\tau\frac{\mathrm{d}G_6}{\mathrm{d}x} - 3\sqrt{5}G_4\tau\frac{\mathrm{d}\theta}{\mathrm{d}x}, \\ G_7 &= -\sqrt{7}\theta\tau\frac{\mathrm{d}G_6}{\mathrm{d}x} - \sqrt{\frac{105}{2}}\theta G_4\tau\frac{\mathrm{d}\theta}{\mathrm{d}x} - 4\sqrt{7}G_6\tau\frac{\mathrm{d}\theta}{\mathrm{d}x}, \\ G_9 &= -3\sqrt{14}\theta G_6\tau\frac{\mathrm{d}\theta}{\mathrm{d}x}, \end{aligned}\right\} \text{(B)} \tag{8}$$

where $\theta \equiv T/m_1$ and G's are G-moments defined in equation (15) below. From these equations, we obtain a key equation for our analysis:

$$G_3 = -\sqrt{\frac{3}{2}}P\tau\frac{\mathrm{d}\theta}{\mathrm{d}x} + 10G_3\tau^2\frac{\mathrm{d}^2\theta}{\mathrm{d}x^2} \tag{9}$$

with

$$G_3 = const., \quad P = const., \tag{10}$$

where G_3 is expressed in terms of the heat flux q such that $G_3 = \sqrt{2/3}q$.

Equation (9) can be regarded as an ordinary differential equation with constant coefficients for the temperature field θ. Its solution consistent to the approximation of the 3rd-order COET can be obtained in a power series form as follows [3,11,12]:

$$\begin{aligned}\theta(x) = \theta_0 + \theta_1\left(x - \frac{L}{2}\right) + \frac{G_3 + \sqrt{\frac{3}{2}}P\tau\theta_1}{20G_3\tau^2}\left(x - \frac{L}{2}\right)^2 \\ + \frac{\frac{3}{2}P^2\tau\theta_1 + \sqrt{\frac{3}{2}}G_3P}{600G_3^2\tau^3}\left(x - \frac{L}{2}\right)^3,\end{aligned} \tag{11}$$

where two constants θ_0 and θ_1 can be specified by the gas temperatures at the boundaries $\theta(x = 0) = \theta_\mathrm{L}$ and $\theta(x = L) = \theta_\mathrm{R}$. Note that the temperatures θ_L and θ_R are, in general, different from the wall temperatures T_L and T_R. In our phenomenological analysis, the values of P and G_3 will be taken as those estimated independently in the numerical simulations. There remains, therefore, only one parameter, that is, the relaxation time τ in the equation (11).

Temperature jumps at the boundaries with the Maxwell boundary condition (1) can be obtained from the following relations, which are derived from the analysis of the conservation laws of mass and energy at the boundaries [3,11–16]:

$$\frac{\frac{T_\mathrm{L}}{m_1} - \theta_\mathrm{L}}{\theta_\mathrm{L}} = \frac{\sqrt{3\pi}}{2}\frac{G_3}{P\sqrt{\theta_\mathrm{L}}} + \left(\frac{T_\mathrm{L}}{3m_1\theta_\mathrm{L}} + 2\right)\frac{G_3^2}{P^2\theta_\mathrm{L}}, \tag{12}$$

$$\frac{\frac{T_\mathrm{R}}{m_1} - \theta_\mathrm{R}}{\theta_\mathrm{R}} = -\frac{\sqrt{3\pi}}{2}\frac{G_3}{P\sqrt{\theta_\mathrm{R}}} + \left(\frac{T_\mathrm{R}}{3m_1\theta_\mathrm{R}} + 2\right)\frac{G_3^2}{P^2\theta_\mathrm{R}}. \tag{13}$$

5. Comparison between Analytical and Simulation Results

5.1. *Temperature Profile*

Temperature profiles obtained by MD simulations for several values of the total particle number N are shown in Fig.3. Theoretical curves based on COET and the curve of the classical Fourier's law are also shown for comparison in the figure. The values of the quantities N, τ, P and G_3 adopted in the analysis by COET are listed in Table 1.

We can see that the agreement between the simulation data of the temperature profile and its theoretical prediction is fairly well. Especially we notice that the 3rd-order COET can quantitatively explain the S-curved temperature profile.

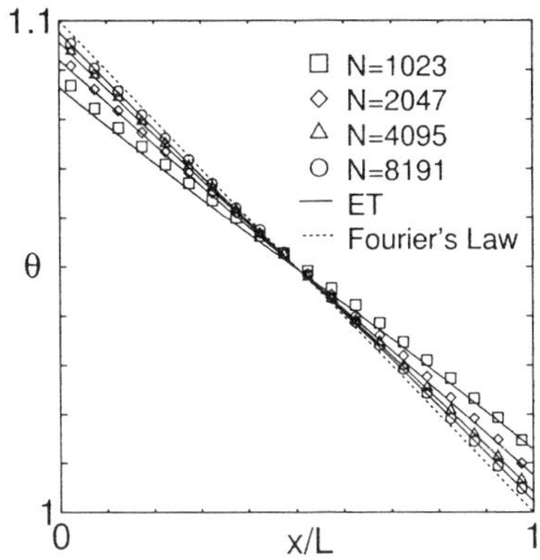

Fig. 3. Temperature profiles obtained by MD simulations. Theoretical curves based on COET (solid curves) and the curve of the classical Fourier's law (dotted line) are also shown for comparison.

Table 1. Values of N, τ, P and G_3 adopted in the analysis by COET.

N	1023	2047	4095	8191
τ	98.7	98.5	101	103
P	0.525	0.525	0.525	0.525
G_3	0.00439	0.00259	0.00141	0.00075

5.2. *Temperature Jumps at the Boundaries*

Gas temperatures at the boundaries of the system obtained by MD simulations and predicted by COET (equations (12) and (13)) for several values of the total particle number N are listed in Table 2. Here Kn is the Knudsen number defined by

$$\mathrm{Kn} \equiv \frac{\tau}{L}\sqrt{\frac{T_{\mathrm{L}} + T_{\mathrm{R}}}{2m_1}}. \tag{14}$$

We can see in the Table 2 that the agreement between the simulation data of the temperature jumps and their theoretical prediction by COET is fairly well.

Now we have confirmed that COET can explain both the S-shaped temperature profile and temperature jumps at the boundaries in a consistent way. We understand the usefulness of COET explicitly.

Table 2. Gas temperatures at the boundaries obtained by MD simulations and predicted by COET. Kn is the Knudsen number.

N	1023	2047	4095	8191
Kn	0.0987	0.0493	0.0252	0.0128
$\theta_{\rm L}$(MD)	1.090	1.094	1.096	1.098
$\theta_{\rm L}$(COET)	1.0865	1.0920	1.0957	1.0977
$\theta_{\rm R}$(MD)	1.012	1.007	1.004	1.002
$\theta_{\rm R}$(COET)	1.0128	1.0075	1.0041	1.0022

5.3. *Heat Conductivity*

From the present analysis based on COET, we recognize that the well-defined heat conductivity is given by $\kappa \sim P\tau$ (equation (9)), and that the relaxation time τ increases slowly with the increase of the total particle number N (Table 2). From this observation, it seems to follow an assertion that the heat conductivity diverges very slowly in the thermodynamic limit. Therefore the value of the exponent α is small.

However, here, we have shown the results only in one special case with a fixed value of the mass ratio $r = 1.1$. Our detailed study, which is omitted here for simplicity, reveals the r-dependence of the heat conductivity κ. [3] The exponent α increases with the increase of the mass ratio $r(> 1)$. This fact may be one of the reasons why the values of the exponent α estimated in previous works are scattered among researchers. Further study of this subject is highly expected.

6. Higher-Order Moments

Higher-order moments also play an important role in COET, but it is usually difficult to measure them experimentally. Therefore it seems to be interesting to show explicitly such higher-order moments obtained by numerical simulations of a hard-point gas and to compare them with those derived from the 3rd-order COET. These numerical data may give us an important information when we try to study higher-order COET than the third one. In fact, we have already made some preliminary studies of higher-order COET, which, we hope, will soon be reported. [?]

For completeness, we firstly give the definition of the G-moments:

$$G_i = m_1 \sqrt{\theta}^i \int \psi_i f \mathrm{d}c, \tag{15}$$

where ψ_i is the i-th orthonormal Hermite polynomial:

$$\psi_i = \frac{(-1)^i}{\sqrt{2^i i!}} e^{\chi^2} \frac{d^n}{d\chi^n} \left(e^{-\chi^2}\right), \left(\chi \equiv \frac{c}{\sqrt{2\theta}}\right). \tag{16}$$

6.1. *Second-order G-Moments*

The 2nd-order G-moments are G_4 and G_6-moments, which correspond to Hermite polynomials ψ_4 and ψ_6. Figure 4 shows the profiles of G_4 and G_6-moments, and their theoretical predictions based on the 3rd-order COET.

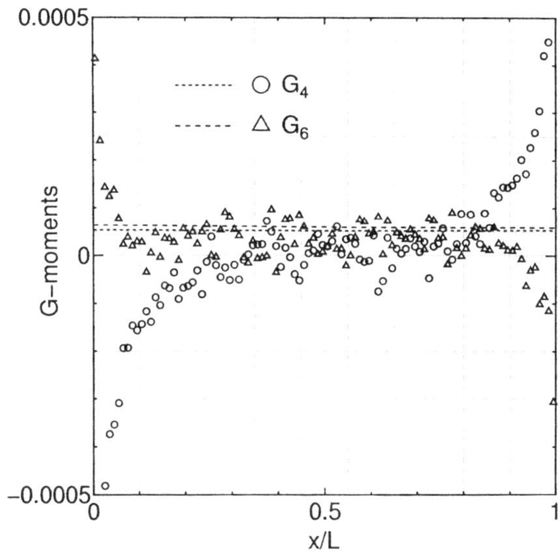

Fig. 4. Second-order G-moments G_4 and G_6 obtained by MD simulations. Their theoretical predictions from the 3rd-order COET are also shown (broken lines). $N = 2047$.

We can see from Fig.4 that the agreement between numerical results and theoretical prediction is good in the bulk region, but not so good near the boundaries because of the boudary layer effect. In order to predict such boundary effect properly, we need an analysis based on higher-order COET than the third-order COET.

6.2. *Third-Order G-Moments*

The 3rd-order G-moments are G_5, G_7 and G_9-moments, which correspond to Hermite polynomials ψ_5, ψ_7 and ψ_9. Figure 5 shows the profiles of G_5, G_7

and G_9-moments, and their theoretical predictions based on the 3rd-order COET. In the figure, the predicted values are very near the horizontal line with the value 0.

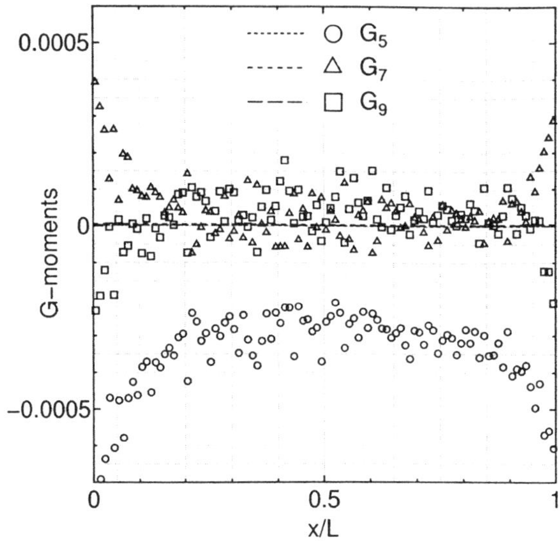

Fig. 5. Third-order G-moments G_5, G_7 and G_9 obtained by MD simulations. Their theoretical predictions from the 3rd-order COET are also shown (broken lines). $N = 2047$.

We notice some discrepancy between numerical data and predicted ones for the G_5-moment probably due to the boundary layer effect. The effect is more remarkable in a system with smaller size. We need careful analysis of this problem.

7. Concluding Remarks

The case that the value of the mass ratio r is near unity is also interesting. If $r = 1$ the energy transport is in a ballistic type. As the heat conductivity is a measure of the intensity of the diffusive energy transport, we can analyze the transition from diffusive energy transport to ballistic energy transport in terms of the heat conductivity. We have found that heat conductivity diverges with a power-law relation as r tends to unity. [3]

In conclusion we have shown that extended thermodynamics is a good phenomenological theory for analyzing heat conduction phenomena in a hard-point gas in highly nonequilibrium.

Acknowledgments

This paper is dedicated to Tommaso Ruggeri with deep esteem and friendship. M. S. and N. Z. were supported by Grant-in-Aids from the Japan Society for the Promotion of Science (No. 1604076) and by Daiko Foundation (No. 10100).

References

1. I. Müller and T. Ruggeri, *Rational Extended Thermodynamics*, (Springer-Verlag, New York, 1998).
2. I. Müller, D. Reitebuch and W. Weiss, *Continuum Mech. Thermodyn.* **15**, 113 (2003).
3. S. Taniguchi, M. Nakamura, M. Sugiyama, M. Isobe and N. Zhao, *J. Phys. Soc. Jpn.* **77**, (2008) (to be published).
4. A. Dhar, *Phys. Rev. Lett.* **86**, 3554 (2001).
5. M. Isobe, *Int. J. Mod. Phys. C* **10**, 1281 (1999).
6. S. Lepri, R. Livi and A. Politi, *Phys. Rep.* **377**, 1 (2003).
7. P. Grassberger, W. Nadler and L. Yang, *Phys. Rev. Lett.* **89**, 180601 (2002).
8. G. Casati and T. Prosen, *Phys. Rev. E* **67**, 015203(R) (2003).
9. P. L. Bhatnagar, E. P. Gross and M. Krook, *Phys. Rev.* **94**, 511 (1954).
10. S. Chapman and T. G. Cowling, *The Mathematical Theory of Non-Uniform Gases* (Cambridge University Press, Cambridge, 1970).
11. M. Sugiyama and N. Zhao, *J. Phys. Soc. Jpn.* **74**, 1899 (2005).
12. M. Sugiyama and N. Zhao, *Rend. Circ. Mat. Palermo, Series II, Suppl.* **78**, 333 (2006).
13. H. Struchtrup and W. Weiss, *Continuum Mech. Thermodyn.* **12**, 1 (2000).
14. H. Struchtrup, *Phys. Rev. E* **65**, 041204 (2002).
15. H. Struchtrup, *Macroscopic Transport Equations for Rarefied Gas Flows, Approximation Methods in Kinetic Theory* (Springer-Verlag, Berlin, Heidelberg, 2005).
16. N. Zhao and M. Sugiyama, *Contiuum Mech. Thermodyn.* **18**, 367 (2007).
17. N. Zhao, M. Sugiyama and S. Taniguchi, In preparation.

EQUIVALENCE TRANSFORMATIONS AND DIFFERENTIAL INVARIANTS FOR GENERALIZED WAVE EQUATIONS

R. TRACINÀ

Dipartimento di Matematica e Informatica, Università di Catania
Viale A. Doria 6, 95125 Catania, Italy
E-mail: tracina@dmi.unict.it

C. SOPHOCLEOUS

Department of Mathematics and Statistics, University of Cyprus,
CY 1678 Nicosia, Cyprus
E-mail: christod@ucy.ac.cy

In this paper we consider the class of equations $u_{tt} = f(x,t,u,u_x,u_t)u_{xx} + g(x,t,u,u_x,u_t)$. We derive equivalence transformations and we classify differential invariants.

Keywords: Wave equations; Equivalence transformations; Invariants

1. Introduction

The differential invariants of Lie groups of continuous transformations can be used in wide fields: classification of invariant differential equations and variational problems arising in the construction of physical theories, solution methods for ordinary and partial differential equations, equivalence problems for geometric structures.

S. Lie [1] showed that every invariant system of differential equations [2] and every variational problem [3] could be directly expressed in terms of differential invariants. Lie also used [2] to integrate ordinary differential equations, and succeeded in completely classifying all the differential invariants for all possible finite-dimensional Lie groups of point transformations in the case of one independent and one dependent variable. Lie's preliminary results on invariant differentiations and existence of finite bases of differential invariants were generalized by Tresse [4] and Ovsiannikov [5].

The general theory of differential invariants of Lie groups including algorithms of construction of differential invariants can be found in Ref. 5,6.

A simple method for constructing invariants of families of linear and nonlinear differential equations admitting infinite equivalence transformation groups was developed in Ref. 7 (see also Ref. 8). This method was then applied to several linear and nonlinear equations [9–19] .

In the present work, in the spirit of Ibragimov's work [9] , we consider the class of nonlinear one-dimensional wave equations of the form

$$u_{tt} = f(x, t, u, u_t, u_x)u_{xx} + g(x, t, u, u_t, u_x). \tag{1}$$

We derive the equivalence transformations for equations (1) and we construct differential invariants and differential equations with the employment of the derived equivalence transformations.

Hyperbolic type second-order nonlinear PDEs in two independent variables are used in modern mathematical physics. They can describe various types of wave propagation and model several phenomena in various fields of hydro and gas dynamics, chemical technology, super conductivity, crystal dislocation.

In the sections 2 and 3, using the infinitesimal method, we find equivalence transformations and differential invariants of the first order for the class (1), respectively. Finally some applications of the invariants and invariant equations appear in the last section.

2. Equivalence transformations

In order to find continuous group of equivalence transformations of the class (1) we consider the arbitrary functions f and g that appear in our equation as dependent variables and we apply the Lie infinitesimal invariance criterion [5] , that is, we look for the equivalent operator in the following form:

$$Y = \xi^1\partial_t + \xi^2\partial_x + \eta\partial_u + \mu\partial_f + \nu\partial_g \tag{2}$$

that suitable prolongation applied to the equation (1) leaves it invariant and where the unknowns infinitesimal coordinates ξ^1, ξ^2 and η depend on t, x and u, while μ and ν depend on t, x, u, u_x, u_t, f and g.

This means that we require the invariance of the equation (1) with respect the following prolongation of operator Y

$$Y^{(2)} = Y + \zeta^{11}\partial_{u_{tt}} + \zeta^{22}\partial_{u_{xx}} \tag{3}$$

under the constraint that the equation (1) is satisfied. Taking into account the expressions of the coordinates ζ^{11} and ζ^{22} that can be found in Ref. 5,20, we get the *determining system* in the unknowns ξ^1, ξ^2, η, μ and ν. After having solved the *determining system* obtained from invariance conditions, we get the equivalence generator.

By applying the above procedure, we find that Eqs. (1) admit an infinite continuous group of equivalence transformations generated by the Lie algebra $\mathcal{L}_{\mathcal{E}}$ spanned by the following infinitesimal operators

$$\begin{aligned} Y_\phi &= \phi(x)\partial_x + 2f\phi'\partial_f + u_x f\phi''\partial_g - u_x\phi'\partial_{u_x}, \\ Y_\tau &= \tau(t)\partial_t - 2f\tau'\partial_f - (2g\tau' + u_t\tau'')\partial_g - u_t\tau'\partial_{u_t}, \\ Y_\psi &= \psi(x,t,u)\partial_u + (\psi_x + u_x\psi_u)\partial_{u_x} + (\psi_t + u_t\psi_u)\partial_{u_t} + \\ &\quad +[g\psi_u + \psi_{tt} + 2u_t\psi_{tu} + u_t^2\psi_{uu} - f(\psi_{xx} + 2u_x\psi_{xu} + u_x^2\psi_{uu})]\partial_g, \end{aligned} \tag{4}$$

where ϕ, τ and ψ are arbitrary functions of the indicated variables. Here ϕ', ϕ'' and τ', τ'' denote the derived functions of ϕ and τ, respectively. Equivalence transformations (4) can be written in the finite form

$$x' = \alpha(x), \quad t' = \beta(t), \quad u' = \gamma(x,t,u). \tag{5}$$

3. Differential invariants and invariant equations

An *invariant* of equations (1) is a function of the form

$$J(x,t,u,u_x,u_t,f,g) \tag{6}$$

which is invariant under the equivalence transformation $\mathcal{E}$. Likewise, differential invariants of $\mathcal{E}$,

$$J(x,t,u,u_x,u_t,f,g,f_t,f_x,f_u,f_{u_x},f_{u_t},g_t,g_x,g_u,g_{u_x},g_{u_t},\dots) \tag{7}$$

are called *differential invariants of order s* of equations (1), where s denotes the maximal order derivative of f and/or g. We call (6) differential invariant of order zero.

In order to find the differential invariants of order zero, we apply the operators (4) on the function J defined by (6) and obtain 3 linear first-order partial differential equations for J :

$$Y_\phi(J) = 0, \quad Y_\tau(J) = 0, \quad Y_\psi(J) = 0. \tag{8}$$

that provide the *invariant criterion* under the transformations (5). Solution of the system (8) gives the required invariants. Since $\phi(x)$, $\tau(t)$ and $\psi(x,t,u)$ are arbitrary functions, these identities lead to 16 linear first order pdes for J. We obtain the solution J = constant. Hence, equations (1) do not admit differential invariants of order zero.

Similarly, the first prolongation of the operators (4) lead to the invariant criterion

$$Y_\phi^{(1)}(J) = 0, \quad Y_\tau^{(1)}(J) = 0, \quad Y_\psi^{(1)}(J) = 0 \tag{9}$$

which lead to a system of first order linear partial differential equations. This system has one integral. Therefore equations (1) admit one differential invariant of order one. This invariant has the form

$$J = \frac{f_{u_x}^2}{f\,f_{u_t}^2}. \tag{10}$$

Furthermore we obtain the two invariant equations

$$f_{u_x} = 0, \quad f_{u_t} = 0. \tag{11}$$

That is,

$$Y_\phi^{(1)}(f_{u_x})\Big|_{f_{u_x}=0} = 0, \quad Y_\tau^{(1)}(f_{u_x})\Big|_{f_{u_x}=0} = 0, \quad Y_\psi^{(1)}(f_{u_x})\Big|_{f_{u_x}=0} = 0$$

and similarly for the second equation.

Finally, we present the results in the case where we consider only the third equation of the invariance criterion (9). That is, we assume that $t' = t$ and $x' = x$. We obtain

$$\begin{aligned} J_1 &= \frac{f_{u_t}}{f_{u_x}}, \\ J_2 &= (f f_{u_t}^2 - f_{u_x}^2)\,[f_u(2f + u_x f_{u_x} - u_t f_{u_t}) \\ &+ f_{u_t}(f g_{u_t} - f_t - g f_{u_t}) + f_{u_x}(f_x - g_{u_x})]\, f_{u_x}^{-3}. \end{aligned} \tag{12}$$

J_1 and J_2 are known as *semi-invariants*. Furthermore we obtain the semi-invariant equations

$$\begin{aligned} &f_u(2f + u_x f_{u_x} - u_t f_{u_t}) + f_{u_t}(f g_{u_t} - f_t - g f_{u_t}) + f_{u_x}(f_x - g_{u_x}) = 0, \\ &f f_{u_t}^2 - f_{u_x}^2 = 0. \end{aligned} \tag{13}$$

4. Example

Differential invariants and invariant equations can be employed to find equations that can be mapped to each other. More generally, we can classify families of equations that can be transformed into simpler, may be linear, equations. For example, consider the problem of finding those forms of (1) that can be mapped into linear wave equation $u_{tt} = u_{xx}$, which is a member of (1) with $f = 1$ and $g = 0$. We note that this linear equation satisfies the invariant equations (11). Hence, any equation of the class (1) which can be linked with this equation must satisfy the invariant equations (11). That is, it must be of the form

$$u_{tt} = f(x,t,u)u_{xx} + g(x,t,u,u_x,u_t).$$

This class of equations was considered in Ref. 19. In this work the form of all linearizable equations of the above class were presented.

As an example, we consider the equation

$$u_{tt} + u_x u_{xx} = 0$$

which is known as the equation of stationary transonic gas flow. This equation is a member of (1) with $f = -u_x$ and $g = 0$. We note that the semi-invariants J_1 and J_2 given by (12) are zero. We also note that corresponding forms of f and g in

$$u_{tt} = \frac{xu_x - u}{x^2}u_{xx} - \frac{2}{x^4}(xu_x - u)^2$$

have the semi-invariants (12) equal to zero. It can be shown that these two latter equations are connected by the transformation

$$u' = \frac{u}{x},$$

where the equation of stationary transonic gas flow is assumed to be written in primed variables.

5. Conclusion

In the present paper we have derived differential invariants of the first order for the class (1) using the infinitesimal method. In order to calculate the differential invariants of the second order, we need the second prolongation of the operators (4). We state that the differential invariants of this order are long expressions. For this reason the presentation of second order differential invariants will be the subject of a longer paper.

Acknowledgments

This paper is dedicated to Prof. Tommaso Ruggeri on the occasion of his 60th birthday. R.T. acknowledges the financial support from P.R.A. (ex 60%) of University of Catania. R.T. also expresses her gratitude to the hospitality shown by Prof. C. Sophocleous during her visit to the University of Cyprus.

References

1. S. Lie, *Math. Ann.* **24**, 537 (1884).
2. S. Lie, *Math.Ann.* **32**, 213 (1888).
3. S. Lie, *Leipz. Berichte* **4**, 369 (1897).
4. A. Tresse, *Acta Math.* **18**, 1 (1894).
5. L.V. Ovsiannikov, *Group analysis of differential equations*, (Academic Press, New York, 1982).
6. P.J. Olver, *Equivalence, invariants and symmetry*, (Cambridge University Press, Cambridge, 1995).
7. N.H. Ibragimov, *Not. South African Math. Soc.*, **29**, 61, (1997).
8. N.H. Ibragimov, *Elementary Lie Group Analysis and Ordinary Differential Equations*, (Wiley, New York, 1999).
9. N.H. Ibragimov, *Nonlinear Dynamics*, **28**, 125 (2002).
10. N.H. Ibragimov, M. Torrisi and A. Valenti, *Commun. Nonlinear Sci. Numer. Simul.* **9**, 69 (2004).
11. N.H. Ibragimov and C. Sophocleous, *Proc. Inst. Math. of NAS of Ukraine* **50**, 142 (2004).
12. N.H. Ibragimov and S.V. Meleshko, *J. Math. Anal. Appl.* **308**, 266 (2005).
13. N.H. Ibragimov and C. Sophocleous, *Commun. Nonlinear Sci. Numer. Simul.* **12**, 1133 (2007).
14. I.K. Johnpillai and F.M. Mahomed, *J. Phys. A:Math.Gen.* **28**, 11033 (2001).
15. M. Torrisi, R. Tracinà and A. Valenti, *Nonlinear Dynamics* **36**, 97 (2004).
16. R. Tracinà, *Commun. Nonlinear Sci. Numer. Simul.* **9**, 127 (2004).
17. M. Torrisi and R. Tracinà, *J. Phys. A:Math. Gen.* **38**, 7519 (2005).
18. C. Sophocleous and N.M. Ivanova, Differential invariants of semilinear wave equations, *Proceedings of Tenth International Conference in Modern Group Analysis*, 198-206, (Larnaca, Cyprus, 2004).
19. C. Sophocleous and R. Tracinà, Differential Invariants for quasi-linear and semi-linear wave-type equations, to appear, (2008).
20. N. H. Ibragimov, M. Torrisi and A. Valenti, *J. Math. Phys.*, **32**, 2988 (1991).

ON THE FORMULATION OF THE QUANTUM EXTENDED THERMODYNAMICS BY USING THE SEMICLASSICAL INTERPRETATION OF THE WIGNER FUNCTION

M. TROVATO

Dipartimento di Matematica e Informatica,
Città Universitaria, Viale A. Doria, 6, 95125, Catania, Italy
E-mail: trovato@dmi.unict.it

We consider an appropriate semiclassical limit to such a nonlinear Schrodinger equation, using a semiclassical interpretation of the Wigner function. We derive a set of quantum hydrodynamic balance equations for an arbitrary number of moments in the framework of the Extended Thermodynamics. These equations can be used as the hydrodynamic limit of the generalized Gross-Pitaevskii. We analyze the closure problem for the hydrodynamic system by using the Maximum Entropy Principle (MEP) and the Moyal expansion for the semiclassical Wigner function.

Keywords: Extended Thermodynamics; Wigner function; Moyal expansion.

1. Introduction

We consider the nonlinear Schrodinger equation

$$i\hbar\frac{\partial\varphi(\mathbf{r},t)}{\partial t} = \left[-\frac{\hbar^2}{2m}\Delta + V(\mathbf{r},t)\right]\varphi(\mathbf{r},t) + \left[\sum_{j=1}^{L} u_j\,|\varphi(\mathbf{r},t)|^{2j}\right]\varphi(\mathbf{r},t) \quad (1)$$

where $\varphi(\mathbf{r},\mathbf{t})$ is the wavefunction and u_j the strength of the each nonlinear term. This equation is the generalization of the Gross-Pitaevskii equation[1] (that is obtained for $j = 1$) with nonlinear terms of the odd type. We can consider a consistent expansion around $\hbar$, in this way we separate classical from quantum dynamics, and we obtain order by order corrections terms. We will do this by considering the dynamics of the Wigner function[2,3] $\mathcal{F}_W$, which is exactly equivalent to the wavefunction φ in the sense that all information about the wavefunction is contained within its Wigner representation. It is well known that introducing the Weyl-Wigner transform W and the inverse Weyl-Wigner transform W^{-1} (Weyl quantization) we obtain

$$W(\widehat{\varrho}) = \varrho(\mathbf{r},\mathbf{p}) = (2\pi\hbar)^3\, \mathcal{F}_W(\mathbf{r},\mathbf{p})\,, \tag{2}$$

$$W^{-1}(\varrho) = \int d^3p\, \mathcal{F}_W\left(\frac{\mathbf{r}''+\mathbf{r}'}{2},\mathbf{p}\right) e^{-\frac{i}{\hbar}\mathbf{p}\cdot(\mathbf{r}''-\mathbf{r}')}\,, \tag{3}$$

where $\widehat{\varrho} = \widehat{\varrho}(\widehat{\mathbf{r}},\widehat{\mathbf{p}})$ is the density matrix operator, $\varrho(\mathbf{r},\mathbf{p})$ the corresponding phase-space function in the space of phase and

$$\mathcal{F}_W(\mathbf{r},\mathbf{p}) = \frac{1}{(2\pi\hbar)^3}\int d^3\tau\, e^{-\frac{i}{\hbar}\tau\cdot\mathbf{p}}\, \varphi^*(\mathbf{r}-\tau/2)\,\varphi(\mathbf{r}+\tau/2) \tag{4}$$

the Wigner function. In order to obtain the *generalized Wigner equation*, starting by definition of Wigner function (4), we calculate the derivative

$$\frac{\partial\mathcal{F}_W}{\partial t} = \frac{1}{(2\pi\hbar)^3}\int d^3\tau\, e^{-\frac{i}{\hbar}\tau\cdot\mathbf{p}}\left[\frac{\partial\varphi^*(\mathbf{r}-\tau/\mathbf{2})}{\partial t}\,\varphi(\mathbf{r}+\tau/\mathbf{2})+ \right. \tag{5}$$
$$\left. \varphi^*(\mathbf{r}-\tau/\mathbf{2})\,\frac{\partial\varphi(\mathbf{r}+\tau/\mathbf{2})}{\partial \mathbf{t}}\right]$$

and we use the eq. (1) to separate the linear and nonlinear contributes

$$\frac{\partial\varphi(\mathbf{r},t)}{\partial t} = \frac{1}{i\hbar}\langle\mathbf{r}|\widehat{H}_L|\varphi\rangle + \frac{1}{i\hbar}\sum_{j=1}^{L} u_j\,|\varphi(\mathbf{r},t)|^{2j}\,\varphi(\mathbf{r},t)$$

where

$$\widehat{H}_L = \frac{\widehat{p}^2}{2m} + \widehat{V}\,.$$

Thus, after some calculations and by using the Moyal formalism,[4] we obtain

$$\frac{\partial\mathcal{F}_w}{\partial t} + \frac{p_k}{m}\frac{\partial\mathcal{F}_w}{\partial x_k} = \frac{\partial V_{eff}}{\partial x_k}\frac{\partial\mathcal{F}_w}{\partial p_k} + \tag{6}$$
$$\sum_{s=1}^{\infty}(-1)^s\left(\frac{\hbar}{2}\right)^{2s}\sum_{|\alpha|=2s+1}\frac{1}{\alpha_1!\alpha_2!\alpha_3!}\left[\frac{\partial^{\alpha_1+\alpha_2+\alpha_3}V_{eff}}{\partial x_1^{\alpha_1}\partial x_2^{\alpha_2}\partial x_3^{\alpha_3}}\right]\left[\frac{\partial^{\alpha_1+\alpha_2+\alpha_3}\mathcal{F}_W}{\partial p_1^{\alpha_1}\partial p_2^{\alpha_2}\partial p_3^{\alpha_3}}\right]$$

where $|\alpha| = \alpha_1+\alpha_2+\alpha_3$, and the nonlinear terms have been included by defining the quantity

$$V_{eff} = V(\mathbf{r},t) + \sum_{j=1}^{L} u_j\,[n(\mathbf{r},t)]^j \quad \text{with} \quad n(\mathbf{r},t) = \int d^3p\,\mathcal{F}_W$$

It should be noted that the quantity V_{eff} represent an *effective potential* in which has been included an additional potential that depends from the density $n(\mathbf{r},t)$ (in the particular case of Gross-Pitaevskii equation is proportional to the density). Therefore, this additional potential can be interpreted as a large number of particles interacting with one another , i.e. as a kind of non-ideal gas.

2. The Extended quantum hydrodynamic (QHD) system

We consider an arbitrary operator $\widehat{\Psi}(\widehat{\mathbf{r}},\widehat{\mathbf{p}})$ in terms of $\widehat{\mathbf{r}}$ and $\widehat{\mathbf{p}}$, and the corresponding space-phase function $\Psi(\mathbf{r},\mathbf{p})$. The macroscopic expectation value of $\widehat{\Psi}(\widehat{\mathbf{r}},\widehat{\mathbf{p}})$ is define by the *global* quantity

$$\langle\widehat{\Psi}(\widehat{\mathbf{r}},\widehat{\mathbf{p}})\rangle = Tr(\widehat{\varrho}\,\widehat{\Psi}) = \int\!\!\int d^3p\; d^3r\;\Psi(\mathbf{r},\mathbf{p})\,\mathcal{F}_{\mathbf{W}}(\mathbf{r},\mathbf{p}) \tag{7}$$

while the macroscopic *local moment* $F(\mathbf{r},t)$ of $\widehat{\Psi}(\widehat{\mathbf{r}},\widehat{\mathbf{p}})$ can be defined by means of the *local* relation

$$F(\mathbf{r},t) = \int d^3p\;\Psi(\mathbf{r},\mathbf{p})\,\mathcal{F}_W(\mathbf{r},\mathbf{p},t) \tag{8}$$

In particular, by introducing the quantity $\varepsilon = p^2/2m$ we can consider as set of phase-space functions the following set of traceless kinetic fields

$$\Psi_A(\mathbf{p}) = \{\varepsilon^s,\;\varepsilon^s p_{i_1},\;\varepsilon^s p_{\langle i_1}\,p_{i_2\rangle},\;\ldots,\;\varepsilon^s p_{\langle i_1}\,p_{i_2}\cdots p_{i_r\rangle}\}\,, \tag{9}$$

and by using the (8) we define the corresponding set of *local moments*

$$F_A(\mathbf{r},t) = \{F_{(s)},\;F_{(s)|i_1},\;F_{(s)|\langle i_1 i_2\rangle},\;\ldots,\;F_{(s)|\langle i_1\cdots i_r\rangle}\} \tag{10}$$

where $s = 0,1,\ldots N$ and $r = 1,2,\ldots M$.
As in the classic Extended Thermodynamics,[5] it is possible to decompose the local moments into their *convective* and *central* parts, respectively. Thus we introduce the mean velocity v_i, the peculiar momentum $\widetilde{p}_i = p_i - m\,v_i$, the quantity $\widetilde{\varepsilon} = \widetilde{p}^2/2m$, and we define the new set of kinetic fields

$$\Psi_A(\widetilde{\mathbf{p}}) = \{\widetilde{\varepsilon}^s,\;\widetilde{\varepsilon}^s\widetilde{p}_{i_1},\;\widetilde{\varepsilon}^s\widetilde{p}_{\langle i_1}\,\widetilde{p}_{i_2\rangle},\;\ldots,\;\widetilde{\varepsilon}^s\widetilde{p}_{\langle i_1}\,\widetilde{p}_{i_2}\cdots\widetilde{p}_{i_r\rangle}\}\,, \tag{11}$$

and in correspondence the new set of *central moments*

$$M_A(\mathbf{r},t) = \{M_{(s)},\;M_{(s)|i_1},\;M_{(s)|\langle i_1 i_2\rangle},\;\ldots,\;M_{(s)|\langle i_1\cdots i_r\rangle}\} \tag{12}$$

being

$$M_{(0)|i_1} = 0\,,\qquad M_{(s)|\langle i_1 i_2\cdots i_r\rangle} = \int d^3p\;\widetilde{\varepsilon}^s\,\widetilde{p}_{\langle i_1}\,\widetilde{p}_{i_2}\cdots\widetilde{p}_{i_r\rangle}\,\mathcal{F}_W\,. \tag{13}$$

Multiplying Eq.(6) by $\Psi_A(\widetilde{\mathbf{p}})$, integrating over the $\mathbf{p}$ space we obtained the quantum balance equations for the moments of the Wigner functions. In particular, by considering only an arbitrary number of scalar and vectorial moments $\{M_{(s)}, M_{(s)|i}\}$, we can determine the corresponding set of balance equations, representing in the right-hand side the first quantum correction

$$\dot{n} + n\,\frac{\partial v_k}{\partial x_k} = 0\,, \tag{14}$$

$$\dot{v}_i + \frac{1}{nm^2}\,\frac{\partial M_{(0)|ik}}{\partial x_k} + \frac{1}{m}\,\frac{\partial V_{eff}}{\partial x_i} = 0\,, \tag{15}$$

$$\dot{M}_{(s)} + M_{(s)} \frac{\partial v_k}{\partial x_k} + \frac{s}{m} M_{(s-1)|ik} \frac{\partial v_i}{\partial x_k} - \frac{s}{n\, m^2} M_{(s-1)|i} \frac{\partial M_{(0)|ik}}{\partial x_k}$$
$$+\frac{1}{m} \frac{\partial M_{(s)|k}}{\partial x_k} = \frac{\hbar^2}{24} \frac{s(s-1)}{m^2} \left\{ \frac{(s-2)}{m} \frac{\partial^3 V_{eff}}{\partial x_{\langle i} \partial x_j \partial x_{k\rangle}} M_{(s-3)|\langle ijk\rangle} \right.$$
$$\left. + \frac{3}{5}(1+2s) \frac{\partial^3 V_{eff}}{\partial x_r \partial x_r \partial x_k} M_{(s-2)|k} \right\}, \tag{16}$$

$$\dot{M}_{(s)|i} + M_{(s)|i} \frac{\partial v_k}{\partial x_k} + \frac{s}{m} M_{(s-1)|ijk} \frac{\partial v_j}{\partial x_k} + \frac{1}{m} \frac{\partial M_{(s)|ik}}{\partial x_k}$$
$$+M_{(s)|k} \frac{\partial v_i}{\partial x_k} - \frac{1}{n\, m} M_{(s)} \frac{\partial M_{(0)|ik}}{\partial x_k} - \frac{s}{n\, m^2} M_{(s-1)|ij} \frac{\partial M_{(0)|jk}}{\partial x_k} =$$
$$\frac{\hbar^2}{24} \frac{s}{m} \left\{ \frac{(s-1)(s-2)}{m^2} \frac{\partial^3 V_{eff}}{\partial x_r \partial x_j \partial x_k} M_{(s-3)|\langle rjki\rangle} + \right.$$
$$\frac{3(s-1)}{m} \frac{(3+2s)}{7} \left(\frac{\partial^3 V_{eff}}{\partial x_r \partial x_r \partial x_k} M_{(s-2)|\langle ki\rangle} + \frac{\partial^3 V_{eff}}{\partial x_{\langle k} \partial x_{r\rangle} \partial x_i} M_{(s-2)|\langle kr\rangle} \right)$$
$$\left. + \frac{(4s^2+8s+3)}{5} \frac{\partial^3 V_{eff}}{\partial x_r \partial x_r \partial x_i} M_{(s-1)} \right\} \qquad \text{with} \quad s = 1, \cdots N. \tag{17}$$

where in (16) and (17) all time derivatives $\dot{v}_i$ were eliminated by means of the balance equation (15) and for $\hbar \to 0$ we obtain the balance equations of the usual classic Extended Thermodynamics. The previous set of equations contain unknown constitutive functions represented by some moments of higher order $H_A = \{M_{N+1}, M_{(l)|\langle ij\rangle}, M_{(r)|\langle ijk\rangle}, M_{(s)|\langle ijqk\rangle}\}$ with $l = 0, \cdots N$; $r = 0, \cdots N-1$, $s = 0, \cdots N-3$, and, in general, the problem of closure can be tackled using the Quantum Maximum Entropy formalism.[6–9]

3. The Quantum Maximum Entropy Principle (QMEP)

We consider as entropy, in quantum mechanics, the *global* quantity $S(\widehat{\varrho}) = -\langle \ln(\widehat{\varrho}) - \widehat{I}\rangle$, and we search the extremal value of entropy subject to the constrain that the information of the physical system is described by the local quantities $M_A(\mathbf{r}, t)$ expressed by (13). Thus we consider the new *global* functional[6–9]

$$\widetilde{S} = S - \int d^3r \left\{ \sum_{A=1}^{N} \lambda_A(\mathbf{r}, t) \left[M_A(\mathbf{r}, t) - \int d^3p\, \Psi_A \mathcal{F}_W \right] \right\}, \tag{18}$$

where $\lambda_A(\mathbf{r}, t)$ are the *local* Lagrange multipliers. Its possible to show that the solution of problem is given by the density operator

$$\widehat{\varrho} = \exp\left\{ W^{-1} \left(- \sum_{A=1}^{N} \lambda_A(\mathbf{r}, t) \Psi_A \right) \right\}. \tag{19}$$

By using the Moyal formalism,[4] its possible to prove that both the Wigner function and the central moments can be obtained as expansion in even power of $\hbar$

$$\mathcal{F}_W = \mathcal{F}_W^{(0)} + \sum_{k=1}^{\infty} \hbar^{2k} \mathcal{F}_W^{(2k)}, \qquad M_A = M_A^{(0)} + \sum_{k=1}^{\infty} \hbar^{2k} M_A^{(2k)}, \tag{20}$$

where the quantities $\{\mathcal{F}_W^{(0)}, M_A^{(0)}\}$ represent the classic contributes, while the remaining terms represent the quantum corrections of order k, respectively. Thus, for $\hbar \to 0$ the quantum Extended Thermodynamics $\to$ classic Extended Thermodynamics. Following this approach we consider the Weyl-Wigner transform (2) of the density operator (19) and, by using the Moyal expansion,[4] we obtain for the first quantum correction the relation

$$\mathcal{F}_W = \exp\left\{ -\alpha - \beta\tilde{\varepsilon} - \sum_{r=0}^{M}\sum_{l=0}^{N} \Lambda_{(l)|\langle i_1 i_2 \cdots i_r\rangle}\, \tilde{\varepsilon}^l\, \tilde{p}_{\langle i_1} \cdots \tilde{p}_{i_r\rangle} \right\} \left[1 + \hbar^2 Q_c\right]$$

where α and β are the Lagrange multipliers of local equilibrium, $\Lambda_A = \{\Lambda_{(l)|\langle i_1 i_2 \cdots i_r\rangle}\}$ are the non-equilibrium lagrange multipliers, and by assuming that $\frac{\partial \Lambda_A}{\partial x_k} \ll \frac{\partial \alpha}{\partial x_k}, \frac{\partial \beta}{\partial x_k}$ the quantum correction term Q_c will be expressed explicitly as $Q_c \approx Q_c\left(\beta, \frac{\partial \alpha}{\partial x_k}, \frac{\partial \beta}{\partial x_k}, \frac{\partial^2 \alpha}{\partial x_i \partial x_k}, \frac{\partial^2 \beta}{\partial x_i \partial x_k}, \widetilde{\mathbf{p}}\right)$.
If we expand the exponential in terms of the nonequilibrium Lagrange multipliers Λ_A, we obtain the *equilibrium* and the linear *nonequilibrium* contributions of the Wigner function $\mathcal{F}_W$

$$\mathcal{F}_W|_E = e^{-\alpha}\, e^{-\beta\tilde{\varepsilon}} \left\{1 + \hbar^2 Q_c\right\} \tag{21}$$

$$\mathcal{F}_W|_{NE} = -e^{-\alpha} e^{-\beta\tilde{\varepsilon}} \left\{1 + \hbar^2 Q_c\right\} \sum_{r=0}^{M}\sum_{l=0}^{N} \Lambda_{(l)|\langle i_1 \cdots i_r\rangle} \tilde{\varepsilon}^l \tilde{p}_{\langle i_1} \cdots \tilde{p}_{i_r\rangle} \tag{22}$$

By inserting these expansions of the distribution function into the definition of the macroscopic quantities (13) we obtain the Lagrange multipliers and, all constitutive functions will be expressed analytically up to the first quantum correction respect to the classic expressions. Thus, by using only an arbitrary set of scalar and vectorial moments, we obtain for the constitutive functions, in the local equilibrium conditions, the relations

$$M_{(N+1)}|_E = n\,(2N+3)!! \left(\frac{K_B T}{2}\right)^{N+1} \left[1 + \hbar^2 Q^{(N+1)}\right]$$

$$M_{(s)|\langle ij\rangle}|_E = \hbar^2 \left\{\Gamma_s^{(1)} A_{\langle ij\rangle}^{(1)} + \Gamma_s^{(2)} A_{\langle ij\rangle}^{(2)}\right\}$$

$$M_{(s)|\langle i_1 i_2 \cdots i_r\rangle}|_E = 0 \quad \text{for} \quad r \geq 3\,.$$

Analogously, by introducing the nonequilibrium variables $\Delta_{(s)} = M_{(s)} - M_{(s)}|_E$ it possible to calculate, the nonequilibrium constitutive functions

$$M_{(N+1)}|_{NE} = \sum_{r=2}^{N} \left[\Phi^{(0)}_{(N+1)\,r} + \hbar^2\, \Phi^{(2)}_{(N+1)\,r}\right] \Delta_{(r)}\,,$$

$$M_{(s)|\langle ij\rangle}|_{NE} = \hbar^2 \left\{ \left[\sum_{r=2}^{N} \Gamma^{(1)}_{sr}\, \Delta_{(r)}\right] A^{(1)}_{\langle ij\rangle} + \left[\sum_{r=2}^{N} \Gamma^{(2)}_{sr}\, \Delta_{(r)}\right] A^{(2)}_{\langle ij\rangle} \right\},$$

$$M_{(s)|\langle ijk\rangle}|_{NE} = \hbar^2 \left\{ \sum_{r=1}^{N} \chi^{(1)}_{sr}\, M_{(r)|\langle i}\, A^{(1)}_{jk\rangle\rangle} + \sum_{r=1}^{N} \chi^{(2)}_{sr}\, M_{(r)|\langle i}\, A^{(2)}_{jk\rangle\rangle} \right\},$$

$$M_{(s)|\langle i_1 i_2 \cdots i_r\rangle}|_{NE} = 0 \quad \text{for} \quad r \geq 4\,.$$

where the coefficients

$$G_A = \{\Gamma^{(1)}_s,\, \Gamma^{(2)}_s,\, \Phi^{(0)}_{(N+1)\,r},\, \Gamma^{(1)}_{sr},\, \Gamma^{(2)}_{sr},\, \chi^{(1)}_{sr},\, \chi^{(2)}_{sr}\}$$

are functions known[9] of $\{n, T\}$ and all the quantities $H_A = \{Q^{N+1}, \Phi^{(2)}_{(N+1)\,r}, A^{(1)}_{\langle ij\rangle}, A^{(2)}_{\langle ij\rangle}\}$ are complicated functions[9]

$$H_A = H_A\left(n, T, \frac{\partial n}{\partial x_k}, \frac{\partial T}{\partial x_k}, \frac{\partial^2 n}{\partial x_i \partial x_k}, \frac{\partial^2 T}{\partial x_i \partial x_k}\right)$$

which we do not show here. We conclude emphasizing that for $\hbar \to 0$ we obtain the usual results obtained in the classic Extended Thermodynamics.

Acknowledgment:
This paper is dedicated to Tommaso Ruggeri on the occasion of his birthday. The paper has been supported by INDAM, by GNFM project *Young Researchers* (2007) and by MIUR-PRIN project *Nonlinear Propagation and Stability in Thermodynamical Processes of Continuous Media* (2005).

References

1. F. Dalfovo, S. Giorgini, L.P. Pitaevskii, S. Stringari, *Rev. Mod. Phys.* **71**,463 (1999)
2. E.P. Wigner, *Phys. Rev.* **40**, 749 (1932).
3. M. Hillery, R. O'Conell, M. Scully, E.P. Wigner, *Phys. Rep.* **106**, 121 (1984).
4. J.E. Moyal, *Proc. Cambridge Phil. Soc.* **45**, 99 (1949).
5. I. Müller, T. Ruggeri, *Rational Extended Thermodynamics.*: Springer Tracts in Natural Philosophy, Vol. **37**, Springer-Verlag New York (1998).
6. E.T. Jaynes, *Phys. Rev.* **106**, 620 (1957). *Phys. Rev.* **108**, 171 (1957).
7. R. Luzzi, A.R. Vasconcellos, J.G. Ramos, *Int. Journal of Modern Phys. B* **14**, 3189 (2000).
8. P. Degond, C. Ringhofer, *J. Stat. Phys.* **112**, 587 (2003).
9. M. Trovato *On the formulation of the Quantum Maximum Entropy Principle by using the Wigner function* (2007).

APPROXIMATE SYMMETRIES OF A VISCOELASTIC MODEL

ANTONINO VALENTI

Dipartimento di Matematica e Informatica,
Università di Catania,
viale A. Doria 6, 95125 Catania, Italy

Approximate symmetries of a mathematical model describing one-dimensional motion in a viscoelastic medium with a small viscosity coefficient are studied. An approximate invariant solution is obtained through the approximate generator of the first-order approximate symmetries.

Keywords: Lie groups, approximate symmetries, dissipative media.

1. Introduction

We consider the third order partial differential equation

$$w_{tt} = \left[\sigma(w_x) + \lambda_0\, w_{tx}\right]_x , \tag{1}$$

where σ is an arbitrary function of its argument, λ_0 is a positive real parameter, $w(t,x)$ is the dependent variable and subscripts denote partial derivative with respect to the independent variables t and x.

A physical prototype of the problem studied here arises when we consider purely longitudinal motions of a homogeneous viscoelastic bar of uniform cross-section and we assume that the material is a nonlinear Kelvin solid. That is, we assume a stress-strain relation of the following form:

$$\tau = \sigma(w_x) + \lambda_0\, w_{xt}, \tag{2}$$

where τ is the stress, x the position of a cross-section (which is assumed to move as a vertical plane section) in the homogeneous rest configuration of the bar, $w(t,x)$ the displacement at time t of the section from its rest position, $\sigma(w_x)$ the elastic tension (w_x is the strain), λ_0 the viscosity positive coefficient.

Some mathematical questions related to (1), as the existence, uniqueness and stability of weak solutions can be found in Ref. 1, moreover a study related to a generalized "shock structure" is showed in Ref. 2 and shear wave solutions are found in Ref. 3, where some explicit examples of blow up for boundary-values problems with smooth initial data are shown. Besides a symmetry analysis is performed in Refs. 4,5.

Making use of the change of variable $w_x = u$, equation (1) can be put in the form

$$u_{tt} = [f(u)u_x]_x + \varepsilon\, u_{txx}, \tag{3}$$

where we have set $\sigma' = f$ and $\lambda_0 = \varepsilon$, with $\varepsilon \ll 1$ a small parameter. Here and in what follows, primes denote derivative of a function with respect to the only variable upon which it depends.

As it is well known, a small dissipation is able to prevent the breaking of the wave profile allowing to study the so called "far field" and a technique widely used is the perturbation analysis performed by expanding the dependent variables in power series of a small parameter (may be a physical parameter or often artificially introduced). For $\varepsilon = 0$, we recover the nonlinear wave equation

$$u_{tt} = [f(u)u_x]_x, \tag{4}$$

which symmetries were widely studied in Ref. 6.

Combination of the Lie group theory and the perturbation analysis gives rise to the so-called approximate symmetry theories. The first paper on this subject is due to Baikov, Gazizov and Ibragimov.[7] Successively another method for finding approximate symmetries was proposed by Fushchich and Shtelen.[8] In the method proposed by Baikov, Gazizov and Ibragimov, the Lie operator is expanded in a perturbation series so that an approximate operator can be found. But, the approximate operator does not reflect well an approximation in the perturbation sense; in fact, even if one uses a first order approximate operator, the corresponding approximate solution could contain higher order terms. In the method proposed by Fushchich and Shtelen the dependent variables are expanded in a perturbation series; equations are separated at each order of approximation and the approximate symmetries of the original equations are defined to be the exact symmetries of the system coming out from equating to zero the coefficients of the smallness parameter. This method is consistent with the perturbation theory and yields correct terms for the approximate solutions but a "drawback" is present: it is impossible to work in hierarchy, i. e. in the searching of symmetries there is a coupled system between the equations at several order of approximation, therefore the algebra can increase enormously.

In this paper we work in the framework of the approximate method proposed in Ref. 9, which removes the "drawback" of the impossibility to work in hierarchy appearing in the the method proposed by Fushchich and Shtelen. We obtain the symmetry classification of the function $f(u)$ through which equation (3) is approximately invariant and, in an application, we obtain an approximate solution.

2. Approximate symmetry method

In general, any solution of (3) will be of the form $u = u(t, x, \varepsilon)$ and the one-parameter Lie group of infinitesimal transformations in the (t, x, u)-space of the equation (1), can be considered in the following form:

$$\hat{t} = t + a\,\xi^1(t, x, u(t, x, \varepsilon), \varepsilon) + \mathcal{O}(a^2), \tag{5}$$

$$\hat{x} = x + a\,\xi^2(t, x, u(t, x, \varepsilon), \varepsilon) + \mathcal{O}(a^2), \tag{6}$$

$$\hat{u} = u + a\,\eta(t, x, u(t, x, \varepsilon), \varepsilon) + \mathcal{O}(a^2), \tag{7}$$

where a is the group parameter.

Let us suppose that $w(t, x, \varepsilon)$ and $\hat{u}(\hat{t}, \hat{x}, \varepsilon)$, analytic in ε, can be expanded in power series of ε, i.e.

$$u(t, x, \varepsilon) = u_0(t, x) + \varepsilon\, u_1(t, x) + \mathcal{O}(\varepsilon^2), \tag{8}$$

$$\hat{u}(\hat{t}, \hat{x}, \varepsilon) = \hat{u}_0(\hat{t}, \hat{x}) + \varepsilon\, \hat{u}_1(\hat{t}, \hat{x}) + \mathcal{O}(\varepsilon^2), \tag{9}$$

where: u_0 and u_1 are some smooth functions of t and x; $\hat{u}_0$ and $\hat{u}_1$, are some smooth functions of $\hat{t}$ and $\hat{x}$.

Upon formal substitution of (8) in (1), equating to zero the coefficients of zero and first degree powers of ε we arrive at the following system of PDEs

$$L_0 := u_{0tt} - f(u_0)\, u_{0xx} - f'(u_0)\, u_{0x}^2 = 0, \tag{10}$$

$$L_1 := u_{1tt} - f(u_0)\, u_{1xx} - f'(u_0)\, u_{0xx}\, u_1$$
$$- 2\, f'(u_0)\, u_{0x}\, u_{1x} - f''(u_0)\, u_{0x}^2\, u_1 - u_{0txx} = 0, \tag{11}$$

where we have set

$$f^{(i)}(u_0) = f^{(i)}(u(t, x, \varepsilon))\,|_{\varepsilon=0}, \quad i = 0, 1, 2.$$

Hence, u_0 is solution of the nonlinear wave equation (10) (or (4)) which we call *unperturbed equation*.

In order to have an one-parameter Lie group of infinitesimal transformations of the system (10)-(11), which is consistent with the expansions of the dependent variables (8) and (9), we introduce these expansions in the infinitesimal transformations (5)-(7). Upon formal substitution, equating to zero the coefficients of zero and first degree powers of ε, we get the following one-parameter Lie group of infinitesimal transformations in the (t, x, u_0, u_1)-space

$$\hat{t} = t + a\, \xi_0^1(t, x, u_0) + \mathcal{O}(a^2), \tag{12}$$

$$\hat{x} = x + a\, \xi_0^2(t, x, u_0) + \mathcal{O}(a^2), \tag{13}$$

$$\hat{u}_0 = u_0 + a\, \eta_0(t, x, u_0) + \mathcal{O}(a^2), \tag{14}$$

$$\hat{u}_1 = u_1 + a\, [\eta_{10}(t, x, u_0) + \eta_{11}(t, x, u_0)\, u_1] + \mathcal{O}(a^2), \tag{15}$$

where we have set

$$\xi_0^i(t, x, u_0) = \xi^i(t, x, u(t, x, \varepsilon), \varepsilon)\mid_{\varepsilon=0}, \quad i = 1, 2 \tag{16}$$

$$\eta_0(t, x, u_0) = \eta(t, x, u(t, x, \varepsilon), \varepsilon)\mid_{\varepsilon=0}, \tag{17}$$

$$\eta_{10}(t, x, u_0) + \eta_{11}(t, x, u_0)\, u_1 = \frac{d\eta}{d\varepsilon}\mid_{\varepsilon=0}. \tag{18}$$

Definition: We call *approximate symmetries* of equation (3) the (exact) symmetries of the system (10)-(11) through the one-parameter Lie group of infinitesimal transformations (12)-(15). Consequently, the one-parameter Lie group of infinitesimal transformations (12)-(15), the associated Lie algebra and the corresponding infinitesimal operator

$$X = \xi_0^1(t, x, u_0)\frac{\partial}{\partial t} + \xi_0^2(t, x, u_0)\frac{\partial}{\partial x} + \eta_0(t, x, u_0)\frac{\partial}{\partial u_0} + [\eta_{10}(t, x, u_0) + \eta_{11}(t, x, u_0)\, u_1]\frac{\partial}{\partial u_1}, \tag{19}$$

are called the approximate Lie group, the approximate Lie algebra and the approximate Lie operator of the equation (3), respectively.

Moreover, after setting

$$X_0 = \xi_0^1(t, x, u_0)\frac{\partial}{\partial t} + \xi_0^2(t, x, u_0)\frac{\partial}{\partial x} + \eta_0(t, x, u_0)\frac{\partial}{\partial u_0}, \tag{20}$$

the approximate Lie operator (19) can be rewritten as

$$X = X_0 + [\eta_{10}(t, x, u_0) + \eta_{11}(t, x, u_0)\, u_1]\frac{\partial}{\partial u_1} \tag{21}$$

and X_0 can be regarded as the infinitesimal operator of the unperturbed equation (10) (or (4)).

It is worthwhile noticing that, thanks to the functional dependencies of the coordinates of the approximate Lie operator (19) (or (21)), now we are able to work in hierarchy in finding the invariance conditions of the system (10)-(11): firstly, by classifying the unperturbed equation (10) through the operator (20) and after by determining η_{10} and η_{11} from the invariance condition that follows by applying the operator (21) to the linear equation (11). In fact the invariance condition of the system (10)-(11) reads:

$$X_0^{(2)}(L_0)\Big|_{L_0=0} = 0, \qquad X^{(3)}(L_1)\Big|_{L_0=0,\ L_1=0} = 0, \tag{22}$$

where $X_0^{(2)}$ and $X^{(3)}$ are the second and third extensions of the operators X_0 and X, respectively.

3. Group classification via approximate symmetries

The classification of the equation (10) is well known (see for details Ref. 6) and through $(22)_1$, we obtain

$$\xi_0^1 = a_5\, t^2 + a_3\, t + a_1, \tag{23}$$

$$\xi_0^2 = a_6\, x^2 + a_4\, x + a_2, \tag{24}$$

$$\eta_0 = (a_5\, t - 3\, a_6\, x + a_7) u_0 + a_8\, t + a_{10}\, x + a_9, \tag{25}$$

$$(a_7\, u_0 + a_9)\; f'(u_0) - 2\,(a_4 - a_3)\, f(u_0) = 0, \tag{26}$$

$$(a_5\, u_0 + a_8)\, f'(u_0) + 4\, a_5\, f(u_0) = 0, \tag{27}$$

$$(3\, a_6\, u_0 - a_{10})\, f'(u_0) + 4\, a_6\, f(u_0) = 0, \tag{28}$$

where a_i, $i = 1, 2, \ldots, 10$ are constants.

Taking (23)-(27) into account, from $(22)_2$ we obtain the following additional conditions:

$$a_5 = a_6 = a_8 = a_{10} = 0, \tag{29}$$

$$\eta_{10} = 0, \qquad \eta_{11} = a_3 + a_7 - 2\, a_4. \tag{30}$$

After observing that conditions (29) impose restrictions upon to X_0, summarizing we have to manage the following relations:

$$\xi_0^1 = a_3\, t + a_1, \tag{31}$$

$$\xi_0^2 = a_4\, x + a_2, \tag{32}$$

$$\eta_0 = a_7\, u_0 + a_9, \tag{33}$$

$$\eta_1 = (a_3 + a_7 - 2\, a_4)\, u_1, \tag{34}$$

$$(a_7\, u_0 + a_9)\; f'(u_0) - 2\,(a_4 - a_3)\, f(u_0) = 0. \tag{35}$$

When f is an arbitrary function, we obtain

$$a_7 = a_9 = 0, \qquad a_4 = a_3 \tag{36}$$

and we call the associate three-dimensional Lie algebra the *Approximate Principal Lie Algebra* of equation (3). We denote it by $Approx\mathcal{L}_\mathcal{P}$ and it is spanned by the three operators

$$X_1 = \frac{\partial}{\partial t}, \qquad X_2 = \frac{\partial}{\partial x}, \qquad X_3 = t\,\frac{\partial}{\partial t} + x\,\frac{\partial}{\partial x} - u_1\,\frac{\partial}{\partial u_1}. \tag{37}$$

The classification of f with the corresponding extensions of $Approx\mathcal{L}_\mathcal{P}$ arising from (31)-(35), are reported in Table 1.

Table 1. Classification of $f(u_0)$ with the corresponding extensions of $Approx\mathcal{L}_\mathcal{P}$. f_0, p and q are constitutive constants with $p \neq 0$.

Case	Forms of $f(u_0)$	Extensions of $Approx\mathcal{L}_\mathcal{P}$
I	$f = f_0\, e^{\frac{u_0}{p}}$	$X_4 = x\,\frac{\partial}{\partial x} + 2p\,\frac{\partial}{\partial u_0} - 2\,u_1\,\frac{\partial}{\partial u_1}$
II	$f = f_0\,(u_0+q)^{\frac{1}{p}}$	$X_4 = x\,\frac{\partial}{\partial x} + 2p\,(u_0+q)\,\frac{\partial}{\partial u_0} + 2\,(p-1)\,u_1\,\frac{\partial}{\partial u_1}$

4. An application

Let us consider the following form of the tension $\sigma(w_x)$:

$$\sigma(w_x) = \sigma_0\,\log(w_x + 1), \tag{38}$$

which was suggested by Capriz.[10] Taking into account that

$$u = w_x, \qquad f = \sigma' = \frac{\sigma_0}{u+1}, \tag{39}$$

we fall in the Case *II* of Table 1 with the following identifications:

$$f_0 = \sigma_0, \quad p = -1, \quad q = 1. \tag{40}$$

In this case, the approximate Lie operator X_4 assumes the form

$$X_4 = x\,\frac{\partial}{\partial x} - 2\,(u_0+1)\,\frac{\partial}{\partial u_0} - 4\,u_1\,\frac{\partial}{\partial u_1} \tag{41}$$

and from the corresponding invariant surface conditions we obtain the following representation for the different terms in the expansion of u:

$$u_0 = \frac{\psi(t)}{x^2} - 1, \qquad u_1 = \frac{\chi(t)}{x^4}, \tag{42}$$

which gives the form of an invariant solution approximated at the first order in ε. The functions ψ and χ must satisfy the following system of ODEs to which, after (38), the system (10)-(11) is reduced through (42):

$$\psi_{tt} - 2\,\sigma_0 = 0, \qquad \chi_{tt} - \frac{6}{t^2}\,\chi - 12\,\sigma_0\,t = 0. \tag{43}$$

After solving (43) and taking (42) into account, we have

$$u_0 = \sigma_0\left(\frac{t}{x}\right)^2 - 1, \qquad u_1 = 3\,\frac{(40\,\sigma_0\,\log t - 8\,\sigma_0 - 25)\,t^5 - 25}{50\,t^2\,x^4}. \tag{44}$$

Therefore, the invariant solution up to the first order in ε is

$$u(t,x,\varepsilon) = \sigma_0\left(\frac{t}{x}\right)^2 - 1 + 3\,\varepsilon\,\frac{(40\,\sigma_0\,\log t - 8\,\sigma_0 - 25)\,t^5 - 25}{50\,t^2\,x^4}. \tag{45}$$

We have an unperturbed state represented by a stretching modified by the viscosity effect.

Acknowledgements

The Author acknowledges the support by M.I.U.R. through P.R.I.N. 2005-2007: *NonLinear Propagation and Stability in Thermodynamical Processes of Continuous Media*, national coordinator Prof. T. Ruggeri, to whom this article is dedicated, on the occasion of his 60th birthday.

References

1. J. M. Greenberg, R. C. Mac Camy, V. J. Mizel, *J. Math. Mech.* **17**, 707 (1968).
2. A. Donato, G. Vermiglio, *J. de Mécanique Théor. Appl.* **1**, 359 (1982).
3. K. R. Rajagopal and G. Saccomandi, *Q. Jl Mech. Appl. Math.* **56**, 311 (2003).
4. M. Ruggieri and A. Valenti, *Proceedings of MOGRAN X*, N. H. Ibragimov et al. Eds., 175 (2005).
5. M. Ruggieri and A. Valenti, *Proceedings of WASCOM 2005*, R. Monaco, G. Mulone, S. Rionero and T. Ruggeri Eds., World Sc. Pub., Singapore, 481 (2006).
6. W. F. Ames, R. J. Lohner, E. Adams, *Int. J. Non-Linear Mech.* **16**, 439 (1981).
7. V. A. Baikov, R. K. Gazizov and N. H. Ibragimov, *Mat. Sb.* **136**, 435 (1988) (English Transl. in: *Math USSR Sb.* **64**, 427 (1989)).
8. W. I. Fushchich and W. M. Shtelen, *J. Phys. A: Math. Gen.* **22**, L887 (1989).
9. A. Valenti, *Proceedings of MOGRAN X*, N. H. Ibragimov et al. Eds., 236 (2005).
10. G. Capriz, Waves in strings with non-local response, in *Mathematical Problems in continuum Mechanics* , Trento 12-17 January (1981).

ON WAVES IN WEAKLY NONLINEAR POROELASTIC MATERIALS MODELING IMPACTS OF METEORITES*

K. WILMANSKI

University of Zielona Góra, Poland
E-mail: krzysztof_wilmanski@t-online.de
www.mech-wilmanski.de

We present a weakly nonlinear model of poroelastic materials in which deformations of both components are assumed to be small and simultaneously material properties depend on the current porosity. Changes of the latter are described by the balance equation. The model is used in the description of nonlinear waves (one-dimensional Riemann problem) created by the impact of meteorite of a moderate size. The asymptotic method of analysis is applied.

Keywords: Nonlinear waves, Poroelastic media, Impacts of meteorites.

1. Introduction

The paper is devoted to the construction of a weakly nonlinear model of a two-component porous material. It is assumed that deformations are small but material parameters depend on changes of porosity. The latter are described by a balance equation. This model is used in the description of propagation of strong discontinuity waves which may appear in soils after an impact of the meteorite of a moderate size. These are meteorites listed in the second and third lines of the Table 1 below.

With a few exceptions the propagation of shock waves in soils has been modelled by the application of one-component models (e.g. [1]). Some results based on the asymptotic analysis of a two-component model have been published under the smallness assumption of the porosity relaxation time (e.g. [2]). In this work, we present an approach in which this assumption is not satisfied.

*dedicated to prof. Tommaso Ruggeri (Bologna) on the occasion of his 60th Anniversary

Table 1. Consequences of impacts of meteorites of various sizes.

Impactor diameter	**Yield** (megatons)	**Interval** (years)	**Consequences**
< 50	< 10	< 1	meteors in upper atmosphere most do not reach surface
75	10-100	1000	irons make craters like Barringer Meteor Crater (USA), stones produce airburst like Tunguska, land impacts destroy area of the size of cities
160	100-1000	5000	irons, stones hit the ground, comets produce airbursts, land impacts destroy area of the size of large urban area (New York, Tokyo)
350	1000-10000	15000	land impacts destroy area of the size of small states, ocean impacts produce mild tsunamis
700	10000-100000	65000	land impacts destroy area of the size of moderate states like Virginia, ocean impacts makes big tsunamis
1700	100000-1000000	250000	land impact raises dust with global implication, it destroys area of the size of large states like California or France

2. Field equations

The fundamental fields describing the mechanical behaviour of the porous material are as follows

– partial mass densities, ρ^S, ρ^f, of the soil and of the water, respectively,
– partial velocities, v_k^S, v_k^F, where we use Cartesian coordinates,
– porosity n.

In addition, as we construct the set of the first order field equations, we consider the deformation tensor of the solid component, e_{kl}^S, as the field as well.

These quantities have to satisfy the following set of balance equations

$$\frac{\partial \rho^S}{\partial t} + \frac{\partial \rho^S v_k^S}{\partial x^k} = 0, \quad \frac{\partial \rho^F}{\partial t} + \frac{\partial \rho^F v_k^F}{\partial x^k} = 0,$$

$$\rho^S \left(\frac{\partial v_k^S}{\partial t} + v_l^S \frac{\partial v_k^S}{\partial x^l} \right) - \frac{\partial \sigma_{kl}^S}{\partial x^l} - \pi \left(v_k^F - v_k^S \right) = 0,$$
$$\rho^F \left(\frac{\partial v_k^F}{\partial t} + v_l^F \frac{\partial v_k^F}{\partial x^l} \right) - \frac{\partial \sigma_{kl}^F}{\partial x^l} + \pi \left(v_k^F - v_k^S \right) = 0, \qquad (1)$$

$$\frac{\partial \Delta_n}{\partial t} + v_k^S \frac{\partial \Delta_n}{\partial x^k} + \Phi \frac{\partial}{\partial x^k} \left(v_k^F - v_k^S \right) + \frac{\Delta_n}{\tau} = 0, \quad \Delta_n = n - n_E,$$

$$\frac{\partial e_{kl}^S}{\partial t} - \frac{1}{2} \left(\frac{\partial v_k^S}{\partial x_l} + \frac{\partial v_l^S}{\partial x_k} \right) = 0,$$

where we have used the assumption on smallness of deformation of both components

$$\max \left(\left\| e_{kl}^S \right\|, |\varepsilon| \right) \ll 1, \quad \left\| e_{kl}^S \right\| = \sup_{\alpha=1,2,3} \left| \lambda^{(\alpha)} \right|, \tag{2}$$

$$\det \left(e_{kl}^S - \lambda^{(\alpha)} \delta_{kl} \right) = 0, \quad |\varepsilon| \ll 1, \quad \varepsilon = \frac{\rho_0^F - \rho^F}{\rho_0^F}.$$

The last equation in the set (1) is the linearized form of the integrability condition.

We have used already the linear constitutive relations for the momentum source: $\pi \left(v_k^F - v_k^S \right)$, and for the porosity source: Δ_n / τ. In these relations π is the permeability coefficient, and τ is the relaxation time of porosity. The flux of porosity has been linearized as well: $\Phi \left(v_k^F - v_k^S \right)$, where Φ is the transport coefficient for porosity.

In order to construct field equations we have to add constitutive relations for the partial stresses, σ_{kl}^S, in the soil, the partial pressure in the fluid, p^F, and the equilibrium porosity, n_E. We assume them to have the following form

$$\sigma_{kl}^S = \sigma_{0kl}^S + \lambda^S (n) I \delta_{kl} + 2\mu^S (n) e_{kl}^S + \tag{3}$$
$$+ \left[Q(n) \varepsilon - N(n)(n - n_0) - \beta \Delta_n \right] \delta_{kl},$$

$$p^F = p_0^F - \rho_0^F \kappa (n) \varepsilon - Q(n) I - N(n)(n - n_0) - \beta \Delta_n, \tag{4}$$

$$I = e_{mm}^S, \quad \sigma_{kl}^F = -p^F \delta_{kl}, \quad n_E = n_0 \left(1 + \delta(n) I \right).$$

Consequently, we have to specify the following material parameters

$$\left\{ \lambda^S, \mu^S, \kappa, Q, N, \delta, \Phi, \beta, \pi, \tau \right\}. \tag{5}$$

They may all be functions of the current porosity, n, which makes the model weakly nonlinear.

In constitutive relations (4) we have included interactions between components through volume changes – the parameter Q introduced by M. Biot, as well as through the porosity changes – the parameter N, introduced on the basis of thermodynamical considerations [3]. However, we have neglected a second order contribution introduced by Signorini and relating stresses to the second invariants of deformations. This has been argued elsewhere to be negligible.

3. 1-D problem

In order to investigate the structure of the field equations we simplify the problem to one spacial dimension. In principle, a similar analysis can be also performed for three dimensions but the problem becomes technically much more involved.

The 1-D problem is described by the set of the following fields

$$\left\{e^S, \varepsilon, v^S, v^F, \Delta_n\right\}, \tag{6}$$

where e^S is the extension/compression of the skeleton (soil) in the x-direction, ε is the relative volume change of the fluid, v^S, v^F are x-components of the velocity, and $\Delta_n = n - n_E$ is the deviation of porosity from its equilibrium value.

Field equations for these quantities follow from (1), (4) and they have the form

$$\begin{gathered}
\frac{\partial e^S}{\partial t} - \frac{\partial v^S}{\partial x} = 0, \quad \frac{\partial \varepsilon}{\partial t} - \frac{\partial v^F}{\partial x} = 0, \\
\rho^S \left(\frac{\partial v^S}{\partial t} + v^S \frac{\partial v^S}{\partial x}\right) - \frac{\partial \sigma^S}{\partial x} - \pi \left(v^F - v^S\right) = 0, \\
\rho^F \left(\frac{\partial v^F}{\partial t} + v^F \frac{\partial v^F}{\partial x}\right) + \frac{\partial p^F}{\partial x} + \pi \left(v^F - v^S\right) = 0, \\
\frac{\partial \Delta_n}{\partial t} + v^S \frac{\partial \Delta_n}{\partial x} + \Phi \frac{\partial}{\partial x} \left(v^F - v^S\right) + \frac{\Delta_n}{\tau} = 0,
\end{gathered} \tag{7}$$

with constitutive relations

$$\begin{gathered}
\Delta_n = n - n_E, \quad n_E = n_0 \left(1 + \delta e^S\right), \\
\sigma^S = \sigma_0^S + \left(\lambda^S + 2\mu^S\right) e^S + Q\varepsilon - N\left(\Delta_n + \delta n_0 e^S\right) - \beta \Delta_n, \\
p^F = p_0^F - \rho_0^F \kappa \varepsilon - Q e^S - N\left(\Delta_n + \delta n_0 e^S\right) - \beta \Delta_n.
\end{gathered} \tag{8}$$

The dependence of compressibilities $\lambda^S + \frac{2}{3}\mu^S, \kappa$, coupling parameters Q, N, and properties δ, Φ of porosity equation can be found from Gedankenexperiments as functions of porosity (see: [3]). On the other hand, the nonequilibrium parameters π, τ, β are assumed in the present work to be constant. Their average values can be estimated (e.g. [4]) but little is known about their behaviour for large changes of porosity. In Table 2, we present these estimates of typical quantities in the case of soils saturated with water.

Bearing these estimates in mind, we can immediately find the orders of magnitude of various contributions to fields equations. These are presented in Table 3. It is seen that the model contains two small parameters: β and $1/\tau$. The last condition is different from that used in the work [2], where it was assumed that the relaxation time τ is small.

Table 2. Typical values of parameters.

macroscopic time scale	$t_0 = 10^{-5}$ [s]
macroscopic length scale ($c = L/t_0 = 10^3$ m/s)	$L = 10^{-2}$ [m]
reference porosity, n_0	0.3
true mass density of solid: ρ_0^{SR} ($\rho_0^S = (1 - n_0)\, \rho_0^{SR}$)	$5 * 10^3$ [kg/m^3]
true mass density of fluid: ρ_0^{FR} ($\rho_0^F = n_0 \rho_0^{FR}$)	10^3 [kg/m^3]
effective Lamé coefficient, λ^S	50 [GPa]
coefficient of fluid compressibility, κ	10^6 [m^2/s^2]
coupling parameter, Q coupling parameter, N	0.2 [GPa] 0.2 [GPa]
coefficient of permeabilty, π	10^8 [kg/m^{3}s]
coefficient of equilibrium changes of porosity, δ	2 [-]
relaxation time of porosity, τ coefficient Φ	1 [s] 0.015 [-]
coupling coefficient, β	0.1 [GPa]
speeds of $P1-$ and $S-$waves	3 [km/s] and 1 [km/s]
changes of partial density, $\rho^F - \rho_0^F$	1 [kg/m^3]
diffusion (seepage) velocity, $v^F - v^S$	0.1 [m/s]
partial pressure in fluid, p_0^F partial stresses in solid, σ_0^{11}	1 [MPa] 150 [MPa]
changes of porosity, $n - n_0$	10^{-3}

In the dimensionless form, these parameters can be defined as follows.

Estimates shown in Table 2 yield the following small parameters of the model

$$\varsigma^2 = \frac{\left[\frac{\Delta_n}{\tau}\right]}{\left[\frac{\partial \Delta_n}{\partial t}\right]} = 10^{-5}, \quad \beta' = \frac{\left[(N + \beta)\, \frac{\partial \Delta_n}{\partial x}\right]}{\left[\rho^F \frac{\partial v^F}{\partial t}\right]} = 0.45 * 10^{-2} \sim \varsigma. \tag{9}$$

Consequently, for our data we can redefine the field equations in terms of a single small parameter ς. They have the following form

– the equations resulting from the mass balance relations

$$\frac{\partial e_s}{\partial t'} - \frac{\partial v_s}{\partial x'} = 0, \quad \frac{\partial \varepsilon_f}{\partial t'} - \frac{\partial v_f}{\partial x'} = 0, \tag{10}$$

Table 3. Orders of magnitude of various contributions to field equations.

$[v^s] = 10$ [m/s]	$[v^F] = 10$ [m/s]
$[\rho^S \partial_t v^S] = 3 * 10^9$ [kg/m^{2}s]	$[\rho^F \partial_t v^F] = 10^9$ [kg/m^{2}s]
$[\rho^S v^S \partial_x v^S] = 3 * 10^8$ [kg/m^{2}s]	$[\rho^F v^F \partial_x v^F] = 10^8$ [kg/m^{2}s]
$[\sigma^S] = 150$ [MPa]	$[p^F] = 1$ [MPa]
$[(\lambda^S + 2\mu^S) e^S] = 150$ [MPa]	$[\rho_0^F \kappa \varepsilon] = 1$ [MPa]
$[Q\varepsilon] = 0.2$ [MPa]	$[Qe^S] = 0.2$ [MPa]
$[\Delta_n] = 1.5 * 10^{-5}$	$\partial_t \Delta_n = 1.5 * 10^0$ [1/s] $[\Delta_n/\tau] = 1.5 * 10^{-5}$ [1/s]
$[N(\Delta_n + \delta n_0 e^S)] = 0.12$ [MPa]	$[\beta\Delta_n] = 1.5 * 10^{-3}$ [MPa] $[\beta\partial_x\Delta_n] = 1.5 * 10^{-1}$ [kg/m^{2}s]
$[\partial_x \sigma^S] = 1.5 * 10^{10}$ [kg/m^{2}s]	$[\partial_x p^F] = 10^8$ [kg/m^{2}s]

$$e_s = \frac{e^S}{e_0}, \quad \varepsilon_f = \frac{\varepsilon}{e_0}, \quad e_0 = \frac{v_0 t_0}{L} = 1, \quad v_{s,f} = \frac{v^{S,F}}{v_0}, \qquad (11)$$
$$v_0 = \frac{L}{t_0}, \quad x' = \frac{x}{L}, \quad t' = \frac{t}{t_0},$$

– the equations resulting from momentum balance equations

$$\rho_s \left(\frac{\partial v_s}{\partial t'} + v_s \frac{\partial v_s}{\partial x'} \right) - \frac{\partial \sigma_s}{\partial x'} - \pi' (v_f - v_s) = 0,$$

$$\rho_f \left(\frac{\partial v_f}{\partial t'} + v_f \frac{\partial v_f}{\partial x'} \right) + \frac{\partial p_f}{\partial x'} + \pi' (v_f - v_s) = 0, \qquad (12)$$

$$\rho_s = \frac{\rho^S}{\rho_0}, \quad \rho^S = \rho_0^S \left(1 - e^S\right), \quad \rho_f = \frac{\rho^F}{\rho_0}, \quad \pi' = \frac{\pi}{\pi_0}, \quad \pi_0 = \frac{\rho_0}{t_0}, \qquad (13)$$

– porosity balance equation

$$\frac{\partial \Delta_n'}{\partial t'} + v_s \frac{\partial \Delta_n'}{\partial x'} + \frac{\partial}{\partial x'} (v_f - v_s) + \varsigma^2 \Delta_n' = 0, \qquad (14)$$

$$\Delta_n' = \frac{\Delta_n}{\Delta_0}, \quad \Delta_0 = \Phi, \quad \varsigma^2 = \frac{t_0}{\tau}, \qquad (15)$$

where the dimensionless constitutive relations have the form

$$\begin{aligned} &\sigma_s = \sigma_{0s} + E' e_s + Q' \varepsilon_f - \varsigma \Delta_n', \\ &\sigma_s = \frac{\sigma^S}{\sigma_0}, \quad \sigma_{0s} = \frac{\sigma_0^S}{\sigma_0}, \quad \sigma_0 = \rho_0 \frac{L^2}{t_0^2}, \\ &E = \lambda^S + 2\mu^S - N\delta n_0, \quad E' = \frac{E}{\sigma_0}, \quad Q' = \frac{Q}{\sigma_0}, \end{aligned} \tag{16}$$

$$\begin{aligned} &p_f = p_{0f} - K' \varepsilon_f - C' e_s - \varsigma \Delta_n', \quad p_f = \frac{p^F}{\sigma_0}, \quad p_{0f} = \frac{p_0^F}{\sigma_0}, \\ &K = \rho_0^F \kappa, \quad K' = \frac{K}{\sigma_0}, \quad C' = \frac{Q - N\delta n_0}{\sigma_0}. \end{aligned} \tag{17}$$

For the data of Tables 2 and 3 the parameters of the dimensionless quantities have the following values

$$\begin{aligned} &L = 10^{-2}\ [\mathrm{m}], \quad t_0 = 10^{-5}\ [\mathrm{s}], \quad \rho_0 = 10^3\ [\mathrm{kg/m^3}], \quad \Phi = 0.015, \\ &v_0 = 10^3\ [\mathrm{m/s}], \quad \sigma_0 = 1\ [\mathrm{GPa}], \quad \pi_0 = 10^8\ [\mathrm{kg/m^3 s}], \\ &\Delta_0 = 1.5 * 10^{-2}, \quad \varsigma = 3 * 10^{-3}. \end{aligned} \tag{18}$$

4. Governing set of equations

Now we are in the position to write the governing set of equations for the one-dimensional fields $\{e_s, \varepsilon_f, v_s, v_f, \Delta_n\}$. They have the following form

$$\begin{aligned} &\frac{\partial \Delta_n}{\partial t} + v_s \frac{\partial \Delta_n}{\partial x} + \frac{\partial}{\partial x}(v_f - v_s) + \varsigma^2 \Delta_n = 0, \\ &\rho_s \left(\frac{\partial v_s}{\partial t} + v_s \frac{\partial v_s}{\partial x} \right) - \frac{\partial}{\partial x}\left(E(n) e_s + Q(n) \varepsilon_f \right) + \varsigma \frac{\partial \Delta_n}{\partial x} - \pi (v_f - v_s) = 0, \\ &\rho_f \left(\frac{\partial v_f}{\partial t} + v_f \frac{\partial v_f}{\partial x} \right) - \frac{\partial}{\partial x}\left(K(n) \varepsilon_f + C(n) e_s \right) - \varsigma \frac{\partial \Delta_n}{\partial x} + \pi (v_f - v_s) = 0, \\ &\frac{\partial e_s}{\partial t} - \frac{\partial v_s}{\partial x} = 0, \quad \frac{\partial \varepsilon_f}{\partial t} - \frac{\partial v_f}{\partial x} = 0, \end{aligned} \tag{19}$$

where

$$\begin{aligned} &\rho_s = \rho_{0s}(1 - e_s), \quad \rho_f = \rho_{0f}(1 - \varepsilon_f), \\ &\frac{dE}{dn} < 0, \quad \frac{d^2E}{dn^2} > 0, \quad \frac{dK}{dn} < 0, \quad \frac{d^2K}{dn^2} > 0, \\ &\frac{dQ}{dn} > 0, \quad \frac{d^2Q}{dn^2} < 0, \quad \frac{dC}{dn} > 0, \quad \frac{d^2C}{dn^2} < 0, \end{aligned} \tag{20}$$

and ρ_{0s}, ρ_{0f} are initial values of dimensionless partial mass densities.

For typographical reasons, we have skipped the prime for the dimensionless quantities. The inequalities (20) follow from micro/macro considerations presented in the paper [3]. The structure of this set makes clear that, for reasons of consistency of orders of magnitude, a weak discontinuity wave P1 or P2, i.e. for waves in which velocities are continuous on the wave front but one of the accelerations may be not, is accompanied by a kink-like solution for the porosity. This follows easily from momentum balance equations. Simultaneously, for strong discontinuities of velocity fields (i.e. kink-like solutions) the porosity must be expressed by the soliton-like solution. This structure has been already indicated for another asymptotic problem in the paper [2].

5. Structure of asymptotic solutions

We construct the asymptotic solution of the Riemann problem. Technical details concerning the asymptotic analysis can be found in the papers [5]. In such a solution, we have a regular part which is a power expansion with respect to the small parameter ς and the singular part related to the "boundary layer" effect near the front of the wave. This contribution is written in the form of dependence on the so-called fast variable $\sigma = \left(x - \overset{0}{x}(t)\right)/\varsigma^2$, where $\overset{0}{x}(t)$ is the position of the wave front. For a typical quantity Π this expression have the following form

$$\Pi_{as} = \Gamma_0(t) + \Pi_0(\sigma, t) + \sum_{j=1}^{N} \varsigma^j \left(\Gamma_1(x,t) + \Gamma_j^p(\sigma, x, t)\right),$$

$$\Gamma_j^p(\sigma, x, t) = \Pi_j(\sigma, t) + H_j^p z_0(\sigma, t), \quad j \geq 1, \tag{21}$$

$$\overset{0}{x}(t) = X_0(t) - \varsigma X_1(t) - \varsigma^2 X_2(t),$$

where $\Gamma_j, \Pi_j H_j^p, z_0$ are smooth bounded functions, and, simultaneously, Π_j is kink-like, while z_0 – soliton-like. They are stabilized in infinity, i.e.

$$\begin{aligned} \forall k, i, j, l : \quad & \sigma^k \frac{d^j}{d\sigma^j} \frac{d^l}{dt^l} \frac{d^i}{dx^i} \left(y - y^{\pm}\right) = 0 \text{ if } \sigma \to \infty, \\ & z_0^- = \lim_{\sigma \to -\infty} z_0 = 0, \quad z_0^+ = \lim_{\sigma \to \infty} z_0 = 0, \text{ or conversely,} \\ & \Pi_j^{\pm} = 0 \text{ (solitons),} \end{aligned} \tag{22}$$

where y is either z_0 or Π_j. Then

Proposition 1: *Strong discontinuities cannot exist simultaneously for v_s and v_f, i.e. the strong discontinuity of v_s yields the weak discontinuity of v_f, or conversely. The propagation velocities of fronts of these two discontinuities satisfy the condition*

$$\dot{X}_0^2 = \frac{E(n)}{\rho_{0s}} \text{ or } \dot{X}_0^2 = \kappa(n). \tag{23}$$

Proposition 2: *Let $\Delta_n = v_s = v_f = 0$ and $\rho_{0s} > 0, \rho_{0f} > 0, e_s$ are constant for $t = 0$. Then asymptotic solutions of any accuracy with respect to ς of the system* (19) *exist on a finite time interval, and they possess the following properties. They are smooth approximations with respect to ς of order $O(\varsigma)$ of strong discontinuities either for v_s, ρ_s, e_s or for $v_f, \rho_f, \varepsilon_f$ and the infinitely thin soliton function of order $O(\varsigma)$ for Δ_n along a small perturbation of characteristics of the linearized problem to* (7).

Details of these asymptotic considerations are the subject of the forthcoming paper.

6. Final remarks

The model presented in this work is able to describe the propagation of strong discontinuity waves which propagate in the intermediate stage of the impact problem. It enables the estimation of the time and the distance from the site of impact before the dynamics of the problem is described by the usual acoustic waves in the two-component poroelastic medium.

References

1. K. Garg, D. H. Brownell, Jr., J. W. Pritchett and R. G. Herrmann, *J. Appl. Phys.* **46**, 702 (1975).
2. E. Radkevich and K. Wilmanski, *Jour. Math. Sci.* **114**, 1431 (2003).
3. K. Wilmanski, *Géotechnique* **54**, 593 (2004).
4. B. Albers and K. Wilmanski, *Arch. Mech.* **58**, 313 (2006).
5. E. Radkevich and K. Wilmanski, *A Riemann problem for Poroelastic Materials with the Balance Equation for Porosity, Part I and II*, Preprint 593, 594, WIAS (2000).